TILAPIA

TILAPIA

Aquaculture, Biology and Health Management

Edited by Kim Thompson, Craig Shoemaker and David C. Little

CABI

CABI is a trading name of CAB International

CABI
Nosworthy Way
Wallingford
Oxfordshire OX10 8DE
UK

Tel: +44 (0)1491 832111
E-mail: info@cabi.org
Website: www.cabi.org

CABI
200 Portland Street
Boston
MA 02114
USA

T: +1 (617)682-9015
E-mail: cabi-nao@cabi.org

A catalogue record for this book is available from the British Library, London, UK.
Library of Congress Control Number: 2024951566.

ISBN-13: 9781800629387 (hardback)
9781800629462 (ePDF)
9781800629455 (ePub)

DOI: 10.1079/9781800629455.0000

Typeset by Cheshire Typesetting Ltd, Cuddington, Cheshire and Exeter Premedia Services Ltd, Chennai, India
Printed and bound in the UK by CPI Group (UK) Ltd, Croydon, CR0 4YY

Contents

Aquatic Animal Diseases Series

EAFP book series editor
Dr. Annette Simone Boerlage, EAFP Publications Officer, Scotland's Rural College (SRUC), Scotland, UK Annette.Boerlage@sruc.ac.uk

When culturing aquatic species, the proper maintenance of health is essential to the sustainability of production. Disease epidemics, for example outbreaks of infectious salmon anaemia (ISA) in Chile, and early mortality syndrome (EMS) in shrimp in South-East Asia have caused widespread and devastating effects on the aquaculture industries in the affected countries. In order for aquatic animal diseases to be properly prevented, identified, monitored, treated and managed, there is a constant need for high quality and authoritative books to inform the range of personnel working in the aquaculture industry.

The EAFP/CABI Books Series encompasses professional-level reference books and practical guides aimed at fish farm operatives and technicians, fish biologists and pathologists, veterinarians, students who are studying aquaculture and fish health, and the many personnel involved in supplying the aquaculture industry with pharmaceuticals, nutraceuticals, feed and equipment.

The first titles in the series include:

Ariadna Sitjà-Bobadilla, James E. Bron, Geert F. Wiegertjes and M. Carla Piazzon (eds) (2021) *Fish Parasites: A Handbook of Protocols for their Isolation, Culture and Transmission*

Albert Tacon (2022) *A Handbook of Nutritional Disorders and Diseases of Fish and Shrimp*

Kim Thompson, Craig Shoemaker and David C. Little (eds) (2024) *Tilapia: Aquwaculture, Biology and Health Management*

The European Association of Fish Pathologists (EAFP) exists to promote the exchange of knowledge and coordination of research related to fish and shellfish pathology. Membership of the EAFP is open to all fish and shellfish pathologists, students of fish pathology and organisations, agencies or other associations which have an interest in maintaining the objectives of the EAFP. https://eafp.org

CABI shares knowledge and science to help tackle global issues like poverty, hunger, education, equality, sustainability, climate change and biodiversity and our aquaculture, fish and fisheries books contribute to that work. www.cabidigitallibrary.org/books

Contributors

Alastair Hamilton – alastair.hamilton@hendrix-genetics.com

Alvin C. Camus – Department of Pathology, College of Veterinary Medicine, University of Georgia, Athens, Georgia, USA, camus@uga.edu

Andrew P. Shinn – INVE (Thailand), Bangpood, Pakkred, Nonthaburi, 11120, Thailand; Institute of Aquaculture, University of Stirling, Stirling FK9 4LA, Scotland, UK, a.shinn@inveaquaculture.com

Annette Simone Boerlage – Centre for Epidemiology and Planetary Health (CEPH), SRUC School of Veterinary Medicine, Inverness, UK, annette.boerlage@sruc.ac.uk

Benjamin R. LaFrentz – Aquatic Animal Health Research Unit, United States Department of Agriculture – Agricultural Research Service (USDA–ARS), Auburn, AL, USA, benjamin.lafrentz@usda.gov

Carlos Garcia de Leaniz – Centre for Sustainable Aquatic Research, Swansea University, Swansea, UK, c.garciadeleaniz@swansea.ac.uk

Channarong Rodkhum – Center of Excellence in Fish Infectious Diseases (CE FID) and Department of Veterinary Microbiology, Faculty of Veterinary Science, Chulalongkorn University, Bangkok, Thailand, channarong.r@chula.ac.th

Craig A. Shoemaker – United States Department of Agriculture – Agricultural Research Service (USDA–ARS), Aquatic Animal Health Research Unit, Auburn, AL, USA, craig.shoemaker@usda.gov

D. Allen Davis – School of Fisheries, Aquaculture, and Aquatic Science, Auburn University, Auburn, AL, USA, davisda@auburn.edu

Darci Carlos Fornari – AQUAM, Animal Science Research Program, Federal University of Rio Grande do Sul – UFRGS, Porto Alegre-RS, Brazil, darci.peixegen@gmail.com

David C. Little – IoA, University of Stirling, UK, d.c.little@stir.ac.uk

De-Hai Xu – United States Department of Agriculture – Agricultural Research Service (USDA–ARS), Aquatic Animal Health Research Unit, Auburn, AL, USA, dehaixu@gmail.com

Divya Rose – Thad Cochran National Warmwater Aquaculture Center, College of Veterinary Medicine, Mississippi State University, Stoneville, MS, USA, dj1371@msstate.edu

Esteban Soto – Department of Medicine and Epidemiology, University of California, School of Veterinary Medicine, Davis, CA, USA, sotomartinez@ucdavis.edu

Francis Murray – Institute of Aquaculture, University of Stirling, FK7 9LA, UK. fm73@stir.ac.uk/francismurray65@gmail.com

Gustavo Ramirez Paredes – Ridgeway Biologicals Ltd (a Ceva Santé Animale company), Compton, UK, gus.ramirez@ceva.com

Ha Thanh Dong – Aquaculture and Aquatic Resources Management Program, School of Environment, Resources and Development, Asian Institute of Technology, Pathum Thani, Thailand, htdong@ait.asia

Halina Sobolewska – halina@noahgene.com

Hetron M. Munang'andu – Faculty of Biosciences and Aquaculture, Nord University, Bodø, Norway, hetron.m.munangandu@nord.no

Ian Goulding – Megapesca Lda., Alfeizerão, Portugal, iangoulding@megapesca.com

Janina Costa – Aquaculture Research Group, Moredun Research Institute, Penicuik, UK, Janina.costa@moredun.ac.uk

Joke van der Giessen – National Institute for Public Health and the Environment, Bilthoven, The Netherlands, joke.van.der.giessen@rivm.nl

Jorge del-Pozo – Easter Bush Pathology, Royal (Dick) School of Veterinary Studies, Edinburgh, UK, jorge.del.pozo@ed.ac.uk

Junning Cai – Aquaculture Officer, Fisheries and Aquaculture Department (FIRA), FAO, Rome, Italy, Junning.Cai@fao.org

Kim Thompson – Aquaculture Research Group, Moredun Research Institute, Penicuik, UK, kim.thompson@moredun.ac.uk

Matt J. Griffin – Thad Cochran National Warmwater Aquaculture Center, College of Veterinary Medicine, Mississippi State University, Stoneville, MS, USA, griffin@cvm.msstate.edu

Mona Dverdal Jansen – Norwegian Veterinary Institute, Ås, Norway, currently Norwegian Food Safety Authority, Sandefjord, Norway, monadjansen@hotmail.com

Nopadon Pirarat – Center of Excellence in Wildlife, Exotic and Aquatic Animal Pathology, Department of Pathology, Faculty of Veterinary Science, Chulalongkorn University, Bangkok, Thailand, nopadonpirarat@gmail.com

Olga L. M. Haenen – NRL for Fish Diseases, Wageningen Bioveterinary Research, Lelystad, The Netherlands, olga.haenen@wur.nl

Partho Pratim Debnath – Center of Excellence in Fish Infectious Diseases (CE FID) and Department of Veterinary Microbiology, Faculty of Veterinary Science, Chulalongkorn University, Bangkok, Thailand, parthoku2004@yahoo.com

Ram C. Bhujel – Aqua-Centre, Asian Institute of Technology, Thailand, bhujel@ait.asia

Roger S. V. Pullin – pullin.roger@gmail.com

Ruth Zadoks – Faculty of Science, Sydney School of Veterinary Science, The University of Sydney, Camden, NSW, Australia, ruth.zadoks@sydney.edu.au

Saengchan Senapin – National Center for Genetic Engineering and Biotechnology (BIOTEC), National Science and Technology Development Agency (NSTDA), Pathum Thani, Thailand, saengchan@biotec.or.th

Sonia Rey Planellas – Institute of Aquaculture, University of Stirling, Stirling, UK, sonia.reyplanellas@stir.ac.uk

Taylor I. Heckman – Thad Cochran National Warmwater Aquaculture Center, Mississippi Agriculture and Forestry

Experiment Station and Department of Wildlife, Fisheries and Aquaculture, College of Forest Resources, Mississippi State University, Stoneville, MS, USA, tih33@msstate.edu

Thomas P. Loch – Department of Fisheries and Wildlife, College of Agriculture and Natural Resources and Department of Pathobiology and Diagnostic Investigation, College of Veterinary Medicine, Michigan State University, East Lansing, MI, USA, lochthom@msu.edu

Waldemar Rossi, Jr – Kentucky State University Aquaculture Research Center, Frankfort, KY, USA, Waldemar.Rossi@kysu.edu

Warren A. Turner – Nam Sai Farms Co. Ltd, Prachinburi, Thailand, warrennamsai@gmail.com

Preface

Tilapia is a group of cichlids, endemic to tropical freshwater in Africa, Jordan and Israel, that are extremely tolerant of challenging aquatic environments. They have a broad temperature range, can reproduce under various physical and environmental conditions, are generally disease resistant, are herbivorous by nature so their nutritional requirements are not as robust and demanding as a carnivorous fish, can be affordably grown without the need for specialized equipment and are extremely nutritious, being high in protein and low in fat while containing many vitamins, minerals and essential fatty acids. For these reasons, tilapia are widely preferred by fish farmers and consumers, making this species the second-most cultured freshwater fish in the world (behind carp). These qualities allow tilapia to serve as a vital source of protein for many countries but, since global demand is high, they can also be a high-value, upscale export commodity.

To illustrate the global importance of tilapia aquaculture, in 2023 the global market value of cultured tilapia was US$13.9 billion and is projected to be US$21.6 billion within the next decade. Following a pause in 2020, global tilapia supply growth has rebounded and is expected to reach 7 million metric tonnes (mmt) in 2024. Global production of farmed tilapia in 2023 is projected to be 6.7 mmt (up approximately 5% from 2022 levels) and 7.03 mmt in 2024 (up a further 5% from 2023 levels). The main producers of farmed tilapia in 2023 were China, Indonesia, Egypt and Brazil. The first two of these countries are both projected to produce over 1.6 mmt in 2023, with Egypt's production estimated at over 1.1 mmt. In China, tilapia appears to be a mature sector with modest expansion; however, Indonesia is still on a rapid growth path with expected growth of 5.0 and 3.7% in 2023 and 2024, respectively. There are some strong growth volumes among other Asian tilapia producers, including Bangladesh, Philippines, Thailand and Vietnam, while in Latin America the top producer is Brazil with around 400,000 mt in 2023, followed by Colombia and Mexico.

Since tilapia have begun to be produced globally in large quantities, challenges have arisen in the industry. While there is still much to learn regarding nutrition, there are diets currently available for tilapia that are economical, nutritious and environmentally responsible. Advances in both genetics and breeding are continuing to assist producers to remain profitable. Diseases, however, have become a serious concern to the industry and more research is required.

With production costs increasing and profit margins becoming tighter, it is imperative that new information be published and disseminated so that the continued growth and success of the global tilapia aquaculture industry can be assured. Considerable technological and production advances have been made in recent years (e.g. recirculating aquaculture systems, bio-floc technology, genetic improvements and sustainable nutrient-dense diet formulations), and this book brings together the latest information on tilapia aquaculture into a single, essential volume.

The book is organized into three distinct sections. Section 1 discusses tilapia species and includes chapters on biotaxonomic and physicochemical requirements; their importance as an aquaculture species; reproduction and genetics; production systems; harvest and post-harvest procedures; nutritional requirements; immunity and functional feeds; and fish welfare. Section 2 presents information on pathogens and includes chapters on bacterial and viral diseases, disease diagnosis, concurrent infections, zoonotics and histology. Section 3 is devoted to health interventions and includes chapters on farm health management, vaccines, certification for disease management and future developments.

This book will be invaluable for students, aquaculturists, scientists, extension personnel, producers, government officials and anyone with an interest in tilapia aquaculture. It is hoped that this book will provide state-of-the-art information on many aspects of tilapia culture, in particular relating to health management as the title implies. The book will contribute to the long-term, sustainable growth of the global tilapia aquaculture industry.

Dr Carl Webster

Chapter 1

An introduction to the tilapias

Roger S. V. Pullin

Introduction

The inclusion of personal experiences in this introductory chapter was agreed with the editors. My own introduction to the tilapias began in 1979 when I joined the International Center for Living Aquatic Resources Management (ICLARM). I was tasked to convene the 1980 Bellagio Conference on the Biology and Culture of Tilapias, with Ziad Shehadeh as mentor. Its proceedings (Pullin & Lowe-McConnell, 1982) became a key reference work, still freely available from the World Fish Center. At the First International Symposium on Tilapia in Aquaculture (ISTA I), I suggested keeping the focus on Nile tilapia (*Oreochromis niloticus*), blue tilapia (*O. aureus*) and *Oreochromis* hybrids, and exploring the potential of *Tilapia rendalli* (now *Coptodon rendalli*) in polyculture (Pullin, 1983).

Farmed tilapias are increasingly important contributors to world fish supply. According to the Food and Agriculture Organization (FAO, 2019a), the global production of farmed tilapias, from 140 countries, grew from 2.83 million mt in 2008 to 5.88 million mt in 2017. *O. niloticus*, *Oreochromis* hybrids and tilapias 'not elsewhere identified' contributed together about 95%, with *O. niloticus* accounting for over 70%. Tilapia farming, especially in Africa, could make increased use of diverse species, such as the three spotted tilapia (*O. andersonii*), *O. aureus* and the blackchin tilapia (*Sarotherodon melanotheron*) (Pullin, 2017). The growth of farmed tilapia production in 2020 was forecast at only 0.1%, to about 6.93 million mt, but with formerly high annual growth rates (typically 5 to 6%) returning after the Covid-19 pandemic (Harkell, 2020).

Kim Thompson, Craig Shoemaker and David C. Little (eds) **Tilapia**
DOI: 10.1079/9781800629455.0001, © CAB International, 2025

Tilapia biology has presented challenges for tilapia farming, especially their capacity for prolific reproduction at small sizes and batch spawning, with low fecundities per spawning compared with most other farmed fish. The growth of tilapia farming has been facilitated by highly effective research to ensure reliable and safe mass production of all-male seed, synchronous egg collection and artificial incubation, and genetic improvement. Control of tilapia diseases is now urgent – hence the importance of this volume.

This introductory chapter summarizes tilapia nomenclature, biology and ecology, their status as food fish and sources of further information. Its main message is to know your fish well in pursuit of their good health. This means consulting key works and keeping up with research findings and advances in husbandry, including disease threats and countermeasures.

Tilapia nomenclature

According to Trewavas (1982), the name 'Tilapia' derives from A. Smith, author of the genus *Tilapia* in 1840, who based it on 'the Bushman (sic) word for 'fish', which began with a click, rendered 'Til'.' 'Fish' in Khoekhoe, a Khoisan language, is written as '//au-p'. 'Tilapia' is now used globally as a common name. FishBase (Froese & Pauly, 2018) lists many other common names for tilapias, by species and in multiple languages.

All tilapias belong to the family Cichlidae, in which the Tribe Tilapiini can be 'tentatively recognized', based on structural differences in the apophysis at the base of the skull (Trewavas, 1983). The genus *Tilapia* used to contain all the substrate spawning and mouthbrooding tilapias (e.g. Thys van den Audenaerde, 1968). Trewavas (1982) put all the mouthbrooders into the genus *Sarotherodon* before finally settling on *Oreochromis* for maternal mouthbrooders and *Sarotherodon* for male and biparental ones (Trewavas, 1983).

The taxonomy of the substrate spawners has also been revised: for example, the former *T. rendalli* and *T. zillii* are now in the genus *Coptodon*. FishBase (Froese & Pauly, 2018) gives the valid names and synonyms of all tilapias by species. The main genera contain the following numbers of species: *Oreochromis*, ten; *Sarotherodon*, 13; *Coptodon*, 31; and *Tilapia*, seven. Trewavas (1983) described multiple subspecies, including seven for *O. niloticus* and five each for *S. galilaeus* and *S. melanotheron*.

All farmed tilapias should be called 'farmed types', which is FAO's term to cover any farmed aquatic organism (Mair & Lucente, 2020). The wild relatives of farmed tilapias are populations of conspecifics and related species, in their native African waters and/or in gene banks. Feral tilapia populations are also widespread, following introductions for aquaculture and the aquarium trade and subsequent escapes and discards. Checking the identity of tilapias is becoming easier through diagnostic SNP markers (e.g. Van Bers *et al.*, 2012; Lind *et al.*, 2017; Hamilton *et al.*, 2020).

Tilapia farmed types are often named based on their local origins. Descendants of the first introduction of *O. niloticus* to Thailand are called the Chitralada I strain, after their initial establishment at the Chitralada Palace, Bangkok. In Indonesia, *O. mossambicus* was named *ikan mudjair* (Mudjair's fish) after Bapak Mudjair, who discovered it as a small feral population in 1939. Some red tilapias are named after the places where they were bred: for example, the Florida Red and the Taiwan Red. Others have attractive names, such as *pla tap tim* (ruby fish) in Thailand and 'king fish' in the Philippines (Guerrero & Guerrero, 2020). Tilapia strain names often indicate high performance: for example, 'BEST', FaST', GET-EXCEL, 'GenoMar Supreme' and 'SaltUno' (Obrero and Racaza, 2020).

Tilapia biology

Feeding and digestion

Tilapia are opportunistic detritivores and herbivores, equipped with gill rakers for filter feeding, teeth for browsing and a pharyngeal mill to reduce food particle size. Mouthbrooders filter feed on phytoplankton and browse on benthic detrital aggregates, epiphytic growths and epilithic algae (Trewavas, 1983). Substrate spawners consume more filamentous algae, aquatic macrophytes and miscellaneous plant material. These low trophic levels facilitate the formulation of cost-effective feeds in tilapia farming. In common with many herbivorous and detritivorous freshwater fish, tilapia fry are more carnivorous than juveniles and adults and consume more zooplankton and zoobenthos.

Gerking (1994), citing various authors, gave examples of tilapia gut contents as follows: wild Lake Chad *S. galilaeus* – 'fine plant debris, dead segmented algae, crustacean plankton, red clay... bacteria, benthic diatoms, protozoans and rotifers'; feral *O. mossambicus* in Hong Kong – 'mud-sand-detritus, 61% (up to 85% in some individuals), *Oscillatoria*, 0.8%; diatoms, 17%; *Chlorella*, 9%; *Eudorina*, 9.8%; and Cladocera, 0.3%'. Feral *O. mossambicus* in Lake Valencia, Venezuela feed mainly on detrital aggregates (Bowen, 1980, 1981).

Bowen (1982) described the dual nature of tilapia digestion, based on gastric acid secretion at pH levels below 2.0 and exceptionally long intestines with ratios to fish standard length of between 7:1 and 10:1. Tilapias can lyse the cell walls of bacteria, as protein-rich food sources less used by other species. In Lake George, Uganda, wild *O. niloticus* feed mainly on cyanobacteria and assimilate 70 to 80% of the carbon provided (Moriarty & Moriarty, 1973). Pauly (1976) described wild *S. melanotheron* in the Sakumo Lagoon, Ghana, eating 'bottom mud' with 30% organic content, and estimated a daily consumption of 1.5g with assimilation of at least 900 calories for a 20g fish.

Growth

Growth varies among and within tilapia species and hybrids and with food availability and quality and environmental conditions, especially temperature and dissolved oxygen (DO). Prein *et al.* (1993) compiled an impressive set of experimental and on-farm tilapia growth studies. Tilapias are farmed mostly as all-male tilapia populations, to avoid reproduction and because males grow faster than females. The fast growth of farmed tilapias enables short production cycles, typically two or more per year in the tropics.

Moreau *et al.* (1986) applied to tilapias the growth performance index Ø' (lower case phi prime), where Ø' = $\log_{10}K + 2 \log_{10}L_\infty$; and K ($year^{-1}$) and L_∞ (cm) are parameters of the von Bertalanffy growth equation. FishBase (Froese and Pauly, 2018) lists Ø' values for many tilapias, by species. Pauly *et al.* (1988) compared Ø' values in wild and farmed populations of seven tilapia species. Farmed *O. niloticus* had the highest mean Ø' (3.21, range 2.64–3.72) compared with *O. aureus* (3.15, range 2.83–3.15) and *O. mossambicus* (3.14, range 2.63–3.50). Overlaps between wild and farmed Ø's were higher for *O. mossambicus* (0.66) and O. *aureus* (0.65) than for *O. niloticus* (0.56) and lowest for *C. rendalli* (0.15), which suggests scope for further domestication in all species.

Reproduction

The reproductive strategies of tilapias have posed problems for tilapia farming, particularly their capacity for precocious maturation and prolific reproduction at small sizes when kept as mixed sex populations and, paradoxically, their low fecundity per spawning, which is several orders of magnitude less than that of many farmed fish. Tilapia farming has overcome these problems by innovative brood stock management that maximizes reproductive output (e.g. Bhujel, 2000; Little *et al.*, 1993; Little & Hulata, 2000), artificial incubation that mimics mouthbrooding (Rana, 1988) and all-male production stocks (see below).

The tilapias include substrate spawners and maternal, male and biparental mouthbrooders (Jalabert and Zohar, 1982; Noakes & Balon, 1982). Tilapias typically display lek courtship behaviour and spawn in shallow waters over soft substrates. Maturation and spawning are determined by interactive environmental factors including the available area, population density, sex ratio, temperature, salinity and particularly DO (see below).

Tilapia oogenesis provides for batch spawning with limited clutch sizes (Peters, 1983). The reproductive output for a given tilapia population and environment is therefore hard to predict. In general, the substrate spawners have smaller eggs than mouthbrooders: for example, 1.3–1.6 mm for *C. tholloni* and 2.7–2.9 mm for *O. mossambicus*

(Peters, 1983). Coward and Bromage (2000) found overlapping fecundities among and within substrate spawners and mouthbrooders: for example, 2,990–12,344 for wild *C. zillii* of 8–24 cm and 2,412–2,964 for wild *O. niloticus* of 29–32 cm. Balarin and Hatton (1979) tabulated wide ranges for tilapia fecundities – for example, as eggs per year and fry per brood, respectively: *O. aureus*, 1,500–4,300 and 64–2,000; *O. mossambicus*, 2,000–10,000 and 75–1,000.

Sex determination

Cnaani (2013) summarized five decades of research on the genetics of sex determination in tilapias, which involves female or male heterogamety and different linkage groups (LGs) within species, for example: *O. niloticus*, XX-XY, LGs 1 and 23; *O. aureus*, ZZ-WZ, LGs 1 and 3. Taslima *et al.* (2020) found that sex determination in GIFT strain *O. niloticus* involved LG1, compared with LG23 for a Stirling strain. Many interspecific *Oreochromis* hybrid crosses yield predominantly male progeny (Hickling, 1963; Wohlfarth & Hulata, 1981).

Wohlfarth (1994) stressed 'broodstock purity' for tilapia hybridization and the preferred use of hormonal sex reversal to obtain all-male progeny. Macintosh (2008) confirmed the safety of using methyl testosterone for masculinization of early tilapia fry. Beardmore *et al.* (2001) recommended the development of YY genetically male tilapia (GMT) brood stocks to provide all-male farmed types without hormone treatment. The success of this approach can vary, as described by Tuan *et al.* (1999) for the Chitralada strain of *O. niloticus*.

Genetic improvement

At the First International Symposium on Genetics in Aquaculture (ISGA I) (Wilkins & Gosling, 1983), I became convinced that selection was a better approach to tilapia genetic improvement than hybridization. Hybrids can provide all-male progeny and/or better cold or salt tolerance, but selection can achieve genetic gain in each successive generation.

With Trygve Gjedrem as mentor, the Genetic Improvement of Farmed Tilapia (GIFT) project began in 1988. Selection was applied to a base population built from four African wild and four Asian farmed types of *O. niloticus* (World Fish, 2004; Asian Development Bank, 2005a). This approach was criticized, as was the Green Revolution, on grounds that it might lead to farmers having to rely on seed of only one or a few improved strains. However, a review in *Seedling* endorsed ICLARM as 'having taken into account the role of genetic resources, both to obtain improved performance and to avoid genetic resource erosion' (Martinez-i-Prat, 1995).

Khaw *et al.* (2007) compared growth between a GIFT strain population bred using cryopreserved sperm from the GIFT base population and another using fresh sperm from the GIFT ninth generation of selection and showed an accumulated gain of at least 63%. Ponzoni *et al.* (2008) reported sustained genetic gains of 10–15% per GIFT generation. The GIFT project's methods were also an important contribution (World Fish, 2004).

GIFT and GIFT-derived farmed types have continued to figure in research. For example, de Verdal *et al.* (2018) showed genetic control of feed conversion ratio (FCR) for GIFT *O. niloticus* juveniles and suggested using weight loss at fasting for selection to improve FCR. Barría *et al.* (2020) found highly heritable resistance to Tilapia Lake Virus (TiLV) in a GIFT breeding population. This might reflect the diverse composition of the GIFT base population.

The GIFT project was only one among many beneficial applications of genetic research in tilapia farming. Komen and Benzie (2015) found genetic gains of 5–10% per generation in tilapia breeding programmes. Shen and Yue (2019) concluded that conventional selective breeding, assisted by quantitative genetics, would continue as the main approach for genetic improvement of farmed fish, including tilapias, with marker assisted selection, genome editing and genomic selection as additional aids.

Natural selection of tilapias to particular farm environments, also called domestication (or inadvertent) selection, proceeds in addition to any purposeful selection. The performance of genetically improved tilapias will differ in different farm environments. Agha *et al.* (2018) showed strong genotype × environment interaction (G × E) for

GIFT *O. niloticus* bred in Malaysia and grown in China, India and Malaysia, and proposed that G × E and micro-environmental sensitivity information be used to select 'robust' tilapia genotypes for high performance across environments and resilience to environmental fluctuations.

Diseases

Roberts and Sommerville (1982) considered tilapias more disease resistant than many other farmed fish and listed mainly their bacterial pathogens and ectoparasites. Up to the mid-1990s, most tilapia disease problems were reported from cooler waters: for example, in Israel, South Africa and the USA (Paperna *et al.*, 1983). Pullin and Maclean (1992) reviewed over 2800 tilapia publications and found only 4% covering health and pathology. Plumb (1997) reviewed tilapia diseases in the Americas, emphasizing stressful environments. ISTAs I, II and III had, respectively, 66, 99 and 81 contributed papers, with only two, two and none on diseases (Fishelson & Yaron, 1983; Pullin *et al.*, 1988, 1996).

By 2000, however, the disease-free honeymoon was over. A Web of Science search, for which I thank David Little, found the following counts of articles citing tilapia and disease and their percentage contributions to all tilapia articles: 1990–2000, 442 (11.5%); 2000–2010, 1066 (15.9%); 2010–2020, 3675 (24.8%). Wendover and Aguirre (2011) noted four major bacterial tilapia pathogens (*Streptococcus agalactiae*, *Streptococcus iniae*, *Flavobacterium columnare* and *Francisella* spp.), with *Streptococcus* spp. accounting for over half of all tilapia bacterial diseases and countermeasures. Sood *et al.* (2020b), described serious infections of *O. niloticus* with *Shewanella putrefaciens* in India.

TiLV is currently the most serious tilapia disease and was the subject of a Global Special Alert (FAO, 2017). Fertilized eggs from infected tilapia broodstock have tested positive for TiLV (Dong *et al.*, 2020). Coinfections of tilapias with TiLV and bacterial pathogens can be more serious than single infections with either: for example, TiLV with *Aeromonas hydrophila* (Nicholson *et al.*, 2020) and with *Aeromonas veronii* (Amal *et al.*, 2018). Sood *et al.* (2020a) reviewed the history of TiLV, noting the lack of treatments and vaccines and placing best hopes on effective biosecurity and breeding for genetic resistance.

The welfare of farmed fish has particular importance in the context of controlling diseases and will figure increasingly in research, codes of conduct and regulations. The general reviews by Ashley (2007) and Segner *et al.* (2019) have much of relevance to the welfare of farmed tilapias. Training manuals tailored to local circumstances are particularly important, as tilapia husbandry advances and disease risks and countermeasures evolve (e.g. Din & Subasinghe, 2017).

Tilapia ecology

Natural distribution

Tilapias are native to Africa and the Levant and occupy diverse warm-water habitats, including rivers, lakes, estuaries and coastal wetlands, typically at depths of less than 15m and up to 30–40m in some lakes (Philippart & Ruwet, 1982; Trewavas, 1983). Some tilapias have wide native ranges; for example, *S. galilaeus* and *O. niloticus* are found from Sénégal to Egypt. Others have discontinuous or narrow ranges, reflecting complex histories of floods and droughts, desertification, seismic shifts and volcanism. For example, *C. zillii* has a relict type of distribution across sub-Saharan Africa, with scattered populations further north, including Israel (Thys van den Audenaerde, 1988).

Tilapia speciation has been allopatric through geographical isolation and/or sympatric through 'ecological diversification' in the same water body (Trewavas, 1982). Many allopatric tilapia species and subspecies are native to only one or a few rivers or parts of rivers. For example, *O. urolepis urolepis* and *O. urolepis hornorum* are native only to Tanzania's Rufigi and Wami rivers, respectively. Some allopatric species can hybridize when they meet: for example, *O. mortimeri* and *O. mossambicus* in the lower Athi (Thys van den Audenaerde, 1988). Wild populations of sympatric tilapias do not hybridize under natural conditions: for example, *O. aureus* and *O. niloticus* in Egypt (Payne and Collinson, 1983).

Environmental tolerances

The wide environmental tolerance limits of tilapias have enabled tilapia farming throughout much of the tropics and subtropics, using ponds, cages, tanks and raceways. The health and performance of any tilapia in any given environment are affected by DO, temperature, salinity and their interactions. DO decreases with increasing temperature and elevated salinities increase osmotic work. De Silva *et al.* (1986) found extremely high oxygen uptake in *O. niloticus* larvae at 17 ppt salinity. Tine *et al.* (2010) showed that wild *S. melanotheron* from estuarine habitats overexpressed the heat shock protein Hsp70, which was taken as an indicator of 'chronic hyper-osmotic stress . . . linked to impaired growth performance'.

All tilapias are thermophilic. Low temperatures limit their geographical ranges and sites for viable farming. Shelton and Popma (2006) give 28–35°C as optimal for tilapia growth, with reproduction only above 20°C and optimal at 25–30°C for *O. niloticus. O. aureus* and *O. niloticus* × *O. aureus* hybrids tolerate slightly lower temperatures than *O. niloticus*. Philippart and Ruwet (1982), citing multiple authors, described most tilapias as eurythermal and some as stenothermal; for example, *C. zillii* survives at 6.5–42.5°C and *O. niloticus* at 8–42°C, whereas *S. melanotheron* inhabits West African lagoons at 18–33°C.

Philippart and Ruwet (1982) also described most tilapias as euryhaline. Tilapias that are normally found in freshwater, including *O. andersonii*, *O. macrochir* and *C. rendalli*, can tolerate salinities up to 20 ppt. Some *O. niloticus* populations tolerate up to 30 ppt. *O. aureus* tolerates up to 45 ppt and *O. mossambicus* can adapt to rapid salinity changes from zero to 120 ppt. For farming in seawater, Watanabe *et al.* (2006) emphasized the suitability of the Florida Red hybrid and discussed the potential of other tilapias.

Balarin and Hatton (1979) recognized the paramount importance of adequate DO for appetency and growth and listed low

DO tolerance limits of about 1.0 mg/L for seven *Oreochromis* spp., with survival for short periods down to 0.1 mg/L, and at least 2.0–3.0 mg/L recommended on farms. Ross and Ross (1983) advised that DO be kept above 3.0 mg/L for farmed *O. niloticus* and found that handling increased their respiratory rate by 150–300% above normal. Tilapias gulp air at the water surface and respire through the buccal cavity when DO becomes critical.

The gill-oxygen limitation theory (GOLT) (Pauly, 2021) holds that fish growth and reproduction are controlled by internal oxygen levels, set by gill surface area relative to fish size. Kolding *et al.* (2008) found that the GOLT explained variations in growth, maturation size, egg size and fecundity in *O. niloticus*. Environmental stressors increase oxygen consumption, constrain growth and ultimately bring on reproduction (Amarasinghe & Pauly, 2021).

In theory, genetic improvement of growth in farmed tilapias could be sought by selection for greater gill surface area, thinner gill epithelia, haemoglobins saturating at lower oxygen tensions, etc. Selection for quiet behaviour would be more feasible, as in the domestication of livestock. Selection for growth per se can also be fortuitous selection for quiet behaviour. For example, Bozynski (1998) found that 'fast-growing GIFT fish performed less locomotory activity and agonistic behaviour than slow-growing control fish'.

As alien species and farmed types

Tilapias have been spread widely beyond their native ranges, as alien species and farmed types, including genetically improved strains and hybrids, and have established feral populations that have impacts on native species and their habitats. Farmed tilapia hybrids are usually fertile (Bartley *et al.*, 2001). Tilapias escape from farms and can establish feral populations, some of which hybridize with wild populations. Tilapia introductions can also spread pathogens and parasites.

All tilapias are potentially invasive as alien species and those with very wide adaptability, such as *O. mossambicus* and *C. zillii*, have the worst track records (Pullin *et al.*, 1997; Canonico *et al.*,

2005). De Silva *et al.* (2004) concluded that tilapias, including *O. niloticus*, have had no major adverse impacts on biodiversity in Asia and the Pacific but recognized the difficulty of determining whether an introduced tilapia had been the primary cause of biodiversity decline in a water body that had experienced other human interventions and consequent ecological change: for example, Lake Buhi, Philippines (Gindelberger, 1981).

In Africa, the genetic diversity and abundance of wild tilapias have long been subject to modification and loss by introductions of alien tilapias for inland fisheries and farming. Lowe-McConnell (1987) attributed the precipitous declines of *O. esculentus* and *O. variabilis* in Lake Victoria to their displacement by introduced *O. niloticus*, *O. leucostictus* and *C. zillii*. Thys van den Audenaerde (1988) warned that 'aquaculturists are undoing the work of natural evolution by transfers and bringing species together'. During the collection of wild *O. niloticus* in Sénégal for the GIFT project, Dirk Thys van den Audenaerde told me that some fish around us looked like hybrids with introduced *O. aureus*. Introductions for tilapia farming are continuing to pose huge risks to wild tilapias and their habitats.

Tilapia farmers everywhere require farmed types that perform well, and it is not surprising that some introductions of alien tilapia species and farmed types have been made unofficially. GIFT *O. niloticus* are now in the lower Volta Basin and some contain *O. mossambicus* genes (Anane-Taabeah *et al.*, 2019) – probably from the Philippine *O. niloticus* farmed type in the GIFT base population (Taniguchi *et al.*, 1985). In Tanzania, important wild populations of *O. korogwe* are threatened by hybridization with *O. niloticus* (Blackwell *et al.*, 2020). However, hybridization between wild and introduced tilapias is sometimes limited and wild refugia can be identified as, for example, in Mozambique for *O. mossambicus* (Firmat *et al.*, 2013).

The challenge is to reconcile tilapia farming with conservation of biodiversity, especially wild tilapias. Tilapia wild types will be needed by breeders for pursuing genetic resistance to diseases. Thorough risk analysis prior to tilapia introductions is essential, as outlined by Lind *et al.* (2015) with a worked example of a hypothetical introduction of GIFT *O. niloticus* to Zambia. 'Twinning', which is the combined planning, governance and financing of aquaculture and biodiversity conservation, with the latter isolated from farmed fish and farm waters, is recommended, as endorsed by FAO (Bartley *et al.*, 2006; FAO, 2008; Pullin, 2008). Disease control is an added reason for recommending it.

Tilapias as food fish

Tilapias from inland fisheries and aquaculture have contributed to food security and nutrition for millennia. From the 1940s, following its wide but largely unsuccessful introductions for aquaculture, somc feral *O. mossambicus* populations became important local sources of food fish: for example, in Sri Lanka (De Silva & Senaratne, 1988) and the Philippines and Indonesia (Costa-Pierce *et al.*, 1988). From the mid-1960s, *O. niloticus* farming, especially in Asia, established tilapia as a popular food fish (e.g. Dey *et al.*, 2000; Asian Development Bank, 2005b).

Young and Muir (2002) described tilapia as 'prized for the neutrality of its taste', but recognition of tilapias for their good taste also has a long history. Edgar Barton Worthington, recalling field trips on African lakes undertaken from 1927 to 1933 (Barton Worthington, 1983), once told me: 'We ate mostly tilapias . . . they were delicious and we never got tired of them.' *O. esculentus* and the Malawian tilapias *O. karongae*, *O. lidole* and *O. squamippinis* have particularly high reputations for taste.

Tilapias are often cooked and served whole. Their lack of intramuscular bones also enables easy production of boneless fillets. Fillets of *O. niloticus* and *Oreochromis* hybrids, including red tilapias, are used worldwide in home cooking and restaurants. On a KLM flight out of Malawi I was served 'Tilapia au Beurre Blanc' from a top chef's recipe. During an ICLARM project in Kuwait in the 1980s, *O. spilurus* fillets were served up unidentified at a prestigious dinner and later praised as delicious *hamoor* (grouper), the customary luxury fish.

Erroneous information is sometimes spread about alleged hazards of eating tilapias (e.g. www.twistedfood.co.uk) and negative opinions can arise where tilapias are regarded as nuisance species. I was told that a comment in Australia about tilapias as food fish received the following response: 'Tilapia? You can't eat them. They're vermin!'

In truth, tilapias contribute much to global fish supply, responsibly and sustainably, and their reputation is growing. I once checked in my fishing rods at London City airport and the baggage handler, who came from Trinidad, said: 'Ah, you are going fishing! My best fish is tilapia. Do you know tilapia?' 'Yes, but I need to know more.'

Information sources

Important sources of information on tilapias include the following, among many: Fryer and Iles (1972); Balarin and Hatton (1979); Pullin and Lowe-McConnell (1982); Trewavas (1983); Lowe-McConnell (1987); Keenleyside (1991); Beveridge and McAndrew (2000); and the proceedings of ISTA, held about every 3 years since ISTA I (Fishelson & Yaron, 1983). Ro Lowe-McConnell's tilapia field guide, based on Trewavas (1983), was published by ICLARM (Pullin, 1988). Her autobiography (Lowe-McConnell, 2006) is based on many tilapia field studies and includes a review of tilapia farming. Shelton and Popma (2006) reviewed tilapia biology at length in the wide-ranging volume compiled by Lim and Webster (2006). Guerrero and Guerrero (2020) have related tilapia biology to current farming practices. The manual by Romana-Eguia *et al.* (2020) is also recommended.

Most of the ICLARM and World Fish publications on tilapias cited here are freely available online from the World Fish Center (www.worldfishcenter.org). Extensive free information on tilapias is also available from many other sources, among which the following have particular importance and/or lead to others: FAO (www.fao.org); NACA (www.enaca.org); FishBase (www.fishbase.org); national, regional and international tilapia research networks and farming associations; and general help sites (e.g. www.thefishsite.

com). Among FAO's many tilapia publications, the review of aquatic genetic resources (FAO, 2019b) is particularly important as it reflects country reports that include tilapias as farmed types and wild relatives, with further updating expected (http://www.fao.org/aquatic-genetic-resources/activities/sow/countryreports/en/).

Acknowledgements

Among the many who provided help for this chapter, I thank in particular Devin Bartley, Barry Costa-Pierce, Peter Edwards, David Little, Graham Mair, Daniel Pauly and Warren Turner. I thank also everyone who taught me about the tilapias over the past 40 years.

References

Agha, S., Mekkawy, W., Ibanez-Escriche, N., Lind, C.E., Kumar, J. *et al.* (2018) Breeding for robustness: Investigating the genotype-by-environment interaction and micro-environmental sensitivity of genetically improved farmed Tilapia (*Oreochromis niloticus*). *Animal Genetics* 49(5), 421–427. DOI: 10.1111/age.12680.

Amal, M.N.A., Koh, C.B., Nurliyana, M., Suhaiba, M., Nor-Amalina, Z. *et al.* (2018) A case of natural co-infection of Tilapia lake virus and *Aeromonas veronii* in a *Malaysian* red hybrid Tilapia (*Oreochromis niloticus* x *O. mossambicus*) farm experiencing high mortality. *Aquaculture* 485, 12–16. DOI: 10.1016/j.aquaculture.2017.11.019.

Amarasinghe, U.S. and Pauly, D. (2021) The relationship between size at maturity and maximum size in cichlid populations corroborates the gill-oxygen limitation theory (GOLT). *Asian Fisheries Science* 34(1), 14–22. DOI: 10.33997/j.afs.2021.34.1.002.

Anane-Taabeah, G., Frimpong, E.A. and Hallerman, E. (2019) Aquaculture-mediated invasion of the genetically improved Tilapia (Gift) into the lower Volta Basin of Ghana. *Diversity* 11(10), 188–210. DOI: 10.3390/d11100188.

Ashley, P.J. (2007) Fish welfare: Current issues in aquaculture. *Applied Animal Behaviour Science* 104(3–4), 199–235. DOI: 10.1016/j.applanim.2006.09.001.

Asian Development Bank (2005a) An Impact Evaluation of the Development of Genetically Improved Farmed Tilapia and Their Dissemination in Selected Countries. Operations Evaluation Department, Asian Development Bank. Available at: http://www.adb.org

Asian Development Bank (2005b) An Evaluation of Small-Scale Freshwater Rural Aquaculture Development for Poverty Reduction. Operations Evaluation Department, Asian Development Bank. Available at: http://www.adb.org

Balarin, J.D. and Hatton, J.P. (1979) *Tilapia. A Guide to Their Biology & Culture in Africa.* Unit of Aquatic Pathobiology, University of Stirling, Scotland.

Barría, A., Trinh, T.Q., Mahmuddin, M., Benzie, J.A.H., Chadag, V.M. *et al.* (2020) Genetic parameters for resistance to Tilapia Lake Virus (TiLV) in Nile tilapia (*Oreochromis niloticus*). *Aquaculture 522*, Article 735126 . DOI: 10.1016/j.aquaculture.2020.735126.

Bartley, D.M., Rana, K. and Immink, A.J. (2001) The use of interspecific hybrids in aquaculture and fisheries. *Reviews in Fish Biology and Fisheries* 10, 325–337.

Bartley, D.M., Harvey, B.J. and Pullin, R.S.V. (eds) (2006) Workshop on status and trends in aquatic genetic resources: a basis for international policy. In: *FAO Fisheries Proceedings, May 8–10, 2006, Victoria, British Columbia, Canada.* Rome, Italy, p. 16 (Vol. 5).

Barton Worthington, E. (1983) *The Ecological Century: A Personal Appraisal.* Clarendon Press, Oxford, pp. 5–27.

Beardmore, J.A., Mair, G.C. and Lewis, R.I. (2001) Monosex male production in finfish as exemplified by tilapia: applications, problems, and prospects. *Aquaculture* 197(1–4), 283–301. DOI: 10.1016/S0044-8486(01)00590-7.

Beveridge, M.C.M. and McAndrew, B.J. (eds) (2000) *Tilapias: Biology and Exploitation*, Kluwer Academic Publishers, Dordrecht, Netherlands.

Bhujel, R.C. (2000) A review of strategies for the management of Nile tilapia (*Oreochromis niloticus*) broodfish in seed production systems, especially hapa-based systems. *Aquaculture* 181(1–2), 37–59. DOI: 10.1016/S0044-8486(99)00217-3.

Blackwell, T., Ford, A.G.P., Ciezarek, A.G., Bradbeer, S.J., Gracida Juarez, C.A. *et al.* (2020) Newly discovered cichlid fish biodiversity threatened by hybridization with non-native species. *Molecular Ecology* 30(4), 895–911. DOI: 10.1111/mec.15638.

Bowen, S.H. (1980) Detrital nonprotein amino acids are the key to rapid growth of Tilapia in Lake Valencia, Venezuela. *Science* 207(4436), 1216–1218. DOI: 10.1126/science.207.4436.1216.

Bowen, S.H. (1981) Digestion and assimilation of periphytic detrital aggregate by *Tilapia mossambica. Transactions of the American Fisheries Society* 110(2), 239–245. DOI: 10.1577/1548-8659(1981)1102.0.CO;2.

Bowen, S.H. (1982) Feeding, digestion and growth – Qualitative considerations. In: Pullin, R.S.V. and Lowe-McConnell, R.H. (eds) *The Biology and Culture of Tilapias. ICLARM Conference Proceedings, Vol. 7.* International Center for Living Aquatic Resources Management, Manila, Philippines, pp. 141–156.

Bozynski, C.C. (1998) Interactions between growth, reproduction, and activity levels in control and fast-growing strains of Nile tilapia *Oreochromis niloticus*. [MSc Thesis], University of British Columbia.

Canonico, G.C., Arthington, A., McCrary, J.K. and Thieme, M.L. (2005) The effects of introduced tilapias on native biodiversity. *Aquatic Conservation: Marine and Freshwater Ecosystems* 15(5), 463–483.

Cnaani, A. (2013) The tilapias' chromosomes influencing sex determination. *Cytogenetics and Genomic Research* 141(2–3), 195–205. DOI: 10.1159/000355304.

Costa-Pierce, B. A., Zainal, S., & Effendi, P. (1988). ICLARM and South–South Technology Transfer: Philippine Aquaculture Technology in Indonesia. *Naga. The ICLARM Quarterly* 11(4), pp. 10–11.

Coward, K. and Bromage, N.R. (2000) Reproductive physiology of female tilapia broodstock. *Reviews in Fish Biology and Fisheries* 10, 1–25. DOI: 10.1023/A:1008942318272.

De Silva, C.D., Premawansa, S. and Keembiyahetty, C.N. (1986) Oxygen consumption in *Oreochromis niloticus* (L.) in relation to salinity, temperature and time of day. *Journal of Fish Biology* 29(2), 267–277. DOI: 10.1111/j.1095-8649.1986.tb04944.x.

De Silva, S.S. and Senaratne, K.A.D.W. (1988) *Oreochromis mossambicus* is not universally a nuisance species: The Sri Lankan experience. In: Pullin, R.S.V., Bhukaswan, T., Tonguthai, K. and Maclean, J.L. (eds) *ICLARM Conference Proceedings 15. The Second International*

Symposium on Tilapia in Aquaculture, International Center for Living Aquatic Resources Management, Manila, Philippines, pp. 445–450.

De Silva, S.S., Subasinghe, R.P., Bartley, D.M. and Lowther, A. (2004) Tilapias as alien aquatics in Asia and the Pacific. *FAO Fisheries Technical Paper* 453. Rome.

de Verdal, H., Vandeputte, M., Mekkawy, W., Chatain, B. and Benzie, J.A.H. (2018) Quantifying the genetic parameters of feed efficiency in juvenile Nile tilapia *Oreochromis niloticus*. *BMC Genetics* 19(1), 105. DOI: 10.1186/s12863-018-0691-y.

Dey, M.M., Bimbao, G.P., Young, L., Regaspi, P., Kohinoor, A.H.M. *et al.* (2000) Current status of production and consumption of tilapia in selected Asian countries. *Aquaculture, Economics and Management* 4(1–2), 13–30.

Din, M. and Subasinghe, R. (2017) Basic Biosecurity Manual for Tilapia Hatchery Technicians in Bangladesh. CGIAR Research Program on Fish Agri-Food Systems, FISH-2017–FISH-2010 pp.

Dong, H.T., Senapin, S., Gangnonngiw, W., Nguyen, V.V., Rodkhum, C. *et al.* (2020) Experimental infection reveals transmission of Tilapia Lake Virus (TiLV) from Tilapia broodstock to their reproductive organs and fertilized eggs. *Aquaculture* 515, article 734541. DOI: 10.1016/j.aquaculture.2019.734541.

FAO (2008) Aquaculture development 5. Genetic resource management. FAO technical guidelines for responsible fisheries No. 5, Suppl. 3, Rome.

FAO (2017) Outbreaks of Tilapia Lake Virus (TiLV) threaten the livelihoods and food security of millions of people dependent on Tilapia farming. *Global Information and Early Warning System. Region: Global. Special Alert* No. 338. Rome.

FAO (2019a) *FAO yearbook. Fisheries and aquaculture statistics, 2017.* Table B12. Rome.

FAO (2019b) *The state of the world's aquatic genetic resources for food and agriculture.* Food and Agriculture Organization Commission on Genetic Resources for Food and Agriculture.

Firmat, C., Alibert, P., Losseau, M., Baroiller, J.F. and Schliewen, U.K. (2013) Successive invasion-mediated interspecific hybridizations and population structure in the endangered cichlid *Oreochromis mossambicus*. *PloS One* 8(5), e63880. DOI: 10.1371/journal.pone.0063880.

Fishelson, L. and Yaron, Z. (eds) (1983) In: *International Symposium on Tilapia in Aquaculture, Nazareth, Israel, May 8–13, 1983*, Tel Aviv University, Israel.

Froese, R. and Pauly, D. (eds) (2018) FishBase. World Wide Web electronic publication. Available at: http://www.fishbase.org

Fryer, G. and Iles, T.D. (1972) *The Cichlid Fishes of the Great Lakes of Africa: Their Biology and Evolution.* Oliver & Boyd, Edinburgh.

Gerking, S.D. (1994) *Feeding Ecology of Fish.* Academic Press, New York.

Gindelberger, B. (1981) Why sinarapan almost disappeared from Lake Buhi. ICLARM Newsletter, 4, 3–5.

Guerrero, R.D.III. and Guerrero, L.A. (2020) *Nile Tilapia Farming: Scientific Studies, Tested Technologies, Profitable Practices.* Kindle Edition.

Hamilton, M.G., Lind, C.E., Barman, B.K., Velasco, R.R., Danting, M.J.C. *et al.* (2020) Distinguishing between Nile tilapia strains using a low-density single-nucleotide polymorphism panel. *Frontiers in Genetics* 11, 594722. DOI: 10.3389/fgene.2020.594722.

Harkell, L. (2020) Global tilapia production positive despite pandemic. *Undercurrent News*, October 5, 2020.

Hickling, C.L. (1963) The cultivation of Tilapia. *Scientific American*, 208, 143–153.

Jalabert, B. and Zohar, Y. (1982) Reproductive physiology in cichlid fishes, with particular reference to *Tilapia* and *Sarotherodon.* In: Pullin, R.S.V. and Lowe-McConnell, R.H. (eds) *The Biology and Culture of Tilapias. ICLARM Conference Proceedings*, 7, International Center for Living Aquatic Resources Management, Manila, Philippines, pp. 129–140.

Keenleyside, M.H.A. (1991) *Cichlid Fishes: Behaviour, Ecology and Evolution.* Chapman & Hall, London.

Khaw, H.L., Ponzoni, R.W. and Danting, M.J.C. (2007) Estimation of genetic change in the GIFT strain by comparing contemporary progeny produced by males born in 1991 or 2003. *Proceedings of the Association for the Advancement of Animal Breeding and Genetics* 17, 111–114.

Kolding, J., Haug, L. and Stefansson, S. (2008) Effect of ambient oxygen on growth and reproduction in Nile tilapia (*Oreochromis niloticus*). *Canadian Journal of Fisheries and Aquatic Sciences* 65(7), 1413–1424. DOI: 10.1139/F08-059.

Komen, H. and Benzie, J. (2015) Selective breeding in Nile tilapia: Achievements and future directions. In: *Proceedings of the International Symposium on Genetics in Aquaculture XII*, Santiago de Compostela, Spain, p. 38.

Lim, C. and Webster, C.D. (2006) *Tilapia. Biology, culture, and nutrition.* Food Products Press. An Imprint of the Haworth Press, New York.

Lind, C.E., Dana, G.V., Perera, R.P. and Phillips, M.J. (2015) Risk analysis in aquaculture: A step-by-step introduction with worked examples. WorldFish, Penang, Malaysia.

Lind, C.E., Kilian, A. and Benzie, J.A.H. (2017) Development of diversity array technology markers as a tool for rapid genomic assessment in Nile tilapia, *Oreochromis niloticus*. *Animal Genetics* 48(3), 362–364. DOI: 10.1111/age.12536.

Little, D.C. and Hulata, G. (2000) Strategies for tilapia seed production. In: Beveridge, M.C.M. and McAndrew, B.J. (eds) *Tilapias: Biology and Exploitation*. Kluwer Academic Publishers, pp. 267–326.

Little, D.C., Macintosh, D.J. and Edwards, P. (1993) Improving spawning synchrony in the Nile tilapia, *Oreochromis niloticus* (L.). *Aquaculture Research* 24(3), 399–405. DOI: 10.1111/j.1365-2109.1993.tb00563.x.

Lowe-McConnell, R.H. (1987) *Ecological Studies in Fish Communities*. Cambridge University Press, pp. 100 and 302 .

Lowe-McConnell, R. (2006) *The Tilapia Trail – The Life Story of a Fish Biologist*. MPM Publishing, Ascot, UK.

Macintosh, D.J. (2008) Risks associated with using methyl testosterone in Tilapia farming. Available at: https://www.sustainablefish.org/Publications?year=2008–01–01T00:00:00Z

Mair, G.C. and Lucente, D. (2020) What are "Farmed types" in aquaculture and why do they matter?. *FAO Aquaculture Newsletter* 61, 40–42.

Martinez-i-Prat, A.-R. (1995) Fishing out aquatic diversity. *Seedling, the Quarterly Newsletter of Genetic Resources Action International* 12(2), 2–13.

Moreau, J., Bambino, C. and Pauly, D. Maclean, J.L., Dizon, L.B. and Hosillos, L.V. (eds) (1986) Indices for overall growth performance of 100 tilapia (Cichlidae) populations. In: Maclean, J.L., Dizon, L.B. and Hosillos, L.V. (eds) *The First Asian Fisheries Forum*. Asian Fisheries Society, Manila, Philippines, pp. 201–206.

Moriarty, D.J.W. and Moriarty, C.M. (1973) The assimilation of carbon from phytoplankton by the two herbivorous fishes: *Tilapia nilotica* and *Haplochromis nigripinnis*. *Journal of Zoology* 171, 41–56.

Nicholson, P., Mon-on, N., Jaemwimol, P., Tattiyapong, P. and Surachetpong, W. (2020) Coinfection of tilapia lake virus and *Aeromonas hydrophila* synergistically increased mortality and worsened the disease severity in Tilapia (*Oreochromis* spp.). *Aquaculture* 520, article 734746. DOI: 10.1016/j.aquaculture.2019.734746.

Noakes, D.L.G. and Balon, E.K. (1982) Life histories of tilapias. In: Pullin, R.S.V. and Lowe-McConnell, R.H. (eds) *The Biology and Culture of Tilapias. ICLARM Conference Proceedings 7*, International Center for Living Aquatic Resources Management, Manila, Philippines, pp. 61–82.

Obrero, T. and Racaza, J.L. (2020) Genetic development of Nile tilapia strains in the Philippines for sustainable and efficient farming. *World Aquaculture Communications* 51(4), 60–65.

Paperna, I., Van As, J.G. and Basson, L. (1983) Review of diseases affecting cultured cichlids. In: Fishelson, L. and Yaron, Z. (eds) *International Symposium on Tilapia in Aquaculture, Nazareth, Israel*, Tel Aviv University, pp. 174–184.

Pauly, D. (1976) The biology, fishery and potential for aquaculture of *Tilapia Melanotheron* in a small West African lagoon. *Aquaculture* 7(1), 33–49. DOI: 10.1016/0044-8486(76)90030-2.

Pauly, D. (2021) The Gill-Oxygen Limitation Theory (GOLT) and its critics. *Science Advances* 7(2), eabc6050. DOI: 10.1126/sciadv.abc6050.

Pauly, D., Moreau, J. and Prein, M. (1988) A comparison of overall growth performance of tilapia in open waters and aquaculture. In: Pullin, R.S.V., Bhukaswan, T., Tonguthai, K. and Maclean, J.L. (eds) *The Second International Symposium on Tilapia in Aquaculture, ICLARM Conference Proceedings 15*, International Center for Living Aquatic Resources Management, Manila, Philippines, pp. 469–479.

Payne, A.I. and Collinson, R.I. (1983) A comparison of the biological characteristics of *Sarotherodon niloticus* (L.) with those of *S. aureus* (Steindachner) and other tilapia of the delta and lower Nile. *Aquaculture* 30(1–4), 335–351. DOI: 10.1016/0044-8486(83)90174-6.

Peters, H.M. (1983) Fecundity, egg weight, and oocyte development in tilapias (Cichlidae, Teleostei). *ICLARM Translations* 2. International Center for Living Aquatic Resources Management, Manila, Philippines. Translated and edited by D. Pauly from Peters, H.M. (1963) Eizal, Eigewicht und Gelegeentwicklung in der Gattung *Tilapia* (Cichlidae, Teleostei). *International Revue der gesampten Hydrobiologie und Hydrographie* 48, pp. 547-576.

Philippart, J.Cl. and Ruwet, J.Cl. (1982) Ecology and distribution of tilapias. In: Pullin, R.S.V. and Lowe-McConnell, R.H. (eds) *The Biology and Culture of Tilapias. ICLARM Conference Proceedings 7*, International Center for Living Aquatic Resources Management, Manila, Philippines, pp. 15–59.

Plumb, J.A. (1997) Infectious diseases of tilapias. In: Costa-Pierce, B.A. and Rackocy, J.E. (eds) *Tilapia Aquaculture in the Americas, I.* World Aquaculture Society, pp. 212–228.

Ponzoni, R.W., Nguyen, N.H., Khaw, H.L., Kamaruzzaman, N., Hamzah, A. *et al.* (2008) Genetic improvement of Nile tilapia (*Oreochromis niloticus*) – present and future. In: *Proceedings of the 8th International Symposium on Tilapia in Aquaculture*, pp. 33–52.

Prein, M., Hulata, G., & Pauly, D. (Eds.). (1993) *Multivariate methods in aquaculture research: Case studies of tilapias in experimental and commercial systems.* ICLARM studies and Reviews 20. International Center for Living Aquatic Resources Management

Pullin, R.S.V. (1983) Choice of tilapia species for aquaculture. In: Fishelson, L. and Yaron, Z. (eds) *International Symposium on Tilapia in Aquaculture, Nazareth, Israel,* Tel Aviv University, pp. 64–76.

Pullin, R.S.V. (ed.) (1988) *Tilapia Genetic Resources for Aquaculture. ICLARM Conference Proceedings 16,* International Center for Living Aquatic Resources Management, Manila, Philippines. (Also available in French.)

Pullin, R.S.V. (2008) Aquaculture and conservation of fish genetic resources: twinning objectives and opportunities. In: Bartley, D.M. (ed.) *Workshop on Pioneering Fish Genetic Resource Management and Seed Dissemination Programs for Africa. CIFA Occasional Paper* No. 29. Food and Agriculture Organization, Rome, pp. 24–29.

Pullin, R.S.V. (2017) Diversification in aquaculture: Species, farmed types and culture systems. In: Harvey, B., Soto, D., Carolsfeld, J., Beveridge, M. and Bartley, D.M. (eds) *Planning for Aquaculture Diversification: The Importance of Climate Change and Other Drivers,* 15–36. FAO Technical Workshop, 23–25 June 2016. *FAO Fisheries and Aquaculture Proceedings* No. 47, Rome, pp. 15–36.

Pullin, R.S.V. and Lowe-McConnell, R.H. (eds) (1982) The Biology and Culture of Tilapias. In: *ICLARM Conference Proceedings 7,* International Center for Living Aquatic Resources Management, Manila, Philippines.

Pullin, R.S.V. and Maclean, J.L. (1992) Analysis of research for the development of tilapia farming: An interdisciplinary approach is lacking. *Netherlands Journal of Zoology* 42(2–3), 512–522.

Pullin, R.S.V., Bhukaswan, T., Tonguthai, K. and Maclean, J.L. (eds) (1988) In: *The Second International Symposium on Tilapia in Aquaculture, ICLARM Conference Proceedings 15,* International Center for Living Aquatic Resources Management, Manila, Philippines.

Pullin, R.S.V., Lazard, J., Legendre, M., Amon-Kothias, J.B. and Pauly, D. (eds) (1996) In: *The Third International Symposium on Tilapia in Aquaculture, ICLARM Conference Proceedings 41,* International Center for Living Aquatic Resources Management, Manila, Philippines. (Also available in French.)

Pullin, R.S.V., Palomares, M.L., Casal, C.M.V., Dey, M.M. and Pauly, D. (1997) Environmental impacts of tilapias. In: Fitzsimmons, K. (ed.) *Proceedings of the from the Fourth International Symposium on Tilapia in Aquaculture, 2,* Northeast Regional Agricultural Engineering Service (NREAS), Cooperative Extension. Ithaca, New York, pp. 554–570.

Rana, K. (1988) Reproductive biology and the hatchery rearing of tilapia eggs and fry. In: Muir, J. and Roberts, R.J. (eds) *Recent Advances in Aquaculture.* Springer, pp. 343–406.

Roberts, R.J. and Sommerville, C. (1982) Diseases of tilapias. In: Pullin, R.S.V. and Lowe-McConnell, R.H. (eds) *The Biology and Culture of Tilapias. ICLARM Conference Proceedings 7,* International Center for Living Aquatic Resources Management, Manila, Philippines, pp. 247–263.

Romana-Eguia, M.R., Eguia, R.V. and Pakingking, R.V., Jr (2020) *Tilapia culture: The Basics. Aquaculture Culture Extension Manual No. 66. July 2020.* Aquaculture Department. Available at: http://www.seafdec.org.ph

Ross, B. and Ross, L.G. (1983) The oxygen requirements of *Oreochromis niloticus* under adverse conditions. In: Fishelson, L. and Yaron, Z. (eds) *International Symposium on Tilapia in Aquaculture, Nazareth, Israel, May 8–13, 1983,* Tel Aviv University, pp. 134–143.

Segner, H., Reiser, S., Ruane, N., Rösch, R., Steinhagen, D. *et al.* (2019) *Welfare of Fishes in Aquaculture. FAO Fisheries and Aquaculture Circular no. 1189.* Budapest.

Shelton, W.L. and Popma, T.J. (2006) Biology. In: Lim, C.E. and Webster, C.D. (eds) *Tilapia: Biology, Culture, and Nutrition.* Food Products Press. An Imprint of the Haworth Press, New York, pp. 1–49.

Shen, Y.B. and Yue, G.H. (2019) Current status of research on aquaculture genetics and genomics information from ISGA XIII. *Aquaculture and Fisheries* 4(2), 43–47. DOI: 10.1016/j.aaf.2018.11.001.

Sood, N., Pradhan, P., Ravindra, V., Yadav, D.K., Mishra, M.K. *et al.* (2020a) Tilapia lake virus: The story so far. *Journal of Fish Diseases* 43(10), 1115–1132. DOI: 10.1111/jfd.13237.

Sood, N., Pradhan, P.K., Ravindra, Verma, D.K., Yadav, M.K., Mishra, R.K. *et al.* (2020b) Large-scale mortality in cultured tilapia

Oreochromis niloticus due to infection with *Shewanella putrefaciens* in India. *Journal of the World Aquaculture Society* 51(2), 563–570. DOI: 10.1111/jwas.12686.

Taniguchi, N., Macaranas, J.M. and Pullin, R.S.V. (1985) Introgressive hybridization in cultured tilapia stocks in the Philippines. *Bulletin of the Japanese Society of Scientific Fisheries* 51(8), 1219–1224. DOI: 10.2331/suisan.51.1219.

Taslima, K., Wehner, S., Taggart, J.B., de Verdal, H., Benzie, J.A.H. *et al.* (2020) Sex determination in the GIFT strain of tilapia is controlled by a locus in linkage group 23. *BMC Genetics* 21(1), 49. DOI: 10.1186/s12863-020-00853-3.

Thys van den, D.F.E. (1968) An Annotated Bibliography of *Tilapia* (Pisces: Cichlidae). *Musée Royale de l'Afrique Centrale Documentation Zoologique* No 14, Tervuren, Belgium.

Thys van den Audenaerde, D.F.E. (1988) Natural distribution of tilapias and its consequences for the possible protection of genetic resources. In: Pullin, R.S.V. (ed.) *Tilapia genetic resources for aquaculture. ICLARM Conference Proceedings 16,* International Center for Living Aquatic Resources Management, Manila, Philippines, pp. 1–12.

Tine, M., Bonhomme, F., McKenzie, D.J. and Durand, J.D. (2010) Differential expression of the heat shock protein Hsp70 in natural populations of the tilapia *Sarotherodon melanotheron,* acclimatized to a range of environmental salinities. *BMC Ecology* 10, 11. DOI: 10.1186/1472-6785-10-11.

Trewavas, E. (1982) Tilapia: Taxonomy and speciation. In: Pullin, R.S.V. and Lowe-McConnell, R.H. (eds) *The Biology and Culture of Tilapias. ICLARM Conference Proceedings 7,* International Center for Living Aquatic Resources Management, Manila, Philippines, pp. 3–13.

Trewavas, E. (1983) *Tilapiine Fishes of the Genera Sarotherodon, Oreochromis and Danakilia.* British Museum (Natural History), London.

Tuan, P.A., Mair, G.C., Little, D.C. and Beardmore, J.A. (1999) Sex determination and the feasibility of genetically male tilapia production in the Thai-Chitralada strain of *Oreochromis niloticus* (L.). *Aquaculture* 173(1–4), 257–269. DOI: 10.1016/S0044-8486(98)00450-5.

Van Bers, N.E.M., Crooijmans, R.P.M.A., Groenen, M.A.M., Dibbits, B.W. and Komen, J. (2012) SNP marker detection and genotyping in tilapia. *Molecular Ecology Resources* 12(5), 932–941. DOI: 10.1111/j.1755-0998.2012.03144.x.

Watanabe, W.O., Fitzsimmons, K. and Yi, Y. (2006) Farming tilapia in saline waters. In: Lim, C.E. and Webster, C.D. (eds) *Tilapia. Biology, Culture, and Nutrition*. Food Products Press. An Imprint of the Haworth Press, New York, pp. 347–447.

Wendover, N. and Aguirre, M. (2011) *Streptococcus* in tilapia: vaccine development, field experiences in Asia. *Global Aquaculture Advocate*, September/October 2011, 58–60.

Wilkins, N.P. and Gosling, E.M. (eds) (1983) Genetics in aquaculture. In: *Proceedings of an International Symposium held in University College, Galway, Ireland. Aquaculture 33, 20 March to 2 April 1982, v–x, 1–426.*

Wohlfarth, G.W. (1994) The unexploited potential of tilapia hybrids in aquaculture. *Aquaculture Research* 25(8), 781–788. DOI: 10.1111/j.1365-2109.1994.tb00743.x.

Wohlfarth, G.W. and Hulata, G.I. (1981) Applied Genetics of Tilapias. *ICLARM studies and Reviews 6*. International Center for Living Aquatic Resources Management, Manila, Philippines.

World Fish (2004) *GIFT technology manual: An Aid to Tilapia Selective Breeding*. World Fish Center, Penang, Malaysia.

Young, J.A. and Muir, J.F. (2002) Tilapia: Both fish and fowl?. *Marine Resource Economics* 17(2), 163–173. DOI: 10.1086/mre.17.2.42629359.

Chapter 2

Tilapia as an aquaculture species

Junning Cai

Introduction

The world fisheries and aquaculture production of tilapias and other cichlids (tilapias in short) increased 100 times, from 70,000 to 7 million tonnes between 1950 and 2018, with the share of farmed tilapias increasing from 8 to 88% (Food and Agriculture Organization (FAO), 2020 a).

Tilapias are one of the most popular aquaculture species, cultivated in over 120 countries (including non-sovereign territories). In 2018, the 6 million tonnes of world tilapia aquaculture production made it the fourth-largest species group accounting for 5.27% of the production of all species, which was only lower than brown seaweeds (13.04%), red seaweeds (15.15%) and carps (25.52%) (FAO, 2020b). Tilapias were the fifth-largest species group in terms of value, with US$11 billion farm-gate value (4.26% of all species) following crayfish (5.49%), salmon/trout/smelts (8.65%), marine shrimps and prawns (14.6%) and carp (23.67%) (FAO, 2020b).

Tilapias are also a popular fish commodity, imported in over 140 countries in 2018 with US$1.5 billion of import value accounting for nearly 30% of the world import of freshwater finfish products and much higher than the US$256 million world import of carp products (FAO, 2020c).

This chapter provides an overview of tilapia aquaculture production, trade and consumption at the global, regional and national levels based primarily on official statistics. Unless specified otherwise: (1) tilapias referred to in this chapter broadly include all cichlid species in the family Cichlidae; (2) countries in this chapter

Kim Thompson, Craig Shoemaker and David C. Little (eds) **Tilapia**
DOI: 10.1079/9781800629455.0002, © CAB International, 2025

include non-sovereign territories; (3) product statistics are derived from the FAO Global Fishery and Aquaculture Production Statistics v2020.1.0, published through FishStatJ (FAO, 2020a); (4) trade statistics are primarily from the United Nations Comtrade database (UN, 2020), with some complementary statistics from the FAO Global Fishery and Aquaculture Commodities Statistics v2020.1.0 published through FishStatJ (FAO, 2020c) used to compare tilapia and other fish products in international trade.

Tilapia aquaculture production

World tilapia aquaculture production increased from 1.2 to 6.0 million tonnes between 2000 and 2018; the 9.4% annual growth was higher than that in the world aquaculture production of freshwater fish (5.5%), finfish (5.5%) and all species (5.9%). Accordingly, the share of tilapia in the world aquaculture production of freshwater fish increased from 6.8 to 13%; its share in finfish increased from 5.7 to 11%; and its share in all species from 2.8 to 5.3%.

Tilapia aquaculture production by region

The 9.4% annual growth in world tilapia aquaculture production between 2000 and 2018 was entirely attributed to developing regions whereas production in developed regions declined from 8665 to 6827 tonnes, which mostly reflected the situation in North America or, specifically, the USA (Table 2.1).

In 2018, tilapia aquaculture production in developing regions accounted for 99.89% of the world total, and the share of least developed countries has increased from 2.38 to 9.52% between 2000 and 2018 thanks to its much higher growth in tilapia aquaculture production (18.21% per year) compared with the world average of 9.44% (Table 2.1). In this regard we may say that tilapias tend to be a poor people's fish.

Landlocked developing countries accounted for 2.69% of the world's tilapia aquaculture production in 2018, and its 10.56% annual growth between 2000 and 2018 was slightly higher than

Table 2.1 Status and trends of tilapia farming in global and regional aquaculture.

Region	Tilapia aquaculture production, 2000–2018										Year 2018	
	Production (tonnes)				Share of world tilapia aquaculture production (%)			Share of each region's total aquaculture production (%)			Number of tilapia-farming countries in each region	Number of tilapia species farmed in each region
	2000	2018	±	Annual growth (%)	2000	2018	±	2000	2018	±		
World	1,189,926	6,031,432	4.841,506	9.44	100.00	100.00	0.00	2.77	5.2 7	2.50	124	17
Developed regions	8,665	6,827	−1,838	-1.32	0.73	0.11	-0.61	0.21	0.14	-0.08	9	2
Developing regions	1,181,261	6,024,605	4,843,344	9.47	99.27	99.89	0.61	3.03	5.50	2.47	115	17
Least developed countries	28,267	574,087	545,820	18.21	2.38	9.52	7.14	3.11	13.39	10.28	35	11
Landlocked developing countries	26,666	162,450	135,784	10.56	2.24	2.69	0.45	34.68	36.39	1.71	19	9
Small island developing states	7,324	6,002	−1,322	-1.10	0.62	0.10	-0.52	10.29	9.39	-0.90	28	4
Africa	176,963	1,316,409	1,139,446	11.79	14.87	21.83	6.95	39.21	57.02	17.81	46	13
Sub-Saharan Africa	19,528	264,619	245,091	15.58	1.64	4.39	2.75	18.19	36.80	18.61	41	13
Eastern Africa	8,040	145,122	137,082	17.44	0.68	2.41	1.73	11.76	47.57	35.80	14	8
Middle Africa	2,837	6,523	3,686	4.73	0.24	0.11	-0.13	98.68	80.66	-18.01	8	3
Southern Africa	148	603	455	8.11	0.01	0.01	-0.00	4.99	5.50	0.51	4	4
Western Africa	7,503	104,371	96,868	15.75	0.63	1.73	1.10	23.34	27.12	3.78	14	6
Northern Africa	158,435	1,059,790	901,355	11.14	13.31	17.57	4.26	45.93	66.25	20.32	6	2

Americas	100,527	561,540	461,013	10.03	8.45	9.31	0.86	6.90	14.70	7.80	31	7
LAC	92,476	554,992	462,516	10.47	7.77	9.20	1.43	10.60	17.55	6.96	30	7
Caribbean	6,284	4,033	−2,251	−2.43	0.53	0.07	−0.46	15.83	11.74	−4.08	12	4
Central America	20,067	122,766	102,699	10.59	1.69	2.04	0.35	22.61	29.91	7.30	8	5
South America	66,125	428,193	362,068	10.94	5.56	7.10	1.54	8.89	15.76	6.87	10	3
Northern America	8,051	6,548	−1,503	−1.14	0.68	0.11	−0.57	1.38	0.99	−0.38	1	1
Asia	911,795	4,151,658	3,239,863	8.79	76.63	68.83	-7.79	2.34	3.95	1.61	31	6
Southern Asia	60	360,360	360,300	62.15	0.01	5.97	5.97	0.00	3.54	3.54	4	2
Eastern Asia	599,330	1,687,617	1,088,287	5.92	50.37	27.98	−22.39	1.85	2.40	0.55	5	3
South-Eastern Asia	298,159	2,086,770	1,788,611	11.42	25.06	34.60	9.54	8.11	8.69	0.58	11	3
Western Asia	14,246	16,910	2,664	0.96	1.20	0.28	−0.92	12.07	3.66	−8.41	11	5
Europe	180	279	99	2.47	0.02	0.00	-0.01	0.01	0.01	0.00	7	2
Eastern Europe		53				0.00			0.01		2	1
Northern Europe		1				0.00			0.00		1	1
Southern Europe		23				0.00			0.00		2	2
Western Europe	180	202	22	0.64	0.02	0.00	−0.01	0.04	0.07	0.03	2	2
Oceania	461	1,546	1,085	6.95	0.04	0.03	−0.01	0.33	0.70	0.37	9	3
Pacific islands	461	1,546	1,085	6.95	0.04	0.03	−0.01	2.16	8.06	5.89	9	3

Unless noted otherwise, country grouping follows the United Nations M49 standard. According to the original 1996 definition under the M49 standard, developed regions include Europe, Northern America, Japan, Australia and New Zealand; other countries are deemed developing regions. Sub-Saharan Africa includes Sudan. LAC = Latin America and the Caribbean.

the developing regions' average. In contrast, tilapia aquaculture production in small island developing states declined from 7324 to 6002 tonnes between 2000 and 2018, which mostly reflected the situation in the Caribbean, whereas tilapia aquaculture production in the Pacific Islands has increased (Table 2.1).

Africa accounted for around one-fifth of the 6 million tonnes of world tilapia aquaculture production in 2018. Tilapias are a major aquaculture species in Africa, farmed in 46 African countries with 1.3 million tonnes of production accounting for nearly 60% of the region's total aquaculture production in 2018 (Table 2.1). Most of tilapia aquaculture production in Africa came from northern Africa (thanks primarily to Egypt), whereas tilapia aquaculture in sub-Saharan Africa grew rapidly from <20,000 tonnes in 2000 to >250,000 tonnes in 2018 (Table 2.1).

The Americas accounted for a little under 10% of the 6 million tonnes of world tilapia aquaculture production in 2018, most of which came from South America (7%) and Central America (2%), whereas northern America and the Caribbean each had a small share (around 0.1%). Indeed, northern America and the Caribbean were the only two geographic subregions in the world with a decline in tilapia aquaculture production between 2000 and 2018 (Table 2.1).

Asia accounted for around 70% of the 6 million tonnes of world tilapia aquaculture production in 2018, mostly from South-eastern Asia (35%) and Eastern Asia (28%). Southern Asia had the fastest tilapia aquaculture growth between 2000 and 2018 (>60% growth per year), with production increasing from <100 to >350,000 tonnes (Table 2.1).

Europe cultivated around 300 tonnes of tilapia in 2018, mostly in Western Europe. Oceania's 1500 tonnes of tilapia aquaculture production in 2018 was entirely from the Pacific Islands. Central Asia, which comprises five landlocked developing countries, had no tilapia production in 2018, nor did the subregion of Australia and New Zealand.

Tilapia aquaculture production by country

In 2018, the top 30 largest tilapia farming countries included 13 countries in Asia, ten in Africa and seven in Latin America and the

Caribbean (Table 2.2). China, Indonesia and Egypt were the only three countries with over 1 million tonnes of tilapia aquaculture production, and together they accounted for two-thirds of the world tilapia aquaculture production in 2018 (26.93, 20.27 and 17.43%, respectively) (Table 2.2).

Tilapia aquaculture production in all top 30 countries increased between 2000 and 2018, with 18 recording average annual growth >10% (Table 2.2), including nine countries in Africa (numbers refer to list position in Table 2.2: Egypt (2), Ghana (10), Uganda (11), Zambia (18), Nigeria (20), Kenya (23), Tanzania (24), Malawi (27) and Sudan (28); six countries in Asia: Indonesia (2), Bangladesh (4), Vietnam (7), Myanmar (14), Sri Lanka (22) and Cambodia (29); and three countries in Latin America and the Caribbean (LAC): (Brazil (5), Mexico (13) and Honduras (15).

Among the top 30 largest tilapia-farming countries, tilapia are the major species, accounting for over half of the total aquaculture production in nine of the ten countries in Africa (except for Nigeria), four of the seven countries in LAC (Brazil (5), Colombia (9), Honduras (15) and Costa Rica (21)) and none of the 13 in Asia (Table 2.2).

While the share of tilapia in total aquaculture production increased between 2000 and 2018 in most of the top 30 countries it decreased in nine of them, including, notably, Saudi Arabia (30; from 66.09 to 10.81%) and Guatemala (25; from 59.58 to 38.53%). In both countries, the dominance of tilapia was taken over by whiteleg shrimp (*Litopenaeus vannamei*), another major farming species and international seafood commodity.

Tilapia aquaculture production by species

In total, 23 tilapia species items were recorded in the FAO statistics on global aquaculture production for the period 1950–2018 (Table 2.3), of which 17 contributed to the 6 million tonnes of production in 2018. The top three species items accounted for nearly the entire production in 2018, primarily Nile tilapia (*Oreochromis niloticus*) and its hybrids.[1]

Table 2.2 Top 30 tilapia-farming countries, 2018.

Region	Tilapia aquaculture production										Year 2018
	Tonnes				Share of world tilapia production (%)			Share of each region's total aquaculture production (%)			Number of farmed tilapia species items
	2000	2018	±	Annual growth (%)	2000	2018	±	2000	2018	±	
1. China (AS)	548,261	1,624,547	1,076,286	6.22	46.08	26.93	−19.14	1.84	2.46	0.61	2
2. Indonesia (AS)	85,179	1,222,741	1,137,562	15.95	7.16	20.27	13.11	8.57	8.28	−0.29	2
3. Egypt (AF)	157,425	1,051,444	894,019	11.13	13.23	17.43	4.20	46.29	67.34	21.05	1
4. Bangladesh (AS)		344,784				5.72			14.33		1
5. Brazil (LAC)	32,459	317,080	284,621	13.50	2.73	5.26	2.53	18.82	52.35	33.52	2
6. Philippines (AS)	92,579	277,006	184,427	6.28	7.78	4.59	-3.19	8.41	12.02	3.61	2
7. Vietnam (AS)		260,000				4.31			6.26		1
8. Thailand (AS)	82,581	211,441	128,860	5.36	6.94	3.51	-3.43	11.19	23.73	12.55	2
9. Colombia (LAC)	22,870	77,933	55,063	7.05	1.92	1.29	-0.63	37.01	58.70	21.69	2
10. Ghana (AF)	3,712	70,628	66,916	17.78	0.31	1.17	0.86	74.24	92.17	17.93	1
11. Uganda (AF)	600	70,095	69,495	30.28	0.05	1.16	1.11	73.17	67.57	-5.60	1
12. Taiwan Province of China (AS)	49,235	62,633	13,398	1.35	4.14	1.04	-3.10	19.20	22.06	2.86	1
13. Mexico (LAC)	6,726	52,748	46,022	12.12	0.57	0.87	0.31	12.47	21.34	8.86	1
14. Myanmar (AS)		44,511				0.74			3.93		1
15. Honduras (LAC)	1,900	33,500	31,600	17.28	0.16	0.56	0.40	18.90	51.54	32.64	1
16. Malaysia (AS)	18,471	31,766	13,295	3.06	1.55	0.53	-1.03	11.00	8.10	-2.90	2

17. Lao People's Democratic Republic (AS)	18,928	31,000	12,072	2.78	1.59	0.51	−1.08	45.00	28.65	−16.35	1
18. Zambia (AF)	4,020	24,190	20,170	10.48	0.34	0.40	0.06	94.81	99.55	4.74	5
19. Ecuador (LAC)	9,201	23,050	13,849	5.23	0.77	0.38	-0.39	15.01	4.27	-10.74	1
20. Nigeria (AF)	2,705	21,734	19,029	12.27	0.23	0.36	0.13	10.52	7.46	-3.06	1
21. Costa Rica (LAC)	8,100	16,200	8,100	3.93	0.68	0.27	-0.41	83.44	77.81	-5.63	2
22. Sri Lanka (AS)	60	14,730	14,670	35.76	0.01	0.24	0.24	1.36	47.64	46.28	1
23. Kenya (AF)	222	12,160	11,938	24.91	0.02	0.20	0.18	43.36	78.33	34.97	1
24. United Republic of Tanzania (AF)	210	11,000	10,790	24.60	0.02	0.18	0.16	17.36	65.27	47.92	1
25. Guatemala (LAC)	2,361	10,910	8,549	8.88	0.20	0.18	-0.02	59.58	38.53	-21.05	1
26. Zimbabwe (AF)	2,041	10,500	8,459	9.53	0.17	0.17	0.00	94.89	99.20	4.31	1
27. Malawi (AF)	500	8,544	8,044	17.08	0.04	0.14	0.10	94.34	94.79	0.45	4
28. Sudan (AF)		8,000				0.13			80.00		1
29. Cambodia (AS)	370	8,000	7,630	18.62	0.03	0.13	0.10	2.56	3.15	0.58	1
30. Saudi Arabia (AS)	3,968	7,780	3,812	3.81	0.33	0.13	-0.20	66.09	10.81	-55.28	2

AF, Africa; AS, Asia; LAC, Latin America and the Caribbean;

Generally growing faster than other tilapia species, Nile tilapia has become the dominant species in global tilapia aquaculture. Many years of research and development have resulted in various genetically improved farmed tilapia (GIFT) strains that dominate tilapia farming in most countries, except for a few countries (e.g. Malawi) that do not allow the introduction of non-native Nile tilapia on account of conservation of biodiversity.

In 2018, nearly the entire 400,000-tonne aquaculture production of Blue Nile tilapia (*O. Oreochromis* × *O. aureus*) (Table 2.3) came from China, which has been farming the hybrid primarily for all-male fingerlings. While growing not as rapidly as GIFT strains, Blue Nile tilapia is favoured by some farmers because it tends to be more stress/disease resistant and can tolerate lower temperatures.

Red tilapia (*O. niloticus* × *O. mossambicus*) is another hybrid tilapia that has become commercially established in many countries and territories (e.g. China, Taiwan Province of China and Thailand), although its production has not been clearly reported in FAO statistics, but may be embedded in 'Tilapia nei' or 'Nile tilapia'.

None of the 20 minority tilapia species items in Table 2.3 exceeded 1% of the total production in 2018. Mozambique tilapia (*O. mossambicus*) had the highest production among them, yet its 54,000-tonne production in 2018 was much lower than the 81,000 tonnes in 2017 (the maximum in 1950–2018), and this decline reflected the 27,000-tonne production drop in Indonesia. In 2018, Mozambique tilapia was farmed in ten countries globally, with Indonesia, Malawi, South Africa, Singapore and Eswatini being the top five producers. This species' high salinity tolerance makes it a suitable species in coastal lagoons and other high-salinity farming environments, yet its relatively small size and dark colour could be a marketing disadvantage. Mozambique tilapia was the first tilapia species introduced in China and was marketed as the 'African crucian carp'. The image of relatively cheap fish has led to a marketing disadvantage for tilapia in general that has had a long-lasting and profound impact on the development of the domestic market for tilapia in China.

Besides Nile and Mozambique tilapia, blue tilapia (*O. aureus*) is another global tilapia species (Table 2.3), with 3000 tonnes of

Table 2.3 Tilapia aquaculture production by species.

ASFIS species items (ranked by production in 2018)		Year 2018				Maximum annual production, 1950–2018			
Common name	Scientific name	2018 production (tonnes)	Share of all tilapia species (%)	Number of countries farming the species	Top five producers	Maximum annual production (tonnes)	Year	Number of countries farming the species	Top five producers
1. Nile tilapia	*Oreochromis niloticus*	4,525,431	75.03	78	China; Indonesia; Egypt; Brazil; Thailand	4,525,431	2018	78	China; Indonesia; Egypt; Brazil; Thailand
2. Tilapias nei	*Oreochromis* (=Tilapia) spp.	1,030,004	17.08	49	Bangladesh; Vietnam; Philippines; Taiwan Province of China; Colombia	1,030,004	2018	49	Bangladesh; Vietnam; Philippines; Taiwan Province of China; Colombia
3. Blue Nile tilapia, hybrid	*Oreochromis aureus* x *O. niloticus*	406,048	6.73	2	China; Panama	406,048	2018	2	China; Panama
4. Mozambique tilapia	*Oreochromis mossambicus*	53,754	0.89	10	Indonesia; Malawi; South Africa; Singapore; Eswatini	80,703	2017	10	Indonesia; Malawi; South Africa; Singapore; Eswatini
5. Tilapia shiranus	*Oreochromis shiranus*	5,036	0.08	1	Malawi	5,311	2017	1	Malawi

Continued

Table 2.3 Continued

ASFIS species items (ranked by production in 2018)		Year 2018				Maximum annual production, 1950–2018			
Common name	Scientific name	2018 production (tonnes)	Share of all tilapia species (%)	Number of countries farming the species	Top five producers	Maximum annual production (tonnes)	Year	Number of countries farming the species	Top five producers
6. Blue tilapia	*Oreochromis aureus*	3,182	0.05	4	Costa Rica; Côte d'Ivoire; Syrian Arab Republic; Cuba	5,786	2008	5	Syrian Arab Republic; Costa Rica; Cuba; Côte d'Ivoire; Panama
7. Three-spotted tilapia	*Oreochromis andersonii*	2,147	0.04	3	Zambia; Namibia; Botswana	4,149	2013	2	Zambia; Namibia
8. Redbreast tilapia	*Coptodon rendalli*	1,903	0.03	3	Zambia; Malawi; Namibia	3,760	2015	2	Zambia; Malawi
9. Longfin tilapia	*Oreochromis macrochir*	1,800	0.03	1	Zambia	2,147	2013	1	Zambia
10. (Oreochromis tanganicae)	*Oreochromis tanganicae*	1,690	0.03	1	Zambia	1,690	2018	1	Zambia
11. Sabaki tilapia	*Oreochromis spilurus*	280	0.00	1	Saudi Arabia	400	2013	1	Saudi Arabia
12. (Cichla spp.)	*Cichla* spp.	80	0.00	1	Brazil	1,415	2012	1	Brazil
13. Blackchin tilapia	*Sarotherodon melanotheron*	56	0.00	1	Senegal	264	2014	1	Senegal
14. Mango tilapia	*Sarotherodon galilaeus*	18	0.00	2	Burkina Faso; Liberia	18	2018	2	Burkina Faso; Liberia

15. Jaguar guapote	*Cichlasoma managuense*	2	0.00	2	Panama; El Salvador	168	1997	1	Guatemala
16. Redbelly tilapia	*Coptodon zillii*	1	0.00	1	Liberia	201	2000	2	Uganda; Liberia
17. Banded jewelfish	*Hemichromis fasciatus*	0	0.00	1	Cameroon	5	2012	1	Cameroon
18. Cichlasoma nei	*Cichlasoma* spp.					1,330	2002	2	Dominican Republic; Guatemala
19. Blackbelt cichlid	*Cichlasoma maculicauda*					140	1995	1	Guatemala
20. Green terror	*Aequidens rivulatus*					3	1990	1	Ecuador
21. Yellow-belly bream	*Serranochromis robustus*					1	1998	1	Eswatini
22. Velvety cichlids	*Astronotus* spp.					1	1986	1	Peru
23. Peacock cichlid	*Cichla ocellaris*					0.01	2017	1	Peru
All tilapia species		6,031,432	100	124					

ASFIS, Aquatic Sciences and Fisheries Information System (www.fao.org/fishery/collection/asfis/en). An ASFIS species item could refer to either an individual species, hybrid or group of related species, such as family (when identification to species is impossible).

production in 2018 contributed by four countries worldwide (Costa Rica in Central America, Côte d'Ivoire in Western Africa, Syrian Arab Republic in Western Asia and Cuba in the Caribbean). The lower production in 2018 compared with the maximum production in 2008 (nearly 6000 tonnes) primarily reflected the production drop in the Syrian Arab Republic from 3000 to 600 tonnes. One key advantage of blue tilapia is its relatively high resistance to cold temperatures, which makes it a suitable farming species in high-latitude and/or high-altitude areas.

Besides Nile tilapia, Mozambique tilapia and blue tilapia, other *Oreochromis* tilapia species that have been farmed in regional aquaculture include (1) tilapia shiranus (*O. shiranus*), three-spotted tilapia (*Oreochromis andersonii*), longfin tilapia (*O. macrochir*) and *O. tanganicae* farmed in Southern Africa and (2) sabaki tilapia (*O. spilurus*) farmed in Saudi Arabia (Table 2.3).

Besides *Oreochromis*, other cichlid species have been farmed mostly as niched species, including, among others (1) *Coptodon*[2] and *Sarotherodon* spp. farmed in Africa and (2) *Chichla* and *Cichlasoma* spp. farmed in Latin America (Table 2.3). It is worth noting that while the aquaculture production of cichlid *Cichlasoma* exceeded 1000 tonnes in 2002 thanks primarily to the production in the Dominican Republic (no. 18 in Table 2.3), the FAO statistics recorded only 2 tonnes of *Cichlasoma managuense* cultured by Panama and El Salvado in 2018 (no. 15 in Table 2.3). However, while not reflected in FAO statistics as a distinctive species item, *C. managuense* has been farmed and marketed in China as 'freshwater grouper' which, as the name suggests, is a high-value niche species.

Tilapia trade and consumption

Tilapia have become increasingly popular in domestic and international markets thanks to desirable characteristics (e.g. no small bones, mild taste and firm texture), affordability and availability. However, the market expansion of tilapia products has been limited by quality issues (e.g. off-flavour), consumer preference (e.g. lack of consumers' acceptance of freshwater fishes and/or farmed fish in general) and trade barriers (e.g. attempts to ban tilapia import in multiple countries in Africa).

Tilapia international trade

According to FAO statistics (FAO, 2020c), world tilapia trade (measured in import value) increased from US$116 million to US$1.5 billion between 2000 and 2018; the 15% annual growth was higher than that of freshwater fishes (11%), finfish (6.1%) and aquatic products in general (5.6%). Accordingly, the share of tilapia in the import value of freshwater fish products increased from 14 to 29%, that of finfish increased from 0.3 to 1.4% and the share in the import value of all fish and fishery products increased from 0.2 to 0.9%.

Data and methodology

Analysis of the international trade of tilapia products in the remainder of this section is based on trade statistics from the UN Comtrade Database (United Nations, 2020). Import statistics are used as the primary data supplemented by export statistics – see the methodology details in FAO, (2020d, pp. 60–61). The mirror-data approach helps compile more comprehensive trade data and avoid discrepancies between export and import statistics that commonly occur in trade statistics.

The trade statistics used here cover four major tilapia products (fresh whole tilapia HS030271, frozen whole tilapia HS030323, fresh tilapia fillets HS030431 and frozen tilapia fillets HS030461). Other tilapia products are not included because of the lack of disaggregate data in the UN Comtrade database.

Four indicators are used to examine tilapia trade patterns, including (1) market size (measured by trade volume and trade value), (2) market price, (3) per capita import and (4) number of suppliers. The first three indicators are straightforward while the last includes two measures: one is the actual number of suppliers, which counts how many countries or territories export tilapia products to the market. The second is the effective number of suppliers that measures the diversity of suppliers (i.e. the evenness of market share distribution among suppliers). The effective number of suppliers in a market is calculated by $e^{-\sum_i [s_i/\ln(s_i)]}$, where s_i denotes exporter *i*'s share in the market (equal to its export to the market divided by the total export of all suppliers to the market). For example, while 11 countries/territories exported tilapia products to Mexico in 2018 – i.e. the actual number of suppliers being 11 – the effective number of suppliers was only 1.05, which reflects the situation that Mexico's tilapia import market in 2018 was dominated by one supplier, i.e. China (99.2% market share).

Global overview

In 2018, more than half a million tonnes of tilapia products were exported from 99 countries to 164 countries, with US$1.5 billion of total value and US$2.87/kg average price. The products comprised US$918 million of frozen tilapia fillets (61% of the total), US$330 million of frozen whole tilapias (22%), US$209 million of fresh tilapia fillets (14%) and US$42 million of fresh whole tilapias (3%) (Tables 2.4 and 2.5 and Table 2.6).

Two-thirds of the US$1.5 billion of tilapia products were exported to developed regions and the average price (US$3.91) was much higher than the US$1.93/kg average price of tilapia export to developing regions (Table 2.4). This reflects a much higher share of fillets in the tilapia import of developed regions (87%) than that of developing regions (53%).

In 2018, the number of countries exporting tilapia products to developed and developing Regions (76 and 75, respectively) was nearly the same (Table 2.4). However, the effective number of exporters to developed regions (5.19) was much higher than to developing countries (2.35), which indicates that the former have more diverse sources of tilapia import.

Northern America

In 2018 Northern America accounted for over half of the world's tilapia import (Table 2.4), thanks primarily to the United States of

Table 2.4. Regional markets of tilapia import, 2018[a].

Regional markets[b]	Import volume (tonnes)	Import price[d] (US$/kg)	Import value		Number of countries exporting tilapia products to the region		Per capita import	
			US$1000	Share of the world market (%)	Actual	Effective[e]	US$/person	kg/person
World[c]	522,397	2.87	1,499,516	100.00	99	4.35	0.20	0.07
Developed regions	248,545	3.91	97,140	64.76	76	5.19	0.76	0.19
Developing regions	273,778	1.93	528,122	35.22	75	2.35	0.08	0.04
Least-developed countries	44,041	1.30	57,397	3.83	30	2.07	0.06	0.04
Landlocked developing countries	19,644	1.89	37,100	2.47	26	1.99	0.07	0.04
Small island developing states	3,515	2.69	9,447	0.63	30	6.86	0.13	0.05
Africa	125,235	1.51	189,616	12.65	34	1.49	0.15	0.10
Sub-Saharan Africa	125,081	1.51	188,303	12.56	32	1.44	0.17	0.12
Eastern Africa	25,760	1.95	50,187	3.35	16	1.83	0.12	0.06
Middle Africa	33,739	1.60	54,176	3.61	17	1.34	0.32	0.20
Southern Africa	10,297	1.05	10,849	0.72	16	1.85	0.17	0.16
Western Africa	55,285	1.32	73,092	4.87	13	1.04	0.19	0.15
Northern Africa	154	7.56	1,313	0.09	7	1.56	0.01	0.00
Americas	293,520	3.47	1,017,219	67.84	39	3.48	1.01	0.29
LAC	107,465	2.41	259,248	17.29	27	1.30	0.40	0.17
Caribbean	1,122	2.82	3,159	0.21	13	2.51	0.07	0.03
Central America	96,352	2.36	227,584	15.18	24	1.21	1.30	0.55
South America	9,991	2.85	28,505	1.90	13	1.56	0.07	0.02
Northern America	186,055	4.07	757,971	50.55	34	4.28	2.08	0.51

Continued

Table 2.4 Continued

Regional markets[b]	Import volume (tonnes)	Import price[d] (US$/kg)	Import value		Number of countries exporting tilapia products to the region		Per capita import	
			US$1000	Share of the world market (%)	Actual	Effective[e]	US$/person	kg/person
Asia	55,178	2.55	141,130	9.41	53	5.87	0.03	0.01
Central Asia	129	2.46	318	0.02	3	1.96	0.00	0.00
Southern Asia	4,320	3.48	15,032	1.00	13	1.07	0.01	0.00
Eastern Asia	5,839	3.36	19,637	1.31	26	6.37	0.01	0.00
South-Eastern Asia	8,002	1.75	13,997	0.93	22	6.03	0.02	0.01
Western Asia	36,887	2.49	92,146	6.15	41	4.82	0.34	0.14
Europe	46,387	3.17	147,050	9.81	62	7.00	0.20	0.06
Eastern Europe	15,799	2.63	41,607	2.77	32	2.14	0.14	0.05
Northern Europe	5,558	3.45	19,165	1.28	36	7.73	0.18	0.05
Southern Europe	7,418	3.26	24,204	1.61	34	6.77	0.16	0.05
Western Europe	17,611	3.52	62,075	4.14	54	8.03	0.32	0.09
Oceania	2,004	2.12	4,247	0.28	14	2.39	0.10	0.05
Australia and New Zealand	1,829	2.11	3,868	0.26	9	2.27	0.13	0.06
Pacific islands	176	2.16	380	0.03	7	2.13	0.03	0.01

[a]Based on mirror-trade data constructed from UN Comtrade statistics – see FAO (2020d, pp. 60–61) for the methodology.
[b]Country grouping based on the UN M49 standard.
[c]Regional trade may not add up exactly to the world trade that includes that in unspecified areas.
[d]A few countries that have data on trade value but not on trade volume are included in the calculation of the regional aggregates for volume and value but not used in the calculation of regional price; hence the price may not be exactly equal to the value divided by the volume.
[e]The effective number of countries measuring the evenness of the market share distribution among exporting countries – see FAO (2020d, p. 60) for the formula for calculating the indicator.

America, which is the largest tilapia international market, accounting for 53 and 80% of the world import of frozen tilapia fillets and fresh tilapia fillets, respectively (Table 2.5). USA was also a major importer of frozen whole and fresh whole tilapia, and Canada was a major importer of tilapia fillets (Table 2.5). The 2.08 US$ per capita import of tilapia products in Northern America was the highest among all regions or subregions (Table 2.4).

Central America

In 2018, Central America accounted for 15% of the world tilapia import value (Table 2.4). The low effective number of exporters to the subregion (1.21) reflects the fact that a single country, China, supplied 97% of tilapia import in the subregion (in terms of value).

In 2018, frozen tilapia fillets accounted for 90% of the tilapia import value of Central America, followed by frozen whole tilapia (7.4%). Mexico and Costa Rica in the subregion were among the top ten frozen tilapia fillets importers in the world, and Costa Rica was also a major importer of fresh tilapia fillets (Table 2.5). Mexico and Honduras were two Central American countries among the top ten exporters of frozen tilapia fillets, yet neither country's share in the world market was more than 1% (Table 2.6).

Seven LAC countries, including four in Central America (Honduras, Costa Rica, Mexico and Guatemala) and three in South America (Colombia, Ecuador and Brazil) were among the top ten exporters of fresh tilapia fillets (Table 2.6). This reflects the competitive advantage of LAC countries in exporting fresh tilapia fillets thanks to their geographic proximity to the USA and Canada that accounted for nearly 90% of the international market of fresh tilapia fillets (Table 2.5).

Africa

Africa accounted for 13% of world tilapia import value in 2018 (Table 2.4). Similar to Central America, the low effective number of exporters to Africa (1.49) reflects the fact that 93% of tilapia import

Table 2.5 Top ten tilapia importers by value, 2018[a].

Country (ranked by import value)[b]	Volume (tonnes)	Price (US$/kg)	Value US$1000	Share of world total (%)	Country (ranked by import value)[b]	Volume (tonnes)	Price (US$/kg)	Value US$1000	Share of world total (%)
Top ten importers of frozen tilapia fillets					**Top ten importers of frozen whole tilapia**				
1. United States of America (NA)	123,738	3.93	486,397	52.97	1. United States of America (NA)	31,665	1.93	60,963	18.46
2. Mexico (LAC)	75,751	2.42	183,649	20.00	2. Côte d'Ivoire (AF)	39,544	1.15	45,599	13.81
3. Israel (AS)	12,242	4.58	56,052	6.10	3. Cameroon (AF)	13,707	2.48	34,037	10.31
4. Russian Federation (ER)	8,973	2.66	23,871	2.60	4. Saudi Arabia (AS)	12,305	1.51	18,526	5.61
5. Costa Rica (LAC)	7,437	2.57	19,078	2.08	5. Kenya (AF)	7,711	2.24	17,283	5.23
6. Iran (Islamic Republic of) (Asia)	4,181	3.45	14,437	1.57	6. Zambia (AF)	7,808	2.20	17,204	5.21
7. Canada (NA)	2,282	5.56	12,679	1.38	7. Mexico (LAC)	10,513	1.58	16,574	5.02
8. Peru (LAC)	4,253	2.91	12,367	1.35	8. Ghana (AF)	7,141	1.38	9,848	2.98
9. Netherlands (ER)	2,542	4.61	11,714	1.28	9. Netherlands (ER)	3,611	1.99	7,179	2.17
10. Spain (ER)	3,714	2.88	10,710	1.17	10. Gabon (AF)	3,118	2.30	7,169	2.17
Rest of the world	24,510	3.56	87,232	9.50	Rest of the world	65,738	1.46	95,899	29.04
World	**269,623**	**3.41**	**918,186**	**100.00**	**World**	**202,861**	**1.63**	**330,282**	**100.00**

Top ten importers of fresh tilapia fillets					Top ten importers of fresh whole tilapia				
1. United States of America (NA)	21,621	7.78	168,185	80.39	1. United States of America (NA)	1,966	3.76	7,394	17.67
2. Canada (NA)	2,018	7.88	15,912	7.61	2. Belgium (ER)	431	14.00	6,036	14.43
3. Costa Rica (LAC)	897	4.65	4,172	1.99	3. Zambia (AF)	5,688	0.94	5,369	12.84
4. United Kingdom (ER)	477	6.89	3,285	1.57	4. Spain (ER)	597	6.03	3,601	8.61
5. Netherlands (ER)	322	9.24	2,974	1.42	5. United Arab Emirates (AS)	3,613	0.95	3,435	8.21
6. Peru (LAC)	540	4.88	2,636	1.26	6. Netherlands (ER)	216	9.40	2,026	4.84
7. Egypt (AF)	143	7.95	1,140	0.54	7. Qatar (AS)	538	3.09	1,662	3.97
8. Italy (ER)	139	7.59	1,058	0.51	8. Ireland (ER)	214	6.69	1,430	3.42
9. Namibia (AF)	93	7.88	731	0.35	9. Cambodia (AS)	3,646	0.39	1,428	3.41
10. France (ER)	64	7.99	510	0.24	10. Kenya (AF)	520	2.43	1,262	3.02
Rest of the world	2,286	3.77	8,614	4.12	Rest of the world	3,885	2.11	8,188	19.57
World	**28,600**	**7.31**	**209,217**	**100.00**	**World**	**21,313**	**1.94**	**41,832**	**100.00**

[a]Based on mirror trade data constructed from UN Comtrade statistics – see FAO (2020d, pp. 60–61) for the methodology.
[b]AF = Africa; AS = Asia; ER = Europe; LAC = Latin America and the Caribbean; NA = Northern America.

Table 2.6 Top ten tilapia exporters by value, 2018[a].

Country (ranked by export value)[b]	Volume (tonnes)	Price (US$/kg)	Value US$1000	Share of world total (%)	Country (ranked by export value)[b]	Volume (tonnes)	Price (US$/kg)	Value US$1000	Share of world total (%)
Top ten exporters of frozen tilapia fillets					**Top ten exporters of frozen whole tilapia**				
1. China (AS)	246,219	3.19	786,319	85.64	1. China (AS)	159,421	1.59	253,471	76.74
2. Indonesia (AS)	9,413	6.49	61,120	6.66	2. Taiwan Province of China (AS)	15,514	1.9	29,420	8.91
3. Taiwan Province of China (AS)	2,760	7.69	21,220	2.31	3. Vietnam (AS)	7,143	1.79	12,776	3.87
4. Mexico (LAC)	1,632	5.11	8,348	0.91	4. Thailand (AS)	5,974	1.42	8,474	2.57
5. Netherlands (ER)	1,153	4.62	5,327	0.58	5. Zimbabwe (AF)	1,625	2.49	4,045	1.22
6. Vietnam (AS)	2,093	2.5	5,240	0.57	6. Myanmar (AS)	2,460	1.23	3,030	0.92
7. China, Hong Kong SAR (AS)	1,151	4.07	4,683	0.51	7. Netherlands (ER)	1,140	2.4	2,734	0.83
8. Honduras (LAC)	811	5.08	4,121	0.45	8. India (AS)	2,340	1.02	2,387	0.72
9. Malaysia (AS)	643	5.89	3,788	0.41	9. Belgium (ER)	847	2.19	1,852	0.56
10. Belgium (ER)	298	8.64	2,572	0.28	10. Indonesia (AS)	528	3.19	1,685	0.51
Rest of the world	3,452	4.48	15,448	1.68	Rest of the world	5,868	1.77	10,408	3.15
World	**269,623**	**3.41**	**918,186**	**100**	**World**	**202,861**	**1.63**	**330,282**	**100**

Top ten exporters of fresh tilapia fillets					Top ten exporters of fresh whole tilapia				
1. Colombia (LAC)	7,271	8.9	64,716	30.93	1. Netherlands (ER)	503	12.65	6,364	15.21
2. Honduras (LAC)	8,079	6.9	55,749	26.65	2. China (AS)	5,985	0.87	5,194	12.42
3. Costa Rica (LAC)	4,809	8.23	39,598	18.93	3. Egypt (AF)	3,936	1	4,417	10.56
4. Mexico (LAC)	1,037	7.91	8,205	3.92	4. Myanmar (AS)	871	3.04	2,646	6.32
5. Ecuador (LAC)	841	6.82	5,736	2.74	5. Bangladesh (AS)	691	3.43	2,371	5.67
6. Netherlands (ER)	790	6.33	5,003	2.39	6. Greece (ER)	338	5.55	1,876	4.49
7. Indonesia (AS)	934	5.12	4,784	2.29	7. Colombia (LAC)	194	8.59	1,666	3.98
8. Brazil (LAC)	497	7.05	3,504	1.67	8. Morocco (AF)	239	6.89	1,645	3.93
9. China (AS)	995	3.05	3,032	1.45	9. Thailand (AS)	3,753	0.41	1,529	3.65
10. Guatemala (LAC)	251	8.13	2,040	0.98	10. France (ER)	193	6.89	1,330	3.18
Rest of the world	3,095	5.44	16,849	8.05	Rest of the world	4,611	2.77	12,794	30.58
World	**28,600**	**7.31**	**209,217**	**100**	**World**	**21,313**	**1.94**	**41,832**	**100**

[a]Based on mirror-trade data constructed from UN Comtrade statistics – see FAO (2020d, pp. 60–61) for the methodology.
[b]AF = Africa; AS = Asia; ER = Europe; LAC = Latin America and the Caribbean; NA = Northern America.

in the region was supplied by China. Nearly all tilapia import in Africa went to sub-Saharan Africa.

By contrast, the import markets in Northern America and Central America were dominated by tilapia fillets (90 and 92%, respectively) while frozen and fresh whole tilapia accounted for, respectively, 94 and 4.1% of Africa's total tilapia import value. This explains the relatively low tilapia import price in Africa (Table 2.4).

Africa imported 54% of frozen whole tilapias traded internationally, and six countries in the subregion (Côte d'Ivoire, Cameroon, Kenya, Zambia, Ghana and Gabon) were among the top ten importers of frozen whole tilapia (Table 2.5). Egypt and Namibia were the only two African countries among the top ten importers of fresh tilapia fillets; Zambia and Kenya were the only two among the top ten importers of fresh whole tilapia. None of the countries in Africa was among the top ten importers of frozen tilapia fillets (Table 2.5).

Contributing to 22% of world tilapia aquaculture production (Table 2.1) notwithstanding, Africa contributed to only 5% of world tilapia export value (Table 2.4). Zimbabwe was the only African country among the top ten exporters of frozen whole tilapia. Egypt and Morocco were the only two among the top ten exporters of fresh whole tilapias. There were no African countries among the top ten exporters of either frozen or fresh tilapia fillets (Table 2.6).

Europe

Europe accounted for 9.8% of world tilapia import value with a relatively high effective number of exporters (7.00), reflecting more diverse sources of tilapia exports to the region (Table 2.4). With little tilapia aquaculture production notwithstanding (Table 2.1), there was substantial intra-regional tilapia trade within Europe because of its sophisticated seafood supply chain. Indeed, the Netherlands was among both the top ten importers (Table 2.5) and top ten exporters (Table 2.6) for all four tilapia products, which may primarily reflect repackaging and/or trans-shipment of tilapia products originating from non-European countries.

The Russian Federation was the largest tilapia importer in Europe, accounting for 16% of the region's tilapia import in 2018. That country imported primarily small-size frozen tilapia fillets with a relatively low average price (Table 2.5). Besides the Netherlands and the Russian Federation, other major tilapia importers and/or exporters in Europe include (1) Spain (for frozen tilapia fillet import), (2) the United Kingdom, Italy and France (for fresh tilapia fillet import), (3) Belgium, Spain and Ireland (for fresh whole tilapia import), (4) Greece and France (for fresh whole tilapia export) and (5) Belgium (for frozen whole tilapia export).

Asia

In 2018 Asia accounted for 69% of world tilapia aquaculture production (Table 2.1) and 82% of world tilapia export. With a 27% share in world tilapia aquaculture production notwithstanding (Table 2.2), China accounted for 70% of world tilapia export and was the top exporter of all four tilapia products (Table 2.6). China exported tilapia products to 99 countries with >50% of the market share in 62 of these, and it accounted for 85% of tilapia export to developing regions, including 93% of tilapia export to Africa and 96% of that to LAC.

Other major tilapia exporters in Asia include (1) Indonesia (among the top ten lists for three products excepting fresh whole tilapia); (2) Taiwan Province of China and Vietnam (among the top ten lists for frozen tilapia fillets and frozen whole tilapia); (3) Thailand and Myanmar (among the top ten lists for fresh tilapia fillets and fresh whole tilapia); (4) Malaysia (among the top ten list for frozen tilapia fillets); (5) Bangladesh (among the top ten list for fresh whole tilapia); and (6) India (among the top ten list for fresh tilapia fillets).

Asia accounted for 9.4% of world tilapia import value, most of which (6.2%) went to Western Asia (Table 2.4). Israel and Iran were among the top ten importers of frozen tilapia fillets, Saudi Arabia was one of the top ten importers of fresh fillets and United Arab Emirates, Qatar and Cambodia were among the top ten importers of fresh whole tilapia (Table 2.5).

Tilapia consumption

Nearly all tilapia production is for direct human consumption. Thus, world average per capita tilapia consumption can be estimated by world tilapia production divided by world population. Tilapia consumption at the regional or national level can be estimated by the food balance approach, i.e. consumption = production + import – export. While applying the food balance approach, the following conversion factors are used to estimate the live weight equivalent of tilapia products: 2.86 for frozen tilapia fillet, 2.48 for fresh tilapia fillet, 1.12 for frozen whole tilapia and 1.00 for fresh whole tilapia (EUFOMA, 2019).

In 2018 the world average per capita tilapia consumption was 0.9 kg (live weight), and regional and subregional consumption varied from <0.005 kg in Central Asia to >5 kg in Northern Africa (Table 2.7). The 0.47 kg average per capita tilapia consumption in developed regions was half of the world average, and 98% of their tilapia supply was from import. By contrast, the 0.99 kg average consumption in developing regions was higher than the world average and 93% of their tilapia supply was from domestic production. The average consumption in landlocked developing countries (0.75 kg) was slightly lower than the world average, and that in small island developing states (0.24 kg) was much lower (Table 2.7).

Africa

The 1.52 kg of per capita tilapia consumption in Africa was higher than the world average. This nevertheless reflects the extraordinarily high consumption in Northern Africa, whereas the consumption in other African subregions was lower than the world average. The consumption in Middle Africa and Southern Africa was especially low, and >80% of their tilapia supply was from import (Table 2.7).

Americas

The Americas was another region with above-average per capita tilapia consumption (1.39 kg), as was consumption in Central America, (2.68 kg), South America (1.08 kg) and Northern America

(1.27 kg), whereas that in the Caribbean (0.2 kg) was much lower (Table 2.7). While most tilapia supply in South America came from its own production (with only 5.9% of import), the import share was 99, 49 and 20%, respectively, in Northern America, Central America and the Caribbean (Table 2.7).

Asia

The average 0.76 kg of per capita tilapia consumption in Asia was below the world average. Nevertheless, consumption in South-eastern Asia was >3 kg thanks to its 2.2 million tonnes of tilapia production (the highest among all subregions). Eastern Asia's 1.7 million tonnes of tilapia production was the second highest, yet its 0.47 kg per capita consumption was around half of the world average because most of its tilapia utilization went to export (Table 2.7). The per capita tilapia consumption in Southern and Western Asia was 0.22 and 0.3 kg, respectively. Tilapia supply in Southern Asia was mostly its own production whereas that in Western Asia was mostly from import (Table 2.7).

Europe and Oceania

Despite its 100,000 tonnes (live weight) tilapia import being the second highest among all regions, Europe's 0.11 kg average per capita tilapia consumption was the lowest (Table 2.7). At the subregional level, the consumption in Eastern Europe (0.14 kg) was the highest, that in Southern and Western Europe (0.1 kg each) was in the middle and that in Northern Europe (0.07) was the lowest. Per capita tilapia consumption was 0.15 kg in Oceania, and the 0.34 kg consumption in Pacific Islands was primarily from own production, whereas the 0.07 kg in Australia and New Zealand was entirely supplied by import (Table 2.7).

Selected countries

Around one-fourth of the world population live in 31 countries with per capita tilapia consumption >1 kg, including ten countries (9.5%

Table 2.7 Tilapia supply and utilization, 2018.

Region[a]	Tilapia supply				Tilapia utilization				
	Production		Import		Export		Consumption		
	Live weight (tonnes)	Share of supply (%)	Live weight equivalent (tonnes)[b]	Share of supply (%)	Live weight equivalent (tonnes)[b]	Share of utilization (%)	Live weight equivalent (tonnes)[c]	Share of utilization (%)	Per capita consumption (kg/capita)[d]
World[c]	6,882,202	86.3	1,090,568	13.7	1,090,568	13.7	6,882,202	86.3	0.90
Developed regions	14,438	2.3	602,658	97.7	21,692	3.5	595,405	96.5	0.47
Developing regions	6,867,764	93.4	487,739	6.6	1,065,744	14.5	6,289,759	85.5	0.99
Least-developed countries	863,084	94.7	48,614	5.3	6,431	0.7	905,267	99.3	0.90
Landlocked developing countries	361,326	94.2	22,403	5.8	3,173	0.8	380,556	99.2	0.75
Small island developing states	11,346	66.8	5,629	33.2	162	1.0	16,813	99.0	0.24
Africa	1,809,088	92.8	141,218	7.2	10,077	0.5	1 940,230	99.5	1.52
Sub-Saharan Africa	636,204	81.9	140,851	18.1	5,796	0.7	771,259	99.3	0.71
Eastern Africa	339,331	92.1	29,022	7.9	3,278	0.9	365,074	99.1	0.86
Middle Africa	6,996	15.6	37,924	84.4	0	-	44,920	100.0	0.27
Southern Africa	638	5.1	11,971	94.9	2,229	17.7	10,379	82.3	0.16
Western Africa	256,239	80.5	61,934	19.5	213	0.1	317,961	99.9	0.83
Northern Africa	1,205,884	100.0	368	0.0	4,356	0.4	1,201,895	99.6	5.07
Americas	726,171	49.4	744,426	50.6	71,138	4.8	1,399,459	95.2	1.39
LAC	719,623	71.6	284,852	28.4	68,125	6.8	936,350	93.2	1.46
Caribbean	7,057	80.0	1,768	20.0	0	-	8,825	100.0	0.20
Central America	260,939	50.6	254,822	49.4	44,789	8.7	470,972	91.3	2.68

South America	451,627	94.1	28,261	5.9	23,336	4.9	456,553	95.1	1.08
Northern America	6,548	1.4	459,574	98.6	3,014	0.6	463,109	99.4	1.27
Asia	4,342,798	97.8	99,623	2.2	987,596	22.2	3,454,825	77.8	0.76
Central Asia	-	-	363	100.0	4	1.1	360	98.9	0.00
Southern Asia	412,170	97.1	12,331	2.9	5,173	1.2	419,328	98.8	0.22
Eastern Asia	1,689,421	99.4	11,043	0.6	920,448	54.1	780,016	45.9	0.47
South-Eastern Asia	2,221,991	99.4	12,331	0.6	61,672	2.8	2,172,651	97.2	3.32
Western Asia	19,215	23.2	63,555	76.8	300	0.4	82,470	99.6	0.30
Europe	279	0.3	102,681	99.7	18,453	17.9	84,507	82.1	0.11
Eastern Europe	53	0.1	44,172	99.9	3,282	7.4	40,944	92.6	0.14
Northern Europe	1	0.0	10,074	100.0	2,444	24.3	7,631	75.7	0.07
Southern Europe	23	0.1	17,732	99.9	2,108	11.9	15,647	88.1	0.10
Western Europe	202	0.7	30,702	99.3	10,618	34.4	20,286	65.6	0.10
Oceania	3,866	61.2	2,448	38.8	171	2.7	6,143	97.3	0.15
Australia and New Zealand	-	-	2,245	100.0	162	7.2	2,083	92.8	0.07
Pacific islands	3,866	95.0	204	5.0	9	0.2	4,060	99.8	0.34

[a]Country grouping based on the UN M49 standard.
[b]Based on UN Comtrade data – see notes in Table 2.4 and conversion ratios from EUFOMA (2019); 2.86 for frozen tilapia fillets, 2.48 for fresh tilapia fillets, 1.12 for frozen whole tilapia and 1.00 for fresh whole tilapia.
[c]Consumption = Production + Import – Export.
[d]Based on population data from United Nations (2019).

Table 2.8 Per capita tilapia consumption, 2018.

Selected countries	Per capita tilapia consumption (kg)	Population (×1000)
I. Per capita consumption > 3 kg, ten countries, 9.5% of world population		
1. Egypt (AF)	11.88	98,424
2. Costa Rica (LAC)	5.42	4,999
3. Israel (AS)	5.11	8,382
4. Indonesia (AS)	4.71	267,671
5. Lao People's Democratic Republic (AS)	4.39	7,062
6. Malawi (AF)	3.43	18,143
7. Mexico (LAC)	3.26	126,191
8. Thailand (AS)	3.25	69,428
9. Sri Lanka (AS)	3.14	21,229
10. Philippines (AS)	3.01	106,651
II. Per capita consumption 1–3 kg, 21 countries, 14.2% of world population		
Vietnam (AS)	2.57	95,546
Bangladesh (AS)	2.13	161,377
Brazil (LAC)	1.61	209,469
United States of America (NA)	1.37	327,096
III. Per capita consumption 0.5–1.0 kg, 23 countries, 24.3 of world population		
United Republic of Tanzania (AF)	0.98	56,313
Myanmar (AS)	0.76	53,708
Kenya (AF)	0.61	51,393
China (AS)	0.51	1,427,648
IV. Per capita consumption 0.1–0.5 kg, 46 countries, 12.5% of world population		
Nigeria (AF)	0.48	195,875
Ethiopia (AF)	0.28	109,224
Russian Federation (ER)	0.17	145,734
Iran (Islamic Republic of) (AS)	0.15	81,800
South Africa (AF)	0.15	57,793
V. Per capita consumption 0.01–0.10 kg, 57 countries, 13.3% of world population		
Democratic Republic of the Congo (AF)	0.09	84,068
United Kingdom (ER)	0.08	67,142
Germany (ER)	0.08	83,124
Italy (ER)	0.07	60,627
France (ER)	0.06	64,991
Republic of Korea (AS)	0.06	51,172
Japan (AS)	0.02	127,202
Türkiye (AS)	0.01	82,340
VI. Per capita consumption between < 0.01 kg, 78 countries, 26.3% of world population		
Pakistan (AS)	0.00	212,228
India (AS)	0.00	1.352,642

Calculation of per capita tilapia consumption is explained in the notes in Table 2.7.

of the world population) with consumption >3 kg and 21 countries (14.2%) with consumption between 1 and 3 kg (Table 2.8). The top ten countries include two in Africa, six in Asia and two in LAC; Indonesia, Thailand and the Philippines in Asia and Mexico in LAC are four countries among those with a population of >50 million. Vietnam and Bangladesh in Asia, Brazil in LAC and the USA in Northern America are four countries in the 1–3 kg category with a population of >50 million (Table 2.8).

Around one-fourth of the world's population live in 23 countries with consumption between 0.5 and 1.0 kg. Four countries in this category have a population of >50 million: the United Republic of Tanzania and Kenya in Africa and Myanmar and China in Asia (Table 2.8).

Around one-fourth of the world's population live in 103 countries where consumption is between 0.01 and 0.50 kg, including 46 countries (12.5% of the world's population) between 0.1 and 0.5 kg and 57 countries (13.3%) between 0.01 and 0.10 kg (Table 2.8). Five countries in the 0.1–0.5 kg range have a population of >50 million: Nigeria, Ethiopia and South Africa in Africa, Iran in Asia and the Russian Federation in Europe. Eight countries are in the 0.01–0.10 kg category with a population of >50 million: one in Africa (Congo, DRC), three in Asia (Republic of Korea, Japan and Türkiye) and four in Europe (United Kingdom, Germany, Italy and France).

Around one-fourth of the world's population lives in 78 countries with consumption <0.01 kg; Pakistan and India are the only two countries in this category with a population >50 million (Table 2.8).

Prospects

While tilapias belong to Labroidei (a suborder under Perciformes; or Cichliformes in more recent taxonomic classifications), they used to be categorized under Percoidei (a suborder under Perciformes) that includes many commercially popular carnivore species (e.g. snappers, groupers, basses and perches). Indeed, tilapia products are sometimes marketed as high-valued species (e.g. seabreams, snappers or groupers) thanks to their desirable characteristics such as absence of small bones, mild and sweet taste, medium-firm and flaky texture, white flesh etc.

While the world population consumed on average 0.9 kg of tilapia per person in 2018, per capita tilapia consumption in countries with a long history of tilapia consumption tends to be much higher: e.g. 12 and 5 kg, respectively, in Egypt and Israel, where Nile tilapia is a native species (Table 2.8). Mostly relying on imports, the 1.37 kg per capita tilapia consumption in the USA was above the world average. Although China (the largest tilapia-farming country) exported over half of its tilapia production and consumed an average of only 0.51 kg of tilapia per capita (much lower than the world average), tilapias have become a well-established and increasingly popular food fish in some provinces (e.g. Guangxi and Yunnan, which are two of the five major tilapia-farming provinces in China).

In 2018, per capita tilapia consumption was <0.5 kg in over 180 countries that accounted for over half of the world's population, and between 0.5 and 1.0 kg (compared with 0.9 kg of world average) in 23 countries accounting for one-fourth of the world's population (Table 2.8). This indicates a large growth potential in per capita tilapia consumption.

Suppose that per capita tilapia consumption in 2018 < 0.5 kg increases to 0.5 kg, that between 0.5 and 1.0 kg increases to 1.0 kg and that >1 kg remain the same as the baseline level: world tilapia demand could reach 10.6 million tonnes in 2030, which is 3.7 million tonnes higher than the baseline level in 2018 (Table 2.9).

Nearly 90% of the 3.7 million tonnes of tilapia demand growth potential is contributed by developing regions, primarily Asia (61.68%) and Africa (23.84%). In seven of the top ten countries (Table 2.9), the demand growth potential primarily reflects the increase in per capita tilapia demand whereas that in the other three countries (Egypt, Indonesia and Mexico) reflects their population growth.

Tilapia aquaculture production in sub-Saharan Africa (SSA) and LAC needs to reach, respectively, 1.3 and 1.1 million tonnes in 2030 in order to fully cover the estimated tilapia demand in these two subregions driven by population growth and higher per capita consumption (Table 2.9). These seem to be achievable targets, in

Table 2.9 Prospects for tilapia demand at the global, regional and national levels.

Country or region	Total consumption (tonnes)						Per capita consumption (kg)	
	Baseline 2018[a]	Population growth only[b]		Population growth and higher per capita consumption[c]				
		2030	Growth between 2018 and 2030	2030	Growth between 2018 and 2030	Share of world growth (%)	2018	2030[d]
World	**6,882,202**	**7,872,621**	**990,418**	**10,591,662**	**3,709,460**	**100.00**		
Developed regions	595,405	636,177	40,772	999,529	404,124	10.89		
Developing regions	6,289,759	7,236,444	946,685	9,592,133	3,302,374	89.03		
Least-developed countries	905,267	1,137,462	232,195	1,431,725	526,458	14.19		
Landlocked developing countries	380,556	517,074	136,518	696,579	316,023	8.52		
Small island developing states	16,813	18,725	1,912	39,541	22,728	0.61		
Asia	3,454,825	3,776,685	321,860	5,742,711	2,287,887	61.68		
Africa	1,940,230	2,480,411	540,181	2,824,678	884,448	23.84		
Sub-Saharan Africa	771,259	1,045,312	274,053	1,333,521	562,262	15.16		
Europe	84,507	84,562	55	370,720	286,213	7.72		
LAC	936,350	1,028,105	91,755	1,129,225	192,875	5.20		
Northern America	463,109	495,488	32,379	500,169	37,060	1.00		
Oceania	6,143	7,370	1,227	24,159	18,016	0.49		
Top ten countries with the greatest demand growth potential								
1. India	~0	~0	~0	751,821	751,821	20.27	~0	0.50
2. China	733,835	752,696	18,861	1,464,340	730,505	19.69	0.51	1.00
3. Egypt	1,168,869	1,434,983	266,115	1,434,983	266,115	7.17	11.88	11.88

Continued

Table 2.9 Continued

Country or region	Total consumption (tonnes)						Per capita consumption (kg)	
	Baseline 2018[a]	Population growth only[b]		Population growth and higher per capita consumption[c]				
		2030	Growth between 2018 and 2030	2030	Growth between 2018 and 2030	Share of world growth (%)	2018	2030[d]
4. Indonesia	1,261,593	1,410,191	148,598	1,410,191	148,598	4.01	4.71	4.71
5. Pakistan	~0	~0	~0	131,479	131,479	3.54	~0	0.50
6. Japan	2,872	27,26	−145	60,379	57,507	1.55	0.02	0.50
7. Congo, DRC	7,845	11,202	3,357	60,023	52,179	1.41	0.09	0.50
8. Philippines	321,194	372,532	51,338	372,532	51,338	1.38	3.01	3.01
9. Mexico	410,897	458,713	47,817	458,713	47,817	1.29	3.26	3.26
10. Russian Federation	25,168	24,756	−412	71,674	46,506	1.25	0.17	0.50
Other countries	*2,949,930*	*3,404,821*	*454,891*	*4,375,525*	*1,425,595*	*38.43*		

[a]Calculated in Table 2.7.
[b]Assuming every country's per capita tilapia consumption in 2030 remains at its 2018 level.
[c]Assuming per capita tilapia consumption <0.5 kg in 2018 increased to 0.5 kg in 2030; that between 0.5 kg and 1.0 kg increased to 1.0 kg; and that no less than 1 kg remaining the same as the baseline level in 2018.
[d]Based on assumption in [c].

light of the encouraging experience of Egypt, which cultivated 1.1 million tonnes of tilapia in 2018 despite its limited land and water resources, as well as its lack of key aquatic feed ingredients (El-Sayed, 2017). However, many technical issues need to be addressed in order to achieve these targets. For example, should countries in SSA and LAC learn from the experience in Asia and Egypt to establish earthen ponds as the main tilapia-farming system despite their lack of experience in pond culture? Or should they try to follow the experience of salmon farming in marine areas to focus on improving and expanding tilapia cage culture in inland waters? In-depth assessments are needed to delineate suitable development paths for tilapia aquaculture in the two subregions, as well as in other places with less developed tilapia aquaculture.

Notes

1 Some Nile tilapia hybrids (e.g. red tilapia *O. niloticus* × *O. mossambicus*) may be recorded under the first item Nile tilapia, whereas the second item, Tilapia nei tends to include mostly Nile tilapia.

2 *Coptodon* is a more updated scientific name for the genus *Tilapia*, which is still used in Aquatic Sciences and Fisheries Information System (ASFIS) and hence the FAO statistics.

References

El-Sayed, A.-F.M. (2017) Social and economic performance of tilapia farming in Egypt. In: Cai, J., Quagrainie, K.K. and Hishamunda, N. (eds) *Social and Economic Performance of Tilapia Farming in Africa.* FAO Fisheries and Aquaculture Circular No. 1130, Rome, pp. 1–48.

EUFOMA (European Market Observatory for Fisheries and Aquaculture Products) (2019) *Conversion factors by CN-8 codes from 2001 to 2019. Metadata 2 - Data management (Annex 7).*

Food and Agriculture Organization (2020a) FAO Global Fishery and Aquaculture Production Statistics v2020.1.0, published through FishStatJ Retrieved March 2020. Available at: http://www.fao.org/fishery/statistics/software/fishstatj/en

Food and Agriculture Organization (2020b) Top 10 species groups in global aquaculture 2018. World Aquaculture Performance Indicators (WAPI) factsheet. Food and Agriculture Organization. Available at: http://www.fao.org/3/ca9245en/ca9245en.pdf and supplementary materials at. http://www.fao.org/3/ca9245en/ca9245en.pdf

Food and Agriculture Organization (2020c) FAO Global Fishery and Aquaculture Commodities Statistics v2020.1.0, published through FishStatJ Retrieved August 2020. Available at: http://www.fao.org/fishery/statistics/software/fishstatj/en

Food and Agriculture Organization (2020d) Tilapia production and trade with a focus on India. World Aquaculture Performance Indicators (WAPI) factsheet. Food and Agriculture Organization. Available at: http://www.fao.org/3/ca9224en/ca9224en.pdf

United Nations (2019) *World Population Prospects* (2019) *Revision.* Available at: https://population.un.org/wpp2019/

United Nations (2020) *UN Comtrade* Database. Retrieved February 14, 2020. Available at: https://comtrade.un.org/

Chapter 3

Reproduction and juvenile production

David C. Little, Warren A. Turner, Halina Sobolewska, Alastair Hamilton and Andrew P. Shinn

Introduction

Tilapias stand out as a group of farmed fish for which the availability of juveniles to stock a production system has rarely been considered a major constraint – the characteristic of effortless natural breeding and recruitment in ponds is a major factor in its global popularity. Swingle (1960) remarked on the 'efficiency' of the Java tilapia, then named *Tilapia mossambica* as part of polycultures with largemouth bass (*Micropterus salmoides*), stocked to control reproduction allowing production of larger individual fish in early trials at Auburn, Alabama, USA. In contrast, the production of a large biomass of individually small, self-recruited fish was widely perceived to be a benefit in many food-insecure countries in the tropics and sub-tropics. By the 1980s the realization that this trait was an impediment to its potential as a major commercial species had been accepted, and the era of hormone-induced sex control to produce monosex male stocks had become demonstrated scientifically and commercially established. Whereas other approaches to producing monosex stocks have been researched (e.g. triploidy through cold shock; Pradeep *et al.*, 2014), genetically male tilapia (GMT) (Pham *et al.*, 1999) have proved to have limited commercial applicability largely because the natural asynchronous, multiple production of

Kim Thompson, Craig Shoemaker and David C. Little (eds) **Tilapia**
DOI: 10.1079/9781800629455.0003, © CAB International, 2025

small batches of eggs makes scaling up using these techniques unviable. Nevertheless, there remains a wide range of strategies in use to produce juveniles to replace fish harvested from farmed systems (Little & Hulata, 2000), and the combination of intensive management and selectively improved strains means that even mixed-sex fish can produce individually large fish consistently (Bostock *et al.*, 2022). Although large commercial hatcheries producing selectively improved monosex fish are a reality in many parts of Asia and elsewhere in the tropics, where the fish has become established commercially, the range of scale and type of hatchery technology remain diverse as do the qualities of fish produced and the implications for health management through the value chain (Fig. 3.1).

Thus, a small hatchery retaining and reusing broodstock to service local seasonal demand for mixed-sex juveniles, in addition to producing food fish, is likely to have different health issues to that facing a large, specialized hatchery competing in national and/or international markets for improved monosex juveniles. Integrated operations producing seed for cage-based lake operations will face different challenges again, especially if multiple producers are present in the lake or reservoir.

Health and welfare over the life cycle of farmed tilapia start in the 'hatchery'. Early environments are known to affect health and welfare outcomes later in the life cycle across a range of biological organisms. This includes pre-birth and early life for humans and other mammals (Veenendaal *et al.*, 2013), insects and fish (Overturf and Huggett, 2015). Trade-offs in hatchery system design and management (e.g. biofloc systems, Deng *et al.*, 2021), selective breeding (Mengistu *et al.*, 2021) and early nutrition (Srisakultiew *et al.*, 2022) have all been found to impact on later performance in tilapia.

Until relatively recently, tilapias were considered robust and to have no major disease issues. The control of trans-boundary fish diseases has been on the radar of scientists and policymakers in Asia for a long time, but the absence of specific mentions of tilapia in work published more than 15 years ago (Lavilla-Pitogo & Nagasawa, 2004; Arthur *et al.*, 2008; Shinn *et al.*, 2023a, b) suggests that attitudes were relaxed until more recently. Reports of the rising economic

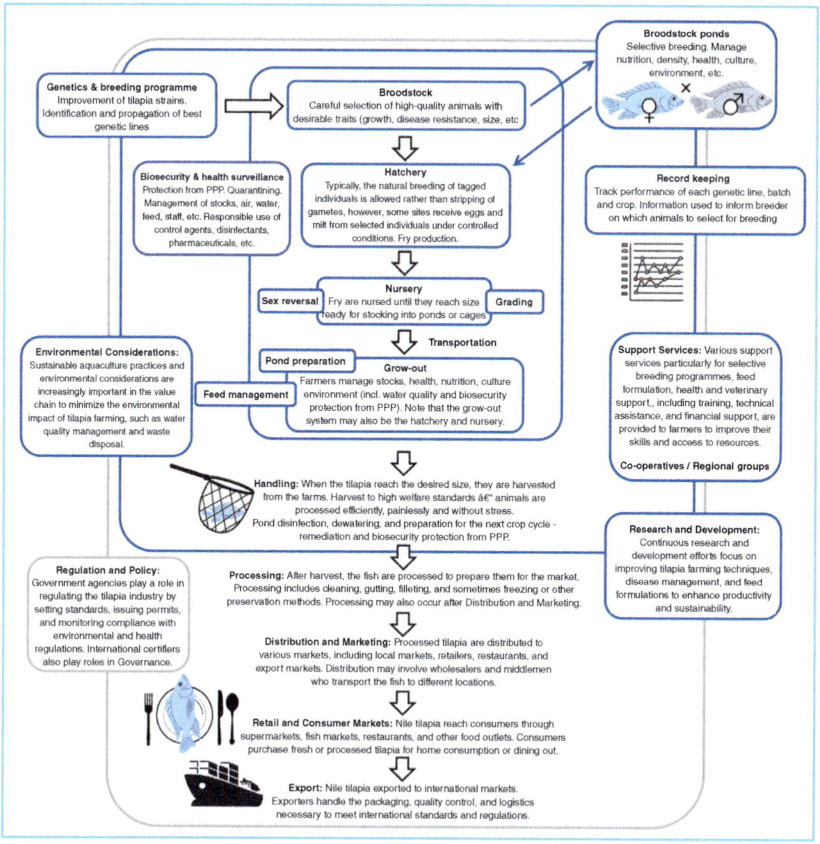

Figure 3.1 Implications for health management through the tilapia value chain – simplified. The boxes enclosed within a blue line cover key issues at the farm level while those enclosed by a grey line cover issues across the entire value chain

significance of *Streptococcus* Group B were becoming common and causing an estimated disease loss of around 300,000 million tonnes valued at $500 million in lost value by 2018 in Asia (Tveteras *et al.*, 2019) and, in the last ten years, prevention and control of disease has become a major focus for the sector.

We will address approaches to tilapia hatchery health that lead to good outcomes for producing fish for consumption. Beginning with natural reproduction strategies and how understanding these can lead to improved hatchery practices, we move to considering decisions around sourcing and replacing broodstock and implications for intensification of broodstock and juvenile management. We then turn to assessing the implications for isolating or protecting tilapia from pathogens, parasites and pests (PPP) or building their resilience and robustness against them through selective breeding. We then reflect on the rapid developments in genomic tools that support such approaches. Hatchery location, design and maintenance are then considered as key elements in ensuring successful and consistent healthy juvenile production before considering in detail prophylactic management steps (e.g. quarantine) and husbandry techniques for broodfish and juveniles that reduce stress and trauma. The specific issue around the use of masculinization hormones to achieve monosex stocks on juvenile health is discussed, before we finally turn to the implications for health management associated with post-hatchery nursery practice and distribution to the final location of grow-out.

Natural breeding and natural environments: implications for high health and welfare-controlled reproduction

Understanding the natural behaviour and breeding of oreochromids, the female mouthbrooding tilapias, is essential to recognizing and overcoming the challenges associated with intensification of juvenile production under controlled hatchery conditions. Producing large numbers of the same age and same size fry, for example, is rarely a problem for fecund species of farmed fish but is more of a challenge for *Oreochromis* that produce small batches of eggs asynchronously and, often, aseasonally. This is even more of a challenge if monosex fish are required, produced through hormone treatment of sexually undifferentiated fry using exogenous hormones in first feeds; large numbers of same-age, same-size eggs and larvae are required for any efficient process. Key to this is the management of large numbers of breeding fish spawning as synchronously as possible. This requires fish to be maintained under optimal breeding conditions in an environment where natural spawning and fertilization of eggs can occur, and their harvest managed, as efficiently and completely as possible, limiting stress and trauma at any stage. It is known from observational studies that synchronization of spawning is a natural phenomenon (Fryer & Iles, 1972) and occurs under captive conditions (e.g. Oliveira & Almada, 1998). The behaviours around lek-based breeding continue to be informed by analysis of pheromones used by males to exert dominance and induce female readiness to spawn (Barata *et al.*, 2007; Li & Buchinger, 2014). Sound production is also known to be important in courtship competition between dominant and subordinate males (Amorim & Almada, 2005). Aggression between males courting females and between males and females whilst mediated in low-density environments where subordinates can escape can result in physical injury and even death in high-density hatchery environments. Mozambique tilapia (*Oreochromis mossambicus*) breeding females are also aggressive on conspecifics, particularly in the late incubation/fry-release stage (Oliveira and Almada, 1998). Early harvest of eggs and yolk sac fry is therefore likely to reduce stress within the broodstock. Ensuring that system design and management are optimized improves performance (see below), but clearly optimal stocking densities and sex ratios are a key management requirement.

Spawning periodicity of maternal mouthbrooding tilapias, as indicated by brood and inter-brood intervals under natural conditions, can be impacted by factors such as lunar cycles, but female health and water quality are critical under captive conditions (Ripley & Lobel, 2005). The maintenance of female condition under intensive breeding management and even brief periods of oral incubation when feeding is restricted is a challenge. The functional trade-offs between successful mouthbrooding and feeding performance have been demonstrated in various species of cichlids (Tkint *et al.*, 2012). The energetic costs of mouthbrooding were significant (about 15% more than for females without their brood) in another group of maternal mouthbrooders (Mrowka and Schierwater, 1988), suggesting the importance of nutrition during the period prior to spawning.

Pathogen-free status or building resistance?

The concept of freedom from disease through designing and managing stringent biosecurity may be possible in some contexts, but managing the microbiome within the tilapia and its environment to avoid disease outbreaks may be a more cost-effective approach than trying to eliminate every pathogen, especially in the open systems that dominate world production (Stentiford *et al.*, 2017). Improving our understanding of the fundamental mechanisms behind pathogen tolerance in wild hosts with higher genetic diversity (Van Houte *et al.*, 2016) and inherited resistance resulting from past exposure to pathogens (Medzhitov *et al.*, 2012; Verbruggen *et al.*, 2015) can provide a basis for controlling the progression of infection to disease in farmed hosts. In addition to supplying specific-pathogen-free (SPF) larvae, which are confirmed to be free of certain pathogens but may not be tolerant to the microbiome or pathobiome of the receiving farm, the provision of more diverse and resilient lines produced through well-managed selective breeding programmes and emerging genetic technologies (such as SNP arrays; Tsai *et al.*, 2016a, b) can augment hatchery supply. This will help mitigate non-listed production diseases (Nir, 2003) and directly benefit farm yield and profit, which is crucial (Cock *et al.*, 2017).

Approaches to breeding for greater resilience and robustness to current and emerging PPP for genetic stocks likely to be farmed in a range of farming systems will involve trade-offs but are likely to benefit from the application of a range of conventional and genomic tools. Progressive improvement in key production traits, most especially in disease resistance through selective breeding programmes, has great potential to help address this challenge. Currently the proportion of aquaculture production derived from selectively bred stocks remains significantly lower than in terrestrial animal- and plant-farming industries (Yáñez *et al.*, 2015; Robledo *et al.*, 2018; Houston *et al.*, 2022). Many aquaculture species, however, are particularly amenable to genetic improvement, due to high fecundity, generally external fertilization and relatively high within-species genetic diversity through retained close relatedness to their wild forebears. Genetic gains through the introduction of selective breeding in aquaculture are generally greater than in terrestrial farm animals (Gjedrem & Rye, 2018), with the widely distributed genetically improved farmed tilapia (GIFT) strain representing a striking exemplar (Nguyen, 2016).

Impaired growth and high mortality levels due to infection remains a significant constraint on aquaculture production, and the possibility of cumulative and permanent increases in disease resilience and its component traits of resistance and tolerance is driving accelerating development of breeding programmes in aquaculture. Nevertheless, it remains the case that uptake of advanced breeding technologies across the industry, in tilapia production most especially, is very variable, ranging from the use of wild seed stocks through to family-based selection utilizing genomic tools (Yáñez *et al.*, 2015; Houston *et al.*, 2022).

Many studies of the genetic basis of disease resistance in aquaculture species have detected significant heritability for these traits (e.g. Yáñez *et al.*, 2014; Gjedrem, 2015), confirming that identification of individual fish with naturally high resistance to a particular disease may be expected to produce offspring with enhanced resistance through selective breeding.

In its broadest sense, resistance includes any genetic mechanism through which improvement will lead to a reduction in disease incidence or severity. Most commonly, this is thought of as an ability to resist *de novo* infection (Bishop & Woolliams, 2014), although it is important to recognize that resistance can comprise other traits: for example, through either reduced impact of the pathogen on performance through tolerance (Doeschl-Wilson *et al.*, 2012) or reduced transmission of the infection to uninfected cohabitees (Lipschutz-Powell *et al.*, 2012). This can complicate identification of breeding goals and capture of accurate disease resistance measurement.

Defining resistance as a binary trait through survival/mortality of individuals during a dedicated challenge simplifies the model, with moderate to high heritability demonstrated for a range of infectious diseases (Ødegård *et al.*, 2011; Yáñez *et al.*, 2014; Gjedrem, 2015). To avoid compromising biosecurity within a breeding nucleus, such data are rarely collected directly from breeding candidates but through the challenge of close relatives (Ødegård *et al.*,

2011). An illustrative study demonstrating the power of a controlled *Streptococcus iniae* challenge in Nile tilapia (*Oreochromis niloticus*) is reported in LaFrentz *et al.* (2016).

Disease data and samples from field outbreaks may be used to determine breeding values under conditions more representative of a commercial production environment, although data of suitable quality may be harder to gather and, in practice, high genetic correlations are often observed between both environments (e.g. Ødegård *et al.*, 2007).

The requirement to use data collected from a challenged population to direct selective breeding decisions in a closely related candidate broodstock population is commonly tackled via family selection. This requires establishing a high number of genetically diverse families with known pedigree relationships tracked via physical tags, genetic markers or both. Breeding values calculated in the experimental population are used to select the best-performing full-sib families among candidates.

A disadvantage with this approach to selection is that the performance of individual fish may be strongly influenced by environment, leading to re-ranking of breeding values between diverse production sites which can significantly reduce the response to selection across the programme. This is particularly applicable to tilapia breeding, with breeding programmes widely disseminating stocks to diverse production environments (Omasaki *et al.*, 2016; Sae-Lim *et al.*, 2016). More generally, progress through family-based selection is inherently rate limited by capture of only the genetic variation between families: the genetic variation within families remains invisible. Since families will necessarily include a range of phenotypes, selection of high-performing families will inevitably include individuals with lower performance and, more seriously in the long term, can also result in the deselection of high-performing individuals within intermediate families.

For a subset of traits this problem may be overcome through marker-assisted selection (MAS), which enables identification of the more resistant selection candidates within full-sibling families, improving on the accuracy of family selection through capture of some of the within-family variation (Sonesson, 2007). MAS requires identification of specific genetic variants linked to a measurable effect on a trait (quantitative trait loci, or QTL) and selection of individuals that carry the favourable allele at the QTL. The relatively recent domestication of aquaculture species, coupled with the novel disease pressures on farmed fish, raises the likelihood that loci with significant effect segregate within farmed populations, and focused efforts to identify these have delivered notable successes in aquaculture. Undoubtedly the most successful example of application of MAS in selective breeding was the discovery that a single QTL explains much of the genetic variance for resistance to infectious pancreatic necrosis virus in salmonids (Houston *et al.*, 2008, 2010; Moen *et al.*, 2009). Subsequent application of MAS has all but eliminated this disease in salmonid aquaculture worldwide (Moen *et al.*, 2015; Norris, 2017). Other notable examples include resistance to *Flavobacterium psychrophilum* in rainbow trout, *Oncorhynchus mykiss* (see Vallejo *et al.*, 2014; Liu *et al.*, 2018), salmonid alphavirus in Atlantic salmon, *Salmo salar* (see Gonen *et al.*, 2015) and lymphocystis disease in Japanese flounder, *Paralichthys olivaceus* (see Fuji *et al.*, 2006).

A significant constraint on progress through MAS in animal breeding generally is that most economically important traits are controlled not by a single gene but by many, each exerting a small effect and thus identified QTLs individually contribute little to the trait (Meuwissen *et al.*, 2013).

A relatively new technology, genomic selection (GS), overcomes this problem and is widely considered to be the state of the art for selective breeding schemes in aquaculture. In GS a population with known phenotypes, such as disease resistance, is genotyped using a large number (thousands to hundreds of thousands) of genetic markers distributed across the genome. Association between phenotype and genotype via a population whose performance has been measured is used as 'training' data to develop a model that can predict genomic breeding values in selection candidates based on their genotypes alone. This has twin benefits: the combined effect of linkage between many markers and QTL, each individually too small to measure, collectively contributes a very significant influence on the trait (Meuwissen *et al.*, 2001).

Secondly, such high-resolution genotypes enable a far more accurate estimate of the genetic relationship between individuals than could be deduced from pedigree records alone, particularly within families (Meuwissen *et al.*, 2013).

Application of GS in aquaculture is currently limited, however all reports to date demonstrate higher prediction accuracy of breeding values than the use of pedigree information alone (Ødegård *et al.*, 2014; Tsai *et al.*, 2015; Dou *et al.*, 2016). Genomic selection remains an expensive technology, although with expanding use costs have fallen, accelerating uptake further. Genotyping arrays suitable for GS have been developed in an expanding range of important aquaculture species, including tilapia (Joshi *et al.*, 2018), Atlantic salmon (Houston *et al.*, 2014; Yáñez *et al.*, 2016; Tsai *et al.*, 2016a, b), rainbow trout (Palti *et al.*, 2015), common carp, *Cyprinus carpio* (see Xu *et al.*, 2014) and catfish, *Ictalurus punctatus* (see Liu *et al.*, 2014).

A compelling demonstration of the ability of genomic technologies to boost prediction accuracy for survival in tilapia exposed to *Streptococcus agalactiae* infection is described in Joshi *et al.* (2021). More recently, Kayansamruaj *et al.* (2023) explored the application of genomics technology in both advanced breeding programmes and simultaneously as tools to develop a next-generation vaccine against streptococcosis in farmed finfish.

Hatchery design to reduce or control PPP

Mouthbrooding tilapias attain breeding condition and spawn naturally under a wide range of managed and unmanaged conditions. The behaviours of breeding *Oreochromis* are adapted to ensure optimal conditions for courtship, spawning and mouthbrooding, which are difficult to replicate in the same 'hatchery'. For example, favoured sites for communal courtship and nest building, leks, are typically in shallow, well-lit environments (Fig. 3.2) ensuring that cues for final maturation, ovulation and spawning are visual and explaining the importance of intensified changes in skin colour and patterns in breeding fish (Turner & Falter, 1989). In contrast, once the female

Figure 3.2 Tilapia spawning pits: active breeding of mixed-sex *Oreochromis* illustrated by the numerous shallow depressions excavated by males in earthen ponds. Optimization of such lek-based breeding behaviour is critical for any hatchery system (courtesy of A. Shinn)

has a mouth full of eggs, hopefully fertilized, she seeks solitude and cover from conspecifics and predators.

'Controlled' hatchery conditions suggests that water quality can be maintained at levels that allow high densities of fish to perform without major losses to PPP. Our understanding of the inter-relationship between specific pathogens and disease, the importance of co-infection by more than one disease agent (Abdel-Latif *et al.*, 2020) and the concept of the pathobiome have all advanced in recent years suggests that approaches to optimizing the production of healthy juveniles cannot rely only on exclusion of PPP but rather is aimed at ensuring high levels of breeder resilience through an optimal environment (particularly diet and water quality).

Maintaining the health and performance of broodstock requires a hatchery design that builds on the understanding of the natural reproductive strategies described above and ensures efficient and trauma-free handling of fish and progeny. Scale and seasonality of

Figure 3.3 Rice fields can make very good breeding systems for tilapias. Low densities of broodfish can result in high individual productivity in such shallow systems, and frequent draining and drying can prevent the build-up of pathogens (courtesy of D. Little)

juvenile production are critical factors affecting hatchery design, and market demand should always be considered. For example, is the demand for frequent sales of small numbers of seed or for larger, more irregular orders? Is there demand year round or is it strongly seasonal? Designs aiming to meet the needs for in-house production within vertically integrated systems have quite different needs to hatcheries competing to sell to a large diversity of customers through time and space. Soil, land elevation and slope, water supply, climate, infrastructure, availability of materials and labour, etc. are further factors that will affect hatchery design.

Health management of broodstock and juveniles in small, dispersed facilities is likely to be quite different to the challenges faced by large centralized systems. At one extreme, the use of irrigated rice fields for low-density spawning and nursing of Nile tilapia, for example, is unlikely to result in significant health issues, even though such systems are 'open' to entry of PPP (Fig. 3.3). There are several reasons for this; rice fields are typically drained and dried between crops, a practice that can eliminate or reduce build-up in

Figure 3.4 Feeding through mesh nets erected to deter bird predation (courtesy of Symrise and W. Turner)

pathogens and parasites. Also, disease outbreaks are less common in natural, polyculture systems using low-density stocking, as this most closely resembles a naturally balanced ecosystem with a fully functioning food web. Disease problems tend to be more economically impactful in large, centralized monoculture systems managed at higher stocking densities.

As tilapia are a very low fecund fish species (producing typically 200–1200 eggs per spawning), most pond space in the production of tilapia juveniles is used for spawning broodfish. For example, a hatchery that produces 5 million tilapia fry per month may cover an area of 15 hectares or more. It is not economically possible to contain such large areas in biosecure buildings to exclude PPP. A better alternative is to build modular units, each with a separate water supply, that give a certain level of biosecurity but, more importantly, facilitate rapid and effective removal of most PPPs as and when they occur. Multiple units that can be isolated and independently filled, drained and sanitized, whether they be tanks, incubators or entire recycle systems, are characteristic of good design. An exception to this is airborne pests such as birds (Fig. 3.4) and flying insects that may need other forms of deterrence or physical exclusion. Greater complexity of design and management also requires more, and a greater range of, skills and capacities in the workforce. Good staff training is essential. Productivity, both in terms of quantity and quality, can be improved by giving staff productivity incentives or even an ownership stake. Training of staff to ensure overt linkages and lines of responsibility between husbandry techniques used and health and productivity outcomes is essential. Ensuring that health status is monitored between production stages is critical. For example, improvements in the hatchery to improve incubation survival and hatchling quality would positively affect outcomes during later larval nursery.

Large businesses might be expected to have more technical capacity to prevent or treat PPP and/or investment in data-driven IT systems to monitor and predict performance. Again, such investments demand higher financial benefits that in turn need to be underwritten by higher productivity. This, in turn, can lead to greater efficiencies (e.g. encouraging higher fish densities) but can result in

Figure 3.5 Mesh barriers to prevent entry of predators such as frogs, air-breathing fish and snakes (courtesy of D. Little)

compromising the health and welfare of stocks. Larger-scale systems with high production will also tend to have greater distribution and delivery challenges, typically serving a more distant, dispersed customer base. Smaller, household-managed enterprises that develop incrementally over time, serving more local demand, may avoid such pressures.

In general, close proximity to customers should be the priority factor in site selection because it reduces fry transportation time and cost. Other important considerations are good, all-year-round road access, consistent water supply and reliable mains electricity. Access to high-quality, perennial water should be a clear aspiration for any new enterprise although in practice many commercial entities operate at less-than-optimal conditions. Proximity to, and extraction

Figure 3.6 Fine mesh hapas can exclude PPP in environments where they are otherwise difficult or impossible to exclude (courtesy of Symrise)

from, perennial surface water can be a double-edged opportunity, with the water body being a reservoir of PPP that is difficult to exclude. Water bugs (*Diplonychus indicus*; see Robert & Venkatesan, 1997), dragonfly nymphs (Tave *et al.*, 1990), snakes (Trape *et al.*, 2008) and birds (e.g. Fitzsimons & Thomas, 2011) have all been connected to predation of tilapia juveniles or to the transfer of pathogens. One study found that, although growth was not impacted by high predation pressure by the predatory water bug *Belostoma anurum*, both monosex and mixed-sex Nile tilapia fry indicated significant stress that was likely to result in poor health outcomes. The 'sit-and-wait' predatory nymph and adult phases were found to grow significantly faster on fish diets than when consuming other invertebrates (Valbon *et al.*, 2019), suggesting that high densities of juvenile tilapias may be at particular risk even in environments with alternative feeds.

Losses from piscivorous fish species and amphibia are also known to be potentially important. The Asian common snakehead (*Channa striata*), and a range of other piscivores, can completely control tilapia recruitment in free-breeding, mixed-sex Nile tilapia ponds. In practice, the complete exclusion of such flying and 'walking' predators in

Figure 3.7 Treatment of pond bottom with lime (courtesy of D. Little)

open-pond systems from neighbouring, unmanaged wetlands is problematic, although physical barriers and control of vegetation in and around the pond periphery and dykes can potentially reduce impacts (Fig. 3.5).

Continuous immersion in water can result in a build-up of PPP in any system. Rapid and complete drying of units holding breeding or juvenile tilapias to remove residual animals and organic waste is a key element of good management. Disease outbreaks in aquaculture can be caused by a wide range of agents, including viruses, bacteria, fungi and parasites (Moreira *et al.*, 2021). Within the ciliates, the trichodinids are among those most encountered in the tilapia culture industry (Abdel-Baki *et al.*, 2017) and their infection pressure is linked to high levels of organic matter in the water (Madsen *et al.*, 2000). Maintaining broodfish and juveniles in hapa-in-pond systems has long been an approach to optimizing performance and excluding predators and pests (Little *et al.*, 1991) in environments that are otherwise open. Fine mesh, UV-resistant, polyethylene net material can exclude major pests and predators allowing high survival, but it must be kept free of major fouling for effective operation and allow essential water exchange (Fig. 3.6). The earthen pond design itself is critical to providing a managed environment in which regular drying is an effective approach to managing many water-quality issues and associated pathogens. This can be very effective if the pond bottom can be completely dried, but high groundwater, seepage from surrounding water or fresh rainwater may prevent complete drying. Thus, drying is often combined with the use of toxicants and/or liming products to raise pH to 11–12 (Boyd, 2017; Fig. 3.7).

Our understanding of the substrate–water interface and the importance of managing the microbiome therein is an increasing area of research stimulated by the ongoing disease challenges faced by the global shrimp industry. Fan *et al.* (2017) characterized tilapia pond sediments from intensively fed systems and found that microbial diversity in the tilapia gut reflects that present in the sediment.

Research into substrate-related pathogen issues in the Pacific whiteleg shrimp, *Penaeus vannamei*, has established the importance of the water–sediment interface and controlling the accumulation of organic solids. The pandemic of acute hepatopancreatic necrosis disease (AHPND)-causing *Vibrio parahaemolyticus* strains and their impacts on shrimp production in South-east Asia have been close to catastrophic, particularly in Thailand, where intensification of pond-based production had peaked. Large-scale implementation of disinfection using chlorine-based products has been linked to a depauperation of the sediment microbiome and its greater vulnerability to dominance by opportunistic pathogenic strains.

Designs of culture units that result in rapid concentration and removal of solid wastes are, therefore, an important attribute in maintaining good water quality and reducing undesirable pathogens. Additionally, adaptations that allow for rapid and complete drainage of water to support sun-drying sanitation of earthen systems are critical. Central sumps that support frequent flushing and removal of sediments can prevent the development of anaerobic conditions, and constrain carry-over of undesirable organisms. In Pacific whiteleg shrimp culture, a central drain or 'shrimp toilet' to facilitate the collection and removal of organic wastes from culture ponds is now commonplace (Kawahigashi, 2018) and is present in approximately 60–85% of ponds from a recent survey of >600 ponds in Thailand (Shinn, unpublished data).

Control of suspended solids and temperature are key factors in maintaining high health in any tilapia hatchery. Ensuring that any water source has low suspended solids and is free of PPP, their eggs/cysts or intermediate hosts is critical to protecting operations. Investment in groundwater that is free of PPP or storage of surface water that allows disinfection and solids settlement may be costly but essential.

Nutrition and diets

Under natural conditions the omnivorous Nile tilapia thrives on microalgae and their decomposition products (periphyton detritus aggregate, PDA). Most pond-based production relies on the dietary contributions of both natural and supplementary feed. Nile tilapia are repeat spawners requiring only short periods of recovery; the challenges to maintaining the nutritional health and productivity of Nile tilapia broodfish are linked to diets supplying key dietary nutrients (Izquierdo *et al.*, 2001). In clearwater systems in which natural food is limited, dietary nutrients need to be higher. Crude protein (CP) should be in the range of 38%, and the digestible energy obtained per gram of CP should be 9.5 (de Oliveira *et al.*, 2014). Under conditions in which natural feeds are available/dominant over several generations, Lu and Takeuchi (2004) demonstrated the sufficiency of such diets in terms of broodstock performance and the higher quality of eggs produced. High levels of n-6 fatty acids are more important than absolute amounts of n-3 lipids, optimizing reproductive performance (Santiago & Reyes, 1993) in freshwater fish.

Maintaining high reproductive health has moved increasingly to the use of pre- and probiotics in broodstock diets through dietary fortification (Hai, 2015). Products based on yeasts and yeast by-products can improve antioxidant defence systems, immune response and disease resistance (Abu-Elala *et al.*, 2021). Dietary inclusion of commercial probiotics has also led to improvements in performance in fingerling production (El-Haroun *et al.*, 2006; Xia *et al.*, 2020).

There are also increasing health and performance claims around the application of biofloc technology to broodfish and juvenile systems. Biofloc, the maintenance of high levels of micro-organisms feeding on organic waste generated within intensive aquaculture systems that have zero or very limited water exchange but aim to treat waste and retain nutrients within the system, have demonstrated prophylactic health benefits under certain conditions. The mechanism for this preventative effect is probably linked to the diversity and concentration of beneficial microbes, even when pathogens are present, probably related to crowding out. Additionally, more stable water quality and high levels of immune-boosting compounds beta-glucans, peptidoglycans and lipopolysaccharides, together with antioxidants contained in biofloc, may explain greater fish immunity and survival (Supono *et al.*, 2014; Kim *et al.*, 2020; Ulloa Walker *et al.*, 2020). At low densities, such as those used for broodfish, growth, digestive enzyme activity and immune response of a GIFT strain were found to be higher in biofloc than in a controlled recirculating aquaculture system (RAS) (Long *et al.*, 2015). In other commercial pilot studies, the high level of suspended solids and unpredictable microorganism stability were issues that limited carrying capacity way below those possible in conventional RAS systems (Little *et al.*, 2008).

Ironically, the expanding literature around the use of pro- and prebiotics comes from countries in which pond-based Nile tilapia culture predominates. Most often these fail to mention that in earthen ponds, microplankton dominate and they are beneficial in removing nitrogenous waste, producing oxygen and providing a rich source of very cheap, natural feed to filter-feeding tilapia. One industry view is that the commercial world focuses on promoting products that can be packaged and sold rather than what the farmer needs.

Much of the emerging nutrition-related disease observed in Nile tilapia is probably an outcome of intensification of culture resulting in loss of natural feed as part of the diet. Research conducted in clearwater within RAS systems can result in research artefacts as a result.

Management of hatcheries and nurseries to reduce or control PPP

Sourcing, holding/quarantine of broodfish prior to entering the hatchery

Ensuring that founder or new stocks of breeding animals introduced into the hatchery are disease free is an important step to ensuring high health of the broodstock and progeny thereafter. Best-practice procedures for the quarantine of stocks entering hatcheries have been framed around three levels of risk for aquatic animals in general by Arthur *et al.* (2008) and Subasinghe and Shinn (2023), and then more specifically for tilapia by Diake *et al.* (2019): (1) high risk for the context of introductions and transfers of fish between regions of different or unknown health status where escape of fish or associated PPP is likely. For hatcheries with poor biosecurity, this can mean not only problems affecting their own hatchery productivity but also problems that can impact fish health and environments in the regions in which they are located; (2) risk is reflected where trading of healthy hatchery juveniles is well established, and quarantine is designed to improve biosecurity and confidence among the trading partners; and (3) routine quarantine of stocks where biosecurity and stock management and health status is high in hatchery-produced or local wild-captured broodstock and juveniles. The general principles of quarantine listed by Arthur *et al.* (2008), therefore, should be followed for high-risk introductions of tilapia, notably for stocks from regions where certain pathogens are known to be present and have the potential to transmit both horizontally and vertically (see Tables 3.1 and 3.2).

A progressive management pathway (PMP) approach for improving aquatic biosecurity aims to manage risks from harmful agents in aquaculture through a comprehensive and collaborative approach involving public and private sectors. The key components of an effective biosecurity management strategy include: (1) animal management to ensure healthy stocks and promote immunity through good practices; (2) pathogen management to prevent, reduce or eliminate pathogens; (3) people management to educate and engage

Table 3.1 Commonly encountered bacterial pathogens of tilapia with comments on their potential for vertical transmission. H, horizontal transmission; *Om*, *O. mossambicus*; *On*, *O. niloticus*; *Osp*, *Oreochromis* sp.; V, vertical transmission.

Name	Classification	Species	Transmission	Comments	Reference
Aeromonas hydrophila (motile *Aeromonas* septicaemia)	Aeromonadaceae	*On*	H	The vertical transmission of *Aeromonas* spp. is not yet elucidated.	Dong *et al.* (2015); Shinn *et al.* (2023b)
Chlamydia spp.	Chlamydiaceae	*On*	H, V?	The 16S rRNA detection of *Chlamydia*-like organisms in pre-hatch eggs in farmed *Lates calcarifer* alludes to the possibility of their vertical transmission.	Stride *et al.* (2013, 2014)
Edwardsiella ictaluri	Enterobacteriaceae	*On* ×*Om*	H, V?	The presence of the bacterium in the gonads of *Ictalurus punctatus* suggests the potential for vertical transmission.	Mqolomba and Plumb (1992); Dong *et al.* (2019); Nhinh *et al.* (2022)
Edwardsiella tarda	Enterobacteriaceae	*On*, *On* ×*Om*	H	No known record of vertical transmission in other fish species.	Clavijo *et al.* (2002)
Enterococcus faecalis	Enterococcaceae	*On*	H	No record of vertical transmission in other fish species.	Abu-Elala *et al.* (2020)
Flavobacterium columnare	Flavobacteriaceae	*On*, *On* ×*Om*	H, V?	Not clear whether transmission is vertical. Presence of a bacterium in the reproductive organs of asymptomatic broodstock, fertilized eggs and newly hatched fry alludes to the possibility of maternal transmission.	Suebsing *et al.* (2015); Haenen *et al.* (2023)
Francisella noatunensis subsp. *orientalis*	Francisellaceae	*On* ×*Om*	H, V	Asymptomatic carriers can vertically transmit bacteria to fertilized eggs. Bacteria detected in ovaries, testes, milt, in both unfertilized and fertilized eggs, and in 1-, 10- and 30-day old larvae.	Pradeep *et al.* (2017); Nguyen *et al.* (2019)

Hahella chejuensis	Hahellaceae	*O*sp	H, V?	*In situ* DNA hybridization showed positive signals in the ovaries and testes of tilapia broodstock.	Senapin *et al.* (2016)
Lactococcus garvieae	Streptococcaceae	*O*sp	H	No record of vertical transmission in other fish species.	Abu-Elala *et al.* (2020)
Micrococcus luteus	Micrococcaceae	*On*	H	Bacteria were isolated from the gonads.	Abd El-Rhman *et al.* (2009)
Mycobacterium marinum	Mycobacteriaceae	*O*sp	H	The vertical transmission of *Mycobacterium* species remains to be elucidated. Granulomas are recorded from the ovary of several fish species.	Mori *et al.* (1986); Harms *et al.* (2003); Kent *et al.* (2016)
Nocardia asteroides, N. seriolae	Nocardiaceae	*O*sp	H	Vertical transmission is unknown; granulomas have been recorded in the ovary of *Micropterus salmoides*	Chen (1992); Labrie *et al.* (2008)
Pseudomonas fluorescens, P. spp.	Pseudomonadaceae	*On*	H	Bacteria were isolated from the gonads.	Abd El-Rhman *et al.* (2009)
Shewanella putrefaciens	Shewanellaceae	*On* × *Om*	H, V	Bacteria detected in ovaries, testes, milt, in both unfertilized and fertilized eggs, and in 1-, 10- and 30-day old larvae.	Pradeep *et al.* (2017)
Streptococcus agalactiae	Streptococcaceae	*O*sp	H, V	Although milt, unfertilized and fertilized eggs were negative when tested by LAMP, 40% of 10-day-old larval offspring tested positive; 55% of the reproductive organs of the asymptomatic parents were positive.	Pradeep *et al.* (2016)
Streptococcus iniae	Streptococcaceae	*O*sp	H, V	90% of milt samples and 70% of unfertilized eggs collected from the broodstock were infected at the time of spawning and transmitted to their offspring.	Pradeep *et al.* (2016)

stakeholders; (4) relevant research; and (5) supportive policies. The objectives are reducing disease burdens, improving farm and national health, limiting global disease spread, maximizing socio-economic benefits, attracting investment and promoting One Health goals (Subasinghe *et al.*, 2023).

One key issue is that sub-clinical levels of PPP in introduced tilapia emerge as a problem only when the fish are stressed. This suggests the importance of containment facilities in which various stressors can be applied or carefully controlled. Co-culture with native species, or the sharing of effluent water holding the quarantine fish, can be used to assess the presence of PPP on imported stocks and the susceptibility of such species. However, negative results depending on the diagnostic tests applied do not conclusively mean that PPP are absent but may mean that asymptomatic carrier fish may harbour infections below detectable levels for the diagnostic used.

In the shrimp industry, some stocks referred to as 'all-pathogen exposed' (APE) stocks may result from 'uncharacterized, selected survivor stocks', i.e. stocks that have been created by selection of surviving shrimp based on their size and overall health appearance over multiple generations. This selection process may take place under farming conditions with less than rigorous biosecurity measures in place against the presence of various known and unknown pathogens in a specific region (Alday-Sanz *et al.*, 2020). The term, however, is misleading because these shrimp stocks have not been, and will never be, exposed to all possible pathogens. While it is suggested that such populations may exhibit some level of resilience when naturally challenged by specific pathogens in ponds, it is crucial to dispel this dangerous myth. These animals can potentially serve as carriers of the pathogens to which they have been exposed, thus acting as reservoirs of infection. Consequently they pose a significant risk for the trans-boundary movement of diseases. The same critical biosecurity considerations should be applied to all aquaculture species and, in the context of this chapter, to tilapias.

The need for large numbers of broodfish and the practice of regularly upgrading broodstock through the introduction of improved lines from external sources make the implementation of effective quarantine and biosecurity procedures critical in commercial hatcheries. The common practice of buying back fish produced at different sites for use as broodstock creates a high risk of introduction of PPP and is a major incentive for hatcheries to keep broodstock production 'in house' and in conditions as biosecure as possible. One perspective is the downside of successful biosecurity measures, in that there will be no development of resistance to local pathogens experienced by farmers, and fish may not perform well as a result. A parallel risk is that the same fish may also show greater resistance to control agents in the event they become infected and require treatment. Under strict biosecurity practices, while animals from the pond cannot return to the broodstock/multiplication centre because of the potential disease risk they pose, the selection for resistance to pathogens can be achieved through carefully managed breeding programmes of selection.

Commercial-scale hatcheries require significant capacity to produce the numbers of broodstock they need, and complete exclusion of PPP is likely to be costly/impractical. For example, cage culture in open water bodies exposes potential broodfish to infection, particularly if such water bodies are shared with numerous independent producers. Repeat introductions to the hatchery of improved strains from external suppliers also represent an additional risk of pathogen introduction to the hatchery. This has motivated larger enterprises to retain in-house genetic improvement programmes, but the success of these in the longer term, compared with the performance of fish from specialized commercial breeding companies, is unknown. The overall benefits of breeding for local conditions compared with global strains developed with cutting-edge technologies at scale have yet to mature, as demonstrated by the chicken industry. Limited availability of diagnostic capacity for even common diseases is likely to be a severe constraint for private hatcheries.

Vertical transmission

Due to the potential for certain viral and bacterial pathogens to transmit vertically from parent fish to their offspring through eggs or larvae, representative samples should be included in regular health assessments as part of passive and active health surveillance.

Table 3.2 Viral pathogens of tilapia with comments on their potential for vertical transmission. *Cz, Coptodon zillii; Hap, Haplochromis; Oamp, O. amphimelas; Oau, Oreochromis aureus; Oe, O. esculentus; Ov, O. variabilis; Sg, Sarotherodon galilaeus.*

Name	Classification	Species	Transmission	Comments	Reference
Bohle iridovirus (BIV)	Ranavirus: Iridoviridae	*Om*	H, V	Recurrent mortalities in fry collected from mouthbrooding females a year after surviving an initial epizootic suggests possible vertical transmission of the virus.	Ariel and Owens (1997); Machimbirike *et al.* (2019)
Infectious pancreatic necrosis virus (IPNV)	Aquabirnavirus: Birnaviridae	*Om, On*	H, V	It is not fully clear how the virus spreads among tilapia, but it is known that the virus can persist in a latent state without causing any symptoms. Later, the virus can be transmitted vertically from infected broodfish to fertilized eggs.	Bootland *et al.* (1991); Garcia *et al.* (2010)
Infectious spleen and kidney necrosis virus (ISKNV)	Megalocytivirus: Iridoviridae	*On, On* × *Om*	H, V	ISKNV detected in 10% of ovaries (ovarian follicles but not oocytes) using a colorimetric LAMP assay with a pre-addition of hydroxynapthol blue. Virus not detected in the testes. An infection of 2.45% found in *On* eggs and 44.2% of *On* × *Om* eggs.	Suebsing *et al.* (2016)
Iridovirus-like agent	Iridoviridae	*On*	H	Systemic disease described in juvenile, 3–6 cm-sized, fish; no evaluation of the developing ovaries or testes.	McGrogan *et al.* (1998)

Lymphocystis disease virus (LCDV)	Lymphocystivirus: Iridoviridae	*Hap, Oamp, Oe, Ov*	H	Case report of lymphocystis on the tails of fish.	Paperna (1973)
Tilapia lake virus (TiLV)	Tilapinevirus (Orthomyxovirus-like): Tilapia tilapinevirus	*Cz, Oau, On, Sg*	H, V	The detection of TiLV genomic RNA and viable virus in the testes and ovaries indicates the potential transmission of this virus from infected broodstock to their offspring.	Yamkasem *et al.* (2019); Dong *et al.* (2020)
Tilapia parvovirus (TiPV)	Chapparvovirus: Parvoviridae	*On*	H, V?	TiPV has been found within the oocytes of infected *O. niloticus* suggesting that vertical transmission is likely.	Dong *et al.* (2022)
Tilapia larvae encephalitis virus (TLEV)	Herpesvirus-like virus: Herpesviridae	*Oau, On, Sg*	H, V	Both sexually mature males and females can be carriers (61.9% males, 56.2% females). 18/25 *Oau* were positive; 4/12 *On* were positive.	Shlapobersky *et al.* (2010); Sinyakov *et al.* (2011)
Viral nervous necrosis (VNN)	Betanodavirus: Noraviridae	*On, Osp*	H, V?	Vertical transmission suspected. Virus has been detected in the developing gonads, eggs, and seminal fluids of several species.	Bigarré *et al.* (2009); Keawcharoen *et al.* (2015); Prihartini *et al.* ((2015); OIE (2019)

Tables 3.1 and 3.2 provide a summary of commonly encountered viral and bacterial pathogens of cultured tilapias and highlight that a significant number have the potential for vertical transmission. However, the mechanism and precise tissue location of many vertically transmitted pathogens remain to be established and, in many cases, it is believed that infection arises from surface contamination of the eggs with the pathogen in ovarian fluid. For instance, white colonies of *Francisella noatunensis* subsp. *orientalis* (*Fno*) were observed in the ovaries of hybrid red tilapia but not within the testes (Nguyen *et al.*, 2019). No infection of the testes was seen on histopathology, but infection was confirmed by PCR and *in situ* hybridization (ISH) in the ovary, testes, fertilized eggs, yolk sac fry and in 5- and 30-day-old fish. The study found *Fno* located in the connective tissue of the testis and the oocyte membrane but not within the oocyte, suggesting that it might be possible to disinfect eggs from infected broodstock to have offspring that are infection free. It is, however, unknown whether the bacterium can be found within milt and ovarian fluid. Additional research is required to investigate whether pathogens can be identified in both fertilized and unfertilized eggs. Moreover, it needs to be determined whether disinfecting eggs can lead to a decrease in the incidence of diseases caused by pathogens that are transmitted not only horizontally but can also be present in ovarian fluid, milt and those that inhabit the external surface of eggs.

Although the application of specific substances (such as chloramine-T, formalin, potassium permanganate, etc.) to sanitize the eggs of *O. niloticus* can lead to reduced survival rates following a bacterial challenge by pathogens such as *Flavobacterium davisii*, a bacterium responsible for columnaris disease, research conducted by Guan *et al.* (2022) demonstrated that infection resulted from diseased eggs and, therefore, that disinfection is warranted. Microscopic examination of eggs confirmed the presence of bacteria on the egg surface; their ability to attach to biological surfaces has been demonstrated to play a critical role in the pathogenesis of *F. columnare* (see Olivares-Fuster *et al.*, 2011). For a review of agents and regimes used for the disinfection of eggs, including those of tilapias, see De Swaef *et al.* (2016).

Impact of temperature regimes on hatchery management

The optimal temperature range for Nile tilapia is 28–30°C, but major areas of tilapia production occur in the sub-tropics where seasonal temperatures dip below or above the preferred range and pose challenges to maintaining both breeding animals and juveniles in high health. Change in temperature outside the normal physiological range can be a stress challenge, reducing immune response and lowering resistance to pathogens (Karvonen *et al.*, 2010). Such conditions make handling, grading and transporting fish riskier and can result in poor outcomes. Temperature regimes have also been used as a form of prophylactic health management, which will be reviewed in this section.

Concerns about climate change and the frequency and duration of elevated water temperatures (>35°C) are increasing. In Southern China changes in gill morphology, reduced hypoxia tolerance and inhibited aerobic metabolism have been demonstrated in response to such high temperatures, with implications for tilapia health and survival (Zhou *et al.*, 2022). In contrast, low water temperature leads to other health challenges for tilapias and ectoparasites becoming more problematic. The specific profiles of sub-optimal temperature regimes are highly variable and may favour different mitigation or remedial approaches. In Northern Vietnam (17–23° 22' N) for example), water temperatures can drop to 10–18°C for up to four months, occasionally dropping to <10°C resulting in high mortalities (Dan and Little, 2000). Such conditions, i.e. an extended 'cool season', the severity of which may vary considerably year to year, are then followed by a rapid return to more optimal temperatures. The strategic siting of ponds used to overwinter both broodstock and juveniles, to protect the water surface from cold prevailing winds and the holding of fish in deep hapas (3 m) within deep (3.5 m) earthen ponds, was identified as a cost-effective mitigation approach. Sheltering ponds behind windbreaks is believed to moderate water temperatures and improve growth rates. Farmers had observed that tilapia burrow into sediments at low temperatures and typically suffer high levels of *Saprolegnia* infection; restricting access to the sediment by holding them in deep hapas improved survival.

Other cool-temperature regimes include those prevalent at high-altitude locations that experience low median temperatures (Green *et al.*, 1997) or for which there is a prolonged winter period of low temperatures and extreme diurnal temperature variation (e.g. Kuwait; Ridha *et al.*, 1998). Polytunnels and heaters are used in Egypt to maintain higher ambient water temperatures (3–4°C higher than ambient) to stimulate spawning in concrete tank systems early in the season (Nasr-Allah *et al.*, 2021; see Fig. 3.8).

Enhanced nutrition has also been demonstrated to improve the resilience of tilapia under lower-temperature regimes. Huang *et al.* (2022) found that Nile tilapia fingerlings fed a diet high in linolenic acid prior to cool temperatures (15–20°C for 8 weeks) had better survival following a challenge test with *Aeromonas hydrophila*. High temperatures are also a recognized risk factor: for example, tilapia in Thailand become more susceptible to *S. agalactiae* infection (Kayansamruaj *et al.*, 2014). In addition to reducing immune

Figure 3.8 Polytunnels may be effective at moderating temperature changes during cold periods in the hatchery or nursery (courtesy of D. Little)

Figure 3.9 Shading of hapas containing spawning fish can improve productivity by moderating high daytime temperatures (courtesy of A. Shinn)

response, it is known to cause changes in the gene expression of *Streptococcus*, and in turn morphology, metabolism and virulence (Mereghetti *et al.*, 2008). The disease is known to particularly impact on large individual fish, making broodstock particularly vulnerable. The susceptibility of tilapia to *S. agalactiae* grown in cages in lake systems in Brazil is also co-related to temperature (Delphino *et al.*, 2019). In contrast, small fish were most impacted by *Fno* during seasonal declines in temperature. Bhujel and Perera (2017) demonstrated the negative impacts of high ambient temperatures on broodstock productivity and used simple net shading to reduce water temperatures by around 2°C (Fig. 3.9). The level of difference in seed output suggests the importance of temperature to the productivity and underlying health and condition of broodfish.

Since tilapia, like most fish and all invertebrates, are poikilothermic, it is not surprising that thermal preference of aquatic animals has been demonstrated for a range of species and not just in Nile tilapia (Boltaña *et al.*, 2013; Cerqueira *et al.*, 2016; Huntingford *et al.*, 2020).

The concept of behavioural prophylaxis stems from the innate behaviour of poikilotherms, including fish, to 'self-treat' when infected by seeking higher external temperatures to expose themselves to 'fever' conditions that disfavour infectious pathogens. The basis of this behaviour appears to be a response to pathogen recognition. Given a thermal gradient, pathogen-challenged individuals will seek higher temperatures and demonstrate higher survival than conspecifics. Proven under controlled laboratory conditions with pathogen-challenged animals, the concept has also been trialled in the field with modified ponds in Egypt.

Under general conditions it would be expected that optimal water quality, including temperature, throughout a culture system would be a positive design feature, thereby reducing exposure to environmental stressors. In practice it is problematic to optimize water quality throughout a pond system exposed to diurnal and seasonal change; designed temperature gradients within such systems may therefore have practical value in minimizing losses to disease.

An alternative approach is to try and provide ideal temperature conditions that reduce stress and disease and improve productivity, but any investments to do so must be commercially viable. Pond-based adaptations include shading and water spraying to reduce temperature, and polytunnels and other greenhouse systems for elevating temperature. Tilapia are even grown in cold temperate regions of the world in heated, insulated buildings. Although there are large costs to temperature control systems, there will be returns in terms of improved production and the reduction or elimination of the use of chemotherapeutants.

Handling, grading and distribution

Handling and transportation are known as stressful nodes, both within farms and the wider value chain. Crowding and confinement are known to raise stress hormones. Vijayan *et al.* (1997) found elevated levels of cortisol, glucose and prolactin hormones (Auperin *et al.*, 1997) in wild-caught fish and reduced levels of growth hormones following confinement. The release of such chemicals is believed to be part of chemical communication between fish; even non-handled fish showed elevated levels of stress hormones when they received effluent water from tanks holding stressed fish (Barcellos *et al.*, 2011). It is well known that salt (NaCl) is commonly used during handling/grading operations to reduce stress and could be related to this phenomenon.

Carneiro and Urbinati (2001) demonstrated the use of salt to mitigate stress during transport that Hegazi *et al.* (2014) applied to Nile tilapia (8 ppt), and found similar beneficial impacts of tolerance to higher density and lower metabolic cost.

Combinations of stressors during handling, confinement and transportation are normal. Evans *et al.* (2006) demonstrated the impacts of high unionized ammonia (UIA) concentrations on Nile tilapia and the acute stress reactions and high levels of mortality that occurred when UIA exceeded 2 mg/l. He also noted that exposure to sub-lethal concentrations (0.32–0.37 mg/l UIA) for 24 hours did not exacerbate mortalities from a pathogen challenge (*S. agalactiae*). Long duration of exposure to sub-lethal low dissolved oxygen (DO) concentrations (Evans *et al.*, 2003) and bacterial challenge at elevated temperatures (Evans *et al.*, 2004) were shown to increase disease susceptibility to *S. agalactiae*, suggesting that stressors can compound to reduce survival under conditions common in hatcheries.

Handling for vaccination is another hatchery-level stressor that can be alleviated by intra-peritoneal multivalent vaccines (Shoemaker *et al.*, 2012), which protect against more than one pathogen. A single injection is less stressful and labour-intensive than multiple injection events and was found to provide good protection against both *V. vulnificus* and *S. iniae*.

Swifter, less physically traumatic, handling and sedation are approaches increasingly used to reduce handling stress. The use of soft gloves and soft netting material for netting adult and juvenile tilapias is advised. The synthetic anaesthetic tricaine methanesulfonate (MS-222) has been the mainstream product used for fish and the only US FDA-approved drug for such purposes, but it has a long chemical withdrawal time (Durhack *et al.*, 2020) and has genotoxic properties (Dos Santos Nascimento *et al.*, 2020). Considerable research has been made into plant-based alternatives such as clove oil extracted from *Eugenia caryophyllata*, the main chemical components of which are eugenol and isoeugenol (Cooke *et al.*, 2004; Javahery *et al.*, 2012). Bicarbonate of soda ('baking soda') is a cheap and readily available product for anaesthesia used on Nile tilapia juveniles. Baking soda works through the release of carbon dioxide, which was effective for both induction and recovery when used at dosages of 50–100 g/l in both fresh and saline water conditions (Avillanosa and Caipang, 2019).

Anaesthesia is commonly used for fish biometric analysis on farms (artificial spawning, weighing, markings, classifications, blood sampling and surgical procedures; Anderson *et al.*, 1997), critically reducing trauma during handling that leads to stress and disease. Anaesthetics are also used as sedatives to reduce stress levels during transport. In one simulated transport model for tilapia fingerlings, benzocaine and MS222 were used at 35 and 75 mg/l, respectively and valerian extract from *Valeriana officinalis* at 200 mg/l. No reductions in stress were observed, and the result was interpreted as due to the dosages being too low (Montañez Calero *et al.*, 2022).

Sedation based on essential oils of natural plants, such as *Hesperozygis ringens*, has been used successfully to sedate juvenile Nile tilapia and reduce handling and transportation stress, as indicated by a range of haematological and biochemical parameters. Nile tilapia (ca. 35 g) subjected to simulated transport (4.5 hour) performed best when sedated using 20 µl/l, but even 10 µl/l caused a decrease in protein, triglycerides and cholesterol values immediately following transport of *O. niloticus* (see Ferreira *et al.*, 2022).

Grading

Although grading for size is stressful and traumatic to fish, it is essential to maintaining good size uniformity. During nursery of tilapia, fry fish become uneven in size within 5–7 days, mainly because the larger fish are more dominant and get more feed. Without grading, many of the small fish would perish. Gregory (1987) described high levels of cannibalism among conspecifics under 30 days of age that were exacerbated by size differences. Grading is most usually used on juvenile fish prior to stocking in grow-out. There is a huge variety of grader types, ranging from simple boxes with plastic mesh to very expensive commercial graders that can separate large volumes of fish into many sizes. These types of graders are often supplied to fish at high speed by fish pumps and connected to fish counters so that stress and labour are reduced to a bare minimum.

Grading of tilapia can be roughly divided into grading in the pond and grading on land. Grading in ponds is carried out using hand-held baskets of various designs (Fig. 3.10) and hapas to contain fish. The big advantage is that fish do not have to be moved from the environment in which they are contained, graders are cheap and it is a quick and efficient process. The disadvantage is that staff have to work in the water, aand in temperatures below 30°C they feel uncomfortable working in the water for prolonged periods. At certain times, particularly when dissolved oxygen and/or temperature is low, fish are inactive and grading takes longer. Some farmers move fish into tanks for grading, but two-way transport adds another level of stress to the fish and there may be some shock to fish when they are stocked in a different environment. Grading in tanks is easier for staff, however, and fish are not lost during handling.

Large commercial graders (Fig. 3.11) work well with large fish down to fingerling size but are not yet available for fry. Larger fish are heavier and so warrant investment in this type of equipment, particularly if labour is expensive and in short supply. Tilapia fry producers are mostly limited to using home-made graders using various meshes such as netting, plastic net sheets, vertical bars and even hand-made graders made by punching round holes in aluminium washing-up bowls. These graders generally divide fish into two sizes, but they can be stacked to grade into three or four sizes at the same time. The graders still need separating, however, to remove the graded fry from each container.

Netting, mesh and/or graders with any sharp edges will cause physical abrasion to fish; many fish will get stuck in the holes and suffer damage. The choice and quality of the grading system, therefore, is an important consideration, and tilapia fry need to be rested for 2–4 days after grading and before they are sold or stocked.

Figure 3.10 Grading can be critical to maintaining consistent batches of juveniles but can be stressful and result in physical trauma; in-hapa grading of fry (courtesy of W. Turner)

Figure 3.11 Mechanical grader for fingerlings (courtesy of W. Turner)

Transportation

Many factors influence the geography of juvenile and food fish tilapia, leading to their co-location or the need for short- or long-distance transportation of juveniles. Variable temperature regimes between northern and southern Vietnam, for example, lead to a concentration of hatcheries in the south but significant grow-out capacity in the North and a subsequent demand for (air) transportation of juveniles between them. Productive hatcheries, even small-scale, can quickly out-produce local demand, necessitating transportation to customers at distance. This is particularly the case if clusters of production develop, such as in areas of Northwest Bangladesh. Rice field-based hatcheries, previously dependent on itinerant fry traders to distribute small consignments, have come to rely on wholesale fingerling markets to distribute the very large volumes of fingerlings that they collectively produce over much larger areas.

The main key to successful transportation of tilapia is to starve fish prior to any movement that will take an hour or more. Because tilapia are filter feeders, starving fish can be problematic especially if cultured in water rich in natural food. Ideally, tilapia fry transport is best if maintained in filtered water (i.e. all planktonic food removed) for a minimum period of 6 hours prior to packing. Concrete tanks with hapas installed are most suitable for holding and starving fish, and enable simple and stress-free netting of all the fish in a tank. This approach also facilitates the removal of dead or weak fish prior to packing. The packing of fish is done by congregating fry at one side of the hapa and allowing the fry to swim out against a flow of water (Fig. 3.12). Weak fish will not be able to swim against the current and can be removed with the dead fish. Such 'swimming out' of fish is a critical step to removal of weak fish that would otherwise transport poorly.

It is important that packing tanks are covered with a roof to prevent direct sunlight and provide good aeration via a blower or water fountain system. The latter is good in very hot climates because the

Figure 3.12 Grading of red tilapia juveniles in concrete tanks prior to transportation (courtesy of D. Little)

moving water allows for rapid cooling due to the increased evaporation of water created by the water fountain. During cold weather, however, it may be necessary to create a greenhouse using clear plastic sheets to surround the packing tank area to stabilize the temperature and prevent drops in temperature.

Tilapia fry are most commonly packed for transport in plastic bags with oxygen (Fig. 3.13). Usually, 30% of the bag's volume is water and the remainder is pure oxygen. Although this does not allow the maximum number of fish to be transported per bag, it does allow the maximum number of fry to be transported per weight of water. Trucks are limited in how much weight they can carry. Most pick-up trucks, for example, can only carry 1 tonne of weight. If a 10-l bag is filled with 3l of water, the truck can potentially carry 300 bags of fish but, if 5l of water is added per bag, only 200 bags can be carried. Double plastic bags are used because these provide a level of insurance against losing fish if one bag has a hole. Thin plastic material is preferable because it facilitates twisting of the bag neck and sealing with an elastic band. The plastic material that the bags are made of is also a very important consideration because brittle plastic shatters when dropped while overly soft, malleable plastic will stretch and bubble up under pressure.

The advantage of packing fry in bags, compared with using tanks with pure oxygen, is that the oxygen gradually diffuses into the water and gas bubble disease is not a problem. Also, it is very easy to divide shipments between different customers and ponds, and customers can easily check a single bag to confirm the quality and quantity of fish. The disadvantage of bags is that carbon dioxide, a byproduct of respiration, cannot escape to the atmosphere and can become toxic to the fish. It also causes a significant reduction in water pH, which most likely stresses the fish, but it does have the benefit of reducing the proportion of ammonia in the unionized and toxic state.

It is important when transporting fish in bags to put down some sort of insulation on the floor of the truck to prevent heat transfer from the engine. Bags should also be covered with material such as jute bags to prevent direct sunlight and wetted down to maintain a 'low' temperature. Fish can survive for longer distances at higher density if the temperature can be reduced to an ideal of 23°C; transportation of fish should be completely avoided during spells of very high temperatures. Anaesthetics (such as clove oil, MS-222 and phenoxyethanol) can be used to reduce metabolism during transport, but reducing the temperature, if possible, is preferable when transporting tilapia. Another significant factor to consider is the timing of fish transportation, with a preference for transfer during the early morning hours, arriving at the destination before the day becomes hotter when there is an increased probability of heat stress.

Open transportation of juvenile tilapias is still practised as for carp in Bangladesh, i.e. in simple aluminum containers on bicycles and motorized rickshaws, sometimes over significant distances and time periods, with aeration achieved through splashing against the container during transportation (Fig. 3.14). Tanks can be used for transporting tilapia fry, but these are more commonly used for transporting fingerlings, broodstock and grown-out fish sent live to market. Transport tanks come in all sizes and shapes, but tall, narrow tanks create less of a problem from water shift and spilling (caused by movement of the vehicle) than low, wide tanks. These are very useful for transport on farms and can be fitted with hapas to allow for the easy removal of fish. Insulated tanks take up more space but are useful for maintaining a 'low' temperature during transport. The advantage of tanks is that carbon dioxide is lost to the atmosphere through diffusion, and the control of temperature is much easier than bags. Given that the oxygen is kept compressed in bottles or air is supplied by a blower, less space is used in the truck.

The stocking density will depend on the size of fish, transport time, temperature and whether using air, oxygen or a combination. For most long-distance transportation, pure oxygen is used because it allows a higher density of fish to be transported. It is important to use fine-bubble diffusers to maximize the amount of oxygen that is dissolved in the water before the bubbles rise to the surface and are lost to the atmosphere, which extends the time that a cylinder of oxygen will last. Ceramic diffusers are ideal for this but are expensive. In Thailand, aquarium aeration tubing with tiny holes made with a needle is used. The tubing is attached to a metal frame which sits on the bottom of the tank (Fig. 3.15). A certain amount

Figure 3.13 Transportation of juveniles in oxygenated plastic bags (courtesy of W. Turner)

Figure 3.14 Transportation of tilapia juveniles in open tanks in Bangladesh (courtesy of D. Little)

of skill is required because the fish will use more oxygen following initial stocking, but this will reduce after the fish settle down. It is important to adjust the amount of oxygen supplied to prevent oxygen super-saturation, because gas bubble disease (oxygen comes out of solution in blood to form bubbles) happens very quickly and is lethal. This problem could be overcome in the future using nano-bubble generators, which produce bubbles of <130 nm, smaller than many viruses. The bubbles are so small that they are not buoyant, and they repel each other due to being negatively charged. High levels of dissolved oxygen can be achieved without the problem of gas bubble disease.

Although fewer fish can be transported by truck than when using oxygen, the use of a blower is much safer for the fish because the air is not under high pressure and cannot lead to gas bubble disease. For short distances with small tanks, 12 V blowers that connect to a car battery are a good option, but petrol-driven

Figure 3.15 Aeration supplied in open tanks (courtesy of W. Turner)

blowers are recommended for large tanks and long-distance travel. Pure oxygen can be added to the air from a bottle to enable more fish to be transported per trip, with only a low risk of gas bubble disease.

There are many chemicals that can be added to water during transport to reduce stress and improve performance. Salt at a concentration of 1–2 ppt and calcium chloride at 50 ppm (although not used for water with a high calcium content) is used to 'harden' the fish, to reduce slime formation and to reduce osmoregulatory stress (Dupree and Huner, 1984). Oxygen levels can be raised using powered products such as hydrogen peroxide, calcium peroxide and potassium permanganate. These products can be useful for on-farm tank transport, where they can perform the dual function of killing ectoparasites. Bacteriocides, including the judicious use of antibiotics such as oxytetracycline and enrofloxacin, and chemicals, such as acriflavine and methylene blue, are sometimes added to transport water to reduce bacterial levels. Foam in tanks can be a problem due to fish slime and high aeration. Antifoam chemicals, however, can be used to reduce this problem. Removal of ammonia is possible with certain carbon compounds and zeolite, but usually this is not necessary if the fish are starved prior to packing.

The condition of roads is a key element in transporting live fish. Moving tilapia fingerlings over poor roads significantly increases metabolic rate (MO_2) and total ammonia nitrogen (TAN) excretion rates and induces significantly higher post-transportation mortality rates compared with fish moved over better-quality roads (Obirikorang *et al.*, 2022).

Once fish have been transported to the destination, it is important to gradually acclimatize them to the water in which they are being stocked. Bags are first left to float for 10–15 min to allow the temperature to change before releasing the fish. Generally this is done by gradually allowing the stocking water to enter the bags before emptying them. In tanks it is more difficult to achieve this but it is possible by pumping water into the tank and fitting an overflow system that does not allow the fish to escape. Figure 3.16 illustrates some of the health conditions discussed above.

Figure 3.16 Various health conditions in tilapia. (A,B) Physical deformity resulting from fusion of the vertebrae, most probably as a consequence of air supersaturation and the formation of emboli during a transport event when juveniles. This physical trauma was consistent within a significant proportion of conspecifics in the same population. (C) An individual with skin lesions. (D). A sub-acute infection of the Gram-positive bacterium *Streptococcus agalactiae*, resulting in bilateral exophthalmia, dark coloration, erratic swimming and a distended abdomen full of ascitic fluid. (E) Nile tilapia 14 hours after an experimental infection with the Gram-negative bacterium *Aeromonas hydrophila*. Infection resulted in changes in pigmentation and behaviour; in severe cases, systemic infections can occur, leading to high mortality rates. (Photos courtesy of A. Shinn)

Impacts of hormone treatment on health

The use and addition of steroids, principally 17-alpha methyltestosterone, to first feeds for tilapia has become standard commercial practice to ensure high levels of phenotypic males to control in-system breeding and consistency of growth and performance. An understanding of the window of sensitivity for exogenous hormone use and an effective approach to incorporating the hormone into feed were critical to establishing the range of effective dosages to achieve functional sex reversal by Clemens and Inslee (1968) in *O. mossambicus*. Commercial methods require the growing of first feeding fry at a high density (6/l water) for 21–28 days, depending on temperature, and being fed five or six times per day to ensure consistent delivery of hormone dose. Clearwater conditions, in which natural feed is limited or absent, require a lower dose (40 mg/kg feed) than fry raised where natural food is abundant, such as hapas in earthen ponds. Such high density results in individually stunted fry after 3 weeks (individual weight 0.1−0.25 g) but is most economical in terms of feed and hormone cost. Most importantly, it reduces the amount of natural food the fish can eat so that all individuals consume the hormone-treated feed, and it is possible to obtain >99% male fish consistently. Although abnormal gonads have been observed in fish receiving sub-optimal dosages of hormone (MacIntosh *et al.*, 1988), in general, survival and growth of monosex fish to 30g fingerling size is higher than in mixed-sex stocks.

High temperature has been observed to impact the sex ratio of Nile tilapia, with male-dominant ratios occurring at higher temperatures (32.0–36.5°C; Nivelle *et al.*, 2019), and has been suggested as a method for creating all-male populations despite such temperatures outside the optimal range for growth (28–30°C; Baroiller *et al.*, 1995a, b) and health. It appears there is a strong genetic component to this. Thermosensitive lines produced 90% male populations following two generations of selection (Baroiller *et al.*, 2009), which is below the level required to prevent impacts of early breeding and recruitment but might still be highly beneficial when juveniles are to be stocked in high-density systems that restrict natural breeding behaviour.

Conclusions

1. Recent history of tilapia aquaculture has led to both extensification and intensification of juvenile production in line with the animal's reproductive biology and behaviour, but posing different challenges to health management both at that stage of production and for later life stages.
2. Demand for more productive and consistent grow-out performance stimulated the application of both monosex technology and, increasingly, the development of selectively superior stocks, both of which require intensive hatchery approaches.
3. Optimizing egg and fry harvest from tilapia broodstock tends to require more handling, and higher densities of animals that can produce relatively small numbers of fertile eggs after complex, density-dependent courtship and spawning, followed naturally by extended periods of maternal mouthbrooding and parental care, can have implications for health and welfare outcome of both breeding and juvenile fish.
4. Economies of scale associated with commercial production and servicing geographically dispersed food fish production also impose costs and challenges associated with transport and distribution.
5. The benefits and practicality of establishing procedures to produce pathogen-free or -resilient fish through the design and management of hatcheries were discussed, together with the role of selective breeding for disease-resistant traits. Particularly while tilapia production in open systems continues to be the dominant production approach, the value of animals that adapt to challenges related to both environment and PPP remains strong.
6. High heritability for both resistance to specific agents and the impacts of PPP make selective breeding a core approach to improving health and welfare in the hatchery. Opportunities for using data from challenged populations to direct selective breeding have been transformed from reliance on family selection using marker-assisted selection. Genomic selection promises rapid advances in selection given the multi-gene basis for resistance to many PPP.

7. The importance of site selection and the capacity to optimize environmental, mainly water, quality and the specific pathobiome in which juvenile tilapias are produced is critical and impacted by its design. Modular designs that can ensure rapid drainage and replacement with PPP- free water of any unit and removal of particulate wastes are key, as is the physical exclusion of PPP, be they predatory insects, reptiles, or birds. Hatchery/nursery location will also affect likely stress associated with the distribution of juveniles.
8. Optimizing environments also includes ensuring temperatures are within a suitable range as are dissolved oxygen and other key water quality parameters. Open environments make seasonality effects difficult or uneconomic to control although polytunnels, use of passive solar and wind breaks, and other simple adaptations can be effective. In contrast behavioural fever approaches can support self-treatment of fish *in situ* prophylactically.
9. Reducing stress and enhancing immune systems is also a function of high nutritional status of both breeding and juvenile fish. The balance of natural and fed diets is a key issue to control costs and optimize fish condition, especially given the competition among promoters of pre- and probiotics. Low margins in tilapia seed production make this balance critical and necessary to focus their use strategically during periods of particular stress – for example, the use of such products during periods of extreme temperatures on valuable broodfish or prior to transportation of juveniles.
10. Ensuring high levels of husbandry during handling of fish, effective grader design, reliable oxygenation/aeration and, above all, avoidance of physical and physiological trauma are critical for reducing stress and resultant health issues. Simple salt and a range of anaesthetics and sedatives are increasingly used during handling and transportation to this end.
11. The roles of quarantine during transfers to control the movement of PPP are often underappreciated. The design and implementation of quarantine needs to be able to identify not only virulent PPP but also sub-clinical infections. Developing biosecurity

protocols that are risk based is essential: a key element of this is to understand modes of infection and transfer. Understanding vertical transmission of pathogens can be critical for reducing risks, and developing simple diagnostic tests is a key priority.

References

Abd El-Rhman, A.M., Khattab, Y.A. and Shalaby, A.M. (2009) *Micrococcus luteus* and *Pseudomonas* species as probiotics for promoting the growth performance and health of Nile tilapia, *Oreochromis niloticus. Fish and Shellfish Immunology* 27(2), 175–180. DOI: 10.1016/j.fsi.2009.03.020.

Abdel-Baki, A.S., Al Ghamdi, A. and Al-Quraishy, S. (2017) First record of three African trichodinids (Ciliophora: Peritrichida) in cultured Nile tilapia (*Oreochromis niloticus*) in Saudi Arabia with re-evaluation of their host specificity. *Parasitology Research* 116(4), 1285–1291. DOI: 10.1007/s00436-017-5407-0.

Abdel-Latif, H.M.R., Dawood, M.A.O., Menanteau-Ledouble, S. and El-Matbouli, M. (2020) The nature and consequences of co-infections in tilapia: A review. *Journal of Fish Diseases* 43(6), 651–664. DOI: 10.1111/jfd.13164.

Abu-Elala, N.M., Abd-Elsalam, R.M. and Younis, N.A. (2020) Streptococcosis, lactococcosis and enterococcosis are potential threats facing cultured Nile tilapia (*Oreochomis niloticus*) production. *Aquaculture Research* 51(10), 4183–4195. DOI: 10.1111/are.14760.

Abu-Elala, N.M., Ali, T.E., Ragaa, N.M., Ali, S.E., Abd-Elsalam, R.M. *et al.* (2021) Analysis of the productivity, immunity, and health performance of Nile tilapia (*Oreochromis niloticus*) broodstock-fed dietary fermented extracts sourced from *Saccharomyces cerevisiae* (Hilyses): A field trial. *Animals* 11(3), 815. DOI: 10.3390/ani11030815.

Alday-Sanz, V., Brock, J., Flegel, T.W., McIntosh, R., Bondad-Reantaso, M.G. *et al.* (2020) Facts, truths and myths about SPF shrimp in aquaculture. *Reviews in Aquaculture* 12(1), 76–84. DOI: 10.1111/raq.12305.

Amorim, M.C.P. and Almada, V.C. (2005) The outcome of male–male encounters affects subsequent sound production during courtship in the cichlid fish *Oreochromis mossambicus. Animal Behaviour* 69(3), 595–601. DOI: 10.1016/j.anbehav.2004.06.016.

Anderson, W.G., McKinley, R.S. and Colavecchia, M. (1997) The use of clove oil as an anesthetic for rainbow trout and its effects on swimming performance. *North American Journal of Fisheries Management* 17(2), 301–307. https://doi.org/10.1577/1548-8675(1997)017<0301:TUOCOA>2.3.CO;2

Ariel, E. and Owens, L. (1997) Epizootic mortalities in tilapia *Oreochromis mossambicus*. *Diseases of Aquatic Organisms* 29(1), 1–6. DOI: 10.3354/dao029001.

Arthur, J.R., Bondad-Reantaso, M.G. and Subasinghe, R.P. (2008) Procedures for the quarantine of live aquatic animals: A manual. *FAO Fisheries Technical Paper.* Food and Agriculture Organization 502, 74.

Auperin, B., Baroiller, J.F., Ricordel, M.J., Fostier, A. and Prunet, P. (1997) Effect of confinement stress on circulating levels of growth hormone and two prolactins in freshwater-adapted tilapia (*Oreochromis niloticus*). *General and Comparative Endocrinology* 108(1), 35–44. DOI: 10.1006/gcen.1997.6938.

Avillanosa, A.L. and Caipang, C.M.A. (2019) Use of sodium bicarbonate as an inexpensive general anesthetic for juvenile red tilapia hybrids. *International Aquatic Research* 11(3), 287–294. DOI: 10.1007/s40071-019-00235-1.

Barata, E.N., Hubbard, P.C., Almeida, O.G., Miranda, A. and Canário, A.V.M. (2007) Male urine signals social rank in the Mozambique tilapia (*Oreochromis mossambicus*). *BMC Biology* 5, 54. DOI: 10.1186/1741-7007-5-54.

Barcellos, L.J.G., Volpato, G.L., Barreto, R.E., Coldebella, I. and Ferreira, D. (2011) Chemical communication of handling stress in fish. *Physiology and Behavior* 103(3–4), 372–375. DOI: 10.1016/j.physbeh.2011.03.009.

Baroiller, J.F., Chourrout, D., Fostier, A. and Jalabert, B. (1995a) Temperature and sex chromosomes govern sex ratios of the mouthbrooding cichlid fish *Oreochromis niloticus. Journal of Experimental Zoology* 273(3), 216–223. DOI: 10.1002/jez.1402730306.

Baroiller, J.F., Clota, F. and Geraz, E. (1995b) Temperature sex determination in two tilapias species, *Oreochromis niloticus* and the red tilapia (Red florida strain): Effect of high or low temperatures. In: Goetz, F.W. and Thomas, P. (eds) *Proceedings of the 5th International Symposium on the Reproductive Physiology of Fish*, University of Texas, Austin, pp. 158–160.

Baroiller, J.F., D'Cotta, H., Bezault, E., Wessels, S. and Hoerstgen-Schwark, G. (2009) Tilapia sex determination: Where temperature and genetics meet. *Comparative Biochemistry and Physiology. Part A, Molecular and Integrative Physiology* 153(1), 30–38. DOI: 10.1016/j.cbpa.2008.11.018.

Bhujel, R.C. and Perera, A. (2017) Shading of breeding hapas enhances reproductive performance of Nile tilapia (*Oreochromis niloticus*) and seed output. *Journal of Aquaculture in the Tropics* 32(3–4), 187–197.

Bigarré, L., Cabon, J., Baud, M., Heimann, M., Body, A. *et al.* (2009) Outbreak of betanodavirus infection in tilapia, *Oreochromis niloticus* (L.), in fresh water. *Journal of Fish Diseases* 32(8), 667–673. DOI: 10.1111/j.1365-2761.2009.01037.x.

Bishop, S.C. and Woolliams, J.A. (2014) Genomics and disease resistance studies in livestock. *Livestock Science* 166, 190–198. DOI: 10.1016/j.livsci.2014.04.034.

Boltaña, S., Rey, S., Roher, N., Vargas, R., Huerta, M. *et al.* (2013) Behavioural fever is a synergic signal amplifying the innate immune response. *Proceedings. Biological Sciences* 280(1766), 20131381. DOI: 10.1098/rspb.2013.1381.

Bootland, L., Dobos, P. and Stevenson, R. (1991) The IPNV carrier state and demonstration of vertical transmission in experimentally infected brook trout. *Diseases of Aquatic Organisms* 10, 13–21. DOI: 10.3354/dao010013.

Bostock, J., Albalat, A., Bunting, S., Turner, W.A., Mensah, A.D. *et al.* (2022) Mixed-sex Nile tilapia (*Oreochromis niloticus*) can perform competitively with mono-sex stocks in cage production. *Aquaculture* 557, 738315. DOI: 10.1016/j.aquaculture.2022.738315.

Boyd, C. (2017) Lime plays crucial role in aquaculture pond management, Retrieved July 10, 2017. Global Seafood Alliance. Available at: https://www.globalseafood.org/advocate/lime-aquaculture/

Carneiro, P.C.F. and Urbinati, E.C. (2001) Salt as a stress response mitigator of matrinxã, *Brycon cephalus* (Gunther), during transport. *Aquaculture Research* 32(4), 297–304. DOI: 10.1046/j.1365-2109.2001.00558.x.

Cerqueira, M., Rey, S., Silva, T., Featherstone, Z., Crumlish, M. *et al.* (2016) Thermal preference predicts animal personality in Nile tilapia *Oreochromis niloticus*. *Journal of Animal Ecology* 85(5), 1389–1400. DOI: 10.1111/1365-2656.12555.

Chen, S.C. (1992) The study on the pathogenicity of *Nocardia asteroides* to largemouth bass *Micropterus salmoides* Lacepede. *Fish Pathology* 27(1), 1–5. DOI: 10.3147/jsfp.27.1.

Clavijo, A.M., Conroy, G., Conroy, D.A., Santander, J. and Aponte, F. (2002) First report of *Edwardsiella tarda* from tilapias in Venezuela. *Bulletin of the European Association of Fish Pathologists* 22(4), 280–282.

Clemens, H.P. and Inslee, T. (1968) The production of unisexual broods by *Tilapia mossambica* sex reversed with methyltestosterone. *Transactions of the American Fisheries Society* 97(1), 18–21. DOI: 10.1577/1548-8659(1968)97[18:TPOUBB]2.0.CO;2.

Cock, J., Salazar, M. and Rye, M. (2017) Strategies for managing disease in non-native shrimp populations. *Reviews in Aquaculture* 9(3), 211–226. DOI: 10.1111/raq.12132.

Cooke, S.J., Suski, C.D., Ostrand, K.G., Tufts, B.L. and Wahl, D.H. (2004) Behavioral and physiological assessment of low concentrations of clove oil anaesthetic for handling and transporting largemouth bass (*Micropterus salmoides*). *Aquaculture* 239(1–4), 509–529. DOI: 10.1016/j.aquaculture.2004.06.028.

Dan, N.C. and Little, D.C. (2000) Overwintering performance of Nile tilapia *Oreochromis niloticus* (L.) broodfish and seed at ambient temperatures in northern Vietnam. *Aquaculture Research* 31(6), 485–493. DOI: 10.1046/j.1365-2109.2000.00466.x.

Delphino, M.K.V.C., Leal, C.A.G., Gardner, I.A., Assis, G.B.N., Roriz, G.D. *et al.* (2019) Seasonal dynamics of bacterial pathogens of Nile tilapia farmed in a Brazilian reservoir. *Aquaculture* 498, 100–108. DOI: 10.1016/j.aquaculture.2018.08.023.

Deng, Y., Kokou, F., Eding, E.H. and Verdegem, M.C.J. (2021) Impact of early-life rearing history on gut microbiome succession and performance of Nile tilapia. *Animal Microbiome* 3(1), 81. DOI: 10.1186/s42523-021-00145-w.

de Oliveira, M.M., Ribeiro, T., Orlando, T.M., de Oliveira, D.G.S., Drumond, M.M. *et al.* (2014) Effects crude protein levels on female Nile tilapia (*Oreochromis niloticus*) reproductive performance parameters. *Animal Reproduction Science* 150(1–2), 62–69. DOI: 10.1016/j.anireprosci.2014.08.006.

De Swaef, E., Van den Broeck, W., Dierckens, K. and Decostere, A. (2016) Disinfection of teleost eggs: A review. *Reviews in Aquaculture* 8(4), 321–341. DOI: 10.1111/raq.12096.

Diake, S., Jr, Pickering, T. and Garcia-Gomez, R. (2019) *Technical guidelines for quarantine, broodstock management and hatchery operations of introduced GIFT Nile tilapia.* Pacific Community (SPC).

Doeschl-Wilson, A.B., Bishop, S.C., Kyriazakis, I. and Villanueva, B. (2012) Novel methods for quantifying individual host response to infectious pathogens for genetic analyses. *Frontiers in Genetics* 3, 266. DOI: 10.3389/fgene.2012.00266.

Dong, H.T., Nguyen, V.V., Le, H.D., Sangsuriya, P., Jitrakorn, S. *et al.* (2015) Naturally concurrent infections of bacterial and viral pathogens in disease outbreaks in cultured Nile tilapia (*Oreochromis niloticus*) farms. *Aquaculture* 448, 427–435. DOI: 10.1016/j.aquaculture.2015.06.027.

Dong, H.T., Senapin, S., Jeamkunakorn, C., Nguyen, V.V., Nguyen, N.T. *et al.* (2019) Natural occurrence of edwardsiellosis caused by *Edwardsiella ictaluri* in farmed hybrid red tilapia (*Oreochromis* sp.) in Southeast Asia. *Aquaculture* 499, 17–23. DOI: 10.1016/j.aquaculture.2018.09.007.

Dong, H.T., Senapin, S., Gangnonngiw, W., Nguyen, V.V., Rodkhum, C. *et al.* (2020) Experimental infection reveals transmission of tilapia lake virus (TiLV) from tilapia broodstock to their reproductive organs and fertilized eggs. *Aquaculture* 515, 734541. DOI: 10.1016/j.aquaculture.2019.734541.

Dong, H.T., Sangpo, P., Dien, L.T., Mai, T.T., Linh, N.V. *et al.* (2022) Usefulness of the pancreas as a prime target for histopathological diagnosis of tilapia parvovirus (TiPV) infection in Nile tilapia, *Oreochromis niloticus*. *Journal of Fish Diseases* 45(9), 1323–1331. DOI: 10.1111/jfd.13663.

Dos Santos Nascimento, H., Crispim, B.D., Francisco, L.F.V., Merey, F.M., Kummrow, F. *et al.* (2020) Genotoxicity evaluation of three anesthetics commonly employed in aquaculture using *Oreochromis niloticus* and *Astyanax lacustris*. *Aquaculture Reports* 17, 100357. DOI: 10.1016/j.aqrep.2020.100357.

Dou, J., Li, X., Fu, Q., Jiao, W., Li, Y. *et al.* (2016) Evaluation of the 2b-RAD method for genomic selection in scallop breeding. *Scientific Reports* 6, 19244. DOI: 10.1038/srep19244.

Dupree, H.K. and Huner, J.V. (1984) *The status of warmwater fish farming and progress in fish farming research.* United States Fish and Wildlife Service, pp. 165–176.

Durhack, T.C., Jeffrey, J.D. and Enders, E.C. (2020) In search of an anaesthesia alternative for field-based research. *Aquaculture* 525, 735285. DOI: 10.1016/j.aquaculture.2020.735285.

El-Haroun, E.R., Goda, A.M.A. and Kabir Chowdhury, M.A. (2006) Effect of dietary probiotic Biogens supplementation as- a growth promoter on growth performance and feed utilization of Nile tilapia *Oreochromis niloticus* (L.). *Aquaculture Research* 37(14), 1473–1480. DOI: 10.1111/j.1365-2109.2006.01584.x.

Evans, J.J., Shoemaker, C.A. and Klesius, P.H. (2003) Effects of sublethal dissolved oxygen stress on blood glucose and susceptibility to *Streptococcus agalactiae* in Nile tilapia, *Oreochromis niloticus*. *Journal of Aquatic Animal Health* 15(3), 202–208. DOI: 10.1577/H03-024.

Evans, J.J., Klesius, P.H., Shoemaker, C.A. and Fitzpatrick, B.T. (2004) *Streptococcus agalactiae* vaccination and infection stress in Nile tilapia, *Oreochromis niloticus*. *Journal of Applied Aquaculture* 16(3/4), 105–115.

Evans, J.J., Pasnik, D.J., Brill, G.C. and Klesius, P.H. (2006) Un-ionized ammonia exposure in Nile tilapia: Toxicity, stress response, and susceptibility to *Streptococcus agalactiae*. *North American Journal of Aquaculture* 68(1), 23–33. DOI: 10.1577/A05-032.1.

Fan, L., Chen, J., Meng, S., Song, C., Qiu, L. *et al.* (2017) Characterization of microbial communities in intensive GIFT tilapia (*Oreochromis niloticus*) pond systems during the peak period of breeding. *Aquaculture Research* 48(2), 459–472. DOI: 10.1111/are.12894.

Ferreira, A.L., Dos Santos, F.A.C., de Sena Souza, A., Favero, G.C., Pinheiro, C.G. *et al.* (2022) Anesthetic and sedative efficacy of essential oil of *Hesperozygis ringens* and the physiological responses of *Oreochromis niloticus* after biometric handling and simulated transport. *Fish Physiology and Biochemistry* 48(5), 1155–1166. DOI: 10.1007/s10695-022-01109-w.

Fitzsimons, J.A. and Thomas, J.L. (2011) Fiji's collared kingfishers (*Todiramphus chloris vitensis*) do hunt for fish in inland waters. *Notornis* 58, 163–164.

Fryer, G. and Iles, T.D. (1972) *The cichlid fishes of the Great Lakes of Africa. Their biology and evolution*. Oliver & Boyd.

Fuji, K., Kobayashi, K., Hasegawa, O., Coimbra, M.R.M., Sakamoto, T. *et al.* (2006) Identification of a single major genetic locus controlling the resistance to lymphocystis disease in Japanese flounder (*Paralichthys olivaceus*). *Aquaculture* 254(1–4), 203–210. DOI: 10.1016/j.aquaculture.2005.11.024.

Garcia, J., Munro, E.S., Monte, M.M., Fourrier, M.C.S., Whitelaw, J. *et al.* (2010) Atlantic salmon (*Salmo salar* L.) serum vitellogenin neutralises infectivity of infectious pancreatic necrosis virus (IPNV). *Fish and Shellfish Immunology* 29(2), 293–297. DOI: 10.1016/j.fsi.2010.04.010.

Gjedrem, T. (2015) Disease resistant fish and shellfish are within reach: A review. *Journal of Marine Science and Engineering* 3(1), 146–153. DOI: 10.3390/jmse3010146.

Gjedrem, T. and Rye, M. (2018) Selection response in fish and shellfish: A review. *Reviews in Aquaculture* 10(1), 168–179. DOI: 10.1111/raq.12154.

Gonen, S., Baranski, M., Thorland, I., Norris, A., Grove, H. *et al.* (2015) Mapping and validation of a major QTL affecting resistance to pancreas disease (salmonid Alphavirus) in Atlantic salmon (*Salmo salar*). *Heredity* 115(5), 405–414. DOI: 10.1038/hdy.2015.37.

Green, B.W., Veverica, K.L. and Fitzpatrick, M.S. (1997) Fry and fingerling production. In: Egna, H.S. and Boyd, C.E. (eds) *Dynamics of Pond Aquaculture.* CRC Press.

Gregory, R.G. (1987) *Con-specific predation in* Oreochromis niloticus *juveniles and its possible significance in the commercial production of fingerlings.* [MSc Thesis], Institute of Aquaculture, University of Stirling.

Guan, B., Cai, Y., Zhou, Y., Zhao, Z., Wang, Y. *et al.* (2022) Pathogen identification, risk factor and preventive measure of a columnaris disease outbreak in tilapia (*Oreochromis niloticus*) eggs and larvae from a tilapia hatchery. *Aquaculture* 561, 738718. DOI: 10.1016/j.aquaculture.2022.738718.

Haenen, O.L.M., Dong, H.T., Hoai, T.D., Crumlish, M., Karunasagar, I. *et al.* (2023) Bacterial diseases of tilapia, their zoonotic potential and risk of antimicrobial resistance. *Reviews in Aquaculture* 15(S1 (Suppl. 1)), 154–185. DOI: 10.1111/raq.12743.

Hai, N.V. (2015) Research findings from the use of probiotics in tilapia aquaculture: A review. *Fish and Shellfish Immunology* 45(2), 592–597. DOI: 10.1016/j.fsi.2015.05.026.

Harms, C.A., Howard, K.E., Wolf, J.C., Smith, S.A. and Kennedy-Stoskopf, S. (2003) Transforming growth factor-beta response to mycobacterial infection in striped bass *Morone saxatilis* and hybrid tilapia *Oreochromis* spp. *Veterinary Immunology and Immunopathology* 95(3–4), 155–163. DOI: 10.1016/s0165-2427(03)00138-7.

Hegazi, M.A.M., Attia, Z.I. and Hegazi, M.M. (2014) Salinity lessens the impact of high stocking densities and metabolic cost in white muscle and liver of Nile tilapia fingerlings. *Aquaculture Research* 45(3), 566–570. DOI: 10.1111/j.1365-2109.2012.03241.x.

Houston, R.D., Haley, C.S., Hamilton, A., Guy, D.R., Tinch, A.E. *et al.* (2008) Major quantitative trait loci affect resistance to infectious pancreatic necrosis in Atlantic salmon (*Salmo salar*). *Genetics* 178(2), 1109–1115. DOI: 10.1534/genetics.107.082974.

Houston, R.D., Haley, C.S., Hamilton, A., Guy, D.R., Mota-Velasco, J.C. *et al.* (2010) The susceptibility of Atlantic salmon fry to freshwater infectious pancreatic necrosis is largely explained by a major QTL. *Heredity* 105(3), 318–327. DOI: 10.1038/hdy.2009.171.

Houston, R.D., Taggart, J.B., Cézard, T., Bekaert, M., Lowe, N.R. *et al.* (2014) Development and validation of a high density SNP genotyping array for Atlantic salmon (*Salmo salar*). *BMC Genomics* 15, 90. DOI: 10.1186/1471-2164-15-90.

Houston, R.D., Kriaridou, C. and Robledo, D. (2022) Animal board invited review: Widespread adoption of genetic technologies is key to sustainable expansion of global aquaculture. *Animal* 16(10), 100642. DOI: 10.1016/j.animal.2022.100642.

Huang, X., Chen, F., Guan, J., Xu, C., Li, Y. *et al.* (2022) Beneficial effects of re-feeding high α-linolenic acid diets on the muscle quality, cold temperature and disease resistance of tilapia. *Fish and Shellfish Immunology* 126, 303–310. DOI: 10.1016/j.fsi.2022.05.053.

Huntingford, F., Rey, S. and Quaggiotto, M.M. (2020) Behavioural fever, fish welfare and what farmers and fishers know. *Applied Animal Behaviour Science* 231, 105090. DOI: 10.1016/j.applanim.2020.105090.

Izquierdo, M.S., Fernández-Palacios, H. and Tacon, A.G.J. (2001) Effect of broodstock nutrition on reproductive performance of fish. *Aquaculture* 197(1–4), 25–42. DOI: 10.1016/S0044-8486(01)00581-6.

Javahery, S., Nekoubin, H. and Moradlu, A.H. (2012) Effect of anaesthesia with clove oil in fish (review). *Fish Physiology and Biochemistry* 38(6), 1545–1552. DOI: 10.1007/s10695-012-9682-5.

Joshi, R., Árnyasi, M., Lien, S., Gjøen, H.M., Alvarez, A.T. *et al.* (2018) Development and validation of 58K SNP-array and high-density linkage map in Nile tilapia (*O. niloticus*). *Frontiers in Genetics* 9, 472. DOI: 10.3389/fgene.2018.00472.

Joshi, R., Skaarud, A., Alvarez, A.T., Moen, T. and Ødegård, J. (2021) Bayesian genomic models boost prediction accuracy for survival to *Streptococcus agalactiae* infection in Nile tilapia (*Oreochromus niloticus*). *Genetics, Selection, Evolution* 53(1), 37. DOI: 10.1186/s12711-021-00629-y.

Karvonen, A., Rintamäki, P., Jokela, J. and Valtonen, E.T. (2010) Increasing water temperature and disease risks in aquatic systems: Climate change increases the risk of some, but not all, diseases. *International Journal for Parasitology* 40(13), 1483–1488. DOI: 10.1016/j.ijpara.2010.04.015.

Kawahigashi, D. (2018) New paradigm for controlling EMS/APHNS in intensive *P. vannamei* boone 1931 culture ponds. *Asian Fisheries Science* 31S, 182–193. DOI: 10.33997/j.afs.2018.31.S1.013.

Kayansamruaj, P., Pirarat, N., Hirono, I. and Rodkhum, C. (2014) Increasing of temperature induces pathogenicity of *Streptococcus agalactiae* and the up-regulation of inflammatory related genes in infected Nile tilapia (*Oreochromis niloticus*). *Veterinary Microbiology* 172(1–2), 265–271. DOI: 10.1016/j.vetmic.2014.04.013.

Kayansamruaj, P., Dinh-Hung, N., Srisapoome, P., Na-Nakorn, U. and Chatchaiphan, S. (2023) Genomics-driven prophylactic measures to increase Streptococcosis resistance in tilapia. *Journal of Fish Diseases* 46(6), 597–610. DOI: 10.1111/jfd.13763.

Keawcharoen, J., Techangamsuwan, S., Ponpornpisit, A., Lombardini, E.D., Patchimasiri, T. *et al.* (2015) Genetic characterization of a Betanodavirus isolated from a clinical disease outbreak in farm-raised tilapia *Oreochromis niloticus* (L.) in Thailand. *Journal of Fish Diseases* 38(1), 49–54. DOI: 10.1111/jfd.12200.

Kent, M.L., Watral, V.G., Kirchoff, N.S., Spagnoli, S.T. and Sharpton, T.J. (2016) Effects of subclinical *Mycobacterium chelonae* infections on fecundity and embryo survival in zebrafish. *Zebrafish* 13(Suppl. 1), S88–S95. DOI: 10.1089/zeb.2015.1204.

Kim, J.H., Sohn, S., Kim, S.K. and Hur, Y.B. (2020) Effects on hematological parameters, antioxidant and immune responses, AChE, and stress indicators of olive flounders, *Paralichthys olivaceus*, raised in biofloc and seawater challenged by *Edwardsiella tarda. Fish and Shellfish Immunology* 97, 194–203. DOI: 10.1016/j.fsi.2019.12.011.

Labrie, L., Ng, J., Tan, Z., Komar, C., Ho, E, *et al.* (2008) Nocardial infections in fish: An emerging problem in both freshwater and marine aquaculture systems in Asia [Retrieved April 19, 2023]. In: Bondad-Reantaso, M.G., Mohan, C.V., Crumlish, M. and Subasinghe, R.P. (eds) *Diseases in Asian Aquaculture VI.* Fish Health Section, Asian Fisheries Society, pp. 297–312. Available at: http://www.fhsafs.net/daa_vi_files/22.pdf

LaFrentz, B.R., Lozano, C.A., Shoemaker, C.A., García, J.C., Xu, D.-H. *et al.* (2016) Controlled challenge experiment demonstrates substantial additive genetic variation in resistance of Nile tilapia (*Oreochromis niloticus*) to *Streptococcus iniae. Aquaculture* 458, 134–139. DOI: 10.1016/j.aquaculture.2016.02.034.

Lavilla-Pitogo, C.R. and Nagasawa, K. (eds) (2004) Transboundary fish diseases in Southeast Asia: Occurrence, surveillance, research and training. SEAFDEC Aquaculture Department. Available at: https://repository.seafdec.org.ph/handle/10862/1676

Li, W. and Buchinger, T. (2014) Pheromone signaling: A pissing contest in tilapia. *Current Biology* 24(18), R843–R845. DOI: 10.1016/j.cub.2014.08.001.

Lipschutz-Powell, D., Woolliams, J.A., Bijma, P. and Doeschl-Wilson, A.B. (2012) Indirect genetic effects and the spread of infectious disease: Are we capturing the full heritable variation underlying disease prevalence?. *PLOS ONE* 7(6), e39551. DOI: 10.1371/journal.pone.0039551.

Little, D.C. and Hulata, G. (2000) Strategies for tilapia seed production. In: Beveridge, M.C.M. and McAndrew, B.J. (eds) *Tilapias: Biology and Exploitation.* Fish and fisheries series, Springer, p. 25.

Little, D.C., Innes Taylor, N.L., Turongruang, D. and Komolmarl, S. (1991) Large seed for small-scale aquaculture. *Aquabyte* 4(2), 2–3.

Little, D.C., Murray, F.J., Azim, E., Leschen, W., Boyd, K. *et al.* (2008) Options for producing a warm-water fish in the UK: Limits to "Green growth. *Trends in Food Science and Technology* 19(5), 255–264. DOI: 10.1016/j.tifs.2007.12.003.

Liu, S., Sun, L., Li, Y., Sun, F., Jiang, Y. *et al.* (2014) Development of the catfish 250K SNP array for genome-wide association studies. *BMC Research Notes* 7, 135. DOI: 10.1186/1756-0500-7-135.

Liu, S., Vallejo, R.L., Evenhuis, J.P., Martin, K.E., Hamilton, A. *et al.* (2018) Retrospective evaluation of marker-assisted selection for resistance to bacterial cold water disease in three generations of a commercial rainbow trout breeding population. *Frontiers in Genetics* 9, 286. DOI: 10.3389/fgene.2018.00286.

Long, L., Yang, J., Li, Y., Guan, C. and Wu, F. (2015) Effect of biofloc technology on growth, digestive enzyme activity, hematology, and immune response of genetically improved farmed tilapia (*Oreochromis niloticus*). *Aquaculture* 448, 135–141. DOI: 10.1016/j.aquaculture.2015.05.017.

Lu, J. and Takeuchi, T. (2004) Spawning and egg quality of the tilapia *Oreochromis niloticus* fed solely on raw Spirulina throughout three generations. *Aquaculture* 234(1–4), 625–640. DOI: 10.1016/j.aquaculture.2003.12.027.

Machimbirike, V.I., Jansen, M.D., Senapin, S., Khunrae, P., Rattanarojpong, T. *et al.* (2019) Viral infections in tilapines: More than just tilapia lake virus. *Aquaculture* 503, 508–518. DOI: 10.1016/j.aquaculture.2019.01.036.

MacIntosh, D.J., Singh, T.B., Little, D.C. and Edwards, P. (1988) Growth and sexual development of 17 alpha-methyltestosterone and progesterone-treated Nile tilapia (*Oreochromis niloticus*) reared in earthen ponds. In: Pullin, R.S.V., Bhukaswan, T., Tonguthai, K. and Maclean, J.L. (eds) *ICLARM Conference Proceedings The Second International Symposium on Tilapia in Aquaculture*, Department of Fisheries, and International Center for Living Aquatic Resources Management, pp. 457–463 (Vol. 15).

Madsen, H.C.K., Buchmann, K. and Mellergaard, S. (2000) Association between trichodiniasis in eel (*Anguilla anguilla*) and water quality in recirculation systems. *Aquaculture* 187(3–4), 275–281. DOI: 10.1016/S0044-8486(00)00323-9.

McGrogan, D.G., Ostland, V.E., Byrne, P.J. and Ferguson, H.W. (1998) Systemic disease involving an iridovirus-like agent in cultured tilapia, *Oreochromis niloticus* L. – A case report. *Journal of Fish Diseases* 21(2), 149–152. DOI: 10.1046/j.1365-2761.1998.00082.x.

Medzhitov, R., Schneider, D.S. and Soares, M.P. (2012) Disease tolerance as a defense strategy. *Science* 335(6071), 936–941. DOI: 10.1126/science.1214935.

Mengistu, S.B., Palstra, A.P., Mulder, H.A., Benzie, J.A.H., Trinh, T.Q. *et al.* (2021) Heritable variation in swimming performance in Nile tilapia (*Oreochromis niloticus*) and negative genetic correlations with growth and harvest weight. *Scientific Reports* 11(1), 11018. DOI: 10.1038/s41598-021-90418-w.

Mereghetti, L., Sitkiewicz, I., Green, N.M. and Musser, J.M. (2008) Remodeling of the *Streptococcus agalactiae* transcriptome in response to growth temperature. *PLOS ONE* 3(7), e2785. DOI: 10.1371/journal.pone.0002785.

Meuwissen, T.H.E., Hayes, B.J. and Goddard, M.E. (2001) Prediction of total genetic value using genome-wide dense marker maps. *Genetics* 157(4), 1819–1829. DOI: 10.1093/genetics/157.4.1819.

Meuwissen, T., Hayes, B. and Goddard, M. (2013) Accelerating improvement of livestock with genomic selection. *Annual Review of Animal Biosciences* 1, 221–237. DOI: 10.1146/annurev-animal-031412-103705.

Moen, T., Baranski, M., Sonesson, A.K. and Kjøglum, S. (2009) Confirmation and fine-mapping of a major QTL for resistance to infectious pancreatic necrosis in Atlantic salmon (*Salmo salar*): Population-level associations between markers and trait. *BMC Genomics* 10, 368. DOI: 10.1186/1471-2164-10-368.

Moen, T., Torgersen, J., Santi, N., Davidson, W.S., Baranski, M. *et al.* (2015) Epithelial cadherin determines resistance to infectious pancreatic necrosis virus in Atlantic salmon. *Genetics* 200(4), 1313–1326. DOI: 10.1534/genetics.115.175406.

Montañez Calero, T., Díaz Coahila, D., Angulo Herrera, P., Cerro Temoche, L., Cruz-Castellón, C. *et al.* (2022) Sedative effect of an alcoholic extract of *Valeriana* sp in tilapia fingerlings (*Oreochromis niloticus*) to reduce stress during simulated transport. *Revista de Investigaciones Veterinarias del Perú* 33(3), e21063. DOI: 10.15381/rivep.v33i3.21063.

Moreira, M., Schrama, D., Farinha, A.P., Cerqueira, M., Raposo de Magalhães, C. *et al.* (2021) Fish pathology research and diagnosis in aquaculture of farmed fish; a proteomics perspective. *Animals: An Open Access Journal from MDPI* 11(1), 125. DOI: 10.3390/ani11010125.

Mori, T., Hati, K. and Kubota, S. (1986) Histopathological findings of long-snouted butterfly fish *Forcipiger longirostris* naturally infected with acid-fast bacterium. *Bulletin of the Nippon Veterinary and Animal Science University* 35, 141–145.

Mqolomba, T.N. and Plumb, J.A. (1992) Longevity of *Edwardsiella ictaluri* in the organs of experimentally infected channel catfish, *Ictalurus punctatus*. *Aquaculture* 101(1–2), 1–6. DOI: 10.1016/0044-8486(92)90226-B.

Mrowka, W. and Schierwater, B. (1988) Energy expenditure for mouthbrooding in a cichlid fish. *Behavioral Ecology and Sociobiology* 22(3), 161–164. DOI: 10.1007/BF00300565.

Nasr-Allah, A., Dickson, M., Al-Kenawy, D.A., Ali, S.E. and Charo-Karisa, H. (2021) [Manual (pp. FISH-2021–FISH-2004)]. *Better management practices for tilapia hatcheries in Egypt.* CGIAR Research Program on Fish Agri-Food Systems.

Nguyen, N.H. (2016) Genetic improvement for important farmed aquaculture species with a reference to carp, tilapia and prawns in Asia: Achievements, lessons and challenges. *Fish and Fisheries* 17(2), 483–506. DOI: 10.1111/faf.12122.

Nguyen, V.V., Dong, H.T., Senapin, S., Gangnonngiw, W., Pirarat, N. *et al.* (2019) Transmission of *Francisella noatuensis* subsp. orientalis from subclinically infected hybrid red tilapia broodstock (*Oreochromis* sp.) to their offspring. *Microbial Pathogenesis* 136, 103670. DOI: 10.1016/j.micpath.2019.103670.

Nhinh, D.T., Giang, N.T.H., Van Van, K., Dang, L.T., Dong, H.T. *et al.* (2022) Widespread presence of a highly virulent *Edwardsiella ictaluri* strain in farmed tilapia, *Oreochromis* spp. *Transboundary and Emerging Diseases* 69(5), e2276–e2290. DOI: 10.1111/tbed.14568.

Nir, O. (2003) What are production diseases, and how do we manage them?. *Acta Veterinaria Scandinavica, Supplement* 98, 21–32.

Nivelle, R., Gennotte, V., Kalala, E.J.K., Ngoc, N.B., Muller, M. *et al.* (2019) Correction: temperature preference of Nile tilapia (*Oreochromis niloticus*) juveniles induces spontaneous sex reversal. *PLOS ONE* 14(3), e0214689. DOI: 10.1371/journal.pone.0214689.

Norris, A. (2017) Application of genomics in salmon aquaculture breeding programs by Ashie Norris: Who knows where the genomic revolution will lead us?. *Marine Genomics* 36, 13–15. DOI: 10.1016/j.margen.2017.11.013.

Obirikorang, K.A., Yeboah, E.A., Gyampoh, B.A. and Amanie-Adjei, S.K. (2022) Effect of road conditions on physiological stress responses and post-transportation growth and survival of Nile tilapia (*Oreochromis niloticus*) fingerlings. *Journal of Applied Aquaculture* 34(1), 180–196. DOI: 10.1080/10454438.2020.1825269.

Ødegård, J., Olesen, I., Gjerde, B. and Klemetsdal, G. (2007) Positive genetic correlation between resistance to bacterial (furunculosis) and viral (infectious salmon anaemia) diseases in farmed Atlantic salmon (*Salmo salar*). *Aquaculture* 271(1–4), 173–177. DOI: 10.1016/j.aquaculture.2007.06.006.

Ødegård, J., Baranski, M., Gjerde, B. and Gjedrem, T. (2011) Methodology for genetic evaluation of disease resistance in aquaculture species: Challenges and future prospects. *Aquaculture Research* 42, 103–114. DOI: 10.1111/j.1365-2109.2010.02669.x.

Ødegård, J., Moen, T., Santi, N., Korsvoll, S.A., Kjøglum, S. *et al.* (2014) Genomic prediction in an admixed population of Atlantic salmon (*Salmo salar*). *Frontiers in Genetics* 5, 402. DOI: 10.3389/fgene.2014.00402.

OIE (2019) Chapter 2.3.12. – viral encephalopathy and retinopathy.. In: *OIE – Manual of Diagnostic Tests for Aquatic Animals* (14th Version).

Olivares-Fuster, O., Bullard, S.A., McElwain, A., Llosa, M.J. and Arias, C.R. (2011) Adhesion dynamics of *Flavobacterium columnare* to channel catfish *Ictalurus punctatus* and zebrafish *Danio rerio* after immersion challenge. *Diseases of Aquatic Organisms* 96(3), 221–227. DOI: 10.3354/dao02371.

Oliveira, R.F. and Almada, V.C. (1998) Maternal aggression during the mouthbrooding cycle in the cichlid fish, *Oreochromis mossambicus*. *Aggressive Behavior* 24(3), 187–196. DOI: 10.1002/(SICI)1098-2337(1998)24:33.0.CO;2-I.

Omasaki, S.K., Charo-Karisa, H., Kahi, A.K. and Komen, H. (2016) Genotype by environment interaction for harvest weight, growth rate and shape between monosex and mixed sex Nile tilapia (*Oreochromis niloticus*). *Aquaculture* 458, 75–81. DOI: 10.1016/j.aquaculture.2016.02.033.

Overturf, M.D. and Huggett, D.B. (2015) Responses to various exposure durations of levonorgestrel during early-life stages of fathead minnows (*Pimephales promelas*). *Aquatic Toxicology* 161, 33–40. DOI: 10.1016/j.aquatox.2015.01.029.

Palti, Y., Gao, G., Liu, S., Kent, M.P., Lien, S. *et al.* (2015) The development and characterization of a 57-K single nucleotide polymorphism array for rainbow trout. *Molecular Ecology Resources* 15(3), 662–672. DOI: 10.1111/1755-0998.12337.

Paperna, I. (1973) Lymphocystis in fish from East African lakes. *Journal of Wildlife Diseases* 9(4), 331–335. DOI: 10.7589/0090-3558-9.4.331.

Pradeep, P.J., Srijaya, T.C., Hassan, A., Chatterji, A.K., Withyachumnarnkul, B. *et al.* (2014) Optimal conditions for cold-shock induction of triploidy in red tilapia. *Aquaculture International* 22(3), 1163–1174. DOI: 10.1007/s10499-013-9736-4.

Pradeep, P.J., Suebsing, R., Sirthammajak, S., Kampeera, J., Jitrakorn, S. *et al.* (2016) Evidence of vertical transmission and tissue tropism of streptococcosis from naturally infected red tilapia (*Oreochromis* spp). *Aquaculture Reports* 3, 58–66. DOI: 10.1016/j.aqrep.2015.12.002.

Pradeep, P.J., Suebsing, R., Sirithammajak, S., Kampeera, J., Turner, W. *et al.* (2017) Vertical transmission and concurrent infection of multiple bacterial pathogens in naturally infected red tilapia (*Oreochromis* spp). *Aquaculture Research* 48(6), 2706–2717. DOI: 10.1111/are.13102.

Prihartini, N.C., Yanuhar, U. and Maftuch(2015) Betanodavirus infections in tilapia seed (*Oreochromis* sp.), in Indonesia. *Journal of Life Science and Biomedicine* 5, 106–109.

Ridha, M.T., Cruz, E.M., Al-Ameeri, A.A. *et al.* (1998) Effects of controlling temperature and light duration on seed production in tilapia, *Oreochromis spilurus* (Günther). *Aquaculture Research* 29, 403–410. DOI: 10.1046/j.1365-2109.1998.00216.x.

Ripley, J.L. and Lobel, P.S. (2005) Reproductive behavior of the lake Malawi cichlid fish, *Tramitichromis intermedius*. *Environmental Biology of Fishes* 73(2), 171–180. DOI: 10.1007/s10641-004-5567-5.

Robert, N. and Venkatesan, P. (1997) Prey preference and predatory efficiency of the water bug, *Diplonychus indicus* Venk. and Rao (Hemiptera: Belostomatidae), an effective biocontrol agent for mosquitoes. *Journal of Entomological Research* 21, 267–272.

Robledo, D., Palaiokostas, C., Bargelloni, L., Martínez, P. and Houston, R.D. (2018) Applications of genotyping by sequencing in aquaculture breeding and genetics. *Reviews in Aquaculture* 10(3), 670–682. DOI: 10.1111/raq.12193.

Sae-Lim, P., Gjerde, B., Nielsen, H.M., Mulder, H. and Kause, A. (2016) A review of genotype-by-environment interaction and microenvironmental sensitivity in aquaculture species. *Reviews in Aquaculture* 8(4), 369–393. DOI: 10.1111/raq.12098.

Santiago, C.B. and Reyes, O.S. (1993) Effects of dietary lipid source on reproductive performance and tissue lipid levels of Nile tilapia *Oreochromis niloticus* (Linnaeus) broodstock. *Journal of Applied Ichthyology* 9(1), 33–40. DOI: 10.1111/j.1439-0426.1993.tb00385.x.

Senapin, S., Dong, H.T., Meemetta, W., Siriphongphaew, A., Charoensapsri, W. *et al.* (2016) *Hahella chejuensis* is the etiological agent of a novel red egg disease in tilapia (*Oreochromis* spp.) hatcheries in Thailand. *Aquaculture* 454, 1–7. DOI: 10.1016/j.aquaculture.2015.12.013.

Shinn, A.P., Avenant-Oldewage, A., Bondad-Reantaso, M.G., Cruz-Laufer, A.J., García-Vásquez, A. *et al.* (2023a) A global review of problematic and pathogenic parasites of farmed tilapia. *Reviews in Aquaculture* 15(S1 (Suppl. 1)), 92–153. DOI: 10.1111/raq.12742.

Shinn, A.P., Dong, H.T., Vinh, N.T., Wongwaradechkul, R. and Lio-Po, G.D. (2023b) Chapter 6. infectious diseases of warmwater fish in fresh water. In: Woo, P.T.K. and Subasinghe, R.P. (eds) *Climate Change on Diseases and Disorders of Finfish in Cage Culture*, 3rd edn. CA B International, pp. 202–277.

Shlapobersky, M., Sinyakov, M.S., Katzenellenbogen, M., Sarid, R., Don, J. *et al.* (2010) Viral encephalitis of tilapia larvae: Primary characterization of a novel herpes-like virus. *Virology* 399(2), 239–247. DOI: 10.1016/j.virol.2010.01.001.

Shoemaker, C.A., LaFrentz, B.R. and Klesius, P.H. (2012) Bivalent vaccination of sex reversed hybrid tilapia against *Streptococcus iniae* and *Vibrio vulnificus*. *Aquaculture* 354–355, 45–49. DOI: 10.1016/j.aquaculture.2012.04.033.

Sinyakov, M.S., Belotsky, S., Shlapobersky, M. and Avtalion, R.R. (2011) Vertical and horizontal transmission of tilapia larvae encephalitis virus:

The bad and the ugly. *Virology* 410(1), 228–233. DOI: 10.1016/j.virol.2010.11.004.

Sonesson, A.K. (2007) Within-family marker-assisted selection for aquaculture species. *Genetics, Selection, Evolution* 39(3), 301–317. DOI: 10.1186/1297-9686-39-3-301.

Srisakultiew, N., Kumkhong, S., Marandel, L., Plagnes-Juan, E., Panserat, S. *et al.* (2022) Short initial period of high carbohydrate feeding improves nutrient utilisation in juvenile Nile tilapia (*Oreochromis niloticus*) fed a high carbohydrate diet. *Aquaculture* 561, 738661. DOI: 10.1016/j.aquaculture.2022.738661.

Stentiford, G.D., Sritunyalucksana, K., Flegel, T.W., Williams, B.A.P., Withyachumnarnkul, B. *et al.* (2017) New paradigms to help solve the global aquaculture disease crisis. *PLOS Pathogens* 13(2), e1006160. DOI: 10.1371/journal.ppat.1006160.

Stride, M.C., Polkinghorne, A., Powell, M.D. and Nowak, B.F. (2013) *'Candidatus similichlamydia laticola'*, a novel *Chlamydia*-like agent of epitheliocystis in seven consecutive cohorts of farmed Australian barramundi, *Lates calcarifer* (Bloch). *PLOS ONE* 8(12), e82889. DOI: 10.1371/journal.pone.0082889.

Stride, M.C., Polkinghorne, A. and Nowak, B.F. (2014) Chlamydial infections of fish: Diverse pathogens and emerging causes of disease in aquaculture species. *Veterinary Microbiology* 170(1–2), 19–27. DOI: 10.1016/j.vetmic.2014.01.022.

Subasinghe, R.P. and Shinn, A.P. (2023) Chapter 11. Biosecurity: Current and future strategies. In: Woo, P.T.K. and Subasinghe, R.P. (eds) *Climate Change on Diseases and Disorders of Finfish in Cage Culture*, 3rd edn. CA B International, pp. 430–461.

Subasinghe, R., Alday-Sanz, V., Bondad-Reantaso, M.G., Jie, H., Shinn, A.P. *et al.* (2023) Biosecurity: Reducing the burden of disease. *Journal of the World Aquaculture Society* 54(2), 397–426. DOI: 10.1111/jwas.12966.

Suebsing, R., Kampeera, J., Sirithammajak, S., Withyachumnarnkul, B., Turner, W. *et al.* (2015) Colorimetric method of loop mediated isothermal amplification with the pre-addition of calcein for detecting *Flavobacterium columnare* and its assessment in tilapia farms. *Journal of Aquatic Animal Health* 27(1), 38–44. DOI: 10.1080/08997659.2014.966212.

Suebsing, R., Pradeep, P.J., Jitrakorn, S., Sirithammajak, S., Kampeera, J. *et al.* (2016) Detection of natural infection of infectious spleen and kidney necrosis virus in farmed tilapia by hydroxynapthol blue-loop-mediated isothermal amplification assay. *Journal of Applied Microbiology* 121(1), 55–67. DOI: 10.1111/jam.13165.

Supono, H.J., Prayitno, S.B. and Darmanto, Y.S. (2014) White shrimp (*Litopenaeus vannamei*) culture using heterotrophic aquaculture system on nursery phase. *International Journal of Waste Resources* 4(2), 142–145.

Swingle, H.S. (1960) Comparative evaluation of two tilapias as pond fishes in Alabama. *Transactions of the American Fisheries Society* 89(2), 142–148. DOI: 10.1577/1548-8659(1960)89[142:CEOTTA]2.0.CO;2.

Tave, D., Rezk, M. and Smitherman, R.O. (1990) Effect of body colour of *Oreochromis mossambicus* (Peters) on predation by dragonfly nymphs. *Aquaculture Research* 21(2), 157–162. DOI: 10.1111/j.1365-2109.1990.tb00452.x.

Tkint, T., Verheyen, E., De Kegel, B., Helsen, P. and Adriaens, D. (2012) Dealing with food and eggs in mouthbrooding cichlids: Structural and functional trade-offs in fitness related traits. *PLOS ONE* 7(2), e31117. DOI: 10.1371/journal.pone.0031117.

Trape, S., Guilhaumon, F. and Baldé, C. (2008) Fish predation by the water snake *Afronatrix anoscopus* in a Guinean rainforest stream. *Journal of Freshwater Ecology* 23(3), 495–496. DOI: 10.1080/02705060.2008.9664233.

Tsai, H.Y., Hamilton, A., Tinch, A.E., Guy, D.R., Gharbi, K. *et al.* (2015) Genome wide association and genomic prediction for growth traits in juvenile farmed Atlantic salmon using a high density SNP array. *BMC Genomics* 16, 969. DOI: 10.1186/s12864-015-2117-9.

Tsai, H.Y., Hamilton, A., Tinch, A.E., Guy, D.R., Bron, J.E. *et al.* (2016a) Genomic prediction of host resistance to sea lice in farmed Atlantic salmon populations. *Genetics, Selection, Evolution* 48(1), 47. DOI: 10.1186/s12711-016-0226-9.

Tsai, H.Y., Robledo, D., Lowe, N.R., Bekaert, M., Taggart, J.B. *et al.* (2016b) Construction and annotation of a high density SNP linkage map of the Atlantic salmon (*Salmo salar*) genome. *Genes, Genomes and Genetics* 6(7), 2173–2179. DOI: 10.1534/g3.116.029009.

Turner, G.F. and Falter, U. (1989) On the adoption of "male-typical" breeding colouration by female mouthbrooding cichlids. *Journal of Fish Biology* 34(5), 805–806. DOI: 10.1111/j.1095-8649.1989.tb03361.x.

Tveteras, R., Nystoyl, R. and Jory, D.E. (2019) *GOAL 2019: Global finfish production review and forecast - Responsible Seafood Advocate*. Available at: https://www.globalseafood.org/advocate/goal-2019-global-finfish-production-review-and-forecast/

Ulloa Walker, D.A.U., Morales Suazo, M.C. and Emerenciano, M.G.C. (2020) Biofloc technology: Principles focused on potential species and the case study of Chilean river shrimp *Cryphiops caementarius*. *Reviews in Aquaculture* 12(3), 1759–1782. DOI: 10.1111/raq.12408.

Valbon, W.R., Haddi, K., Gutiérrez, Y., Cruz, F.M., Azevedo, K.E.X. *et al.* (2019) Life history traits and predatory performance of *Belostoma anurum* (Hemiptera: Belostomatidae), a biological control agent of disease vector mosquitoes. *Neotropical Entomology* 48(6), 899–908. DOI: 10.1007/s13744-019-00710-8.

Vallejo, R.L., Palti, Y., Liu, S., Evenhuis, J.P., Gao, G. *et al.* (2014) Detection of QTL in rainbow trout affecting survival when challenged with *Flavobacterium psychrophilum*. *Marine Biotechnology* 16(3), 349–360. DOI: 10.1007/s10126-013-9553-9.

Van Houte, S., Ekroth, A.K.E., Broniewski, J.M., Chabas, H., Ashby, B. *et al.* (2016) The diversity-generating benefits of a prokaryotic adaptive immune system. *Nature* 532(7599), 385–388. DOI: 10.1038/nature17436.

Veenendaal, M.V., Painter, R.C., de Rooij, S.R., Bossuyt, P.M., van der Post, J.A. *et al.* (2013) Transgenerational effects of prenatal exposure to the 1944–45 Dutch famine. *BJOG* 120(5), 548–553. DOI: 10.1111/1471-0528.12136.

Verbruggen, B., Bickley, L.K., Santos, E.M., Tyler, C.R., Stentiford, G.D. *et al.* (2015) *De novo* assembly of the *Carcinus maenas* transcriptome and characterization of innate immune system pathways. *BMC Genomics* 16(1), 458. DOI: 10.1186/s12864-015-1667-1.

Vijayan, M.M., Pereira, C., Grau, E.G. and Iwama, G.K. (1997) Metabolic responses associated with confinement stress in tilapia: The role of cortisol. *Comparative Biochemistry and Physiology Part C* 116(1), 89–95. DOI: 10.1016/S0742-8413(96)00124-7.

Xia, Y., Wang, M., Gao, F., Lu, M. and Chen, G. (2020) Effects of dietary probiotic supplementation on the growth, gut health and disease resistance of juvenile Nile tilapia (*Oreochromis niloticus*). *Animal Nutrition* 6(1), 69–79. DOI: 10.1016/j.aninu.2019.07.002.

Xu, J., Zhao, Z., Zhang, X., Zheng, X., Li, J. *et al.* (2014) Development and evaluation of the first high-throughput SNP array for common carp (*Cyprinus carpio*). *BMC Genomics* 15, 307. DOI: 10.1186/1471-2164-15-307.

Yamkasem, J., Tattiyapong, P., Kamlangdee, A. and Surachetpong, W. (2019) Evidence of potential vertical transmission of tilapia lake virus. *Journal of Fish Diseases* 42(9), 1293–1300. DOI: 10.1111/jfd.13050.

Yáñez, J.M., Houston, R.D. and Newman, S. (2014) Genetics and genomics of disease resistance in salmonid species. *Frontiers in Genetics* 5, 415. DOI: 10.3389/fgene.2014.00415.

Yáñez, J.M., Newman, S. and Houston, R.D. (2015) Genomics in aquaculture to better understand species biology and accelerate genetic progress. *Frontiers in Genetics* 6, 128. DOI: 10.3389/fgene.2015.00128.

Yáñez, J.M., Naswa, S., López, M.E., Bassini, L., Correa, K. *et al.* (2016) Genomewide single nucleotide polymorphism discovery in Atlantic salmon (*Salmo salar*): Validation in wild and farmed American and European populations. *Molecular Ecology Resources* 16(4), 1002–1011. DOI: 10.1111/1755-0998.12503.

Zhou, Y., Zhang, Y., Wei, S., Li, W., Li, W. *et al.* (2022) Reduced hypoxia tolerance and altered gill morphology at elevated temperatures may limit the survival of tilapia (GIFT, *Oreochromis niloticus*) under global warming. *Fishes* 7(5), 216. DOI: 10.3390/fishes7050216.

Chapter 4

Tilapia farming systems

Ram C. Bhujel

Introduction

Tilapias are the most widely farmed fish. Production data have been reported to the Food and Agriculture Organization (FAO) from 145 countries worldwide, and Nile tilapia (*Oreochromis niloticus*) is the dominant species (75%) among the 23 species of farmed tilapia (Miao and Wang, 2020). FAO data show that the annual production of Nile tilapia, along with its hybrids, especially red tilapias, has surpassed 7 million metric tonnes, which is higher than the production of grass carp (*Ctenopharyngodon idella*), which was the highest until 2018. Therefore, Nile tilapia, including its hybrids, has become the number one species amongst the individual species used for aquaculture globally in terms of total tonnage. It plays important roles in food security, employment creation, income generation and economic development. Moreover, as the poor man's fish, tilapias have an important impact on poor people, especially in developing countries (Little, 2000). The market value of tilapia has recently been reported to be US$13.9 billion and is projected to reach US$21.6 billion by 2033 (Choudhary, 2023).

Tilapias are cultured in a wide range of environments and culture systems (AIT, 1994; Bhujel, 2014). Their culture is rapidly growing worldwide. They have adapted to survive and thrive in a wide range of environments, including freshwater to brackish and saline waters, small ditches and tanks to large man-made or natural open waters. They can feed on natural foods, artificial pellets or various kinds of food combinations to grow well and breed. Their

Kim Thompson, Craig Shoemaker and David C. Little (eds) **Tilapia**
DOI: 10.1079/9781800629455.0004, © CAB International, 2025

precocious breeding, short generation cycles, fast growth and the above-mentioned adaptive characteristics are the primary causes of their expansion worldwide, even though they originate from the African continent.

As a tropical fish, they grow well at water temperatures ranging from 24 to 34°C, salinity 0–35 ppt, pH 6.5–9.5, dissolved oxygen (DO) 2–12 mg/L and can tolerate ammonia (unionized NH_3) of up to 0.5 ppm and nitrite (NO_2) 0.2 ppm (Boyd, 1990; Bhujel, 2014). Understanding these values and maintaining the environment in their culture systems, within their comfort zone to keep their health status as high as possible, is key to successful farming of this species. Management of the culture system is to maintain the optimum DO in the water by optimizing plankton growth or effective aeration or water exchange and converting or extracting nitrogenous waste from the system. A large percentage of total ammoniacal nitrogen (TAN) is converted to toxic ammonia (NH_3) at high temperature and pH levels. Within the fish's comfort zone, the growth rate of tilapia is high at high temperatures; therefore, maintaining a high temperature is beneficial but there is an increased chance of ammonia toxicity occurring (Boyd, 1990). It is necessary to convert ammoniacal nitrogen to non-toxic nitrate (NO_3) or remove it before it builds up to harmful levels. The rate of nitrogenous waste production depends on the density of the fish or their biomass. Maximizing the fish biomass, or highest productivity per unit area of space or volume of water, requires careful balancing of these factors in each farming system.

The comfort level for tilapia may vary depending on their life stage and specific activities, such as breeding, egg incubation, larval rearing, fry nursing and grow-out farming. However, in this chapter, the main focus will be on grow-out farming systems. For grow-out, several farming systems are found, practised traditionally or in new systems, that have evolved for culture of tilapias around the globe. Tilapias are mostly cultured in open-earth ponds, cages in lakes, rivers and reservoirs, canals, ditches, rice fields and in tanks with aquaponics, biofloc, water recirculatory aquaculture systems (RAS) and in-pond race systems (IPRS).

The rapidly growing demand for fish produced from sustainable farming and environmental concerns are leading to the development of new and innovative production systems with greater efficiency in terms of resources. A wide range of research and technological advancements has been achieved in farming systems, contributing to the development and expansion of the aquaculture industry worldwide. Culture systems vary in regard to the size and shape of enclosures that have been developed or adapted to suit the local conditions to enhance productivity and profitability while maintaining the excellent health of the fish and the environment. The main objective of this chapter is to describe the systems used for culture of tilapias at all life cycle stages. Some systems are simple while others are more complex, but they might be the only option for various reasons, such as land shortages, water or availability of natural water bodies. This chapter describes the farming systems used by the tilapia aquaculture industry.

Traditional subsistence farming system

Typically, in rural areas, each farmer grows crops, vegetables and fruit and raises animals in the same area, a tradition practised since the earliest human civilization. Subsistence farming is widespread throughout the Asia–Pacific region (Ahmed and Rahman, 1992; Dewar, 2003). By-products of crops, e.g. rice bran, oil cakes, vegetable leaves, fruit peel, etc., are used to feed both livestock and fish. Farm animals consume most of the inputs and produce a large amount of manure, mainly used for fertilization of crops, vegetables and fruit. As aquaculture has emerged as an alternative livelihood option, farmers see an opportunity to use livestock manure as a pond input to produce fish, which is more efficient in regard to nutrient utilization and income generation. Among fish species farmed, tilapia has been their choice species because they breed naturally. The prolific breeder *Oreochromis mossambicus* was introduced to Asian countries to help subsistence farmers. However, when farmers tried to farm it, an overcrowding problem was encountered due to overbreeding and the fish did not grow sufficiently to suit the market. Less precocious Nile tilapia were later introduced, which performed better. Initially, farmers bred fish naturally and used a small number of recruits as seed for new ponds or for sale to earn additional income. Subsistence farmers heavily depend on rainfall or irrigation water for food production, including fish farming (Dewar, 2003). Fish farms where irrigation water is available could harvest more than double the amount of fish, i.e. 2.3 t/ha compared with 1.1 tonnes/ha of drought-prone land (Pant *et al.*, 2004). In many rural areas, tilapia farms utilize unused and unfertile land, swamps and, sometimes, conversion of rice plots into fishponds. Therefore, farm by-products are reused within integrated farming systems (IFS). Farmers need minimal external inputs because they manage resources from within their farms. If a large amount of external resources have to be purchased, they have to invest a lot, which may not be sustainable in the long term. By integrating all the components within their farm, including livestock, fish, crops, fruit and vegetables, farmers try to maximize the productivity of all the components and fulfil their family needs, i.e. hunger and nutrition, and also obtain some revenue by selling surplus amounts of produce to cover the cost of their children's education, buy clothes and other items that they are unable to produce themselves. However, more recently, farmers have had to become more specialized to compete commercially or earn more money. Therefore, they may focus on one or two primary produce that have high income potential so that they can sell to earn a good income and purchase food items when needed. They often grow high-value products and buy cheaper ones for family consumption. For example, some farmers grow fish, converting rice fields into ponds and purchase rice with the income they make from fish.

Traditionally, most farmers in rural areas grow cereals crops, vegetables, fruits and spices and raise animals such as buffalo, cattle, chicken, pigs, goats, sheep and so on, living together with them on a plot of land. They have been taking waste from one component of their farming system to use in another so that outside resources are unnecessary. These traditional systems are naturally sustainable farming systems because nutrients are recycled within the system. However, when commercial farming of monocropping was promoted, the importance of traditional subsistence and IFS still needed to be realized. Recent research has focused on these systems and has attracted global attention because they are often considered to be organic farms.

Because each farmer is doing agriculture differently to support their family's livelihood, generalizing and obtaining detailed explanations about subsistence farming can be challenging. In the mid-1990s, when the economic crisis hit Thailand and South-east Asia, the former King of Thailand started the Royal Initiative of the 'Philosophy of Sufficiency Economy', in which the sustainable agriculture strategy was described. It was all about managing farmers' limited resources to produce most of the things by the farmers themselves so that external inputs were unnecessary. According to this edict, the land should be divided by use, i.e. 30% for fishponds, which also serves as storage for water for agriculture and household purposes, 30% for growing rice, cassava or cereals, 30% for field crops and vegetables and 10% for buildings, animal sheds or residences (Chaipattana Foundation, 2020). The fish suggested for this purpose should be a self-recruiting species (SRS) so that they can breed and produce offspring without the need for external inputs – for example, Nile tilapia, which is considered the king's fish introduced as a gift from a Japanese prince to the then King of Thailand in 1965 (Bangkok Post, 2016).

Rice–fish system

Rice is a staple food of billions. China has a history of about 1700 years of rice–fish farming and rice–fish-farming systems are considered globally important and ingenious agricultural heritage systems (Lu & Li, 2006). In Asia it is said that 'Where there is water, there is fish'. Therefore, Asian society is often called the 'rice–fish' society; people often say, 'rice in the field and fish in the water'. Since the early days of introduction, tilapias have been found in natural water bodies, including rice fields, and people were able to catch them for food.

Tilapias can grow in rice fields without the addition of feed because they consume natural food present in rice fields. Normally, fish are stocked at low densities (e.g. one fish/m^2), which means they have sufficient space and no major health problems. The rice plot has to be modified to house the fish, making it deeper at the corners so that if fish are too small to harvest they can be retained in the system for longer to grow. A trench of about 1m deep and 1–2m wide, along one or all sides of a dyke, is made for shelter so that water remains there even when the rice field needs to be drained to harvest the rice. The species generally used for rice fish culture are common carp (*Cyprinus carpio*) and grass carp. More recently, Nile tilapia has been found to be more suitable as it grows fast and has a short culture period that suits high-yielding varieties of rice, which require only 3–4 months to grow.

Rice–fish culture has long been widely practised in Asia but has not been developed much as a commercial farming system. The leading rice-producing countries are PR China, Indonesia, Japan and the Philippines, where traditional rice farming methods have been refined somewhat over centuries. Having a huge rice production area in India, fish culture has been practised in only limited areas; however, rice–fish farming is one such solution that optimizes the use of resources, maintains a sustainable environment and provides economic benefits to the farmers (Sathoria & Roy, 2022). Therefore, although there is huge scope for culturing tilapia on rice–farms, the use of pesticides needs to stop. Experience shows that fish-farming intervention will become easier when farmers are encouraged to try farming fish in a small corner of a rice plot where the rice does not grow well (Bhujel *et al.*, 2008). The higher income generated from fish compared with the low revenue obtained from rice on partially used land makes it much easier to convince farmers to use this system. Many farmers gradually expanded their ponds to become commercial fish farmers. Therefore, rice–fish culture can be an entry point to commercial fish farming, especially in rural areas.

In the Philippines, Nile tilapia used to be cultured with common carp. Usually, large fingerlings, 15–25 g, are recommended so that they can be harvested within one rice crop of about 3–4 months. If the fingerlings are small (5–10 g), farmers must wait for another rice crop to grow further and harvest the fish in two stages (Dela Cruz *et al.*, 2001). More recently, attempts have been made to culture tilapia at high densities so that the income obtained would be many times greater than when only rice is grown.

Commercial pond system

Tilapias are widely cultured in earthen ponds, mostly in developing countries where land is still available cheaply to rent or purchase. Farming systems are categorized as either extensive, semi-intensive or intensive, which are briefly described in this section.

Extensive system

Traditionally, extensive farming systems have been practised. Mixed-sex fingerlings are stocked with little input and partially harvested several times per year. The size of ponds may vary from 100 m^2 to several hectares. Productivity is normally <1 tonne/ha/year. Traditionally, farmers culture mixed-sex tilapia in green water ponds with or without supplementary feeds. Supplementary feed can be just rice bran, corn meal or some oilcakes, alone or in combination. For feeding there is no hard and fast rule. Some farmers may feed occasionally, i.e. once every few days, while others may feed regularly, such as once or twice per day. Because the fish depend on natural feed, pond fertilization is essential in farm management. The application of manure from chicken, pig, cow or other animals is widespread because these are normally available locally from nearby farms. However, these can be bulky, requiring costly transportation. There are some chemical fertilizers – namely, urea, triple superphosphate (TSP) or diammonium phosphate (DAP) – that are also applied, either in combination with animal manure or replacing them, because these contain high levels of nitrogen and phosphorus and are readily available for agricultural purposes.

Tilapias reproduce naturally in culture systems without the need for hormone injections. This was considered useful in some rural areas where fish farming is mainly carried out for family consumption. Their small fry, 'recruits' produced as a by-product in grow-out ponds, can serve as 'seed' to stock other ponds or even for selling to earn cash, an additional source of income for rural farmers.

One of the major constraints of traditional mixed-tilapia culture is low productivity due to excessive breeding and production of recruits. Female tilapia start breeding at about 60 g or around 6 months of age. They waste energy during reproduction, and overcrowding occurs due to their offspring occupying space and competing for food. Stocked fish do not grow quickly or reach a larger size. When harvesting, almost one quarter or one third of fish are normally too small to be saleable. Small fish have a low price per kilogram, which can be three to four times lower than that for larger ones. Farmers are, therefore, attracted to monosex tilapia fingerlings. However, monosex fry are available in only certain areas, especially rural areas, or are too expensive for the farmer to purchase (three to four times the price of normal seed). Therefore, the inclusion of carnivorous fish such as African catfish (*Clarias gariepinus*) has become a common practice to control unwanted fry in mixed-sex tilapia farming systems.

In addition to using African catfish, the unwanted fry is reduced by rearing tilapia in combination with other predator fish. Several predator fish species are successfully used in various parts of the world, e.g. Asian snakehead (*Channa striata* or *C. micropeltes*), giant snakehead (*C. marulius*), Asian sea bass or barramundi (*Lates calcarifer*), Nile perch (*Lates niloticus*) and African snakehead (*Ophiocephalus obscuris*). These fish are stocked at about 1000 fingerlings/ha whereas tilapias are stocked at >30,000/ha. The most important point to consider is that they are stocked after tilapia start producing recruits, and they should be smaller than the stocked tilapia so that predator fish don't attack or kill them.

Semi-intensive system

When farmers want to produce more fish in their traditional pond system, they start to add inputs such as fertilizers and/or feeds. Ponds are fertilized to make water green, applying animal (chicken, pig, etc.) manure ranging from a few hundred kilograms up to 2 t/ha, enhancing productivity to 1–2 t/ha/year. They may be fed with rice bran, corn meal or oilcake alone, or in combination as supplementary feed. It is possible to double production (4 t/ha/year) by adding 60 kg urea and 30 kg TSP or DAP/ha weekly. Productivity can be further enhanced (6 t/ha/year) through daily supplementary feeding with commercial pellets at 1–2% of biomass. Further increments in productivity, up to 15 t/ha, are possible when monosex fry/fingerlings are stocked at higher densities, e.g. five to seven fingerlings/m^2, and the amount of supplementary feed is increased (Table 4.1). If aeration is provided, stocking density can be >10/m^2, increasing productivity further. The main benefit of semi-intensive systems is that tilapia can utilize plankton, a major source of nutrients that helps reduce feed costs considerably. When tilapia are grown in green waters they may contain higher levels of omega-3 compared with fish grown in cages given complete feed (Karapanagiotidis *et al.*, 2006). This may help boost the image of tilapia, which has been criticised as being a cheap fish containing low levels of omega-3.

Intensive system

When stocking densities are raised above 10/m^2 and feed inputs are increased, emergency aeration is necessary, especially during the early morning, i.e. 03:00–06:00, and then it is regarded as an 'intensive system'. During the daytime, emergency aeration is also needed to mix the water or release DO from the pond water into the atmosphere. Higher productivity, of approximately three to four times, is obtained through aeration where electricity is affordable. Highly intensive farming can be carried out in ponds supplied with 24 hour aeration and high stocking and feeding fish to satiation. The health of the fish is not an issue when they are stocked at low densities or in extensive or semi-intensive systems. Productivity of 40–50 t per fish/ha/year is possible (Edwards & Karim, 2007). Pond productivity also depends on the water's physical and chemical properties (temperature, pH, ammonia, nitrite, alkalinity, salinity, etc.) and types of plankton, microbes, soil and the presence of other fish in the pond.

Table 4.1 Monosex tilapia-farming techniques (modified from Bhujel, 2013).

Stocking density (no./m^2)	Farming systems	Farming technique	Amount of feed required (kg/ha)	Expected production (t/ha)
2.0	Extensive	Only fertilization[a]	-	<2.0
2.5	Semi-intensive	Fertilization +home made feed	-	5.0
3.0	"	Fertilization +pellets for the last 1 month	1,800	6.0
3.5	"	Fertilization +pellets for the last 2 months	3,750	7.5
4.5	"	Fertilization +pellets for the last 3 months	5,600	8.0
5.0	"	Fertilization +pellets for the last 4 months	7,500	12.0
7.5	"	Fertilization +complete pellet feed	10,000	15.0
10.0	Intensive	Fertilization +Complete pellets + aeration	15,000	20.0
>10.0	"	Fertilization +Complete pellets + aeration	>20,000	>20.0

[a]Chicken manure, urea and TSP or DAP

These parameters are more important to monitor and control, especially in intensive systems because fish can become stressed, which favours infection by pathogens. Therefore, intensive systems carry a high risk of having disease outbreaks, which can lead to failure. Intensive farming is necessary because land is expensive, or becoming expensive, in most locations. To increase productivity in tilapia culture systems, different approaches have evolved, such as polyculture, integrated farming, IPRS, cage-cum-pond and tank farming systems, which are often regarded as super-intensive, super-super intensive or hyper-intensive (Gullian-Klanian & Arámburu-Adame, 2013), which are described in the following sections.

Polyculture

Because tilapia consume various micro-organisms, detritus and provided feed, it can be grown in polyculture. Several studies have been performed in which carp were added to tilapia ponds in polyculture (Hossain *et al.*, 2003). Total fish production did not differ between polyculture with carp and monoculture of tilapia only. Polyculture is a complex system, with the farmer stock of many species in the same pond except for very intensively managed ponds. Polyculture is practised mainly in semi-culture systems, and this is considered the best way of reducing the risk of farming single fish species. The farmer can benefit from multiple harvests and varying markets for different fish species.

For polyculture, species with different feeding habits, such as bottom, column and surface feeders, are chosen to be compatible with each other or are less competitive in terms of space and food. The type of feed and its amount depend on the species used alongside the tilapia. For grass carp, grasses from or around the dyke are fed by chopping and spreading onto the pond water surface. Providing fish with grass is carried out by applying it within designated areas covered by PVC pipes, often a square area, so that fish are familiar with the area and grasses remain in one place so that they can be easily removed if necessary. For other species, farmers may use vegetables and succulent plants, banana leaves, water hyacinth, ipil ipil, etc., which are minced and mixed with trash fish, rice bran or other ingredients to produce home-made feed, which is normally in the form of dough or sinking pellets.

Polyculture is usually carried out in semi-intensive farming systems with multiple harvests rather than one final harvest (Edwards *et al.*, 1988). Multiple harvests maximize total fish production and income for farmers. Harvesting is carried out whenever there are reasonable market prices for certain species or they are needed for family consumption, especially for special occasions. In addition to higher production, multiple harvests from polyculture ponds allow more flexibility for farmers to follow market demand, providing them with a continuous cash flow. However, there are some disadvantages to polyculture, including the following.

- Farmers need to know about the culture methods of many species.
- It is relatively more complicated and can be difficult to manage.
- Farmers have to worry about the market for many species, and it is sometimes difficult to find good prices for all of their cultured species.
- Total cost of management and operation can be high because this requires additional feed, facilities and skilled personnel.

Tilapias are omnivorous, and their culture has been trialled in polyculture in traditional Chinse and Indian systems.

Traditional polyculture in China:

1. grass carp (*Ctenopharyngodon idella*) – feed on grass or herbivorous
2. silver carp (*Hypophthalmichthys molitrix*) – phytoplankton feeder
3. bighead carp (*Aristichthys nobilis*) – zooplankton feeder
4. mud carp (*Cirrhinus molitorella*) – detritus feeder.

Similarly, another traditionally practised polyculture, referred to as 'composite culture', is mainly found in India or the Indian subcontinent, which might have existed as early as the polyculture reported in China. The species used in composite aquaculture include:

1. rohu (*Labeo rohita*) – herbivorous column feeder
2. catla (*Catla catla*) – zooplankton feeder
3. mrigal (*Cirrhinus mrigala*) – bottom dweller.

Attempts to introduce Nile tilapia into these systems have been tried where it is the main fish species, and it has shown good growth without affecting the performance of other species when stocked at reasonable densities. Because tilapia performs well at high densities it has been used as the main species, with carp as the minor species. Mouth-brooding tilapia are raised with other fish species (Table 4.2) to utilize natural food within the pond, thereby increasing fish productivity. Tilapias prefer phytoplankton and detritus,

Table 4.2. An example of natural food utilization and stocking ratio of polyculture species per hectare of pond (Bhujel, 2014).

Species	Benefits/purpose	Stocking/ha	
		Per cent	Number
Nile tilapia (*Oreochromis niloticus*)	Consumes mostly phytoplankton	90	27,000
Grass carp (*Ctenopharyngodon idella*)	Grasses and macrophyte grown on the dykes or nearby	5	1,500
Common or mirror carp (*Cyprinus carpio*	Detritus or debris in the pond bottoms/oxygenation	3	900
African catfish (*Clarias gariepinus*)	Controls recruits or unwanted fry of tilapia and other species, and also ill/dead fish	2	600
Total		**100**	**30,000**

and any uneaten food will be eaten by other species and will not be wasted. The addition of one more, or a few other, species does not add to the cost for the farmers but would increase the yield, adding to their net profit when the fish are sold.

The type of species and number of fish stocked are entirely a matter of farmer preference and experience, but the feeding habits of the available fish species have to be kept in mind as well as a market for them. Grass carp is highly preferred as the main species in Chinese polyculture systems because it consumes many different grasses, it produces excreta to fertilize the pond and it supports other species. It is sold for a high price to Chinese communities. Chinese farmers grow tilapia as the main species and silver carp as cleaner fish because they can filter plankton. In tilapia ponds, farmers prefer a small number of carp as indicators of water quality: if water quality deteriorates, silver carp are the first to die, indicating poor water quality. Similarly, silver barb (*Puntius gonionotus*) is used as an indicator in Thailand because it quickly dies through changes in water quality.

More recently, in Bangladesh, farmers have been adding Nile tilapia and *Pangasius* to traditional polyculture containing Indian major carp. Productivity from traditional polyculture systems was 1–3 t/ha (Edwards & Karim, 2007). In some cases Nile tilapia has been the main species farmed with carp, while in other polyculture systems *Pangasius* is used as the main species. Carp are stocked at the rate of 5–10%, only because those newly introduced species can grow well at higher densities and produce greater tonnage, e.g. 40–50 t/ha (Edwards & Karim, 2007) or even more, i.e. 400 t/ha in Vietnam (Phan *et al.*, 2009). In some cases, local species, e.g. tengra (*Mystus* sp.), one of the preferred species available throughout the year (Ara *et al.*, 2006), are added to the polyculture of Nile tilapia, *Pangasius* and carp.

IFS

There are several models of IFS, e.g. fish with rice, poultry, duck, livestock, vegetables, spices and fruits. Rice is the staple food of most developing countries, especially in Asia, and it is the main crop that is grown. Because most rice paddies need water to grow, rice fields have become the breeding ground and shelter for broods, their offspring, juveniles and adult fish. Traditionally fish are caught from the rice field, especially during the summer when fish migrate upstream for breeding in rice fields. However, due to the excessive use of chemicals and pesticides applied to rice, fish may not survive in rice fields. Therefore, some modifications are made to rice fields, such as creating trenches across or around them. At the same time, modifications in farming methods are also needed, such as rice farming without pesticides. Therefore, growing fish in rice fields helps farmers produce organic rice because they must stop using chemicals. At the same time this will help produce fish for family consumption and sale. Fish are grown in trenches (1.0–1.5m deep and 2–3m wide) dug around the rice field, or hapa nets are installed (5–10 m^2 or larger, depending on the need) at corners where rice does not grow well. In addition, vegetables and fruit can be grown on the dykes using pond water or irrigation. At the same time, cattle, buffalo, goats and sheep can graze on grasses on the dykes.

Since the very beginning of farming or primitive agriculture, people around the world have been raising animals along with crops, fruits, vegetables and spices. The knowledge or practices of how a farmer effectively utilizes/reutilizes their available resources, e.g. land, water, by-products or wastes, to produce sufficient food and many varieties of items to support their family living is termed 'integrated farming'. The commercialization of agriculture monocrops has become popular due to high productivity and profitability, resulting from the development of specialized knowledge and technologies. However, the sustainability of commercial monocrop systems has been questioned. Revisiting and reviewing the traditional systems, academicians and researchers have tried to understand the linkages or interactions among the subsectors within the farming system. Interestingly, the system of growing many crops, vegetables and fruits along with raising animals is more sustainable and more efficient in recycling available resources. In an IFS, an output from one subsystem becomes an input to another, resulting in greater output efficiency e.g. rice bran is used as animal feed and animal manures are used to fertilize rice fields, vegetable gardens, fruit orchards and, more recently, fishponds.

In Vietnam, a popular model of IFS called the VAC system (FAO, 2001; Luu, 2001) is found; V denotes *vuon* (garden), A denotes *ao* (pond) and C denotes *chuong* (livestock house). In this system the pond plays a central role. Farmers grow vegetables using water from the pond and raise animals, especially pigs. Pig manure is used to fertilize both the pond and the vegetables. This system is a traditional IFS approach practised by mostly low-income families for their livelihoods in rural Vietnam. According to rural farmers, the most profitable activity is aquaculture and the second is agriculture, followed by horticulture.

In addition to traditional types of IFS, commercially practised integrated livestock–fish farming systems began in the early 1980s. Livestock farming is the main source of income, and much animal manure is produced as a farm by-product. Manure is used mainly to fertilize crop fields, vegetable gardens and fruit orchards. Since that time farmers have started applying manures in their ponds to

Figure 4.1 Tilapia grown in a pond while pumpkin are grown on the dyke using trellis above the pond; pumpkins are seen hanging above the pond (photo by R. Bhujel)

enhance phytoplankton growth and provide substrates for bacteria and microfauna, which the fish eat.

Ponds are often used for water storage and can be used for livestock, cleaning utensils, washing clothes and bathing. Animal manures, from such as chicken, cow and pig, etc., are used as inputs for fishponds to develop green water. Chicken manure contains 1.5–8.5% nitrogen whereas the content in pig excreta ranges 1.8–7.4%. Similarly, in some countries, cattle faeces/manure are also used, which normally contain 0.5–4.4% nitrogen. Because these excreta have very low levels, farmers must use large amounts to develop green water and support good fish growth. Ponds should not be overloaded with manure or organic matter because this creates an anoxic environment at the pond bottom. Maximum dry matter loading should not exceed 100 kg/ha/day.

In China and other parts of Asia, construction of pig sheds on the pond dyke is common, with urine and litter being flushed out directly to the ponds. This system has become very popular because it provides ample pond input as pigs consume a lot but digest less and produce a lot of faeces, enhancing phytoplankton growth in the pond. As a result, tilapia can grow rapidly, consuming plankton without needing pelleted feed. Cheap tilapia were produced using this system but eventually this system was discouraged, especially if the farmer is planning to export. Pellet feeding was compulsory on those farms rearing fish for export. On the other hand, poultry houses are built over fishponds so that faeces and uneaten feed can drop into the pond directly as pond input. The introduction of mixed-sex tilapia with catfish used to be common practice to control tilapia recruits.

On the dykes in the semi-intensive culture system of tilapia in Bangladesh, vegetables are grown such as tomatoes, potatoes, pumpkin, snake gourds and many others, to better utilize the space. Farmers usually raise the dyke, adding fertile sediments or pond mud, which serve as fertilizer and a good source of nutrients. They often use bamboo or wooden poles (or trellis) above the fishponds to expand the space available for vegetables. As a result, vegetables can be seen hanging above the pond (Fig. 4.1). Pond water is used to irrigate the vegetables when there is no rainfall.

Cage systems

Cage culture has long been practised. However, in the case of tilapia it started as an alternative solution to the off-flavour or muddy smell that is often found with pond-cultured tilapia. Tilapia has the characteristics of absorbing both the colour and smell of its environment. The muddy smell in fish is caused by geosmin (GSM) and 2-methylisoborneol (MIB), formed as secondary by-products of bacterial metabolism (Tucker, 2000); GSM and other chemicals are produced by blue-green algae or cyanobacteria in high-nutrient-load tilapia ponds. The most prevalent genera of cyanobacteria found in Thailand are *Anabaena* sp., *Oscillatoria* sp., *Pseudanabaena* sp., *Lyngbya* sp., *Phormidium* sp., *Planktolyngbya* sp. and *Synchococcus* sp. However, GSM is associated with abundance of *Anabaena* sp. and *Oscillatoria* sp., and MIB with *Pseudanabaena* sp. and the culture period (Pimolrat *et al.*, 2015). These compounds can be detected at 5–10 ppb, so even minor traces must be removed. Because these compounds are volatile and degrade quickly, it is possible to purge them from tilapia by placing the fish in clean water for 2–3 days (Fitzsimmons, 2008). This depuration system is common in farms whose products are bound for export, although this is not always necessary in Asia because fish identified as having an off-flavour are considered lower grade and are sent to the domestic market and cooked with herbs and spices to mask the off-flavour.

More advanced methods could be tried to remove or decrease GSM and MIB in tilapia, including ozonation, advanced oxidation processes (AOPs) and adsorption removal from water using activated carbon and/or zeolites. Purging with clean water followed by three ppt saline water and keeping them for at least 16 days has been recommended as an efficient treatment by some farmers (Yamprayoon & Noomhorm, 2008). An anoxic situation and heavily nutrient-loaded ponds also cause stress and can result in poor fish health, leading to fish kills. Therefore, ponds should be free of high levels of manure and feed to avoid excessive green water, especially cyanobacteria. Once off-flavour is noticed, the fish should be harvested from the pond and the pond completely drained, dried for 2–3 weeks and limed before restocking.

Alternative methods for tilapia culture using cages in running water such as rivers, canals or lakes and reservoirs are being increasing used globally. Various cage systems are described below.

Cages in lakes and reservoirs

Fishing in lakes and rivers is a tradition worldwide, but fish catch is declining due to pollution, overfishing and other reasons. In many countries, governments or communities are trying to manage the natural stock by restocking natural water bodies with hatchery-produced fingerlings, restricting fishing during the breeding season or setting harvest limits or quotas. However, the productivity of natural water bodies is very low, e.g. 82 kg/ha/year for lakes and 273 kg/ha/year for rivers (Randall *et al.*, 1995). Even from the productive reservoirs of Africa, fish yields were up to 329 kg/ha/year, in Latin America and the Caribbean to 125 kg/ha/year and in Asia up to 650 kg/ha/year (Kolding & van Zwieten, 2006). To better utilize water space, various recommendations have been made.

One suggestion is cage culture, which started in China around 800 years ago (Hu, 1994). The highest productivity of 187.5 kg/m^3 was reached when the Freshwater Fisheries Research Centre introduced small-volume (1 m^3 or 2 × 2 × 1 m^3), high-density culture techniques from the USA starting in the 1990s (Zhou *et al.*, 2012). The productivity of cage culture per square meter is almost the same as that from 1 ha of high-yielding reservoirs.

In cages, Chinese carp, common carp, eels, bream, rainbow trout and other species were commonly cultured in previous times (Hu & Min, 1984). This type of farming spread worldwide, including in South-east Asia, where the earliest records date back to the late 1800s, particularly for freshwater lakes and river systems (Gopakumar, 2009). Cage culture in sea cages of many marine species was also trialled, such as the yellow tail, grouper, sea bream, milkfish and others. Similarly, the culture of many freshwater species, such as carp, tilapias and others has been trialled with some successes in cages in lakes, reservoirs, rivers, canals, etc. However, cage culture either grew slowly or did not take off, except for a few species, due to several issues such as the declining natural fertility of water, pollution

Figure 4.2 Cages in the Mekong River, Thailand (photo by R. Bhujel)

Figure 4.3 Tilapia cages alongside a residence in a lake in Thailand (photo by R. Bhujel)

and environmental concerns. Among others, salmon is a major cage culture species, especially in Norway, Scotland, Chile and some other countries in the northern hemisphere. In the tropics, tilapia has emerged as the leading species for cage culture (Bhujel, 2019).

The mass-scale production technologies developed to produce high-quality monosex tilapia fingerlings have contributed to the expanding cage culture of tilapia globally. The tilapia industry has taken off in Asia, especially during the past three decades. Hundreds of tilapia hatcheries have emerged in Thailand and then Bangladesh; several of these can supply >20 million high-quality monosex fry each month (Bhujel, 2011; 2013). Cage culture was further boosted due to the production and promotion of red tilapia (known locally as Thapthim) initiated and led by the Charoen Pokaphand Co. Ltd, headquartered in Thailand. Branding and marketing boosted its demand, resulting in a huge expansion of cage culture in Thailand and is now being expanded to many other countries. Standard cage sizes are either $5 \times 5\,m^2$ or $6 \times 4\,m^2$, installed at a depth of 2m in rivers, canals or lakes (Figs. 4.2 and 4.3). Although red tilapia cage farmers number fewer than pond farmers, production is up to

1/3–1/2 of total tilapia produced by volume (Belton *et al.*, 2009). Farmers are organized in clusters as a group under contract farming. They are given guidance on feeding: 5–8% of biomass per day of good-quality floating pellets containing 30–35% crude protein fed twice or thrice daily. Normally they stock 30–40 g juveniles and harvest about 1 t of tilapia per cage – i.e. 20 kg/m^3 of size 0.6–1.0 kg.

Tilapia farming is becoming specialized and segmented by hatchery, nursing and grow-out carried out by different groups of farmers, which is a sign of a mature industry. The estimated net profit can be about US$1000 per cage of standard size 50 m^3. Similar systems are found in Cambodia, Laos and Vietnam along the Mekong River and its tributaries. The major advantage of cage culture is that tilapia do not develop a muddy flavour, which occurs when they are grown in some greenwater ponds.

Although efforts have been made to introduce cage culture in Bangladesh, commercial cage culture of tilapia really started following the success of monosex seed production, becoming popular in rivers around Chandpur and Lakhipur (Baqui & Bhujel, 2011).

Figure 4.4 Tilapia culture in cages installed in a lake in Mali (photo by R. Bhujel)

Cage culture was adopted by over 6000 families, including those living on public land along the riverbanks but who were considered as landless (Pervez, 2018). The standard cage size is $6 \times 3 \times 1.5\,m^3$, stocking size for tilapia is 20g and density is 35–40 fish/m^3. Harvest size was around 400 g. Tilapia are fed with floating pellets having a food conversion ratio (FCR) of 1.75. Productivity was found to be 350 kg per cage, i.e. 13 kg/m^3. Net profit showed approximately US$200 per cage over 6 months. Tilapia farming boomed in early 2000 (Baqui and Bhujel, 2011).

In Vietnam, tilapia are grown in cages in the Mekong River delta, along the river where *Pangasius* is the dominant commercial species. Some of the largest farms have over 100 cages. Each cage is 16 m × 8 m × 4 m (512 m^3) in size. Stocking at about 80–100 fish/m^3, they produce about 20 tonnes per cage, i.e. 39 kg/m^3. The Ministry of Agriculture and Rural Development, Vietnam, has passed the Development Scheme on large-scale tilapia cultivation by 2020 with a vision towards 2030 (VOV, 2016). They planned for 1.5 million m^3 cage volume on rivers and reservoirs by 2020 and 1.8 million m^3 cage volume by 2030 to create over 50,000 jobs, with the target of 0.4 million mt by 2030 (VOV, 2016). However, the main obstacle for the tilapia industry was the requirement for high-quality monosex seeds. There are about 200 smaller tilapia hatcheries producing about 0.5 billion per year, but large quantities of broodstock and monosex fry still need to be imported.

In the Philippines, tilapia used to be produced in Lakes Laguna and Taal using small to large cages, i.e. $10 \times 10\,m^2$ with a depth of 3–4 m. Millions of monosex fingerlings used to be produced by large hatcheries, but many of those have collapsed due to natural disasters and other reasons. Tilapia produced in the cages in those lakes were grown without feeding because the lakes used to be very fertile. Nowadays, tilapia are fed with pellets and other types of feed. An excessive number of cages and overstocking of biomass have created environmental problems, resulting in large-scale fish kills.

In India, tilapia cage farming has been booming because many state governments are providing cage culture facilities to the private sector, mainly for monosex tilapia farming. Similarly, in Africa, tilapia is the main species. Many private companies and public agencies promote monosex tilapia farming in cages in several countries, including Ghana, Ivory Coast, Mali, Nigeria, Rwanda, Uganda, Zambia and Zimbabwe (Fig. 4.4). Most of them use large circular cages commonly used for salmon farming. Cage sizes in these countries range from 10 to 20m in diameter and 4–10m deep, some of which can accommodate 100,000 fingerlings in a single crop, similar to growing them in a pond bigger than a hectare. However, most farmers need more good quality monosex seed, and efforts are being made to establish mono-sex tilapia hatcheries in most countries.

Cage culture and pollution issues

Cage culture is considered a polluter of water because nutrients are added to natural water bodies. Pollution is enrichment of nutrients, especially nitrogen and phosphorus. However, the enrichment of these nutrients from cage aquaculture is negligible and can be quantified. For example, for nitrogen, if we have 600 cages producing about 600 tonnes of fish in a lake of area 100 ha and the feed gives a FCR of 1.5: the lake contains around 30% crude protein (CP) and the protein contains about 16% nitrogen (N), according to Osti *et al.* (2018), and 26% of N input through the feed is harvested as fish (in the case of Nile tilapia). Using this information, N enrichment can be estimated by dividing the volume of the water body, as shown below:

The total amount of N added to the system as feed =
$= 600 \text{ t fish} \times 1.5 \text{ FCR} \times 30\% \text{ CP} \times 16\% \text{ N}$
$= 43.2 \text{ t N}$

The total amount of N expected to be harvested in the form of fish (26% wet weight basis):

$= 43.2 \text{ t N} \times 26\%$
$= 11.232 \text{ t N}$
Total N remaining in the system $= 43.2 – 11.232 = 31.968 \text{ t}$

Figure 4.5 Tilapia grown in cages (5 × 5 m^2 or 6 × 4 m^2, at a depth of 2 m) installed in ponds, Chiang Mai, Thailand; other fish species are farmed outside the cages with no extra feeding (photo by R. Bhujel)

If a lake covers 100 ha (or 1,000m long and 1,000m wide), the total volume of water would be:

$$= 100\,\text{ha} \times 10{,}000\ \text{m}^2 \times 1{,}000\ \text{l}$$
$$= 1{,}000{,}000{,}000\ \text{l of water}$$

Increased N concentration due to cage culture = (31.968 × 1000 × 1000 × 1000 mg)/(1,000,000,000 l)

= 1.00 ppm (the limit given by Xu *et al.*, 2014)

This estimation shows that the nitrogen enrichment level due to having cages, assuming all nitrogen remains in the water for a whole week, is almost equivalent to the benchmark level of pollution, i.e. 1.0 ppm or 1.0 mg/l (Xu *et al.*, 2014). Because the proportion of the area covered by 600 cages would be approximately 3%, cages covering over 3% of the water surface area could be set as the maximum allowable or highest carrying capacity for cage culture in any water body. However, a study from Bangladesh (Hossain *et al.*, 2017) has reported that local fish consume remaining feed outside the cages and are utilized by phyto- and zooplankton; therefore, the actual carrying capacity of any lake is much higher. Moreover, if a lake receives water continuously from upstream areas and drains, it will have a higher carrying capacity.

On the other hand, lakes normally receive run-off water from adjacent areas, especially from agricultural fields where pesticides are heavily used, and the waste from industries and municipalities drains directly into water bodies. Although these are the major cause of water pollution, man-made structures in pristine water, especially in lakes, reservoirs, rivers or seas, may destroy the natural beauty. In such cases, zoning can be carried out so that cage culture can be located away from tourist areas.

Cages in the pond

Cages are installed mostly in natural lakes or man-made reservoirs where water is stored for irrigation or electricity generation by the construction of dams. They are also found in slow-flowing rivers in plain areas. Almost all of these belong to the public. Communities use these water bodies for irrigation and transportation. In some rural villages, water is used by animals and even by humans for bathing, washing clothes, cleaning utensils and even drinking. Therefore, having cages often creates conflicts with those activities. Those activities may also disturb the cage culture and sometimes cause water pollution, affecting the health of the fish stock in cages. Mass-scale fish kill can occur due to excessive pollution or nutrient enrichment. For example, two incidents have happened in Thailand.

Tonnes of tilapia died in thousands of cages along Thailand's Chaophraya and Bang Pakong rivers (Bhujel, 2013). The fish reared in cages died downstream of the Chaophraya river because a large amount of molasses poured into the river when a barge transporting thousands of tonnes of molasses capsized. In the Bang Pakong river, the death of fish in cages was believed to have been caused by a sluice gate suddenly opening and releasing acidic humus material and different types of chemicals, including pesticides that had accumulated with run-off water from surrounding agricultural land. To avoid these types of risk, some farmers/companies started culturing fish in cages installed in deeper (>2 m) and relatively larger ponds (>0.5 ha) belonging to them, with no others having access to cause these sorts of problems. Fig. 4.5 shows an example of cages installed in ponds in Chiang Mai, Thailand. Tilapia are grown in cages at either high feeding rates (3–5% biomass daily) or *ad libitum*. The most common cage sizes are $5 \times 5\,m^2$ or $6 \times 4\,m^2$, at a depth of 2 m. Outside of the cages other fish species, in particular carp, are grown in polyculture but without feeding. The excess feed and excreta of heavily fed tilapia are sufficient to create plankton growth in ponds, which serves as a natural food for carp or smaller tilapia.

Some farmers also grow *Pangasius* in cages at high densities with extra feeding to achieve fast growth. Uneaten feed and excreta become fertilizers for the pond water. Cages are installed as required but should cover at most one third of the pond area. Farmers stock fingerlings in the pond (1–5 g) and, when they are 30–40 g, they collect them and use them to stock cages, using high feeding rates so that the fish can rapidly grow to a size that will fetch a high price for their harvest. Farmers grade fingerlings based on their size and stock these in different cages, which they sell as a full cage when the fish are sufficiently well grown. This allows fish to be sold continuously on a regular basis, rather than waiting for the harvest of large pond stock. This helps farmers have a regular cash flow and a regular food source for the family if they choose to consume the fish, which helps enhance the health status of the farmer's family.

Modified pond system

Hapa in the pond

In order to utilize pond space and enhance productivity per unit area of land and volume of water, hapas are also used, which helps maintain the fish stock of different species and different sizes within a pond. Initial hapas were constructed from mosquito nets placed upside down to keep alive surplus wild-caught fish for a short period.

These are made of cotton or nylon nets. Gradually, they became very popular with farmers, and nowadays special UV-resistant net materials are produced of varying mesh size, such as nos. 16–28. The larger the number, the smaller the mesh holes, e.g. no. 16 contains 16 holes per inch, meaning larger holes compared with no. 24, for example, which has 24 holes per inch.

Most monosex tilapia hatcheries use hapas for breeding, sex reversal, nursing and even for holding them for conditioning in tanks. Grow-out farmers may also use hapas for nursing the seed in a corner of a grow-out site before releasing them into the pond so that they become large enough to be protected from predation by other fish, birds, insects, snakes and other predators. The size of the hapa may differ based on its purpose: for example, $24 \times 5\,m^2$, $60 \times 5\,m^2$ and $8 \times 5\,m^2$ are the most common sizes of hapa used for breeding, $10 \times 2\,m^2$, $5 \times 4\,m^2$ or $3 \times 2\,m^2$ for nursing from fry to fingerlings for 1–3 months and sex-reversal treatment for 21 days.

Hapas make fish gathering or harvesting easier for handling and allow complete harvesting, but the installation of hapas requires the additional arrangement of either bamboo, wood or concrete poles for supports because they need to be hung tied by ropes on top and at the bottom to avoid touching the mud of the pond bottom. Hapas are installed in series in shallow ponds (up to 1 m), either along the dykes or across the ponds. Some tilapia hatcheries even arrange hapas spread all over the pond, which requires personnel to go into the pond for feeding of the fish.

Stocking density may depend on the size of the fish. For example, broodfish are stocked at six fish/m^2 of hapa space, whereas higher densities are applied for smaller fish (fry and fingerlings). For fry during sex reversal, the density of fish may range from 5000 to 6000 /m^2 but may reduce to 500–1000 /m^2 for subsequent nursing.

More recently, hapas have also been used to nurse tilapia fingerlings in the middle of a pond where shrimp larvae are reared. This system is useful for cleaning and maintenance of water quality, helping to improve the health of shrimp and reducing the chances of disease. Although tilapia may grow poorly in this system there is scope to develop it, but more research is needed to determine the densities and type of feed suitable for both shrimp and tilapia.

In addition, hapas are also used for both grow-out farming of tilapia and installion in deeper ponds (2.5 m) to accommodate more fish and allow better water exchange through wider-mesh nets. Typically, $10 \times 10\,m^2$ hapas at 2m depth are used, which is four times larger than normal cage size, i.e. $5 \times 5\,m^2$.

IPRS

The productivity of the pond can be further enhanced by creating a raceway system at one corner. Recently an IPRS was designed for channel catfish and has been used for tilapia in various countries including Bangladesh, Brazil, Colombia, Egypt, Indonesia, PR China, Thailand and Vietnam (Cremer *et al.*, 2014; Nasr-Allah, 2019; Gharti *et al.*, 2023; Nugroho *et al.*, 2023). In this system the water is recirculated through large bodies of water that assimilate wastes from the small, cultivated tank areas where fish are stocked at high densities.

IPRS involves the construction of a series of long, narrow raceway tanks at one corner of the same pond or lake to create a water flow-through system through the use of a strong pump; fish are stocked at high density and fed with high-quality feed to achieve rapid growth (Arana *et al.*, 2018). In this system, usually a larger pond (0.4–2.0 ha) is chosen. A series of four or more narrow, parallel concrete tanks are constructed, depending on the target production. Normally a small pond of 0.4 ha will have four raceway tanks of size $12.2 \times 3.1 \times 1.2\,m^3$, which receive the water from a pump of 2.5 horsepower (HP) or higher for each raceway tank. Two more 1.5-HP pumps are needed to move the water at the two corners of the pond. Arana *et al.* (2018) reported a trial showing that 57 kg/m^3 in a raceway tank is possible in around 125 days by stocking 17,000 tilapia fingerlings of 18g per concrete tank of size 150 m^3 ($25 \times 5 \times 1.2\,m^3$). In China, grass carp stocked at 750g reached a stocking density of 127 kg/m^3 at a harvest size of 2.6 kg after 182 days, with a FCR of 1.89 (Cremer *et al.*, 2014). IPRS is described in more detail by Chappell (2017).

The major advantages of this system are that it creates a water current within each tank similar to that of rivers or small streams, which fish like as it supplies more oxygen through mixing with the atmosphere and, more importantly, flushes out nitrogenous waste from the pond. It also facilitates feeding, close monitoring, sampling and harvesting of the fish. As a result, the system creates a healthy environment for fish so that high-density culture is possible in the tank. These tanks can be used for depuration before harvest by having a clean water supply system where the fish are placed for a few days before being sold. However, the key issue is the high initial investment cost. Some farmers in Asia have pointed out that it requires at least US$20,000 to construct four concrete tanks as raceways. Another challenge is that it requires an uninterrupted 24-hour electricity supply, and the daily cost of running the pump using electricity can be very high.

A recent study in Pakistan has shown the high productivity and economical viability of IPRS, in which the average harvested biomass was 57.3 kg/m^3 with an average daily weight gain per fish of 4.47g (Fatima *et al.*, 2021). Similarly, in Indonesia, stocking 75–80 juveniles of 33 g Nile tilapia, they were able to harvest 13.0–14.7 kg/m^3 with almost 80–90% survival (Nugroho et al., 2023).

Tank farming systems

Tilapia are also produced in more intensive systems in tanks at high densities using various models such as RAS, bio-RAS, aquaponics, the biofloc system (BFS) and integrated multitrophic aquaculture (IMTA). Recent concerns about the environment, pollution, climate change and global warming have led the aquaculture industry to evolve more sustainable aquaculture systems aimed at preserving finite natural resources, especially land and water, and many schemes have been trialled to use less, reuse and conserve resources.

RAS

Water for aquaculture is sourced from various water bodies, e.g. pumping from rivers, lakes or underground aquifers. RAS are seen as the solution where water and land space are scarce. Water is reused repeatedly in these systems, requiring efficient cleaning and filtration, which adds significantly to costs. Normally, high-value species are stocked at high density in concrete tanks or tarpaulins. However, fish species must tolerate stress due to the high density. Tilapia is one of the best species for such conditions because it has strong innate and adaptive immune responses that are sufficient to resist significant alterations in somatic growth and haematological status. However, in RAS, complete feed is given to the fish to fulfil its nutritional requirements, which is normally costly. Under prolonged culture periods, fish growth might be impaired in systems with low water-flow rates (Obirikorang *et al.*, 2022). Chronic exposure of fish to poor water quality under low water-exchange conditions predisposes them to a significantly higher risk of disease despite their hardy nature. Attempts should be made to minimize stress, and to make the system function well.

Recirculatory aquaculture systems, including types and components, are well described by Bregnballe (2015). Filtration systems involve the separation of coarse and heavy materials by settlement under gravity, filtration by gravel and coarse sand or drum filtration, etc. For dissolved nitrogenous wastes, on the other hand, conversion of toxic ammonia (NH_3) to nitrite (NO_2) and then to non-toxic nitrate (NO_3) has to be carried out with the help of the bacterial genera *Nitrosomonas* and *Nitrobacter*, respectively. Bioballs, bottle caps, oyster shells and nylon nets, etc. are used to create a separate compartment to provide shelter for these bacteria, also called a biological filter. Charcoal is also used to absorb chemical waste. Because it is not possible to make this conversion very efficient, nitrogen waste builds up at high levels in the entire system and dirty water must then be replaced by fresh clean water at specific intervals. This means that all nitrogen remains as waste, unutilized and thrown away as sludge. High water-flow rates of 100% water exchange/h may be needed for Nile tilapia to maintain its health and welfare status. One of the most important benefits of RAS is that biosecurity is better than in other systems.

Huge amounts of sludge may be produced from RAS and bio-RAS systems, which can be reused as either predigested ingredients (highly digestible) in rations for aquatic and terrestrial animals or fertilizers for plant crops. However, RAS should help conserve water and generate less effluent, so the surrounding environment does not become polluted. Attempts have been made to devise a proper use for this effluent or, on the other hand, to design more intensive RAS with a series of structures and equipment, such as settlers, mechanical filters, biological filters, UV disinfection, water pumps, air blowers, power generators, emergency aeration or ozone generation, etc.

For RAS, high investment is the key issue in regard to building structures, buying and installing equipment and high operating costs, i.e. electricity, maintenance and depreciation. This is partly compensated for by the flexibility allowed to locate production facilities near large markets, complete and convenient harvesting and quick and efficient disease control.

Moreover, technical skills and understanding are crucial to running the whole system. Due to a lack of fine-tuning in managing and handling RAS systems, there have been many failures of commercial RAS operations. RAS has been widely used for tilapia hatcheries and nurseries without grow-out, and there are many vertically integrated and commercially successful farms worldwide, including hatcheries, nurseries and grow-outs.

Aquaponics system

Tilapia have been most widely used for aquaponics systems in which fish and vegetables are grown together, recycling both water and waste. Nitrogenous waste produced by tilapia in tanks is transferred, along with the water, to vegetable-growing beds serving as sources of nutrients (Somerville *et al.*, 2014). An aquaponics system is one of the best solutions in the production of various food items, including fish, i.e. mainly as a source of high-quality protein and minerals and vegetables rich in vitamins. Its importance was realized during the COVID-19 pandemic when travel was restricted. Many families started with aquaponics to produce food; at the same time, others became engaged in aquaponics and backyard gardening. This system uses fish nitrogenous waste to grow vegetables. NH_3 is converted to nitrate, which plants take up as a source of nitrogen; therefore, fish waste can serve as a fertilizer. Plants act as filtering agents and clean the water, which is then circulated repeatedly to the fish tanks. Normally at least 20 kg of tilapia from 1000 l of water and an average of 2–3 kg of vegetables/m^2 can be produced from an aquaponics system (Rakocy *et al.*, 2003; Pantanella and Bhujel, 2014). The optimum fish species for this system is Nile tilapia and its hybrids, such as red tilapia. Other suitable species include *Pangasius*, catfish, common carp, trout and shrimp/prawn, which can also be grown with tilapia. However, the system's productivity depends on its management and vegetable type used.

Over 100 vegetable species can be grown, including sweet basil, eggplant, Chinese kale, tomatoes, lettuce and many immunity-boosting spices such as ginger, garlic, mint and coriander (Somerville *et al.*, 2014). It has many benefits in addition to family nutrition: it can be installed in a small space on a balcony, on the rooftop or in the backyard garden. One of the constraints is that it requires 24-hour electrical supply along with a back-up DC generator. Simple tap water can be used to fill the tanks initially, with topping up by about 10% weekly or as needed to compensate for evaporative or leakage losses. To run the system successfully it is necessary to understand how to grow a variety of vegetables, their requirements and tilapia farming.

Flow-through aquaculture system

The flow-through aquaculture system (FAS) involves the culture of fish in tanks or raceways with flowing water; these are artificial channels used to grow aquatic organisms and are among the earliest aquaculture systems. Typically raceways are long, rectangular, concrete tanks or canals with inlets and outlets for massive amounts of water to pass through. Continuous flow-through for large quantities of water is arranged to allow a high level of water exchange to maintain good water quality so that fish can be stocked at high densities. Raceways have been used for coldwater fish species such as trout, which requires very high water quality levels and is a high-value fish. Other freshwater species that have been used include catfish, tilapia and some marine species, especially juveniles, which require a constant water flow, especially salmon, brackish-water sea bass, sea bream and others.

High-density culture of tilapia using raceways was used in Taiwan in the 1970s to export fish to the USA, but eventually they became unsustainable when tilapia were produced more cheaply in ponds.

FAS is suitable for areas where clean water is plentiful, but the initial investment required for the construction of concrete raceways is the biggest problem.

BFS/aqua/biomimicry

Biofloc systems have been developed to utilize nitrogenous waste rather than removing it from the system as in RAS, to produce microbial protein by the addition of starch or carbohydrates such as molasses or rice bran (Avnimelech *et al.*, 2014). The addition of a small amount of carbon (C:N ratio 1:10–1:20) helps enhance the growth of bacteria and other microbes. As they clump together in the water forming flocs, this is called biofloc, which serves as the natural food available for the fish in the culture system. A biofloc is the aggregation of heterotrophic, nitrifying and cyanobacteria, also including various algae and fungi. It does not require filtration systems with many compartments but still requires a simple aerator or blower.

The biofloc system is considered cost effective by enhancement of natural food in the form of microbial protein converted from nitrogenous wastes, which serves as the main source of nutrients without having to introduce feed. In this system, water contains the biofloc and therefore it should not be changed. Hence the requirement for water is minimal.

According to Avnimelech *et al.* (2014), 20–30 kg of tilapia/m^2 can maintain high-density stocking. Several invividuals in Asia have started using biofloc as a small business, but with limited knowledge and resources, as these can be set up in small backyard gardens in limited spaces. These are high-tech systems with high production capacity per unit area, minimum water exchange and less land area required. In these systems, water is used efficiently and waste is reused. However, the system is highly vulnerable because millions of microbes and living organisms have to survive, dependent on reduced space or water volume and a limited amount of oxygen.

The hierarchical food web has to be maintained in a balanced way; the population of one microbe may affect others. Maintaining this balance is not easy, and a 24-hour supply of oxygen is needed so that all living organisms can use it for their survival and growth. The system is so complex that the farmer must understand all aspects of the water's microbes and its physical and chemical characteristics. As the density of living organisms is high, maintaining their health and that of the cultured fish is not easy.

Other systems

Brackish/saline water

Nile tilapia, especially its hybrids (red varieties), can adapt to brackish water and seawater (up to 30 ppt). However, high salinity may cause stress, which can predispose the fish to disease. Tilapia has been cultured in the marine environment, with FAO (2022) data showing 107.4 tonnes of production in marine and coastal aquaculture. It can include both cages and pond culture. Vast areas of coastal lands and seas are available for aquaculture, and its widespread adoption may not be too far off: the prospects for tilapia culture in brackish and saline water are huge (Suresh & Lin, 1992). Among aquaculture species, saline water shrimp occupies the majority of coastal areas; however, shrimp culture has its pros and cons. Whenever shrimp culture diminishes, tilapia has been the alternative for farmers. Nile tilapia has been trialled with sea bass (ratio 5 - 6:1) in some coastal areas of Bangladesh, also known as brownfields, that shrimp farmers abandoned after disease devasted the shrimp industry. At low densities, i.e. one fish/m^2, productivity was achieved with approximately 1 t/ha (Monwar *et al.*, 2013). More recently the culture of tilapia with shrimp and prawn in the same pond at low densities has been one of the simple solutions introduced to reduce the risk of mortality in shrimp cultured in brackish or freshwater ponds in Bangladesh, Thailand and other countries of Asia (Jewel *et al.*, 2021).

IMTA

Two or more species with different trophic niches are managed in this farming system, with different food items produced in IMTA. Fish and other aquatic and/or terrestrial species, such as plants, pigs and poultry, can also be maintained with multiple relationships among resources to better utilize space and reuse water and nutrients. This integration facilitates the sharing of resources among different species with greater potential for technical and economic efficiency. Therefore, it has been promoted as a suitable system for sustainable food production (Chopin, 2013).

In some areas tilapia have been cultured with shrimp, bivalves or seaweed in brackish water. In other cases, fish, terrestrial animals and/or vegetables (hydroponics) can be produced in the same production system as recirculation with single or multiple loops. In a broad sense, IMTA combines the farming of fish that are fed, e.g. tilapia, with grazers and filter feeders, providing both organic matter (echinoderms, molluscs, especially bivalves, microcrustaceans and worms, other herbivorous fish) and inorganic matter utilizing species such as phytoplankton, macroalgae or hydroponic vegetables.

Aquaponics is a type of IMTA because it combines fish farming and vegetable gardening in which a crop benefits from the waste of another system, reducing the need for inputs, nutrients and effluent production. The main challenge in combining many organisms in IMTA is finding the correct proportions to create a balanced system that is environmentally and financially sustainable and socially acceptable.

Bio-RAS

Combinations of certain unique systems are evolving. For example, a recirculation system with bioflocs in more than one compartment is bio-RAS, the combination of RAS and BFS (Zimmermann *et al.*, 2023). The advantages of BFS over traditional RAS became apparent when different systems of bioflocs were developed. A trend can be seen when merging these two low-water demand systems to optimize crops with reduced production costs (especially food and electricity). The bio-RAS strategy uses the best and most efficient of the previous technologies, with cost reduction combined with maximization of technological, zootechnical and animal welfare efficiency and crop sustainability.

Bio-RAS has been used in the past decade for low-cost aquaculture projects (Zimmermann *et al.*, 2023). In bio-RAS, bioflocs can form in either one or more compartments or the entire circulating water.

Comparison among farming systems

Amongst the tilapia farming systems described above, extensive or semi-intensive pond systems are the cheapest and most sustainable because these can be carried out using low-cost inputs and have a low risk of fish health problems or diseases. However, land may be available only in some countries and, if available, it can be costly to rent or purchase large areas. At the same time, good water availability can be another constraint to filling a large land area. Land and water may be available in only a few convenient places and, if available, prices can be prohibitively high. Normally adequate land and water are available in rural areas, far away from roads, requiring costly transportation to bring in resources and take products to market for selling. There is also a shortage of labour and technical managers and staff who are unwilling to live and work in remote areas. Therefore, alternative farming systems that require little land and water and can be recycled or reused have been explored. For example, land is not required for cage culture; this can be carried out in communal ponds, lakes, reservoirs and slow-flowing rivers. Therefore, landless people can also raise tilapia by installing cages in lakes, rivers and reservoirs. However, the farmer must obtain a licence or permit from the government. Each farming system has its pros and cons. The relative advantages and disadvantages of each system are shown in Table 4.3.

The production capacity, profitability and applicability of those systems need to be tested and verified by partnering with the public

Table 4.3 Comparison of tilapia culture systems (Bhujel and Suharman, 2021).

Systems	Land required	Water use	Feed use	Waste production	Cost	Productivity	Risk level	Power needed	Profit potential
Lake/sea	-	++++	-	-	+	+	-	-	+
Pond	+++	+++	+	+	+	+	+	+	+
Aquaponics	+	+	++	+	+++	++	++	++	+++
Cages	-	++++	+++	+	++	+++	++	-	+++
RAS	+	+	+++	++	+++	+++	+++	++	++
Biofloc	+	+	++	+	++	+++	++++	+++	+++
Aquamimicry	+	+	++	+	++	+++	+++	+++	+++
Raceway	++	+++	+++	++	++	++++	++	+	++
IPRS	++	++	+++	++	+++	++++	+++	+++	++
Rice + fish	++++	+++	+	+	+	+	+	+	+
Rice + animals	+++	++	+	+	++	++	++	+	+++

and private sectors before promoting them for expansion (Bhujel & Woollard, 2011). More importantly, comparison among systems should be based on the actual potential of production volume, productivity per unit area/water volume, production costs per tonne or kilogram and also initial capital investment and day-to-day running costs. In addition, systems are also evaluated based on their sustainability and whether they are environmentally sound.

Trends in tilapia farming systems

Tilapia farming systems are moving from traditional open ponds and cages to IPRS, RAS and biofloc systems; aquaponics and IMTA are fully integrated systems. The traditional aquaculture systems produce waste either deposited in the system itself or released directly into the environment, raising nitrogen and phosphorus concentrations in soil and water. The waste deposited is harmful to the welfare of fish, and waste released into the environment threatens both human health and the environment.

These systems include a series of structural changes in aquaculture activity that consider water and waste treatment through closed land-based RAS and the reuse of waste as nutrients. They are evolving along with trends in sustainable food production systems that consider the minimization of input and waste, utilization of waste and effluent and cleaning and reuse of the same water in recirculation systems, which contributes to the 'circular economy' (Zimmermann *et al.*, 2023).

The circular economy is a pathway towards further green and sustainable development that aims to reduce inputs, energy consumption and waste production, close economic and ecological resource flows or links, decentralize production systems (local production and consumption) and question the tools used for measuring economic performance and the role of money and finance in building natural and social economic capital.

Along the path of circular economy, tilapia farming systems should also reduce the use of land, water, energy, labour and feed ingredients, especially fishmeal and oil, and minimize waste production to maintain fish health and environmental sustainability. Wherever possible we should maximize the use of by-products and their derivatives, e.g. rice bran, oil cake and feather meal, as ingredients to reduce the cost of production and maximize profit for financial sustainability. More importantly, feed should be supplementary to the natural food. Emphasis should be given to creating a cultural environment where live food such as phytoplankton (*Chlorella*, *Microcystis*, diatoms, etc.) and zooplanktons (e.g. *Moina*, rotifers, copepods and others) can be grown as needed.

More research on feeding management specific to the farming systems, and based on fish gut health, functional food or specific purposes, is important, such as supplementation of deficient micronutrients, enzymes to increase digestibility and natural products to reduce stress and improve immunity. Because it is difficult to avoid the presence of bacteria and viruses in aquaculture systems, developing a specific vaccine is also important; however, this is lagging well behind in the case of aquatic species compared with the livestock and poultry sectors.

The aims of implementing the farming systems described above for aquaculture by 2030 are to reduce production and processing costs tenfold, utilize 90% fewer natural resources and decrease waste production by a factor of ten, as stated by Zimmermann *et al.* (2023).

Conclusions and Recommendations

There are a number of aquaculture systems used for growing and producing fish. Newer systems are evolving mainly to enhance productivity, maximize the utilization of land area or water volume and inputs and minimize wastes and their effects so that fish health can be maintained at the highest possible level. Each system has its pros and cons, and a particular system may fit well as the optimum in the local context. Profitability is the ultimate goal, and long-term sustainability depends on selecting the right system and overall management.

The pond culture system is the most accessible form of tilapia culture. It still dominates tilapia farming around the globe and is characterized mostly by low inputs and low productivity, but is the cheapest way of producing fish because it utilizes natural food and health management is normally not a major issue. Therefore it is the most suitable system for small and poor farmers, especially in rural areas. It also utilizes large ponds where land is plentiful and cheap to rent or purchase. However, these large ponds need to be intensified to enhance productivity and increase overall fish production to feed the increasing population.

Polyculture is another method of enhancing the productivity of fish per unit area of water area or volume. Various forms of traditional polyculture, such as Chinese major carp and Indian major carp, still exist in Asia. Recently, however, the inclusion of tilapia in carp polyculture has changed. Tilapia is often stocked as the main species because it grows rapidly and can be stocked at high densities. Carp are often used as an indicator of water quality. If water quality deteriorates, carp show symptoms or may even die, during which time farmers can do something to save tilapia. Tilapia is cultured with shrimp, either in polyculture or co-cultured with tilapia in cages/pens/hapas and other species in ponds to solve the problem of shrimp disease. Therefore, the optimal utilization of natural foods can be obtained from polyculture if the appropriate species are chosen. Although more research is needed, this system should be further promoted as a sustainable aquaculture method.

To utilize pond space and enhance productivity, cages, hapas and pens are installed, which helps maintain fish stock of varying size from different species, makes harvesting easier and allows complete harvesting, but the installation of cages requires deeper ponds of around 2 m. Hapas and pens are installed in shallow ponds (up to 1.5 m). Monosex tilapia hatcheries are becoming popular and are an excellent example of hapa-in-pond systems.

Cage culture has become increasingly important, where artificial water bodies (canals and reservoirs) and natural ones (lakes and rivers) are available, and their use has been allowed by government policies and communities. By applying higher stocking densities and complete pellet feed, productivity can be increased; thus, the tilapia produced are of higher quality, accelerating both processing and trade (import + export) and creating more jobs in the value chain. Large volumes of water and their exchange ensure a supply of dissolved oxygen and dilution of nitrogenous wastes, which helps maintain the overall health of fish in cages. However, excessive use of feed has lowered the profit margin and caused fish health problems and environmental concerns where overstocking of fish and an excessive number of cages are installed in some water bodies.

More recently, IPRS has been designed and promoted to enhance pond system productivity and is being tested in several countries. Results have proved, however, that the high cost of electricity has constrained its adoption among farmers. It could be applicable where electricity is free or subsidized for aquaculture or food production businesses. Similarly, the high cost of installation of tanks and water-flow systems is another obstacle for poor farmers, but is suitable for those farmers with access to larger capital investment and who can see profitability in the long term. Traditionally, raceway systems for growing fish are found in some countries and have been applied where abundant water is available, especially for growing high-value species such as trout.

In areas where both the land and water are scarce, RAS using concrete tanks are appropriate. Water is reused repeatedly, passing through various types of cleaning and biofiltration systems for unwanted soluble and insoluble wastes so that fish health can be maintained, even at higher stocking densities, and productivity can be enhanced. However, it is not easy to maintain a balance among stocking density, feed and water flow rates. For this system the initial investment can be another constraint, but it is the only system that can be used in countries where land is limited and expensive and where temperature and other environmental conditions need to be well controlled.

An aquaponic system is one in which water is reused for growing vegetables for additional use. More importantly, vegetables are produced utilizing nitrogenous fish waste as fertilizer (inputs), which converts turbid and polluted water into cleaner water, which is supplied to the tanks where fish can grow and ensures good health. The aquaponic system is suitable for family nutrition because fresh green vegetables supply many vitamins and fish supplies protein and minerals. For larger-scale food supply and better use of water and waste, an IFS has often been the solution and has traditionally been practised in many countries, especially in Asia. Farming fish together with rice, vegetables, fruit and animals such as pigs, ducks, chickens and livestock is the most common system. Farms normally need to be larger in size to accommodate the many animals and plants utilizing the space.

Biofloc systems have been developed and promoted as a potential aquaculture system, practised recently using a tank system at high density producing natural food organisms *in situ*. These require an in-depth understanding of micro-aquatic organisms, their food webs and management of the physical and chemical properties of water affecting fish growth and survival. Therefore, this is an advanced aquaculture system that needs 24-hour aeration to supply oxygen to all the organisms. However, it has been promoted heavily in small areas such as backyards, with no need for large areas of land, which is a constraint in many areas. Because the system provides a wide variety of natural food, fish health should be good and productivity per unit area high. However, the risk of failure is also high.

Over 70% of the world's population resides in coastal areas, but vast land and sea areas globally are either un- or under-utilized; these could be a good source of nutritious food for the increasing population. Brackish/saltwater culture of tilapia has great potential. Red hybrids tolerate higher salinity and black tilapia can be acclimatized at the fry and fingerling stages to grow at higher salinity in coastal land or sea cages. People perceive that tilapia farmed in saline water has a better taste and price and may be considered a genuine seafood item of marine origin. Recently introduced practices of tilapia–shrimp polyculture/co-culture could be expanded in coastal areas where tilapia may enjoy better health, because pathogens from freshwater environments do not survive at higher salinity. Mariculture is facing difficulty in its expansion, mainly due to seed shortages resulting mainly from low survival rates during the hatchling and fry stages. Mass seed production techniques developed for tilapia and then commercialized could play a significant role in expanding coastal aquaculture, thereby compensating for seafood production expected from marine species culture.

References

Ahmed, S. and Rahman, S. (1992) Resource allocation by subsistence farmers in Asia: Some findings. *Asia-Pacific Journal of Rural Development* 2(2), 36–76. DOI: 10.1177/101852911992 0203.

AIT (1994) Partners in development: The promotion of aquaculture. AIT aquaculture. Asian Institute of Technology.

Ara, H., Jesmin, M., Rahman, S.M., Karim, K.M.R. and Rahaman, M.S. (2006) Proximate and mineral composition of four Tengra (Mystus sp.) species of Bangladesh. *Khulna University Studies* 7(2), 51–54.

Arana, E., Jesse, C., Hanson, T., Amezquita, J., Romellon, F. *et al.* (2018) Commercial demonstration of in-pond raceways global aquaculture advocate, May 21, 2018.

Avnimelech, Y., De-Schryver, P., Emmereciano, M., Kuhn, D., Ray, A, *et al.* (2014) *Biofloc technology – A practical guidebook*, 3rd edn. World Aquaculture Society.

Bangkok Post (2016) For the welfare of his people. Benefits to the nation bestowed by His Majesty's Sufficiency Economy Philosophy. Available at: https://www.bangkokpost.com/business/1111081/for-the-welfare-of-his-people

Baqui, M.A. and Bhujel, R.C. (2011) A hands-on training helped proliferation of tilapia culture in Bangladesh p. In: Liping, L. and Fitzsimmons, K. (eds) *Better Science, Better Fish, Better Life proceedings of the ninth international symposium on tilapia in aquaculture*, Aqua-FishCollaborative Research Support Program, pp. 311–322.

Belton, B., Turongruang, D., Bhujel, R. and Little, D.C. (2009) The history, status and future prospects of mono-sex tilapia culture in Thailand. Apr-Jun 2009, *Aquaculture Asia* 16–19.

Bhujel, R.C. (2011) How to produce billions of high-quality tilapia fry p. In: Liping, L. and Fitzsimmons, K. (eds) *Better Science, Better Fish, Better Life proceedings of the ninth international symposium on tilapia in aquaculture*, Aqua-Fish Collaborative Research Support Program, pp. 123–131.

Bhujel, R.C. (2013) On-farm feed management practices for Nile tilapia (*Oreochromis niloticus*) in Thailand. In: Hasan, M.R. and New, M.B. (eds) *On-Farm Feeding and Feed Management in Aquaculture*. FAO fisheries and aquaculture technical Paper no. 583. Food and Agriculture

Organization, pp. 159–189. Available at: https://www.fao.org/fishery/docs/CDrom/T583/root/06.pdf

Bhujel, R.C. (2014) A manual for tilapia business management. CABI Publishing. ISBN (10): 1780641362

Bhujel, R. (2019) Cage culture of monosex tilapia for food and financial security. In: Klinbunga, S., Karoonuthaisiri, N. and Piyapattanakorn, S. (eds) *Proceeding of the Sixth International Symposium on Cage Aquaculture in Asia*, Asian Fisheries Society, pp. 19–28.

Bhujel, R.C. and Suharman, I. (2021) Food and nutrition security in the context of COVID-19 and the potential role of tilapia aquaculture. *IOP Conference Series: Earth and Environmental Science* 695(1), 012023. DOI: 10.1088/1755-1315/695/1/012023.

Bhujel, R.C. and Woollard, M. (2011) Tilapia farming faces expansion issues in Thailand. May/June 2011, *Advocate, Global Aquaculture Alliance* 16–18.

Bhujel, R.C., Shrestha, M.K., Pant, J. and Buranrom, S. (2008) Ethnic women in aquaculture in Nepal. *Development* 51(2), 259–264. DOI: 10.1057/dev.2008.11.

Boyd, C.E. (1990) Water quality in ponds for aquaculture. Auburn University.

Bregnballe, C. (2015) A guide to recirculation aquaculture, an introduction to the new environmentally friendly and highly productive closed fish farming systems. Publication of Food and Agriculture Organization of the United Nations (FAO) and EUROFISH International Organisation. ISBN 978-92-5-108776-3

Chaipattana Foundation (2020) Philosophy of sufficiency Economy. Retrieved September 17, 2020. Available at: https://www.chaipat.or.th/eng/concepts-theories/sufficiency-economy-new-theory.html

Chappell, J. (2017) In-pond raceway system manual for construction and management; principle driven aquaculture production. Technology Publishing. United States Soybean Export Council.

Chopin, T. (2013) Aquaculture, integrated multi-trophic (IMTA). In: Christou, P., Savin, R., Costa-Pierce, B.A., Misztal, I. and Whitelaw, C.B.A. (eds) *Sustainable Food Production*. Springer. DOI: 10.1007/978-1-4614–5797–8_173.

Choudhary, N.R. (2023) Tilapia market outlook 2023-2033. Future market Institute. Available at: https://www.futuremarketinsights.com/reports/tilapia-market

Cremer, M., Chappell, J., Zhang, J. and Zhou, E.H. (2014) New intensive pond aquaculture technology demonstrated in China. *Global Aquaculture Advocate* 6062.

Dela Cruz, C., Sevilleja, R.C. and Torres, J. (2001) Rice-fish systems in Guimba, Nueva Ecija, Philippines. In: *Integrated Agriculture-Aquaculture a Primer.* FAO Fisheries Technical Paper 407. Food and Agriculture Organization of the United Nations. Available at: http://www.fao.org/3/Y1187E/y1187e24.htm#bb

Dewar, R.E. (2003) Rainfall variability and subsistence systems in southeast Asia and the western Pacific. *Current Anthropology* 44(3), 369–388. DOI: 10.1086/368348. (accessed 22 September 2023).

Edwards, P.E. and Karim, M. (2007) Continued evolution: Polyculture in Bangladesh, January/February 2007 print edition of the Global Aquaculture Advocate.

Edwards, P.E., Pullin, R.S.V. and Gartner, J.A. (1988) Research and education for the development of integrated crop-livestock-fish farming systems in the tropics. In: *ICLARM Studies and Reviews 16.* International Center for Living Aquatic Resources Management, p. 53.

Fatima, S., Komal, W., Manzoor, F., Latif, A.A., Liaqat, R. *et al.* (2021) Analysis of the growth performance, stress, profile of fatty acids and amino acids and cortisol in tilapia (*Oreochromis niloticus*), cultured at high stocking density using in-pond raceway system. *Saudi Journal of Biological Sciences* 28(12), 7422–7431. DOI: 10.1016/j.sjbs.2021.08.048.

Fitzsimmons, K. (2008) Food safety, quality control in tilapia products. 2008/January/February, *Global Aquaculture Advocate* 42–44.

Food and Agriculture Organization (2001) Integrated agriculture-aquaculture – A primer. Fisheries Technical Paper no.407. Food and Agriculture Organization/IIRR/WorldFish Center. Available at: https://www.fao.org/3/Y1187E/y1187e00.htm#TopOfPage

Food and Agriculture Organization (2022) Towards blue transformation. State of World Fisheries and Aquaculture. Food and Agriculture Organization. DOI: 10.4060/cc0461en.

Gharti, K., Yan, L., Li, K., Boonpeng, N. and Liu, L. (2023) Growth and muscle quality of grass carp (*Ctenopharyngodon idella*) in in-pond raceway aquaculture and traditional pond culture. *Water* 15(9), 1771. DOI: 10.3390/w15091771.

Gopakumar, G. (2009) History of cage culture, cage culture operations, advantages and disadvantages of cages and current global status of cage farming. In: *Course Manual: National Training on Cage Culture of Sea Bass*. CMFRI and NFDB, Kochi, pp. 8–12.

Gullian-Klanian, G. and Arámburu-Adame, C.A. (2013) Performance of Nile tilapia *Oreochromis niloticus* fingerlings in a hyper-intensive recirculating aquaculture system with low water exchange. *Latin American Journal of Aquatic Research* 41(1), 150–162. DOI: 10.3856/vol41-issue1-fulltext-12.

Hossain, M.A., Little, D.C. and Bhujel, R.C. (2003) Nursing duration and pond fertilization level affects polycultures of Indian major carp; rohu, Labeo rohita and mrigal, (*Cirrhina mrigala*) with monosex Nile tilapia, *Oreochromis niloticus*. *Aquaculture Research* 34(9), 765–775. DOI: 10.1046/j.1365-2109.2003.00883.x.

Hossain, M.R.A., Rahman, Md.A., Akter, S., Hosain, Md.E. and Naser, M.N. (2017) Intervention of tilapia cage culture in the river Dakatia: Threaten or blessed to local fish diversity. *International Journal of Fisheries and Aquatic Studies* 5(1), 228–232.

Hu, B. (1994) Cage culture development and its role in aquaculture. *China Aquaculture and Fisheries Management* 24, 305–310.

Hu, B. and Min, K. (1984) Cage culture techniques adopted for developing reservoir fisheries. FAO Fisheries Report No. 312, 16–19. Food and Agriculture Organization.

Jewel, M.A.S., Haque, M.A., Rahman, M.H., Khatun, M.S., Akter, S. *et al.* (2021) Shrimp polyculture: An economically viable and environmentally friendly farming system in low saline coastal region of Bangladesh. *Iranian Journal of Fisheries Sciences* 20(6), 1649–1663. DOI: 10.22092/ijfs.2021.125434.

Karapanagiotidis, I.T., Bell, M.V., Little, D.C., Yakupitiyage, A. and Rakshit, S.K. (2006) Polyunsaturated fatty acid content of wild and farmed tilapias in Thailand: Effect of aquaculture practices and implications for human nutrition. *Journal of Agricultural and Food Chemistry* 54(12), 4304–4310. DOI: 10.1021/jf0581877.

Kolding, J. and van Zwieten, P.A.M. (2006) Improving productivity in tropical lakes and reservoirs. challenge program on water and food – aquatic ecosystems and fisheries review series 1. Theme 3 of CPWF, C. O WorldFish Center. ISBN: 977-17-3087-8

Little, D.C. (2000) Meeting the needs of the poor in Asia—tilapia in the new millennium. In: Fitzsimmon, K. and Filho, J.C. (eds) *Proceedings of the from the Fifth International Symposium on Tilapia Aquaculture, Rio de Janeiro, Brazil, 3–7 September 2000*, pp. 641–650.

Lu, J.B. and Li, X. (2006) Review of rice-fish-farming systems in China—One of the globally important ingenious agricultural heritage systems (GIAHS). *Aquaculture* 260(1–4), 106–113. DOI: 10.1016/j.aquaculture.2006.05.059.

Luu, T.L. (2001) The VAC system in Northern Viet nam. In: *Food and Agriculture Organization/ICLARM/IIRR. Integrated Agriculture-Aquaculture: A Primer. FAO Fisheries Technical Paper No.407. Food and Agriculture Organization*. FAO fisheries technical Paper no. 407. Food and Agriculture Organization, pp. 29–32.

Miao, W. and Wang, W. (2020) Trends of aquaculture production and trade: Carp, tilapia, and shrimp. *Asian Fisheries Science* 33(Suppl. 1), 1–10. DOI: 10.33997/j.afs.2020.33.S1.001.

Monwar, MD.M., Ruhul Amin Sarker, A.K.M. and Das, N.G. (2013) Polyculture of seabass with tilapia for the utilization of brown fields in the coastal areas of Cox's Bazar, Bangladesh. *African Journal of Fisheries Science* 1(5), 023–027.

Nasr-Allah, A.M. (2019) Tilapia production using climate smart aquaculture system in Egypt, in-pond raceway system (IPRS). In: *Fourth International Conference on Sustainable Development of Livestock Production Systems "Smart and Precise Farming";. November 19–21, 2019*, Alexandria, Egypt.

Nugroho, E., Kristanto, A.H., Pamungkas, W., S Dewi, S., Rifaldi, P. *et al.* (2023) Optimizing tilapia biofloc technology systems and its economic profitability on industrial scale in Indonesia. *IOP Conference Series: Earth and Environmental Science* 1137(1), 012061. DOI: 10.1088/1755-1315/1137/1/012061.

Obirikorang, K.A., Opoku, E.N. and Gyampoh, B.A. (2022) Feed digestion, growth and disease prevalence in Nile tilapia (*Oreochromis niloticus*) cultured at different water exchange rates in a recirculating aquaculture system. *Aquaculture Studies* 22(3), AQUAST565. DOI: 10.4194/AQUAST565.

Osti, J.A.S., Moraes, M.A.B., Carmo, C.F. and Mercante, C.T.J. (2018) Nitrogen and phosphorus flux from the production of Nile tilapia through the application of environmental indicators. *Brazilian Journal of Biology* 78(1), 25–31. DOI: 10.1590/1519-6984.02116.

Pant, J., Demaine, H. and Edwards, P. (2004) Assessment of the aquaculture subsystem in integrated agriculture– aquaculture systems in Northeast Thailand. *Aquaculture Research* 35(3), 289–298. DOI: 10.1111/j.1365-2109.2004.01014.x.

Pantanella, E. and Bhujel, R. (2014) Saline aqua-ponics: Potential player in food, energy production. *Advocate* 34–35.

Pervez, S. (2018) Fish farming in cages spreading out. The Daily Star. Retrieved October 8, 2018. Available at: https://www.thedailystar.net/business/fish-farming-cages-spreading-out-1533949

Phan, L.T., Bui, T.M., Nguyen, T.T.T., Gooley, G.J., Ingram, B.A. *et al.* (2009) Current status of farming practices of stripped catfish, pangasianodon hypophthalamus in the Mekong Delta, Vietnam. *Aquaculture* 296(3–4), 227–236. DOI: 10.1016/j.aquaculture.2009.08.017.

Pimolrat, P., Whangchai, N., Chitmanat, C., Itayama, T. and Lebel, L. (2015) Off-flavor characterization in high-nutrient-load tilapia ponds in Northern Thailand. *Turkish Journal of Fisheries and Aquatic Sciences* 15(2), 275–283. DOI: 10.4194/1303-2712-v15_2_09.

Rakocy, J.E., Shultz, R.C., Bailey, D.S. and Thoman, E.S. (2003) Aquaponic production of tilapia and basil: Comparing a batch and staggered cropping system. *Acta Horticulturae* 648, 63–69. DOI: 10.17660/ActaHortic.2004.648.8.

Randall, R.G., Minns, C.K. and Kelso, J.R.M. (1995) Fish production in freshwaters: Are rivers more productive than lakes?. *Canadian Journal of Fisheries and Aquatic Sciences* 52(3), 631–643. DOI: 10.1139/f95-063.

Sathoria, P. and Roy, B. (2022) Sustainable food production through integrated rice-fish farming in India: A brief review. *Renewable Agriculture and Food Systems* 37(5), 527–535. DOI: 10.1017/S1742170522000126.

Somerville, C., Cohen, M., Pantanella, E., Stankus, A., & Lovatelli, A. (2014). Small-scale aquaponics food production – Integrated fish and plant farming. FAO fisheries and aquaculture technical Paper 589. Food and Agriculture Organization.

Suresh, A.V. and Lin, C.K. (1992) Tilapia culture in saline waters: A review. *Aquaculture* 106(3–4), 201–226. DOI: 10.1016/0044-8486(92)90253-H.

Tucker, C.S. (2000) Off-flavour problems in aquaculture. *Reviews in Fisheries Science* 8(1), 45–88. DOI: 10.1080/10641260091129170.

VOV, voice of Vietnam (2016) Vietnam plans to up tilapia production by 160%. Online– Retrieved on October 8, 2018 from. Available at: https://english.vov.vn/trade/vietnam-plans-to-up-tilapia-production-by-160–328647.vov

Xu, Z., Zhang, X., Xie, J., Yuan, G., Tang, X. *et al.* (2014) Total nitrogen concentrations in surface water of typical agro- and forest ecosystems in China, 2004–2009. *PLOS ONE* 9(3), e92850. DOI: 10.1371/journal.pone.0092850.

Yamprayoon, J. and Noomhorm, A. (2008) Geosmin and off-flavor in Nile tilapia (*Oreochromis niloticus*). *Journal of Aquatic Food Technology.* DOI: 10.1300/J030v09n02_04.

Zhou, E., Zhang, J., Cremer, M.C. and O'Keefe, T. (2012) Tilapia LVHD cage production in hainan. November/12, *Aqua Culture Asia Pacific Magazine* 35–37.

Zimmermann, S., Kiessling, A. and Zhang, J. (2023) The future of intensive tilapia production and the circular bioeconomy without effluents: Biofloc technology, recirculation aquaculture systems, bio-RAS, partitioned aquaculture systems and integrated multitrophic aquaculture. *Reviews in Aquaculture* 15(S1), 22–31. DOI: 10.1111/raq.12744.

Chapter 5

Tilapia as human food

Ian Goulding

Introduction

Aquaculture makes increasingly important contributions to the global supply of fish and seafood. According to the Food and Agriculture Organization (FAO) (2020), global fish production from all sources was about 179 million tonnes in 2018, of which 46% (82 million tonnes) was derived from farm-raised fish, up from about 15 million tonnes annually during the early 1990s. Since approximately 23 million tonnes of fish from capture fisheries is destined for industrial purposes (mostly fishmeal and oil), aquaculture accounted for about 52% of fish consumed as food (Table 5.1).

This general picture is reflected in the characteristics of tilapia production. Whilst wild catches of tilapias and other cichlids, in common with other capture fisheries, are static at about 0.8 million tonnes per year, production from aquaculture has risen considerably, from 3.1 million tonnes in 2010 to 5.6 million tonnes in 2018 (Table 5.2). In comparison, aquaculture production of crustacea was 9.4 million tonnes, and that of molluscs was 17.5 million tonnes.

The tilapia sector is one of the best-performing sectors of finfish aquaculture, with its share of all farmed finfish production rising from 8.3 to 10.2% during the period. Overall, accounting for supplies from both aquaculture and capture fisheries, tilapias account for some 4% of global fish consumption, corresponding to a global average of 830 g per capita, perhaps the equivalent of two portion-sized fish per person.

Kim Thompson, Craig Shoemaker and David C. Little (eds) **Tilapia**
DOI: 10.1079/9781800629455.0005, © CAB International, 2025

Table 5.1 World fisheries and aquaculture production, utilization and trade.

	Average per year					
	1986–1995	1996–2005	2006–2015	2016	2017	2018
Production[a]			(million tonnes live weight)			
Capture						
Inland	6.4	8.3	10.4	11.4	11.9	12
Marine	80.5	83	79.3	78.3	81.2	84.4
Total capture	86.9	91.4	89.8	89.6	93.1	96.4
Aquaculture						
Inland	8.6	19.8	36.8	48	49.6	51.3
Marine	6.3	14.4	22.8	28.5	30	30.8
Total aquaculture	14.9	34.2	59.7	76.5	79.5	82.1
Total fisheries and aquaculture	101.8	125.6	149.5	166.1	172.7	178.5
Utilization[b]			(million tonnes live weight)			
Human consumption	71.8	98.5	129.2	148.2	152.9	156.4
Non-food uses	29.9	27.1	20.3	17.9	19.7	22.2
Population (billions)	5.4	6.2	7	7.5	7.5	7.6
Per capita apparent consumption (kg)	13.4	15.9	18.4	19.9	20.3	20.5
Trade			(million tonnes live weight)			
Fish exports – in quantity	34.9	46.7	56.7	59.5	64.9	67.1
Share of exports in total production (%)	34.3	37.2	37.9	35.8	37.6	37.6
Fish exports – in value (US$ billions)	37	59.6	117.1	142.6	156	164.1

Source: FAO (2020)

[a]Excludes aquatic mammals, reptiles, seaweeds and other aquatic plants. Totals may not match due to rounding.

[b]Utilization data for 2014 –2018 are provisional

A better understanding of tilapia as a food source may lead to improved utilization. This chapter considers tilapia as human food and characterizes its main features in terms of process and products, composition, shelf life and storage conditions, as well as some of the quality and food safety issues that should be considered during the production and distribution of this valuable product.

Handling during harvest

Tilapia are native to areas with warm temperate and tropical climates, conditions that can lead to rapid spoilage if fish are not harvested and treated in the proper way. Post-harvest losses of fish in such countries are especially high. A study by Akande and Diei-Ouadi (2010) in five African countries found that, whilst only about 5% of fish was physically damaged, up to 70% was sold at a discount due to reduced quality. Another study in Ethiopia (Teklu, 2015) found that overall losses of tilapia in distribution were in the region of 7%. Quality losses and spoilage are determined by many factors, including some that occur during production and immediately pre-harvest.

How fish spoils

There are three independent biological and chemical processes that occur during the spoilage of fish. Each may be present to a greater or lesser extent, depending on the environmental conditions and fish species, as well as the season and other factors. Populations of aerobic mesophilic bacteria start to consume fish slime, organs and muscle, leading eventually to putrefaction. Proteolytic enzymes present in the fish stomach and intestines continue to perform their function after death, working to digest the fish from within. Intracellular autolytic enzymes continue to function in a similar way. Fats in fish are often polyunsaturated and highly susceptible to oxidation, leading to oxidative rancidity. This is especially noticeable in fatty fish stored for a long time, e.g. in frozen storage. In fish such as tilapia, where the flesh is typically relatively low in fat, the greatest

Table 5.2 Global aquaculture production of tilapia and finfish (2012–2018).

	Thousand tonnes live weight				
	2010	2012	2014	2016	2018
Nile tilapia, *Oreochromis niloticus*	2,658	3,342	3,758	4,165	4,525
Other Tilapia spp.	473	693	961	973	1,030
All tilapias	3,130	4,036	4,719	5,138	5,555
All finfish	37,745	42,338	47,219	51,078	54,279
% Tilapias	8.3	9.5	10.0	10.1	10.2

Source: FAO (2020)

concern is bacterial spoilage. Bacterial populations must adjust to the new environment (in air), resulting in a lag phase before spoilage starts. Spoilage time can be considerably shortened by exposure to high temperatures, and by contamination with bacterial populations already adapted to the conditions. Cleanliness to avoid cross-contamination and rapid chilling provide the two main means of maintaining fish quality after harvest (Connell, 1995).

If fish are not handled correctly prior to and immediately post-harvest, spoilage can commence rapidly. Once initial quality is lost it cannot be recovered. The aquaculture sector has an advantage over capture fisheries in that it can control harvest conditions to reduce spoilage. For example, it is commonly regarded as good aquaculture practice to restrict or cease feeding for a period prior to harvest, at least for sufficient time for the fish to completely evacuate their guts. This reduces the quantity of active enzymes in the intestinal tract, as well as reducing the likelihood of fish being contaminated with faecal matter after harvest. In salmon, pre-harvest feed withdrawal has also been shown to reduce stress during transport (Waagbø *et al.*, 2017), improving not only welfare but maintaining muscle quality. Of course, if the farming system relies to some extent on natural production in the grow-out pond, then fish will continue to feed. Feed withdrawal is therefore an effective tactic only when using intensive production systems.

Harvesting commonly takes place at night or early morning when ambient temperatures are lowest (Eltholth *et al.*, 2015). The harvesting method chosen will be largely determined by the type of farming system employed, and whether the harvest is partial or full.

Harvesting of earth ponds creates the biggest challenge in keeping fish free from pond substrate (which can contain unconsumed food, dead fish, faeces and soil).

Effect of stress

As well as ensuring optimal cleanliness, harvesting practices should also aim to limit exposure of the fish to stress. There is growing evidence that teleost fish such as tilapia can experience nociception, i.e. pain and suffering (Sneddon, 2009; Sneddon *et al.*, 2014). For example, rainbow trout (*Oncorhynchus mykiss*) shows biochemical responses to stressful stimuli, including increased stress hormones (cortisol), which has implications not only for welfare but also for product quality (Sneddon, 2002).

The World Animal Health Organisation (OIE) Aquatic Code (2019) sets out the principle that the use of fish carries with it an ethical responsibility to optimize welfare to the greatest extent practicable. The OIE Aquatic Code contains specific sections relating to the welfare of farmed fish during transport, stunning and killing. These measures are therefore recognized as international standards under the World Trade Organization (WTO) Sanitary and Phytosanitary Agreement, implying that countries may impose them as trade measures without challenge, subject to the usual conditions of proportionality and non-discrimination. Some jurisdictions (such as the EU and Norway) have adopted legal requirements for welfare of aquaculture fish throughout production and slaughter. In the EU these are set out in Council Directive 98/58/EC concerning the protection of animals kept for farming purposes, Council Regulation (EC) no. 1/2005 on the protection of animals during transport and related operations, and Council Regulation (EC) no. 1099/2009 on the protection of animals at the time of killing. Fish welfare also features in aquaculture certification systems such as Global GAP. Chapter eight considers in more detail the welfare conditions applicable to tilapia production and slaughter.

In addition to improved welfare, reduced stress leads to higher levels of residual muscle glycogen at the time of death (Hoffman *et al.*, 1970). High-stress death is usually accompanied by excessive muscular activity and higher metabolic rates that deplete energy stores. Maintaining muscle glycogen levels delivers a more extensive post-mortem glycolysis, resulting in a lower terminal pH (due to build-up of lactic acid through anaerobic glycolytic pathways), and in the delayed onset and longer duration of rigor mortis. Wilkinson *et al.* (2008) showed that, with barramundi (*Lates calcarifer*), a low-stress death could delay the onset of rigor from 3 to 12 hours, and Matos *et al.* (2010) observed that time to completion of rigor mortis in gilthead seabream (*Sparus aurata*) could be extended from 3 to 32 hours by pre-killing anaesthetization. The pH of fish muscle can fall to <5.5 in some species (such as tunas), but for most fish it does not fall below 6.0–6.2 even under optimal conditions (Daskalova, 2019). Even so, this slight additional fall in pH helps to delay spoilage, thus improving the keeping quality of the resulting product.

This knowledge requires that operators consider the most effective means of reducing stress during harvesting and killing of the 15–20 billion cultured tilapias estimated to be harvested each year. The typical practice of leaving fish to asphyxiate is not considered by animal welfare organizations to provide the best welfare conditions, and was shown to result in a lower-quality product (Poli *et al.*, 2005). Harvesting fish into an ice-water slurry was found to compromise welfare conditions for sea bass (*Dicentrarchus labrax*) (Zampacavallo *et al.*, 2015).

As with mammalian food animals, in terms of welfare and post-harvest quality the most effective killing method involves stunning of fish followed by exsanguination. Stunning can be percussive (manual or automated) or electrical (Roth *et al.*, 2002). Carbon dioxide narcosis is also an option, and may also be combined with chilling to reduce aversive behaviour and handling stress, and to provide a longer pre-rigor period, reportedly increasing fillet quality (Erikson *et al.*, 2006).

Modern aquaculture processing facilities should be equipped with suitable equipment for stunning. Most of the work on implementation of welfare standards for aquaculture finfish has been carried out for species such as Atlantic salmon (*Salmo salar*), trout, European seabass and gilthead seabream, for which commercial systems are now available. Not as much applied research on fish stunning and killing has been undertaken with tilapia. Goes *et al.* (2015) showed that Nile tilapia behave as do other fish species in terms of levels of cortisol during transport and post-harvest pH changes. Lambooij *et al.* (2008) showed the effectiveness of high-welfare electro-stunning of Nile tilapia (*Oreochromis niloticus*), followed by killing the stunned fish in ice water. Oliveira Filho *et al.* (2015) showed that CO_2 narcosis of tilapia resulted in slower onset of rigor mortis, and was significantly less stressful compared with ice-slurry killing. However, there was no clear improvement in product quality with either method: fish remained fit for consumption for 3 weeks on ice. Venturini *et al.* (2018) compared electronarcosis (203 V, ~3.3 A for 15 s) with a chemical anaesthetic control (eugenol) and found significantly elevated plasma lactate levels in tilapia stunned at higher-frequency alternating currents (1200 hz). Wang *et al.* (2018) investigated the use of saturated nitric oxide solution for stunning of *O. niloticus* before slaughter. The impact on animal welfare and quality was also investigated, finding that during refrigerated storage treated fish showed a slower reduction in muscle pH, later onset of rigor mortis post-mortem and less drip-loss than the control. This study also demonstrated an extension of shelf life. On the other hand, de Castro *et al.* (2017) showed that, compared with *O. niloticus* killed immediately following removal from water, a period of hypoxia of 6 min pre-slaughter did not influence the onset of rigor mortis, muscle pH, perceived quality or blood glucose levels, suggesting that short periods of hypoxia do not greatly impact quality. Nevertheless, for killing of farmed fish in general, the Humane Slaughter Association of the UK (2020) does not recommend the use of ice slurry, live chilling or gill cutting without prior stunning or carbon dioxide narcosis. This advice is based on scientific advice from the European Food Safety Authority (EFSA, 2009) in relation to slaughter of farmed trout, which states that 'CO_2, asphyxia on ice, and asphyxia are the methods resulting in the poorest welfare.'

Handling and processing

General approach

The general rule for handling fish after harvest is to keep it clean and keep it cool. This minimizes both bacterial load and bacterial reproduction rate. Fig. 5.1 shows schematically the impact of different temperature and contamination scenarios on the standard bacterial growth curve. Spoilage bacteria are naturally present in the gills, slime and gut of fish. When fish are removed from the aquatic environment the numbers remain constant during a period of adaptation to the new conditions (lag phase). Once adapted, multiplication proceeds rapidly (logarithmic phase) until such time as the available nutrients are consumed. Chilling extends the lag phase and reduces the rate of multiplication. Reducing contamination avoids the introduction of bacteria already in the logarithmic phase of reproduction. Application of both measures ensures that the shelf life benefits from both an extended lag phase and no introduction of actively reproducing bacteria. Clean fresh water or potable water should therefore be used to remove contamination, followed by chilling, ideally to 0°C, as soon as possible after death.

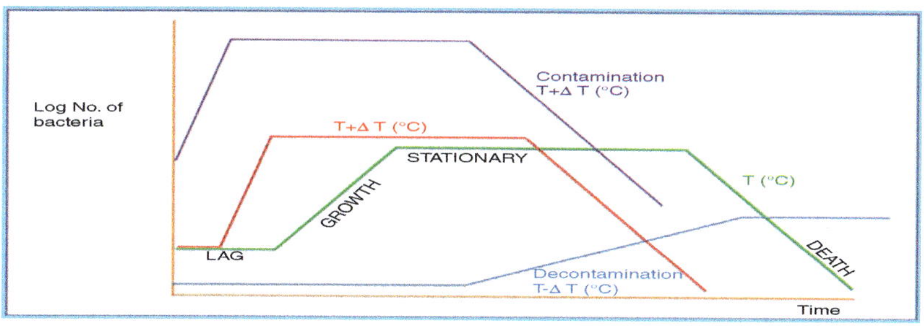

Figure 5.1 Schematic representation of the effect of temperature and hygiene controls on the bacterial growth curve. Source: Author, based on Baranyi and Roberts (1994)

Chilling and use of ice

Melting ice made from potable (i.e. bacteriologically safe) water is the most convenient chilling medium due to its inert nature, constant temperature (i.e. will not freeze the fish) and high latent heat of fusion (~80 kcal/kg), the energy extracted from the fish as the ice melts. In theory, ~0.25 kg of ice is needed to reduce the temperature of 1 kg of lean fish at 25°C to 0°C, but in practice additional ice will be required to account for heat gains from the environment and inefficient heat transfer. Numerous guidelines are published by various international organizations such as FAO, which set out the accepted best practices in terms of icing (as well as general fish handling) and, in general, all are applicable to tilapia. Depending on the circumstances, storage time and temperature, up to 1 kg of ice may be required for each 1 kg of fish to be chilled (FAO, 1992).

The key principles to consider are having sufficient ice to both chill fish from ambient temperatures to 0°C, and to maintain this temperature by countering heat gains from the exterior, also ensuring that all fish are in contact with ice through alternate layering of fish and ice. Insulation helps to reduce ice consumption during prolonged storage and transport. Fish may need to be washed and re-iced during storage to obtain the maximum shelf life. Fish should not be left in melt water, which should be allowed to drain away (for example, using perforated plastic boxes).

Various types of ice are used in the fishery sector. Block ice (typically 25-kg blocks) can be stored onsite for longer periods (up to several days on-farm without refrigeration, providing that it is protected with insulation) and crushed for use when required; however, if improperly crushed it is less effective and can physically damage the product. Use of mechanical crushers is therefore recommended, to ensure regular small ice particles. Flake or tube ice extracts heat from the fish more quickly but, unless it is to be used quickly, requires refrigerated storage.

Another key consideration in transport and storage is the choice between bulk storage of fish and boxing. The latter reduces the possibility of fish being crushed, and facilitates transfer of product during distribution without handling, but occupies more space as well as requiring attention to box washing and sanitizing. Bulking, especially during transport, can result in loss of quality unless separating boards are used to prevent crushing. The different approaches are described by Waterman (undated).

Hygiene and sanitation

As well as chilling, post-harvest handling should avoid contaminating the product with bacteria already in the growth phase. Thus, any surface contacting the fish should be as clean as possible. This includes nets used to harvest fish, boxes and vehicles, as well as equipment used for handling of ice. It requires two processes – cleaning and sanitizing. Cleaning involves the processes of removal of gross contamination, grease etc. that may trap bacteria, and requires the use of soap or detergent. Sanitizing applies a treatment to ensure that most remaining vegetative bacterial cells are inactivated or killed. Most frequently recommended is a dilute solution of hypochlorite (50–100 ppm) with a surface contact time of about 30 min. However, other chemical solutions are available such as quaternary ammonium compounds. These are cationic surfactants that dissolve bacterial cells walls and have a residual effect. Although more expensive, they have the advantage of detergent action and can be used in a single cleaning and sanitizing step.

Cleanliness includes ensuring that the hands of workers are washed before fish is handled. This is important to minimize spoilage rates, as well as to avoid transmission of potentially pathogenic bacteria. Good hygienic practices therefore demand the provision of handwashing facilities for workers handling fish (Global GAP, 2020). Of course, all other good hygiene-related practices should be in place concerning the provision of hygienic facilities. These should include access to safe toilet facilities for farm workers, personal hygiene rules and hygienic design of equipment.

Value-added processing of tilapia

Currently most tilapia produced for human consumption are consumed whole following removal of the viscera. Many producers seek to deliver a portion-sized tilapia to market, but the size required is highly variable among markets. Natasha (2014) reports a typical market size of 200–250g in Fiji, whereas Clement and Lovell (1994) suggest a target harvest weight for the US market averaging 610 g; these authors report an overall fillet yield for *O. niloticus* of 25.4%. However, Garduño-Lugo *et al.* (2003) obtained a fillet yield of 32.0% for Stirling Nile tilapia (wild-type) *O. niloticus* and 33.4% for a hybrid (Florida red tilapia ×Stirling red *O. niloticus*). Rutten *et al.* (2005) showed that yield is dependent on one phenotype, whilst in the author's experience larger fish of most species typically deliver higher fillet yields.

Significant quantities of tilapia are subject to further processing, either to increase convenience (e.g. by filleting) or to provide shelf life extension such as freezing (Arannilewa *et al.*, 2005). In developing countries, preservation by salting, smoking and/or drying are common processes (Yanar *et al.*, 2006). Tilapia by-products from processing may also be used in the production of fish gelatine (Jamilah and Harvinder, 2002). The author has also observed tanned tilapia skins used to produce leather for fashion items.

Traceability requirements for aquaculture

Traceability is defined by Codex Alimentarius Commission (2007) as 'the ability to follow the movement of a food through specified stage(s) of production, processing, and distribution'. Traceability is a critical tool that provides the means to identify the source and cause of food safety problems and/or to remove unsafe food from the market. Regulations such as EU Regulation 178/2002 (European Union, 2002) may require that all food chain operators should be able to identify any person from whom they have been supplied with an aquaculture product, aquaculture feed or any other item (such as a veterinary medicine) incorporated into the product. They should also be able to identify anyone to whom they have sold their products (Moretti *et al.*, 2003). Good agricultural practices such as Global GAP (2020) also require that traceability of aquaculture products and feeds should be established at all stages of production, processing and distribution.

Borit and Olsen (2016) have set out some of the ideal features of a traceability system applicable to fish. The system should record origins of all inputs received (feed, seed, fertilizers, juveniles, supplements and medicines), along with identification of the input batch (such as date and supplier codes). Different batches should be kept separately; this also helps with stock rotation. As far as possible the operator should preserve a record of batches of fish treated differently during production. Where batches are mixed (e.g. after size grading to ensure an appropriate feeding regime for their size), the mixing should be recorded. Effective records should be kept of each batch of fish grown in each enclosure, to include veterinary drug regimes, feeding methods and quantities, pond fertilizers added and any results of water quality parameters. Each batch of fish leaving the farm should be allocated a number, and should be labelled so as to identify the origin, date of harvest and batch number. The operators should keep records and have in place a functional plan detailing the procedures to be followed in case a batch of aquaculture products subsequently needs to be withdrawn or recalled from distribution.

Quality considerations

Nutritional quality

Aquaculture delivers one of best feed conversion ratios of all animal protein foods and, as with other farmed fish, tilapia can provide consumers with a sustainable and efficient source of high-quality dietary protein. The protein content of wild tilapia fillets from Ethiopia was found by Alemu *et al.* (2013) to be between 13.3 and 15.8%, and fat content between 0.5 and 0.7%. Jim *et al.* (2017) found that *O. niloticus* grown in more polluted lake waters in Zimbabwe had a significantly lower protein content, suggesting a direct linkage between water and nutritional quality. The muscle of farmed tilapia typically has a significantly higher fat content due to the increased fat levels in formulated feeds. In one Nigerian study, Arannilewa *et al.* (2005) reported a fat content of farmed *O. niloticus* of 9.72%. Clement and Lovell (1994) found a fat content of 5.7% in US farmed *O. niloticus* fillet, although this was significantly less than for channel catfish (*Ictalurus punctatus*) fed to satiation on the same diet (fat content 7.4%). Tonial *et al.* (2009) found that a compound feed supplemented with flaxseed oil resulted in a muscle lipid content of 9%, as well as significant increase in omega-3 as a proportion of total fatty acids (from 1.5% of fatty acids with feed based on soya oil, to 10.5%), with docosahexaenoic acid (DHA) peaking at 1.2%. Omega-3 fatty acids, in particular essential DHA and eicosapentaenoic acid (EPA), are known to provide significant health benefits in the human diet, and especially so in terms of foetal development and maternal health (Broadhurst *et al.*, 2002).

Al-Souti *et al.* (2012) fed red hybrid tilapia on diets containing cod liver oil (in place of corn oil) and found no significant differences in feed conversion or proximate composition, but the total *n*-3 content in the fatty acid profiles of dorsal muscle increased from 7.6 to 18.6% when the diet contained 12% fish oil. The essential *n*-3 fatty acid DHA also increased from 5.8 to 16.1% of total fatty acids. The evidence suggests that farmed tilapia can have a significantly superior nutritional quality than captured and, as Al-Souti *et al.* (2012) suggest, there appears to be potential to enhance the total *n*-3 fatty acids and DHA in tilapia fillets by manipulating the nature and content of dietary oils. However, tilapias with increased fat content are more susceptible to spoilage due to oxidative rancidity, especially during frozen storage (Arannilewa *et al.*, 2005). Although not reported, it is expected that fat content will also impact on the eating quality (flavour and texture). Preparation and cooking methods are also influential, with frying, one of the most common cooking methods, leading to significant reductions in unsaturated fatty acid content (Mekonnen *et al.*, 2020).

Spoilage mechanisms and shelf life

Spoilage of fish is characterized through sensory, chemical and microbiological means, with a wide variation between species in the nature and rate of spoilage. Cyprian *et al.* (2014) developed a sensory vocabulary for cooked tilapia fillets for use in shelf life trials (Table 5.3), and measured these organoleptic characteristics through sensory evaluation by trained taste panels.

Potentially these constructs could be developed into technical grading scales to reflect the progressive change in these variables throughout storage, similar to the ten-point Torry scale (Torry Research Station, 1989) or commercial grading schemes (E-Extra, A-, B- and C-unfit), such as employed in EU regulations (European Union, 1996).

Bacterial growth is the major cause of spoilage in fresh fish, as indicated in Fig. 5.1. This is primarily manifested in changes in its sensory features (appearance, odour, taste and texture), which ultimately render the product unfit for consumption. However, microbiological indicators are not regarded as accurate predictors of spoilage due to the wide range of bacterial species and the fact that total numbers of bacteria can reflect recent contamination rather than spoilage over time. In Malawi, Kapute *et al.* (2013) found *O. niloticus* sampled from local open markets had bacterial counts averaging 9.5×10^8 colony-forming units (CFU)/g, considered to be above acceptable limits, whereas fish supermarkets had a much more hygienic product (mean 2.7×10^5 CFU/g). Most of the spoilage bacteria isolated were of the genera *Corynebacterium*, *Micrococcus*, *Pseudomonas*, *Bacillus* and *Flavobacterium*. *Escherichia coli* (an indicator of faecal contamination) was also detected.

Chemical indicators provide more effective measures of spoilage, since they are a direct consequence of the biochemical processes involved. Rodrigues *et al.* (2016) and Dergal *et al.* (2013) have successfully used total volatile bases nitrogen (TVB-N) to demonstrate progressive quality deterioration in studies of shelf life and quality of *Oreochromis* spp. TVB-N is a common fish freshness measure, since biogenic amines (including ultimately ammonia) are released from bacterial metabolism of muscle proteins, rendering the typical odours and flavours of decomposed flesh. The EU sets TVB-N limits between 25 and 35 mg/100g in fish muscle for several European marine species (European Union, 2019). Operators should check whether their own national regulations set limits to TVB-N (or other chemical parameters) in freshwater fish in general, or in tilapia.

Trimethylamine oxide (TMAO) is implicated in the maintenance of osmotic balance and, when present, is reduced by post-mortem bacterial actions to trimethylamine (TMA), the level of which is often used as an indicator of quality in certain fish species. Although TMAO is generally not substantially present in fish grown in freshwater (Hebard *et al.*, 1982), tilapias are euryhaline and, when grown in sea or brackish water, may contain TMAO in their flesh. Waliszewski and Avalos (2001) also suggest that the presence of TMAO could also be related to the presence of fishmeal of marine origin in the feed. Cyprian *et al.* (2014) found that the amount of TMA in tilapia fillets during storage time of up to 25 days was very low (<1 mg/100 g). However, other studies have indicated that TMA can develop in tilapia grown in freshwater (Reddy *et al.*, 1995). The use of TMA as a chemical indicator of quality in this species may not be reliable, except in restricted circumstances such as marine environments and in diets with a high choline content (Niizeki *et al.*, 2002).

Another range of indicators are based on the breakdown of nucleotides. A composite *K*-value based on adenosine triphosphate (ATP) degradation products (such as inosine and hypoxanthine) has been found to correlate well with the freshness of fish, especially

Table 5.3 Characteristics of sensory attributes of cooked tilapia fillets for shelf life determination.

Sensory factor	Sensory attribute	Description of attribute
Odour	Boiled potatoes	Whole newly boiled potatoes
	Boiled milk	Hot milk, fruity odour
	Earth	Fresh earth
	Musty	Mouldy
	Rancid	Rancidity
Appearance	Light/dark colour	Light, white or dark, yellowish colour
	Dark brown/light grey	Dark brownish or light grey colour
	Black threads	Black threads in flesh
Flavour	Arctic charr	Arctic charr, new trout
	Sweet	Typical sweet flavour of fresh fish
	Metallic	Metallic flavour
	Earthy	Fresh earth

Source: Cyprian *et al.* (2014)

useful in freshwater species where TMA may not be present. Ryder (1985) and Laly *et al.* (2017) have both reported the use of *K*-values to demonstrate quality changes in *O. niloticus* stored at chilled temperatures, values that correlated well with sensory and microbiological values.

Whilst the oxidative rancidity of lipids is not likely to be a limiting factor in the spoilage of wild tilapia, the elevated lipid content of the muscle of farmed fish suggests that this kind of spoilage could play a greater role, especially during extended frozen storage. This feature emphasizes the need for good packaging practices (using an effective oxygen barrier) for frozen farmed tilapia.

Numerous authors have studied the shelf life of tilapias under different conditions. Whole fresh tropical tilapia stored in melting ice have been found to retain fitness for human consumption for between 16 and 18 days in the case of tilapia from Lake Malawi (Kapute *et al.*, 2013). The study by Laly *et al.* (2017) considered the effect of gutting on a range of quality indicators of farmed *O. niloticus* in India, finding that the shelf life of whole and gutted samples during iced storage at 4 ± 2°C was 24 and 21 days, respectively. The limiting factor was the more rapid development of biogenic amines (such as spermine, spermidine and cadaverine) in gutted fish. This suggests that recommendations for immediate post-harvest gutting may not always deliver optimal shelf life, although the study did not record the state of starvation at death, which may also impact on different spoilage rates between whole and gutted fish.

Filleting reduces shelf life (as expected, since there is greater exposure of the muscle to bacterial penetration). He *et al.* (2016) found that tilapia fillets held at 0°C had a shelf life of only 11 days, compared with 14 days for fillets stored in a modified atmosphere of 30% CO_2 and 70% N_2. Cyprian *et al.* (2014) developed a quality index methodology for the determination of shelf life of tilapia fillets, finding that the maximum, as determined by an index comprising sensory and microbial count data, was 16 days at 1°C and 19 days at −1°C. Overall, it would appear that filleting under ideal conditions reduces shelf life by about 1 week compared with storage of whole fish.

Some authors have reported on the effect of application of preservation technologies to extend the shelf life of fillets, seeking to redress the reduction in shelf life. The effects of various combinations of citric acid, ascorbic acid and sorbic acid and their salts were investigated by Waliszewski and Avalos (2001), who recommended an optimal combination of citric acid (0.15%) and potassium sorbate (0.1%). Other authors found essential oils of rosemary (*Rosmarinus officinalis* L.) and thyme (*Thymus vulgaris*) to be effective at extending the shelf life of *O. niloticus* (Albarracín *et al.*, 2012; Khalafalla *et al.*, 2015). The results suggested that rosemary extract has only an antioxidant effect and does not impact greatly on shelf life, whereas thyme extract has both strong antimicrobial and antioxidant activities, resulting in a shelf life extension for fillets of up to 9 days. It should be noted that the use of chemical preservatives in fish may be subject to regulatory controls that differ depending on the country.

Off-flavours

An important factor in the quality of farmed tilapia is the presence of earthy, muddy or musty off-flavours, a common occurrence that can affect market quality and thus price (Howgate, 2004). These flavours are attributed to the presence of geosmin (GSM) and 2-methylisoborneol (MIB). Both are compounds produced by blue-green algae and cyanobacteria, for which humans have a detection threshold as low as five parts per trillion (Polak & Provasi, 1992). Such off-flavours are a common problem encountered in farmed fish (Lovell *et al.*, 1986), but especially in freshwater fish such as carp, catfish and tilapia (Vallod *et al.*, 2007) and particularly in recirculating aquaculture systems (Azaria & Van Rijn, 2018). Prevention requires attention to water quality to minimize algal blooms. When these do occur, the typical solution is to place the fish in clean water without feeding. Yamprayoon and Noomhorm (2003) reported that under such conditions GSM flavours in *O. niloticus* fell and were not detectable after 16 days. However, Schram *et al.* (2021) showed that tilapia fed during depuration eliminated GSM from the ovaries and fillets more quickly than starved fish. Under acidic conditions GSM decomposes rapidly, and consumers in some regions are accustomed to applying vinegar or citrus juices to deodorize the product.

Food safety hazards and controls

As with all aquaculture products, tilapia are subject to a range of food safety hazards whether derived from their environment, feed, chemotherapeutic interventions or subsequent post-harvest treatments. Aquaculture operators should be aware of the origins of hazards and the possible availability of mechanisms to eliminate or limit their presence. In some cases, regulations may be in place with a view to protecting consumers, and official control activities of competent authorities are directed to the monitoring of the presence of the hazards and enforcement of the regulatory requirements. This section reviews some of the most frequently encountered food safety hazards associated with the consumption of freshwater farmed fish in general, and tilapia in particular. These include parasites and residues of veterinary treatments, pesticides and aflatoxins. Other hazards such as heavy metals and organic pollutants, including dioxins and polychlorinated biphenyls (PCBs), may be present in feeds and accumulate in farmed fish, but the risks associated with tilapia are considered to be minimal if grown in unpolluted waters, due to their rapid growth. In fact, as feed suppliers seek to reduce the content of fishmeal and oil in tilapia feeds, the risks are likely to be further reduced.

Parasites transmissible to humans

Zoonotic parasites infecting aquaculture-produced fish are reviewed by dos Santos and Howgate (2011). In aquaculture species grown in freshwater, helminths in the families Opisthorchiidae and Heterophyidae (class Trematodea, subclass Digenea) are considered to be the major hazards, whereas zoonotic nematoda such as Anisakidae and Gnathostomidae are more frequently associated with marine species (Myers, 1970). Diphyllobothridae of the class Cestoda (tapeworms) are limited to temperate regions (Meyer, 1970) and are not often encountered in aquaculture systems. Humans acquire infection through the consumption of undercooked infected raw fish, or by cross-contamination from raw to cooked products. Fish-borne zoonotic trematodes (FBZT) derived from the consumption of raw, wild-caught freshwater fish and crustacea have long been associated with morbidity and death, especially in South-east Asia. In this region a substantial proportion of common species shows the presence of infective metacercaria (Khoa *et al.*, 2020). The most important trematodes of teleosts from a public health point of view appear to be liver flukes of the family Opisthorchiidae, especially *Clonorchis sinensis*, *Opisthorchis (Op.) viverrini* and *Op. felineus* (dos Santos & Howgate, 2011). Keiser and Utzinger (2005) estimated that 601 million people globally were at risk of infection with *C. sinensis* and a further 79.8 million were exposed to *Opisthorchis* spp. Lun *et al.* (2005) estimated that 35 million individuals could be infected by *C. sinensis,* including some 15 million in China. The life cycle of *C. sinsensis* is typical of the digenetic flukes, with hosts in aquatic gastropoda, fish and mammals that can include rats, humans and domestic carnivorous animals (Fig. 5.2).

More recently, awareness has increased of the public health importance of intestinal heterophyid flukes, which are much smaller and linked to a significant pathology, often fatal, in the heart, brain and spinal cord of humans (Chai, 2007; Chai & Jung, 2017).

Chi *et al.* (2008), in a survey of 53 Vietnamese fish farms, found that the overall prevalence of FBZT metacercariae in aquaculture fish (*O. niloticus* and six species of carp) was 44.6%, ranging from 12.5 to 61.0%. However, the problem is not limited to South-east Asia. *Op. felineus* is prevalent in freshwater fish in Russia and countries of Central Europe (Yossepowitch *et al.*, 2004) and could be present in aquaculture systems in these regions. More recently, the presence of FBZTs was demonstrated in tilapia from some of the most important regions of production. Satour *et al.* (2019) found that 57% of *O. niloticus* from the Egyptian delta demonstrated encysted metacercariae, with six species of trematode identified, predominantly zoonotic *Heterophyes* spp. (*H. heterophyes* and *H. equalis*) encountered in 37% of samples. El-Seify *et al.* (2020) found that 65% of *O. niloticus* collected from local markets in Egypt showed the presence

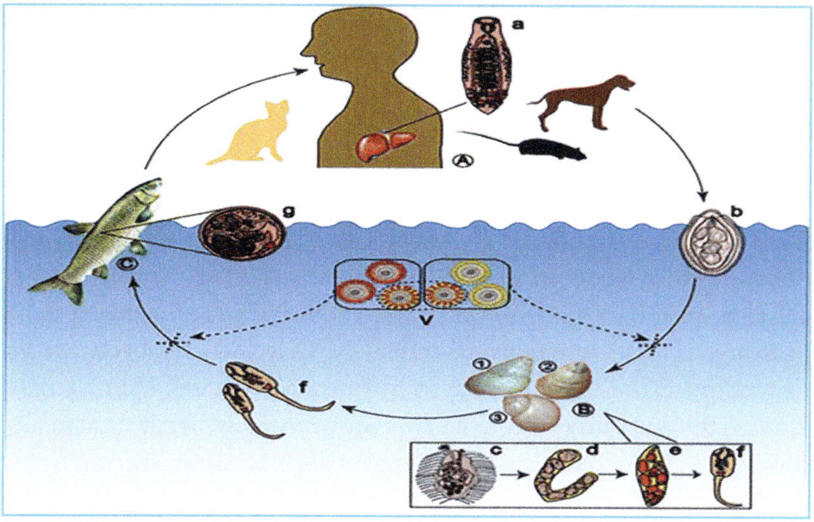

Figure 5.2 The life cycle of *C. sinensis*. The adult worms of *C. sinensis* (a) live mainly in the bile ducts of the definitive host (A) (human beings, dogs, cats, mice, etc.). The eggs (b) are discharged from the definitive host with the faeces. Freshwater snails, mainly *Parafossarulus striatulus* (1), *Bithynia fuchsianus* (2) or *Alocinma longicornis* (3), can serve as the first intermediate host (B). Eggs develop into miracidia (c) in water and then to sporocysts (d), rediae (e) and cercariae (f) after they are swallowed by snails. The mature cercariae are shed from the snails, swim freely in water and invade into the second intermediate host (C) (freshwater fish) through the skin and then form metacercariae (g) in the fish musculature. Humans or carnivorous mammals can become infected due to the ingestion of raw or undercooked fish. Intervention in the formation of cercaria or metacercaria by effective vaccines (V) will block the transmission of *C. sinensis* and fundamentally control clonorchiasis. Source: Tang *et al.* (2016)

of infective encysted metacercariae of FBZT, with *Heterophyes* spp. being the most common. Given the widespread prevalence of FBZT in freshwater systems in tropical and even some temperate regions, and the difficulty in ensuring that no infective stages of the life cycle are present in aquaculture systems (unless perhaps grown in a recirculatory aquaculture system (RAS) with controlled conditions), the assumption should be that fresh tilapia may be a source of infection to consumers and should therefore be adequately treated before consumption, with care taken to avoid cross-contamination from raw fish to cooked foods.

One control method available is the use of molluscicides in fishponds to reduce snail populations, but these are expensive, break down quickly and risk leaving contaminating residues in the product (Slootweg *et al.*, 1993). Other post-harvest treatments used to eliminate the FBZT hazard are described by Franssen *et al.* (2019). These include cooking at a core temperature of 60–75°C for 15–30 min or freezing at -21°C for 1–7 days, which is also effective but cannot be relied upon in the domestic situation. Parasitic stages are also sensitive to 2–5% NaCl, and reduced pH, for example by pickling. Low-dose radiation under 1.0 kGy has also been shown to inactivate the infective stages of most trematodes (Abdussalam *et al.*, 1995). Where fish is to be consumed raw, such treatments should be undertaken prior to preparation. In some jurisdictions, such as the EU, the requirement for treatment of susceptible species when the product may be consumed raw is mandated by regulations (European Union, 2004). However, these measures specify only time and temperature combinations designed to eliminate parasites other than trematodes. European Food Safety Authority (2010) has provided a scientific opinion, indicating that a temperature of 70°C for 30 min is necessary to ensure the inactivation of metacercariae of *Clonorchis* spp. and *Opisthorchis* spp. In relation to freezing, EFSA refers to WHO statements that the metacercaria of *Opisthorchis* spp. and *Clonorchis* spp. are killed by freezing at −10°C for 5 days. It also refers to temperature–time parameters set out by FAO (2003), which suggest that 3–4 days at −20°C are required to inactivate the larvae of *C. sinensis*, and 32 hours at −28°C to kill the larvae of *Op. felinus*. Based on these studies, additional guidance concerning parasite safety in fishery products is provided by the European Commission (2011).

However, in relation to farmed tilapia, awareness of this hazard is not widely known outside South-east Asia. Many consumers are therefore likely to remain exposed to the risk of serious illness, or worse, due to the consumption of raw or undercooked tilapia, which is sometimes used fraudulently as a substitute for higher-value species such as red snapper in raw fish dishes such as sushi and sashimi (Warner *et al.*, 2013).

Bacterial food poisoning

Aquaculture water supplies may also be contaminated with human and animal faeces and other biohazardous substances, with associated risks of contamination with pathogenic bacteria such as *Salmonella* spp. and verotoxic *E. coli* (Begadiya *et al.*, 2018). Additionally, in some cases ponds are fertilized intentionally with manure (a practice known as septage), especially in developing countries where artificial fertilizers may be unavailable or too expensive. Although Eves *et al.* (1995) found that there were no consistent significant differences in sensory or microbiological quality of fish samples during iced storage, there would be a food safety risk if the fish were to be consumed raw or allowed to cross-contaminate ready-to-eat foods.

One such risk was identified for the first time in a 2015 outbreak of food poisoning reported in Singapore in 2015 (Barkham *et al.*, 2019). This involved the first and only reported outbreak of a food-borne invasive disease caused by GBS (*Streptococcus agalactiae*), in which over 160 people were hospitalized suffering from septic arthritis and meningitis, acquired from eating raw freshwater tilapia. The authors report that 'although GBS sepsis is well-described in neonates and older adults with co-morbidities, this outbreak affected non-pregnant and younger adults with fewer co-morbidities, suggesting greater virulence.' Whilst GBS is reportedly widespread in South-east Asia, where it is a common pathogen of freshwater fish (Sun *et al.*, 2016), and is implicated as a leading cause of bovine mastitis (Botelho *et al.*, 2018), this is the first reporting of a food-borne transmission to humans. Barkham *et al.* (2019) hypothesize that the incident was caused by the emergence of a new strain, identified as belonging to serotype III-4 and multilocus sequence type (MLST) 283 (ST283), and suggest that this could be evidence of a wider regional, but as yet unidentified, outbreak in South-east Asia. Leal *et al.* (2019) have also identified GBS ST283 in tilapia in Brazil. There is evidence reported by Chau *et al.* (2017) that ST283 might not be pathogenic for freshwater fish, since affected fish samples recovered from a survey in Singapore did not exhibit typical GBS symptoms such as corneal opacity and exophthalmia. Da Cunha *et al.* (2014) have hypothesized that the emergence of new GBS strains able to cause human infections are associated with the acquisition of antimicrobial resistance (AMR).

As with the risks of FBZT, the common use of tilapia for raw consumption, both as a named product and in fraudulent substitution for other fish, can result in the transmission of this infection. Consumers and catering suppliers should always be advised to thoroughly cook freshwater tilapia to eliminate both parasitic and bacterial food safety hazards. The Government of Singapore has banned the serving of ready-to-eat raw tilapia in catering establishments, but other governments in the region have not yet reacted. There may also be an occupational health risk to workers handling fresh tilapia from GBS, because they may be subject to cutaneous infections.

Residues of veterinary medicines

As with all farmed fish, tilapia may be subject to various parasitic, bacterial, virus and fungal infections requiring preventative or therapeutic treatments, usually involving chemical interventions (as well as other actions). In addition, they may be subject to various chemical treatments during routine production processes such as sex-reversal hormones or disinfectants during handling. The use of veterinary medicines and medicated feeds is usually regulated in most countries, with restrictions on registration, prescription, distribution and application of compounds, as well as lists of banned substances. It should be noted that fish used for non-food purposes (e.g. broodstock or for ornamental purposes) may not be subject to such stringent limitations on veterinary interventions as fish produced for human consumption. The regulatory requirements will vary among countries, especially so since the animal health status and risks of potential AMR will vary from region to region, requiring the application of different veterinary regimes. The approach to ensuring an appropriate and WTO-compliant regime for veterinary medicine controls is set out in the OIE Aquatic Animal Health Guide, which addresses antimicrobial use in aquatic animals.

A modern regulatory regime for drug use in food animals will typically contain two lists. The first will be of banned substances, whose use in food animals including fish is prohibited due to either either the toxicity of their residues or the risk of development of resistant strains of bacteria in the human population. Several common compounds, including the antibiotics chloramphenicol and nitrofuran, are not generally permitted due to the risk of resistance developing in bacterial populations, which would reduce the effectiveness of human medical applications. The second list usually specifies permitted applications. Codex Alimentarius Standards recognize only one permitted veterinary medicine for general aquaculture use (oxytetracycline) and two for finfish (amoxicillin and ampicillin), although several others are also listed for specific species such as trout and salmon (Codex Alimentarius, 2020).

Dyes used as pharmacologically active substances in aquaculture are comprehensively reviewed by Verdon and Andersen (2017). Malachite green, gentian violet and methylene blue are all effective anti-protozoan and anti-fungal compounds commonly used in disinfectant baths for ova and juveniles. However, the widespread use of malachite green raises concerns for consumer health, as highlighted in reviews of the toxicity of its residues in muscle by Culp and Beland (1996) and Srivastava *et al.* (2004). It exhibits a long residence time in fatty muscle tissue and is shown (along with its metabolite leucomalachite green) to be a potent mutagen and to cause an increase in cancerous tumour formation in female rats (Culp *et al.*, 2006). For this reason, the use of such chemicals has been banned in several jurisdictions (including the EU, Canada and the USA). Other banned substances (in the EU) include dimetridazole, metronidazole and anabolic substances administered for growth-promotion purposes. However, other applications of anabolic substances, such as sex reversal, are permitted. This allows use of the synthetic androgen 17α-methyl testosterone for sex reversal in tilapia hatchery operations. The US Food and Drug Administration requires a 120-day withdrawal period following such treatment. Green and Teichert-Coddington (2000) showed that this might be excessive in tilapia, since methyl testosterone and metabolites levels in this species fell to <10 pg/g at 6–50 days following withdrawal.

Regulations setting the permitted veterinary medicines usually also establish the safe dosages and maximum residue levels permitted in the final product sold for consumption, as well as any conditions of use (for example, restricted to certain species or application methods such as immersion, injection or oral delivery). The registration decision should also specify the mandatory withdrawal period (the minimum time to elapse between the last treatment and harvesting for human consumption). These parameters are determined by experimentation. An example is provided by Xu *et al.* (2005), who measured the changes in the muscle concentration of residues of enrofloxacin and furazolidone and their metabolites in *O. niloticus* following a single oral dose event, with a view to recommending the withdrawal period. Many drugs are not harmful to humans (they may be identical to those used in human medicine), and they may also be degraded during cooking (Dilna *et al.*, 2017). However, there are two important concerns. One is that some compounds, or their metabolites in humans, may be carcinogenic or mutagenic. The other is that constant exposure to low dosages may give rise to disruption of intestinal flora in consumers and the development of AMR.

For the EU market, operators should refer to Commission Regulation (EU) no. 37/2010 of 22 December 2009 on pharmacologically active substances, and their classification regarding maximum residue limits in foodstuffs of animal origin. A summary of relevant substances permitted for use in aquaculture is shown in Table 5.4. For aquaculture operators supplying the US market, the list of approved aquaculture drugs by the U.S. Food and Drug Administration is published in an online database.

The on-farm management of the application of veterinary drugs should follow good agricultural practices (see Box 5.1).

The national competent authority responsible for food safety should implement a national annual residue monitoring programme, with the objective of assessing the extent of the compliance of aquaculture establishments with the regulatory provisions regarding veterinary medicines. Such a programme comprises a series of sampling and testing activities, followed by any necessary corrective actions that provide information regarding the effectiveness of the controls

Table 5.4 Some typical MRLs for veterinary medicines used in aquaculture.

Pharmacologically active substance	Marker residue	Animal species	MRL (XIIg/kg)	Target tissues	Other provisions	Therapeutic classification
Cloxacillin	Cloxacillin	All food-producing species	300	Muscle	For finfish the muscle MRL relates to 'muscle and skin in natural proportions'. Not for use in animals from which eggs are produced for human consumption	Anti-infectious agents/ antibiotics
Emamectin	Emamectin B1a	Finfish	100	Muscle and skin in natural proportions		Antiparasitic agents/agents acting against endo- and ectoparasites
Deltamethrin	Deltamethrin	Finfish	10	Muscle and skin in natural proportions		Antiparasitic agents/agents acting against ectoparasites
Erythromycin	Erythromycin A	All food-producing species	200	Muscle	For finfish, the muscle MRL relates to 'muscle and skin in natural proportions'	Anti-infectious agents/ antibiotics
Flumequine	Flumequine	Finfish	600	Muscle and skin in natural proportions		Anti-infectious agents/ antibiotics
Oxolinic acid	Oxolinic acid	All food-producing species	100	Muscle	For finfish, the muscle MRL relates to 'muscle and skin in natural proportions'. Not for use in animals from which milk or eggs are produced for human consumption	Anti-infectious agents/ antibiotics
Oxytetracycline	Sum of parent drug and its 4-epimer	All food-producing species	100	Muscle	For finfish, the muscle MRL relates to 'muscle and skin in natural proportions'	Anti-infectious agents/ antibiotics

Source: Commission *Regulation* (EU) No 37/2010 of 22 December 2009, European Union

Box 5.1 Good agricultural practices for the management of veterinary drugs applied to food animals

Veterinary medicines should be obtained only from authorized veterinary pharmacies or suppliers. Veterinary medicines should be prescribed and administered only by the persons permitted do so under national legislation, such as a qualified veterinarian or fish health specialist. Any storage of veterinary medicines on the farm should be in locked facilities, with a named key-holder. A written record of stock control should be maintained. Fish which are undergoing a veterinary drug treatment should be kept separate from those which are not (or which are subject to a different regime). Note should be taken of all warning statements and contra-indications for use, particularly in relation to withdrawal periods. A written record of veterinary applications should be made. This should record every application of drugs and other chemicals, and should include the treatment start date, treatment stop date, compound used, reason(s) for use, dose, identity of ponds or cages where the drug was applied, and harvest date for the treated ponds or cages. The entries in the register should be signed by the veterinarian, or fish disease specialist responsible for administering the drug programme. Aquaculture products must not be harvested before the end of the withdrawal period. At the end of the withdrawal period, and before harvest, consideration should be given to drawing a sample of treated fish for analysis of the residue level, to check that it does not exceed the permitted limit (the maximum residue level – MRL). For those fish tested with drug residue concentrations above the MRL, harvest of the batch should be postponed until the batch complies with the appropriate MRL. If aquaculture products which are treated with a veterinary medicine are sold live for on-growing before the end of the withdrawal period, then the buyer must be informed in writing of the treatment applied, the last date of treatment and the date of termination of the withdrawal period.

applied (Subasinghe, 2009). In the case of exports to the EU market, this regime must follow the requirements set out in Commission Delegated Regulation (EU) 2022/1644 on specific requirements for the performance of official controls on the use of pharmacologically active substances authorised as veterinary medicinal products or as feed additives and of prohibited or unauthorised pharmacologically active substances and residues thereof.

For aquaculture animals, residue monitoring has two objectives. The first is to establish whether there has been any use of banned substances, with samples drawn throughout the production cycle (fish feed, eggs, larvae, juveniles, grow-out and market). Samples may include organs (liver and kidney), and tests are conducted with

Box 5.2 Groups A and B compounds

GROUP A: substances having anabolic effects and unauthorized substances

1. Stilbenes, stilbene derivatives and their salts and esters
2. Antithyroid agents
3. Steroids
4. Resorcylic acid lactones, including zeranol
5. Beta-agonists
6. Compounds included in Annex IV to Council Regulation (EEC) No 2377/90 of 26 June 1990

GROUP B: veterinary drugs and contaminants

1. Antibacterial substances, including sulphonamides and quinolones
2. Other veterinary drugs
   (a) Anthelmintics
   (b) Anticoccidials, including nitroimidazoles
   (c) Carbamates and pyrethroids
   (d) Sedatives
   (e) Non-steroidal anti-inflammatory drugs (NSAIDs)
   (f) Other pharmacologically active substances
3. Other substances and environmental contaminants
   (a) Organochlorine compounds, including PCBs
   (b) Organophosphorus compounds
   (c) Chemical elements
   (d) Mycotoxins
   (e) Dyes
   (f) Others

Source: Commission Delegated Regulation (EU) 2022/1644 on specific requirements for the performance of official controls on the use of pharmacologically active substances authorised as veterinary medicinal products or as feed additives and of prohibited or unauthorised pharmacologically active substances and residues thereof.

a view to detecting the presence of any banned substance at any level. The second objective is to check that permitted substances have been applied correctly (for example, observance of withdrawal periods) and therefore samples are drawn from market-size and harvested fish, and tests conducted on edible portions (skin and muscle), to establish compliance with the MRL. In the EU regime the sampling targets those compounds falling within Groups A and B as shown in Box 5.2. For official sampling for a national Residue

Monitoring Programme, Commission Delegated Regulation (EU) 2022/1644 requires at least 1 fish per 100 tonnes of annual production, with one third of the total samples for Group A compounds (taken at farm level, on fish at all stages of farming), and two thirds of the total samples for Group B compounds (taken on fish ready to be placed on the market for consumption).

Non-compliant results from residue monitoring indicate a failure in the control system. In each case these should be followed up to determine the cause, and the conditions giving rise to the non-compliance should be corrected. If non-compliant products are detected and have not been released to the market, appropriate control measures should be applied. In the case of detection of banned substances, the affected batch(es) should be destroyed and cannot be used for animals. In the case of residues exceeding the MRL, if the fish are alive the withdrawal period can be extended to allow the residues to be metabolized and the fish harvested at a later date. If the fish have already been harvested, they should not be used for human consumption but may be used for animal feed (e.g. fishmeal). If non-compliant aquaculture products have been released to the market, they should be the subject of a recall or withdrawal process.

Pesticides

Concerns are often expressed regarding water quality and its potential impact on product safety, as well as the effects on fish health and growth. This is an important issue, especially in countries such as Egypt where regulations prohibit the use of irrigation supply water for aquaculture, thus confining activities to the use of water from drainage canals (Goulding & Kamel, 2013). When freshwater is contaminated with agriculture run-off, it has been shown to increase the risk of pesticide residues being present in the muscle (Ibrahim & Ramzy, 2013). Yahia and Elsharkawy (2014) also found a wide range of different types of organophosphorus (OPs), organochlorine (OC), polychlorinated biphenyl (PCBs), hexachlorobenzene (HCB) and trifluralin pesticides in samples of *O. niloticus* and African catfish (*Clarias gariepinus*) from Egyptian aquaculture farms; such compounds are known to accumulate in adipose tissues (Bräuner *et al.*, 2012). These findings raise concerns, because exposure assessments used to determine dosage rates during registration decisions are usually based on the target crops rather than incidental contamination from run-off. There have been no extensive specific human health risk assessments for pesticides in fishery products and, in their absence, the EU applies a general default MRL of 0.01 mg/kg under Regulation (EC) No 396/2005 on maximum residue levels of pesticides in or on food and feed of plant and animal origin. Certainly, where tilapia is farmed in regions of intensive agriculture, the risk of pesticide contamination should be monitored and managed by the farm operator. There is a particular additional risk where aquaculture and crop production are integrated, for example in rice–fish culture systems, although Berg (2001) observed that Vietnamese farmers growing fish in rice paddies were aware of this risk and used significantly less pesticide as a result.

Aflatoxins

There are substantial and mounting concerns regarding the human health impacts of consumption of fungal toxins, a group of contaminants particularly associated with hepatocellular carcinoma (Liu & Wu, 2010) and stunting in children (Mosites *et al.*, 2017). Aflatoxins are a group of toxins produced by *Aspergillus* spp. and other filamentous fungi that are found on agricultural crops such as maize and other grains, groundnuts and some dried products. Human exposure in the diet occurs via a wide range of sources, including products of farmed animals that may have consumed contaminated animal feeds. Maize is one of the crops in which aflatoxin content is frequently found at high levels (Wu, 2015) and is a common component of fish feeds.

There is clear evidence that aquaculture feeds can be a source of aflatoxins in the human diet. Marijani *et al.* (2017) found *Asperigillus flavus* in 54% and aflatoxins in 64% of fish feeds sampled in Kenya, Tanzania, Rwanda and Uganda. All affected samples were locally manufactured feed and no contamination was detected in imported feed. The study highlighted the potential risks to both fish performance and the health of consumers. Fallah *et al.* (2014) found that 67% of aquaculture feed samples from Iranian fish farms were contaminated, with levels of aflatoxin B1 up to 68.5 μg/kg. As a result of awareness of the risks, regulators often establish maximum limits for aflatoxins in animal feed, and reputable animal feed manufacturers monitor their raw materials to ensure that levels are within acceptable limits. In the EU the regulatory limits are set by Directive 2002/32/EC 32/EC of the European Parliament and of the Council on undesirable substances in animal feed. This sets a maximum content of aflatoxin B1 in feed ingredients of 0.02 mg/kg and in finished feeds of 0.01 mg/kg (relative to a feed with a moisture content of 12%). The associated sampling and testing requirements are set out in Commission Regulation (EC) No 152/2009 of 27 January 2009, 'laying down the methods of sampling and analysis for the official control of feed'. The EU has also published guidance notes on sampling and testing of animal feeds to check compliance (European Commission, 2014). A useful code of practice is available, Codex Alimentarius (1997) which, whilst directed at the dairy sector, provides good advice for managing this important hazard in all animal feed production. In all cases, farmers should ensure dry and well-ventilated, pest-secure on-farm storage conditions for feed and feed ingredients.

Fish health and human food safety

It is worth noting that many of the food safety hazards described in this section also impact negatively on fish health, and therefore on farm productivity. Jiraungkoorskul *et al.* (2003) and Ayoola (2008) found adverse histopathological and biochemical alterations in tilapia exposed to glyphosate pesticides. Almeida *et al.* (2001) found that tilapia exposed to sublethal concentrations of cadmium in water for as little as 7 days led to metabolic effects, similar to those observed in starvation. Deng *et al.* (2010) noted hepatic disorders in *O. niloticus* and *O. aureus* fed a long-term diet contaminated with aflatoxin B1. It is therefore in the interests of producers, not only on public health grounds but also to ensure good productivity, to ensure that their fish are grown with healthy diets and environment. As with many aspects of animal production, a healthy diet and environment meet the needs of farmers, animals and consumers.

Conclusion

The increasing production and consumption of tilapia makes this group of fishes one of the most important aquaculture species in the world, in terms not only of production and economics (where other aquaculture products may assume importance) but also as a source of human food, contributing substantially to food security, especially in populous Asian, African and South American countries. The potential is clearly demonstrated by countries such as Egypt, where fish from aquaculture (mostly tilapia) accounts for 68% of the 19.4 kg per capita annual consumption of fish, despite less than ideal production conditions (Goulding & Kamel, 2016). This chapter has demonstrated the importance of tilapia, from both capture fisheries and aquaculture, in providing food security and nutritional benefits to consumers.

Fish farmers often make production decisions on issues such as selection of breed, diet, production conditions and environment on the exclusive basis of production criteria such as yields, feed conversion and growth rates. However, this chapter shows that decisions made during production, even in terms of choice of feed and hatchery disinfectants, can have negative consequences on both the product and consumers. Whilst much of aquaculture research is directed at improving production technologies, to ensure that maximum societal benefits are derived there is a need to ensure that fish welfare, quality and food safety considerations are also addressed.

To this end, the author therefore recommends that policy makers, aquaculture producers and researchers should ensure that the following important issues are addressed in the future:

- exploring the potential for nutritional fortification of tilapia feeds with *n*-3 and *n*-6 oils (which can include the essential fatty acids DHA and EPA if the supplements are derived from fish or algal sources)

- managing water quality to prevent off-flavours and risks of contamination with bacterial pathogens of faecal origin
- ensuring that commonly used aquaculture chemicals that leave toxic residues in the product (such as malachite green) are not used at any stage of production
- ensuring that intensification to obtain efficiency gains does not lead to increased hazards from residues of veterinary medicines
- ensuring that, as feed formulations for tilapia progressively replace fish protein with vegetable sources, this does not lead to increased risk of aflatoxins and other chemical hazards (although some hazards such as heavy metals and dioxins may be reduced by this trend)
- improving fish welfare and quality during slaughter by ensuring humane stunning prior to killing
- improving hygiene and temperature control, to reduce post-harvest losses and improve spatial and temporal reach of marketing chains
- ensuring that proper advice is provided to consumers regarding handling, preparation and consumption, to eliminate the hazards presented by pathogenic bacterial and, perhaps most importantly of all, fish-borne zoonotic trematodes.

References

Abdussalam, M., Käferstein, F.K. and Mott, K.E. (1995) Food safety measures for the control of foodborne trematode infections. *Food Control* 6(2), 71–79. Available at: DOI: https://doi.org/10.1016/0956-7135(95)98910-S.

Akande, G., & Diei-Ouadi, Y. (2010). Postharvest losses in small-scale fisheries: Case studies in five sub-Saharan African countries. FAO fisheries and aquaculture technical Paper no.550Rome

Albarracín, H.W., Alfonso, A.C. and Sánchez, B. (2012) Application of essential oils as a preservative to improve the shelf life of Nile tilapia (oreochoromis niloticus). *Vitae* 19(1), 34–40.

Alemu, L.A., Melese, A.Y. and Gulelat, D.H. (2013) Effect of endogenous factors on proximate composition of Nile tilapia (*Oreochromis niloticus*

L.) fillet from lake zeway. *American Journal of Research Communication* 1(11), 405–410.

Almeida, J.A., Novelli, E.L.B., Dal Pai Silva, M.D.P. and Júnior, R.A. (2001) Environmental cadmium exposure and metabolic responses of the Nile tilapia, *Oreochromis niloticus*. *Environmental Pollution* 114(2), 169–175. DOI: 10.1016/s0269-7491(00)00221-9.

Al-Souti, A., Al-Sabahi, J., Soussi, B. and Goddard, S. (2012) The effects of fish oil-enriched diets on growth, feed conversion and fatty acid content of red hybrid tilapia, oreochromis sp. *Food Chemistry* 133(3), 723–727. DOI: 10.1016/j.foodchem.2012.01.080.

Arannilewa, S.T., Salawu, S.O., Sorungbe, A.A. and Ola-Salawu, B.B. (2005) Effect of frozen period on the chemical, microbiological and sensory quality of frozen tilapia fish (sarotherodun galiaenus). *African Journal of Biotechnology* 4(8), 852–855.

Ayoola, S.O. (2008) Toxicity of herbicide, glyphosate on Nile tilapia (*Oreochromis niloticus*) juvenile.

Azaria, S. and Van Rijn, J. (2018) Off-flavor compounds in recirculating aquaculture systems (RAS): Production and removal processes. *Aquacultural Engineering* 83, 57–64. DOI: 10.1016/j.aquaeng.2018.09.004.

Baranyi, J. and Roberts, T.A. (1994) A dynamic approach to predicting bacterial growth in food. *International Journal of Food Microbiology* 23(3–4), 277–294. DOI: 10.1016/0168-1605(94)90157-0.

Barkham, T., Zadoks, R.N., Azmai, M.N.A., Baker, S., Bich, V.T.N. *et al.* (2019) One hypervirulent clone, sequence type 283, accounts for a large proportion of invasive *Streptococcus agalactiae* isolated from humans and diseased tilapia in southeast Asia. *PLOS Neglected Tropical Diseases* 13(6), e0007421. DOI: 10.1371/journal.pntd.0007421.

Begadiya, H., Nayak, J., Mathakiya, R., Parmar, B. and Solanki, K. (2018) Isolation, biotyping and serotyping of escherichia coli isolated from fresh water fish. *Biosciences* 3572.

Berg, H. (2001) Pesticide use in rice and rice–fish farms in the mekong delta, vietnam. *Crop Protection* 20(10), 897–905. DOI: 10.1016/S0261-2194(01)00039-4.

Borit, M. and Olsen, P. (2016) Seafood traceability systems: Gap analysis of inconsistencies in standards and norms, Vol. C1123. *FAO Fisheries and Aquaculture Circular*.

Botelho, A.C.N., Ferreira, A.F.M., Fracalanzza, S.E.L., Teixeira, L.M. and Pinto, T.C.A. (2018) A perspective on the potential zoonotic role of *Streptococcus agalactiae*: Searching for A missing link in alternative transmission routes. *Frontiers in Microbiology* 9, 608. DOI: 10.3389/fmicb.2018.00608.

Bräuner, E.V., Raaschou-Nielsen, O., Gaudreau, E., Leblanc, A., Tjønneland, A, *et al.* (2012) Predictors of adipose tissue concentrations of organochlorine pesticides in a general danish population. *Journal of Exposure Science and Environmental Epidemiology* 22(1), 52–59. DOI: 10.1038/jes.2011.39.

Broadhurst, C.L., Wang, Y., Crawford, M.A., Cunnane, S.C., Parkington, J.E, *et al.* (2002) Brain-specific lipids from marine, lacustrine, or terrestrial food resources: potential impact on early African homo sapiens. *Comparative Biochemistry and Physiology. Part B, Biochemistry and Molecular Biology* 131(4), 653–673. DOI: 10.1016/s1096-4959(02)00002-7.

Chai, J.Y. (2007) Intestinal flukes. In: *Food-Borne Parasitic Zoonoses.* Springer, pp. 53–115. DOI: 10.1007/978-0-387-71358-8_2.

Chai, J.Y. and Jung, B.K. (2017) Fishborne zoonotic heterophyid infections: an update. *Food and Waterborne Parasitology* 8–9, 33–63. DOI: 10.1016/j.fawpar.2017.09.001.

Chau, M.L., Chen, S.L., Yap, M., Hartantyo, S.H.P., Chiew, P.K.T, *et al.* (2017) Group B streptococcus infections caused by improper sourcing and handling of fish for raw consumption, singapore, 2015–2016. *Emerging Infectious Diseases* 23(12), 2002–2010. DOI: 10.3201/eid2312.170596.

Chi, T.T., Dalsgaard, A., Turnbull, J.F., Tuan, P.A. and Murrell, K.D. (2008) Prevalence of zoonotic trematodes in fish from a vietnamese fish-farming community. *Journal of Parasitology* 94(2), 423–428. DOI: 10.1645/GE-1389.1.

Clement, S. and Lovell, R.T. (1994) Comparison of processing yield and nutrient composition of cultured Nile tilapia (*Oreochromis niloticus*) and channel catfish (*Ictalurus punctatus*). *Aquaculture* 119(2–3), 299–310. Available at: https://doi.org/10.1016/0044-8486(94)90184-8

Codex Alimentarius commission code of practice for the reduction of aflatoxin b1 (1997) In Raw materials and supplemental feedingstuffs for milk producing animals (CAC/RCP 45–1997). Retrieved October 27, 2020. Available at: http://www.ipcinfo.org/fileadmin/user_upload/animalfeeding/docs/CXP_045e.pdf

Codex Alimentarius (2020) Codex veterinary drug residue in food online database. Retrieved October 19, 2020. Available at: http://www.fao.org/fao-who-codexalimentarius/codex-texts/dbs/vetdrugs/en/

Codex Alimentarius Commission (2007) Joint FAO/WHO food standards programme and world health organization. *Codex Alimentarius Commission: Procedural Manual.Food & Agriculture Org.*

Commission Delegated Regulation (EU) 2022/1644 of 7 July 2022 supplementing Regulation (EU) 2017/625 of the European Parliament and of the Council with specific requirements for the performance of official controls on the use of pharmacologically active substances authorised as veterinary medicinal products or as feed additives and of prohibited or unauthorised pharmacologically active substances and residues thereof. Available at: https://eur-lex.europa.eu/homepage.html

Commission regulation (EC) no 152/2009 of 27 January (2009) Laying down the methods of sampling and analysis for the official control of feed. Current consolidated version available at. Retrieved October 27, 2020. Available at: https://eur-lex.europa.eu/homepage.htm

Connell, J.J. (1995) *Control of Fish quality*, 4th ed. Fishing Books.

Council Directive 98/58/EC of (1998) Concerning the protection of animals kept for farming purposes. Available at: https://eur-lex.europa.eu/homepage.html

Council Regulation (EC) no. 1099/2009 of 24 September 2009 on the protection of animals at the time of killing. https://eur-lex.europa.eu/homepage.html

Council Regulation (EC) (2004) No 1/2005 of 22 December 2004 on the protection of animals during transport and related operations and amending Directives 64/432/EEC and 93/119/EC and Regulation (EC)) No 1255/97. https://eur-lex.europa.eu/homepage.html

Culp, S.J. and Beland, F.A. (1996) Malachite green: A toxicological review. *Journal of the American College of Toxicology* 15(3), 219–238. Available at: DOI: https://doi.org/10.3109/10915819609008715.

Culp, S.J., Mellick, P.W., Trotter, R.W., Greenlees, K.J., Kodell, R.L. *et al.* (2006) Carcinogenicity of malachite green chloride and leucomalachite green in B6C3F1 mice and F344 rats. *Food and Chemical Toxicology* 44(8), 1204–1212. Available at: DOI: https://doi.org/10.1016/j.fct.2006.01.016.

Cunha, V., Davies, M.R., Douarre, P.E., Rosinski-Chupin, I., Margarit, I. *et al.* (2014) *Streptococcus agalactiae* clones infecting humans were selected and fixed through the extensive use of tetracycline. *Nature Communications* 5, 4544. DOI: 10.1038/ncomms5544.

Cyprian, O.O., Sveinsdóttir, K., Magnússon, H., Arason, S., Jóhannsson, R. *et al.* (2014) Development of Quality Index Method (QIM) scheme for farmed tilapia fillets and its application in shelf life study. *Journal of*

Aquatic Food Product Technology 23(3), 278–290. Available at: DOI: https://doi.org/10.1080/10498850.2012.713088.

Daskalova, A. (2019) Farmed fish welfare: stress, post-mortem muscle metabolism, and stress-related meat quality changes. *International Aquatic Research* 11(2), 113–124. DOI: 10.1007/s40071-019-0230-0.

de Castro, P.L.D., Lewandowski, V., Souza, M.L.R.D., Coradini, M.F., Alexandre, A.A.D.C, *et al.* (2017) Effect of different periods of pre-slaughter stress on the quality of the Nile tilapia meat. *Food Science and Technology* 37(1), 52–58. DOI: 10.1590/1678-457x.05616.

Deng, S., Tian, L., Liu, F., Jin, S., Liang, G, *et al.* (2010) Toxic effects and residue of aflatoxin b1 in tilapia (*oreochromis niloticus* × O. aureus) during long-term dietary exposure. *Aquaculture* 307(3–4), 233–240. DOI: 10.1016/j.aquaculture.2010.07.029.

Dergal, N.B., Abi-Ayad, S.M.E.A., Douny, C., Brose, F., Daube, G, *et al.* (2013) Microbial, biochemical and sensorial quality assessment of algerian farmed tilapia (*Oreochromis niloticus*) stored at 4 and 30C. *African Journal of Food Science* 7(12), 498–507. DOI: 10.5897/AJFS2013.1063.

Dilna, S.V., Nanitha Krishna, E.K., Panda, S.K. and Ashok Kumar, K. (2017) Impact of different processing methods on retention of oxytetracycline residues in treated tilapia. 11th Indian Fisheries and Aquaculture Forum, 406, *Book of Abstracts* 406.

Directive 2002/32/EC of the European Parliament and of the Council of 7 May (2002) On undesirable substances in animal feed. Available at: https://eur-lex.europa.eu/homepage.html

dos Santos, C.A.M. and Howgate, P. (2011) Fishborne zoonotic parasites and aquaculture: A review. *Aquaculture* 318(3–4), 253–261. DOI: 10.1016/j.aquaculture.2011.05.046.

El-Seify, M.A., Sultan, K., Elhawary, N.M., Satour, N.S. and Marey, N.M. (2020) Prevalence of heterophyid infection in tilapia fish "*Orechromas niloticus*" with emphasize of cats role as neglected reservoir for zoonotic heterophyes heterophyes in Egypt. *Journal of Parasitic Diseases* 1–8.

Eltholth, M., Fornace, K., Grace, D., Rushton, J. and Häsler, B. (2015) Characterisation of production, marketing and consumption patterns of farmed tilapia in the nile delta of egypt. *Food Policy* 51, 131–143. Available at: DOI: https://doi.org/10.1016/j.foodpol.2015.01.002.

Erikson, U., Hultmann, L. and Steen, J.E. (2006) Live chilling of atlantic salmon (*Salmo salar*) combined with mild carbon dioxide anaesthesia: I. establishing a method for large-scale processing of farmed fish. *Aquaculture* 252(2–4), 183–198.

European Commission (2011) Guidance on viable parasites in fishery products that may represent A risk to the health of the consumer, endorsed 16 November 2011 by the standing committee on the food chain and animal health, section biological safety of the food chain, DG health and consumer protection. Retrieved October 27, 2020. Available at: https://ec.europa.eu/food/sites/food/files/safety/docs/biosafety_fh_eu_food_establishments-20111214_scfcah_guidance_parasites_en.pdf

European Commission (2014) Guidance document for the implementation of Commission Regulation (EU). No 152/2009 as regards methods of sampling and analysis endorsed by the Standing Committee on the food chain and animal health section animal nutrition at its meeting on 16–17. Retrieved October 27, 2020, 691/2013 of 19 July 2013 amending regulation (EC. Available at: https://ec.europa.eu/food/sites/food/files/safety/docs/animal-feed-guidance_documents_691_2013_en.pdf

European Food Safety Authority (2009) Scientific opinion of the panel on animal health and welfare on a request from the European commission on species-specific welfare aspects of the main systems of stunning and killing of farmed rainbow trout. *EFSA Journal* 1013, 1–55. Available at: http://www.efsa.europa.eu/en/efsajournal/pub/1012.htm

European Food Safety Authority (2010) Opinion of the Scientific Panel on Biological hazards (BIOHAZ) on parasites in fishery products adopted on 11 March 2010. Retrieved October 27, 2020. Available at: https://efsa.onlinelibrary.wiley.com/doi/epdf/10.2903/j.efsa.2010.1543

European Union (2002), Regulation (EC) No 178/2002 of the European Parliament and of the Council of 28 January 2002 laying down the general principles and requirements of food law, establishing the European Food Safety Authority and laying down procedures in matters of food safety. https://eur-lex.europa.eu/homepage.html

Eves, A., Turner, C., Yakupitiyage, A., Tongdee, N. and Ponza, S. (1995) The microbiological and sensory quality of septage-raised Nile tilapia (*Oreochromis niloticus*). *Aquaculture* 132(3–4), 261–272. Available at: DOI: https://doi.org/10.1016/0044-8486(94)00328-L.

Fallah, A.A., Pirali-Kheirabadi, E., Rahnama, M., Saei-Dehkordi, S.S. and Pirali-Kheirabadi, K. (2014) Mycoflora, aflatoxigenic strains of aspergillus section flavi and aflatoxins in fish feed. *Quality Assurance and Safety of Crops and Foods* 6(4), 419–424. Available at: DOI: https://doi.org/10.3920/QAS2012.0186.

Food and Agriculture Organization (FAO) (1992) Ice in fisheries. In: Graham, J., Johnston, W.A. and Nicholson, F.J. (eds) *FAO Fisheries Technical Paper, 331*. Food and Agriculture Organization.

Food and Agriculture Organization (FAO) (2003) Fisheries technical paper 444 huss, HH and embarek, PKB, chapter 5.1.4 [Retrieved October 27, 2020. Parasites in Assessment and management of seafood safety and quality, Huss, H. H. Ababouch, L, and Gram, L]. In: *Food and Agriculture Organization of the United Nations*. Available at: http://www.fao.org/docrep/006/y4743e/y4743e0c.htm

Food and Agriculture Organization (FAO) (2020) Sustainability in action. state of world fisheries and aquaculture. *Food and Agriculture Organization of the United Nations.*

Franssen, F., Gerard, C., Cozma-Petruţ, A., Vieira-Pinto, M., Jambrak, A.R. *et al.* (2019) Inactivation of parasite transmission stages: Efficacy of treatments on food of animal origin. *Trends in Food Science and Technology* 83, 114–128. Available at: DOI: https://doi.org/10.1016/j.tifs.2018.11.009.

Garduño-Lugo, M., Granados-Alvarez, I., Olvera-Novoa, M.A. and Muñoz-Córdova, G. (2003) Comparison of growth, fillet yield and proximate composition between stirling Nile tilapia (wild type) (*Oreochromis niloticus*, linnaeus) and red hybrid tilapia (florida red tilapia× stirling red O. niloticus) males. *Aquaculture Research* 34(12), 1023–1028. Available at: DOI: https://doi.org/10.1046/j.1365-2109.2003.00904.x.

Global, G.A.P. (2020) Integrated Farm Assurance All Farm base – Aquaculture Checklist (ed 5.1–1)_AQ_NOV2017. Retrieved October 23, 2020. Available at: https://www.globalgap.org/uk_en/for-producers/globalg.a.p./integrated-farm-assurance-ifa/aquaculture/

Goes, E.S., Lara, J.A., Gasparino, E., Del Vesco, A.P., Goes, M.D. *et al.* (2015) Pre-slaughter stress affects ryanodine receptor protein gene expression and the water-holding capacity in fillets of the Nile tilapia. *PLOS ONE* 10(6), e0129145. Available at: DOI: https://doi.org/10.1371/journal.pone.0129145.

Goulding, I.C. and Kamel, M. (2013) *Institutional, policy and regulatory framework for sustainable development of the Egyptian aquaculture sector.* WorldFish.

Goulding, I. C., & Kamel, M. (2016). Improved marketing of Egyptian aquaculture products, Consultancy report for US Soya bean Export Council Chesterfield, MO, USA

Green, B.W. and Teichert-Coddington, D.R. (2000) Human food safety and environmental assessment of the use of 17α-methyltestosterone to produce male tilapia in the united states. *Journal of the World*

Aquaculture Society 31(3), 337–357. Available at: DOI: https://doi.org/10.1111/j.1749-7345.2000.tb00885.x.

He, Y., Huan, H., Li, L., Yang, X., Zhao, Y. *et al.* (2016) Freshness and shelf life of air packaged and modified atmosphere packaged fresh tilapia fillets during freezing-point storage. *Journal of Nutrition and Food Sciences* 06(6). Available at: DOI: https://doi.org/10.4172/2155-9600.1000564.

Hebard, C. (1982) *Occurrence and significance of trimethylamine oxide and its derivatives in fish and shellfish.* Chemistry & Biochemistry of Marine Food Products.

Hoffman, A., Disney, J.G., Grimwood, B.E. and Jones, N.R. (1970) Glucose levels in fresh tilapia muscle. *Journal of the Fisheries Research Board of Canada* 27(4), 801–803. Available at: DOI: https://doi.org/10.1139/f70-082.

Howgate, P. (2004) Tainting of farmed fish by geosmin and 2-methyl-iso-borneol: A review of sensory aspects and of uptake/depuration. *Aquaculture* 234(1–4), 155–181. Available at: DOI: https://doi.org/10.1016/j.aquaculture.2003.09.032.

Humane Slaughter Association. (2020). Humane harvesting of fish. https://www.hsa.org.uk/humane-harvesting-of-fish-introduction/introduction-6. Retrieved October 27, 2020

Ibrahim, L.A. and Ramzy, E.M. (2013) Water quality and its impact on *Tilapia zilli* (case study) qarun lake-egypt. *International Water Technology Journal* 3(4), 170–191.

Jamilah, B. and Harvinder, K.G. (2002) Properties of gelatins from skins of fish—black tilapia (*Oreochromis mossambicus*) and red tilapia (*Oreochromis nilotica*). *Food Chemistry* 77(1), 81–84. Available at: DOI: https://doi.org/10.1016/S0308-8146(01)00328-4.

Jim, F., Garamumhango, P. and Musara, C. (2017) Comparative analysis of nutritional value of *Oreochromis niloticus* (linnaeus), Nile tilapia, meat from three different ecosystems. *Journal of Food Quality* 1–8. DOI: 10.1155/2017/6714347.

Jiraungkoorskul, W., Upatham, E.S., Kruatrachue, M., Sahaphong, S., Vichasri-Grams, S. *et al.* (2003) Biochemical and histopathological effects of glyphosate herbicide on Nile tilapia (*Oreochromis niloticus*). *Environmental Toxicology* 18(4), 260–267. Available at: DOI: https://doi.org/10.1002/tox.10123.

Kapute, F., Likongwe, J., Kang'ombe, J. and Kiiyukia, C. (2013) Shelf life of whole fresh lake malawi tilapia (oreochromis species–chambo) stored in ice. *African Journal of Food, Agriculture, Nutrition and Development* 13(1), 7138–7156. Available at: DOI: https://doi.org/10.18697/ajfand.56.12590.

Keiser, J. and Utzinger, J. (2005) Emerging foodborne trematodiasis. *Emerging Infectious Diseases* 11(10), 1507–1514. Available at: DOI: https://doi.org/10.3201/eid1110.050614.

Khalafalla, F.A., Ali, F.H.M. and Hassan, A.H.A. (2015) Quality improvement and shelf-life extension of refrigerated Nile tilapia (*Oreochromis niloticus*) fillets using natural herbs. *Beni-Suef University Journal of Basic and Applied Sciences* 4(1), 33–40. Available at: DOI: https://doi.org/10.1016/j.bjbas.2015.02.005.

Khoa, D.V., Hoa, D.T., Anh, D.N., Van, N.T., Dung, D.T. *et al.* (2020) Fish-borne trematode metacercariae detected in fish commonly used for raw consumption in ninh binh province. *Vietnam. Tropical Biomedicine* 37(2), 443–451.

Laly, S.J., Ashok Kumar, K.N., Sankar, T.V., Lalitha, K.V. and Ninan, G. (2017) Quality of monosex tilapia under ice storage: gutting effects on the formation of biogenic amines, biochemical, and microbiological characteristics. *International Journal of Food Properties* 20(6), 1368–1377. Available at: DOI: https://doi.org/10.1080/10942912.2016.1209777.

Lambooij, E., Gerritzen, M.A., Reimert, H., Burggraaf, D. and Van de Vis, J.W. (2008) A humane protocol for electro-stunning and killing of Nile tilapia in fresh water. *Aquaculture* 275(1–4), 88–95. DOI: 10.1016/j.aquaculture.2008.01.026.

Leal, C.A.G., Queiroz, G.A., Pereira, F.L., Tavares, G.C. and Figueiredo, H.C.P. (2019) *Streptococcus agalactiae* sequence type 283 in farmed fish. *Brazil. Emerging Infectious Diseases* 25(4), 776–779. Available at: DOI: https://doi.org/10.3201/eid2504.180543.

Liu, Y. and Wu, F. (2010) Global burden of aflatoxin-induced hepatocellular carcinoma: A risk assessment. *Environmental Health Perspectives* 118(6), 818–824. Available at: DOI: https://doi.org/10.1289/ehp.0901388.

Lovell, R.T., Lelana, I.Y., Boyd, C.E. and Armstrong, M.S. (1986) Geosmin and musty-muddy flavors in pond-raised channel catfish. *Transactions of the American Fisheries Society* 115(3), 485–489. DOI: 10.1577/1548-8659(1986)1152.0.CO;2.

Lun, Z.R., Gasser, R.B., Lai, D.H., Li, A.X., Zhu, X.Q, *et al.* (2005) Clonorchiasis: A key foodborne zoonosis in China. *Lancet. Infectious Diseases* 5(1), 31–41. DOI: 10.1016/S1473-3099(04)01252-6.

Marijani, E., Wainaina, J.M., Charo-Karisa, H., Nzayisenga, L., Munguti, J, *et al.* (2017) Mycoflora and mycotoxins in finished fish feed and feed ingredients from smallholder farms in east africa. *Egyptian Journal of Aquatic Research* 43(2), 169–176. DOI: 10.1016/j.ejar.2017.07.001.

Matos, E., Gonçalves, A., Nunes, M.L., Dinis, M.T. and Dias, J. (2010) Effect of harvesting stress and slaughter conditions on selected flesh quality criteria of gilthead seabream (*Sparus aurata*). *Aquaculture* 305(1–4), 66–72. DOI: 10.1016/j.aquaculture.2010.04.020.

Mekonnen, M.F., Desta, D.T., Alemayehu, F.R., Kelikay, G.N. and Daba, A.K. (2020) Evaluation of fatty acid-related nutritional quality indices in fried and raw Nile tilapia, (*Oreochromis niloticus*), fish muscles. *Food Science and Nutrition* 8(9), 4814–4821. DOI: 10.1002/fsn3.1760.

Meyer, M.C. (1970) Cestode zoonoses of aquatic animals. *Journal of Wildlife Diseases* 6(4), 249–254. DOI: 10.7589/0090-3558-6.4.249.

Moretti, V.M., Turchini, G.M., Bellagamba, F. and Caprino, F. (2003) Traceability issues in fishery and aquaculture products. *Veterinary Research Communications* 27(1), 497–505. DOI: 10.1023/b:verc.0000014207.01900.5c.

Mosites, E., Dawson-Hahn, E., Walson, J., Rowhani-Rahbar, A. and Neuhouser, M.L. (2017) Piecing together the stunting puzzle: A framework for attributable factors of child stunting. *Paediatrics and International Child Health* 37(3), 158–165. DOI: 10.1080/20469047.2016.1230952.

Myers, B.J. (1970) Nematodes transmitted to man by fish and aquatic mammals. *Journal of Wildlife Diseases* 6(4), 266–271. DOI: 10.7589/0090-3558-6.4.266.

Natasha, J. (2014) Postharvest processing of farmed tilapia (*Oreochromis niloticus*) for potential commercialisation in fiji ([doctoral dissertation] [master's thesis]. University of the South Pacific. Faculty of Science Technology and Environment, School of Marine Studies 50–51.

Niizeki, N., Daikoku, T., Hirata, T., El-Shourbagy, I., Song, X, *et al.* (2002) Mechanism of biosynthesis of trimethylamine oxide from choline in the teleost tilapia, *Oreochromis niloticus*, under freshwater conditions. *Comparative Biochemistry and Physiology. Part B, Biochemistry and Molecular Biology* 131(3), 371–386. DOI: 10.1016/s1096-4959(01)00508-5.

Oliveira Filho, P.R.C., Oliveira, C.A.F., Sobral, P.J.A., Balieiro, J.C.C., Natori, M.M. *et al.* (2015) How stunning methods affect the quality of Nile tilapia meat. *CyTA – Journal of Food* 13(1), 56–62. DOI: https://doi.org/10.1080/19476337.2014.911211.

Polak, E.H. and Provasi, J. (1992) Odor sensitivity to geosmin enantiomers. *Chemical Senses* 17(1), 23–26. Available at: https://doi.org/10.1093/chemse/17.1.23

Poli, B.M., Parisi, G., Scappini, F. and Zampacavallo, G. (2005) Fish welfare and quality as affected by pre-slaughter and slaughter management. *Aquaculture International* 13(1–2), 29–49. DOI: 10.1007/s10499-004-9035-1.

Reddy, N.R., Villanueva, M. and Kautter, D.A. (1995) Shelf life of modified-atmosphere-packaged fresh tilapia fillets stored under refrigeration and temperature-abuse conditions. *Journal of Food Protection* 58(8), 908–914. DOI: 10.4315/0362-028X-58.8.908.

Rodrigues, T.P., Eliane, T.M., Robson, M.F., Pereira Mello Silvia, S.I.C.R.P., Ivone, C.S, *et al.* (2016) Quality index method (QIM) and quantitative descriptive analysis (QDA) of Nile tilapia (*Oreochromis niloticus*) quality indices. *African Journal of Agricultural Research* 11(3), 209–216. DOI: 10.5897/AJAR2015.9565.

Roth, B., Moeller, D., Veland, J.O., Imsland, A. and Slinde, E. (2002) The effect of stunning methods on rigor mortis and texture properties of atlantic salmon (*Salmo salar*). *Journal of Food Science* 67(4), 1462–1466. DOI: 10.1111/j.1365-2621.2002.tb10306.x.

Rutten, M.J.M., Bovenhuis, H. and Komen, H. (2005) Genetic parameters for fillet traits and body measurements in Nile tilapia (*Oreochromis niloticus* L. *Aquaculture* 246(1–4), 125–132. DOI: 10.1016/j.aquaculture.2005.01.006.

Ryder, J.M. (1985) Determination of adenosine triphosphate and its breakdown products in fish muscle by high-performance liquid chromatography. *Journal of Agricultural and Food Chemistry* 33(4), 678–680. DOI: 10.1021/jf00064a027.

Satour, N.S., Zayed, A.F. and Rahman, M.A. (2019) Occurrence of encysted metacercariae in tilapia nilotica (oreochromusniloticus) in alexandria province and their public health significance. *Alexandria Journal of Veterinary Sciences* 61(2). DOI: 10.5455/ajvs.40272.

Schram, E., Kwadijk, C., Hofman, A., Blanco, A., Murk, A, *et al.* (2021) Effect of feeding during off-flavour depuration on geosmin excretion by Nile tilapia (*Oreochromis niloticus*). *Aquaculture* 531, 735883. DOI: 10.1016/j.aquaculture.2020.735883.

Slootweg, R., Vroeg, P.A. and Wiersma, S.J. (1993) Effects of molluscivorous fish, water quality and pond management on the development of schistosomiasis vector snails in aquaculture ponds. *Aquaculture Research* 24(1), 123–128. DOI: 10.1111/j.1365-2109.1993.tb00835.x.

Sneddon, L.U. (2002) Anatomical and electrophysiological analysis of the trigeminal nerve in a teleost fish, oncorhynchus mykiss. *Neuroscience Letters* 319(3), 167–171. DOI: 10.1016/s0304-3940(01)02584-8.

Sneddon, L.U. (2009) Pain perception in fish: Indicators and endpoints. *ILAR Journal* 50(4), 338–342. DOI: 10.1093/ilar.50.4.338.

Sneddon, L.U., Elwood, R.W., Adamo, S.A. and Leach, M.C. (2014) Defining and assessing animal pain. *Animal Behaviour* 97, 201–212. DOI: https://doi.org/10.1016/j.anbehav.2014.09.007.

Srivastava, S., Sinha, R. and Roy, D. (2004) Toxicological effects of malachite green. *Aquatic Toxicology* 66(3), 319–329. DOI: 10.1016/j.aquatox.2003.09.008.

Subasinghe, R. (2009) Disease control in aquaculture and the responsible use of veterinary drugs and vaccines: the issues, prospects and challenges. *Options Méditerranéennes* 86, 5–11.

Sun, J., Fang, W., Ke, B., He, D., Liang, Y, *et al.* (2016) Inapparent *Streptococcus agalactiae* infection in adult/commercial tilapia. *Scientific Reports* 6, 26319. DOI: 10.1038/srep26319.

Tang, Z.L., Huang, Y. and Yu, X.B. (2016) Current status and perspectives of clonorchis sinensis and clonorchiasis: epidemiology, pathogenesis, omics, prevention and control. *Infectious Diseases of Poverty* 5(1), 71. DOI: 10.1186/s40249-016-0166-1.

Teklu, D. (2015) Determinant factors for wasted fish during harvesting at amerti and fichawa reservoirs oromia/ethiopia. *Fisheries Science* 9(4), 12.

Tonial, I.B., Stevanato, F.B., Matsushita, M., De Souza, N.E., Furuya, W.M. *et al.* (2009) Optimization of flaxseed oil feeding time length in adult Nile tilapia (*Oreochromis niloticus*) as a function of muscle omega-3 fatty acids composition. *Aquaculture Nutrition* 15(6), 564–568. DOI: 10.1111/j.1365-2095.2008.00623.x.

Torry Research Station (1989) Sensory assessment of fish quality, Torry Advisory Note no.91. http://www.fao.org/3/x5989e/x5989e00.htm Retrieved October 27, 2020.

Union, E. (2004) of the European Parliament and of the Council of 29 April 2004 laying down specific hygiene rules for food of animal origin. Consolidated version. Retrieved October 27, 2020, 853/2004, Annex III to Regulation (EC). Available at: https://eur-lex.europa.eu/legal-content/EN/TXT/PDF/?uri=CELEX:02004R0853–20190726&from=EN

Vallod, D., Cravedi, J.P., Hillenweck, A. and Robin, J. (2007) Analysis of the off-flavor risk in carp production in ponds in dombes and forez

(France). *Aquaculture International* 15(3–4), 287–298. Available at: https://doi.org/10.1007/s10499-007-9080-7

Venturini, F.P., Vargas Baldi, S.C., Parisi, G., Costa, T.D., Rucinque, D.S, *et al.* (2018) Effects of different stunning methods on blood markers and enzymatic activity of stress responses of tilapia (*Oreochromis niloticus*). *Italian Journal of Animal Science* 17(4), 1094–1098. DOI: 10.1080/1828051X.2018.1426396.

Verdon, E., & Andersen, W. C. (2017). Certain dyes as pharmacologically active substances in fish farming and other aquaculture products. Food and Drug Administration Papers. https://digitalcommons.unl.edu/usfda/13/Retrieved October 28, 202013.

Waagbø, R., Jørgensen, S.M., Timmerhaus, G., Breck, O. and Olsvik, P.A. (2017) Short-term starvation at low temperature prior to harvest does not impact the health and acute stress response of adult atlantic salmon. *PeerJ* 5, e3273. DOI: 10.7717/peerj.3273.

Waliszewski, K.N. and Avalos, M. (2001) Shelf-life extension of tilapia in crushed ice with additives at- 3C. *Journal of Food Processing and Preservation* 25(4), 299–307. DOI: 10.1111/j.1745-4549.2001.tb00462.x.

Wang, Z.C., Yan, Y., Su, P., Zhao, M.M., Xia, N, *et al.* (2018) Treatments of tilapia (*Oreochromis niloticus*) using nitric oxide for quality improvement: establishing a potential method for large-scale processing of farmed fish. *Nitric Oxide: Biology and Chemistry* 77, 19–25. DOI: 10.1016/j.niox.2018.04.003.

Warner, K., Timme, W., Lowell, B. and Hirschfield, M. (2013) *Oceana study reveals seafood fraud nationwide.* Oceana Publications, pp. 1–69.

Waterman, J.J. (undated) Ministry of agriculture, fisheries and food, torry research station. Retrieved October 27, 2020, *Torry Advisory Note No.15 (Revised)*. Available at: http://www.fao.org/3/x5896e/x5896e00.htm

Wilkinson, R.J., Paton, N. and Porter, M.J.R. (2008) The effects of pre-harvest stress and harvest method on the stress response, rigor onset, muscle ph and drip loss in barramundi (lates calcarifer). *Aquaculture* 282(1–4), 26–32. DOI: 10.1016/j.aquaculture.2008.05.032.

World Animal Health Organisation. (2019). Aquatic animal health code. https://www.oie.int/en/standard-setting/aquatic-code/access-online/ Retrieved October 27, 2020.

Wu, F. (2015) Global impacts of aflatoxin in maize: trade and human health. *World Mycotoxin Journal* 8(2), 137–142. DOI: 10.3920/WMJ2014.1737.

Xu, W., Zhu, X., Wang, X., Deng, L. and Zhang, G. (2006) Residues of enrofloxacin, furazolidone and their metabolites in Nile tilapia (*Oreochromis niloticus*). *Aquaculture* 254(1–4), 1–8. DOI: 10.1016/j.aquaculture.2005.10.030.

Yahia, D. and Elsharkawy, E.E. (2014) Multi pesticide and PCB residues in Nile tilapia and catfish in assiut city, egypt. *Science of the Total Environment* 466–467, 306–314. DOI: 10.1016/j.scitotenv.2013.07.002.

Yamprayoon, J. and Noomhorm, A. (2003) *Off-flavor in Nile tilapia (Oreochromis iloticus). Off-Flavors in Aquaculture.* American Chemical Society.

Yanar, Y., Çelik, M. and Akamca, E. (2006) Effects of brine concentration on shelf-life of hot-smoked tilapia (*Oreochromis niloticus*) stored at 4 C. *Food Chemistry* 97(2), 244–247. Available at: DOI: https://doi.org/10.1016/j.foodchem.2005.03.043.

Yossepowitch, O., Gotesman, T., Assous, M., Marva, E., Zimlichman, R, *et al.* (2004) Opisthorchiasis from imported raw fish. *Emerging Infectious Diseases* 10(12), 2122–2126. DOI: 10.3201/eid1012.040410.

Zampacavallo, G., Parisi, G., Mecatti, M., Lupi, P., Giorgi, G, *et al.* (2015) Evaluation of different methods of stunning/killing sea bass (*Dicentrarchus labrax*) by tissue stress/quality indicators. *Journal of Food Science and Technology* 52(5), 2585–2597. DOI: 10.1007/s13197-014-1324-8.

Chapter 6

Nutrient requirements of tilapia

D. Allen Davis, Darci Carlos Fornari and Waldemar Rossi Jr

Introduction

In recent years production of tilapia has increased by 70%, which has positively affected global freshwater aquaculture (FAO, 2022). Tilapia is the third-most farmed fish species in the world (FAO, 2022), reflected in a significant increase in per capita consumption during the last 7 years (FAO, 2022). Paralleling the increase in feed-based animal production is an increased need for feed and feed ingredients. The cost of production fluctuates from year to year but, in general, feed ingredient prices increase over time. Industries must balance cost increases through the utilization of less costly or more efficient protein and energy sources. This drives the need for testing of new ingredients, as well as the use of feed supplements such as probiotics, exogenous enzymes, mycotoxin adsorbents and immunostimulants. Formulations need to be fine-tuned and manufacturing practices optimized. Inappropriate changes in feed formulations adversely interfere not only with the growth of fish but also with their health, and thus nutrition is key to success in producing fish of high quality and at competitive prices.

Nutritional research on tilapia has been conducted globally under a range of rearing conditions and with different perspectives regarding feed optimization. Information from various studies, however,

Kim Thompson, Craig Shoemaker and David C. Little (eds) **Tilapia**
DOI: 10.1079/9781800629455.0006,

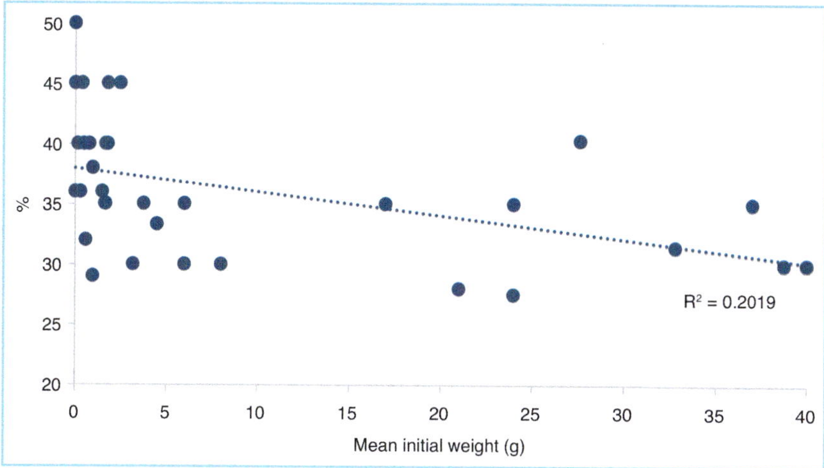

Figure 6.1 Dietary crude protein requirements of tilapia, *Oreochromis* spp. (%; *n*=33). Bahnasawy *et al.* (2009); Abdel-Tawwab *et al.* (2010); Abdel-Tawwab (2012); Kpundeh *et al.* (2015); Carneiro *et al.* (2017); Ogungi and Wirth (1999); Shiau (2002); Sumi *et al.* (2011); Qiang *et al.* (2012); Zeng et al. (2009). Lowest mean initial weight = 0.012 g.

converges into the concept of nutritionally balanced diets for the efficient production and optimum health of farmed fish. The aim of this chapter is to provide an updated summary of the nutritional requirements of tilapia. Here, the term 'tilapia' refers collectively to species and crosses within the genus *Oreochromis: O. niloticus* (Nile tilapia), *O. aureus* (blue tilapia), *O. mossambicus* (Mozambique tilapia), *O. niloticus* × *O. aureus* (hybrid tilapia) and (*O. mossambicus* × *O. niloticus* (red hybrid tilapia). For the sake of simplification, the collective and common names above are used hereafter.

This chapter is not exhaustive in summarizing existing information on nutritional requirements, functional additives, feeds and feeding of farmed tilapia. Additional information can be found in the reviews of Shiau (2002), Ng and Romano (2013) and Maas *et al.* (2020), among others.

Protein and amino acids

Second to water, protein constitutes the largest nutrient in the animal's body and the most expensive component of aquaculture feeds. In the diet, proteins are the sources of amino acids (AAs) that have a suite of critical roles in metabolism. Fish require dietary protein in the form of both essential and non-essential AAs that should be supplied through a balanced diet (Tacon and Cowey, 1985).

Several studies have assessed the dietary protein requirements of tilapia and have shown requirements as high as 50% crude protein (CP) in fry of Mozambique tilapia (Jauncey & Ross, 1982) and 28% CP in juvenile (21 g) hybrid tilapia (Twibell & Brown, 1998). The available literature reveals protein requirements (mean ± standard difference (SD)) of tilapia ranging from 38 ± 5.3% CP for fish of ≤5g initial weight to 32 ± 3.6% CP for fish of initial weight ranging from ~6 to 40 g. A higher variation in reported requirements is observed when initial sizes were ≤2.5g (Fig. 6.1).

The genetic potential for growth of different tilapia strains has been shown to influence dietary protein requirements. Ng and Hanim (2007) fed genetically improved farmed tilapia (GIFT) and red hybrid tilapia iso-energetic diets containing 25 or 35% CP and observed superior growth performance of the GIFT tilapia at both protein levels, whereas the increase in dietary CP from 25 to 35% did not affect growth rates of red hybrid tilapia. While most studies assessing the nutritional requirements of tilapia were conducted with young fish, Koch *et al.* (2017) observed improved growth performance of Nile tilapia grown from 450 to 800g under intensive conditions when fed a practical diet containing 27% digestible protein (DP) and 3.0 Mcal digestible energy (DE)/kg. Adequate protein nutrition not only supports optimum growth performance but also affects egg and fry production, as shown by El-Sayed and Kawanna (2008). Improved spawning performance of Nile tilapia was observed when the fish were fed 40% versus 30 or 35% CP diets, indicating that tilapia broodstock benefit from high dietary protein.

Tilapia tolerate a wide range of salinities and temperatures. Inverse relationships between salinity and protein requirements of tilapia have been reported (Shiau, 2002; Daudpota *et al.*, 2014; Singha *et al.*, 2020, Singha *et al.*, 2021; Wu *et al.*, 2021). Compiled information indicates that tilapia require 34 ± 3.6% CP at salinities ranging from 0 to 15 g/L and 23 ± 2.0% CP when reared at salinities ranging from 32 to 37 g/L. By contrast, Larumbe-Morán *et al.* (2010) observed no effect of salinity (from 1.2 to 25 g/L) on the dietary protein requirement of Nile tilapia. Under sub-optimum water temperature (22°C), GIFT tilapia displayed reduced digestibility of dry matter and protein retention efficiency compared with fish reared at 28°C. Maximum growth performance was observed in fish fed ~37 versus 30% dietary CP (Zeng *et al.*, 2021).

Growing animals need a constant supply of protein and energy to sustain maximum growth rates. While in laboratory studies fish are fed at or near satiety, feed restrictions may occur under intensive farming conditions. Under conditions of restricted feed allowance, the use of balanced feeds with higher protein content than that recommended by laboratory studies may support faster growth rates, while protein retention efficiency generally decreases with increased protein intake.

AAs

Amino acids are organic compounds that contain an amine, a carboxyl group and a side chain specific to each AA. Amino acids have critical roles in protein synthesis, immune function and fish health (Furuya and Furuya, 2010; National Research Council (NRC), 2011). Among the 20 proteinogenic AAs, arginine, histidine, isoleucine, leucine, lysine, methionine, phenylalanine, threonine, tryptophan and valine are not synthesized by animals (including fish) and are defined as dietary essential AAs without which animals would not survive. As with most animals, fish can synthesize alanine, asparagine, aspartate, cysteine, glutamate, glutamine, glycine, proline, serine and tyrosine, which are classified as non-essential AAs (NRC, 2011). The compiled essential AA requirements of tilapia are presented in Table 6.1.

Arginine (Arg)

Arginine is a precursor of nitric oxide and creatine, promotes anabolic processes by stimulating insulin and growth hormone (NRC, 2011) and has been shown to improve the production performance and immune responses of tilapia (Yue *et al.*, 2015; Vianna *et al.*, 2020). Although Arg can be limiting in practical feed formulations, limitations are uncommon. Studies have shown that tilapia of initial weight ranging from 0.015 to 159g require Arg at 5.3 ± 0.82% of dietary CP based on production performance and nitrogen (N) deposition.

Histidine (His)

Histidine is essential in the synthesis of haemoglobin, is the direct precursor of histamine, serves as a substrate in the synthesis of purines and has both antioxidant and anti-inflammatory properties (Wade and Tucker, 1998; NRC, 2011). Different studies using tilapia of initial weight ranging from 0.075 to 159g have determined His requirements of 2.2 ± 0.61% of dietary CP based on production performance, N deposition, and myogenin- and MyoD-gene expression.

Isoleucine (Ile), leucine (Leu) and valine (Val)

Isoleucine, Leu and Val comprise the branched-chain proteinogenic AAs (BCAAs) having non-linear aliphatic side chains. They are hydrophobic AAs and their functions, including structure and conformation of proteins, are mediated by interactions with other hydrophobic molecules (NRC, 2011). Leucine stimulates muscle protein synthesis as an activator of mTOR (Andersen *et al.*, 2016), while Ile and Val have critical roles in energy and glucose metabolism (Ahmad *et al.*, 2021). Dietary Leu supplementation promoted intestinal mucosal development and increased gastrointestinal digestion capacity of GIFT tilapia (Shi *et al.*, 2014). The dietary requirements for BCAAs in tilapia have been assessed in fish of initial weight ranging from <1 to 159g and were reported based on dietary concentrations maximizing weight gain and N/amino acid deposition.

Table 6.1 Dietary essential AA requirements of tilapia, *Oreochromis* spp.

Amino acid	Fish initial weight (g)	Dietary protein (% of dry diet)	AA requirement (% of crude protein)	Response criteria	References
Arginine					
	0.015	28	4.2	WG	Santiago and Lovell (1988)
	3.0	28	4.9	WG, FCR, PRE	Neu *et al.* (2016)
	6.0	28	6.2	WG	Yue *et al.* (2015)
	20	32	6.3[a]	ND[a]	Diógenes *et al.* (2016)
	81.5	30	5.0–5.3	WG, FCR, PER	Wu *et al.* (2016)
	159	28	4.9[a]	ND[a]	Nascimento *et al.* (2020)
Mean ± SD			5.3 ± 0.82		
Histidine					
	0.075	28	1.72	WG	Santiago and Lovell (1988)
	4.8	26	3.1	WG; FCR; PER	Michelato *et al.* (2017)
	20	32	1.72[a]	ND[a]	Diógenes *et al.* (2016)
	64	28	2.8	WG, MD, MyoD-gene	Zaminhan-Hassemer *et al.* (2020)
	159	28	1.73[a]	ND[a]	Nascimento *et al.* (2020)
Mean ± SD			2.2 ± 0.61		
Isoleucine					
	0.087	28	3.11	WG	Santiago and Lovell (1988)
	2.5	27.5	5.0	WG	Araújo *et al.* (2021)
	20	32	2.84[a]	ND[a]	Diógenes *et al.* (2016)
	159	28	3.26[a]	ND[a]	Nascimento *et al.* (2020)
Mean ± SD			3.6 ± 0.85		
Leucine					
	0.057	28	3.39	WG	Santiago and Lovell (1988)

Continued

Table 6.1 Continued

Amino acid	Fish initial weight (g)	Dietary protein (% of dry diet)	AA requirement (% of crude protein)	Response criteria	References
	0.65	35	4.5	WG, Leu gain, ND	Zehra and Khan (2017)
	2.0	29	4.31	WG	Gan *et al.* (2016)
	20	32	4.83[a]	ND[a]	Diógenes *et al.* (2016)
	159	28	4.82[a]	ND[a]	Nascimento *et al.* (2020)
Mean ± SD			4.4 ± 0.53		
Lysine					
	0.041	28	5.12	WG	Santiago and Lovell (1988)
	5.2	32	5.6–6.0	WG, ND	Prabu et al. (2020)
	5.7	30	5.67	WG	Furuya *et al.* (2006)
	9	28	7.12	GR; WG	Ovie and Eze. (2013)
	20	32	5[a]	ND[a]	Diógenes *et al.* (2016)
	159	28	5.87[a]	ND[a]	Nascimento *et al.* (2020)
	275	25	5.8	FY	Michelato et al. (2016)
Mean ± SD			5.8 ± 0.61		
Methionine					
	0.062	28	2.68 (with 0.54% cystine) TSAA = 3.2	WG	Santiago and Lovell (1988)
	2.3	28	3.3 (with 0.3% cysteine) TSAA = 3.6	WG	He *et al.* (2017)
	5.62	28	1.75 (with 1.6% cystine) TSAA = 3.35	WG	Nguyen and Davis (2009)
	20	32	TSAA = 3.2[a]	ND[a]	Diógenes *et al.* (2016)

Mean ± SD			3.3 ± 0.16		
Phenylalanine					
	0.013	28	3.75 (with 1.79% Tyr) TAAA = 5.54	WG	Santiago and Lovell (1988)
	5.6	29	3.03 (with 3.45% Tyr) TAAA = 6.48	WG; TGC; PER	Xiao *et al.* (2020)
	20	32	TAAA = 5.03[a]	ND[a]	Diógenes *et al.* (2016)
Mean ± SD			5.7 ± 0.60		
Threonine					
	0.054	28	3.75	WG	Santiago and Lovell (1988)
	3.0	28	4.74	WG	Yue *et al.* (2014)
	20	32	4.67[a*]	ND[a]	Diógenes *et al.* (2016)
	159	28	6.02[a]	ND[a]	Nascimento *et al.* (2020)
	563	29	4.1	WG	Michelato et al. (2016)
Mean ± SD			4.7 ± 0.8		
Tryptophan					
	0.056	28	1	WG	Santiago and Lovell (1988)
	1.4	28	1	WG	Wu *et al.* (2018)
	3.4	30	1.1	WG	Zaminhan *et al.* (2018)
	5.3	32	1.2	WG, ND	Prabu et al. (2020)
	7.9	30	0.83–1.1	TGC, FE, PRE, Trp gain	Nguyen *et al.* (2019)
	20	32	1.19[a]	ND[a]	Diógenes *et al.* (2016)
	38	28	1.1	WG, FCR, PER	Zaminhan *et al.* (2017)
	159	28	0.87[a]	ND[a]	Nascimento *et al.* (2020)

Continued

Table 6.1 Continued

Amino acid	Fish initial weight (g)	Dietary protein (% of dry diet)	AA requirement (% of crude protein)	Response criteria	References
Mean ± SD			1.0 ± 0.1		
Valine					
	0.05	28	2.8	WG	Santiago and Lovell (1988)
	6.4	28	4.1–4.5	WG, PRE	Xiao *et al.* (2018)
	20	32	3.8	ND[a]	Diógenes *et al.* (2016)
	159	28	3.5	ND[a]	Nascimento *et al.* (2020)
Mean ± SD			3.74 ± 0.57		

[a]Deletion method

WG, weight gain; FCR, feed conversion ratio; PRE, protein retention efficiency; PER, protein efficiency ratio; MD, muscle development; ND, nitrogen/protein deposition; GR, growth rate; FY, fillet yield; TGC, thermal-unit growth coefficient; FE, feed efficiency; SD, standard deviation.

Compiled requirement values from different studies indicated that tilapia require Ile, Leu and Val at 3.6 ± 0.85%, 4.4 ± 0.53% and 3.74 ± 0.57% of dietary CP, respectively.

Lysine (Lys)

Lysine is among the first limiting AA in practical feed formulations based on plant and rendered animal protein ingredients (Nunes *et al.*, 2014). Lysine is found in high concentrations in fish muscle and is the precursor of hydroxylysine in regard to the formation of collagen (NRC, 2011). Studies assessing dietary Lys requirements of tilapia have used fish of initial weight ranging from 0.041 to 275g and indicated a Lys requirement of 5.8 ± 0.61% of dietary CP based on weight gain, N deposition and fillet yield.

Methionine (Met)

Along with Lys, supplementation of Met in crystalline forms in tilapia feeds may be necessary to prevent deficiencies due to high inclusion rates of plant-based protein feedstuffs, notably soybean meal. Fish fed diets deficient in Met display poor growth and feed efficiency, and may develop lenticular cataracts (NRC, 2011).

In addition to serving as a proteinogenic AA, Met also participates in two main metabolic pathways as (1) a source of sulfur for the synthesis of other sulfur-containing compounds, including the AAs cysteine and taurine, and (2) as a donor of methyl groups in methylation reactions (Mehler, 1986). Because cysteine is derived from Met, dietary cysteine has been demonstrated to spare ~50% of Met requirement on an equimolar basis in several animals, including fish (NRC, 2011). Consequently, because some cysteine is always present in basal diets used in Met requirement studies, the designation of total sulfur AA (TSAA) requirement is adopted to account for the sparing value of cysteine. Compiled information from different studies using tilapia of initial weight ranging from 0.062 to 20g indicates a TSAA requirement of 3.3 ± 0.16% of dietary CP based on weight gain and N retention. In addition, recent studies with Nile tilapia have demonstrated that some Met can be spared by dietary taurine (Michelato *et al.*, 2018) and that ~1% taurine is required in soybean meal-based diets to improve reproductive performance and larval quality (Al-Feky *et al.*, 2016a, b).

Phenylalanine (Phe)

Known as an aromatic AA, Phe is an essential component of body proteins and an important precursor of aromatic compounds required for normal body functions. The conversion of Phe to tyrosine that occurs primarily in the liver and kidney mediates the production of hormones and neurotransmitters such as dopamine, epinephrine, norepinephrine, thyroxine and triiodothyronine (Li *et al.*, 2009). Because tyrosine is a proteinogenic AA and a derivative of Phe, it can partially satisfy the dietary Phe requirement of animals, including fish (NRC, 2011). Consequently, the term total aromatic AA (TAAA) requirement is adopted in studies on Phe requirement to reflect basal levels of tyrosine, which has been shown to spare 37% of the Phe requirement in Nile tilapia on an equimolar basis (Zehra & Yousif, 2021). Studies in tilapia of initial weight ranging from 0.013 to 20g showed a TAAA requirement of 5.7 ± 0.6% of dietary CP based on weight gain, protein efficiency ratio and N deposition.

Threonine (Thr)

Threonine is abundant in mucins, serves as precursor for the synthesis of glycine and has been shown to modulate immune responses and antioxidant status in fish (NRC, 2011; Habte-Tsion *et al.*, 2016). Threonine typically becomes the third-most limiting AA in plant-based formulations and is the third-most supplemented crystalline AA in commercial feeds (Nunes *et al.*, 2014). The dietary Thr requirement of tilapia, compiled from different studies using fish of initial weight ranging from 0.054 to 563g, was 4.7 ± 0.8% of dietary CP based on weight gain and N deposition.

Tryptophan (Trp)

Tryptophan is a precursor of the neurotransmitter serotonin and neurohormone melatonin (NRC, 2011). In addition to improvement in production performance, an adequate dietary supply of Trp prevents scoliosis and adverse effects on mineral metabolism, and improves intestinal functions, including immune responses, barrier function and antioxidant status in fish (NRC, 2011; Wen *et al.*, 2014). Compiled information from different studies indicates that Trp is required in the diet of tilapia (initial weight 0.056–159 g) at 1.0 ± 0.1% of dietary CP based on production performance, N and Trp deposition.

Lipids and fatty acids

Lipids occur in nature in highly reduced states as fats, oils, fatty acids, phospholipids, cholesterol and derivatives. Lipids have several physiological roles including supply of chemical energy, cell membrane structure, mediation of immune responses and neural functions, comprising essential macronutrients in feeds.

Lipids comprise concentrated sources of energy in feeds by providing ~2.5 times the energy supplied by the oxidation of an equal amount of proteins or carbohydrates. Tilapia utilize digestible carbohydrates efficiently as energy, and hence their dietary lipid requirement is relatively low. Lipid levels in tilapia feeds will vary depending on the nutritional density of the ingredients and nutritional specifics of formulations. Generally, grow-out diets for tilapia contain <8% lipid which, besides supplying essential fatty acids, allows formulations to attain optimum energy:protein ratios of ~10–12 kcal DE/g DP (Winfree and Stickney, 1981; Koch *et al.*, 2017). Using purified diets containing 29% CP, Chou and Shiau (1996) observed maximum growth of hybrid tilapia fed 10% dietary lipid, but the performance of the fish was not statistically different from groups fed diets with either five or 15% lipid, indicating that a minimum of 5% dietary lipid was adequate. Likewise, the increase in dietary lipid from 6 to 9% in diets containing 20 or 28% CP did not affect the growth, feed efficiency, energy and protein retention efficiency of Nile tilapia (Li *et al.*, 2013), indicating that 6% dietary lipid was adequate. Generally, studies evaluating optimum carbohydrate (CHO):lipid ratios in tilapia diets have observed satisfactory production performance with ≤7% dietary lipid (Xie *et al.*, 2017; Liu *et al.*, 2018; Coutinho *et al.*, 2018; Kabir *et al.*, 2020).

Fatty acids

Tilapia, like other vertebrates, are unable to synthesize omega-6 (*n*-6) and omega-3 (*n*-3) fatty acids (FAs) due to the lack of Δ−12 and Δ−15 fatty acid desaturases required in the lipogenesis pathway. Consequently, the supply of *n*-6 and *n*-3 FAs in tilapia diets is necessary. On the other hand, tilapia can synthesize the biologically active long-chain polyunsaturated FAs (LC-PUFAs) arachidonic (ARA; 20:4*n*-6), eicosapentaenoic (EPA; 20:5*n*-3) and docosahexaenoic (DHA; 22:6*n*-3) acids by elongating and desaturating linoleic (LA; 18:2*n*-6) and α-linolenic (ALA: 18:3*n*-3) acids commonly present in plant-based oils. Therefore, although dietary provision of LC-PUFAs generally leads to higher concentrations of these FAs in edible tissue (Stoneham *et al.*, 2018), dietary requirements have not been demonstrated in tilapia.

Tilapia have been shown to require both *n*-6 and *n*-3 FAs in diets. Dietary requirements for LA or ARA at ~0.5−1.0% (Kanazawa *et al.*, 1980; Takeuchi *et al.*, 1983) and ALA at 0.45−0.64% (Chen *et al.*, 2013) have been demonstrated. Therefore, the provision of both *n*-6 and *n*-3 FAs at ~1% in formulated feeds would satisfy the EFA requirements of tilapia when the supply of FAs from the rearing environment is negligible.

Carbohydrates

To date, no dietary requirement for carbohydrates (CHO) has been determined in fish, including tilapia. The glucose needed in their metabolism can be synthesized from non-glucose precursors, especially glucogenic AAs, through gluconeogenesis (NRC, 2011).

Digestible CHOs, however, comprise the least expensive source of dietary energy for animals, including fish, and whose dietary provision prevents the increased catabolism of protein and lipids for energy and synthesis of biologically important compounds. In addition, key properties of aquaculture feeds such as water stability, density and retention of water-soluble nutrients are influenced by CHO gelatinization during feed manufacture.

Starch is the main digestible CHO in aquaculture feeds and is highly digestible in tilapia in both native and gelatinized form (Amirkolaie *et al.*, 2006). The main carbohydrate sources for fish feed are corn, sorghum, millet, rice and wheat, and the choice of input is usually related to availability and price.

Although fish endogenous enzymes cannot hydrolyse β-glycosidic bonds of non-starch polysaccharides (NSP), digestibility of soluble NSP, probably mediated by gut fermentation, has been demonstrated in tilapia, suggesting that NSP can contribute digestible energy (Haidar *et al.*, 2016; Maas *et al.*, 2020). Insoluble NSP (primarily cellulose and insoluble hemicellulose) commonly present in tilapia feeds are virtually indigestible (Amirkolaie *et al.*, 2005), and dietary levels exceeding ~8% (as crude fibre) should be avoided for best performance (Dioundick and Stom, 1990; Meurer *et al.*, 2003). Some indigestible carbohydrates, such as fructo- and mannan-oligosaccharides, inulin and β-glucan, have been shown to have prebiotic functions in fish, including tilapia (Song *et al.*, 2014), indicating their applicability as functional additives in commercial feeds.

Following digestion and absorption, CHOs meet the immediate energy needs of cells and are converted to glycogen and fat, which supply energy in post-absorptive states. Tilapia utilize digestible CHOs well, and adequate growth and feed efficiency can be achieved using diets containing >40% digestible CHOs (Wang *et al.*, 2005; Xie *et al.*, 2017; Coutinho *et al.*, 2018). Recent findings indicate that when above ~35% in the diet, digestible CHOs can cause metabolic disorders and fatty liver syndrome in tilapia (Jiang *et al.*, 2013; Xie *et al.*, 2017). Supplementation of myo-inositol at 400 mg/kg in diets containing 30 or 45% starch appears to attenuate such effects (Zhu *et al.*, 2020). Intestinal inflammation in Nile tilapia fed a high-starch diet (45% maize starch) was also observed but was alleviated by the addition of dietary sodium acetate at 1.85 g/kg (Li *et al.*, 2020).

An adequate supply of non-protein energy in feeds maximizes protein retention and feed efficiency, thus reducing nutrient losses to receiving waters and production costs related to feeds and feeding. Digestible CHOs can contribute to the sparing of dietary protein by providing metabolic energy and directing more protein to anabolic routes (Wang *et al.*, 2005; Azaza *et al.*, 2015; Boonanuntanasarn *et al.*, 2018). For instance, Shiau and Peng (1993) observed similar weight gain and feed efficiency of hybrid tilapia when dietary protein was decreased from 28 to 24% if replaced by dextrin or starch in closely isoenergetic diets.

Digestible CHOs typically comprise one of the main sources of non-protein energy and dietary lipids in tilapia feeds. Coutinho *et al.* (2018) demonstrated that a CHO:lipid ratio of 6.14 (NFE = 43%) in 28% CP diets supported the best performance of Nile tilapia reared in net cages, indicating that the proper use of digestible CHOs in nutritionally balanced formulations aids in reducing feed costs by sparing both protein and lipid.

Minerals

Fish satisfy their mineral requirements by absorbing essential elements from the diet during digestion and from the aquatic environment through the gills (NRC, 2011). Independently of absorption route, minerals must be present in sufficient quantities to support normal physiological functions in animals such as the formation of bone and other hard tissues, electron transfer and enzyme function (Furuya, 2010; NRC, 2011).

Of the ~29 naturally-occurring inorganic elements considered essential for all farmed animals, including fish (Lall & Milley, 2008), 13 are considered most relevant with respect to dietary requirements. Of these, eight are cations: calcium, copper, iron, magnesium, manganese, potassium, sodium and zinc; and five are anions: phosphorus, chloride, iodide, molybdate and selenium (Scott *et al.*,

Table 6.2 Dietary mineral requirements of tilapia, *Oreochromis* spp.

Mineral	Fish initial weight (g)	Dietary requirement	Response criteria	Reference
Calcium (Ca)	0.52	0.35–0.43% (Ca in water = 27–33 mg Ca/l)	WG, bone and scale Ca	Shiau and Tseng (2007)
	0.8	0.7% (Ca-free water)	WG, FCR	Robinson *et al.* (1987)
	1.2	0.65% (Ca-free water)	Bone Ca and Ash	Robinson *et al.* (1984)
Phosphorus (P)	0.27	0.75%[a]	WG, FCR, WB minerals	Pezzato *et al.* (2006)
	0.41	0.59–0.64%[b]	FCR, PER, bone P	Furuya *et al.* (2008a)
	1.5	0.5% (P in water = 0.52 mg/L)	Bone ash, P, Ca	Robinson *et al.* (1987)
	25.2	0.76–0.79%[b]	WB and bone P	Phromkunthong and Udom (2008)
	35	0.53%[a]	FCR, PER	Furuya *et al.* (2008b)
	46	0.86%[b]	SGR, WB P	Yao *et al.* (2014)
	146	0.61%[b]	WG, WG and bone minerals	Schamber *et al.* (2014)
Potassium (K)	0.77	0.2–0.3% (K in water = 1.08–3.62 mg/L)	WG, gill NaK-ATPase, K retention	Shiau and Hsieh (2001)
Sodium (Na)	0.61	0.15% (FW)	WG, FE, gill NaK-ATPase	Shiau and Lu (2004)
Magnesium (Mg)	0.5	0.05% (Mg in water = 0.1 mg/L)	WG, FE, tissue minerals	Reigh *et al.* (1991)
	0.88	0.02%	WG, ALP, tissue Mg	Lin *et al.* (2013)
	10	0.059–0.077% (Mg in water = 9.1 mg/L)	WG, FE, WB minerals	Dabrowska *et al.* (1989)
Manganese (Mn)	0.64	7.0 mg/kg	Liver Mn-SOD, Mn retention	Lin et al. (2008)

Continued

Table 6.2 Continued

Mineral	Fish initial weight (g)	Dietary requirement	Response criteria	Reference
Iron (Fe)	0.6	85 mg/kg (from ferric sulfate); 150–160 mg/kg (from ferric citrate) (Fe in water = 61.4 μg/L)	WG, liver Fe, Hb	Shiau and Su (2003)
	175	121 mg/kg[b]	Hb	Cheng *et al.* (2019)
Copper (Cu)	0.79	4 mg/kg	WG, Cu retention	Shiau and Ning (2003)
Zinc (Zn)	0.55	26–29 mg/kg (Zn in water = 8.1 μg/L)	ALP, Zn retention	Lin et al. (2008)
	3.1	20 mg/kg (Zn in water = 4.0 μg/L)	Scale and bone Zn	McClain and Gatlin III (1988)
	8.4	30 mg/kg (Zn in water = 15 μg/L)	WG, FE, serum and bone Zn	Eid and Ghonim (1994)
	13.3	44.5–79.5 mg/kg[c] (Zn in water = 7.95 μg/L)	Bone Zn	Sá *et al.* (2004)
	167	37.2, 52.1 mg/kg (Zn in water = 1.4–1.7 μg/L)	WG, bone Zn	Huang *et al.* (2015)
Selenium (SE)	1.85	1.06 mg/kg	WG	Lee *et al.* (2016)
	3.0	0.57 mg/kg (Se-free water)	WG and liver GPx	Ning *et al.* (2020)

[a]Total non-phytate P;
[b]available;
[c]plant-based basal diet containing 65% soybean meal.
WG, weight gain; FCR, feed conversion ratio; WB, whole body; PER, protein efficiency ratio; SGR, specific growth rate; FE, feed efficiency; ALP, alkaline phosphatase; SOD, superoxide dismutase; Hb, haemoglobin; GPx, glutathione peroxidase.

1982). At the time of this review, dietary requirements for ten of the above minerals had been reported for tilapia (Table 6.2).

Calcium (Ca) and phosphorus (P)

Calcium and P comprise the major inorganic constituents in the diets and hard tissues of animals. Calcium is essential in several physiological processes including blood clotting, muscle contraction, osmoregulation and as a co-factor in enzyme processes (NRC, 2011). Tilapia cultured in Ca-free water required ~0.7% Ca in the diet for optimum growth performance, bone Ca and ash (Robinson *et al.*, 1984, 1987). Likewise, O'Connell & Gatlin, 1994 observed adequate growth and tissue mineralization in blue tilapia reared in low-Ca water (<1.0 mg/L) and fed a purified diet supplemented with 7.5% Ca. A lower dietary Ca requirement (0.35–0.43%) was observed by Shiau and Tseng (2007) when the rearing water contained ~30 mg Ca/l. Albeit that a Ca requirement is presented, this is only in 'Ca-free' water which typically does not occur under production conditions. When typical levels of Ca are present there is no need to supplement Ca in most feed formulations.

Phosphorus plays essential physiological roles as a component of phospholipids, co-enzymes and nucleic acids (NRC, 2011). Although absorption of soluble P from water has been demonstrated in *Tilapia zillii* (Al-Kholy *et al.*, 1970), soluble P is generally present at much lower concentrations in traditional aquaculture production systems compared with Ca. Consequently, most of the required P must be supplied in the diet and supplementation of inorganic P to plant-based feeds is generally required due to the low availability of phytate P. The reported dietary requirements for available P in tilapia range from 0.59 to 0.86%. Tilapia fry (0.41 g) required 0.59–0.64% available P for optimum production performance and bone P (Furuya *et al.*, 2008a), whereas 0.61% available P was required in the diet of 146 g fish (Schamber *et al.*, 2014). Yao *et al.* (2014) observed maximum specific growth rate and whole-body P of Nile tilapia fed 0.86% available P/kg of diet.

Potassium (K) and sodium (Na)

Along with chloride, K and Na play essential roles in the control of osmotic pressure and acid-based equilibrium of intracellular and extracellular fluids. Dietary deficiency of K and Na reportedly cause convulsions, tetany and impaired growth in fish (NRC, 2011). The dietary K requirement of hybrid tilapia has been reported to be 0.2–0.3 g/kg based on growth, gill NaK-ATPase activity and K retention (Shiau and Hsieh, 2001). In hybrid tilapia, dietary supplementation of Na was unnecessary in seawater; 0.15% dietary Na improved growth, feed efficiency and NaK-ATPase activity while ≥0.3% dietary Na tended to reduce fish performance in freshwater (Shiau and Lu, 2004).

Magnesium (Mg)

In vertebrates, Mg exerts primarily structural roles because ~60% of the body's Mg is found in the skeleton (NRC, 2011). Magnesium is a major intracellular cation participating in acid–base balance and an essential co-factor in enzymatic reactions, including hydrolysis and transfer of phosphate groups, FA oxidation and AA synthesis. A dietary deficiency of Mg may lead to tetany and muscle flaccidity in fish (NRC, 2011), whereas 0.32% Mg in low- but not high-CP diets affected production performance and blood parameters of Nile tilapia (Dabrowska *et al.*, 1989). The reported dietary requirement for Mg in tilapia has ranged from 0.05 to 0.077% in waters with Mg concentrations of 0.1 and 9.1 mg/, respectively (Dabrowska *et al.*, 1989; Reigh *et al.*, 1991).

Manganese (Mn)

In vertebrates, Mn is found primarily in bone and mitochondria. Manganese functions as a co-factor in enzyme activation and as a component of metallo-enzymes involved in carbohydrate, lipid and protein metabolism. Signs of Mn deficiency in fish include skeletal abnormalities and cataracts (NRC, 2011). In hybrid tilapia, optimum Mn-dependent superoxide dismutase (Mn-SOD) activity and whole-body Mn retention were observed at a dietary Mn concentration of 7.0 mg/kg (Lin *et al.*, 2008b).

Iron (Fe)

Iron is a structural component of haemoglobin and is essential in cellular respiration through its participation in oxidation-reduction and electron-transfer processes. Iron deficiency signs in fish include impaired growth and anaemia (NRC, 2011). The reported dietary Fe requirement for optimum growth, hepatic Fe and haemoglobin concentrations of hybrid tilapia reared in 61.4 μg Fe/l water was 85 and ~160 mg/kg when supplied by ferric sulfate and ferric citrate, respectively (Shiau and Su, 2003). In Nile tilapia, 121 mg of available Fe/kg diet was required to maximize haemoglobin concentration (Cheng *et al.*, 2019).

Copper (Cu)

Copper is essential due to its role in physiological processes, including bone formation, cellular respiration, connective tissue development, keratinization and tissue pigmentation. Copper functions as a co-factor of several physiologically important metallo-enzymes, including cytochrome oxidase, lysyl oxidase and Cu/Zinc-SOD (NRC, 2011). Not surprisingly, signs of Cu deficiency are normally associated with reduced activities of Cu-dependent enzymes. Hybrid tilapia was shown to require 4 mg Cu/kg of diet based on weight gain and Cu retention (Shiau and Ning, 2003), while Damasceno *et al.* (2016) found no improvement in the performance of Nile tilapia in response to supplemental Cu in a basal diet containing 6.8 mg available Cu/kg.

Zinc (Zn)

Zinc is an essential co-factor in several enzyme systems and an integral component of many metallo-enzymes, some of which are involved in carbohydrate, lipid and protein metabolism. Most published studies on the Zn requirement of tilapia reared in waters containing <2–15 μg Zn/l observed dietary requirements between 20 and ~50 mg/kg. However, when a practical basal diet containing 65% soybean meal was used, 79.5 mg Zn/kg was required to maximize bone Zn concentration (Sá *et al.*, 2005). This higher dietary requirement might reflect the antagonistic effect of phytate on Zn availability and should be considered in the formulation of plant-based feeds.

Selenium (Se)

Selenium is an integral component of glutathione peroxidases (GPx), an enzyme family that converts hydrogen peroxide and fatty acid hydroperoxides to water and lipid alcohols, respectively, preventing oxidative damage to cellular membranes. The activity of GPx in tissues has been shown to correlate positively with Se intake in animals (NRC, 2011), underscoring the importance of an adequate dietary supply of Se. When Se-Met was the dietary source of Se for Nile tilapia, Lee *et al.* (2016) observed maximum weight gain in fish fed 1.06 mg Se/kg of diet, whereas Ning *et al.* (2020) found that 0.57 mg Se/kg of diet maximized weight gain and liver GPx activity.

Chromium (Cr)

Chromium appears to be conditionally essential for tilapia when glucose is the main source of dietary CHO. Chromic oxide (Cr_2O_3) supplementation at levels as low as 2 mg/kg was shown to improve dietary glucose utilization in tilapia, but it appears to have little effect on the utilization of dietary starch (Shiau & Lin, 1993). Maximum weight gain of hybrid tilapia fed a diet containing 40% glucose was observed when dietary Cr_2O_3 was 204 mg/kg (Shiau & Shy, 1998).

Iodine (I)

Iodine is important for the proper functioning of the thyroid gland, requirements ranging from 100 to 300 μg/kg of diet (Furuya, 2010).

Vitamins

Vitamins are organic compounds that animals require at relatively minor amounts but are essential for normal growth, reproduction and health. Vitamins are classified according to their solubility into either water-soluble (ascorbic acid, biotin, choline, cyanocobalamin (B_{12}), folic acid, myo-inositol, niacin, pantothenic acid, pyridoxine (B_6), riboflavin and thiamine) or fat-soluble (A, D, E and K) vitamins. Of the 15 vitamins known to be required by animals, ascorbic acid, choline, myo-inositol and vitamins A, D and E have roles other than co-enzymes (NRC, 2011). With the exception of vitamin B_{12}, which has been demonstrated as being synthesized by gut microbiota in many fish including tilapia (Lovell and Limsuwan, 1982; Shiau and Lin, 1993), all 14 remaining vitamins have been demonstrated to be required in the diet of tilapia (Table 6.3).

Water-soluble vitamins

Ascorbic acid (vitamin C)

Specific roles of vitamin C include (1) participation in collagen synthesis by the hydroxylation of proline and lysine to procollagen; (2) being a reducing agent aiding in iron absorption; and (3) its paired role with vitamin E as a non-enzymatic antioxidant. While synthesis of vitamin C occurs in many animal species, the lack of gulonolactone oxidase activity has been indicated as the limiting step preventing its endogenous synthesis from D-glucose in teleosts (Moreau and Dabrowski, 2001). Optimum growth and survival of tilapia have been observed when dietary levels of vitamin C ranged from ~20 to 96 mg/kg being supplied by stable forms of the vitamin, whereas maximum concentration in the liver of Nile tilapia was observed when 115 mg/kg vitamin C was supplied by L-ascorbyl-2-polyphophate (C2PP) (Huang *et al.*, 2016).

Table 6.3 Dietary vitamin requirements of tilapia, *Oreochromis* spp.

Vitamin	Dietary requirement (mg/kg; otherwise noted)	Response criteria	Reference
Water-soluble vitamins			
Ascorbic acid (vit. C)	41–48 (C2S; 48%), 40.5 (C2MP-Mg; 46.5%) and 63.4 (C2MP-Na; 25%); 180, 460* (C2PP; 25%); 100–200 (C2S; 48%)	Growth, vertebral collagen, *liver vit. C	Shiau and Hsu (1995, 1999); Huang *et al.*, 2016; Al-Amoudi et al. (1992)
Biotin	0.06	Growth	Shiau and Chin (1999)
Choline	CE, 1000	Growth, body and liver concentrations	Kasper *et al.* (2000, 2002); Shiau and Lo (2000)
Cyanocobalamin (B_{12})	Dispensable	Growth, blood and hepatosomatic indices, liver B_{12}	Lovell and Limsuwan (1982); Shiau and Lung (1993);
Folic acid	0.82	Growth, liver concentration	Shiau and Huang (2001)
Myo-inositol	400	Growth, liver myo-inositol, liver and muscle lipid	Peres *et al.* (2004); Shiau and Su (2005), Bu *et al.* (2021)
Niacin	20.4, 84.6; 121	Growth, liver niacin	Shiau and Suen (1992); Jiang *et al.* (2014)
Pantothenic acid	10 (Ca-d-pantothenate; 92%)	Growth, FE, survival, blood indices.	Roem *et al.* (1991); Soliman and Wilson (1992a)
Pyridoxine (B_6)	3	Growth, survival, health	Lim *et al.* (1995)
	1.7/9.5 (28% dietary CP), 16.5/15 (36% dietary CP)	Growth/AST	Shiau and Hsieh (1997)
	10	Growth, health	Teixeira *et al.* (2012)
Riboflavin	5	Growth, survival	Lim *et al.* (1993)
	6	FE, liver D-AA oxidase	Soliman and Wilson (1992b)
Thiamin	3.5; 1.16, 2.06	Growth, FE, health, liver thiamine	Lim *et al.* (2011); Chun et al. (2015)

Fat-soluble vitamins			
Retinol (A)	5,000; 5,850–6970 IU/kg	Growth, FE, survival, health	Saleh *et al.* (1995); Hu *et al.* (2006)
Colecalcipherol (D)	374.8 IU/kg	Growth	Shiau and Suen (1992)
Tocopherol (E)	5–10 (5% dietary lipid); 10, 25 (3, 6% dietary lipid); 42–44, 60–66 (5, 12% dietary lipid)	Growth, FE, health, liver MDA	Satoh *et al.* (1987); Roem *et al.* (1990); Shiau and Shiau (2001);
Phyllokinone (K)	5.2	Growth, plasma prothrombin	Lee (2003)

CE, conditionally essential; FE, feed efficiency; AST, aspartate transaminase; AA, amino acid; MDA, malondialdehyde.
*connects data in two columns (reg and resp) indicating this response is due to liver levels.

Biotin

Biotin serves as co-enzyme in carboxylation reactions and is covalently bound to carboxylases such as acetyl-CoA carboxylase and pyruvate carboxylase (NRC, 2011; Zempleni & Kuroishi, 2012). Biotin participates in key metabolic steps in fatty acid synthesis and gluconeogenesis. The hepatic activity of the above carboxylases in hybrid tilapia was augmented in response to supplemental biotin, which was required at 0.06 mg/kg for maximum weight gain (Shiau & Chin, 1999).

Choline

Choline is an essential component of membrane phospholipids, a precursor in the biosynthesis of acetylcholine and a key component in methylation reactions. In animals, choline can be supplied by the diet and via *de novo* synthesis by methylation of phosphatidylethanolamine to phosphatidylcholine. In fish, endogenous synthesis alone may not meet physiological needs for choline, which must be supplied in the diet (NRC, 2011). Although the sparing effects of Met and betaine on the dietary choline requirement of Nile tilapia have been shown (Kasper *et al.*, 2000, 2002, respectively), Shiau and Lo, 2000 demonstrated that ~1000 mg of choline/kg of diet was required to maximize growth and body choline concentration of hybrid tilapia.

Cyanocobalamin (B_{12})

Vitamin B_{12} is the generic descriptor for all compounds containing a corrin nucleus exhibiting the biological activity of cyanocobalamin. It serves as a co-enzyme for methionine synthase and methylmalonyl-CoA mutase that play critical roles in the supply of methyl groups in methylation processes and the formation of succinyl-CoA, an intermediate in the tricarboxylic acid cycle, respectively. Although B_{12} is synthesized only by bacteria, supplementation of the vitamin in animal diets is not required because its provision is mediated by the gut microbiota in the presence of cobalt, a constituent of B_{12}.

As shown in other fish species (NRC, 2011), studies in Nile tilapia (Lovell & Limsuwan, 1982) and hybrid tilapia (Shiau & Lung, 1993) have confirmed the dietary dispensability of B_{12}.

Folic acid

Reflecting its co-enzyme role, most of the folic acid present in the body of animals is found tightly bound to enzymes. In its active form as tetrahydrofolic acid, it participates in single carbon transfer reactions in AA metabolism and the biosynthesis of purines, pyrimidines and nucleotides (NRC, 2011). In fish, signs of folic acid deficiency have included reduced growth and feed efficiency, anorexia, abnormal skin coloration and impaired haematopoiesis (NRC, 2011). Reduced weight gain and feed efficiency have been observed in hybrid tilapia fed folic acid-deficient diets, and its provision at 0.82 mg/kg in the diet was found to maximize growth (Shiau and Huang, 2001).

Myo-inositol

Myo-inositol is the biologically active form of inositol. In the form of phosphatidylinositol, it affects the structure and function of biological membranes by supporting synthesis of eicosanoids and cell signalling in the regulation of metabolism. In fish, deficiency of inositol has been shown to cause adverse effects including reduced appetite and growth, anaemia, fin erosion and dark skin coloration (NRC, 2011). Evidence from previous studies suggests that endogenous synthesis of myo-inositol occurs in some fish, including common carp (Aoe & Masuda, 1967) and channel catfish (Burtle & Lovell, 1989). In Nile tilapia, Peres *et al.* (2004) indicated that myo-inositol supplementation was not required in purified diets to support adequate growth, feed efficiency and erythropoiesis, but was required to prevent lipotropic effects in liver and muscle. Conversely, myo-inositol at 400 mg/kg was required to maximize growth and liver myo-inositol concentration in hybrid tilapia (Shiau & Su, 2005), and to maximize growth and alleviate salinity stress in Mozambique tilapia (Bu *et al.*, 2021).

Niacin

Niacin is required for the biosynthesis of nicotinamide adenine dinucleotide (NADH) and NAD phosphate (NADPH), both of which are essential co-enzymes in several oxidation-reduction reactions in carbohydrate, lipid and amino acid metabolism (NRC, 2011). Reduction in growth, feed efficiency and hepatosomatic index was observed in Nile tilapia fed niacin-deficient diets, and 20.4 and 84.6 mg/kg dietary niacin was required to maximize growth and liver niacin concentration, respectively (Jiang *et al.*, 2014). A slightly higher dietary niacin requirement of 121 mg/kg was observed in hybrid tilapia when dextrin rather than glucose was included in the diet at 38% (Shiau & Suen, 1992), suggesting that CHO complexity may affect niacin requirement.

Pantothenic acid

Pantothenic acid functions as the precursor of co-enzyme A and acyl carrier protein, which play essential roles in acyl transfer reactions in the metabolism of FAs, AAs and CHOs. Deficiency of pantothenic acid directly affects mitochondria-rich cells, with clubbed gills a classical deficiency sign of the vitamin in fish (NRC, 2011). Blue tilapia displayed reduced growth when dietary Ca-d-pantothenate was <6 mg/kg, and a minimum of 10 mg/kg was required in the diet for optimum feed efficiency and survival (Roem *et al.*, 1991). Likewise, Soliman and Wilson, 1992a, b determined that a minimum of 10 mg/kg dietary Ca-d-pantothenate was required for optimum production performance and health of blue tilapia.

Pyridoxine (B_6)

Pyridoxine is one of the three naturally occurring forms of vitamin B_6, with pyridoxal and pyridoxamine constituting the other two forms, also metabolized to a common co-enzyme form, pyridoxal phosphate. Pyridoxal phosphate is required for many enzymatic reactions involving AAs including transamination, decarboxylation and dehydration, porphyrin biosynthesis and glycogen catabolism (NRC, 2011). In red hybrid tilapia the absence of dietary pyridoxine-HCl resulted in poor growth and feed efficiency, low haematocrit and several adverse effects including anorexia, convulsions, muscle spasms and lethargy compared with fish fed diets containing ≥3 mg of pyridoxine-HCl/kg (Lim *et al.*, 1995). Probably due to the key role of B_6 in AA metabolism, its requirement appears to be affected by changes in dietary protein level. Shiau and Hsieh (1997) observed the B_6 requirement for maximum growth and hepatic alanine aminotransferase activity of hybrid tilapia to increase from 1.7 to 16.5 mg/kg and from 9.5 to 15 mg/kg as dietary protein increased from 28 to 36%. In Nile tilapia, Teixeira *et al.* (2012) found that 10 mg of pyridoxine-HCl/kg of diet was required for normal growth and health.

Riboflavin

Riboflavin is the essential component of the co-enzymes flavin mononucleotide and flavin adenine dinucleotide, both co-factors of flavoenzymes with key roles in the intermediary metabolism of CHOs, AAs and lipids and the activation of pyridoxine and folic acid to their functional forms (Combs & McClung, 2017). Red hybrid tilapia fed a riboflavin-deficient diet (<5 mg riboflavin/kg) displayed anorexia, reduced growth and nervous symptoms after 4–6 weeks of feeding (Lim *et al.*, 1993). Likewise, improved feed efficiency and maximum liver D-AA oxidase activity were observed in blue tilapia fed 6 mg of riboflavin/kg of diet (Soliman and Wilson, 1992a, b).

Thiamine

Thiamine pyrophosphate, the co-enzyme form of thiamine, is essential in CHO metabolism and neural function and is required to produce ATP, ribose, NAD and DNA (Combs & McClung, 2017). Lim *et al.* (2011) reported that dietary thiamine at 3.5 mg/kg was adequate for optimum production performance and health of Nile tilapia, whereas Ren *et al.* (2015) found that maximum growth and hepatic thiamine concentrations occurred when dietary thiamine was 1.16 and 2.06 mg/kg, respectively.

Lipid-soluble vitamins

Vitamin A

Vitamin A is the generic descriptor for compounds with the biological activity of retinol. In addition to retinol (alcohol), other basic forms of vitamin A also include retinal (aldehyde) and retinoic acid (acid). Vitamin A is essential for the maintenance of several physiological processes in animals including growth and development, vision, immunocompetence and reproduction. In fish, poor growth, high mortality, anaemia and eye lesions, including haemorrhages and retinal degeneration, are among the reported signs of vitamin A deficiency (NRC, 2011). Nile tilapia fed purified diets without vitamin A supplementation developed typical signs of deficiency including blindness, exophthalmia and reduced feed consumption and growth, which were not observed in fish fed a minimum of 5000 IU of vitamin A/kg of diet (Saleh *et al.*, 1995). These authors also noted signs of hypervitaminosis in fish receiving diets containing 10,000 IU of vitamin A. As observed in other fish species, tilapia are able to bioconvert β-carotene and other carotenoids into vitamin A in hepatocytes (Kaisuyama and Matsuno, 1988). The ability of dietary β-carotene to fulfil the vitamin A requirement (5,850–6970 IU/kg) of hybrid tilapia at a 19:1 ratio (w/w) was demonstrated by Hu *et al.* (2006).

Vitamin D

Vitamin D is the generic descriptor for all steroids exhibiting the biological activity of cholecalciferol. Natural sources of the vitamin are ergocalciferol (vitamin D_2) and cholecalciferol (vitamin D_3), which are metabolized in the liver and ultimately in the kidney to originate 1,25-(OH)2-vitamin D_3, the biologically active form of vitamin D. In its active form, vitamin D functions in concert with calcitonin and parathyroid hormone to mediate mobilization and utilization of calcium and phosphorus, being therefore critical for normal skeleton formation and development (NRC, 2011). Fish fed vitamin D-deficient diets can display reduced growth, abnormal skeleton formation and development, along with reduced ash content in whole body and skeleton. Supplementation of 2000 IU of D_3/kg in a basal diet containing 7.5 g of calcium/kg was found to improve growth and feed efficiency of blue tilapia after 12 weeks of feeding (O'Connell & Gatlin, 1994). In hybrid tilapia fed diets deficient in vitamin D, reduced growth and feed efficiency along with low haemoglobin, hepatosomatic index (HSI) and plasma alkaline phosphatase activity were observed, whereas supplementation of ~375 IU of vitamin D/kg of diet was required for maximum growth (Shiau & Hwang, 1993).

Vitamin E

Vitamin E is the general descriptor of all tocopherol and tocotrienol derivatives exhibiting the biological activity of α-tocopherol, with d-α-tocopherol having the highest biological activity among derivatives. Vitamin E is required for normal cellular metabolism and plays a critical role as a non-enzymatic antioxidant. The dietary requirement of vitamin E can be affected by antioxidant nutrients such as selenium and vitamin C, as well as by exposure to pro-oxidants such as PUFAs. Satoh *et al.* (1987) observed a decline in vitamin E concentration in the liver and whole body of Nile tilapia as dietary lipid increased from 5 to 15%, and recommended 5–10 mg of α-tocopherol/kg of diet. Roem *et al.* (1990) found 10 mg of dl-α-tocopherol acetate/kg to be required in blue tilapia diets containing 3% lipid, whereas 25 mg of the vitamin/kg was required when dietary lipid was increased to 6%. Likewise, Shiau and Shiau (2001) observed the vitamin E requirement for maximum growth and minimum liver malondialdehyde concentration of hybrid tilapia to increase from 44 to 66 mg/kg in diets containing 5 and 12% lipid, respectively. In Nile tilapia, supplementation of 50 mg of vitamin E/kg to a practical diet formulated to contain 6% lipid was found to support adequate growth and increase the survival of fish challenged with *Streptococcus iniae* (Lim *et al.*, 2011).

Vitamin K

Vitamin K is the generic descriptor for 2-methyl-1,4-naphthoquinone and all its derivatives exhibiting the anti-haemorrhagic activity of phylloquinone (Combs Jr. & McClung, 2017). Animals require vitamin K for blood clotting and bone formation through vitamin K-dependent osteocalcin, among other functions. Haemorrhagic areas in gills, eyes and vascular tissue have been observed in vitamin K-deficient fish (NRC, 2011). Hybrid tilapia fed a vitamin K-deficient diet (<5.2 mg of vitamin K/kg) for 8 weeks displayed poor growth and low plasma prothrombin concentration (Lee, 2003).

Future Perspectives

It is evident that nutrition directly influences both production performance and health of tilapia and that best results are achieved when nutritional requirements are met. However, there are still poorly understood complexities regarding dietary nutrients, antinutrients and adventitious toxins, and the onset of pathologies. With intensive production systems for tilapia becoming increasingly common worldwide, the development and efficacy validation of feed additives capable of boosting disease tolerance and improving feed efficiency are paramount. Examples include additives targeting the modulation of the gut microbiome towards beneficial taxa, immunostimulants to enhance the fish's response to pathogens, exogenous enzymes to improve ingredient/feed nutrient and energy availability and mycotoxin mitigation products to prevent mycotoxicosis from contaminated feeds. The continued efforts of multi-disciplinary researchers will certainly be key to fine-tuning diet formulations given the complexity of factors influencing animal health and performance under intensive aquaculture production.

References

Abdel-Tawwab, M. (2012) Effects of dietary protein levels and rearing density on growth, performance and stress response of Nile tilapia, *Oreochromis niloticus* (L.). *International Aquatic Research* 4(1), 1–13. DOI: 10.1186/2008-6970-4-3.

Abdel-Tawwab, M., Ahmad, M.H., Khattab, Y.A.E. and Shalaby, A.M.E. (2010) Effect of dietary protein level, initial body weight, and their interaction on the growth, feed utilization, and physiological alterations of Nile tilapia, *Oreochromis niloticus* (L.). *Aquaculture* 298(3–4), 267–274. DOI: 10.1016/j.aquaculture.2009.10.027.

Ahmad, I., Ahmed, I., Fatma, S. and Peres, H. (2021) Role of branched-chain amino acids on growth, physiology and metabolism of different fish species: A review. *Aquaculture Nutrition* 27(5), 1270–1289. DOI: 10.1111/anu.13267.

Al-Feky, S.S.A., El-Sayed, A.-F.M. and Ezzat, A.A. (2016a) Dietary taurine improves reproductive performance of Nile tilapia (*Oreochromis niloticus*) broodstock. *Aquaculture Nutrition* 22(2), 392–399. DOI: 10.1111/anu.12256.

Al-Feky, S.S.A., El-Sayed, A.-F.M. and Ezzat, A.A. (2016b) Dietary taurine enhances growth and feed utilization in larval Nile tilapia (*Oreochromis niloticus*) fed soybean meal-based diets. *Aquaculture Nutrition* 22(2), 457–464. DOI: 10.1111/anu.12266.

Al-Kholy, A., Ishak, M.M., Youssef, Y.A. and Khalil, S.R. (1970) Phosphorus uptake from water by *Tilapia zillii* (Gervais). *Hydrobiologia* 36(3–4), 471–478. DOI: 10.1007/BF00039800.

Al-Amoudi, M. A., El-Makhzangy, S. A., Azzam, M. M., & El-Sharawy, A. A. (1992). "The requirement of vitamin C for the growth and health of Nile tilapia, *Oreochromis niloticus*." Aquaculture, 104(1–2), 157–166. https://doi.org/10.1016/0044-8486(92)90118-T

Amirkolaie, A.K., Leenhouwers, J.I., Verreth, J.A.J. and Schrama, J.W. (2005) Type of dietary fibre (soluble versus insoluble) influences digestion, faeces characteristics and faecal waste production in Nile tilapia (*Oreochromis niloticus* L.). *Aquaculture Research* 36(12), 1157–1166. DOI: 10.1111/j.1365-2109.2005.01330.x.

Amirkolaie, A.K., Verreth, J.A.J. and Schrama, J.W. (2006) Effect of gelatinization degree and inclusion level of dietary starch on the characteristics of digesta and faeces in Nile tilapia (*Oreochromis niloticus* (L.). *Aquaculture* 260(1–4), 194–205. DOI: 10.1016/j.aquaculture.2006.06.039.

Andersen, S.M., Waagbø, R. and Espe, M. (2016) Functional amino acids in fish health and welfare. *Frontiers in Bioscience* 8(1), 143–169. DOI: 10.2741/757.

Aoe, H. and Masuda, I. (1967) Water-soluble vitamin requirements of carp-II. *NIPPON SUISAN GAKKAISHI* 33(7), 674–680. DOI: 10.2331/suisan.33.674.

Araújo, F.E., Michelato, M., Schemberger, M.O., Salaro, A.L., Vidal, L.V.O, *et al.* (2021) Assessment of isoleucine requirement of fast-growing Nile tilapia fingerlings based on growth performance, amino acid retention, and expression of muscle growth-related and *mTOR* genes. *Aquaculture* 539, 736645. DOI: 10.1016/j.aquaculture.2021.736645.

Azaza, M.S., Khiari, N., Dhraief, M.N., Aloui, N., Kraïem, M.M, *et al.* (2015) Growth performance, oxidative stress indices and hepatic carbohydrate metabolic enzymes activities of juvenile Nile tilapia, *Oreochromis niloticus* L., in response to dietary starch to protein ratios. *Aquaculture Research* 46(1), 14–27. DOI: 10.1111/are.12153.

Bahnasawy, M., El-Ghobashy, A. and Abdel-Hakim, N. (2009) Culture of the Nile tilapia (*Oreochromis niloticus*) in a recirculating water system using different protein levels. *Egyptian Journal of Aquatic Biology and Fisheries* 13(2), 1–15. DOI: 10.21608/ejabf.2009.2029.

Bu, X., Zhu, J., Liu, S., Wang, C., Xiao, S, *et al.* (2021) Growth, osmotic response and transcriptome response of the euryhaline teleost, *Oreochromis mossambicus* fed different myo-inositol levels under long-term salinity stress. *Aquaculture* 534. DOI: 10.1016/j.aquaculture.2020.736294.

Boonanuntanasarn, S., Kumkhong, S., Yoohat, K., Plagnes-Juan, E., Burel, C, *et al.* (2018) Molecular responses of Nile tilapia (*Oreochromis niloticus*) to different levels of dietary carbohydrates. *Aquaculture* 482, 117–123. DOI: 10.1016/j.aquaculture.2017.09.032.

Burtle, G.J. and Lovell, R.T. (1989) Lack of response of channel catfish (*Ictalurus punctatus*) to dietary myo-inositol. *Canadian Journal of Fisheries and Aquatic Sciences* 46(2), 218–222. DOI: 10.1139/f89-030.

Carneiro, W.F., Pandini, F., Silva, L.C.R, *et al.* (2017) Digestible protein requirement of Nile tilapia fed with rations based on soybean meal and corn. *Acta Scientiarum* 39(4), 343–349.

Chen, C., Sun, B., Li, X., Li, P., Guan, W, *et al.* (2013) N-3 essential fatty acids in Nile tilapia, *Oreochromis niloticus*: Quantification of optimum requirement of dietary linolenic acid in juvenile fish. *Aquaculture* 416–417, 99–104. DOI: 10.1016/j.aquaculture.2013.09.003.

Cheng, X., Tian, J., Yao, Y, *et al.* (2019) Dietary iron requirements of adult GIFT tilapia (*Oreochromis niloticus*). *Israeli Journal of Aquaculture* 71, 1–10.

Chou, B.S. and Shiau, S.Y. (1996) Optimal dietary lipid level for growth of juvenile hybrid tilapia, *Oreochromis niloticus* × *Oreochromis aureus*. *Aquaculture* 143(2), 185–195. DOI: 10.1016/0044-8486(96)01266-5.

Chun, S. S., Kim, J. W., Kim, H. C., & Lee, S. M. (2015). Thiamine requirement of Nile tilapia (*Oreochromis niloticus*) and its effects on growth, survival, and tissue thiamine levels. Aquaculture Nutrition, 21(6), 913–921. https://doi.org/10.1111/anu.12210

Combs, Jr., G.F. and McClung, J.P. (2017) *The vitamins: Fundamental aspects in nutrition and health*, 5th ed. Academic Press.

Coutinho, J.J.O., Neira, L.M., Sandre, L.C.G., Costa, J.I., Martins, M.I.E.G, *et al.* (2018) Carbohydrate-to-lipid ratio in extruded diets for Nile tilapia farmed in net cages. *Aquaculture* 497, 520–525. DOI: 10.1016/j.aquaculture.2018.08.014.

Dabrowska, H., Meyer-Burgdorff, K. and Günther, K.-D. (1989) Interaction between dietary protein and magnesium level in tilapia (*Oreochromis niloticus*). *Aquaculture* 76(3–4), 277–291. DOI: 10.1016/0044-8486(89)90081-1.

Damasceno, F.M., Fleuri, L.F., Sartori, M.M.P., Amorim, R.L., Pezzato, L.E, *et al.* (2016) Effect of dietary inorganic copper on growth performance and hematological profile of Nile tilapia subjected to heat-induced stress. *Aquaculture* 454, 257–264. DOI: 10.1016/j.aquaculture.2015.12.029.

Daudpota, A.M., Siddiqui, P.J.A., Abbas, G, *et al.* (2014) Effect of dietary protein level on growth performance, protein utilization and body composition of Nile tilapia cultured in low salinity water. *International Journal of Interdisciplinary and Multidisciplinary Studies* 2(2), 135–147.

Diógenes, A.F., Fernandes, J.B.K., Dorigam, J.C.P., Sakomura, N.K., Rodrigues, F.H.F, *et al.* (2016) Establishing the optimal essential amino acid ratios in juveniles of Nile tilapia (*Oreochromis niloticus*) by the deletion method. *Aquaculture Nutrition* 22(2), 435–443. DOI: 10.1111/anu.12262.

Dioundick, O.B. and Stom, D.I. (1990) Effects of dietary α-cellulose levels on the juveNile tilapia, *Oreochromis mossambicus* (peters). *Aquaculture* 91(3–4), 311–315. DOI: 10.1016/0044-8486(90)90196-T.

Eid, A.E. and Ghonim, S.I. (1994) Dietary zinc requirement of fingerling *Oreochromis niloticus*. *Aquaculture* 119(2–3), 259–264. DOI: 10.1016/0044-8486(94)90180-5.

El-Sayed, A.M. and Kawanna, M. (2008) Effects of dietary protein and energy levels on spawning performance of Nile tilapia (*Oreochromis niloticus*) broodstock in a recycling system. *Aquaculture* 280(1–4), 179–184. DOI: 10.1016/j.aquaculture.2008.04.030.

FAO (2022) Towards blue transformation [The State of World Fisheries and Aquaculture (SOFIA]. In: *The State of World Fisheries and Aquaculture 2022*. FAO,Rome, Italy, pp. 46–58. DOI: 10.%204060/cc0461en.

Furuya, E.W. (2010) *Tabelas brasileiras para a nutrição de tilápias/T113*. GFM.

Furuya, W.M. and Furuya, V.R.B. (2010) Nutritional innovations on amino acids supplementation in Nile tilapia diets. *Revista Brasileira de Zootecnia, 39 Suppl* 88–94. DOI: 10.1590/S1516-35982010001300010.

Gan, L., Zhou, L.-L., Li, X.-X. and Yue, Y.-R. (2016) Dietary leucine requirement of juvenile Nile tilapia, *Oreochromis niloticus*. *Aquaculture Nutrition* 22(5), 1040–1046. DOI: 10.1111/anu.12353.

Habte-Tsion, H.M., Ren, M., Liu, B., Ge, X., Xie, J, *et al.* (2016) Threonine modulates immune response, antioxidant status, and gene expressions of antioxidant enzymes and antioxidant-cytokine-related signaling molecules in juvenile blunt snout bream (*Megalobrama amblycephala*). *Fish and Shellfish Immunology* 51, 189–199. DOI: 10.1016/j.fsi.2015.11.033.

Haidar, M.N., Petie, M., Heinsbroek, L.T.N., Verreth, J.A.J. and Schrama, J.W. (2016) The effect of type of carbohydrate (starch vs. nonstarch polysaccharides) on nutrients digestibility, energy retention and maintenance requirements in Nile tilapia. *Aquaculture* 463, 241–247. DOI: 10.1016/j.aquaculture.2016.05.036.

Furuya, W.M., Santos, V.G., Silva, L.C.R., Furuya, V.R.B. and Sakaguti, E.S. (2006) Exigências de lisina digestível para juvenis de tilápia-do-nilo. *Revista Brasileira de Zootecnia* 35(3), 937–942. DOI: 10.1590/S1516-35982006000400001.

Furuya, W.M., Fujii, K.M., Santos, L.D., Silva, T.Sd.C., Silva, L.C.Rd, *et al.* (2008a) Exigência de fósforo disponível para juvenis de tilápia-do-nilo. *Revista Brasileira de Zootecnia* 37(9), 1517–1522. DOI: 10.1590/S1516-35982008000900001.

Furuya, W.M., Fujii, K.M., Santos, L.D., Silva, T.Sd.C., Silva, L.C.Rd, *et al.* (2008b) Exigência de fósforo disponível para tilápia-do-nilo (35 a 100 g). *Revista Brasileira de Zootecnia* 37(6), 961–966. DOI: 10.1590/S1516-35982008000600001.

He, J.Y., Long, W.Q., Han, B., Tian, L.X., Yang, H.J, *et al.* (2017) Effect of dietary L-methionine concentrations on growth performance, serum immune and antioxidative responses of juvenile Nile tilapia, *Oreochromis niloticus. Aquaculture Research* 48(2), 665–674. DOI: 10.1111/are.12913.

Hu, C.J., Chen, S.M., Pan, C.H. and Huang, C.H. (2006) Effects of dietary vitamin A and β-carotene concentrations on growth of juvenile hybrid tilapia, *Oreochromis niloticus* × *O. aureus. Aquaculture* 253(1–4), 602–607. DOI: 10.1016/j.aquaculture.2005.09.003.

Huang, F., Jiang, M., Wen, H., Wu, F., Liu, W, *et al.* (2015) Dietary zinc requirement of adult Nile tilapia (*Oreochromis niloticus*) fed semi-purified diets, and effects on tissue mineral composition and antioxidant responses. *Aquaculture* 439, 53–59. DOI: 10.1016/j.aquaculture.2015.01.018.

Huang, F., Jiang, M., Wen, H., Wu, F., Liu, W, *et al.* (2016) Dietary vitamin C requirement of genetically improved farmed tilapia, *Oreochromis niloticus. Aquaculture Research* 47(3), 689–697. DOI: 10.1111/are.12527.

Jauncey, K. and Ross, B. (1982) *A guide to tilapia feeds and feeding.* Institute of Aquaculture, University of Stirling.

Jiang, L.H., Wu, H.Y., Huang, K., Ma, Y., Yang, Q, *et al.* (2013) Effects of dietary carbohydrate levels on growth performance and liver metabolism functions of juvenile tilapia (*Oreochromis niloticus*). *Journal of Fisheries of China* 37(2), 245–255. DOI: 10.3724/SP.J.1231.2013.38329.

Jiang, M., Huang, F., Wen, H., Yang, C., Wu, F, *et al.* (2014) Dietary niacin requirement of GIFT tilapia, *Oreochromis niloticus*, reared in freshwater.

Journal of the World Aquaculture Society 45(3), 333–341. DOI: 10.1111/jwas.12119.

Kabir, K.A., Verdegem, M.C.J., Verreth, J.A.J., Phillips, M.J. and Schrama, J.W. (2020) Effect of dietary carbohydrate to lipid ratio on performance of Nile tilapia and enhancement of natural food in pond aquaculture. *Aquaculture Research* 51(5), 1942–1954. DOI: https://doi.org/10.1111/are.14546.

Kaisuyama, M. and Matsuno, T. (1988) Carotenoid and vitamin A, and metabolism of carotenoids, β-carotene, canthaxanthin, astaxanthin, zeaxanthin, lutein, and tunaxanthin in tilapia tilapia nilotica. *Comparative Biochemistry and Physiology Part B* 90(1), 131–139. DOI: 10.1016/0305-0491(88)90049-1.

Kanazawa, A., Teshima, S.I., Sakamoto, M. and Awal, M.A. (1980) Requirements of *tilapia zillii* for essential fatty acids. *Nippon Suisan Gakkaishi* 46(11), 1353–1356. DOI: 10.2331/suisan.46.1353.

Kasper, C.S., White, M.R. and Brown, P.B. (2000) Choline is required by tilapia when methionine is not in excess. *Journal of Nutrition* 130(2), 238–242. DOI: 10.1093/jn/130.2.238.

Kasper, C.S., White, M.R. and Brown, P.B. (2002) Betaine can replace choline in diets for juvenile Nile tilapia, *Oreochromis niloticus. Aquaculture* 205(1–2), 119–126. DOI: 10.1016/S0044-8486(01)00658-5.

Koch, J.F.A., Barros, M.M., Teixeira, C.P., Carvalho, P.L.P.F., Fernandes Junior, A.C, *et al.* (2017) Protein-to-energy ratio of 21.43 MJ^{-1} improves growth performance of Nile tilapia at the final rearing stage under commercially intensive rearing conditions. *Aquaculture Nutrition* 23(3), 560–570. DOI: 10.1111/anu.12423.

Kpundeh, M.D., Qiang, J., He, J., Yang, H. and Xu, P. (2015) Effects of dietary protein levels on growth performance and haemato-immunological parameters of juvenile Genetically Improved Farmed Tilapia (GIFT). *Oreochromis niloticus. Aquaculture International* 23(5), 1189–1201. DOI: 10.1007/s10499-014-9876-1.

Lall, S.P. and Milley, J.E. (2008) Trace mineral requirements of fish and crustaceans. In: Schlegel, P., Durosoy, S. and Jongbloed, A.W. (eds) *Trace Elements in Animal Production Systems*. Wageningen Academic Publishers, pp. 203–214.

Larumbe-Morán, E., Hernández-Vergara, M.P., Olvera-Novoa, M.A. and Pérez Rostro, C.I.P. (2010) Protein requirements of Nile tilapia (*Oreochromis niloticus*) fry cultured at different salinities. *Aquaculture Research* 41, 1150–1157. DOI: 10.1111/j.1365-2109.2009.02402.x.

Lee, J.Y. (2003) Vitamin K requirements of juvenile hybrid tilapia (*Oreochromis niloticus* × *O. aureus*) and grouper (*Epinephelus malabaricus*). MSc Thesis, National Taiwan Ocean University.

Lee, S., Nambi, R.W., Won, S., Katya, K. and Bai, S.C. (2016) Dietary selenium requirement and toxicity levels in juvenile Nile tilapia, *Oreochromis niloticus. Aquaculture* 464, 153–158. DOI: 10.1016/j.aquaculture.2016.06.027.

Li, M., Hu, F.C., Qiao, F., Du, Z.Y. and Zhang, M.L. (2020) Sodium acetate alleviated high-carbohydrate induced intestinal inflammation by suppressing MAPK and NF-κB signaling pathways in Nile tilapia (*Oreochromis niloticus*). *Fish and Shellfish Immunology* 98, 758–765. DOI: 10.1016/j.fsi.2019.11.024.

Li, P., Mai, K., Trushenski, J. and Wu, G. (2009) New development in fish amino acid nutrition: towards functional and environmentally oriented aquafeeds. *Amino Acids* 37(1), 43–53. DOI: 10.1007/s00726-008-0171-1.

Li, Y., Bordinhon, A.M., Allen Davis, D.A., Zhang, W. and Zhu, X. (2013) Protein: Energy ratio in practical diets for Nile tilapia *Oreochromis niloticus*. *Aquaculture International* 21(5), 1109–1119. DOI: 10.1007/s10499-012-9616-3.

Lim, C., LeaMaster, B.R. and Brock, J.A. (1993) Riboflavin requirement of fingerling red hybrid tilapia grown in seawater. *Journal of the World Aquaculture Society* 24(4), 451–458. DOI: 10.1111/j.1749-7345.1993.tb00573.x.

Lim, C., LeaMaster, B.R. and Brock, J.A. (1995) Pyridoxine requirement of fingerling red hybrid tilapia grown in seawater. *Journal of Applied Aquaculture* 5(2), 49–60. DOI: 10.1300/J028v05n02_05.

Lim, C., Yildirim-Aksoy, M., Barros, M.M. and Klesius, P. (2011) Thiamin requirement of Nile tilapia, *Oreochromis niloticus*. *Journal of the World Aquaculture Society* 42(6), 824–833. DOI: 10.1111/j.1749-7345.2011.00531.x.

Lin, Y. C., Shiau, S. Y., & Hsieh, S. L. (2008). Dietary manganese requirement of hybrid tilapia (*Oreochromis niloticus* × *Oreochromis aureus*) fed practical diets. *Aquaculture*, 278(1-4), 206–210. https://doi.org/10.1016/j.aquaculture.2008.03.009

Lin, Y.H., Lin, S.M. and Shiau, S.Y. (2008b) Dietary manganese requirements of juvenile tilapia, *Oreochromis niloticus* × *Oreochromis aureus*. *Aquaculture* 284(1–4), 207–210. DOI: 10.1016/j.aquaculture.2008.07.049.

Lin, Y.H., Ku, C.Y. and Shiau, S.Y. (2013) Estimation of dietary magnesium requirements of juvenile tilapia, *Oreochromis niloticus* × *Oreochromis aureus*, reared in freshwater and seawater. *Aquaculture* 380–383, 47–51. DOI: 10.1016/j.aquaculture.2012.11.034.

Liu, H.Y., Chen, Q., Tan, B.P., Dong, X., Chi, S, *et al.* (2018) Effects of dietary carbohydrate levels on growth, glucose tolerance, glucose homeostasis and GLUT4 gene expression in tilapia nilotica. *Aquaculture Research* 49(12), 3735–3745. DOI: 10.1111/are.13841.

Lovell, R.T. and Limsuwan, T. (1982) Intestinal synthesis and dietary nonessentiality of vitamin B_{12} for tilapia nilotica. *Transactions of the American Fisheries Society* 111(4), 485–490. DOI: 10.1577/1548-8659(1982)1112.0.CO;2.

Maas, R.M., Verdegem, M.C.J., Wiegertjes, G.F. and Schrama, J.W. (2020) Carbohydrate utilisation by tilapia: A meta-analytical approach. *Reviews in Aquaculture* 12(3), 1851–1866. DOI: 10.1111/raq.12413.

McClain, W.R. and Gatlin III, D.M. (1988) Dietary zinc requirement of *Oreochromis aureus* and effects of dietary calcium and phytate on zinc bioavailability. *Journal of the World Aquaculture Society* 19(3), 103–108. DOI: 10.1111/j.1749-7345.1988.tb00936.x.

Mehler, A.H. (1986) Amino acid metabolism II: Metabolism of the individual amino acids. In: Devlin, T.M. (ed.) *Textbook of Biochemistry: With Clinical Correlations*. John Wiley & Sons Inc., USA, pp. 462–464.

Meurer, F., Hayashi, C. and Boscolo, W.R. (2003) Fibra bruta para alevinos de tilápia do nilo (*Oreochromis niloticus* L.). *Revista Brasileira de Zootecnia* 32(2), 256–261. DOI: 10.1590/S1516-35982003000200002.

Michelato, M., Dantas, P. E. L., & Kessler, A. M. (2016). Lysine in tilapia diets: Effects on fillet yield and protein retention. *Aquaculture*, 450, 68–75. https://doi.org/10.1016/j.aquaculture.2015.07.030

Michelato, M., Vidal, L.V.O., Xavier, T.O., Graciano, T.S., Moura, L.B, *et al.* (2016b) Dietary threonine requirement to optimize protein retention and fillet production of fast-growing Nile tilapia. *Aquaculture Nutrition* 22(4), 759–766. DOI: 10.1111/anu.12293.

Michelato, M., Zaminhan, M., Boscolo, W.R., Nogaroto, V., Vicari, M, *et al.* (2017) Dietary histidine requirement of Nile tilapia juveniles based on growth performance, expression of muscle-growth-related genes and haematological responses. *Aquaculture* 467, 63–70. DOI: 10.1016/j.aquaculture.2016.06.038.

Michelato, M., Furuya, W.M. and Gatlin III, D.M. (2018) Metabolic responses of Nile tilapia *Oreochromis niloticus* to methionine and taurine supplementation. *Aquaculture* 485, 66–72. DOI: 10.1016/j.aquaculture.2017.11.003.

Moreau, R. and Dabrowski, K. (2001) Gulonolactone oxidade presense in fishes: activity and significance. In: Dabrowski, K. (ed.) *Ascorbic Acid in Aquatic Organisms – Status and Perspectives*. CRC Press, pp. 13–95.

Nascimento, T.M.T., Mansano, C.F.M., Peres, H, *et al.* (2020) Determination of the optimum dietary essential amino acid profile for growing phase of Nile tilapia by the deletion method. *Aquaculture* 523, 735204.

National Research Council (NRC). (2011) *Nutrient requirements of fish and shrimp*. National Academies Press.

Neu, D., Boscolo, W., Zaminhan, M., Almeida, F., Sary, C, *et al.* (2016) Growth performance, biochemical responses, and skeletal muscle development of juvenile Nile tilapia, *Oreochromis niloticus*, fed with increasing levels of arginine. *Journal of the World Aquaculture Society* 47(2), 248–259. DOI: 10.1111/jwas.12262.

Ng, W.K. and Hanim, R. (2007) Performance of genetically improved nile tilapia compared with red hybrid tilapia fed diets containing two protein levels. *Aquaculture Research* 38(9), 965–972. DOI: 10.1111/j.1365-2109.2007.01758.x.

Ng, W.K. and Romano, N. (2013) A review of the nutrition and feeding management of farmed tilapia throughout the culture cycle. *Reviews in Aquaculture* 5(4), 220–254. DOI: 10.1111/raq.12014.

Nguyen, T.N. and Davis, D.A. (2009) Re-evaluation of total sulphur amino acid requirement and determination of replacement value of cysteine for methionine in semi-purified diets of juvenile Nile tilapia, *Oreochromis niloticus*. *Aquaculture Nutrition* 15(3), 247–253. DOI: 10.1111/j.1365-2095.2008.00589.x.

Nguyen, L., Salem, S.M.R., Salze, G.P., Dinh, H. and Allen Davis, D. (2019) Tryptophan requirement in semi-purified diets of juvenile Nile tilapia *Oreochromis niloticus*. *Aquaculture* 502, 258–267. DOI: 10.1016/j.aquaculture.2018.12.049.

Ning, L., Tan, Y., Wang, W.X., Wu, S., Chen, F, *et al.* (2020) Optimum selenium requirement of juvenile Nile tilapia, *Oreochromis niloticus*. *Aquaculture Nutrition* 26(2), 528–535. DOI: 10.1111/anu.13014.

Nunes, A.J.P., Sá, M.V.C., Browdy, C.L. and Vazquez-Anon, M.V. (2014) Practical supplementation of shrimp and fish feeds with crystalline amino acids. *Aquaculture* 431, 20–27. DOI: 10.1016/j.aquaculture.2014.04.003.

O'Connell, J. P., & Gatlin III, D. M (1994) Effects of dietary calcium and vitamin D_3 on weight gain and mineral composition of the blue tilapia (*Oreochromis aureus*) in low-calcium water. *Aquaculture* 125(1–2), 107–117. DOI: 10.1016/0044-8486(94)90287-9.

Ogungi, O., & Wirth, E. (1999). Protein requirements of Nile tilapia (*Oreochromis niloticus*) in intensive culture systems. Aquaculture International, 7(1), 25-34. https://doi.org/10.1023/A:1009244902823

Ovie, S. I., & Eze, J. U. (2013). Lysine requirement for growth and protein efficiency ratio in tilapia: A comparative study. Journal of Aquaculture Research & Development, 4(2), 256–265

Peres, H., Lim, C. and Klesius, P.H. (2004) Growth, chemical composition and resistance to streptococcus iniae challenge of juvenile Nile tilapia (*Oreochromis niloticus*) fed graded levels of dietary inositol. *Aquaculture* 235(1–4), 423–432. DOI: 10.1016/j.aquaculture.2003.09.021.

Pezzato, L.E., Rosa, M.J.S., Barros, M.M. and Guimarães, I.G. (2006) Exigência de fósforus disponível para alevinos de tilápia do nilo. *Ciência Rural* 36(5), 1600–1605. DOI: 10.1590/S0103-84782006000500041.

Phromkunthong, W. and Udom, U. (2008) Phosphorus requirement of sex-reversed red tilapia fed all-plant diets. *Songklanakarin Journal of Science and Technology* 30(1), 7–16.

Prabu, P., Sahoo, S. K., Kumar, V., & Chakraborty, S. (2020a). Dietary lysine levels for optimal growth and nitrogen retention in tilapia. *Aquaculture Research*, 51(5), 1697–1705. https://doi.org/10.1111/are.14665

Qiang, J., Yang, H., Wang, H., Kpundeh, M.D. and Xu, P. (2012) Growth and IGF-I response of juvenile Nile tilapia (*Oreochromis niloticus*) to changes in water temperature and dietary protein level. *Journal of Thermal Biology* 37(8), 686–695. DOI: 10.1016/j.jtherbio.2012.07.009.

Reigh, R.C., Robinson, E.H. and Brown, P.B. (1991) Effects of dietary magnesium on growth and tissue magnesium content of blue tilapia, *Oreochromis aureus*. *Journal of the World Aquaculture Society* 22(3), 192–200. DOI: 10.1111/j.1749-7345.1991.tb00734.x.

Ren, C., Wen, H., Huang, F., Jiang, M., Wu, F. *et al.* (2015) Dietary thiamine requirement of GIFT *Oreochromis niloticus*. *Journal of Fisheries of China* 39(4), 539–546.

Robinson, E.H., Rawles, S.D., Yette, H.E. and Greene, L.W. (1984) An estimate of the dietary calcium requirement of fingerling tilapia aurea reared in calcium-free water. *Aquaculture* 41(4), 389–393. DOI: 10.1016/0044-8486(84)90206-0.

Robinson, E.H., Labomascus, D., Brown, P.B. and Linton, T.L. (1987) Dietary calcium and phosphorus requirement of *Oreochromis aureus* reared in calcium-free water. *Aquaculture* 64(4), 267–276. DOI: 10.1016/0044-8486(87)90189-X.

Roem, A.J., Kohler, C.C. and Stickney, R.R. (1990) Vitamin E requirements of the blue tilapia, *Oreochromis aureus* (*Steindachner*), in relation to dietary lipid level. *Aquaculture* 87(2), 155–164. DOI: 10.1016/0044-8486(90)90272-O.

Roem, A.J., Stickney, R.R. and Kohler, C.C. (1991) Dietary pantothenic acid requirement of blue tilapia. *Progressive Fish-Culturist* 53(4), 216–219. DOI: 10.1577/1548-8640(1991)0532.3.CO;2.

Sá, M. V., Furuya, W. M., & Oliveira, L. G. (2005). Zinc requirement of Nile tilapia (*Oreochromis niloticus*) fed soybean meal-based diets. Aquaculture, 243(1-4), 145–151. https://doi.org/10.1016/j.aquaculture.2004.09.014

Sá, M.V., Pezzato, L.E., Ferreira Lima, M.M.B. and Magalhães Padilha, P. (2004) Optimum zinc supplementation level in Nile tilapia *Oreochromis niloticus* juveniles diets. *Aquaculture* 238(1–4), 385–401. DOI: 10.1016/j.aquaculture.2004.06.011.

Saleh, G., Eleraky, W. and Gropp, J.M. (1995) A short note on the effects of vitamin A hypervitaminosis and hypovitaminosis on health and growth of tilapia nilotica (*Oreochromis niloticus*). *Journal of Applied Ichthyology* 11(3–4), 382–385. DOI: 10.1111/j.1439-0426.1995.tb00046.x.

Santiago, C.B. and Lovell, R.T. (1988) Amino acid requirements for growth of Nile tilapia. *Journal of Nutrition* 118(12), 1540–1546. DOI: 10.1093/jn/118.12.1540.

Satoh, S., Takeuchi, T. and Watanabe, T. (1987) Requirement of tilapia for α-tocopherol. *Nippon Suisan Gakkaishi* 53(1), 119–124. DOI: 10.2331/suisan.53.119.

Schamber, C.R., Boscolo, W.R., Natali, M.R.M., Michelato, M., Furuya, V.R.B, *et al.* (2014) Growth performance and bone mineralization of large Nile tilapia (*Oreochromis niloticus*) fed graded levels of available phosphorus. *Aquaculture International* 22(5), 1711–1721. DOI: 10.1007/s10499-014-9776-4.

Scott, M.L., Nesheim, M.C. and Young, R.J. (1982) *Nutrition of the chicken.* Scott and Associates.

Shiau, S.-Y. and Tseng, H.-C. (2007) Dietary calcium requirements of juvenile tilapia, *Oreochromis niloticus* × *O. aureus*, reared in fresh water. *Aquaculture Nutrition* 13(4), 298–303. DOI: 10.1111/j.1365-2095.2007.00481.x.

Shiau, S.Y. and Hsieh, H.L. (1997) Vitamin B_6 requirements of tilapia *Oreochromis niloticus* × O. aureus fed two dietary protein concentrations. *Fisheries Science* 63(6), 1002–1007. DOI: 10.2331/fishsci.63.1002.

Shiau, S.Y. (2002) Tilapia, *Oreochromis* spp. In: Webster, C.D. and Lim, C. (eds) *Nutrient Requirements and Feeding of Finfish for Aquaculture.* CABI Publishing, pp. 273–292.

Shiau, S.Y. and Chin, Y.H. (1999) Estimation of the dietary biotin requirement of juvenile hybrid tilapia, *Oreochromis niloticus* × O. aureus. *Aquaculture* 170(1), 71–78. DOI: 10.1016/S0044-8486(98)00391-3.

Shiau, S. Y., & Lung, S. C. (1993). Vitamin B12 requirement for the growth of tilapia, *Oreochromis niloticus* × O. aureus. *Aquaculture,* 115(1–2), 181–185. https://doi.org/10.1016/0044-8486(93)90222-O

Shiau, S.Y. and Hsieh, J.F. (2001) Quantifying the dietary potassium requirement of juvenile hybrid tilapia (*Oreochromis niloticus* × O. aureus). *British Journal of Nutrition* 85(2), 213–218. DOI: 10.1079/bjn2000245.

Shiau, S.Y. and Huang, S.Y. (2001) Dietary folic acid requirement for maximum growth of juvenile tilapia *Oreochromis niloticus* × O. aureus. *Fisheries Science* 67(4), 655–659. DOI: 10.1046/j.1444-2906.2001.00302.x.

Shiau, S.Y. and Hwang, J.Y. (1993) Vitamin D requirements of juvenile hybrid tilapia *Oreochromis niloticus* × O. aureus. *Nippon Suisan Gakkaishi* 59(3), 553–558. DOI: 10.2331/suisan.59.553.

Shiau, S.Y. and Lin, S.F. (1993) Effect of supplemental dietary chromium and vanadium on the utilization of different carbohydrates in tilapia, *Oreochromis niloticus* × O. aureus. *Aquaculture* 110(3–4), 321–330. DOI: 10.1016/0044-8486(93)90379-D.

Shiau, S.Y. and Lo, P.S. (2000) Dietary choline requirements of juvenile hybrid tilapia, *Oreochromis niloticus* × O. aureus. *Journal of Nutrition* 130(1), 100–103. DOI: 10.1093/jn/130.1.100.

Shiau, S.Y. and Lu, L.S. (2004) Dietary sodium requirement determined for juvenile hybrid tilapia (*Oreochromis niloticus* × O. aureus) reared in fresh water and seawater. *British Journal of Nutrition* 91(4), 585–590. DOI: 10.1079/BJN20041091.

Shiau, S.Y. and Ning, Y.C. (2003) Estimation of dietary copper requirements of juvenile tilapia, *Oreochromis niloticus* × *Oreochromis aureus. Animal Science* 77(2), 287–292. DOI: 10.1017/S1357729800059026.

Shiau, S.Y. and Peng, C.Y. (1993) Protein-sparing effect by carbohydrates in diets for tilapia, *Oreochromis niloticus* × O. aureus. *Aquaculture* 117(3–4), 327–334. DOI: 10.1016/0044-8486(93)90329-W.

Shiau, S.Y. and Shiau, L.F. (2001) Re-evaluation of the vitamin E requirements of juvenile tilapia (*Oreochromis niloticus* × O. aureus). *Animal Science* 72(3), 529–534. DOI: 10.1017/S135772980005205X.

Shiau, S.Y. and Shy, S.M. (1998) Dietary chromic oxide inclusion level required to maximize glucose utilization in hybrid tilapia, *Oreochromis niloticus* × O. aureus. *Aquaculture* 161(1–4), 357–364. DOI: 10.1016/S0044-8486(97)00283-4.

Shiau, S.Y. and Su, L.W. (2003) Ferric citrate is half as effective as ferrous sulfate in meeting the iron requirement of juvenile tilapia, *Oreochromis niloticus* × *Oreochromis aureus*. *Journal of Nutrition* 133(2), 483–488. DOI: 10.1093/jn/133.2.483.

Shiau, S.Y. and Su, S.L. (2005) Juvenile tilapia (*Oreochromis niloticus* × *Oreochromis aureus*) requires dietary myo-inositol for maximal growth. *Aquaculture* 243(1–4), 273–277. DOI: 10.1016/j.aquaculture.2004.10.002.

Shiau, S.Y. and Suen, G.S. (1992) Estimation of the niacin requirement of tilapia fed diets containing glucose or dextrin. *Journal of Nutrition* 122(10), 2030–2036. DOI: 10.1093/jn/122.10.2030.

Shi, W.Q., Sun, Y.X., Luo, L., Jing, T., Chen, Y.J. *et al.* (2014) Dietary leucine requirement of tilapia (GIFT *Oreochromis niloticus*). *In Chinese). Journal of Fisheries of China* 10, 1778–1786.

Singha, K.P., Shamna, N., Sahu, N.P., Sardar, P., HariKrishna, V, *et al.* (2020) Feeding graded levels of protein to genetically improved farmed tilapia (GIFT) juveniles reared in inland saline water: Effects on growth and gene expression of IGF-I, IGF-IR, and IGF-BPI. *Aquaculture* 525, 735306. DOI: 10.1016/j.aquaculture.2020.735306.

Singha, K.P., Shamna, N., Sahu, N.P., Sardar, P., Harikrishna, V, *et al.* (2021) Optimal dietary protein for culture of Genetically Improved Farmed Tilapia (GIFT), *Oreochromis niloticus* (linnaeus, 1758) juveniles in low saline water: Effects on growth, metabolism and gene expression. *Animal Feed Science and Technology* 271, 114713. DOI: 10.1016/j.anifeedsci.2020.114713.

Soliman, A.K. and Wilson, R.P. (1992a) Water-soluble vitamin requirements of tilapia. 1 pantothenic acid requirement of blue tilapia, *Oreochromis aureus*. *Aquaculture* 104(1–2), 121–126. DOI: 10.1016/0044-8486(92)90142-8.

Soliman, A.K. and Wilson, R.P. (1992b) Water-soluble vitamin requirements of tilapia. 2. riboflavin requirement of blue tilapia, *Oreochromis aureus*. *Aquaculture* 104(3–4), 309–314. DOI: 10.1016/0044-8486(92)90212-4.

Song, S.K., Beck, B.R., Kim, D., Park, J., Kim, J, *et al.* (2014) Prebiotics as immunostimulants in aquaculture. A review. *Fish and Shellfish Immunology* 40(1), 40–48. DOI: 10.1016/j.fsi.2014.06.016.

Stoneham, T.R., Kuhn, D.D., Taylor, D.P., Neilson, A.P., Smith, S.A, *et al.* (2018) Production of omega-3 enriched tilapia through the dietary use of algae meal or fish oil: improved nutrient value of fillet and offal. *PLOS ONE* 13(4), e0194241. DOI: 10.1371/journal.pone.0194241.

Sumi, K., Das, M. and Siddika, I. (2011) Effect of different protein levels of fry feed on the production of quality tilapia (*Oreochromis niloticus*) fry. *Journal of the Bangladesh Agricultural University* 9(2), 365–374. DOI: 10.3329/jbau.v9i2.11053.

Tacon, A.G.J. and Cowey, C.B. (1985) Protein and amino acid requirements. In: Tytler, P. and Calow, P. (eds) *Fish Energetics*. Springer, pp. 155–183.

Takeuchi, T., Satoh, S. and Watanabe, T. (1983) Requirement of tilapia nilotica for essential fatty acids. *Nippon Suisan Gakkaishi* 49(7), 1127–1134. DOI: 10.2331/suisan.49.1127.

Teixeira, C.P., Barros, M.M., Pezzato, L.E., Fernandes, A.C., Koch, J.F.A, *et al.* (2012) Growth performance of Nile tilapia, *Oreochromis niloticus*, fed diets containing levels of pyridoxine and haematological response under heat stress. *Aquaculture Research* 43(8), 1081–1088. DOI: 10.1111/j.1365-2109.2011.02911.x.

Twibell, R.G. and Brown, P.B. (1998) Optimal dietary protein concentration for hybrid tilapia (*Oreochromis niloticus* × *Oreochromis aureus*) fed all-plant diets. *Journal of the World Aquaculture Society* 29(1), 9–16. DOI: 10.1111/j.1749-7345.1998.tb00294.x.

Vianna, R.A., Chideroli, R.T., Costa, A.R., Ribeiro Filho, O.P., Oliveira, L.L, *et al.* (2020) Effect of experimental arginine supplementation on the growth, immunity, and resistance of tilapia fingerlings to *Streptococcus agalactiae*. *Aquaculture Research* 51(3), 1276–1283. DOI: 10.1111/are.14478.

Wade, A.M. and Tucker, H.N. (1998) Antioxidant characteristics of L-histidine 11the work described in this manuscript was partially sponsored and funded by cytos pharmaceuticals, LLC. *Journal of Nutritional Biochemistry* 9(6), 308–315. DOI: 10.1016/S0955-2863(98)00022-9.

Wang, Y., Liu, Y.J., Tian, L.X., Du, Z., Wang, J, *et al.* (2005) Effects of dietary carbohydrate level on growth and body composition of juvenile tilapia, *Oreochromis* × *O. aureus*. *Aquaculture Research* 36(14), 1408–1413. DOI: 10.1111/j.1365-2109.2005.01361.x.

Wen, H., Feng, F., Jiang., W, *et al.* (2014) Dietary tryptophan modulates intestinal immune response, barrier function, antioxidant status and gene expression of TOR and nrf2 in young grass carp (*Ctenopharyngodon idella*). *Fish & Shellfish Immunology* 40, 1–12. DOI: 10.1016/j.fsi.2014.07.004.

Winfree, R.A. and Stickney, R.R. (1981) Effects of dietary protein and energy on growth feed conversion efficiency and body composition of tilapia aurea. *Journal of Nutrition* 111(6), 1001–1012. DOI: 10.1093/jn/111.6.1001.

Wu, L., Liang, H., Hamunjo, C.M.K., Ge, X., Ji, K, *et al.* (2021) Culture salinity alters dietary protein requirement, whole body composition and nutrients metabolism related genes expression in juvenile genetically improved farmed tilapia (GIFT) (*Oreochromis niloticus*). *Aquaculture* 531, 735961. DOI: 10.1016/j.aquaculture.2020.735961.

Wu, W.Y., Jiang, M., Liu, W, *et al.* (2016) Dietary arginine requirement of genetically improved farmed tilapia (*Oreochromis niloticus*). *Chinese Journal of Animal Nutrition* 28(5), 1412–1424.

Wu, Y., Zhang, L., Teng, T, *et al.* (2018) Effect of dietary tryptophan levels on growth and hemato-immunological parameters in juvenile Nile tilapia *Oreochromis niloticus*. *Israeli Journal of Aquaculture* 70, 1–10.

Xiao, W., Li, D.Y., Zhu, J.L., Zou, Z.Y., Yue, Y.R, *et al.* (2018) Dietary valine requirement of juvenile Nile tilapia, *Oreochromis niloticus*. *Aquaculture Nutrition* 24(1), 315–323. DOI: 10.1111/anu.12562.

Xiao, W., Zou, Z., Li, D., Zhu, J., Yue, Y, *et al.* (2020) Effect of dietary phenylalanine level on growth performance, body composition, and biochemical parameters in plasma of juvenile hybrid tilapia, oreochromis × *Oreochromis aureus*. *Journal of the World Aquaculture Society* 51(2), 437–451. DOI: 10.1111/jwas.12641.

Xie, D., Yang, L., Yu, R., Chen, F., Lu, R, *et al.* (2017) Effects of dietary carbohydrate and lipid levels on growth and hepatic lipid deposition of juvenile tilapia, *Oreochromis niloticus*. *Aquaculture* 479, 696–703. DOI: 10.1016/j.aquaculture.2017.07.013.

Yao, Y.F., Jiang, M., Wen, H., Wu, F., Liu, W, *et al.* (2014) Dietary phosphorus requirement of GIFT strain of Nile tilapia *Oreochromis niloticus* reared in freshwater. *Aquaculture Nutrition* 20(3), 273–280. DOI: 10.1111/anu.12075.

Yue, Y., Zou, Z., Zhu, J., Li, D., Xiao, W, *et al.* (2014) Dietary threonine requirement of juvenile Nile tilapia, *Oreochromis niloticus*. *Aquaculture International* 22(4), 1457–1467. DOI: 10.1007/s10499-014-9759-5.

Yue, Y., Zou, Z., Zhu, J., Li, D., Xiao, W, *et al.* (2015) Effects of dietary arginine on growth performance, feed utilization, haematological parameters and non-specific immune responses of juvenile Nile tilapia (*Oreochromis niloticus* L. *Aquaculture Research* 46(8), 1801–1809. DOI: 10.1111/are.12333.

Zaminhan, M., Boscolo, W.R., Neu, D.H., Feiden, A., Barriviera Furuya, V.R, *et al.* (2017) Dietary tryptophan requirements of juvenile Nile tilapia fed corn-soybean meal-based diets. *Animal Feed Science and Technology* 227, 62–67. DOI: 10.1016/j.anifeedsci.2017.03.010.

Zaminhan, M., Michelato, M., Furuya, V.R.B., Boscolo, W.R., Araújo, F.E, *et al.* (2018) Total and available tryptophan requirement of Nile tilapia, *Oreochromis niloticus*, fingerlings. *Aquaculture Nutrition* 24(5), 1553–1562. DOI: 10.1111/anu.12792.

Zaminhan-Hassemer, M., Michelato, M., Boscolo, W.R., Urbich, A.V., Cruz, T.Pd, *et al.* (2020) Dietary histidine requirement of grow-out Nile tilapia (*Oreochromis niloticus*), based on growth performance, muscle development, expression, of muscle-growth-related genes, and blood parameters. *Revista Brasileira de Zootecnia* 49, e20180210. DOI: 10.37496/rbz4920180210.

Zehra, S. and Khan, M.A. (2017) Dietary leucine requirement for growth and maintenance, and its utilization efficiency for fingerling of Nile tilapia,*Oreochromis niloticus* (linnaeus). *Borneo Journal of Marine Science and Aquaculture* 1, 16–24. DOI: 10.51200/bjomsa.v1i0.986.

Zehra, S. and Yousif, R.A. (2021) Dietary total aromatic amino acid requirement and tyrosine replacement value of phenylalanine for fingerling *Oreochromis niloticus* (linnaeus). *Aquaculture Nutrition* 27(4), 1009–1018. DOI: 10.1111/anu.13242.

Zempleni, J. and Kuroishi, T. (2012) Biotin. *Advances in Nutrition* 3(2), 213–214. DOI: 10.3945/an.111.001305.

Zeng, N.N., Jiang, M., Wen, H., Liu, W., Wu, F, *et al.* (2021) Effects of water temperatures and dietary protein levels on growth, body composition and blood biochemistry of juvenile GIFT tilapia (*Oreochromis niloticus*). *Aquaculture Nutrition* 27(1), 240–251. DOI: 10.1111/anu.13181.

Zeng, C., Sun, M., & Liao, R. (2009). Effect of dietary protein levels on the growth and feed utilization of hybrid tilapia (*Oreochromis niloticus* × *Oreochromis aureus*). *Aquaculture*, 292(1-2), 86-90. https://doi.org/10.1016/j.aquaculture.2009.03.020

Zhu, J., Pan, J., Wang, X., Huang, Y., Qin, C. *et al.* (2020) Alleviation of the adverse effect of dietary carbohydrate by supplementation of myo-inositol to the diet of Nile tilapia (*Oreochromis niloticus*). *Animals: An Open Access Journal from MDPI* 10(11), 2190. DOI: 10.3390/ani10112190.

Chapter 7

Tilapia immunology and immunostimulation

Kim D. Thompson

Introduction

Like other teleost species and higher vertebrates, tilapia has a sophisticated immune system comprising two main components: innate and adaptive immunity (Smith *et al.*, 2019). Innate immunity is the first line of defence, offering broad-spectrum protection against various pathogens immediately or relatively soon upon infection (Ellis, 2001; Riera Romo *et al.*, 2016). Following this, the adaptive immune system is activated, targeting specific pathogens through antigen presentation facilitated by the innate response and establishing a memory for quicker and more efficient responses to future exposure by the pathogen. While there is a significant overlap in the components of innate and adaptive immune responses across fish species, certain aspects are unique to each (Smith *et al.*, 2019).

Despite initially being considered highly resilient and adaptable to farming conditions, tilapia is now known to be susceptible to a broad spectrum of infectious agents, including bacteria such as *Streptococcus* spp., *Flavobacterium columnare*, *Francisella orientalis*, *Edwardsiella ictaluri* and *Aeromonas* spp., and viruses like tilapia lake virus, infectious spleen and kidney necrosis virus and tilapia parvovirus. These pathogens pose significant challenges to the sustainability of tilapia aquaculture, causing substantial fish mortality (Wang *et al.*, 2023).

Understanding the immune defences of tilapia is essential for improving their resistance to diseases, a key factor in addressing

Kim Thompson, Craig Shoemaker and David C. Little (eds) **Tilapia**
DOI: 10.1079/9781800629455.0007, © CAB International, 2025

disease challenges in tilapia aquaculture. These include the emergence of new diseases, the need to reduce antibiotic use, the introduction of new strains of tilapia, issues related to larval development, the impacts of climate change and pollution and the use of alternative feed ingredients. Gaining insights into the immune mechanisms and interactions between host and pathogen is vital for creating optimal conditions to prevent immunosuppression, developing effective vaccination strategies, breeding programmes for disease resistance and formulating functional feeds enhanced with immunostimulants and probiotics.

Innate immune response

Constant exposure to pathogens in their aquatic surroundings is challenging for farmed tilapia. The innate immune system of tilapia serves as an important defence against the plethora of potential pathogens present in their aquatic environment, encompassing an intricate network of physical barriers, cellular components and humoral factors (Ellis, 2001; Esteban, 2012; Smith *et al.*, 2019; Dalmo and Bøgwald, 2022) (Table 7.1). These components operate synergistically to detect and neutralize infectious agents. The physical barriers act as the first line of defence, successfully preventing the entry of pathogens. The cellular and humoral responses provide a more direct attack against pathogens that have evaded the physical barriers, using a variety of mechanisms including phagocytosis, inflammation and the production of antimicrobial peptides (AMP) together with other antimicrobial products (Esteban, 2012; Cabillon and Lazado, 2019; Dalmo and Bøgwald, 2022).

Physical barriers

Protective barriers, including the skin, scales and epithelial layers of the gills and gastrointestinal tract, are key locations where pathogens can enter and play an important role in blocking their entry. The mucus that covers these areas also acts as a barrier and contains antimicrobial agents such as lectins, lysozymes, complement proteins and

AMPs, effectively neutralizing and destroying pathogens. Together, the physical barriers and the mucus provide a dual-action defence essential for the fish's survival in environments rich in potential infectious agents (Esteban, 2012; Gomez *et al.*, 2013; Cabillon & Lazado, 2019; Smith *et al.*, 2019).

Innate humoral components

When pathogens overcome the initial physical defences and infiltrate the host, the innate immune system's cellular and humoral mechanisms are activated to impede the progression of the infection. The fish relies on a comprehensive suite of humoral responses within its innate immune system to deal with infections. This includes an assortment of proteins and glycoproteins that can either neutralize or impede the proliferation of pathogens, including AMPs, lytic enzymes such as lysozymes, cathepsin, chitinase and the complement system. Other components include agglutinins, precipitins and natural antibodies that recognize and bind to foreign entities. The immune response is further augmented by cytokines, including interleukin-1 (IL-1) and IL-6, and tumour necrosis factor-alpha (TNFα), which mediate inflammation and immune cell activation. In addition, the presence of growth inhibitors, serum protease inhibitors (including α2 macroglobulin and α1 anti-trypsin), chemokines and acute-phase proteins plays an important role in the robust defence against infections in fish. These agents work either by directly eliminating the pathogen or through inflammation and phagocytosis via the cellular components of the innate immune system (Esteban, 2012; Gomez *et al.*, 2013; Rakers *et al.*, 2013; Salinas, 2015; Cabillon & Lazado, 2019; Smith *et al.*, 2019). Complement, lysozyme and AMPs are the most extensively researched elements of tilapia's humoral innate immune response, discussed below.

Complement system

The complement system is involved in the direct destruction of bacteria, enhancement of inflammation, clearance of damaged cells and promotion of an adaptive immune response (Nakao *et al.*, 2011). In teleost fish the complement system exhibits similarities with that of mammals, possessing homologues to mammalian proteins. This resemblance is partly due to gene or genome duplications, resulting in multiple paralogues for certain components in fish (Nakao *et al.*, 2006). The primary production site for many of the complement proteins is the liver hepatocytes, which release them into the bloodstream. However, some complement proteins are also produced in extra-hepatic tissues (Jorgensen, 2014). These proteins are generally found in an inactive form and become sequentially activated into an enzymatic cascade when a pathogen is identified.

The complement system of fish includes more than 30 proteins that operate through the classical, alternative and lectin pathways. Initiation of the classical pathway occurs through antigen–antibody complex formation whereas the alternative pathway begins with the spontaneous conversion of C3 to C3b following exposure to microorganism surfaces. The lectin pathway is triggered by the binding of mannose-binding lectin to specific carbohydrate patterns on pathogens (Dunkelberger & Song, 2010; Smith *et al.*, 2019). These pathways ultimately lead to the production of C3 convertases, which cleave C3 into C3a and C3b. C3a acts as a chemotactic factor, recruiting leukocytes to the site of infection and triggering inflammation. C3b, meanwhile, enhances the phagocytosis of pathogens by acting as an opsonin. Subsequently, C5 convertase splits C5 into C5a and C5b; C5a is a potent inflammatory agent and immune cell attractant. C5b joins with additional complement components to form the membrane attack complex that directly eliminates pathogens (Dunkelberger and Song, 2010). Some of the studies examining the functions and actions of tilapia's complement system are described by Abdel-Salam *et al.* (2014), Valladão *et al.* (2017), Chen *et al.* (2018, 2019a, b), Galagarza *et al.* (2018), and Dhanarso *et al.* (2021).

Lysozyme

Lysozyme has several immune functions in tilapia. Lysozyme acts as a primary defence against bacterial pathogens by targeting the peptidoglycan layer in the cell walls of Gram-positive bacteria, leading to their destruction. This enzyme can also break down the inner peptidoglycan layer of Gram-negative bacteria once their outer protective wall is compromised by actions of the complement system and various enzymes, allowing for bacterial degradation (Saurabh and Sahoo, 2008; Smith *et al.*, 2019). It also serves as an opsonin, enhancing phagocytosis by marking pathogens for destruction; opsonins bind to the surface of pathogens or other particles, making them more attractive to phagocytes such as neutrophils and macrophages. This binding signals the phagocyte to engulf and destroy the marked particles, facilitating the immune system's ability to clear infections and remove debris (Saurabh & Sahoo, 2008; Smith *et al.*, 2019). Lysozyme can also activate the complement system to promote both pathogen destruction and an elevated inflammatory response that fortifies the fish's immune defence against infections (Saurabh & Sahoo, 2008; Smith *et al.*, 2019). The mucosal layer is enriched with lysozyme, which helps to reduce the microbial load on the fish's skin and prevent potential pathogen entry (Saurabh & Sahoo, 2008; Smith *et al.*, 2019).

The measurement of lysozyme activity has frequently been used as an indicator of the innate immune response in tilapia, with examples of studies provided by Sankaran and Gurnani (1972), Taoka *et al.* (2006), Welker *et al.* (2007), Saurabh and Sahoo (2008), Smith *et al.* (2019), Nurani *et al.* (2020), and Biller *et al.* (2021). These investigations highlight the enzyme's activity levels as crucial indicators of health status and immune competence in response to environmental stressors, pathogenic challenges or immunostimulatory agents.

AMPs

Antimicrobial peptides are short oligopeptides, recognized for their ability to combat bacterial and viral pathogens. Essential amino acids such as arginine, lysine and histidine within the peptide bestow a positive charge on the molecules. This charge, coupled with a balance between hydrophobic and hydrophilic amino acids, imparts an amphipathic nature to AMPs, facilitating their interaction with the lipid membranes of pathogens. AMPs often exert their antimicrobial effect by forming pores in bacterial cell membranes, which can lead to cell death. Additionally, AMPs play a critical role in modulation of immune responses: for example, they can trigger immune cell activation in fish (Hazam and Chen, 2020).

Research into AMPs within teleost fish has identified over 90 distinct AMPs across various molecule families, including β-defensins, cathelicidins, hepcidins, histone-derived peptides and the fish-specific piscidin family. The antimicrobial efficacy of these peptides is significant. β-defensin from Nile tilapia, for instance, has shown inhibitory effects on pathogens such as *Escherichia coli* and *Streptococcus agalactiae*, indicating its therapeutic potential (Dong *et al.*, 2015). The piscidins, including those derived from tilapia, have gained attention for their therapeutic potential, with recent reviews by Hazam and Chen (2020) and Raju *et al.* (2021) highlighting their promise in treating infections.

Beyond their direct antimicrobial action, AMPs are also being explored for their ability to modulate the immune system. This could potentially enhance innate immune responses in fish, leading to innovative treatments for bacterial and viral infections. The discovery of a diverse range of AMPs in teleost fish highlights the potential to uncover new antimicrobial agents that could confront current treatment challenges, particularly antibiotic-resistant infections. The continued study of AMPs from fish and other organisms is expected to contribute substantially to both veterinary and human medicine, offering novel approaches to combat infectious diseases.

Innate cellular components

The innate immune system of fish comprises several cellular components, including various granulocytes such as neutrophils and mast/eosinophilic granule cells, along with monocytes/macrophages, dendritic cells and non-specific cytotoxic cells (Dalmo & Bøgwald, 2022).

Phagocytes

The innate immune response is predominantly mediated by myeloid cells, including mononuclear and polymorphonuclear phagocytes.

Mononuclear phagocytes comprise monocytes that transition from the bloodstream into tissues, differentiating into macrophages and dendritic cells. These cells are pivotal in antigen presentation to T-cells, facilitating the activation of the adaptive immune response and thus acting as a bridge between the innate and adaptive systems. Dendritic cells are highly efficient in this role (Forlenza *et al.*, 2011). In addition, fish possess unique melanomacrophages and phagocytic cells containing melanin within their cytoplasm, found in the spleen, head kidney and inflammation sites; these cells are thought to play a role in antigen presentation (Agius and Roberts, 2003; Koppang *et al.*, 2003).

Neutrophilic granulocytes, the primary polymorphonuclear phagocytes, execute an early and potent antimicrobial response. They engulf pathogens through phagocytosis, deploying toxic granules, reactive oxygen species (ROS) and neutrophil extracellular traps to neutralize invaders (Filho, 2007; Palić *et al.*, 2007; Katzenback & Belosevic, 2009; Rieger *et al.*, 2012; Flerova & Balabanova, 2013; Pijanowski *et al.*, 2013). Neutrophils and macrophages, driven by inflammatory cytokines, are among the first leukocytes to converge on infection sites, playing a role in the development of an inflammatory response in fish (Havixbeck & Barreda, 2015; Havixbeck *et al.*, 2016).

Following pathogen detection, phagocytes utilize pattern-recognition receptors (PRRs) to initiate engulfment. This interaction activates several signalling pathways, including MAPK, NF-XB and JAK-STAT, enhancing the immune response (Chen *et al.*, 2018). The phagosome formed from pathogen engulfment fuses with a lysosome to create a phagolysosome in which the pathogen is eliminated using superoxide radicals, inducible nitric oxide species and antimicrobial peptides, among other mechanisms. This is where the pathogen is destroyed using antimicrobial agents such as ROS and nitric oxide (NO), as documented by Hodgkinson *et al.* (2015) and Lu and Chen (2019). This process culminates in the degradation of the pathogen by hydrolytic enzymes, allowing macrophages and dendritic cells to present the digested material to lymphocytes, thus linking innate and adaptive immunity (Esteban *et al.*, 2015).

Like mammals, monocytes/macrophages and neutrophils are central to phagocytosis in fish. However, teleosts also employ B-1 cells, γIV-T cells and thrombocytes in phagocytic roles (Li *et al.*, 2006; Nagasawa *et al.*, 2014; Wan *et al.*, 2016). Teleost phagocytes, including monocytes/macrophages, demonstrate polarization, a process significantly influencing their functionality (Wiegertjes *et al.*, 2016). Studies such as those by Suzuki (1986) and Fernandes *et al.* (2019) provide examples of the phagocytic capabilities of tilapia's immune cells.

Non-specific cytotoxic cells

Non-specific cytotoxic cells (NCCs) and natural killer (NK)-like cells are two immune cell types implicated in non-specific cell-mediated cytotoxicity (CMC) observed in fish, playing a role in their defence against various pathogens, and they react against xenogeneic and allogenic cells. However, the precise relationship between NCCs and NK-like cells remains unclear. NCCs, identified as small agranular lymphocytes, are distinguished by their expression of a unique type III 34-kDa membrane protein called the NCC receptor protein (Evans *et al.*, 1984), which has also been identified in tilapia (*Oreochromis niloticus*) (Ishimoto *et al.*, 2004; Huang *et al.*, 2022). These cells, primarily sourced from lymphoid tissues, have been posited to act as the evolutionary antecedents to mammalian NK cells (Jorgensen, 2014). Contrarily, NK-like cells are characterized by their more prominent, granular appearance and are predominantly located in the peripheral blood (Shen *et al.*, 2004).

Shen *et al.* (2002) provided an overview of the functions of these cells in channel catfish *Ictalurus punctatus*, although further characterization of CMC in other fish species is required (Jorgensen, 2014). NCCs have also been found in various fish species, including tilapia (*Oreochromis aureus*) (Huang *et al.*, 2022). NCCs are known for their inherent ability to spontaneously eliminate tumour cells, cells infected with viruses and protozoan parasites (Frøystad *et al.*, 1998; Praveen *et al.*, 2004). On the other hand, NK-like cells execute their cytotoxic activity through a mechanism involving perforin- and granzyme-induced apoptosis (Secombes and Wang,

2012), marking a distinct pathway for induction of cell death. The variation in mechanisms between these two types of cell reflects the complexity of the fish immune system, pointing to the need for further characterization of CMC in fish (Shen *et al.*, 2002; Jorgensen, 2014).

PRRs

The elimination of pathogens via phagocytosis is facilitated when phagocytes identify and attach to pathogens using surface PRRs (Ángeles *et al.*, 2015). These receptors engage with pathogen-associated molecular patterns (PAMPs) from a wide range of microbial pathogens – including viruses, bacteria, fungi, parasites and protozoa – and with danger-associated molecular patterns (DAMPs) emitted by stressed or damaged cells, initiating an immune response for their clearance (Smith *et al.*, 2019). Key PRR categories identified in teleosts include Toll-like receptors (TLRs), nucleotide-binding oligomerization domain-like receptors (NLRs), retinoic acid-inducible gene I-like receptors (RLRs), C-type lectin receptors (CLRs) and absent in melanoma-2-like receptors (ALRs) (Uribe *et al.*, 2011; Roberts, 2012; Bruce & Brown, 2017; Li *et al.*, 2017; Smith *et al.*, 2019). These receptors are positioned on the cell surface (e.g. TLRs, CLRs) to detect external threats, within the cytoplasm (e.g. NLRs, RLRs, ALRs) to sense intracellular dangers or in endosomes (TLRs) to identify ingested pathogens. Following recognition of PAMPs or DAMPs, these PRRs activate signalling pathways, activating transcription factors such as NF-XB and interferon (IFN)-response factors (IRFs). This process generates pro-inflammatory and chemotactic cytokines alongside antimicrobial responses (Li *et al.*, 2017).

Seventeen TLRs have been identified in teleosts, of which seven share an orthology with mammalian counterparts, namely TLR1, 2, 3, 5M, 7, 8 and 9. In addition, eight TLRs are unique to fish, specifically TLR5S, 14, 16, 19, 20, 21, 22 and 23 (Palti, 2011). TLRs 1, 2, 4, 5 and 6 are acknowledged for recognizing various pathogens, including bacteria, fungi and protozoa (Kumar *et al.*, 2011). Conversely, the intracellular TLRs 3, 7, 8 and 9 are specialized for detecting viral nucleic acids where TLR7/8 identifies single-strand RNA (ssRNA), TLR3 recognizes double-strand RNA (dsRNA) and TLR9 is responsive to DNA (Gerlier & Lyles, 2011). TLR3, 7, 21 and 22 have been identified in tilapia, with TLR3 exhibiting significant presence in the liver, brain and spleen (Abouelmaatti *et al.*, 2013; Pang *et al.*, 2017).

Aoki *et al.* (2013) highlight the specialized function of RLRs in detecting viral RNA. Among fish, three RLR variants have been identified: RIG-1, sensitive to dsRNA, melanoma differentiation-associated protein 5 (MDA5) and laboratory of genetics and physiology 2 (LGP2). Notably, tilapia lack the genetic coding for RIG-I and LGP2 but possess the genes for MDA5 and the RLR adaptor IPS-1 (Aoki *et al.*, 2013).

Gao *et al.* (2018) have identified the presence of NLRs, specifically NOD1, NOD2 and NLR3, in tilapia. These receptors are crucial for detecting host cell membrane damage and the presence of intracellular pathogens. They trigger the immune response by recognizing DAMPs or PAMPs, producing inflammatory cytokines such as IL-1β. This cytokine production is integral to the formation of inflammasomes (Mogensen, 2009).

Inflammation

Inflammation is a vital host response instigated to eliminate pathogens, repair tissue damage and neutralize harmful substances. It is driven by the release of pro-inflammatory cytokines, which coordinate the movement of cells, including APCs, towards the site of infection for antigen presentation. This activity is significantly influenced by various cytokines that become active during the inflammatory response. The inflammatory process involves the release of several mediators such as vasoactive amines and peptides, eicosanoids, pro-inflammatory cytokines and acute-phase proteins. These substances lead to alterations in vascular permeability, facilitating the recruitment and gathering of leukocytes at the inflammation site. This process is initiated primarily due to microbial products and specific cytokines, namely IL-1β, IL-6 and TNFα. To prevent overzealous cell migration, anti-inflammatory cytokines are subsequently released to control this (Chertov *et al.*, 2000; Ferrero-Miliani *et al.*, 2007).

Studies on tilapia have revealed the presence and variable expression of inflammatory cytokines in response to pathogens. Zhu *et al.* (2017) and Xia *et al.* (2018) documented a rise in IL-1β and TNFα following infection with *Lactobacillus rhamnosus* and *Streptococcus agalactiae*, respectively. Additionally, a comprehensive analysis by Dong *et al.* (2018) on *S. agalactiae* infection in tilapia showed a complex cytokine response, with IL-10 levels increasing while those of TNFα and IL-6 were reduced. This indicates a complex interaction between pro- and anti-inflammatory cytokines in managing tilapia's immune defences.

Adaptive immune response

In tilapia and other teleost species, the adaptive immune system is crucial in managing pathogens persisting within the organism that the innate immune response has not removed. This system comprises cell-mediated immune responses from lymphocytes and the humoral immune response mediated by antibodies. Antigen-presenting cells such as dendritic cells, monocytes/macrophages and B cells bridge the gap between innate and adaptive immunity. They do this by engulfing pathogens and presenting antigens from the pathogen to T cells, thereby initiating T-cell-mediated immune responses (Uribe *et al.*, 2011; Smith *et al.*, 2019).

The adaptive immune system is characterized by its specificity and memory, enabling the rapid targeting and elimination of pathogens following subsequent exposure. This level of precision is achieved through both B- and T-cell receptors (BCR and TCR, respectively), the activation of recombination-activating genes (RAG1 and RAG2) and the involvement of the major histocompatibility complex (MHC I and II). As lymphocytes patrol the organism they identify and attach to pathogens via cell-surface receptors, initiating a process of clonal expansion that results in daughter cells equipped with identical antigen-recognition capabilities.

Table 7.1 Simplified overview of innate immune mechanisms in fish.

Component	Function	Key features	References
Physical Barriers	Prevent the entry of pathogens.	This includes the skin, scales, mucus, epithelial layers of gills, and the gastrointestinal tract. Mucus contains antimicrobial agents like lectins and lysozymes.	Esteban (2012), Gomez *et al.* (2013), Cabillon & Lazado (2019), Smith *et al.* (2019)
Humoral Components	Neutralise and impede the proliferation of pathogens.	Consists of AMPs, lysozymes, cathepsin, chitinase, complement system, cytokines (IL-1, IL-6, TNF-α), serum protease inhibitors, chemokines, and acute phase proteins.	Esteban (2012), Gomez *et al.* (2013), Rakers *et al.* (2013), Salinas (2015), Cabillon & Lazado (2019), Smith *et al.* (2019)
Complement System	Direct destruction of pathogens enhances inflammation and promotes adaptive immune response.	It operates through classical, alternative, and lectin pathways. More than 30 proteins are involved. It is produced mainly in liver hepatocytes.	Nakao *et al.* (20062011), Dunkelberger & Song (2010), Smith *et al.* (2019)
Lysozyme	It breaks down bacterial cell walls, acts as an opsonin, and activates the complement system.	Found in tissues, serum and the mucosal layer, it enhances phagocytosis.	Smith *et al.* (2019), Saurabh & Sahoo (2008)
Antimicrobial Peptides (AMPs)	Combat bacterial and viral pathogens and modulate immune responses.	Includes β-defensins, cathelicidins, and hepcidins. Amphipathic nature facilitates interaction with pathogen membranes.	Dong *et al.* (2015), Hazam & Chen (2020), Raju *et al.* (2021)
Cellular Components	A direct attack on pathogens via phagocytosis and antigen presentation.	Includes monocytes/macrophages, dendritic cells, neutrophils, and non-specific cytotoxic cells (NCCs). Polarisation affects functionality.	Dalmo & Bøgwald (2022), Forlenza *et al.* (2011), Suzuki (1986), Fernandes *et al.* (2019)
Phagocytes	Engulf and destroy pathogens. They are mediated by pattern recognition receptors (PRRs).	Mononuclear and polymorphonuclear phagocytes involve the formation of phagolysosomes.	Esteban *et al.* (2015), Hodgkinson *et al.* (2015), Lu & Chen (2019)

Non-specific Cytotoxic Cells (NCC)	Participate in non-specific cell-mediated cytotoxicity (CMC).	React against xenogeneic, allogenic cells, and pathogens. Express unique type III 34 kDa membrane protein.	Evans *et al.* (1984), Ishimoto *et al.* (2004), Huang *et al.* (2022)
Pattern Recognition Receptors (PRRs)	Identify and bind pathogens using pathogen-associated molecular patterns (PAMPs) and danger-associated molecular patterns (DAMPs).	This includes TLRs, NLRs, RLRs, CLRs, and ALRs. They are located on the cell surface or within the cytoplasm and activate inflammatory and antimicrobial responses.	Uribe *et al.* (2011), Roberts (2012), Bruce & Brown (2017), Li *et al* (2017), Smith *et al.* (2019)
Inflammation	Eliminate pathogens, repair tissue damage, and neutralise harmful substances. Coordinate the movement of cells to the site of infection.	It involves pro-inflammatory cytokines like IL-1β, IL-6, TNF-α and anti-inflammatory cytokines to control the response.	Ferrero-Miliani *et al.* (2007), Chertov *et al.* (2000), Xia *et al.* (2018), Zhu *et al.* (2017), Dong *et al.* (2018)

T lymphocytes

Antigen-presenting cells exhibit processed antigens alongside MHC molecules on their surface as part of the antigen-presentation process, and TCRs recognize these antigen–MHC complexes. This interaction initiates the differentiation of T cells into different subtypes, such as helper T cells ($CD4^+$), cytotoxic T cells ($CD8^+$) and regulatory T cells (T_{regs}) (Uribe *et al.*, 2011; Fischer *et al.*, 2013; Smith *et al.*, 2019). The specificity of the interaction between TCRs and antigens hinges on whether the antigen is presented with MHC class I or II molecules. This distinction guides the development of naïve T cells into either $CD8^+$ or $CD4^+$ T cells (Secombes and Belmonte, 2016).

T helper cells can differentiate into various subsets, including Th1, Th2 and Th17 cells (Mutoloki *et al.*, 2014). Each subset secretes a distinct profile of cytokines that plays a unique role in the immune response. Th1 cells produce cytokines crucial for CMC, enabling the immune system to target cells infected with intracellular pathogens, driven by T-bet and STAT-4 transcription factors and influenced by IL-12 and interferon-gamma (IFNγ). Th2 cells secrete cytokines that support humoral immunity, crucial for defence against extracellular pathogens. These are mediated by transcription factors Gata-3 and STAT-6 and IL-4, IL-5 and IL-13. Cytokines from Th17 cells direct the immune response against bacterial and fungal infections, encouraging their growth by IL-23 (Secombes & Belmonte, 2016; Smith *et al.*, 2019; Zhu and Zhu, 2020).

Cytotoxic T lymphocytes (CTLs), known as $CD8^+$ T cells due to their expression of CD8 glycoprotein, play a pivotal role in CMC, targeting cells compromised by intracellular pathogens or tumour cells. This process is facilitated by MHC class I molecules. These T cells employ three primary strategies to eliminate target cells: the granule exocytosis pathway, the Fas pathway and the deployment of cytolytic cytokines such as TNFα, lymphotoxin-α and IFNγ. With granule exocytosis, CTLs discharge granules that contain granzymes and perforin near the target cell's membrane, leading to the internalization of these molecules and subsequent cell apoptosis. Granzymes further contribute to this process by causing DNA fragmentation within the nucleus. The second mechanism involves the interaction between the Fas ligand on the CTL surface and the Fas receptor on the target cell, triggering apoptosis. Lastly, the release of cytolytic cytokines by CTLs also contributes to the destruction of target cells. The phenomenon of CMC was initially identified in channel catfish, as reported by Miller *et al.*, 1986, and has subsequently been documented in various fish species (Somamoto *et al.*, 2000; Stuge *et al.*, 2000; Nakanishi *et al.*, 2002; Fischer *et al.*, 2006; Utke *et al.*, 2007). Both granzyme (Praveen *et al.*, 2006) and the FasL protein (Ma *et al.*, 2014) have been identified in tilapia.

T-cell receptors are intricate transmembrane glycoproteins comprising either α and β chains or γ and IV chains, which are connected through disulfide bonds. These proteins exhibit a wide range of diversity due to the somatic recombination process of variable (V), diversity (D) and joining (J) gene segments; the TCR β chain loci have been described in Nile tilapia (Nithikulworawong *et al.*, 2012). In the mechanism of T-cell activation, CD4 and CD8 molecules act as crucial co-receptors. CD4 assists in the recognition of antigens presented by the MHC-II molecule. On the other hand, CD8 interacts with antigens presented through MHC-I and can be found as a homodimer of α chains or a heterodimer of α and β chains (Ellmeier *et al.*, 2013; Smith *et al.*, 2019). The genes responsible for CD4 and CD8 have been identified in tilapia (Mai *et al.*, 2021).

MHC

The distinction between MHC class I and II antigens in teleost fish is marked by their chromosomal segregation, a unique feature not seen in cartilaginous fish or other vertebrates (Kuroda *et al.*, 2002). MHC class I molecules, essential for presenting processed antigens to CTLs, consist of three extracellular domains (α1, α2 and α3) and are associated with a β2-microglobulin light chain. On the other hand, MHC class II molecules present antigens to Th2 cells ($CD4^+$ T cells), consist of two α and two β peptide chains and are primarily located on the surface of APCs. The diversity of the MHC class Iα gene has been documented in tilapia (Poonsawat *et al.*, 2009), and the MHC-II molecules MHC-2A and MHC-2B have been identified in Nile tilapia (Pang *et al.*, 2013).

The Class III region of the MHC in teleosts encompasses genes not directly linked to the processing and presentation of antigens. Instead, this diverse group of genes is instrumental in managing the inflammatory response, including those encoding components of the complement system, immune-signalling cytokines and heat shock-proteins that safeguard cells against stress (Deakin *et al.*, 2006). Unlike other vertebrates, the MHC III genes in teleosts are distributed across various chromosomes. This distinct organizational characteristic suggests broad participation in immune responses, although studies on MHC III genes in tilapia are lacking.

B lymphocytes and immunoglobulins

The B lymphocytes phagocytose and act as APCs to stimulate T cells by presenting processed antigens and producing antigen-specific immunoglobulins (Igs) (Parra *et al.*, 2012; Smith *et al.*, 2019). Following antigen recognition and helper T-cell assistance, B cells differentiate into plasma cells, producing antigen-specific Igs (Reyes-Cerpa *et al.*, 2012; Roberts, 2012; Ye *et al.*, 2013). In teleost fish, three antibody classes have been identified, including IgM, IgD and IgZ/T, with IgM being the most significant Ig in the teleost systemic immune response; it can be found in serum, gut and mucus (Ye *et al.*, 2013; Salinas, 2015; Velázquez *et al.*, 2018). The teleost IgM molecular structure differs from mammalian IgM, with light and heavy chains joined non-covalently rather than by disulphide bonds, resulting in monomeric, dimeric and tetrameric antibody forms. This non-covalent linkage is thought to help tetrameric forms bind to epitopes by adjusting their orientation (Solem and Stenvik, 2006). Fish IgM can opsonize pathogens to promote macrophage phagocytosis, activate the classical complement pathway and agglutinate pathogens (Secombes and Fletcher, 1992; Solem and Stenvik, 2006). Despite lower intrinsic affinity and heterogeneity compared with mammalian antibodies (Solem and Stenvik, 2006; Ma *et al.*, 2013), they are still very effective in protecting fish against bacterial and viral infections (Mutoloki *et al.*, 2014).

Immunoglobulin D (IgD), an ancient type of antibody, is widespread across various vertebrates and is present in the immune tissues of fish. However, its exact role remains uncertain (Edholm *et al.*, 2011). The gene for IgD has been cloned in several fish species, including tilapia (Wang *et al.*, 2016). Secreted forms of IgD (sIgD) have been identified in both catfish and rainbow trout *Oncorhynchus mykiss*, with notable structural differences between the two fish species. In catfish, sIgD is missing the CXII1 and variable domains, unlike in trout where these domains are present (Edholm *et al.*, 2011; Ramirez-Gomez *et al.*, 2012; Parra *et al.*, 2013). In addition, sIgD in trout is released as a monomer and exists in at least two forms (Ramirez-Gomez *et al.*, 2012). The ability of sIgD to bind the IgD-binding receptor on granulocytes through their Fc receptor suggests it may have a role as a PRR (Edholm *et al.*, 2011), with a potential function in pattern recognition (Edholm *et al.*, 2011).

IgT/Z appears to be involved in the mucosal immunity of most teleost species, akin to the function of IgA in mammals, and is primarily located on mucosal surfaces like gills, skin, and intestines (Zhang *et al.*, 2010). The concentration of Igs on mucosal surfaces tends to be lower than in the serum (Salinas & Parra, 2015). IgM and IgT are actively transported from the epithelial cells of the lamina propria to the intestinal lumen by the polymeric Ig receptor (pIgR). IgT is found in multimeric forms in mucosal tissues and as a monomer in the serum. The importance of IgT in mucosal immunity is suggested by the significantly higher ratio of IgT/IgM in mucosal secretions compared with plasma in rainbow trout, a phenomenon observed even without antigenic stimulation (Zhang *et al.*, 2010; Xu *et al.*, 2013; Tacchi *et al.*, 2014; Salinas & Parra, 2015). The specific function of IgT/Z in mucosal immunity still requires further investigation.

All three Igs have been characterized in tilapia (Phuyindee *et al.*, 2015; Wang *et al.*, 2016; Velázquez *et al.*, 2018). IgM is frequently used as a measure of systemic antibody responses while IgT is used for mucosal antibody response, including tilapia (Tellez-Bañuelos *et al.*, 2010; Velázquez *et al.*, 2018; Mai *et al.*, 2021).

Immunological tissues

In teleosts the principal lymphoid organs include the thymus, kidney (anterior and middle sections) and spleen, which play a significant role in the immune system. The thymus, located beneath the pharyngeal epithelium in the dorsolateral region of the gills, varies in structure across species and life stages without clear corticomedullary differentiation (Zapata et al., 2006; Secombes & Wang, 2012). It is crucial for the development of T lymphocytes, as in other jawed vertebrates (Alvarez-Pellitero, 2008), with evidence of its role in T-cell production from studies involving T-dependent antigens, monoclonal antibodies (Passer *et al.*, 1996) and functional assays (Ellsaesser *et al.*, 1988), showing lymphocyte migration from the thymus to the spleen and kidney (Tatner & Findlay, 1991).

The anterior kidney, identified as the most important haematopoietic organ (Zapata *et al.*, 1996), is involved in blood cell differentiation and early immune responses. It features maternal microchimeric cells (MMCs), which play a role in antigen trapping and the development of immunologic memory (Secombes *et al.*, 1982; Herraez & Zapata, 1986; Tsujii & Seno, 1990). These centres are aggregates of various immune cells, including macrophages and lymphocytes, highlighting the kidney's role in phagocytosis (Dannevig *et al.*, 1994), antigen processing and IgM production (Kaattari & Irwin, 1985; Brattgjerd & Evensen, 1996).

The spleen functions as a major secondary lymphoid organ and is involved in immune activity and blood cell formation and breakdown (Manning, 1994; Zapata *et al.*, 1996; Galindo-Villegas *et al.*, 2004). Its architecture is less distinct, although it includes areas similar to the red and white pulp found in mammals. The spleen's MMCs and ellipsoids are believed to be involved in antigenic stimulation and long-term retention of antigens, suggesting a role in immunologic memory (Van Muiswinkel *et al.*, 1991). Blood-borne material trapped within ellipsoidal walls is phagocytosed by macrophages, which then migrate into the MMCs (Press and Evensen, 1999).

Mucosal-associated lymphoid tissues (MALTs), including the skin-associated lymphoid tissue (SALT), gut-associated lymphoid tissue (GALT), gill-associated lymphoid tissues (GIALT), nasopharynx-associated lymphoid tissue (NALT) and the organized interbranchial lymphoid tissue (ILT), constitute integral components of the teleost immune defence against infections (Gomez *et al.*, 2013; Rombout *et al.*, 2014; Salinas, 2015; Salinas & Parra, 2015; Smith *et al.*, 2019). These tissues are critical in pathogen resistance, presenting as a diffuse network of immune cells across mucosal surfaces (Salinas, 2015). ILT, found at the base of gill filaments in Atlantic salmon, is noted for its organized structure and has been shown to be involved in antiviral protection (Haugarvoll *et al.*, 2008; Koppang *et al.*, 2010; Aas *et al.*, 2014).

MALTs are composed of innate and adaptive immune cells and molecules, with B and T cells diffusely scattered throughout, showing phenotypic differences from their systemic counterparts. These cells respond to mucosal infection and vaccination, with IgT^{+} B cells being predominant in GALT, SALT and NALT, in contrast to the spleen or head kidney where IgM^{+} B cells are the main subset (Zhang *et al.*, 2010; Xu *et al.*, 2013; Tacchi *et al.*, 2014; Salinas, 2015).

Research into mucosal vaccination, such as the study by Kitiyodom *et al.* (2021a, b), which explored mucosal vaccination against *Fl. columnare* in tilapia, reveals the efficacy of such vaccines in eliciting both mucosal and systemic immune responses.

Gut microbiome

The gut microbiome is essential for fish health, playing a significant role in the development of intestinal lymphoid tissues and GALT across various fish species (Nayak, 2010a), including tilapia. While the influence of gut bacteria on immune cell recruitment and lymphoid tissue formation has been well documented in species such as zebrafish (*Danio rerio*) and ayu (*Plecoglossus altivelis*) (Rawls *et al.*, 2004; Nayak, 2010a; Nie *et al.*, 2017), similar studies in tilapia are still needed. Disruptions in the gut microbiota can lead to excessive inflammation and depletion of lymphoid tissues, critical components of the immune system (Bates *et al.*, 2006, 2007; Nie *et al.*, 2017). Gut bacteria contribute to the immune system by modulation of gene expression related to intestinal homeostasis and epithelial regeneration (Rawls *et al.*, 2004; Gómez & Balcázar, 2008; Pérez *et al.*, 2010).

In addition, gut bacteria are crucial for nutrient metabolism by producing enzymes such as amylase and protease, which enhance the breakdown of complex proteins and other indigestible substances (Ray *et al.*, 2012), thus improving nutrient absorption and overall fish health and growth (Rawls *et al.*, 2004; Bates *et al.*, 2006; Ray *et al.*, 2012; Semova *et al.*, 2012). The presence of beneficial bacteria also prevents the colonization of pathogenic bacteria by competing for attachment sites on the intestinal lining, significantly reducing the risk of infections in the fish gut (Nie *et al.*, 2017; Xiong *et al.*, 2019).

Environmental factors, including seasonal changes, farming locations and aquaculture systems, also influence the diversity of gut microbiota in tilapia. For instance, studies have shown that bacterial communities within tilapia can vary significantly with the season, with different bacterial phyla dominating at different times of the year (Al-Harbi & Naim, 2004; Bereded *et al.*, 2021). The type of aquaculture system also impacts microbial diversity, where fish raised in recirculating systems exhibit a richer microbial environment, potentially enhancing fish survival compared with flow-through systems (Giatsis *et al.*, 2015; Deng *et al.*, 2021).

The widespread use of antibiotics in aquaculture to treat infections can negatively affect the diversity and function of gut microbiota, increasing the susceptibility of fish to diseases (Defoirdt *et al.*, 2011). Alternative strategies, such as dietary supplementation with organic salts and phytogenic compounds, have been employed to mitigate this. These interventions have been successful in maintaining a healthy diversity of gut microbiota, even during bacterial infections, and increasing the resistance of tilapia to diseases (Haygood & Jha, 2018; Wang *et al.*, 2023).

To study and understand these microbial communities, researchers employ molecular techniques such as 16S ribosomal RNA sequencing (Fraher *et al.*, 2012). This method helps identify and quantify bacterial species in the gut, providing insights into their dynamic changes in response to various environmental and therapeutic interventions (Tsuchiya *et al.*, 2008; Bereded *et al.*, 2021).

Effectors of the immune response

The physiological and immunological responses of fish in aquaculture are profoundly influenced by many factors, especially under conditions of high-density farming. The challenges in understanding these influences include a lack of solid baseline data and the inherent variability in fish immune responses, as highlighted by Bowden (2008) and Makrinos and Bowden (2016). Among the various determinants impacting fish immunity, temperature and stress are consistently identified as having a significant effect (Pickering & Pottinger, 1989; Pulsford *et al.*, 1994; Espelid *et al.*, 1996; Wendelaar-Bonga, 1997). In addition, the role of seasonal changes, particularly the effect of photoperiod on the fish immune system, has been explored, highlighting the importance of circadian rhythms and their manipulation through photoperiod adjustment to alter reproductive cycles (Bowden *et al.*, 2007). This practice, common in aquaculture, enables the production of eggs beyond the natural spawning season, facilitating multiple introductions of fish to marine environments annually.

In the realm of fish farming, stressors such as poor water quality, high population density, inadequate nutrition, weather events, temperature fluctuations, handling, transportation and treatment for disease can significantly impact fish health (Wang *et al.*, 2023). Wendelaar-Bonga (1997) defines stress as the physiological response to a threatening situation orchestrated by hormonal systems that activate secondary stress response factors. These factors have a twofold action: they disturb the homeostatic equilibrium and elicit a set of behavioural and physiological responses designed to be compensatory and/or adaptive, aiding the fish in overcoming the threat. However, in the face of intense, chronic stress, this response may become maladaptive, leading to growth inhibition, reproductive failure and a diminished capacity to resist pathogens, thus highlighting the complex interplay between environmental conditions and fish health in aquaculture settings.

Improvement of the immune response

Aquaculture practices are constantly evolving to improve the resilience of the immune response of fish and enhance their ability to withstand disease. Such approaches include optimization of farming conditions to mitigate stress and improve the health status of aquatic populations. Selective breeding strategies have also been instrumental in augmenting disease resistance (Gjedrem, 2015). Breeding programmes for disease resistance in Nile tilapia use genomic technologies to identify fish with genetic traits that confer resistance to diseases such as tilapia lake virus (TiLV) and *Streptococcus iniae*. Significant genomic regions linked to TiLV resistance have been identified, enabling selective breeding for disease-resistant tilapia (Barría *et al.*, 2020). Similarly, genetic markers associated with resistance to *S. iniae* (Chen *et al.*, 2021; Shoemaker *et al.*, 2022) have been found, aiding the selective breeding of more resilient fish against this pathogen. Immunomodulators, which modulate the immune response by initiating, enhancing or suppressing it, can significantly reduce disease-related losses in aquaculture (Falco *et al.*, 2023). Immunostimulants are commonly used immunomodulators that activate or enhance the immune responses of fish. This activation can occur via either specific antigen responses or non-specific routes not involving antigen recognition, such as adjuvants or general immunostimulants. Vaccines and functional feeds incorporating immunostimulants, nucleotides and pre-and probiotics are commercially available and support aquaculture operators in lowering disease-related mortality rates. Vaccines and functional feeds incorporating immunostimulants, nucleotides and pre- and probiotics are commercially available and support aquaculture operators in lowering disease-related mortality rates. Adjuvants enhance the efficacy of vaccines by inducing a more robust immune response to the vaccine's antigens. Cytokines, produced by the cellular immune system, are highlighted as potent immunostimulants that significantly enhance immune functions (Secombes & Wang, 2012).

Probiotics, prebiotics and synbiotics

The benefits of gut flora for animal health have been extensively studied, focusing on their preventive effects against diseases (Holzapfel & Schillinger, 2002). In aquaculture, as with successes seen in livestock, probiotics and prebiotics are increasingly popular for enhancing the health and growth of tilapia. Defined by the FAO/WHO (2001), probiotics are live microorganisms which, if administered in sufficient amounts, can provide health benefits to their host. Introducing a novel probiotic product requires confirmation of its effectiveness and safety *in vivo*. Such a product must be non-harmful to the host, easily ingested and capable of colonizing and proliferating within the host's gastrointestinal tract. It is also crucial that the probiotic lacks any genes that might promote virulence or antibiotic resistance (Kesarcodi-Watson *et al.*, 2008).

Research has explored the probiotic potential of various bacteria in aquaculture, including species such as *Lactobacillus*, *Lactococcus*, *Leuconostoc*, *Enterococcus*, *Carnobacterium*, *Shewanella*, *Bacillus*, *Aeromonas*, *Vibrio*, *Enterobacter*, *Pseudomonas*, *Clostridium*, *Saccharomyces*, *Pediococcus* and *Streptococcus* (Mancuso, 2013; Pandiyan *et al.*, 2013). These probiotics help maintain a balanced gut microbiome by competing for nutrients and attachment sites, thereby inhibiting the growth of pathogenic bacteria through the secretion of antimicrobial substances (Irianto & Austin, 2002). Furthermore, these microorganisms have been shown to bolster the immune response of their fish hosts (Balcázar *et al.*, 2006; Nayak, 2010b).

Studies have demonstrated that specific probiotics, such as *Pseudomonas* sp. M174, can suppress aquaculture pathogens, notably *Flavobacterium psychrophilum*, by producing siderophores with limited iron (Korkea-aho *et al.*, 2011). In addition, some probiotics have been shown to activate the immune system, as indicated by increased levels of serum immunoglobulins and enhanced activity of macrophages and neutrophils (Panigrahi *et al.*, 2005; Balcázar *et al.*, 2007).

Feeding probiotics to tilapia has been shown to boost their innate immune responses, as evidenced by several studies (Martínez Cruz *et al.*, 2012; Zorriehzahra *et al.*, 2016; Xia *et al.*, 2020). Abdel-Tawwab and Ahmad (2009) showed that Nile tilapia fed with diets supplemented with Spirulina (*Arthrospira platensis*) exhibited increased red and white blood cell counts and higher macrophage respiratory burst activity than the control group. In addition, incorporation of *Bacillus subtilis*, *Aspergillus oryzae* and *Saccharomyces cerevisiae* into tilapia diets significantly enhanced macrophage respiratory burst activity, improved erythrocyte robustness and elevated white blood cell counts (P <0.05), while also reducing mortality rates following experimental infections with *A. hydrophila* and *S. iniae* (Iwashita *et al.*, 2015). Lactic acid bacteria (LAB) and other probiotic bacteria have also been noted for their ability to enhance fish survival by modulation of immune functions (Kuebutornye *et al.*, 2019). Aly *et al.* (2008) found significant increases in total leukocyte counts and the populations of lymphocytes and monocytes after feeding *Bacillus pumilus* to fish. *Bacillus subtilis* boosts not only innate immunity in Nile tilapia but also alleviates stress-induced immunosuppression that is common in densely stocked environments, leading to increases in mean corpuscular haemoglobin and enhanced lysozyme and phagocytic activities (Telli *et al.*, 2014). Furthermore, *Lactobacillus plantarum* JCM 1149 was found to modulate the localized immune response in an *ex vivo* anterior sac of hybrid tilapia (*Oreochromis niloticus* ♀ × *Oreochromis aureus* ♂) (Ren *et al.*, 2013). In addition, *P. fluorescens* effectively reduced mortality rates in Nile tilapia when faced with challenges from *A. hydrophila* (Ridha & Azad, 2012; Eissa & Abou-ElGheit, 2014).

On the other hand, prebiotics, which are non-digestible food components, play a crucial role in enhancing host health by selectively promoting the growth and/or activity of beneficial gut microbes (Roberfroid, 2007). Substances such as inulin, fructo-oligosaccharides (FOS) and mannan-oligosaccharides have proved effective in improving gut health by nurturing beneficial bacteria that produce vital short-chain fatty acids (Ringø *et al.*, 2010). These functional saccharides enhance gut morphology and activate immune responses by interacting with pathogen recognition receptors on immune cells (Song *et al.*, 2014).

The synergistic combination of probiotics and prebiotics, known as synbiotics, enhances the beneficial impacts of probiotics on the host's gut health and immune system. Such interactions have been shown to improve tilapia overall health, growth performance and disease resistance (Ringø & Song, 2016). In addition, the inclusion of plant products or β-glucans in synbiotic formulations has been investigated, with findings indicating positive effects on the gut microbiota and immune parameters of tilapia (Rodriguez-Estrada *et al.*, 2009; Ringø & Song, 2016).

The use of probiotics, prebiotics and synbiotics in tilapia aquaculture represents a promising strategy to enhance fish health and productivity and help improve their resistance to disease. These feed additives have been shown to potentially control diseases in tilapia farming, including protection against *A. hydrophila* (Cavalcante *et al.*, 2020). Gaining a deeper understanding of the interactions between these substances and host–microbe dynamics *in vivo* is crucial for optimizing their effectiveness.

Immunostimulants

Immunostimulants are widely used in aquaculture disease control strategies to boost disease resistance and immune response in fish under stress. These agents can be natural and synthetic, including β-glucans, chitin, lactoferrin, levamisole, vitamins B and C, growth hormones and prolactin (Sakai, 1999; Ringo, et al., 2011). Commonly derived from the cell walls of baker's yeast (*Saccharomyces cerevisiae*), β-glucans are also sourced from various plants, other yeasts, seaweeds, mushrooms, fungi and bacteria (Sirimanapong *et al.*, 2015). These glucose polymers consist of a β (1,3)-linked backbone with varying β-(1,6)-linked side chains, influencing their immunomodulation effects (Auinger *et al.*, 2013; Meena *et al.*, 2013).

β-Glucans boost the fish's innate defence systems by improving macrophage phagocytic and bactericidal functions (Ranjan *et al.*,

2012), enhancing complement and lysozyme activities and strengthening antibody responses (Sakai, 1999; Dong *et al.*, 2015). Many studies, including a review by Newaj-Fyzul and Austin (2015), document the efficacy of immunostimulants in enhancing disease resistance in fish. Medicinal plants are also explored as immune modulators, particularly those used in traditional Chinese and Indian medicine (Bulfon *et al.*, 2015; Hai, 2015; Newaj-Fyzul and Austin, 2015). The synergistic effects of combining immunostimulants with vaccines have also been recognized (Tafalla *et al.*, 2013).

Research shows that increasing yeast in the diet of Nile tilapia lowers mortality rates from *A. hydrophila* infections. In addition, the addition of 1.0% of the carotenoid product Lycogen™ from *Rhodobacter sphaeroides* WL-APD911 to feed has been shown to enhance the growth and immune health of tilapia hybrids (*O. mossambicus* ×*O. niloticus*) (Wang *et al.*, 2023). Also, see Chapter 8 for additional reference to immunostimulants.

Nucleotide diets

Dietary nucleotides are low-molecular-weight molecules with a nitrogenous base, a five-carbon sugar and a phosphate group. They are considered conditionally essential nutrients, especially under specific physiological challenges (Kenari *et al.*, 2013). They function as co-factors in cell signalling and metabolism and are vital precursors for DNA replication (Burrells *et al.*, 2001; Peng *et al.*, 2013). As fundamental components of nucleic acids, nucleotides play a key role in various biological processes, including development, cell repair (Trichet, 2010) and immunopotentiation. This latter role enhances the fish's innate and adaptive immune responses, improving its resistance to viral, bacterial and parasitic infections. The benefits of dietary nucleotides extend to higher vertebrates as well, influencing the gastrointestinal tract's physiological, morphological and microbiological aspects, including altering the composition of intestinal microflora. However, the specific impact of exogenous nucleotides on immune expression remains less explored and requires further investigation (Leonardi *et al.*, 2003; Glencross & Rutherford, 2010; Tahmasebi-Kohyani *et al.*, 2011; Ringø *et al.*, 2012).

In aquaculture, nucleotide-supplemented diets have demonstrated positive effects on fish health and growth, improvement in growth performance, enhancement of intestinal morphology and function, support of liver function and modulation of both innate and adaptive immune responses (Burrells *et al.*, 2001; Sakai *et al.*, 2001; Low *et al.*, 2003; Malina *et al.*, 2005; Li & Gatlin III, 2006; Jha *et al.*, 2007; Li *et al.*, 2007; Lin *et al.*, 2009; Tahmasebi-Kohyani *et al.*, 2012; Peng *et al.*, 2013). Relevant studies focusing on tilapia include Shiau *et al.* (2015), Xu *et al.* (2015), De Lima *et al.* (2020) and Selim *et al.* (2020). Given the range of benefits observed, feed formulations must be appropriately balanced with dietary nucleotides to meet the physiological needs of the fish.

Medicinal plants

Medicinal plants, recognized for their broad spectrum of bioactive compounds, have been used for thousands of years in traditional medicine as a source of immunostimulants (Vaseeharan & Thaya, 2014). They contain various phytochemicals such as alkaloids, terpenoids, lectins, polyphenolics, phenolics, quinones and polypeptides (Olusola *et al.*, 2013), which have immunomodulatory, anti-inflammatory, antioxidant, anti-tumour, antimicrobial and antiviral properties (Harikrishnan *et al.*, 2011; Foysal *et al.*, 2019; Kuebutornye and Abarike, 2020; Abd El-Hamid *et al.*, 2021; Ching *et al.*, 2021).

Applied as whole plants, plant parts or extracts, these medicinal components are often added directly to the water or included in the fish's diet, sometimes alone or combined with other bioactive compounds. The integration of herbs as feed additives in tilapia diets has gained significant attention (Wang *et al.*, 2023). While growth performance is an important metric for evaluation of feed additives in tilapia, enhancing the fish's immune system is equally vital for disease prevention. Studies have demonstrated that these additives can boost growth, immunity and overall health in fish (Kuebutornye and Abarike, 2020). Experimental diets supplemented with specific herbal extracts such as cinnamon (*Cinnamomum zeylanicum*) (Abdel-Tawwab *et al.*, 2018), purslane (*Portulaca oleracea*) (Abdelrazek *et al.*,

2019), turmeric (*Curcuma longa*) (Abdelrazek *et al.*, 2017), elephant's foot (*Elephantopus scaber*) (Doan *et al.*, 2019) and garlic (*Allium sativum*) (Zare *et al.*, 2021) have been shown to enhance disease resistance in farmed tilapia. Fish exhibited increased anti-bacterial activity against both Gram-positive and -negative bacteria. The primary bioactive compounds in these herbs also regulate immune system functions, mainly through the proliferation of immune cells (Ardó *et al.*, 2008; Ding *et al.*, 2021), and have elevated antioxidant properties (Subramanian *et al.*, 2014). Furthermore, the mucosal immune response in tilapia is triggered using medicinal plants (Srichaiyo *et al.*, 2020). The potential of herbal medicine as a viable alternative for disease control by boosting immunity, improving disease resistance and mitigating immunosuppression is discussed further in Chapter 8 and in reviews by Reverter *et al.* (2017), Kuebutornye and Abarike (2020), Tadese *et al.* (2022) and Balaganesan *et al.* (2023).

Vaccines

Vaccination is an important strategy in the management of infectious diseases within finfish aquaculture due to its ability to elicit protective immunity and memory responses in vaccinated tilapia (Thompson *et al.*, 2023). Fewer vaccines have been commercialized specifically for tilapia. These include vaccines against streptococcosis from MSD Animal Health – namely AQUAVAC· Strep Sa for *S. agalactiae* serotype Ib, AQUAVAC· Strep Sa1 for *S. agalactiae* serotypes Ia and III and AQUAVAC· Strep Si for *S. iniae* and ALPHA JECT· micro 1 TiLa from Pharmaq. Additionally, MSD Animal Health provides an ISKNV vaccine (AQUAVAC· IridoV), although availability is limited to countries with regulatory approval.

There are extensive reports in the literature on experimental vaccines targeting various pathogens of tilapia. Current vaccines for rapidly spreading TiLV include inactivated (Mai *et al.*, 2021; Zeng *et al.*, 2021a, b), live-attenuated (Bacharach *et al.*, 2016), subunit and DNA formulations (Zeng *et al.*, 2021a, b), with reported survival rates between 58.0 and 86.7%. Similar vaccine technologies are also employed for other key tilapia pathogens such as *S. agalactiae*, *S. iniae*, *A. hydrophila*, *Edwardsiella tarda*, *F. orientalis* and

F. columnare in reviews from Munang'andu *et al.* (2016), Bøgwald & Dalmo (2019), Komar *et al.* (2021), Miccoli *et al.* (2021) and Wang *et al.* (2023).

Despite the availability of these vaccines, only 5% of tilapia globally are vaccinated, with the highest uptake in Latin America at 35% and notably <1% in Asia and Africa (Komar *et al.*, 2021). Vaccines are primarily administered via intraperitoneal injection, providing robust and lasting protection. However, vaccination rates are low during the grow-out phases on farms due to the low economic value of tilapia and tight profit margins, combined with logistical challenges (Komar *et al.*, 2021). The optimal strategy suggests vaccinating tilapia in hatcheries before transfer to grow-out sites, although this method faces constraints including high costs and difficulties in handling small fish. For smaller fish, immersion vaccination is widely utilized although its effectiveness is generally lower due to inadequate antigen uptake through mucosal-associated lymphoid tissues such as skin and gills (Munang'andu *et al.*, 2016). Oral vaccination, favoured for its ease and lower costs, is hindered by poor efficacy issues stemming from antigen degradation in the stomach and inconsistent dosing among individual fish (Dalmo *et al.*, 2016; Embregts & Forlenza, 2016). Adjuvants are often added to injection vaccines to enhance the immune response and activate specific lymphocyte actions (Schijns, 2001; Ribeiro & Schijns, 2010). While commercial adjuvants, such as Montanides from Seppic SA, are commonly used in these vaccines, options for adjuvants in mucosal vaccines (both immersion and oral) remain limited (Adams, 2019; Galindo-Villegas *et al.*, 2019; Hoare *et al.*, 2021; Wangkaghart *et al.*, 2021). Further information about vaccines for tilapia can be found in Chapter 16.

Concluding remarks and future perspectives

Effective disease management is crucial for sustainable aquaculture, emphasizing strategies that prevent disease onset and mitigate losses during outbreaks. Despite advanced molecular tools for identification of immune gene changes, we still need better methods to analyse immune cells and proteins. Innovations such as next-generation sequencing, proteomics and bioinformatics are enhancing our understanding of fish immunity. These technologies aid in developing selective breeding for disease resistance, reducing the experimental use of fish and creating novel vaccines with mucosal delivery systems.

The fish's environment significantly influences its health. Stress from routine husbandry and fluctuating water conditions can suppress the immune system. By understanding these impacts we can predict and counteract immunosuppressive events. Functional feeds, including probiotics, prebiotics and immunostimulants, are also beneficial. These products enhance growth, disease resistance and immune responses, particularly in young tilapia. Probiotics, for instance, modify the gut's microbial community, boosting health outcomes. However, more research is needed on the composition and function of the gut microbiota in fish to better manipulate it for improved health and performance. The concept of synbiotics, combining pre- and probiotics, further optimizes these benefits. A comprehensive approach to disease management in aquaculture should integrate biosecurity, optimal nutrition with functional feeds, effective vaccine, and selective breeding to enhance overall fish well-being.

References

Aas, I.B., Austbø, L., König, M., Syed, M., Falk, K. *et al.* (2014) Transcriptional characterisation of the T cell population within the salmonid interbranchial lymphoid tissue. *Journal of Immunology* 193(7), 3463–3469. DOI: 10.4049/jimmunol.1400797.

Abd El-Hamid, M.I., Ibrahim, S.M., Eldemery, F., El-Mandrawy, S.A.M., Metwally, A.S. *et al.* (2021) Dietary cinnamaldehyde nanoemulsion boosts growth and transcriptomes of antioxidant and immune-related genes to fight *Streptococcus agalactiae* infection in Nile tilapia (*Oreochromis niloticus*). *Fish and Shellfish Immunology* 113, 96–105. DOI: 10.1016/j.fsi.2021.03.021.

Abdelrazek, H.M.A., Tag, H.M., Kilany, O.E., Reddy, P.G. and Hassan, A.M. (2017) Immunomodulatory effect of dietary turmeric supplementation on Nile tilapia (*Oreochromis niloticus*). *Aquaculture Nutrition* 23(5), 1048–1054. DOI: 10.1111/anu.12472.

Abdelrazek, N., Awad, S.M. and Abdel-Tawwab, M. (2019) Effect of dietary purslane (*Portulaca oleracea* L.) leaves powder on growth, immunostimulation, and protection of Nile tilapia, *Oreochromis niloticus* against *Aeromonas hydrophila* infection. *Fish Physiology and Biochemistry* 45(6), 1907–1917. DOI: 10.1007/s10695-019-00685-8.

Abdel-Salam, S.G.R., Tsujikura, M., Kondo, M., Somamoto, T. and Nakao, M. (2014) Purification and functional characterization of complement C3 and a novel zymosan-binding protein in tilapia serum. *Fisheries Science* 80(2), 301–310. DOI: 10.1007/s12562-014-0700-7.

Abdel-Tawwab, M. and Ahmad, M.H. (2009) Live spirulina (*Arthrospira platensis*) as a growth and immunity promoter for Nile tilapia, *Oreochromis niloticus* (L.), challenged with pathogenic *Aeromonas hydrophila*. *Aquaculture Research* 40(9), 1037–1046. DOI: 10.1111/j.1365-2109.2009.02195.x.

Abdel-Tawwab, M., Samir, F., Abd El-Naby, A.S. and Monier, M.N. (2018) Antioxidative and immunostimulatory effect of dietary cinnamon nanoparticles on the performance of Nile tilapia, *Oreochromis niloticus* (L.) and its susceptibility to hypoxia stress and *Aeromonas hydrophila* infection. *Fish and Shellfish Immunology* 74, 19–25. DOI: 10.1016/j.fsi.2017.12.033.

Abouelmaatti, R.R., Algammal, A.M., Li, X., Ma, J., Abdelnaby, E.A. *et al.* (2013) Cloning and analysis of Nile tilapia toll-like receptors type-3 mRNA. *Central European Journal of Immunology* 38, 277–282.

Adams, A. (2019) Progress, challenges and opportunities in fish vaccine development. *Fish and Shellfish Immunology* 90, 210–214. DOI: 10.1016/j.fsi.2019.04.066.

Agius, C. and Roberts, R.J. (2003) Melano-macrophage centres and their role in fish pathology. *Journal of Fish Diseases* 26(9), 499–509. DOI: 10.1046/j.1365-2761.2003.00485.x.

Al-Harbi, A.H. and Naim, U.M. (2004) Seasonal variation in the intestinal bacterial flora of hybrid tilapia (*Oreochromis niloticus* x *Oreochromis aureus*) cultured in earthen ponds in Saudi arabia. *Aquaculture* 229(1), 37–44.

Alvarez-Pellitero, P. (2008) Fish immunity and parasite infections: from innate immunity to immunoprophylactic prospects. *Veterinary Immunology and Immunopathology* 126, 171–198.

Aly, S.M., Mohamed, M.F. and John, G. (2008) Effect of probiotics on the survival, growth, and challenge infection in *Tilapia nilotica*

(*Oreochromis niloticus*). *Aquaculture Research* 39(6), 647–656. DOI: 10.1111/j.1365-2109.2008.01932.x.

Ángeles, M.C., Cuesta, A., Chaves-Pozo, E. and Meseguer, J. (2015) Phagocytosis in teleosts. Implications of the new cells involved. *Biology* 4, 907–922.

Aoki, T., Hikima, J., Hwang, S.D. and Jung, T.S. (2013) Innate immunity of finfish: Primordial conservation and function of viral RNA sensors in teleosts. *Fish and Shellfish Immunology* 35(6), 1689–1702. DOI: 10.1016/j.fsi.2013.02.005.

Ardó, L., Yin, G., Xu, P., Váradi, L., Szigeti, G. *et al.* (2008) Chinese herbs (*Astragalus membranaceus* and *Lonicera japonica*) and boron enhance the non-specific immune response of Nile tilapia (*Oreochromis niloticus*) and resistance against *Aeromonas hydrophila*. *Aquaculture* 275(1), 26–33.

Auinger, A., Riede, L., Bothe, G., Busch, R. and Gruenwald, J. (2013) Yeast (1, 3)-(1, 6)-beta-glucan helps to maintain the body's defence against pathogens: A double-blind, randomized, placebo-controlled, multicentric study in healthy subjects. *European Journal of Nutrition* 52, 1913–1918.

Bacharach, E., Mishra, N., Briese, T., Zody, M.C., Kembou Tsofack, J.E. *et al.* (2016) Characterization of a novel Orthomyxo-like virus causing mass die-offs of tilapia. *mBio* 7(2), e00431–e00416. DOI: 10.1128/mBio.00431-16.

Balaganesan, M., Kaliyamurthi, V. and Binesh, A. (2023) Medicinal plants or plant-derived compounds used in aquaculture. In: Mathew, J., Jose, M.S., Jose, R.E.K. and Kumar, A. (eds) *Recent Advances in Aquaculture Microbial Technology*. Academic Press, pp. 153–207.

Balcázar, J.L., de Blas, I.D., Ruiz-Zarzuela, I., Cunningham, D., Vendrell, D. *et al.* (2006) The role of probiotics in aquaculture. *Veterinary Microbiology* 114(3–4), 173–186. DOI: 10.1016/j.vetmic.2006.01.009.

Balcázar, J.L., de Blas, I., Ruiz-Zarzuela, I., Vendrell, D., Calvo, A.C. *et al.* (2007) Changes in intestinal microbiota and humoral immune response following probiotic administration in brown trout (*Salmo trutta*). *The British Journal of Nutrition* 97(3), 522–527. DOI: 10.1017/S0007114507432986.

Barría, A., Trinh, T.Q., Mahmuddin, M., Benzie, J.A.H., Chadag, V.M. *et al.* (2020) Genetic parameters for resistance to tilapia lake virus (TiLV) in Nile tilapia (*Oreochromis niloticus*). *Aquaculture* 522, 735126. DOI: 10.1016/j.aquaculture.2020.735126.

Bates, J.M., Mittge, E., Kuhlman, J., Baden, K.N., Cheesman, S.E. *et al.* (2006) Distinct signals from the microbiota promote different aspects of zebrafish gut differentiation. *Developmental Biology* 297(2), 374–386. DOI: 10.1016/j.ydbio.2006.05.006.

Bates, J.M., Akerlund, J., Mittge, E. and Guillemin, K. (2007) Intestinal alkaline phosphatase detoxifies lipopolysaccharide and prevents inflammation in zebrafish in response to the gut microbiota. *Cell Host and Microbe* 2(6), 371–382. DOI: 10.1016/j.chom.2007.10.010.

Bereded, N.K., Abebe, G.B., Fanta, S.W., Curto, M., Waidbacher, H. *et al.* (2021) The impact of sampling season and catching site (wild and aquaculture) on gut microbiota composition and diversity of Nile tilapia (*Oreochromis niloticus*). *Biology* 10(3). DOI: 10.3390/biology10030180.

Biller, J.D., Polycarpo, G.V., Moromizato, B.S., Sidekerskis, A.P.D., Silva, T.D. *et al.* (2021) Lysozyme activity as an indicator of innate immunity of tilapia (*Oreochromis niloticus*) when challenged with LPS and *Streptococcus agalactiae*. *Revista Brasileira de Zootecnia* 50, e20210053. DOI: 10.37496/rbz5020210053.

Bøgwald, J. and Dalmo, R.A. (2019) Review on immersion vaccines for fish: An update 2019. *Microorganisms* 7(12), 627. DOI: 10.3390/microorganisms7120627.

Bowden, T.J. (2008) Modulation of the immune system of fish by their environment. *Fish and Shellfish Immunology* 25(4), 373–383. DOI: 10.1016/j.fsi.2008.03.017.

Bowden, T.J., Thompson, K.D., Morgan, A.L., Gratacap, R.M. and Nikoskelainen, S. (2007) Seasonal variation and the immune response: A fish perspective. *Fish and Shellfish Immunology* 22(6), 695–706. DOI: 10.1016/j.fsi.2006.08.016.

Brattgjerd, S. and Evensen, O. (1996) A sequential light microscopic and ultrastructural study on the uptake and handling of *Vibrio salmonicida* in phagocytes of the head kidney in experimentally infected atlantic salmon (*Salmo salar* L.). *Veterinary Pathology* 33(1), 55–65.

Bruce, T.J. and Brown, M.L. (2017) A review of immune system components, cytokines, and immunostimulants in cultured finfish species. *Open Journal of Animal Sciences* 07(3), 267–288. DOI: 10.4236/ojas.2017.73021.

Bulfon, C., Volpatti, D. and Galeotti, M. (2015) Current research on the use of plant-derived products in farmed fish. *Aquaculture Research* 46(3), 513–551. DOI: 10.1111/are.12238.

Burrells, C., Williams, P.D., Southgate, P.J. and Wadsworth, S.L. (2001) Dietary nucleotides: A novel supplement in fish feeds: 2. effects on vaccination, salt water transfer, growth rates and physiology of atlantic salmon (*Salmo salar* L.). *Aquaculture* 199(1–2), 171–184.

Cabillon, N.A.R. and Lazado, C.C. (2019) Mucosal barrier functions of fish under changing environmental conditions. *Fishes* 4(1), 2. DOI: 10.3390/fishes4010002.

Cavalcante, R.B., Telli, G.S., Tachibana, L., Dias, D.D., Oshiro, E. *et al.* (2020) Probiotics, prebiotics, and synbiotics for Nile tilapia: Growth performance and protection against *Aeromonas hydrophila* infection. *Aquaculture Reports* 17, 100343. DOI: 10.1016/j.aqrep.2020.100343.

Chen, C.C., Huang, C.W., Lin, C.Y., Ho, C.H., Pham, H.N. *et al.* (2021) Development of disease-resistance-associated microsatellite DNA markers for selective breeding of tilapia (*Oreochromis* spp.) farmed in taiwan. *Genes* 13(1), 99. DOI: 10.3390/genes13010099.

Chen, J., Zhi, T., Xu, X., Zhang, S., Zheng, Y. *et al.* (2019b) Molecular characterization and dynamic expressions of three Nile tilapia (*Oreochromis niloticus*) complement genes after *Gyrodactylus cichlidarum* (Monogenea) infection. *Aquaculture* 502, 176–188. DOI: 10.1016/j.aquaculture.2018.12.018.

Chen, L., Deng, H., Cui, H., Fang, J., Zuo, Z. *et al.* (2018) Inflammatory responses and inflammation-associated diseases in organs. *Oncotarget* 9(6), 7204–7218. DOI: 10.18632/oncotarget.23208.

Chen, M., Ding, M., Li, Y., Zhong, X., Liu, S. *et al.* (2018) The complement component 1 q (C1q) in Nile tilapia (*Oreochromis niloticus*): Functional characterization in host defense against bacterial infection and effect on cytokine response in macrophages. *Developmental and Comparative Immunology* 87, 98–108. DOI: 10.1016/j.dci.2018.05.023.

Chen, M., Liu, S., Yan, F., Zhou, E., Zhong, X. *et al.* (2019a) Complement 1q-binding protein from Nile tilapia (*Oreochromis niloticus*): Molecular characterization, expression pattern upon bacterial infection, and its binding properties. *Aquaculture* 500, 31–40. DOI: 10.1016/j.aquaculture.2018.09.060.

Chertov, O., Yang, D., Howard, O.M. and Oppenheim, J.J. (2000) Leukocyte granule proteins mobilize innate host defenses and adaptive immune responses. *Immunological Reviews* 177, 68–78. DOI: 10.1034/j.1600-065x.2000.17702.x.

Ching, J.J., Shuib, A.S., Abdullah, N., Majid, N.A., Taufek, N.M. *et al.* (2021) Hot water extract of *Pleurotus pulmonarius* stalk waste enhances innate immune response and immune-related gene expression in red hybrid tilapia *Oreochromis* sp. following challenge with pathogen-associated molecular patterns. *Fish and Shellfish Immunology* 116, 61–73. DOI: 10.1016/j.fsi.2021.06.005.

Dalmo, R.A. and Bøgwald, J. (2022) Innate immunity. In: Buchmann, K. and Secombes, C.J. (eds) *Principles of Fish Immunology*. Springer nature.

Dalmo, R., Bogwald, J. and Tafalla, C. (2016) Adjuvants and delivery methods: Current and novel. In: Adams, A. (ed.) *Advances in Infectious Diseases*. Fish Vaccines. Birkhäuser, Springer, pp. 75–103.

Dannevig, B.H., Lauve, A., Press, C.M. and Landsverk, T. (1994) Receptor-mediated endocytosis and phagocytosis by rainbow trout head kidney sinusoidal cells. *Fish and Shellfish Immunology* 4(1), 3–18. DOI: 10.1006/fsim.1994.1002.

Deakin, J.E., Papenfuss, A.T., Belov, K., Cross, J.G., Coggill, P. *et al.* (2006) Evolution and comparative analysis of the MHC class III inflammatory region. *BMC Genomics* 7, 281. DOI: 10.1186/1471-2164-7-281.

Defoirdt, T., Sorgeloos, P. and Bossier, P. (2011) Alternatives to antibiotics for the control of bacterial disease in aquaculture. *Current Opinion in Microbiology* 14(3), 251–258.

De Lima, S.A., Pedreira, A.C.D.O., De Freitas, J.M.A., Dalmaso, A.C.S., Chiella, R.J. *et al.* (2020) Diets containing purified nucleotides reduce oxidative stress, interfere with reproduction, and promote growth in Nile tilapia females. *Aquaculture* 528, 735509. DOI: 10.1016/j.aquaculture.2020.735509.

Deng, Y., Verdegem, M., Eding, E. and Kokou, F. (2021) Effect of rearing systems and dietary probiotic supplementation on the growth and gut microbiota of Nile tilapia (*Oreochromis niloticus*) larvae. *Aquaculture* 546, 737297.

Dhanarso, P., Yunissa, H., Istiqomah, I. and Isnansetyo, A. (2021) Complement system activation in red tilapia (*Oreochromis* sp.) orally administered with probiotics SEAL. *IOP Conference Series: Earth and Environmental Science* 718(1). DOI: 10.1088/1755-1315/718/1/012055.

Ding, Z.-H., Hong, J.-M., Guo, W.-L., Li, G.-H., Zhao, Z.-C. *et al.* (2021) The screen herbal immunopotentiator and research on its effect on the innate immune system and disease resistance of Nile tilapia (*Oreochromis niloticus*) against *Streptococcus agalactiae*. *Aquaculture* 541, 736778. DOI: 10.1016/j.aquaculture.2021.736778.

Doan, H.V., Hoseinifar, S.H., Sringarm, K., Jaturasitha, S., Khamlor, T. *et al.* (2019) Effects of elephant's foot (*Elephantopus scaber*) extract on growth performance, immune response, and disease resistance of Nile tilapia (*Oreochromis niloticus*) fingerlings. *Fish and Shellfish Immunology* 93, 328–335. DOI: 10.1016/j.fsi.2019.07.061.

Dong, H.T., Nguyen, V.V., Le, H.D., Sangsuriya, P., Jitrakorn, S. *et al.* (2015) Naturally concurrent infections of bacterial and viral pathogens in disease outbreaks in cultured Nile tilapia (*Oreochromis niloticus*) farms. *Aquaculture* 448, 427–435. DOI: 10.1016/j.aquaculture.2015.06.027.

Dong, J., Wei, Y., Sun, C., Tian, Y., Hu, J. *et al.* (2018) Interaction of group B Streptococcus sialylated capsular polysaccharides with host siglec-like molecules dampens the inflammatory response in tilapia. *Molecular Immunology* 103, 182–190. DOI: 10.1016/j.molimm.2018.09.014.

Dong, X.-H., Geng, X., Tan, B.-P., Yang, Q.-H., Chi, S.-Y. *et al.* (2015) Effects of dietary immunostimulant combination on the growth performance, non-specific immunity and disease resistance of cobia, *Rachycentron canadum* (linnaeus). *Aquaculture Research* 46(4), 840–849. DOI: 10.1111/are.12257.

Dunkelberger, J.R. and Song, W.C. (2010) Complement and its role in innate and adaptive immune responses. *Cell Research* 20(1), 34–50. DOI: 10.1038/cr.2009.139.

Edholm, E.S., Bengten, E. and Wilson, M. (2011) Insights into the function of IgD. *Developmental and Comparative Immunology* 35(12), 1309–1316. DOI: 10.1016/j.dci.2011.03.002.

Eissa, N. and Abou-ElGheit, E. (2014) Dietary supplementation impacts of potential non-pathogenic isolates on growth performance, hematological parameters, and disease resistance in Nile tilapia (*Oreochromis niloticus*). *Journal of Veterinary Advances* 4(10), 712–719. DOI: 10.5455/jva.20141025045451.

Ellis, A.E. (2001) Innate host defense mechanisms of fish against viruses and bacteria. *Developmental and Comparative Immunology* 25(8–9), 827–839. DOI: 10.1016/s0145-305x(01)00038-6.

Ellmeier, W., Haust, L. and Tschismarov, R. (2013) Transcriptional control of CD4 and CD8 co-receptor expression during T cell development. *Cellular and Molecular Life Sciences* 70(23), 4537–4553. DOI: 10.1007/s00018-013-1393-2.

Ellsaesser, C.F., Bly, J.E. and Clem, L.W. (1988) Phylogeny of lymphocyte heterogeneity: the thymus of the channel catfish. *Developmental and Comparative Immunology* 12(4), 787–799. DOI: 10.1016/0145-305x(88)90053-5.

Embregts, C.W.E. and Forlenza, M. (2016) Oral vaccination of fish: Lessons from humans and veterinary species. *Developmental and Comparative Immunology* 64, 118–137. DOI: 10.1016/j.dci.2016.03.024.

Espelid, S., Løkken, G.B., Steiro, K. and Bøgwald, J. (1996) Effects of cortisol and stress on the immune system in Atlantic salmon (*Salmo salar* L.). *Fish and Shellfish Immunology* 6(2), 95–110. DOI: 10.1006/fsim.1996.0011.

Esteban, M.Á. (2012) An overview of the immunological defenses in fish skin. *International Scholarly Research Notices* 853470.

Esteban, M.Á., Cuesta, A., Chaves-Pozo, E. and Meseguer, J. (2015) Phagocytosis in teleosts. Implications of the new cells involved. *Biology* 4(4), 907–922. DOI: 10.3390/biology4040907.

Evans, D.L., Hogan, K.T., Graves, S.S., Carlson, R.L., Floyd, E. *et al.* (1984) Nonspecific cytotoxic cells in fish (*Ictalurus punctatus*). III. Biophysical and Biochemical Properties Affecting Cytolysis. *Developmental and Comparative Immunology* 8(3), 599–610. DOI: 10.1016/0145-305x(84)90092-2.

Falco, F.F., Banaee, M., Mauro, M., Faggio, C., Kollath, A. *et al.* (2023) Immunomodulators: An introduction. In: *Immunomodulators in Aquaculture and Fish Health.* eBook, 1st edn. CRC Press ISBN, p. 15. ISBN: 9781003361183

Fernandes, D.C., Eto, S.F., Funnicelli, M.I.G., Fernandes, C.C., Charlie-Silva, I. *et al.* (2019) Immunoglobulin Y in the diagnosis of *Aeromonas hydrophila* infection in Nile tilapia (*Oreochromis niloticus*). *Aquaculture* 500, 576–585. DOI: 10.1016/j.aquaculture.2018.10.045.

Ferrero-Miliani, L., Nielsen, O.H., Andersen, P.S. and Girardin, S.E. (2007) Chronic inflammation: importance of NOD2 and NALP3 in interleukin-1β generation. *Clinical and Experimental Immunology* 147(2), 227–235. DOI: 10.1111/j.1365-2249.2006.03261.x.

Fischer, U., Utke, K., Somamoto, T., Köllner, B., Ototake, M. *et al.* (2006) Cytotoxic activities of fish leucocytes. *Fish and Shellfish Immunology* 20(2), 209–226. DOI: 10.1016/j.fsi.2005.03.013.

Fischer, U., Koppang, E.O. and Nakanishi, T. (2013) Teleost T and NK cell immunity. *Fish and Shellfish Immunology* 35(2), 197–206. DOI: 10.1016/j.fsi.2013.04.018.

Flerova, E.A. and Balabanova, L.V. (2013) Ultrastructure of granulocytes of teleost fish (Salmoniformes, Cypriniformes, Perciformes). *Journal of Evolutionary Biochemistry and Physiology* 49(2), 223–233. DOI: 10.1134/S0022093013020126.

Food and Agriculture Organization/World Health Organization (2001) Health and nutritional properties of probiotics in food including powder milk with live lactic acid bacteria. Report of Joint Food and Agriculture Organization of the United Nations/World Health Expert Consultation Report on Evaluation of Health and Nutritional Properties of Probiotics in Food Including Powder Milk with Live Lactic Acid. The Bacteria, Córdoba, Argentina, October 1-4.

Forlenza, M., Fink, I.R., Raes, G. and Wiegertjes, G.F. (2011) Heterogeneity of macrophage activation in fish. *Developmental and Comparative Immunology* 35(12), 1246–1255. DOI: 10.1016/j.dci.2011.03.008.

Foysal, M.J., Alam, M., Momtaz, F., Chaklader, M.R., Siddik, M.A.B.M. *et al.* (2019) Dietary supplementation of garlic (*Allium sativum*) modulates gut microbiota and health status of tilapia (*Oreochromis niloticus*) against *Streptococcus iniae* infection. *Aquaculture Research* 50(8), 2107–2116. DOI: 10.1111/are.14088.

Fraher, M.H., O'Toole, P.W. and Quigley, E.M. (2012) Techniques used to characterize the gut microbiota: A guide for the clinician. *Nature Reviews. Gastroenterology and Hepatology* 9(6), 312–322. DOI: 10.1038/nrgastro.2012.44.

Frøystad, M.K., Rode, M., Berg, T. and Gjøen, T. (1998) A role for scavenger receptors in phagocytosis of protein-coated particles in rainbow trout head kidney macrophages. *Developmental and Comparative Immunology* 22(5–6), 533–549. DOI: 10.1016/s0145-305x(98)00032-9.

Galagarza, O.A., Smith, S.A., Drahos, D.J., Eifert, J.D., Williams, R.C. *et al.* (2018) Modulation of innate immunity in Nile tilapia (*Oreochromis niloticus*) by dietary supplementation of bacillus subtilis endospores. *Fish and Shellfish Immunology* 83, 171–179. DOI: 10.1016/j.fsi.2018.08.062.

Galindo-Villegas, J., Hosokawa, H, *et al.* (2004) Immunostimulants: Towards temporary prevention of diseases in marine fish. In: Suárez, L.E.C., Ricque, D., Marie, M.G., López, M.G.N. and Villarreal, D. (eds) *Avances en nutrición acuícola VII. Memorias del VII Symposium Internacional de Nutrición Acuícola, 16-19 Noviembre 2004*, Sonara. México, pp. 279–319.

Galindo-Villegas, J., García-Alcazar, A., Meseguer, J. and Mulero, V. (2019) Aluminum adjuvant potentiates gilthead seabream immune responses but induces toxicity in splenic melanomacrophage centers. *Fish and Shellfish Immunology* 85, 31–43. DOI: 10.1016/j.fsi.2018.02.047.

Gao, F.Y., Pang, J.C., Lu, M.X., Yang, X.L., Zhu, H.P. *et al.* (2018) Molecular characterization, expression, and functional analysis of NOD1, NOD2, and NLRC3 in Nile tilapia (*Oreochromis niloticus*). *Fish and Shellfish Immunology* 73, 207–219. DOI: 10.1016/j.fsi.2017.12.012.

Gerlier, D. and Lyles, D.S. (2011) Interplay between innate immunity and negative strand RNA viruses: Towards a rational model. *Microbiology and Molecular Biology Reviews* 75(3), 468–490. DOI: 10.1128/MMBR.00007-11.

Giatsis, C., Sipkema, D., Smidt, H., Heilig, H., Benvenuti, G. *et al.* (2015) The impact of rearing environment on the development of gut microbiota in tilapia larvae. *Scientific Reports* 5(1), 1–15.

Gjedrem, T. (2015) Disease resistant fish and shellfish are within reach: A review. *Journal of Marine Science and Engineering* 3(1), 146–153. DOI: 10.3390/jmse3010146.

Glencross, B. and Rutherford, N. (2010) Dietary strategies to improve the growth and feed utilization of barramundi, lates calcarifer under high water temperature conditions. *Aquaculture Nutrition* 16(4), 343–350. DOI: 10.1111/j.1365-2095.2009.00670.x.

Gomez, D., Sunyer, J.O. and Salinas, I. (2013) The mucosal immune system of fish: The evolution of tolerating commensals while fighting pathogens. *Fish and Shellfish Immunology* 35(6), 1729–1739. DOI: 10.1016/j.fsi.2013.09.032.

Gómez, G.D. and Balcázar, J.L. (2008) A review on the interactions between gut microbiota and innate immunity of fish. *FEMS Immunology and Medical Microbiology* 52(2), 145–154. DOI: 10.1111/j.1574-695X.2007.00343.x.

Hai, N.V. (2015) The use of medicinal plants as immunostimulants in aquaculture: A review. *Aquaculture* 446(1), 88–96.

Harikrishnan, R., Balasundaram, C. and Heo, M.-S. (2011) Impact of plant products on innate and adaptive immune system of cultured finfish and shellfish. *Aquaculture* 317(1–4), 1–15. DOI: 10.1016/j.aquaculture.2011.03.039.

Haugarvoll, E., Bjerkås, I., Nowak, B.F., Hordvik, I. and Koppang, E.O. (2008) Identification and characterization of a novel intraepithelial lymphoid tissue in the gills of Atlantic salmon. *Journal of Anatomy* 213(2), 202–209. DOI: 10.1111/j.1469-7580.2008.00943.x.

Havixbeck, J.J. and Barreda, D.R. (2015) Neutrophil development, migration, and function in teleost fish. *Biology* 4(4), 715–734. DOI: 10.3390/biology4040715.

Havixbeck, J.J., Rieger, A.M., Wong, M.E., Hodgkinson, J.W. and Barreda, D.R. (2016) Neutrophil contributions to the induction and regulation of the acute inflammatory response in teleost fish. *Journal of Leukocyte Biology* 99(2), 241–252. DOI: 10.1189/jlb.3HI0215-064R.

Haygood, A.M. and Jha, R. (2018) Strategies to modulate the intestinal microbiota of tilapia (*Oreochromis* sp.) in aquaculture: A review. *Reviews in Aquaculture* 10(2), 320–333. DOI: 10.1111/raq.12162.

Hazam, P.K. and Chen, J.Y. (2020) Therapeutic utility of the antimicrobial peptide tilapia Piscidin 4 (TP4). *Aquaculture Reports* 17, 100409. DOI: 10.1016/j.aqrep.2020.100409.

Herraez, M.P. and Zapata, A.G. (1986) Structure and function of the melano-macrophage centres of the goldfish *Carassius auratus*. *Veterinary Immunology and Immunopathology* 12(1–4), 117–126.

Hoare, R., Leigh, W., Limakom, T., Wongwaradechkul, R., Metselaar, M. *et al.* (2021) Oral vaccination of Nile tilapia (*Oreochromis niloticus*) against francisellosis elevates specific antibody titres in serum and mucus. *Fish and Shellfish Immunology* 113, 86–88. DOI: 10.1016/j.fsi.2021.03.019.

Hodgkinson, J.W., Grayfer, L. and Belosevic, M. (2015) Biology of bony fish macrophages. *Biology* 4(4), 881–906. DOI: 10.3390/biology4040881.

Holzapfel, W.H. and Schillinger, U. (2002) Introduction to pre- and probiotics. *Food Research International* 35(2–3), 109–116. DOI: 10.1016/S0963-9969(01)00171-5.

Huang, Y., Chen, Z., Xie, R., Wang, P., Zhang, Z. *et al.* (2022) Transferrin mediated NCC killing activity through NCCRP-1 in Nile tilapia (*Oreochromis niloticus*). *Fishes* 7(5), 253. DOI: 10.3390/fishes7050253.

Irianto, A. and Austin, B. (2002) Probiotics in aquaculture. *Journal of Fish Diseases* 25(11), 633–642. DOI: 10.1046/j.1365-2761.2002.00422.x.

Ishimoto, Y., Savan, R., Endo, M. and Sakai, M. (2004) Non-specific cytotoxic cell receptor (NCCRP)-1 type gene in tilapia (*Oreochromis niloticus*): Its cloning and analysis. *Fish and Shellfish Immunology* 16(2), 163–172. DOI: 10.1016/S1050-4648(03)00059-7.

Iwashita, M.K.P., Nakandakare, I.B., Terhune, J.S., Wood, T. and Ranzani-Paiva, M.J.T. (2015) Dietary supplementation with *Bacillus subtilis*, *Saccharomyces cerevisiae* and *Aspergillus oryzae* enhance immunity and disease resistance against *Aeromonas hydrophila* and *Streptococcus iniae* infection in juveNile tilapia *Oreochromis niloticus. Fish and Shellfish Immunology* 43(1), 60–66. DOI: 10.1016/j.fsi.2014.12.008.

Jha, A.K., Pal, A.K., Sahu, N.P., Kumar, S. and Mukherjee, S.C. (2007) Haemato-immunological responses to dietary yeast RNA, ω-3 fatty acid and β-carotene in catla catla juveniles. *Fish and Shellfish Immunology* 23(5), 917–927. DOI: 10.1016/j.fsi.2007.01.011.

Jorgensen, J.B. (2014) Chapter 8. The innate immune response of fish. In: Gudding, R., Lillehaug, A. and Evensen, Ø. (eds) *Fish Vaccination*. John Wiley and Sons, Ltd, pp. 85–103.

Kaattari, S.L. and Irwin, M.J. (1985) Salmonid spleen and anterior kidney harbor populations of lymphocytes with different B cell repertoires. *Developmental and Comparative Immunology* 9(3), 433–444. DOI: 10.1016/0145-305x(85)90006-0.

Katzenback, B.A. and Belosevic, M. (2009) Isolation and functional characterization of neutrophil-like cells, from goldfish (*Carassius auratus* L.) kidney. *Developmental and Comparative Immunology* 33(4), 601–611. DOI: 10.1016/j.dci.2008.10.011.

Kenari, A.A., Mahmoudi, N., Soltani, M. and Abediankenari, S. (2013) Dietary nucleotide supplements influence the growth, haemato-immunological parameters and stress responses in endangered caspian brown trout (*Salmo trutta* caspius kessler, 1877). *Aquaculture Nutrition* 19(1), 54–63. DOI: 10.1111/j.1365-2095.2012.00938.x.

Kesarcodi-Watson, A., Kaspar, H., Lategan, M.J. and Gibson, L. (2008) Probiotics in aquaculture: The need, principles and mechanisms of action and screening processes. *Aquaculture* 274(1), 1–14. DOI: 10.1016/j.aquaculture.2007.11.019.

Kitiyodom, S., Trullàs, C., Rodkhum, C., Thompson, K.D., Katagiri, T. *et al.* (2021a) Modulation of the mucosal immune response of red tilapia (*Oreochromis* sp.) against columnaris disease using a biomimetic-mucoadhesive nanovaccine. *Fish and Shellfish Immunology* 112, 81–91. DOI: 10.1016/j.fsi.2021.02.017.

Kitiyodom, S., Yata, T., Thompson, K.D., Costa, J.Z., Elumalai, P. *et al.* (2021b) Immersion vaccination by a biomimetic-mucoadhesive nanovaccine induces humoral immune response of red tilapia (*Oreochromis* sp.) against Flavobacterium columnare challenge. *Vaccines* 9(11), 1253. DOI: 10.3390/vaccines9111253.

Komar, C., Ariav, R., Thompson, K.D. and Luengnaruemitchai, A. (2021) To vaccinate or not to vaccinate? Tilapia health: *Quo vadis?* Retrieved December 1–3, 2021. Retrieved April 13, 2024. Available at: http://

infofish.org/tilapia/media/attachments/2021/12/06/18_komar_to-vaccinate-or-not.pdf

Koppang, E.O., Hordvik, I., Bjerkås, I., Torvund, J., Aune, L. *et al.* (2003) Production of rabbit antisera against recombinant MHC class II β chain and identification of immunoreactive cells in Atlantic salmon (*Salmo salar*). *Fish and Shellfish Immunology* 14(2), 115–132. DOI: 10.1006/fsim.2002.0424.

Koppang, E.O., Fischer, U., Moore, L., Tranulis, M.A., Dijkstra, J.M. *et al.* (2010) Salmonid T cells assemble in the thymus, spleen, and in novel interbranchial lymphoid tissue. *Journal of Anatomy* 217(6), 728–739. DOI: 10.1111/j.1469-7580.2010.01305.x.

Korkea-aho, T.L., Heikkinen, J., Thompson, K.D., von Wright, A. and Austin, B. (2011) *Pseudomonas* sp. M174 inhibits the fish pathogen *Flavobacterium psychrophilum. Journal of Applied Microbiology* 111(2), 266–277. DOI: 10.1111/j.1365-2672.2011.05044.x.

Kuebutornye, F.K.A. and Abarike, E.D. (2020) The contribution of medicinal plants to tilapia aquaculture: A review. *Aquaculture International* 28(3), 965–983. DOI: 10.1007/s10499-020-00506-3.

Kuebutornye, F.K.A., Abarike, E.D. and Lu, Y. (2019) A review on the application of Bacillus as probiotics in aquaculture. *Fish and Shellfish Immunology* 87, 820–828. DOI: 10.1016/j.fsi.2019.02.010.

Kumar, H., Kawai, T. and Akira, S. (2011) Pathogen recognition by the innate immune system. *International Reviews of Immunology* 30(1), 16–34. DOI: 10.3109/08830185.2010.529976.

Kuroda, N., Figueroa, F., O'hUigin, C. and Klein, J. (2002) Evidence that the separation of MHC class II from class I loci in the zebrafish Danio rerio occurred by translocation. *Immunogenetics* 54(6), 418–430. DOI: 10.1007/s00251-002-0473-5.

Leonardi, M., Sandino, A. and Klempau, A. (2003) Effect of a nucleotide-enriched diet on the immune system, plasma cortisol levels and resistance to infectious pancreatic necrosis (IPN) in juvenile rainbow trout (*Oncorhynchus mykiss*). *Bulletin of the European Association of Fish Pathologists* 23(2), 52–59.

Li, J., Barreda, D.R., Zhang, Y.A., Boshra, H., Gelman, A.E. *et al.* (2006) B lymphocytes from early vertebrates have potent phagocytic and microbicidal abilities. *Nature Immunology* 7(10), 1116–1124. DOI: 10.1038/ni1389.

Li, P. and Gatlin III, D.M. (2006) Nucleotide nutrition in fish: Current knowledge and future applications. *Aquaculture* 251(2–4), 141–152. DOI: 10.1016/j.aquaculture.2005.01.009.

Li, P., Gatlin, D.M. and Neill, W.H. (2007) Dietary supplementation of a purified nucleotide mixture transiently enhanced growth and feed utilization of juvenile red drum, *Sciaenops ocellatus*. *Journal of the World Aquaculture Society* 38(2), 281–286.

Li, Y., Li, Y., Cao, X., Jin, X. and Jin, T. (2017) Pattern recognition receptors in zebrafish provide functional and evolutionary insight into innate immune signaling pathways. *Cellular and Molecular Immunology* 14(1), 80–89. DOI: 10.1038/cmi.2016.50.

Lin, Y.-H., Wang, H. and Shiau, S.-Y. (2009) Dietary nucleotide supplementation enhances growth and immune responses of grouper, *Epinephelus malabaricus*. *Aquaculture Nutrition* 15(2), 117–122. DOI: 10.1111/j.1365-2095.2007.00561.x.

Low, C., Wadsworth, S., Burrells, C. and Secombes, C.J. (2003) Expression of immune genes in turbot (*Scophthalmus maximus*) fed a nucleotide-supplemented diet. *Aquaculture* 221(1–4), 23–40. DOI: 10.1016/S0044-8486(03)00022-X.

Lu, X.J. and Chen, J. (2019) Specific function and modulation of teleost monocytes/macrophages: Polarization and phagocytosis. *Zoological Research* 40(3), 146–150. DOI: 10.24272/j.issn.2095-8137.2019.035.

Ma, C., Ye, J. and Kaattari, S.L. (2013) Differential compartmentalization of memory B cells versus plasma cells in salmonid fish. *European Journal of Immunology* 43(2), 360–370. DOI: 10.1002/eji.201242570.

Ma, T.Y., Wu, J.Y., Gao, X.K., Wang, J.Y., Zhan, X.L. *et al.* (2014) Molecular cloning, functional identification, and expressional analyses of FASL in tilapia, *Oreochromis niloticus*. *Developmental and Comparative Immunology* 46(2), 448–460. DOI: 10.1016/j.dci.2014.06.003.

Mai, T.T., Kayansamruaj, P., Taengphu, S., Senapin, S., Costa, J.Z. *et al.* (2021) Efficacy of heat-killed and formalin-killed vaccines against tilapia tilapinevirus in juvenile Nile tilapia (*Oreochromis niloticus*). *Journal of Fish Diseases* 44(12), 2097–2109. DOI: 10.1111/jfd.13523.

Makrinos, D.L. and Bowden, T.J. (2016) Natural environmental impacts on teleost immune function. *Fish and Shellfish Immunology* 53, 50–57. DOI: 10.1016/j.fsi.2016.03.008.

Malina, A., Tassakka, A. and Sakai, M. (2005) Current research on the immunostimulatory effects of CpG oligodeoxynucleotides in fish. *Aquaculture* 246(1–4), 25–36.

Mancuso, M. (2013) Probiotics in aquaculture. *Journal of Fisheries and Livestock Production* 02(1), e107. DOI: 10.4172/2332-2608.1000e107.

Manning, M.J. (1994) The blood leukocytes of fish: A comparison between species. *Journal of Leukocyte Biology* 55, 483–488.

Martínez Cruz, P., Ibáñez, A.L., Monroy Hermosillo, O.A. and Ramírez Saad, H.C. (2012) Use of probiotics in aquaculture. *ISRN Microbiology* 916845. DOI: 10.5402/2012/916845.

Meena, D.K., Das, P., Kumar, S., Mandal, S.C., Prusty, A.K. *et al.* (2013) Beta-glucan: An ideal immunostimulant in aquaculture (A Review). *Fish Physiology and Biochemistry* 39(3), 431–457. DOI: 10.1007/s10695-012-9710-5.

Miccoli, A., Manni, M., Picchietti, S. and Scapigliati, G. (2021) State-of-the-art vaccine research for aquaculture use: The case of three economically relevant fish species. *Vaccines* 9(2), 140. DOI: 10.3390/vaccines9020140.

Miller, N.W., Deuter, A. and Clem, L.W. (1986) Phylogeny of lymphocyte heterogeneity: The cellular requirements for the mixed leukocyte reaction in channel catfish. *Immunology* 59(1), 123–128.

Mogensen, T.H. (2009) Pathogen recognition and inflammatory signaling in innate immune defenses. *Clinical Microbiology Reviews* 22(2), 240. DOI: 10.1128/CMR.00046-08.

Munang'andu, H.M., Paul, J. and Evensen, Ø. (2016) An overview of vaccination strategies and antigen delivery systems for *Streptococcus agalactiae* vaccines in Nile tilapia (*Oreochromis niloticus*). *Vaccines* 4(4), 48. DOI: 10.3390/vaccines4040048.

Mutoloki, S., Jørgensen, J.B. and Evensen, Ø. (2014) Chapter 9. the adaptive immune response in fish. In: Gudding, R., Lillehaug, A. and Evensen, Ø. (eds) *Fish Vaccination.* John Wiley and Sons, Ltd, pp. 104–115.

Nagasawa, T., Nakayasu, C., Rieger, A.M., Barreda, D.R., Somamoto, T. *et al.* (2014) Phagocytosis by thrombocytes is a conserved innate immune mechanism in lower vertebrates. *Frontiers in Immunology* 5, 445. DOI: 10.3389/fimmu.2014.00445.

Nakanishi, T., Fischer, U., Dijkstra, J.M., Hasegawa, S., Somamoto, T. *et al.* (2002) Cytotoxic T cell function in fish. *Developmental and Comparative Immunology* 26(2), 131–139. DOI: 10.1016/s0145-305x(01)00055-6.

Nakao, M., Kajiya, T., Sato, Y., Somamoto, T., Kato-Unoki, Y. *et al.* (2006) Lectin pathway of bony fish complement: Identification of two homologs of the mannose-binding lectin associated with MASP2 in the common carp (*Cyprinus carpio*). *Journal of Immunology* 177(8), 5471–5479. DOI: 10.4049/jimmunol.177.8.5471.

Nakao, M., Tsujikura, M., Ichiki, S., Vo, T.K. and Somamoto, T. (2011) The complement system in teleost fish: Progress of post-homolog-hunting researches. *Developmental and Comparative Immunology* 35(12), 1296–1308. DOI: 10.1016/j.dci.2011.03.003.

Nayak, S.K. (2010a) Role of gastrointestinal microbiota in fish. *Aquaculture Research* 41(11), 1553–1573. DOI: 10.1111/j.1365-2109.2010.02546.x.

Nayak, S.K. (2010b) Probiotics and immunity: A fish perspective. *Fish and Shellfish Immunology* 29(1), 2–14. DOI: 10.1016/j.fsi.2010.02.017.

Newaj-Fyzul, A. and Austin, B. (2015) Probiotics, immunostimulants, plant products and oral vaccines, and their role as feed supplements in the control of bacterial fish diseases. *Journal of Fish Diseases* 38(11), 937–955. DOI: 10.1111/jfd.12313.

Nie, L., Zhou, Q.J., Qiao, Y. and Chen, J. (2017) Interplay between the gut microbiota and immune responses of ayu (*Plecoglossus altivelis*) during *Vibrio anguillarum* infection. *Fish and Shellfish Immunology* 68, 479–487. DOI: 10.1016/j.fsi.2017.07.054.

Nithikulworawong, N., Yakupitiyage, A., Rakshit, S.K. and Srisapoome, P. (2012) Molecular characterization and increased expression of the Nile tilapia *Oreochromis niloticus* (L.) T-cell receptor beta chain in response to *Streptococcus agalactiae* infection. *Journal of Fish Diseases* 35(5), 343–358. DOI: 10.1111/j.1365-2761.2012.01353.x.

Nurani, F.S., Sukenda, S. and Nuryati, S. (2020) Maternal immunity of tilapia broodstock vaccinated with polyvalent vaccine and resistance of their offspring against *Streptococcus agalactiae*. *Aquaculture Research* 51(4), 1513–1522. DOI: 10.1111/are.14499.

Olusola, S., Emikpe, B.O. and Olaifa, F.E. (2013) The potentials of medicinal plant extracts as bioantimicrobials in aquaculture. *Journal of Medicinal and Aromatic Plant Sciences* 3, 404–412.

Palić, D., Andreasen, C.B., Ostojić, J., Tell, R.M. and Roth, J.A. (2007) Zebrafish (*Danio rerio*) whole kidney assays to measure neutrophil extracellular trap release and degranulation of primary granules. *Journal of Immunological Methods* 319(1–2), 87–97. DOI: 10.1016/j.jim.2006.11.003.

Palti, Y. (2011) Toll-like receptors in bony fish: From genomics to function. *Developmental and Comparative Immunology* 35(12), 1263–1272. DOI: 10.1016/j.dci.2011.03.006.

Pandiyan, P., Balaraman, D., Thirunavukkarasu, R., George, E.G.J., Subaramaniyan, K. *et al.* (2013) Probiotics in aquaculture. *Drug Invention Today* 5(1), 55–59. DOI: 10.1016/j.dit.2013.03.003.

Pang, J.C., Gao, F.Y., Lu, M.X., Ye, X., Zhu, H.P. *et al.* (2013) Major histocompatibility complex class IIA and IIB genes of Nile tilapia *Oreochromis niloticus*: Genomic structure molecular polymorphism and expression patterns. *Fish and Shellfish Immunology* 34(2), 486–496.

Pang, J.C., Gao, F.Y., Wang, M., Zhao, J.L. and Lu, M.X. (2017) Isolation and characterization of toll-like receptor 21 and 22 genes from Nile tilapia, *Oreochromis niloticus* (Linnaeus). *Aquaculture Research* 48(7), 3528–3544. DOI: 10.1111/are.13179.

Panigrahi, A., Kiron, V., Puangkaew, J., Kobayashi, T., Satoh, S. *et al.* (2005) The viability of probiotic bacteria as a factor influencing the immune response in rainbow trout *Oncorhynchus mykiss*. *Aquaculture* 243(1–4), 241–254. DOI: 10.1016/j.aquaculture.2004.09.032.

Parra, D., Rieger, A.M., Li, J., Zhang, Y.A., Randall, L.M. *et al.* (2012) Pivotal advance: peritoneal cavity B-1 B cells have phagocytic and microbicidal capacities and present phagocytosed antigen to CD4+ T cells. *Journal of Leukocyte Biology* 91(4), 525–536. DOI: 10.1189/jlb.0711372.

Parra, D., Takizawa, F. and Sunyer, J.O. (2013) Evolution of B cell immunity. *Annual Review of Animal Biosciences* 1, 65–97. DOI: 10.1146/annurev-animal-031412-103651.

Passer, B.J., Chen, C.H., Miller, N.W. and Cooper, M.D. (1996) Identification of a T lineage antigen in the catfish. *Developmental and Comparative Immunology* 20(6), 441–450.

Peng, M., Xu, W., Ai, Q., Mai, K., Liufu, Z. *et al.* (2013) Effects of nucleotide supplementation on growth, immune responses and intestinal morphology in juvenile turbot fed diets with graded levels of soybean meal (*Scophthalmus maximus* L.). *Aquaculture* 392–395, 51–58. DOI: 10.1016/j.aquaculture.2013.02.002.

Pérez, T., Balcázar, J.L., Ruiz-Zarzuela, I., Halaihel, N., Vendrell, D. *et al.* (2010) Host-microbiota interactions within the fish intestinal ecosystem. *Mucosal Immunology* 3(4), 355–360. DOI: 10.1038/mi.2010.12.

Phuyindee, C., Unajak, S. and Srisapoome, P. (2015) Diversity analysis of the immunoglobulin M heavy chain gene in Nile tilapia *Oreochromis niloticus* (Linnaeus). *African Journal of Biotechnology* 14(29), 2282–2299.

Pickering, A.D. and Pottinger, T.G. (1989) Stress responses and disease resistance in salmonid fish: Effects of chronic elevation of plasma cortisol. *Fish Physiology and Biochemistry* 7(1–6), 253–258. DOI: 10.1007/BF00004714.

Pijanowski, L., Golbach, L., Kolaczkowska, E., Scheer, M., Verburg-van Kemenade, B.M.L. *et al.* (2013) Carp neutrophilic granulocytes form extracellular traps via ROS-dependent and independent pathways. *Fish and Shellfish Immunology* 34(5), 1244–1252. DOI: 10.1016/j.fsi.2013.02.010.

Poonsawat, S., Areechon, N., Srisapoome, P., Maita, M. and Endo, M. (2009) Polymorphism of major histocompatibility complex class Iα gene and resistance against streptococcosis of six strains of Nile tilapia (*Oreochromis niloticus* linnaeus). *Kasetsart Journal (Natl Sci.)* 43, 348–357.

Praveen, K., Evans, D.L. and Jaso-Friedmann, L. (2004) Evidence for the existence of granzyme-like serine proteases in teleost cytotoxic cells. *Journal of Molecular Evolution* 58(4), 449–459. DOI: 10.1007/s00239-003-2566-7.

Praveen, K., Leary, J.H., Evans, D.L. and Jaso-Friedmann, L. (2006) Molecular characterization and expression of a granzyme of an ectothermic vertebrate with chymase-like activity expressed in the cytotoxic cells of Nile tilapia (*Oreochromis niloticus*). *Immunogenetics* 58(1), 41–55. DOI: 10.1007/s00251-005-0063-4.

Press, C.M. and Evensen, Ø. (1999) The morphology of the immune system in teleost fishes. *Fish and Shellfish Immunology* 9, 309–318.

Pulsford, A.L., Lemairegony, S., Tomlinson, M., Collingwood, N. and Glynn, P.J. (1994) Effects of acute stress on the immune system of the dab *Limanda limanda*. *Comparative Biochemistry and Physiology. Part C: Pharmacology, Toxicology and Endocrinology* 109(2), 111–117.

Raju, S.V., Sarkar, P., Kumar, P. and Arockiaraj, J. (2021) Piscidin fish antimicrobial peptide: Structure, classification, properties, mechanism, gene regulation and therapeutical importance. *International Journal of Peptide Research and Therapeutics* 27(1), 91–107. DOI: 10.1007/s10989-020-10068-w.

Rakers, S., Niklasson, L., Steinhagen, D., Kruse, C., Schauber, J. *et al.* (2013) Antimicrobial peptides (AMPs) from fish epidermis: Perspectives for investigative dermatology. *The Journal of Investigative Dermatology* 133(5), 1140–1149. DOI: 10.1038/jid.2012.503.

Ramirez-Gomez, F., Greene, W., Rego, K., Hansen, J.D., Costa, G. *et al.* (2012) Discovery and characterization of secretory IgD in rainbow trout: Secretory IgD is produced through a novel splicing mechanism. *Journal of Immunology* 188(3), 1341–1349. DOI: 10.4049/jimmunol.1101938.

Ranjan, R., Prasad, K.P., Vani, T. and Kumar, R. (2012) Effect of dietary chitosan on haematology, innate immunity and disease resistance of Asian sea bass *Lates calcarifer* (Bloch). *Aquaculture Research* 45(6), 983–993.

Rawls, J.F., Samuel, B.S. and Gordon, J.I. (2004) Gnotobiotic zebrafish reveal evolutionarily conserved responses to the gut microbiota. *Proceedings of the National Academy of Sciences of the United States of America* 101(13), 4596–4601. DOI: 10.1073/pnas.0400706101.

Ray, A.K., Ghosh, K. and Ringø, E. (2012) Enzyme-producing bacteria isolated from fish gut: A review. *Aquaculture Nutrition* 18(5), 465–492. DOI: 10.1111/j.1365-2095.2012.00943.x.

Ren, P., Xu, L., Yang, Y., He, S., Liu, W. *et al.* (2013) *Lactobacillus planarum* subsp. plantarum JCM 1149 vs. *Aeromonas hydrophila* NJ-1 in the anterior intestine and posterior intestine of hybrid tilapia *Oreochromis niloticus* ♀ x *Oreochromis aureus* ♂: An *ex vivo* study. *Fish and Shellfish Immunology* 35(1), 146–153. DOI: 10.1016/j.fsi.2013.04.023.

Reverter, M., Tapissier-Bontemps, N., Sasal, P. and Saulnier, D. (2017) Use of medicinal plants in aquaculture. In: Austin, B. and Newaj-Fyzul, A. (eds) *Diagnosis and Control of Diseases of Fish and Shellfish*, 1st edn. Blackwell Science Publishing, pp. 223–261.

Reyes-Cerpa, S., Maisey, K., Reyes-López, F., Toro-Ascuy, D., María, A. *et al.* (2012) Fish cytokines and immune response. *New Advances and Contributions to Fish Biology* 1–58.

Ribeiro, C.M.S. and Schijns, V.E.J.C. (2010) Immunology of vaccine adjuvants. *Methods in Molecular Biology* 626, 1–14. DOI: 10.1007/978-1-60761-585-9_1.

Ridha, M.T. and Azad, I.S. (2012) Preliminary evaluation of growth performance and immune response of Nile tilapia *Oreochromis niloticus* supplemented with two putative probiotic bacteria. *Aquaculture Research* 43(6), 843–852. DOI: 10.1111/j.1365-2109.2011.02899.x.

Rieger, A.M., Konowalchuk, J.D., Grayfer, L., Katzenback, B.A., Havixbeck, J.J. *et al.* (2012) Fish and mammalian phagocytes differentially regulate pro-inflammatory and homeostatic responses *in vivo*. *PLOS ONE* 7(10), e47070. DOI: 10.1371/journal.pone.0047070.

Riera Romo, M., Pérez-Martínez, D. and Castillo Ferrer, C. (2016) Innate immunity in vertebrates: An overview. *Immunology* 148(2), 125–139. DOI: 10.1111/imm.12597.

Ringø, E. and Song, S.K. (2016) Application of dietary supplements (Synbiotics and probiotics in combination with plant products and β-glucans) in aquaculture. *Aquaculture Nutrition* 22(1), 4–24. DOI: 10.1111/anu.12349.

Ringø, E., Olsen, R.E., Gifstad, T.Ø., Dalmo, R.A., Amlund, H. *et al.* (2010) Prebiotics in aquaculture: A review. *Aquaculture Nutrition* 16(2), 117–136. DOI: 10.1111/j.1365-2095.2009.00731.x.

Ringø, E., Erik Olsen, R., Gonzalez Vecino, J.L. and Wadsworth, S. (2011) Use of immunostimulants and nucleotides in aquaculture: A review. *Journal of Marine Science: Research and Development* 02(1). DOI: 10.4172/2155-9910.1000104.

Ringø, E., Olsen, R.E., Vecino, J.L.G., Wadsworth, S. and Song, S.K. (2012) Use of immunostimulants and nucleotides in aquaculture: A review. *Journal of Marine Science: Research and Development* 1, 104.

Roberfroid, M.B. (2007) Prebiotics: The concept revisited. *The Journal of Nutrition* 137(3 (Suppl. 2)), 830S–837S. DOI: 10.1093/jn/137.3.830S.

Roberts, R.J. (2012) The bacteriology of teleosts. In: Roberts, R.J. (ed.) *Fish Pathology*. Wiley-Blackwell, pp. 339–382.

Rodriguez-Estrada, U., Satoh, S., Haga, Y., Fushimi, H. and Sweetman, J. (2009) Effects of single and combined supplementation of *Enterococcus faecalis*, mannan oligosaccharides and polyhydroxybutyrate acid on growth performance and immune response of rainbow trout, *Oncorhynchus mykiss*. *Aquaculture Science* 57, 609–617.

Rombout, J.H.W.M., Yang, G. and Kiron, V. (2014) Adaptive immune responses at mucosal surfaces of teleost fish. *Fish and Shellfish Immunology* 40(2), 634–643. DOI: 10.1016/j.fsi.2014.08.020.

Sakai, M. (1999) Current research status of fish immunostimulants. *Aquaculture* 172(1–2), 63–92. DOI: 10.1016/S0044-8486(98)00436-0.

Sakai, M., Taniguchi, K., Mamoto, K., Ogawa, H. and Tabata, M. (2001) Immunostimulant effects of nucleotide isolated from yeast RNA on carp, *Cyprinus carpio* L. *Journal of Fish Diseases* 24(8), 433–438. DOI: 10.1046/j.1365-2761.2001.00314.x.

Salinas, I. (2015) The mucosal immune system of teleost fish. *Biology* 4(3), 525–539. DOI: 10.3390/biology4030525.

Salinas, I. and Parra, D. (2015) Fish mucosal immunity: Intestine. In: Beck, B.H. and Peatman, E. (eds) *Mucosal Health in Aquaculture*. pp. 35–170.

Sankaran, K. and Gurnani, S. (1972) On the variation in the catalytic activity of lysozyme in fishes. *Indian Journal of Biochemistry and Biophysics* 9(2), 162–165.

Saurabh, S. and Sahoo, P.K. (2008) Lysozyme: An important defence molecule of fish innate immune system. *Aquaculture Research* 39(3), 223–239. DOI: 10.1111/j.1365-2109.2007.01883.x.

Schijns, V.E.J.C. (2001) Induction and detection of immune responses by vaccine adjuvants. *Critical Reviews in Immunology* 21, 456–463.

Secombes, C.J. and Belmonte, R. (2016) Overview of the fish adaptive immune system. In: Adams, A. (ed.) *Fish Vaccines. Birkhäuser Advances in Infectious Diseases*. Springer.

Secombes, C.J. and Fletcher, T.C. (1992) The role of phagocytes in the protective mechanisms of fish. *Annual Review of Fish Diseases* 2, 53–71. DOI: 10.1016/0959-8030(92)90056-4.

Secombes, C.J. and Wang, T. (2012) The innate and adaptive immune system of fish. In: Austin, B. (ed.) *In Infectious Disease in Aquaculture: Prevention and Control* [Chapter]: 1. Woodhead Publishing Limited, pp. 3–68.

Secombes, C.J., Manning, M.J. and Ellis, A.E. (1982) The effect of primary and secondary immunization on the lymphoid tissue of the carp, *Cyprinus carpio* L. *The Journal of Experimental Zoology* 220(3), 277–287. DOI: 10.1002/jez.1402200303.

Selim, K.M., Reda, R.M., Mahmoud, R. and El-Araby, I.E. (2020) Effects of nucleotides supplemented diets on growth performance and expressions of ghrelin and insulin-like growth factor genes in Nile tilapia, *Oreochromis niloticus*. *Journal of Applied Aquaculture* 32(2), 157–174. DOI: 10.1080/10454438.2019.1696911.

Semova, I., Carten, J.D., Stombaugh, J., Mackey, L.C., Knight, R. *et al.* (2012) Microbiota regulate intestinal absorption and metabolism of fatty acids in the zebrafish. *Cell Host and Microbe* 12(3), 277–288. DOI: 10.1016/j.chom.2012.08.003.

Shen, L., Stuge, T.B., Zhou, H., Khayat, M., Barker, K.S. *et al.* (2002) Channel catfish cytotoxic cells: A mini review. *Developmental and Comparative Immunology* 26(2), 141–149. DOI: 10.1016/s0145-305x(01)00056-8.

Shen, L., Stuge, T.B., Bengtén, E., Wilson, M., Chinchar, V.G. *et al.* (2004) Identification and characterization of clonal NK-like cells from channel catfish (*Ictalurus punctatus*). *Developmental and Comparative Immunology* 28(2), 139–152. DOI: 10.1016/s0145-305x(03)00119-8.

Shiau, S., Gabaudan, J. and Lin, Y. (2015) Dietary nucleotide supplementation enhances immune responses and survival to *Streptococcus iniae* in hybrid tilapia fed diet containing low fish meal. *Aquaculture Reports* 2, 77–81. DOI: 10.1016/j.aqrep.2015.08.002.

Shoemaker, C.A., Lozano, C.A., LaFrentz, B.R., Mumma, P., Vela-Avitúa, S. *et al.* (2022) Additive genetic variation in resistance of Nile tilapia (*Oreochromis niloticus*) to *Francisella orientalis* and its genetic (co)variation to both harvest weight and resistance to *Streptococcus agalactiae* Ib. *Aquaculture* 561, 738736. DOI: 10.1016/j.aquaculture.2022.738736.

Sirimanapong, W., Adams, A., Ooi, E.L., Green, D.M., Nguyen, D.K. *et al.* (2015) The effects of feeding immunostimulant β-glucan on the immune response of *Pangasianodon hypophthalmus. Fish and Shellfish Immunology* 45(2), 357–366. DOI: 10.1016/j.fsi.2015.04.025.

Smith, N.C., Rise, M.L. and Christian, S.L. (2019) A comparison of the innate and adaptive immune systems in cartilaginous fish, ray-finned fish, and lobe-finned fish. *Frontiers in Immunology* 10, 2292. DOI: 10.3389/fimmu.2019.02292.

Solem, S.T. and Stenvik, J. (2006) Antibody repertoire development in teleosts—A review with emphasis on salmonids and *Gadus morhua* L. *Developmental and Comparative Immunology* 30(1–2), 57–76. DOI: 10.1016/j.dci.2005.06.007.

Somamoto, T., Nakanishi, T. and Okamoto, N. (2000) Specific cell-mediated cytotoxicity against a virus-infected syngeneic cell line in isogeneic Ginbuna crucian carp. *Developmental and Comparative Immunology* 24(6–7), 633–640. DOI: 10.1016/s0145-305x(00)00018-5.

Song, S.K., Beck, B.R., Kim, D., Park, J., Kim, J. *et al.* (2014) Prebiotics as immunostimulants in aquaculture: A review. *Fish and Shellfish Immunology* 40(1), 40–48. DOI: 10.1016/j.fsi.2014.06.016.

Srichaiyo, N., Tongsiri, S., Hoseinifar, S.H., Dawood, M.A.O., Esteban, M.Á. *et al.* (2020) The effect of fishwort (*Houttuynia cordata*) on skin mucosal, serum immunities, and growth performance of Nile tilapia. *Fish and Shellfish Immunology* 98, 193–200. DOI: 10.1016/j.fsi.2020.01.013.

Stuge, T.B., Wilson, M.R., Zhou, H., Barker, K.S., Bengtén, E. *et al.* (2000) Development and analysis of various clonal alloantigen-dependent cytotoxic cell lines from channel catfish. *Journal of Immunology* 164(6), 2971–2977. DOI: 10.4049/jimmunol.164.6.2971.

Subramanian, A., Kumar, S.S., Suja, S.K., Sudarshan, M. and Chakraborty, A. (2014) Antioxidant and free radical scavenging activity of annular and seasonal ayurvedic medicinal plants *Elephantopus scaber* L. and *Biophytum sensitivum* DC. *International Journal of Pharmacy and Chemical Biology Sciences* 23(1), 6–17.

Suzuki, K. (1986) Morphological and phagocytic characteristics of peritoneal exudate cells in tilapia, *Oreochromis niloticus* (Trewavas), and carp, *Cyprinus carpio* L. *Journal of Fish Biology* 29(3), 349–364. DOI: 10.1111/j.1095-8649.1986.tb04951.x.

Tacchi, L., Musharrafieh, R., Larragoite, E.T., Crossey, K., Erhardt, E.B. *et al.* (2014) Nasal immunity is an ancient arm of the mucosal immune system of vertebrates. *Nature Communications* 5, 5205. DOI: 10.1038/ncomms6205.

Tadese, D.A., Song, C., Sun, C., Liu, B., Liu, B. *et al.* (2022) The role of currently used medicinal plants in aquaculture and their action mechanisms: A review. *Reviews in Aquaculture* 14(2), 816–847. DOI: 10.1111/raq.12626.

Tafalla, C., Bøgwald, J. and Dalmo, R.A. (2013) Adjuvants and immunostimulants in fish vaccines: Current knowledge and future perspectives. *Fish and Shellfish Immunology* 35(6), 1740–1750. DOI: 10.1016/j.fsi.2013.02.029.

Tahmasebi-Kohyani, A., Keyvanshokooh, S., Nematollahi, A., Mahmoudi, N. and Pasha-Zanoosi, H. (2011) Dietary administration of nucleotides to enhance growth, humoral immune responses, and disease resistance of the rainbow trout (*Oncorhynchus mykiss*) fingerlings. *Fish and Shellfish Immunology* 30(1), 189–193. DOI: 10.1016/j.fsi.2010.10.005.

Tahmasebi-Kohyani, A., Keyvanshokooh, S., Nematollahi, A., Mahmoudi, N. and Pasha-Zanoosi, H. (2012) Effects of dietary nucleotides supplementation on rainbow trout (*Oncorhynchus mykiss*) performance and acute stress response. *Fish Physiology and Biochemistry* 38(2), 431–440.

Taoka, Y., Maeda, H., Jo, J.Y., Kim, S.M., Park, S.I. *et al.* (2006) Use of live and dead probiotic cells in tilapia, *Oreochromis niloticus*. *Fisheries Science* 72(4), 755–766. DOI: 10.1111/j.1444-2906.2006.01215.x.

Tatner, M.F. and Findlay, C. (1991) Lymphocyte migration and localization patterns in rainbow trout, *Onchorhynchus mykiss*, studies using the tracer sample method. *Fish and Shellfish Immunology* 1(2), 107–117. DOI: 10.1016/S1050-4648(06)80011-2.

Tellez-Bañuelos, M.C., Santerre, A., Casas-Solis, J. and Zaitseva, G. (2010) Endosulfan increases seric interleukin-2 like (IL-2L) factor and immunoglobulin M (IgM) of Nile tilapia (*Oreochromis niloticus*) challenged with *Aeromonas hydrophila*. *Fish and Shellfish Immunology* 28(2), 401–405. DOI: 10.1016/j.fsi.2009.11.017.

Telli, G.S., Ranzani-Paiva, M.J.T., Dias, D.C., Sussel, F.R., Ishikawa, C.M. *et al.* (2014) Dietary administration of bacillus subtilis on hematology and non-specific immunity of Nile tilapia *Oreochromis niloticus* raised at different stocking densities. *Fish and Shellfish Immunology* 39(2), 305–311. DOI: 10.1016/j.fsi.2014.05.025.

Thompson, K.D., Rodkhum, C., Bunnoy, A., Thangsunan, P., Kitiyodom, S. *et al.* (2023) Addressing nanovaccine strategies for tilapia. *Vaccines* 11(8), 1356. DOI: 10.3390/vaccines11081356.

Trichet, V.V. (2010) Nutrition and immunity: An update. *Aquaculture Research* 41(3), 356–372.

Tsuchiya, C., Sakata, T. and Sugita, H. (2008) Novel ecological niche of *Cetobacterium somerae*, an anaerobic bacterium in the intestinal tracts of freshwater fish. *Letters in Applied Microbiology* 46(1), 43–48.

Tsujii, T. and Seno, S. (1990) Melano-macrophage centers in the aglomerular kidney of the sea horse (teleosts): Morphologic studies on its formation and possible function. *The Anatomical Record* 226(4), 460–470. DOI: 10.1002/ar.1092260408.

Uribe, C., Folch, H., Enríquez, R. and Moran, G. (2011) Innate and adaptive immunity in teleost fish: A review. *Veterinární Medicína* 56(10), 486–503. DOI: 10.17221/3294-VETMED.

Utke, K., Bergmann, S., Lorenzen, N., Kollner, B, *et al.* (2007) Cell mediated cytotoxicity in rainbow trout, *Oncorhynchus mykiss*, infected with viral haemorrhagic septicaemia virus. *Fish and Shellfish Immunology* 22, 182–196.

Valladão, G.M.R., Gallani, S.U., Pala, G., Jesus, R.B., Kotzent, S. *et al.* (2017) Practical diets with essential oils of plants activate the complement system and alter the intestinal morphology of Nile tilapia. *Aquaculture Research* 48(11), 5640–5649. DOI: 10.1111/are.13386.

Van Muiswinkel, W.B., Lamers, C.H. and Rombout, J.H. (1991) Structural and functional aspects of the spleen in bony fish. *Research in Immunology* 142(4), 362–366. DOI: 10.1016/0923-2494(91)90093-x.

Vaseeharan, B. and Thaya, R. (2014) Medicinal plant derivatives as immunostimulants: An alternative to chemotherapeutics and antibiotics in aquaculture. *Aquaculture International* 22(3), 1079–1091. DOI: 10.1007/s10499-013-9729-3.

Velázquez, J., Acosta, J., Lugo, J.M., Reyes, E., Herrera, F. *et al.* (2018) Discovery of immunoglobulin T in Nile tilapia (*Oreochromis niloticus*): A potential molecular marker to understand mucosal immunity in this species. *Developmental and Comparative Immunology* 88, 124–136. DOI: 10.1016/j.dci.2018.07.013.

Wan, F., Hu, C.B., Ma, J.X., Gao, K., Xiang, L.X. *et al.* (2016) Characterization of γδ T cells from zebrafish provides insights into their important role in adaptive humoral immunity. *Frontiers in Immunology* 7, 675. DOI: 10.3389/fimmu.2016.00675.

Wang, B., Wang, P., Wu, Z.H., Lu, Y.S., Wang, Z.L. *et al.* (2016) Molecular cloning and expression analysis of IgD in Nile tilapia (*Oreochromis niloticus*) in response to *Streptococcus agalactiae* stimulus. *International Journal of Molecular Sciences* 17(3), 348. DOI: 10.3390/ijms17030348.

Wang, B., Thompson, K.D., Wangkahart, E., Yamkasem, J., Bondad-Reantaso, M.G. *et al.* (2023) Strategies to enhance tilapia immunity to improve their health in aquaculture. *Reviews in Aquaculture* 15(S1), 41–56. DOI: 10.1111/raq.12731.

Wangkaghart, E., Deville, S., Wang, B., Srisapoome, P., Wang, T. *et al.* (2021) Immune response and protective efficacy of two new adjuvants, Montanide™ ISA 763B VG and Montanide™ GEL02, administered with a *Streptococcus agalactiae* ghost vaccine in Nile tilapia (*Oreochromis niloticus*). *Fish and Shellfish Immunology* 116, 19–29. DOI: 10.1016/j.fsi.2021.06.014.

Welker, T.L., Lim, C., Yildirim-Aksoy, M. and Klesius, P.H. (2007) Growth, immune function, and disease and stress resistance of juvenile Nile tilapia (*Oreochromis niloticus*) fed graded levels of bovine lactoferrin. *Aquaculture* 262(1), 156–162. DOI: 10.1016/j.aquaculture.2006.09.036.

Wendelaar Bonga, S.E. (1997). The stress response in fish. *Physiological reviews,* 77(3), 591–625. https://doi.org/10.1152/physrev.1997.77.3.591

Wiegertjes, G.F., Wentzel, A.S., Spaink, H.P., Elks, P.M. and Fink, I.R. (2016) Polarization of immune responses in fish: The "macrophages first" point of view. *Molecular Immunology* 69, 146–156. DOI: 10.1016/j.molimm.2015.09.026.

Wilhelm Filho, D.W. (2007) Reactive oxygen species, antioxidants and fish mitochondria. *Frontiers in Bioscience: A Journal and Virtual Library* 12, 1229–1237. DOI: 10.2741/2141.

Xia, Y., Lu, M., Chen, G., Cao, J., Gao, F. *et al.* (2018) Effects of dietary *Lactobacillus rhamnosus* JCM1136 and *Lactococcus lactis* subsp. Lactis JCM5805 on the growth, intestinal microbiota, morphology, immune response, and disease resistance of juvenile Nile tilapia, *Oreochromis niloticus. Fish and Shellfish Immunology* 76, 368–379. DOI: 10.1016/j.fsi.2018.03.020.

Xia, Y., Wang, M., Gao, F., Lu, M. and Chen, G. (2020) Effects of dietary probiotic supplementation on the growth, gut health and disease resistance of juvenile Nile tilapia (*Oreochromis niloticus*). *Animal Nutrition* 6(1), 69–79. DOI: 10.1016/j.aninu.2019.07.002.

Xiong, J.B., Nie, L. and Chen, J. (2019) Current understanding on the roles of gut microbiota in fish disease and immunity. *Zoological Research* 40(2), 70–76. DOI: 10.24272/j.issn.2095-8137.2018.069.

Xu, L., Ran, C., He, S., Zhang, J., Hu, J. *et al.* (2015) Effects of dietary yeast nucleotides on growth, non-specific immunity, intestine growth and intestinal microbiota of juvenile hybrid tilapia *Oreochromis niloticus* ♀ × *Oreochromis aureus* ♂. *Animal Nutrition* 1(3), 244–251. DOI: 10.1016/j.aninu.2015.08.006.

Xu, Z., Parra, D., Gómez, D., Salinas, I., Zhang, Y.A. *et al.* (2013) Teleost skin, an ancient mucosal surface that elicits gut-like immune responses. *Proceedings of the National Academy of Sciences of the United States of America* 110(32), 13097–13102. DOI: 10.1073/pnas.1304319110.

Ye, J., Kaattari, I.M., Ma, C. and Kaattari, S. (2013) The teleost humoral immune response. *Fish and Shellfish Immunology* 35(6), 1719–1728. DOI: 10.1016/j.fsi.2013.10.015.

Zapata, A.G., Chibá, A. and Varas, A. (1996) Cells and tissues of the immune system of fish. In: Iwama, G. and Nakanishi, T. (eds) *The Fish Immune System: Organism, Pathogen, and Environment.* Academic Press, pp. 1–62. DOI: 10.1016/S1546-5098(08)60271-X.

Zapata, A., Diez, B., Cejalvo, T., Gutiérrez-de Frías, C. and Cortés, A. (2006) Ontogeny of the immune system of fish. *Fish and Shellfish Immunology* 20(2), 126–136. DOI: 10.1016/j.fsi.2004.09.005.

Zare, M., Tran, H.Q., Prokešová, M. and Stejskal, V. (2021) Effects of garlic *Allium sativum* powder on nutrient digestibility, haematology, and immune and stress responses in Eurasian Perch *Perca fluviatilis* juveniles. *Animals: An Open Access Journal from MDPI* 11(9), 2735. DOI: 10.3390/ani11092735.

Zeng, W., Wang, Y., Chen, X., Wang, Q., Bergmann, S.M. *et al.* (2021a) Potency and efficacy of VP20-based vaccine against tilapia lake virus using different prime-boost vaccination regimens in tilapia. *Aquaculture* 539, 736654. DOI: 10.1016/j.aquaculture.2021.736654.

Zeng, W., Wang, Y., Hu, H., Wang, Q., Bergmann, S.M. *et al.* (2021b) Cell culture-derived tilapia lake virus-inactivated vaccine containing montanide adjuvant provides high protection against viral challenge for tilapia. *Vaccines* 9(2), 86. DOI: 10.3390/vaccines9020086.

Zhang, Y.A., Salinas, I., Li, J., Parra, D., Bjork, S. *et al.* (2010) IgT, a primitive immunoglobulin class specialized in mucosal immunity. *Nature Immunology* 11(9), 827–835. DOI: 10.1038/ni.1913.

Zhu, J., Fu, Q., Ao, Q., Tan, Y., Luo, Y. *et al.* (2017) Transcriptomic profiling analysis of tilapia (*Oreochromis niloticus*) following *Streptococcus agalactiae* challenge. *Fish and Shellfish Immunology* 62, 202–212. DOI: 10.1016/j.fsi.2017.01.023.

Zhu, X. and Zhu, J. (2020) CD4 T helper cell subsets and related human immunological disorders. *International Journal of Molecular Sciences* 21(21), 8011. DOI: 10.3390/ijms21218011.

Zorriehzahra, M.J., Delshad, S.T., Adel, M., Tiwari, R., Karthik, K. *et al.* (2016) Probiotics as beneficial microbes in aquaculture: An update on their multiple modes of action: A review. *The Veterinary Quarterly* 36(4), 228–241. DOI: 10.1080/01652176.2016.1172132.

Chapter 8

Tilapia welfare

Carlos Garcia de Leaniz and Sonia Rey Planellas

What is fish welfare and why does it matter?

Fish welfare has been notoriously difficult to define (Iwama, 2007; Volpato *et al.*, 2007), particularly in relation to fish farming (Huntingford *et al.*, 2006; Kristiansen & Bracke, 2020; Stien *et al.*, 2020). Good welfare is not the same as good health, although these terms are sometimes confused. To some authors, good health is a prerequisite for high welfare, but good health alone does not guarantee high welfare (Huntingford *et al.*, 2006). Some assess fish welfare in relation to the ability to exercise preferences, i.e. the extent to which fish are able to live under conditions that were 'freely chosen' (Volpato *et al.*, 2007). However, such a definition is not readily applicable in captivity and is virtually impossible to implement in a commercial fish farm. A more pragmatic definition of fish welfare in relation to aquaculture is one that emphasizes freedom of negative experiences (e.g. pain, fear, hunger) and access to positive experiences (e.g. social companionship; Huntingford *et al.*, 2006; Huntingford & Kadri; 2009, Stien *et al.*, 2020).

Regardless of what definition of animal welfare one adopts, consideration of fish welfare is important in fish farming for several practical reasons. Consumers want to make informed choices about the fish they purchase, and good welfare is one criterion that informs consumer choice. For example, half of consumers are willing to pay 25% extra for rainbow trout of high welfare standards (Solgaard & Yang, 2011) and 79% of EU citizens think that the welfare of the fish they eat should be better protected than it is now (Lymbery &

Kim Thompson, Craig Shoemaker and David C. Little (eds) **Tilapia**
DOI: 10.1079/9781800629455.0008, © CAB International, 2025

Hameleers, 2018). Welfare is also becoming an integral part of quality certification schemes (Huntingford *et al.*, 2006, Huntingford & Kadri, 2009, 2014; Royal Society for the Prevention of Cruelty to Animals (RSPCA), 2020, 2021), and promoting good welfare plays an increasingly important role in marketing (Huntingford & Kadri, 2009, 2014; Huntingford *et al.*, 2006). Maintaining good welfare also makes economic sense as it often results in less stress, better growth, lower mortality and higher productivity (Branson, 2008; Huntingford & Kadri, 2009; Segner *et al.*, 2012, 2019). Nevertheless, guidance on good welfare practices typically exists only for the most valuable carnivorous species such as Atlantic salmon (RSPCA, 2018), rainbow trout (RSPCA, 2020), cleaner fish (Garcia de Leaniz *et al.*, 2022; Gutierrez Rabadan *et al.*, 2021), sea bass and sea bream (Pavlidis & Samaras, 2020). Little practical guidance is available on how to measure or improve welfare for low-trophic species such as cyprinids or tilapia. Indeed, manuals on tilapia culture make only a cursory reference to fish welfare (El-Sayed, 2020) or, more often, omit it altogether (Food and Agriculture Organization (FAO), 2022; Fermon, 2006; Lim & Webster, 2006; Quagrainie *et al.*, 2017).

Tilapia diversity, domestication and implications for welfare

Tilapias are a particularly diverse group of fish with a long history of domestication, both of which have implications for welfare. They belong to the family Cichlidae (cichlids), the second largest and most diverse vertebrate family after the cyprinids. More than 1,700 species and 200 genera of cichlids have been described to date, although the number is probably much higher and could be as high as 4000 species (Salzburger, 2018). The tilapiine cichlid tribe includes ~100 species belonging to three genera, the genus *Tilapia* (speckled tilapias, +40 spp.), the genus *Sarotherodon* (brushtooth tilapias, +10 spp.) and the genus *Oreochromis* (blue tilapias, +30 spp.).

True tilapias are native to a small, narrow range in Eastern Africa (El-Sayed, 2020; Nelson, 2006) but they are now being farmed all over the world with 98% of production taking place outside their native range (Champneys *et al.*, 2021). This means that tilapias are now farmed under environmental conditions that may differ widely from those under which the species evolved. Tilapia are considered to be the first farmed fish, with many commercial strains selected for particular commercial traits that affect their environmental requirements and suitability for different rearing systems (El-Sayed, 2020; FAO, 2022). A hieroglyph and artefacts displaying the unmistakable shape of Nile tilapia have been recovered from archaeological sites dating from ancient Egypt (Salzburger, 2018), raising the possibility that some form of proto-aquaculture of Nile tilapia (*Oreochromis niloticus*) may have started ~4,000 years ago.

Nine species of tilapia are commonly used in fish farming. Nile tilapia accounts for ~75% of global production, with Mozambique tilapia (*O. mossambicus*) and blue tilapia (*O. aureaus*) accounting for 5% each. However, there are also several different red hybrids obtained by crossing Mozambique tilapia with Nile tilapia, blue tilapia or Zanzibar (Wami) tilapia (*O. hornorum* syn. *Tilapia urolepis*), which make up 10–20% of global production (El-Sayed, 2020; Gupta & Acosta, 2004).

Tilapias can be farmed in nearly all aquaculture rearing systems, from semi-extensive systems such as earthen ponds to intensive systems including raceways, tanks, floating cages and recirculation aquaculture systems (RAS) (El-Sayed, 2020; Lim & Webster, 2006). They can be reared in monoculture, but also as part of polyculture systems and aquaponics. However, species differ in their suitability for different production systems and in their plasticity and environmental requirements (Table 8.1), all of which can affect their welfare. For example, although most tilapias are omnivorous grazers and filter-feeders, there are also specialized herbivores like the redbreast tilapia (*Tilapia rendalii*) that feed on macrophytes (Gupta & Acosta, 2004; Hlophe & Moyo, 2013). Nile tilapia and red tilapia hybrids are particularly well suited for intensive farming, although they can also be farmed under simpler, less technologically dependent, rearing systems. Some species and hybrids can be farmed in brackish water and even in full sea-water while others are also tolerant to cold water (Lim & Webster, 2006). For example, red belly tilapia and red hybrids are more saline resistant than most other tilapias and are ideally suited for rearing in places where freshwater is scarce. Likewise, blue tilapia, Mozambique tilapia and some red hybrid strains can be safely reared at cool temperatures without impacting their health or welfare (Table 8.1). In addition to species variation, several commercial strains have been developed as a result of artificial selection for particular commercial traits. These include genetically improved farmed tilapia (GIFT), the Chitralada strain developed in Thailand, the Genomar strain developed through Norwegian and Brazilian collaborations, the GIFT EXCEL developed in the Philippines and the genetically male tilapia (GMT) developed by Fishgen, a spin-off of Swansea University and the University of Stirling in the UK. These differ in performance traits and environmental requirements, affecting their suitability for different rearing systems (El-Sayed, 2020; FAO, 2022). Tilapias fall between levels 4 and 5 in the domestication scale suggested by Teletchea and Fontaine (2014), with level 0 being capture fisheries and level 5 when there are selective breeding programmes targeting specific commercial traits.

Feeding behaviour

Stress and disease lower the feeding response of most farmed fish, so feeding behaviour is a good indicator of health and welfare. Most farmed tilapias are omnivorous and feed low in the trophic chain (Table 8.2), making their farming more sustainable than carnivorous species (Edwards, 2000). In nature, tilapia larvae feed mostly on phytoplankton and/or zooplankton (Stickney, 2017), which in culture can be obtained by fertilizing the ponds (Green, 2006). Tilapia also feed on fish pellets, although live feed is preferred and could be considered a form of environmental enrichment that can improve welfare (Ufodikel & Wada, 1991). An ontogenetic shift from feeding on zooplankton to feeding on particulate matter (visual feeder) and filter-feeding (suction) occurs when the fish are 6–7 cm in standard length (Beveridge & Baird, 2000). However, juveniles can continue using both feeding strategies until adults: they can filter-feed on rotifers and actively prey on microcrustaceans using mucus secreted by the gill rakers.

Table 8.1 Suitability of the main tilapine species to different rearing systems and environmental conditions (adapted from Gupta and Acosta, 2004).

Common name	Species	Rearing system				Environmental conditions			
		Ponds	Cages	Raceway	RAS	Freshwater	Brackish	Sea-water	Cool temp.
Nile tilapia	*Oreochromis niloticus*	✓	✓	✓	✓	✓			
Blue tilapia	*Oreochromis aureus*	✓				✓			✓
Mozambique tilapia	*Oreochromis mossambicus*	✓				✓	✓		✓
Spilurus tilapia	*Oreochromis spilurus*		✓			✓	✓	✓	
Zanzibar tilapia	*Oreochromis hornorum* (*Tilapia urolepis*)	✓				✓	✓		
Galilee tilapia	*Sarotherodon galilaeus*	✓				✓	✓		
Black-chinned tilapia	*Sarotherodon melanotheron*	✓	✓			✓	✓	✓	
Red breast tilapia	*Tilapia rendalii*	✓				✓			
Red belly tilapia	*Tilapia zillii*		✓			✓	✓	✓	
Red tilapia hybrid	*O. mossambicus* × (*O. niloticus, O. aureus or O. hornorum*)	✓	✓	✓	✓	✓	✓	✓	✓

Table 8.2 Feeding habits of the main species of farmed tilapia (adapted and updated from Stickney, 2017).

Species	Food items	References
O. niloticus	Phytoplankton and detritus	Wohlfarth and Hulata (1983)
	Phytoplankton	Moriarty and Moriarty (1973), Trewavas (1983), McDonald (1985a,b), de Moor *et al.* (1986), Northcott *et al.*(1991)
	Zooplankton (rotifers, cladocera and microcrustaceans)	Beveridge and Baird (2000), Fatah Ibrahim *et al.* (2015)
O. mossambicus	Benthic algae, macrophytes, phytoplankton, zooplankton, fish eggs and larvae, detritus	Munro (1967), Naik (1973), Man and Hodgkiss (1977), Weatherley and Cogger (1977), Bowen (1979, 1980), Wohlfarth and Hulata (1983), Trewavas (1983), McDonald (1985a,b), de Moor *et al.* (1986), Northcott *et al.* (1991)
O. aureus	Phytoplankton, zooplankton	Trewavas (1983), McDonald (1985a,b), de Moor *et al.* (1986), Northcott *et al.* (1991)
	Detritus and zooplankton	Spataru and Zorn (1978)
	Adults are omnivores. Fry feed on cladocerans, and juveniles larger than 2 cm on rotifers and filamentous algae	Wohlfarth and Hulata (1983), Yowell and Vinyard (1993) Balarin and Hatton (1979)

Tilapias in their natural habitat can endure to short or long starvation periods for various reasons. For example, tilapia of the genus *Oreochromis* are female mouthbreeders and will cease feeding during the incubation period (Turner & Robinson, 2000). Even when the eggs hatch, after some weeks, the larvae remain in the mouth of the female until egg yolk absorption. After several days of free swimming, larvae will still seek refuge in the female mouth if startled. In nature that means that females grow much slower than males. In intensive production systems, eggs are removed from the female mouth just after fertilization and are bred in incubators or hatching jars to avoid the slow growth associated with breeding starvation. This means that, in general, no natural parental care is allowed for tilapias under intensive production systems, which may have welfare implications.

Short periods of starvation can also occur during transport. Starving fish for 24 h is common practice to reduce stress, lower the metabolic rate and reduce waste that could compromise water quality during transportation. Short-term starvation can have significant physiological effects, but fish recover quickly when they resume feeding and may experience compensatory growth. Starvation can also have some benefits on the immune and stress response in tilapia (Sakyi *et al.*, 2020), although the effects on welfare are unknown and merit further study.

Environmental requirements

Tilapias have modest environmental requirements and, because they are sturdy, can tolerate poor-quality water conditions better than most other farmed fish. They can also thrive under crowded conditions, making them ideal for intensive production. However, sub-optimal water quality stocking density values may result in stunted growth and poor welfare. The main environmental requirements for tilapia farming are well known (Beveridge & McAndrew, 2012; Chapman, 2000; El-Sayed, 2020; FAO, 2022; Hussain, 2004; Stickney, 1993) and are summarized in Table 8.3.

Water temperature

Tilapias spawn when the water temperature reaches 24–25°C, typically after 5–6 months of growth, when the fish weigh ~30g (Froese & Pauly, 2019). Most tilapias cannot survive in cold temperatures for a prolonged time, and their preferred water temperatures are between 31 and 36°C, although this is affected by salinity and also depends on the species. Nile tilapia and most red hybrids cease feeding when water temperature falls <16°C, and begin to die at ~13°C, with temperatures in the range 25–32°C being considered acceptable for growth (Chapman, 2000; DeLong *et al.*, 2009) and 27–29°C being optimal (DeLong *et al.*, 2009). On the other hand, blue tilapia,

Table 8.3 Environmental and production requirements of Nile tilapia.

Parameter	Optimal range	Reference
Water quality		
Optimal temperature (°C) for growth	27–29	DeLong *et al.* (2009)
	28–36	DAFF (2018)
	28–32	Chapman (2000)
	31–36	FAO (2022)
Optimal temperature (°C) for reproduction and embryo development	>24	FAO (2022)
	25–30	Chapman (2000)
Dissolved oxygen (mg/l)	5.0–7.5	DeLong *et al.* (2009)
	>3.0	FAO (2022)
	>4.0	Chapman (2000)
		Stickney (1993)
		Ross (2000)
		Hussain (2004)
pH	6.0–9.0	DeLong *et al.* (2009)
	6.0–8.5	Chapman (2000)
	6.5–9.0	FAO (2022)
		Stickney (1993)
Carbon dioxide (CO_2, mg/l)	<40	DeLong *et al.* (2009
Salinity (ppt)	0–28	Chapman (2000)
Total ammonia nitrogen (TAN, mg/l)	<1.0	Chapman (2000)
Unionized ammonia (NH_3, mg/l)	<1.0	DeLong *et al.* (2009)
	<0.2	Hussain (2004)
Nitrites (NO_2-N, mg/l)	<5.0	DAFF (2018)
		DeLong *et al.* (2009)
Nitrates (NO_3-N, mg/l)	<1.0	DeLong *et al.* (2009)
Stocking density		
Raceway/tank (kg/m^3)	160–185	FAO (2022)
	4	Aldon (1998)
	20	DAFF (2018)
RAS (kg/m^3)	60–120	FAO (2022)
	53	DAFF (2018)
Cage (kg/m^3)	50–150	FAO (2022)
	0.8–4.0	Aldon (1998)
	1.5	DAFF (2018)
	0.8–1.1	FAO (2022)
Pond (kg/m^2)	0.08–0.2	Aldon (1998)
Loading		
Raceway (kg/l/min)	1.2–1.5	FAO (2022)

Mozambique tilapia and red tilapia hybrids can withstand cooler temperatures better than other species. For example, Mozambique tilapia can survive cool conditions (11°C) in brackish waters (DAFF, 2018) while blue tilapia can be reared at 8–9°C (Gupta & Acosta, 2004), although their growth might be compromised.Tilapias can show individual differences in thermal preferences depending on their coping styles, which can also impact their behavioural fever response and ability to fight infection (Cerqueira *et al.*, 2016).

Dissolved oxygen

Tilapias have low dissolved oxygen requirements compared with most other farmed fish. Some species, such as Mozambique tilapia, are facultative air-breathers and can survive out of the water for several hours (DAFF, 2018). Dissolved oxygen levels as low as 0.1 mg/l can be tolerated for short periods (Maruyama, 1948), and tilapia can survive dissolved oxygen (DO) concentrations <0.3–0.8 mg/l (DeLong *et al.*, 2009; Popma & Masser, 1999), although recommended DO levels in ponds should not routinely fall below 1 mg/l (DAFF, 2018) and their growth and feed conversion rate may be impacted at DO <3.5 mg/l (DeLong *et al.*, 2009). It is recommended that DO values be kept >3.0 or 4.0 mg/l, with 5.0–7.5 mg/l considered optimal for growth (DeLong *et al.*, 2009).

pH

Tilapias can survive over a wide range of pH values, from acidic (pH = 5.0) to alkaline waters (pH = 10.0; DeLong *et al.*, 2009), although the optimal range for growth and survival is between 6.0 and 9.0 (Chapman, 2000; DeLong *et al.*, 2009; FAO, 2022; Stickney, 1993). A neutral pH of ~7.0 is recommended in RAS to ensure proper functioning of the biofilter (DeLong *et al.*, 2009).

Carbon dioxide (CO_2)

Carbon dioxide should be kept <40 mg/l because elevated values can result in lethargy and listlessness in tilapia (DeLong *et al.*, 2009), which can compromise foraging. In RAS, stripping of carbon dioxide gas may be required to maintain a neutral pH and promote the growth of nitrifying bacteria in the biofilter (DeLong & Losordo, 2012).

Salinity

Tilapias are mainly freshwater species inhabiting shallow streams, ponds, rivers and lakes. However, some species are euryhaline and can be reared in brackish water, or even in full seawater (El-Sayed, 2020; Pullin, 1991; Watanabe *et al.*, 2006). For example, Nile tilapia can be successfully reared at salinities up to 16 g/l (Souza *et al.*, 2019) and Mozambique tilapia at 20 g/l. Some species have an even higher salinity tolerance; red-belly tilapia can be reared in full seawater (35 g/l) and black-chinned tilapia at salinities up to 45 g/l (Gupta & Acosta, 2004). The availability of euryhaline strains and extreme-saline-tolerant tilapia (halophilic) is important for commercial production in dry locations where freshwater is at a premium and where fish can be reared only in saline or hypersaline ponds and lagoons (Suresh & Lin, 1992; Watanabe *et al.*, 2006).

Ammonia (NH_3)

Ammonia is a primary metabolic waste product of fish, excreted through the gills and the urine, although its main source is the fish feed, hence the importance of avoiding overfeeding (Francis-Floyd & Watson, 2006). Tilapia can survive in polluted waters and are more resistant to high values of free (unionized) ammonia (NH_3) than most other farmed fish (Benli & Köksal, 2005; Paracueles, 2015). However, free ammonia is highly toxic to fish and typically lethal when levels reach 2.0 mg/l. Values >0.05 mg/l can be stressful, resulting in stunted growth and compromised welfare in tilapia (da Silva *et al.*, 2013). The recommended concentration of unionized ammonia (NH_3) in fish farms is always zero (Francis-Floyd & Watson, 2006), and should never be allowed to exceed 0.2 mg/l (Hussain, 2004).

Nitrites (NO_2-N)

It is recommended that the concentration of nitrite nitrogen is <5 mg/l in freshwater systems (DAFF, 2018) when the concentration of chloride (Cl^-) is <10 mg/l (DeLong *et al.*, 2009). Increasing chloride concentration to 150–200 mg/l can help reduce nitrite toxicity, which can be achieved by adding common sea salt or calcium chloride (DeLong *et al.*, 2009).

Nitrates (NO_3-N)

Nitrate nitrogen is toxic when it exceeds 300–400 mg/l in RAS, but this can be controlled by water exchanges, filtration and denitrification (DeLong *et al.*, 2009). The recommended level is <1 mg/l to maintain total ammonia nitrogen (TAN) <1.0 mg/l (Chapman (2000).

Welfare indicators for tilapia

To improve the welfare of tilapia, fish farmers need to know what to measure and, for this, they need welfare indicators. Traditionally fish welfare has been measured using 'negative indicators', i.e. points are deducted for deviations from 'normal' or ideal conditions and the more that fish deviate the more points are scored and the worse the welfare of fish is supposed to be (Noble *et al.*, 2018, 2020; Stien *et al.*, 2013, 2020). The advantage of such a system is that it is simple, repeatable and easy to score, but it has been criticized because it does not guarantee that fish are under good welfare: it simply shows when fish are being maintained in a bad state. More recently, attempts have been made to develop 'positive welfare indicators' that can be used to assess 'positive welfare'. These take into account mental states and choice preferences (Kristiansen *et al.*, 2020; Stien *et al.*, 2020), and more sophisticated assays are being trialled for this purpose (Maia & Volpato, 2017); however, these are still in their infancy and are difficult to apply under commercial conditions.

To be useful, welfare indicators (WIs) must be fit for purpose (i.e. they must measure what they are intended to measure), be species and life-stage specific (i.e. one size will rarely fit all (Segner *et al.*, 2019)) and must be easy to use (Noble *et al.*, 2018; Stien *et al.*, 2020). The latter point is crucial because otherwise WIs will not be used by fish farmers. To ensure that WIs are useful, operational welfare indicators (OWIs) have been developed that take into account the expertise and insights of stakeholders involved in the farming and care of fish (typically fish farmers; North *et al.*, 2008; van de Vis *et al.*, 2012), scientists, non-governmental organizations (NGOs) and regulators. For this purpose, a consultation process is recommended and this is best achieved through a Delphi expert approach (Garcia de Leaniz *et al.*, 2022; Gutierrez Rabadan *et al.*, 2021). The defining characteristic of OWIs (as opposed to laboratory-based indicators) is that they can be used on the farm (Gutierrez Rabadan *et al.*, 2021; Noble *et al.*, 2018). OWIs can then be used to benchmark farming operations, identify areas needing improvement and develop best practices (Folkedal *et al.*, 2016; Noble *et al.*, 2019). Typically, the development of OWIs consists of four steps (Gutierrez Rabadan *et al.*, 2021): (1) selection and screening of OWIs; (2) OWI validation; (3) OWI refinement and simplification; and (4) OWI testing at commercial sites.

There are no validated OWIs for tilapia, which is surprising considering this species' global importance as a farmed fish. As with other species, stress has been found to lower disease resistance and compromise the welfare of tilapia (Ellison *et al.*, 2018, 2020). For this reason, stress-related indicators tend to form part of tilapia fish welfare assessments. For example, Azim and Little (2008) examined fin condition, gill histology, proximate composition, blood parameters and plasma cortisol levels as indicators of tilapia welfare, all directly or indirectly related to stress. Likewise, ten welfare metrics were used by Brazilian researchers to asses tilapia welfare in semi-intensive farms (Pedrazzani *et al.*, 2020), including several stress-related indicators. A selection of welfare indicators used to assess the welfare of tilapia is shown in Table 8.4, and these are discussed briefly below. Although all of these metrics provide some useful information, not all can be applied on the farm and more work is needed to validate them and determine how individual welfare metrics can be best combined to remove redundancies.

Eye condition

Eye condition is a useful WI that is easy to score and is related to a number of underlying causes that can impair welfare, growth and the reproduction (Obirikorang *et al.*, 2022; Pedrazzani *et al.*, 2020). Among the eye conditions found in tilapia are cataract and exophthalmia (commonly known as pop-eye). Cataracts can be symptomatic of nutritional deficits but can also be caused by trauma and by some parasites. Exophthalmia is a common, non-specific sign of several infectious diseases.

Injuries and lesions

Injuries and lesions may cause pain and are also a point of entry for secondary infections that affect health and welfare. The most common injuries and lesions found in farmed tilapia include fin damage, lesions in the mouth and skin (including loss of scales), fusion and thickening of lamella, gill haemorrhages, gill hyperplasia and gill splitting. Fin damage (erosion and splitting) is typically caused by aggression and antagonistic competition (fin nipping), but also by poor water quality, trauma and infection, which are exacerbated by stress (Noble *et al.*, 2007). Skin and mouth lesions often also result from stress, infectious diseases, poor water quality or trauma. Gill injuries may reflect the effect of parasites, poor water quality and disease, but also rough handling and exposure to air during transport and grading (Benli & Köksal, 2005; Pedrazzani *et al.*, 2020).

Malformations

Malformations may interfere with normal physiological functioning, including breathing and feeding, swimming and normal behaviours (Segner *et al.*, 2019). Various malformations have been reported for tilapia that can be used as part of a welfare assessment, as done for other species (Gutierrez Rabadan *et al.*, 2021; Noble *et al.*, 2012). Tilapia malformations include deformities of the upper and lower jaw (Pedrazzani *et al.*, 2020), shortening of the operculum

Table 8.4 Welfare indicators for tilapia.

Welfare indicator	Underlying causes and impacts	Example of use
Eye condition		
Cataracts	Nutritional deficiency, trauma	Obirikorang *et al.* (2022)
Exophthalmia	Disease, infection	Pedrazzani *et al.* (2020)
Injuries and lesions		
Fin erosion	Competition, stress, poor water quality, trauma, infection	Obirikorang *et al.* (2022)
Mouth lesions	Disease, poor water quality, infection	Obirikorang *et al.* (2022)
Skin lesions	Competition, stress, poor water quality, trauma, infection	Obirikorang *et al.* (2022)
Gill injuries, lamella fusion, thickening, splitting	Parasites, disease, handling trauma, air exposure, poor water quality	Pedrazzani *et al.* (2020) Benli and Köksal (2005)
Malformations		
Deformed upper or lower jaw	Nutritional deficiency, trauma during embryo development, inbreeding	Pedrazzani *et al.* (2020)
Shortened operculum	Nutritional deficiency, trauma during embryo development, inbreeding	Handwerker and Tave (1994)
Caudal deformity	Nutritional deficiency, high incubation temperature, inbreeding	Mair (1992)
Vertebral deformity (lordosis, scoliosis)	Nutritional deficiency, high incubation temperature, inbreeding	Wang and Tsai (2000)
Fluctuating asymmetry	↑ = ↑ Developmental stress	Fessehaye *et al.* (2007)
Behaviour		
Laterality	↑ = ↑ Reactive coping style	Berlinghieri *et al.* (2021)
Nutritional status and growth		
Condition factor	↓ = ↑ Emaciation/underfeeding	Pedrazzani *et al.* (2020)
Relative weight	↓ = ↑ Emaciation/underfeeding	Pedrazzani *et al.* (2020)
Feed conversion ratio	↑ = ↑ Nutritional disorder ↓ growth	Pedrazzani *et al.* (2020)
Stress indicators		
Plasma cortisol	↑ = ↑ Chronic or acute stress	Pedrazzani *et al.* (2020)
Plasma glucose	↑ = ↑ Stress, nutritional disorder	Pedrazzani *et al.* (2020)
Ventilation frequency	↑ = ↑ Stress/hypoxia	Benli and Köksal (2005)
Eye darkening	↑ = ↑ Stress	Benli and Köksal (2005)
Skin darkening	↑ = ↑ Stress	Paracueles (2015)

Continued

Table 8.4 (continued)

Welfare indicator	Underlying causes and impacts	Example of use
Haematological parameters		
Red blood cells	↓ = ↑ Anaemia, liver and kidney disease	Obirikorang *et al.* (2022)
Haemoglobin	↓ = ↑ Anaemia, nutritional deficiency	Obirikorang *et al.* (2022)
Haematocrit	↓ = ↑ Anaemia, infection, nutritional deficiency	Obirikorang *et al.* (2022)
MCH	↓ = ↑ Anaemia, nutritional deficiency, kidney disease, autoimmune diseases, liver disease, inflammation	Obirikorang *et al.* (2022)
MCV	↓ = ↑, nutritional deficiency, chronic diseases, inflammation	Obirikorang *et al.* (2022)
MCHC	↓ = ↑ Anaemia, nutritional deficiency parasitic infections	Obirikorang *et al.* (2022)
White blood cells	↑ = ↑ Infection, inflammation, stress, immune disease	Obirikorang *et al.* (2022)
Platelets	↓ = ↑ Anaemia, viral infections	Obirikorang *et al.* (2022)
Mortality	Poor welfare suspected when >10%	Pedrazzani *et al.* (2020)
	Poor welfare suspected when >40% for 5–30-g fish	Chapman (2000)
	Poor welfare suspected when >30% for 30–680-g fish	Chapman (2000)

(Handwerker & Tave, 1994), caudal deformity (Mair, 1992) and various forms of vertebral deformities, including lordosis and scoliosis (Wang & Tsai, 2000). These can result from a range of causes, including nutritional deficiencies, high incubation temperatures, trauma during embryo development and inbreeding. One particular condition is fluctuating asymmetry in paired structures (such as fin rays and gill rakers), which may reflect stress during embryonic development (Fessehaye *et al.*, 2007).

Behavioural indicators

Monitoring deviations from 'normal' behaviours may reflect stressful conditions and can be used to assess welfare. For example, stressed tilapia display erratic swimming, combining rapid bursts of swimming with freezing bouts, as well as rapid movements of the fins and operculum and gasping for air at the water's surface (Paracueles, 2015). Other behaviours that could form part of a suite of behavioural welfare indicators include behavioural laterality (handedness), because this may be related to stress-coping styles (Berlinghieri *et al.*, 2021; Nepomnyashchikh & Izvekov, 2006). Both Nile tilapia and Mozambique tilapia display significant laterality (Nepomnyashchikh & Gremyachikh, 1993), with 30% of fish preferring right turns, 20% left turns and 50% being ambidextrous (Gonçalves & Hoshino, 1990). An increase in laterality is thought to reflect a reactive coping style (Berlinghieri *et al.*, 2021), but also an increase in stereotypical behaviours and, possibly, impaired welfare.

Body condition, nutritional status and growth

Stunted growth and loss of weight (emaciation) are common signs of poor welfare in farmed fish (Noble *et al.*, 2020, Stien *et al.*, 2020; Weirup *et al.*, 2022) that can be caused by underfeeding, but also by poor water quality, competition, nutritional deficits, chronic stress and disease. Emaciated fish are more susceptible to disease (Stien *et al.*, 2013) and tend to display abnormal behaviours, often swimming at the surface or perimeter of the tank (Weirup *et al.*, 2022). Two non-invasive, operational indicators that can be used to detect poor nutritional status and assess body condition are Fulton's condition factor and relative weight.

Fulton's condition factor (K) assumes isometric growth (i.e. that body shape does not change with growth) and is calculated as:

$$K = \left(\frac{WT}{TL^3} \right) \times 100$$

where *WT* is the body mass in grams and *TL* is the total length in cm (Blackwell *et al.*, 2000).

Pedrazzani *et al.* (2020) considered that the optimal Fulton's condition factor for Nile tilapia was between 1.6 and 1.9 in semi-extensive systems, and that $K<1.0$ or >2.4 would be indicative of severely impaired welfare. However, the condition factor varies widely among tilapia studies, ranging from $K = 0.4–0.8$ (Paredes-Trujillo *et al.*, 2021) to 1.3–2.3 depending on the season (Keyombe *et al.*, 2017), between 1.0 and 2.5 depending on density (Fatima *et al.*, 2021) and between 1.5 and 2.0 depending on rearing system and diet (Anani & Nunoo, 2016; Ighwela *et al.*, 2011). In red-belly tilapia (*Tilapia zillii*) mean *K* was 0.74 when fish were reared in full seawater and 2.07 in freshwater, while in Zanzibar tilapia (*Oreochromis urolepis*) *K* was 0.53 in seawater and 0.86 in freshwater (Nehemia *et al.*, 2012).

Some studies have found a significantly lower condition factor among parasitized tilapia (Paredes-Trujillo *et al.*, 2021), suggesting that this could be used as a metric of welfare. Other studies, however, have failed to find any relation between parasite loads and condition factor (Zanolo *et al.*, 2009). Thus, while the use of *K* is helpful in detecting changes in body condition over time, or among tanks within a farm, that may be related to welfare, it is not generally possible to give absolute indicative values for welfare assessment because this depends on the season, diet, species and specific farm conditions.

Another popular measure of body condition is relative weight (W_r), which is defined as the ratio between the observed weight of an individual and the predicted weight of an average fish of the same length (Blackwell *et al.*, 2000), the latter being derived empirically from length:weight relationships (Blackwell *et al.*, 2000). Relative weight is expressed as a percentage, 100% being the average relative weight of 'normal' fish and is equivalent to body mass index (BMI) in humans. The use of relative weight has a number of advantages compared with Fulton's condition factor for assessing body condition: (1) it does not assume a constant slope of $b = 3$ for the length:weight relationship (instead it is derived empirically for each population and developmental stage (Gutierrez Rabadan *et al.*, 2021; Rey *et al.*, 2021); (2) it is invariant to measurement units; (3) it compensates for inherent changes in body shape; and (4) it can be used to compare populations and fish of different length (Blackwell *et al.*, 2000). It has been proposed for lumpfish that a relative weight of 90–110% should be considered normal, that 75–90% represents underweight fish and <75% corresponds to emaciated fish (Gutierrez Rabadan *et al.*, 2021). The same classification is probably also useful for tilapia welfare assessment.

One widespread cause of poor growth and loss of weight among farmed fish is inadequate feeding, either because daily feeding rations are too low, aquafeeds are of the wrong size or type for the size of fish, they are not delivered properly or lack essential nutrients (El-Sayed, 2013, 2020; Lim & Webster, 2006). Tilapia are omnivorous fish that can be fed on a wide variety of diets, but are also a voracious species that require a lot of food (Ng & Romano, 2013). To help farmers assess whether their fish are being fed properly, the feed conversion ratio (FCR) can be calculated as part of the general assessment and monitoring. FCR is calculated as the weight of feed given to the fish divided by the weight gained over the same period, typically over a few weeks or months. A low FCR typically means a more efficient system whereas a high FCR may be indicative of overfeeding or poor growth (Fry *et al.*, 2018). Thus, FCR can be used to benchmark the efficiency of an operation and, in combination with data on growth rate and condition factor or relative weight, to reduce the risk of underfeeding and inform the feeding strategy and adjustment of daily feeding rates. In Nile tilapia, recommended daily feeding rates in semi-intensive systems are 15–20% body weight (BW) for post-hatch larvae (0.5 g), 10–15% BW for 5-g juveniles, 3–5% for 18–75-g fish and 1.5–3.0% BW for market-size fish (+150 g), being ~ 2% BW for broodstock (DeLong *et al.*, 2009). However, these depend on specific conditions and should be used only as rough guides. Recommended feeding ratios in intensive systems are typically 50–67% higher than in semi-intensive systems (Bhujel, 2013).

The optimal (i.e. target) FCR for Nile tilapia depends on fish size, feed type, feeding regime and rearing system. Globally, the FCR of tilapia has been found to vary between 1.4 and 2.4 (Tacon & Metian, 2008). However, there is considerable variation between sites. In

Brazil the recommended FRC is 0.8–1.6 in earthen ponds and 0.8–2.0 in cages: the larger the fish the higher the FCR (Pedrazzani *et al.*, 2020). In Thailand the typical FCR is 0.6 in ponds and 1.4–1.8 in cages (Bhujel, 2013), while 1.4–1.8 is the expected FCR in tanks in the USA (DeLong *et al.*, 2009).

Stress indicators

As with other species, stress has been found to lower disease resistance and compromise the welfare of tilapia (Ellison *et al.*, 2018, 2020). For this reason, stress-related indicators tend to form part of most tilapia fish welfare assessments, although not all can be measured at the farm.

Plasma cortisol

Two stress indicators that can be measured as part of welfare assessment are levels of cortisol and glucose. Unlike some other fish species, tilapia are 'stress competent' immediately upon hatching (Pepels & Balm, 2004) and able to respond to stressors without any gradual transition during the juvenile phase.

Unstressed Nile tilapia display a whole-body cortisol concentration of 5.5 ng/g at the end of yolk sac absorption (Hwang *et al.*, 1992) and have a mean blood plasma concentration of 3.9 ± 0.69 ng/ml (SE) when they reach 75 g (Villarroel *et al.*, 2022), whereas Mozambique tilapia display 25 ng/ml when they reach 223 g in weight. In all cases, tilapias are considered to be low-stress-response species, displaying cortisol concentrations in blood plasma typically <40 ng/ml (Moreira & Volpato, 2004).

Although cortisol is typically measured in blood plasma, recent advances make it possible to measure it non-invasively in water, skin, scales and faeces (Ellis *et al.*, 2013; Sadoul & Geffroy, 2019; Uren Webster *et al.*, 2020), and biologgers can also be used to monitor it in free-swimming sentinel fish (Barreto *et al.*, 2022). Such advances will make it possible to use biosensors to monitor stress continuously on the farm, and incorporate welfare assessment into precision fish farming (Føre *et al.*, 2018).

Blood glucose

Recently blood glucose has been proposed as a potential welfare indicator in Nile tilapia (Teodózio de Araújo *et al.*, 2022). The concentration of plasma glucose is directly linked to nutritional stress and increases when cortisol is being mobilized (Barton, 2002; Vijayan *et al.*, 1997), but it is easier and cheaper to analyse than plasma cortisol. Glucose biologgers have also been developed. These have been implanted into Nile tilapia and can be used to monitor blood glucose levels in free-swimming fish in real time (Wu *et al.*, 2019), which can facilitate in-farm welfare assessment.

Eye darkening

Eye darkening represents a simple way of detecting stress in tilapia (Champneys *et al.*, 2018; Freitas *et al.*, 2014; Miyai *et al.*, 2011; Vera Cruz & Brown, 2007; Vera Cruz & Tauli, 2015; Volpato *et al.*, 2003) and can therefore be used as an indirect welfare indicator. In general, eye darkening correlates positively with stress (Champneys *et al.*, 2018; Freitas *et al.*, 2014; Miyai *et al.*, 2011). For example, the eyes and skin of tilapia become darker when the fish are subjected to high levels of ammonia (Benli & Köksal, 2005; Paracueles, 2015). Eye darkening can be assessed from diagrams obtained from photographs (Freitas *et al.*, 2014), but could also be obtained from video images, which might facilitate its use in camera-based monitoring systems. The major shortcoming of the use of eye darkening is that it is very labile and also context dependent (Champneys *et al.*, 2018; Rodriguez-Barreto *et al.*, 2019), and more research is needed to understand the mechanisms and persistence of eye colour changes in fishes (Vanderzwalmen *et al.*, 2021).

Skin darkening

As with eye darkening, skin darkening increases with chronic stress in tilapia (Paracueles, 2015) but is also affected by changes in social dominance and aggression (Champneys *et al.*, 2018), as also seen in other fish (Höglund *et al.*, 2000). In zebrafish, aggressive individuals are darker while fearful individuals become paler (Gerlai *et al.*, 2000). Skin darkening is positively correlated to eye darkening in Nile tilapia (Champneys *et al.*, 2018), but is easier to score and might be amenable to continuous welfare assessment via machine vision (Barreto *et al.*, 2022; Li *et al.*, 2022).

Opercular beat rate (ventilation frequency)

Like many other fish species, when tilapia are stressed they increase their ventilation frequency (Obirikorang *et al.*, 2019), and this can be quantified by measuring opercular beat rate (OBR), which can be used as a welfare indicator (Barreto & Volpato, 2004, 2006). Ventilation frequency is positively correlated with both eye darkening and stress in tilapia and also increases during hypoxia, which could be used to detect changes in dissolved oxygen through machine vision before welfare is compromised (Xu *et al.*, 2006; Zheng *et al.*, 2014).

Haematological parameters

The analysis of haematological parameters can reveal a lot about the health status and well-being of fish. For example, changes in glucose and triglycerides can be used to assess nutritional status, and other parameters to infer reproductive status and the presence of diseases that will affect welfare (Seibel *et al.*, 2021). In addition to cortisol and glucose, the most useful haematological parameters are given in Table 8.4, because these are part of the complete blood count (CBC) used to evaluate overall health and detect various disorders. For example, low counts of red blood cells, haemoglobin and haematocrit can be symptomatic of anaemia, but also liver and kidney diseases and nutritional deficiencies (Obirikorang *et al.*, 2022). Low values of mean corpuscular haemoglobin (MCH), mean corpuscular volume (MCV) and mean corpuscular haemoglobin concentration (MCHC) can be caused by anaemia, kidney and liver disease and nutritional deficiencies, but also by autoimmune diseases, inflammation and parasitic infections (Obirikorang *et al.*, 2022). High values of white blood cells and platelets can indicate viral infections, inflammation, stress or immune diseases (Obirikorang *et al.*, 2022).

Mortality

Increases in mortality above the seasonal norm can be used as part of a welfare assessment (Garcia de Leaniz *et al.*, 2022; Rey Planellas & Garcia de Leaniz, 2024), although what constitutes 'normal' may be difficult to establish because it depends on the stage of development and may also be site specific. For example, in semi-extensive systems in Brazil it has been suggested that poor welfare might be suspected when the mortality of Nile tilapia exceeds 10% (Pedrazzani *et al.*, 2020). Other authors have indicated threshold values for tilapia of 30–40% mortality (Chapman, 2000). In general, monitoring of mortality will allow the farmer to identify unusual events and for this the median and the tenth and 90th percentiles can be used to benchmark historical mortality losses (Soares *et al.*, 2011).

Practical steps to improve and promote positive welfare in Tilapia

Knowing what to measure is the first necessary step to achieving better welfare in farmed fish, but measurements alone are not enough to improve welfare. As mentioned before, the mental state of the fish should be the final welfare indicator. That has been measured with cognitive tests in research studies and could also be improved by the use of enrichment in their environment to mitigate some of the negative impacts of farming practices. Fortunately, there are a number of environmental enrichment strategies that farmers can adopt to improve the welfare of farmed tilapia. Fourteen such practical steps are described below.

Optimal rearing densities

Stocking density is a key production parameter in the fish farm as it affects economic costs and revenue, but also fish health, stress and welfare. The Farm Animal Welfare Council recommends that farmed fish are given 'sufficient space to show most normal behaviours with minimal pain, stress and fear' (Fine Arts Work Center in Provincetown, 2014). The choice of an optimal density is therefore paramount in order to maintain high welfare and maximum production. However, unlike some other farmed fish, a low stocking density does not necessarily result in better welfare in tilapia. Indeed, tilapia show increased aggression and elevated stress at low densities (Champneys *et al.*, 2018; Evans *et al.*, 2008; Rodriguez-Barreto *et al.*, 2019), which results in greater susceptibility to disease and compromised immunocompetence (Ellison *et al.*, 2018). At low densities, social dominance develops and dominant fish may become aggressive and monopolize resources to the detriment of other fish in the tank (Segner *et al.*, 2019). For example, Nile tilapia reared at 48 kg/m^3 were less aggressive than those reared at 13 kg/m^3 (Rodriguez-Barreto *et al.*, 2019).

Maximum growth rate is achieved at a density of 30 kg/m^3, although this can be as high as 60–72 kg/m^3 with vigorous aeration

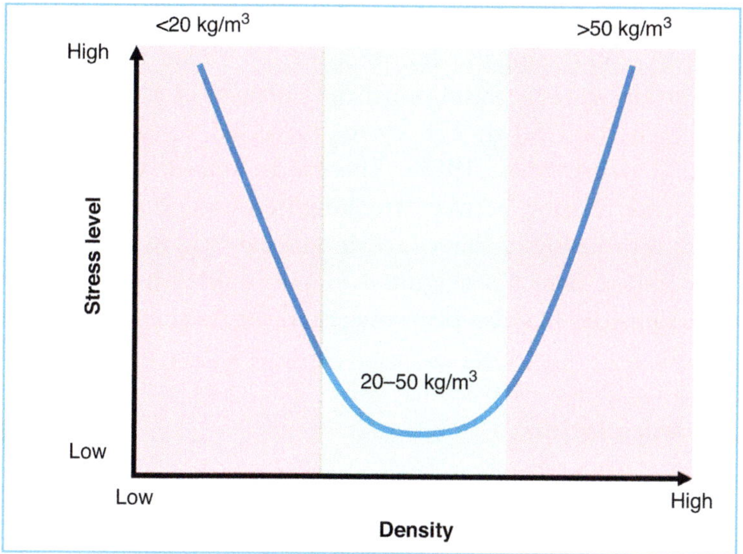

Figure 8.1 Schematic relationship between stocking density (kg/m^3) in RAS and stress levels in farmed tilapia. Stress is reduced at intermediate densities

and can even reach 90–120 kg/m^3 with supplemental oxygen (DeLong *et al.*, 2009). In Thailand, stocking densities of fish of market size (200–250 g) are 40–60 kg/m^3 in raceways (Bhujel, 2013).

From a welfare perspective it is difficult to determine a single optimal density across all contexts because this depends on fish size, the number of social interactions (Gonçalves-de-Freitas *et al.*, 2019), the rearing system (Table 8.3) and tank volume, with larger tanks requiring lower densities. For example, 15-g Nile tilapia can be reared at 11–22 kg/m^3 in 7-l tanks but the density needs to be reduced to 8–12 kg/m^3 in 19-l tanks (Evans *et al.*, 2008).

Nile tilapia can be reared at densities as high as 185 kg/m^3 in raceways (FAO, 2022), but a lower density has been recommended for maintaining high welfare (~34 kg/m^3; Zaki *et al.*, 2020). However, the response to crowding in Nile tilapia depends on social status (Rodriguez-Barreto *et al.*, 2019). At low density the formation of social hierarchies is maintained by agonistic interactions and this results in social stress among subordinate fish, whereas at high density chronic stress is widespread and is caused by crowding (Rodriguez-Barreto *et al.*, 2019). This suggests that there is an optimal density (Figure 8.1). Increasing the density from 2 to 30 kg/m^3 resulted in a significant decrease in plasma cortisol (Villarroel *et al.*, 2022), while increasing it from 50 to 178 kg/m^3 increased cortisol more than fourfold (Vijayan *et al.*, 1997). This lends support to the contention that the relation between stocking density and stress response in tilapia is probably U-shaped, with optimal welfare values reached at intermediate densities (Figure 8.1). An optimal welfare at intermediate densities has also been suggested for Arctic charr (Sevier *et al.*, 2019).

Sex manipulation

Mixed-sex tilapia are easy to rear but, as males outgrow females, this results in small harvest weight, mixed sizes and increased competition and aggression mainly in adults during reproduction, which tends to impact on welfare. The advantages of producing all-male tilapia are a larger weight and more uniform size distribution, which can result in better welfare. The disadvantages of producing monosex tilapia are that it is technically difficult and that it introduces a technological dependency that may curtail development. Production of all-male tilapia is therefore desirable in commercial aquaculture and can be achieved directly by sex reversal of the fry via administration of methyltestosterone in the feed (FAO, 2022; Hiott & Phelps, 1993), or indirectly by the use of YY male broodstock obtained via oestrogen feminization of the parents using the so-called GMT technology (Beardmore *et al.*, 2001). Increasing the water temperature by 4°C at 10–30 days post-fertilization can also increase the proportion of males (Nivelle *et al.*, 2019). However, recent studies with different tilapia strains show that good production and welfare can also be achieved without sex manipulation (Dan & Little 2000).

Grading

Removal of the largest fish (partial harvesting) can be used to lessen competition and provide fish with more space, which may improve welfare (El-Sayed, 2020). However, care must be taken to ensure that density does not become too low because this may encourage aggression and compromise welfare. Passive grading is recommended to avoid excessive handling.

Background colour

Tilapia avoid red backgrounds (Maia *et al.*, 2020) and prefer blue (Maia & Volpato, 2019) or yellow colours (Luchiari *et al.*, 2007), which can inform the choice of rearing tanks to maximize welfare.

Dietary supplementation

There is growing interest in the use of dietary supplements in tilapia aquaculture, particularly in relation to herbal extracts (Gabriel *et al.*, 2015b; Gabriel, 2019) and microalgae (Trevi *et al.* 2023, 2024). Over 40 herbal extracts, minerals and compounds have been shown to have beneficial effects on tilapia, a selection of which are shown in

Table 8.5. In addition to enhanced growth, these include improvements in haematological parameters, protection against oxidative damage, immune stimulation, enhanced disease resistance, lowering of the stress response and bactericidal properties among others. Several essential oils and extracts including sweet orange, lemon, aloe vera, allspice, garlic and tea plant can increase the size and density of red blood cells and concentration of haemoglobin, which can facilitate oxygen uptake and could improve welfare under anoxic or overcrowding conditions. Other dietary supplements, including aloe vera, Indian bael, ashwagandha, ginger and holy basil can modulate the stress response, as evidenced by changes in the concentration of plasma cortisol and glucose (Table 8.5).

One group of dietary supplements that can be particularly useful in improving tilapia welfare includes those rich in L-tryptophan, such as Spirulina and certain other microalgae. Tryptophan is an essential amino acid, required for maintenance of the body's normal functioning, that cannot be synthesized and must be obtained from the diet. Tryptophan is a precursor of serotonin and can improve stress responses by lowering the production of corticosteroids and reduce aggression. Dietary supplementation with tryptophan has been shown to reduce stress and aggression in salmonids (Lepage *et al.*, 2002, 2003; Winberg *et al.*, 2001), Atlantic cod (Höglund *et al.*, 2005) and other species (Hoseini *et al.*, 2019). In Nile tilapia the dietary addition of tryptophan reduced fin erosion (Lall & Tibbetts, 2009) and the frequency of antagonistic interactions, lowered stress and improved welfare (Martins *et al.*, 2013; Neto & Giaquinto, 2020; Vieira *et al.*, 2021) and could be used to modify the stress-coping style (Figure 8.2).

Structural enrichment

Structural enrichment in the form of plastic plants reduces aggression in tilapia (Barreto *et al.*, 2011; Neto & Giaquinto, 2020; Torrezani *et al.*, 2013). Tilapia prefer gravel-enriched over shelter-enriched or non-enriched environments (Delicio *et al.*, 2006). Provision of natural and structured substrates has been shown to reduce neophobia (Tatemoto *et al.*, 2021) and aggression (Galhardo *et al.*,

Table 8.5 Dietary supplements that can help improve health and welfare in tilapia (NT = Nile tilapia; MT = Mozambique tilapia).

Compound	Type	Delivery	Welfare/health benefit	Species	Reference
Sweet kelp *Laminaria japonica*	Seaweed	Dietary	Improves haematological parameters, protects against oxidative damage	NT	Abdel-Daim *et al.* (2020)
Spirulina *Spirulina platensis*	Microalgae	Dietary	Protects against oxidative damage	NT	Abdelkhalek *et al.* (2017)
Spirulina *Spirulina platensis*	Microalgae	Dietary	Immune stimulant, disease resistance	NT	Abdel-Tawwab and Ahmad (2009)
Spirulina *Spirulina platensis Bacillus amyloliquefaciens*	Microalgae+ bacterium	Dietary	Immune stimulant	NT	Al-Deriny *et al.* (2020b)
Spirulina *Spirulina platensis* Selenium nanoparticles	Microalgae+ mineral	Dietary	Immune stimulant	NT	Al-Deriny *et al.* (2020a)
Yucca *Yucca schidigera*	Plant	Bath	Protects against oxidative damage	NT	Abdel-Tawwab *et al.* (2021)
Saccharomyces *Saccharomyces cerevisiae*	Yeast	Bath	Protects against oxidative damage	NT	Abdel-Tawwab *et al.* (2021)
Guava *Psidium guajava*	Plant	Dietary	Improves haematological parameters	NT	Abdel-Tawwab and Hamed (2020)
Chitosan–vitamin E nanocomposite	Polysaccharide + vitamin	Dietary	Protects against oxidative damage; immune stimulant, disease resistance	NT	Ahmed *et al.* (2021)

Continued

Table 8.5 (continued)

Compound	Type	Delivery	Welfare/health benefit	Species	Reference
Thyme *Thymus vulgaris* Fenugreek *Trigonella foenum graecum* Neem *Azadirachta indica*	Plant phytobiotic	Dietary	Lowers cortisol response, immune stimulant	NT	Antache *et al.* (2014)
Cinnamon *Cinnamomum* sp.	Plant	Dietary	Immune stimulant	NT	Dos Santos *et al.* (2016)
Oregano *Origanum vulgare*	Plant	Dietary	Reduces oxidative stress and aggression	NT	El-Hawarry *et al.* (2018)
Linseed *Linum usitatissimum* Soapbark *Quillaja saponaria*	Plant	Dietary	Improves fin damage and immune-oxidative status	NT	Elkaradawy *et al.* (2022)
Sweet orange *Citrus sinensis*	Plant	Dietary	Disease resistance, improves haematological parameters	MT	Acar *et al.* (2015)
Gokshura *Tribulus terrestris*	Plant	Dietary	Disease resistance	MT	Yilmaz *et al.* (2014)
Lemon *Citrus limon*	Plant	Dietary	Immune stimulant, improves haematological parameters	MT	Baba *et al.* (2016)
Aloe vera *Aloe barbadensis*	Plant	Dietary	Improves haematological parameters, stress response	NT	Gabriel *et al.* (2015c) Gabriel *et al.* (2015a)
Blackberry *Rubus* sp.	Plant	Dietary	Immune stimulant, improves haematological parameters	NT	Yilmaz (2019a)
Allspice *Pimenta dioca*	Plant	Dietary	Improves haematological parameters, stress response, immune stimulant	MT	Güllü *et al.* (2016)

Continued

Table 8.5 (continued)

Compound	Type	Delivery	Welfare/health benefit	Species	Reference
Allspice *Pimenta dioca*	Plant	Dietary	Immune stimulant, improves haematological parameters	MT	Yılmaz *et al.* (2015)
Tea plant *Camellia sinensis*	Plant	Dietary	Improves haematological parameters, stress response, immune stimulant	NT	Abdel-Tawwab *et al.* (2010)
Garlic *Allium sativum*	Plant	Dietary	Improves haematological parameters, stress response, immune stimulant	NT	Shalaby *et al.* (2006)
Bright green nori *Ulva clathrata*	Seaweed	Dietary	Improves haematological parameters	NT	del Rocío Quezada-Rodríguez and Fajer-Ávila (2017)
Bermuda grass *Cynodon dactylon*	Plant	Dietary	Improves haematological parameters	MT	Immanuel *et al.* (2009)
Indian bael *Aegle marmelos*	Plant	Dietary	Improves haematological parameters	MT	Immanuel *et al.* (2009)
Ashwagandha *Withania somnifera*	Plant	Dietary	Improves haematological parameters	MT	Immanuel *et al.* (2009)
Ginger *Zingiber officinale*	Plant	Dietary	Improves haematological parameters	MT	Immanuel *et al.* (2009)
Mistletoe *Viscum album coloratum*	Plant	Dietary	Immune stimulant	NT	Park and Choi (2012)
Heal-all *Prunella vulgaris labiatae*	Plant	Injection	Improves haematological parameters	NT	Park and Choi (2014)

Continued

Table 8.5 (continued)

Compound	Type	Delivery	Welfare/health benefit	Species	Reference
Ginseng *Panax ginseng*	Plant	Dietary	Immune stimulant, improves haematological parameters	NT	Goda (2008)
Macsumsuk	Mineral	Dietary	Improves haematological parameters	NT	Shahkar *et al.* (2015)
Rosemary *Rosmarinus officinalis*	Plant	Dietary	Improves haematological parameters, disease resistance	MT	Gurkan *et al.* (2015) Gültepe *et al.* (2014)
Fenugreek *Trigonella foenum graecum*	Plant	Dietary	Improves haematological parameters, disease resistance	MT	Gurkan *et al.* (2015) Gültepe *et al.* (2014)
Thyme *Thymus vulgaris*	Plant	Dietary	Improves haematological parameters, disease resistance	MT	Gurkan *et al.* (2015) Gültepe *et al.* (2014)
Black cumin *Nigella sativum*	Plant	Dietary	Improves haematological parameters, disease resistance	NT	Elkamel and Mosaad (2012)
Caffeic acid	Plant	Dietary	Improves haematological parameters, anti-oxidant, immunostimulant, disease resistance	NT	Yilmaz (2019b)
Purple coneflower *EChinacea purpurea*	Plant	Dietary	Immune stimulant, improves haematological parameters, bactericidal	NT	El-Sayed *et al.* (2014)
Japanese honeysuckle *Lacinera japonica*	Plant	Dietary	Immune stimulant, disease resistance	NT	Yin *et al.* (2008)

Continued

Table 8.5 (continued)

Compound	Type	Delivery	Welfare/health benefit	Species	Reference
Holy basil *Ocimum tenuiflorum*	Plant	Dietary	Stress response	NT	Arenal *et al.* (2012)
Lingzhi *Ganoderma lucidum*	Fungus	Dietary	Immune stimulant, disease resistance	NT	Yin *et al.* (2008)

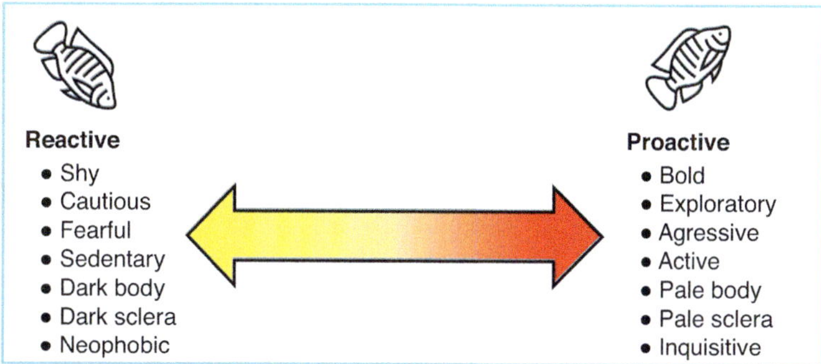

Figure 8.2 Manipulation of rearing density and diet can help modify stress-coping styles in tilapia (adapted from Champneys *et al.*, 2018)

2008, 2009) and may improve welfare. This is particularly important during reproduction because tilapia prefer natural substrates for nesting (Freitas & Volpato, 2013; Mendonça *et al.*, 2010). In some cases thermal gradients for overwintering can be created in indoor ponds or in aquaponics systems, and this may be viewed as a form of structural enrichment and behavioural prophylaxis that allows thermal choice in tilapia (Cerqueira *et al.*, 2016).

Low lighting and provision of overhead cover

Reducing luminosity has been found to reduce aggression in Nile tilapia (Tatemoto & Serra, 2021). Provision of overhead cover is therefore recommended because it can help achieve the low luminosity favoured by the species, but also provides additional benefits because it deters bird predators, provides shading and can reduce heat stress (Bhujel, 2013).

Reduction of anthropogenic noise

Anthropogenic noise can have negative impacts on many fish species (Cox *et al.*, 2018; Popper, 2003) and tilapia are particularly sensitive to noise pollution because they have a closed swim bladder (Halvorsen *et al.*, 2012). Nile tilapia subjected to anthropogenic noise can suffer from several long-term deleterious effects, including stunted growth, altered behaviours and increased stress (Kusku, 2020; Kusku *et al.*, 2020), which can compromise welfare. Fish farms can be very noisy environments, but noise pollution can be reduced by adopting simple and inexpensive techniques, including covering of the tanks and shielding of pumps and standpipes (Barnes *et al.*, 2015).

Prevention of fear contagion

The sight of distressed individuals can make observers more stressed in zebrafish (Fernandes Silva *et al.*, 2019) and the same is also likely to happen with tilapia. Although this may not be a problem in large commercial facilities, where fish are typically reared in large tanks or cages, it can be a problem in smaller tanks and aquaponics facilities. Alarm cues, free cortisol released in the water and other steroid hormones can build up in RAS and may trigger a stress response that can reduce growth and potentially also welfare (Martins *et al.*, 2009; Mota *et al.*, 2014).

Training to reduce crowding stress

Training can reduce stress by allowing tilapia to anticipate negative experiences (Galhardo *et al.*, 2011), and could be used to reduce crowding stress and improve welfare during handling, grading and transportation.

Use of on-demand (self-) feeders

The use of on-demand feeders (operated by the fish, typically through a pendulum system) can reduce competition and cannibalism and result in less wastage, better feed conversion ratio, more uniform growth and enhanced welfare in many species (Attia *et al.*, 2012; Huntingford *et al.*, 2006). In tilapia, on-demand feeders reduce stress and boost immunocompetence (Endo *et al.*, 2002) and are therefore recommended (Benhaïm *et al.*, 2017; McCarthy *et al.*, 1999). Given a choice of diets, Nile tilapia can learn to operate self-feeders that deliver the feed that best meets their nutritional requirements (Attia *et al.*, 2012), and so dietary choice could be used to promote positive aspects of welfare.

Biofloc technology

Biofloc technology (BFT) can be used to reduce waste, improve water quality and recycle nutrients in tilapia fish farming, which can increase survival and net production without any negative impacts on welfare (Azim & Little, 2008; Ekasari *et al.*, 2015; Gullian Klanian *et al.*, 2020). Some studies have also reported better growth and immunocompetence in tilapia reared using BFT (Long *et al.*, 2015), which might result in better welfare.

Use of knotless nets and vacuum pumps during handling and transport

Prolonged crowding, air and light exposure and improper handling are common causes of loss of welfare in tilapia farms that can result in eye injuries, skin lesions and loss of scales. as well as opercular and gill damage if fish are exposed to the air for too long (Pedrazzani *et al.*, 2020). The use of knotless nets and vacuum pumps to move fish around has been recommended (Pedrazzani *et al.*, 2020).

Humane slaughter methods

Humane ways of euthanizing tilapia in the laboratory are reviewed by Close *et al.* (1997) and Rey Planellas and Garcia de Leaniz (2024), but these are rarely used under commercial conditions (Pedrazzani *et al.*, 2020). Most farmed tilapia are killed without prior stunning or loss of consciousness, typically by cold shock or asphyxiation (El-Sayed, 2020). Of the various ways of slaughtering tilapia, asphyxiation in iced water is the most stressful (and longest) procedure, followed by electrical stunning, both of which cause pain and should be avoided. Narcosis by CO_2 has been touted as the preferred slaughter method for tilapia (Campagnoli de Oliveira *et al.*, 2015), but this has also been criticized because loss of consciousness is not immediate and fish may simply become paralysed rather than unconscious (Barkerud, 2021). The use of nitric oxide has been recommended, because this can also improve fillet colour (Wang *et al.*, 2017).

References

Abdel-Daim, M. M., Dawood, M. A. O., Aleya, L., & Alkahtani, S. (2020). Effects of fucoidan on the hematic indicators and antioxidative responses of Nile tilapia (*Oreochromis niloticus*) fed diets contaminated with aflatoxin b1. *Environmental Science and Pollution Research International*, *27*(11), 12579–12586. https://doi.org/10.1007/s11356-020-07854-w

Abdelkhalek, N. K. M., Eissa, I. A. M., Ahmed, E., Kilany, O. E., El-Adl, M., Dawood, M. A. O., Hassan, A. M., & Abdel-Daim, M. M. (2017). Protective role of dietary *Spirulina platensis* against diazinon-induced oxidative damage in Nile tilapia; *Oreochromis niloticus*. *Environmental Toxicology and Pharmacology*, *54*, 99–104. https://doi.org/10.1016/j.etap.2017.07.002

Abdel-Tawwab, M., & Ahmad, M. H. (2009). Live Spirulina (*Arthrospira platensis*) as a growth and immunity promoter for Nile tilapia, *Oreochromis niloticus* (L.), challenged with pathogenic *Aeromonas hydrophila*. *Aquaculture Research*, *40*(9), 1037–1046. https://doi.org/10.1111/j.1365-2109.2009.02195.x

Abdel-Tawwab, M., & Hamed, H. S. (2020). Antagonistic effects of dietary guava (*Psidium guajava*) leaves extract on growth, hemato-biochemical, and immunity response of cypermethrin-intoxicated Nile tilapia, *Oreochromis niloticus*, fingerlings. *Aquaculture, 529*, 735668. https://doi.org/10.1016/j.aquaculture.2020.735668

Abdel-Tawwab, M., Ahmad, M. H., Seden, M. E. A., & Sakr, S. F. M. (2010). Use of green tea, Camellia sinensis L., in practical diet for growth and protection of Nile tilapia, *Oreochromis niloticus* (L.), against *Aeromonas hydrophila* infection. *Journal of the World Aquaculture Society, 41*(s2), 203–213. https://doi.org/10.1111/j.1749-7345.2010.00360.x

Abdel-Tawwab, M., Mounes, H. A. M., Shady, S. H. H., & Ahmed, K. M. (2021). Effects of yucca, *Yucca schidigera*, extract and/or yeast, *Saccharomyces cerevisiae*, as water additives on growth, biochemical, and antioxidants/oxidant biomarkers of Nile tilapia, *Oreochromis niloticus*. *Aquaculture, 533*, 736122. https://doi.org/10.1016/j.aquaculture.2020.736122

Adams, C., & Huntingford, F. (2005). Behavioural syndromes in farmed fish: Implications for production and welfare. *Behaviour, 142*(9–10), 1207–1221. https://doi.org/10.1163/156853905774539382

Ahmed, S. A. A., Ibrahim, R. E., Farroh, K. Y., Moustafa, A. A., Al-Gabri, N. A., Alkafafy, M., & Amer, S. A. (2021). Chitosan vitamin E nanocomposite ameliorates the growth, redox, and immune status of Nile tilapia (*Oreochromis niloticus*) reared under different stocking densities. *Aquaculture, 541*, 736804. https://doi.org/10.1016/j.aquaculture.2021.736804

Al-Deriny, S. H., Dawood, M. A. O., Elbialy, Z. I., El-Tras, W. F., & Mohamed, R. A. (2020a). Selenium nanoparticles and Spirulina alleviate growth performance, hemato-biochemical, immune-related genes, and heat shock protein in Nile tilapia (*Oreochromis niloticus*). *Biological Trace Element Research, 198*(2), 661–668. https://doi.org/10.1007/s12011-020-02096-w

Al-Deriny, S. H., Dawood, M. A. O., Zaid, A. A. A., El-Tras, W. F., Paray, B. A., Van Doan, H., & Mohamed, R. A. (2020b). The synergistic effects of *Spirulina platensis* and *Bacillus amyloliquefaciens* on the growth performance, intestinal histomorphology, and immune response of Nile tilapia (*Oreochromis niloticus*). *Aquaculture Reports, 17*, 100390. https://doi.org/10.1016/j.aqrep.2020.100390

Aldon, E. T. (1998). The culture of tilapia. *SEAFDEC Asian Aquaculture, 20*, 16–17.

Anani, F. K., & Nunoo, F. K. E. (2016). Length-weight relationship and condition factor of Nile tilapia, *Oreochromis niloticus* fed farm-made and commercial tilapia diet. *International Journal of Fisheries and Aquatic Studies, 4*, 647–650.

Antache, ACristea, V., Grecu, I., Dediu, L., Cretu, M., Bocioc, E., & Petrea, S. M. (2014). Effects of dietary supplementation at Nile tilapia with *Thymus vulgaris, Trigonela foenum graecum* and *Azadirachta indica* on welfare status. ANTACHE, A. *Bulletin of University of Agricultural Sciences and Veterinary Medicine Cluj-Napoca. Animal Science and Biotechnologies, 71*.

Araújo, F. C., Ribeiro, R. P., Campos, E. C., Todesco, H., Tsujii, K. M., Mantovani, L. S. C., Ribeiro, R. F., Carvalho, J. C., Casetta, J., Lopera-Barrero, N. M., Gasparino, E., & Oliveira, C. A. Ld. (2022). Could serum glucose be a selection criterion in Nile tilapia breeding programs? *Aquaculture, 548*, 737573.

Arenal, A., Martín, L., Castillo, N. M., De La Torre, D., Torres, U., & González, R. (2012). Aqueous extract of *Ocimum tenuiflorum* decreases levels of blood glucose in induced hyperglycemic tilapia (*Oreochromis niloticus*). *Asian Pacific Journal of Tropical Medicine, 5*(8), 634–637. https://doi.org/10.1016/S1995-7645(12)60130-8

Attia, J., Millot, S., Di-Poï, C., Bégout, M. L., Noble, C., Sanchez-Vazquez, F. J., Terova, G., Saroglia, M., & Damsgård, B. (2012). Demand feeding and welfare in farmed fish. *Fish Physiology and Biochemistry, 38*(1), 107–118. https://doi.org/10.1007/s10695-011-9538-4

Azim, M. E., & Little, D. C. (2008). The biofloc technology (BFT) in indoor tanks: Water quality, biofloc composition, and growth and welfare of Nile tilapia (*Oreochromis niloticus*). *Aquaculture, 283*(1–4), 29–35. https://doi.org/10.1016/j.aquaculture.2008.06.036

Baba, E., Acar, Ü., Öntaş, C., Kesbiç, O. S., & Yılmaz, S. (2016) Evaluation of Citrus limon peels essential oil on growth performance, immune response of Mozambique tilapia *Oreochromis mossambicus* challenged with *Edwardsiella tarda. Aquaculture, 465*, 13–18. https://doi.org/10.1016/j.aquaculture.2016.08.023

Barkerud, R. (2021). *Welfare evaluation of stunning practices for farmed fish in the European Union.* MSc Animal Welfare, Linnaeus University.

Barnes, M. E., Hewitt, C. R., & Parker, T. M. (2015). Fish hatchery noise levels and noise reduction techniques. *Journal of Agricultural Safety and Health, 21*(3), 187–195. https://doi.org/10.13031/jash.21.11072

Barreto, R. E., & Volpato, G. L. (2004). Caution for using ventilatory frequency as an indicator of stress in fish. *Behavioural Processes, 66*(1), 43–51. https://doi.org/10.1016/j.beproc.2004.01.001

Barreto, R. E., & Volpato, G. L. (2006). Ventilatory frequency of Nile tilapia subjected to different stressors. *Journal of Experimental Animal Science, 43*(3), 189–196. https://doi.org/10.1016/j.jeas.2006.05.001

Barreto, R. E., Carvalho, G. G., & Volpato, G. L. (2011). The aggressive behavior of Nile tilapia introduced into novel environments with variation in enrichment. *Zoology, 114*(1), 53–57. https://doi.org/10.1016/j.zool.2010.09.001

Barreto, M. O., Rey Planellas, S., Yang, Y., Phillips, C., & Descovich, K. (2022). Emerging indicators of fish welfare in aquaculture. *Reviews in Aquaculture, 14*(1), 343–361. https://doi.org/10.1111/raq.12601

Barton, B. A. (2002). Stress in fishes: A diversity of responses with particular reference to changes in circulating corticosteroids. *Integrative and Comparative Biology, 42*(3), 517–525. https://doi.org/10.1093/icb/42.3.517

Beardmore, J. A., Mair, G. C., & Lewis, R. I. (2001). Monosex male production in finfish as exemplified by tilapia: Applications, problems, and prospects. *Aquaculture, 197*(1–4), 283–301. https://doi.org/10.1016/S0044-8486(01)00590-7

Benhaïm, D., Akian, D. D., Ramos, M., Ferrari, S., Yao, K., & Bégout, M.-L. (2017). Self-feeding behaviour and personality traits in tilapia: A comparative study between *Oreochromis niloticus* and *Sarotherodon melanotheron*. *Applied Animal Behaviour Science, 187*, 85–92. https://doi.org/10.1016/j.applanim.2016.12.004

Benli, A. C. K., & Köksal, G. (2005). The acute toxicity of ammonia on tilapia (*Oreochromis niloticus* L.) larvae and fingerlings. *Turkish Journal of Veterinary and Animal Sciences, 29*, 339–344.

Berlinghieri, F., Panizzon, P., Penry-Williams, I. L., & Brown, C. (2021). Laterality and fish welfare-a review. *Applied Animal Behaviour Science, 236*, 105239. https://doi.org/10.1016/j.applanim.2021.105239

Beveridge, M. C. M., & Baird, D. J. (2000). Diet, feeding and digestive physiology. *Tilapias: Biology and exploitation*. Springer.

Beveridge, M. C. M., & McAndrew, B. (Eds.). (2012). *Tilapias: Biology and exploitation*. Springer Science+Business Media.

Bhujel, R. C. (2013). On-farm feed management practices for Nile tilapia (*Oreochromis niloticus*) in Thailand. In M. R. Hasan & M. B. New (Eds.), *On-farm feeding and feed management in aquaculture*. Food and Agriculture Organization.

Blackwell, B. G., Brown, M. L., & Willis, D. W. (2000). Relative weight (Wr) status and current use in fisheries assessment and management. *Reviews in Fisheries Science*, *8*(1), 1–44. https://doi.org/10.1080/10641260091129161

Branson, E. J. (Ed.). (2008). *Fish welfare*. Blackwell Publishing Ltd.

Campagnoli De Oliveira, P. R., Girao, P. J. M., Pires De Melo, M., & Viegas, E. M. M. (2015). Indicators of stress in tilapia subjected to different stunning methods. *Boletim do Instituto de Pesca, (São Paulo)*, *41*, 335–343.

Cerqueira, M., Rey, S., Silva, T., Featherstone, Z., Crumlish, M., & Mackenzie, S. (2016). Thermal preference predicts animal personality in Nile tilapia *Oreochromis niloticus*. *Journal of Animal Ecology*, *85*(5), 1389–1400. https://doi.org/10.1111/1365-2656.12555

Champneys, T., Castaldo, G., Consuegra, S., & Garcia De Leaniz, C. (2018). Density-dependent changes in neophobia and stress-coping styles in the world's oldest farmed fish. *Royal Society Open Science*, *5*(12), 181473. https://doi.org/10.1098/rsos.181473

Champneys, T., Genner, M. J., & Ioannou, C. C. (2021). Invasive Nile tilapia dominates a threatened indigenous tilapia in competition over shelter. *Hydrobiologia*, *848*(16), 3747–3762. https://doi.org/10.1007/s10750-020-04341-8

Chapman, F. A. (2000). *Culture of hybrid tilapia: A reference profile. University of Florida cooperative extension service*. Institute of Food and Agricultural Sciences.

Close, B., Banister, K., Baumans, V., Bernoth, E. M., Bromage, N., Bunyan, J., Erhardt, W., Flecknell, P., Gregory, N., Hackbarth, H., Morton, D., & Warwick, C. (1997). Recommendations for euthanAsia of experimental animals: Part 2. DGXT of the European Commission. *Laboratory Animals*, *31*(1), 1–32. https://doi.org/10.1258/002367797780600297

Cox, K., Brennan, L. P., Gerwing, T. G., Dudas, S. E., & Juanes, F. (2018). Sound the alarm: A meta-analysis on the effect of aquatic noise on fish behavior and physiology. *Global Change Biology*, *24*(7), 3105–3116. https://doi.org/10.1111/gcb.14106

DAFF. (2018). *Nile and Mozambique tilapia feasibility study*. Department of Agriculture, Forestry and Fisheries, Chief Directorate: Aquaculture and Economic Development.

Dan, N. C., & Little, D. C. (2000). The culture performance of monosex and mixed-sex new-season and overwintered fry in three strains of Nile tilapia (*Oreochromis niloticus*) in northern Vietnam. *Aquaculture*, *184*(3–4), 221–231. https://doi.org/10.1016/S0044-8486(99)00329-4

da Silva, F. J. R., Dos Santos Lima, F. R., Do Vale, D. A., & Do Carmo, M. V. (2013). High levels of total ammonia nitrogen as NH 4+ are stressful and harmful to the growth of Nile tilapia juveniles. *Acta Scientiarum – Biological Sciences*, *35*, 475–481.

Delicio, H. C., Barreto, R. E., Normandes, E. B., Luchiari, A. C., & Marcondes, A. L. (2006) A place preference test in the fish Nile tilapia. *Journal of Experimental Animal Science*, *43*(3), 141–148. https://doi.org/10.1016/j.jeas.2006.01.001

Delong, D. P., & Losordo, T. (2012). *How to start a biofilter*. Southern Regional Aquaculture Center.

Delong, D. P., Losordo, T., & Rakocy, J. (2009). *Tank culture of tilapia*.

Del Rocío Quezada-Rodríguez, P., & Fajer-Ávila, E. J. (2017). The dietary effect of ulvan from *Ulva clathrata* on hematological-immunological parameters and growth of tilapia (*Oreochromis niloticus*). *Journal of Applied Phycology*, *29*(1), 423–431. https://doi.org/10.1007/s10811-016-0903-7

Dos Santos, W., S de Brito, T., de A Prado, S., G de Oliveira, C., C De Paula, A., C de Melo, D., & A P Ribeiro, P. (2016). Cinnamon (*Cinnamomum* sp.) inclusion in diets for Nile tilapia submitted to acute hypoxic stress. *Fish and Shellfish Immunology*, *54*, 551–555. https://doi.org/10.1016/j.fsi.2016.04.135

Edwards, P. (2000). *Aquaculture, poverty impacts and livelihoods*. Overseas Development Institute.

Ekasari, J., Rivandi, D. R., Firdausi, A. P., Surawidjaja, E. H., Zairin, M., Bossier, P., & De Schryver, P. (2015). Biofloc technology positively affects Nile tilapia (*Oreochromis niloticus*) larvae performance. *Aquaculture*, *441*, 72–77. https://doi.org/10.1016/j.aquaculture.2015.02.019

El-Hawarry, W. N., Mohamed, R. A., & Ibrahim, S. A. (2018). Collaborating effects of rearing density and oregano oil supplementation on growth, behavioral and stress response of Nile tilapia (*Oreochromis niloticus*). *Egyptian Journal of Aquatic Research*, *44*(2), 173–178. https://doi.org/10.1016/j.ejar.2018.06.008

Elkamel, A. A., & Mosaad, G. M. (2012). Immunomodulation of Nile tilapia, *Oreochromis niloticus*, by *Nigella sativa* and *Bacillus subtilis*. *Journal of Aquaculture Research and Development*, *3*.

Elkaradawy, A., Abdel-Rahim, M. M., & Mohamed, R. A. (2022). Quillaja Saponaria and/or linseed oil improved growth performance, water quality, welfare profile and immune-oxidative status of Nile tilapia, *Oreochromis niloticus* fingerlings. *Aquaculture Research*, *53*(2), 576–589. https://doi.org/10.1111/are.15602

Ellis, T., Sanders, M., & Scott, A. (2013). Non-invasive monitoring of steroids in fishes. *Veterinary Medicine, (Austria), 100*, 255–269.

Ellison, A. R., Uren Webster, T. M., Rey, O., Garcia de Leaniz, C., Consuegra, S., Orozco-Terwengel, P., & Cable, J. (2018). Transcriptomic response to parasite infection in Nile tilapia (*Oreochromis niloticus*) depends on rearing density. *BMC Genomics, 19*(1), 723. https://doi.org/10.1186/s12864-018-5098-7

Ellison, A. R., Uren Webster, T. M., Rodriguez-Barreto, D., Garcia de Leaniz, C., Consuegra, S., Orozco-Terwengel, P., & Cable, J. (2020). Comparative transcriptomics reveal conserved impacts of rearing density on immune response of two important aquaculture species. *Fish and Shellfish Immunology, 104*, 192–201. https://doi.org/10.1016/j.fsi.2020.05.043

El-Sayed, A.-F. M. (2013). Tilapia feed management practices in sub-Saharan Africa. *On-farm feeding and feed management in aquaculture. FAO fisheries and aquaculture technical paper*, 377–405.

El-Sayed, A.-F. M. (2020). *Tilapia culture*. Academic Press.

El-Sayed, S. A. A., El-Galil, S. Y. A., & Rashied, N. A. (2014). Effects of some herbal plants as natural feed additives in comparison with antibiotic on growth performance and immune status of Nile tilapia (*Oreochromis niloticus*). *International Journal of Research in Fisheries and Aquaculture, 4*, 22–30.

Endo, M., Kumahara, C., Yoshida, T., & Tabata, M. (2002). Reduced stress and increased immune responses in Nile tilapia kept under self-feeding conditions. *Fisheries Science, 68*(2), 253–257. https://doi.org/10.1046/j.1444-2906.2002.00419.x

Evans, J. J., Pasnik, D. J., Horley, P., Kraeer, K., & Klesius, P. H. (2008). Aggression and mortality among Nile tilapia (*Oreochromis niloticus*) maintained in the laboratory at different densities. *Research Journal of Animal Sciences, 2*, 57–64.

FAO. (2022). *Oreochromis niloticus.* Cultured Aquatic Species Information Programme [Online]. FAO Fisheries and Aquaculture Department.

Fatima, S., Komal, W., Manzoor, F., Latif, A. A., Liaqat, R., Ameen, S., & Janjua, R. S. (2021). Analysis of the growth performance, stress, profile of fatty acids and amino acids and cortisol in tilapia (*Oreochromis niloticus*), cultured at high stocking density using in-pond raceway system. *Saudi Journal of Biological Sciences, 28*(12), 7422–7431. https://doi.org/10.1016/j.sjbs.2021.08.048

Favero Neto, J. F., & Giaquinto, P. C. (2020). Environmental enrichment techniques and tryptophan supplementation used to improve the quality of life and animal welfare of Nile tilapia. *Aquaculture Reports, 17*, 100354. https://doi.org/10.1016/j.aqrep.2020.100354

Fermon, Y. (2006). *Subsistence fish farming in Africa: A technical manual.* ACF International.

Fernandes Silva, P., Garcia de Leaniz, C., & Luchiari, A. C. (2019). Fear contagion in zebrafish: A behaviour affected by familiarity. *Animal Behaviour, 153*, 95–103. https://doi.org/10.1016/j.anbehav.2019.05.004

Fessehaye, Y., Komen, H., Rezk, M. A., Van Arendonk, J. A. M., & Bovenhuis, H. (2007). Effects of inbreeding on survival, body weight and fluctuating asymmetry (FA) in Nile tilapia, *Oreochromis niloticus. Aquaculture, 264*(1–4), 27–35. https://doi.org/10.1016/j.aquaculture.2006.12.038

Fine Arts Work Center in Provincetown. (2014). *Opinion on the welfare of farmed fish.* Farm Animal Welfare Committee.

Folkedal, O., Pettersen, J. M., Bracke, M. B. M., Stien, L. H., Nilsson, J., Martins, C., Breck, O., Midtlyng, P. J., & Kristiansen, T. (2016). On-farm evaluation of the salmon Welfare Index Model (SWIM 1.0): Theoretical and practical considerations. *Animal Welfare, 25*(1), 135–149. https://doi.org/10.7120/09627286.25.1.135

Føre, M., Frank, K., Norton, T., Svendsen, E., Alfredsen, J. A., Dempster, T., Eguiraun, H., Watson, W., Stahl, A., Sunde, L. M., Schellewald, C., Skøien, K. R., Alver, M. O., & Berckmans, D. (2018). Precision fish farming: A new framework to improve production in aquaculture. *Biosystems Engineering, 173*, 176–193. https://doi.org/10.1016/j.biosystemseng.2017.10.014

Francis-Floyd, R., & Watson, C. (2006). Ammonia. *Fish Site, 5m* Publishing.

Freitas, R. H. A., & Volpato, G. L. (2013). Motivation and time of day affect decision-making for substratum granulometry in the Nile tilapia *Oreochromis niloticus. Journal of Applied Ichthyology, 29*(1), 239–241. https://doi.org/10.1111/jai.12026

Freitas, R. H. A., Negrão, C. A., Felício, A. K. C., & Volpato, G. L. (2014). Eye darkening as a reliable, easy and inexpensive indicator of stress in fish. *Zoology, 117*(3), 179–184. https://doi.org/10.1016/j.zool.2013.09.005

Froese, R., & Pauly, D. (2019). *FishBase World Wide Web electronic publication.* 12 (2019 ed).

Fry, J. P., Mailloux, N. A., Love, D. C., Milli, M. C., & Cao, L. (2018). Feed conversion efficiency in aquaculture: Do we measure it correctly? *Environmental Research Letters*, *13*(2), 024017. https://doi.org/10.1088/1748-9326/aaa273

Gabriel, N. N. (2019). Review on the progress in the role of herbal extracts in tilapia culture. *Cogent Food and Agriculture*, *5*(1), 1619651. https://doi.org/10.1080/23311932.2019.1619651

Gabriel, N. N., Qiang, J., He, J. MA, Ma, X. Y., Kpundeh, M. D., & Xu, P. (2015a). Dietary Aloe vera supplementation on growth performance, some haemato-biochemical parameters and disease resistance against *Streptococcus iniae* in tilapia (GIFT). *Fish and Shellfish Immunology*, *44*(2), 504–514. https://doi.org/10.1016/j.fsi.2015.03.002

Gabriel, N. N., Qiang, J., Kpundeh, M. D., & Xu, P. (2015b). Use of herbal extracts for controlling reproduction in tilapia culture: Trends and Prospects-a Review. *Israeli Journal of Aquaculture Bamidgeh*, *67*, 20705.

Gabriel, N. N., Qiang, J. MA, Ma, X. Y., He, J., Xu, P., & Liu, K. (2015c). Dietary Aloe vera improves plasma lipid profile, antioxidant, and hepatoprotective enzyme activities in GIFT-tilapia (*Oreochromis niloticus*) after *Streptococcus iniae* challenge. *Fish Physiology and Biochemistry*, *41*(5), 1321–1332. https://doi.org/10.1007/s10695-015-0088-z

Galhardo, L., Correia, J., & Oliveira, R. (2008). The effect of substrate availability on behavioural and physiological indicators of welfare in the African cichlid (*Oreochromis mossambicus*). *Animal Welfare*, *17*(3), 239–254. https://doi.org/10.1017/S0962728600032164

Galhardo, L., Almeida, O., & Oliveira, R. F. (2009). Preference for the presence of substrate in male cichlid fish: Effects of social dominance and context. *Applied Animal Behaviour Science*, *120*(3–4), 224–230. https://doi.org/10.1016/j.applanim.2009.07.001

Galhardo, L., Vital, J., & Oliveira, R. F. (2011). The role of predictability in the stress response of a cichlid fish. *Physiology and Behavior*, *102*(3–4), 367–372. https://doi.org/10.1016/j.physbeh.2010.11.035

Garcia de Leaniz, C., Gutierrez Rabadan, C., Barrento, S. I., Stringwell, R., Howes, P. N., Whittaker, B. A., Minett, J. F., Smith, R. G., Pooley, C. L., Overland, B. J., Biddiscombe, L., Lloyd, R., Consuegra, S., Maddocks, J. K., Deacon, P. T. J., Jennings, B. T., Rey Planellas, S., Deakin, A., . . . Pavlidis, M. (2022). Addressing the welfare needs of farmed lumpfish: Knowledge gaps, challenges and solutions. *Reviews in Aquaculture*, *14*(1), 139–155. https://doi.org/10.1111/raq.12589

Goda, A. M. A.-S. (2008). Effect of dietary ginseng herb (Ginsana® G115) supplementation on growth, feed utilization, and hematological indices of Nile tilapia, *Oreochromis niloticus* (L.), fingerlings. *Journal of the World Aquaculture Society, 39*(2), 205–214. https://doi.org/10.1111/j.1749-7345.2008.00153.x

Gonçalves, E., & Hoshino, K. (1990). Lateralized behavior of Nile tilapia in natural conditions. Annals of the 5th Annual Meeting of the Federation of Brazilian Societies for Experimental Biology. Caxambu, M.G., Brazil, *423*.

Gonçalves-De-Freitas, E., Bolognesi, M. C., Gauy, A. C. D. S., Brandão, M. L., Giaquinto, P. C., & Fernandes-Castilho, M. (2019). Social behavior and welfare in Nile tilapia. *Fishes, 4*(2), 23. https://doi.org/10.3390/fishes4020023

Green, B. W. (2006). Fingerling production systems. In C. D. Webster & C. E. Lim (Eds.), *Tilapia: Biology, culture, and nutrition.*

Gullian Klanian, M., Delgadillo Díaz, M., Sánchez Solís, M. J., Aranda, J., & Moreno Moral, P. (2020). Effect of the content of microbial proteins and the poly-β-hydroxybutyric acid in biofloc on the performance and health of Nile tilapia (*Oreochromis niloticus*) fingerlings fed on a protein-restricted diet. *Aquaculture, 519*, 734872. https://doi.org/10.1016/j.aquaculture.2019.734872

Güllü, K., Acar, Ü., & Kesbiç, O. S., Yılmaz, S., Ağdamar, S., Ergün, S. & Türker, A. 2016. Beneficial effects of oral Allspice, *Pimenta dioica* powder supplementation on the hemato-immunological and serum biochemical responses of *Oreochromis mossambicus*. *Aquaculture Research,* 47, 2697-2704.

Gupta, M. V., & Acosta, B. O. (2004). A review of global tilapia farming practices. *Aquaculture Asia*, IX, 7–16.

Gurkan, M., Yilmaz, S., Hasan, K., Ergun, S., & Alkan, S. (2015). Influence of three spice powders on the survival and histopathology of *Oreochromis mossambicus* before and after *Streptococcus iniae* infection. *Marine Science and Technology Bulletin, 4*, 1–5.

Gutierrez Rabadan, C., Spreadbury, C., Consuegra, S., & Garcia de Leaniz, C. (2021). Development, validation and testing of an Operational Welfare Score Index for farmed lumpfish *Cyclopterus lumpus* L. *Aquaculture, 531*, 735777. https://doi.org/10.1016/j.aquaculture.2020.735777

Halvorsen, M. B., Casper, B. M., Matthews, F., Carlson, T. J., & Popper, A. N. (2012). Effects of exposure to pile-driving sounds on the lake sturgeon, Nile tilapia and hogchoker. *Proceedings Royal Society Biological Sciences, 279*(1748), 4705–4714. https://doi.org/10.1098/rspb.2012.1544

Handwerker, T. S., & Tave, D. (1994). Semioperculum: A nonheritable deformity in Mozambique tilapia. *Journal of Aquatic Animal Health*, *6*(1), 85–88. https://doi.org/10.1577/1548-8667(1994)006<0085:SANDIM>2.3.CO;2

Hiott, A. E., & Phelps, R. P. (1993). Effects of initial age and size on sex reversal of *Oreochromis niloticus* fry using methyltestosterone. *Aquaculture*, *112*(4), 301–308. https://doi.org/10.1016/0044-8486(93)90391-B

Hlophe, S. N., & Moyo, N. A. G. (2013). The aquaculture potential of Tilapia rendalli in relation to its feeding habits and digestive capabilities. *Physics and Chemistry of the Earth, Parts A/B/C*, *66*, 33–37. https://doi.org/10.1016/j.pce.2013.09.006

Höglund, E., Bakke, M. J., Øverli, Ø., Winberg, S., & Nilsson, G. E. (2005). Suppression of aggressive behaviour in juvenile Atlantic cod (*Gadus morhua*) by l-tryptophan supplementation. *Aquaculture*, *249*(1–4), 525–531. https://doi.org/10.1016/j.aquaculture.2005.04.028

Hoseini, S. M., Pérez-Jiménez, A., Costas, B., Azeredo, R., & Gesto, M. (2019). Physiological roles of tryptophan in teleosts: Current knowledge and perspectives for future studies. *Reviews in Aquaculture*, *11*(1), 3–24. https://doi.org/10.1111/raq.12223

Huntingford, F. A., & Kadri, S. (2009). Taking account of fish welfare: Lessons from aquaculture. *Journal of Fish Biology*, *75*(10), 2862–2867. https://doi.org/10.1111/j.1095-8649.2009.02465.x

Huntingford, F. A., & Kadri, S. (2014). Defining, assessing and promoting the welfare of farmed fish. *Revue Scientifique et Technique*, *33*(1), 233–244. https://doi.org/10.20506/rst.33.1.2286

Huntingford, F. A., Adams, C., Braithwaite, V. A., Kadri, S., Pottinger, T. G., Sandøe, P., & Turnbull, J. F. (2006). Current issues in fish welfare. *Journal of Fish Biology*, *68*(2), 332–372. https://doi.org/10.1111/j.0022-1112.2006.001046.x

Hussain, M. G. (2004). *Farming of tilapia: Breeding plans, mass seed production and aquaculture techniques*. Momin Offset Press.

Hwang, P. P., Wu, S. M., Lin, J. H., & Wu, L. S. (1992). Cortisol content of eggs and larvae of teleosts. *General and Comparative Endocrinology*, *86*(2), 189–196. https://doi.org/10.1016/0016-6480(92)90101-o

Ighwela, K. A., Ahmed, A. B., & Abol-Munafi, A. (2011). Condition factor as an indicator of growth and feeding intensity of Nile tilapia fingerlings (*Oreochromis niloticus*) feed on different levels of maltose. *American-Eurasian Journal of Agricultural and Environmental Sciences*, *11*, 559–563.

Immanuel, G., Uma, R. P., Iyapparaj, P., Citarasu, T., Peter, S. M., Babu, M. M., & Palavesam, A. (2009). Dietary medicinal plant extracts improve growth, immune activity and survival of tilapia *Oreochromis mossambicus*. *Journal of Fish Biology*, *74*(7), 1462–1475. https://doi.org/10.1111/j.1095-8649.2009.02212.x

Iwama, G. K. (2007). The welfare of fish. *Diseases of Aquatic Organisms*, *75*(2), 155–158. https://doi.org/10.3354/dao075155

Keyombe, J., Malala, J., Waithaka, E., Lewo, R., & Obwanga, B. (2017). Seasonal changes in length-weight relationship and condition factor of Nile tilapia, *Oreochromis niloticus* (Linnaeus, 1758) (Cichlidae) in Lake Naivasha, Kenya. *International Journal of Aquatic Biology*, *5*, 7–11.

Kristiansen, T. S., & Bracke, M. (2020). A brief look into the origins of fish welfare science. In T. S. Kristiansen, A. Fernö, M. A. Pavlidis & H. Van De Vis (Eds.), *The welfare of Fish*. Springer.

Kristiansen, T. S., Madaro, A., Stien, L. H., Bracke, M. B. M., & Noble, C. (2020). Theoretical basis and principles for welfare assessment of farmed fish. *Fish Physiology*. Elsevier, 193–236. https://doi.org/10.1016/bs.fp.2020.09.006

Kusku, H. (2020). Acoustic sound–induced stress response of Nile tilapia (*Oreochromis niloticus*) to long-term underwater sound transmissions of urban and shipping noises. *Environmental Science and Pollution Research International*, *27*(29), 36857–36864. https://doi.org/10.1007/s11356-020-09699-9

Kusku, H., Yigit, Ü., Yilmaz, S., Yigit, M., & Ergün, S. (2020). Acoustic effects of underwater drilling and piling noise on growth and physiological response of Nile tilapia (*Oreochromis niloticus*). *Aquaculture Research*, *51*(8), 3166–3174. https://doi.org/10.1111/are.14652

Lall, S. P., & Tibbetts, S. M. (2009). Nutrition, feeding, and behavior of fish. *Veterinary Clinics of North America. Exotic Animal Practice*, *12*(2), 361–372. https://doi.org/10.1016/j.cvex.2009.01.005

Lepage, O., Tottmar, O., & Winberg, S. (2002). Elevated dietary intake of L-tryptophan counteracts the stress-induced elevation of plasma cortisol in rainbow trout (*Oncorhynchus mykiss*). *Journal of Experimental Biology*, *205*(23), 3679–3687. https://doi.org/10.1242/jeb.205.23.3679

Lepage, O., Vílchez, I. M., Pottinger, T. G., & Winberg, S. (2003). Time-course of the effect of dietary L-tryptophan on plasma cortisol levels in rainbow trout Oncorhynchus mykiss. *Journal of Experimental Biology*, *206*(20), 3589–3599. https://doi.org/10.1242/jeb.00614

Li, D., Wang, G., Du, L., Zheng, Y., & Wang, Z. (2022). Recent advances in intelligent recognition methods for fish stress behavior.

Aquacultural Engineering, 96, 102222. https://doi.org/10.1016/j.aquaeng.2021.102222

Lim, C. E., & Webster, C. D. (Eds.). (2006). *Tilapia: Biology, culture, and nutrition*. CRC Press.

Long, L., Yang, J., Li, Y., Guan, C., & Wu, F. (2015). Effect of biofloc technology on growth, digestive enzyme activity, hematology, and immune response of genetically improved farmed tilapia (*Oreochromis niloticus*). *Aquaculture, 448*, 135–141. https://doi.org/10.1016/j.aquaculture.2015.05.017

Luchiari, A. C., Do Amaral Duarte, C. R., De Morais Freire, F. A., & Nissinen, K. (2007). Hierarchical status and colour preference in Nile tilapia (*Oreochromis niloticus*). *Journal of Ethology*, *25*(2), 169–175. https://doi.org/10.1007/s10164-006-0013-0

Lymbery, A. J., & Hameleers, R. (2018). *Time to take fish welfare more seriously*.

Maia, C. M., & Volpato, G. L. (2017). Preference index supported by motivation tests in Nile tilapia. *PLOS ONE*, *12*(4), e0175821. https://doi.org/10.1371/journal.pone.0175821

Maia, C. M., & Volpato, G. L. (2019). What to choose when the best preference is not available: Does the Nile tilapia follow a linear sequence of preferences? *Journal of Zoology*, *307*(4), 274–281. https://doi.org/10.1111/jzo.12646

Maia, C. M., Alves, N. P. C., & Tatemoto, P. (2020). Juvenile Nile tilapia fish avoid red shelters. *Journal of Applied Animal Welfare Science*, 24(1), 98–106. https://doi.org/10.1080/10888705.2020.1848567.

Mair, G. C. (1992). Caudal deformity syndrome (CDS): An autosomal recessive lethal mutation in the tilapia, *Oreochromis niloticus* (L.). *Journal of Fish Diseases*, *15*(1), 71–75. https://doi.org/10.1111/j.1365-2761.1992.tb00638.x

Martins, C. I. M., Ochola, D., Ende, S. S. W., Eding, E. H., & Verreth, J. A. J. (2009). Is growth retardation present in Nile tilapia *Oreochromis niloticus* cultured in low water exchange recirculating aquaculture systems? *Aquaculture*, *298*(1–2), 43–50. https://doi.org/10.1016/j.aquaculture.2009.09.030

Martins, C. I. M., Silva, P. I. M., Costas, B., Larsen, B. K., Santos, G. A., Conceição, L. E. C., Dias, J., Øverli, Ø., Höglund, E., & Schrama, J. W. (2013). The effect of tryptophan supplemented diets on brain serotonergic activity and plasma cortisol under undisturbed and stressed conditions in grouped-housed Nile tilapia *Oreochromis niloticus. Aquaculture, 400–401*, 129–134. https://doi.org/10.1016/j.aquaculture.2013.02.035

Maruyama, T. (1948). An observation on *Tilapia mossambica* in pond referring to diurnal movements with temperature change. *Boll. Freshwater Fish. Res. Lab. Tokyo, 8*, 25–32.

McCarthy, I. D., Gair, D. J., & Houlihan, D. F. (1999). Feeding rank and dominance in Tilapia rendalli under defensible and indefensible patterns of food distribution. *Journal of Fish Biology, 55*(4), 854–867. https://doi.org/10.1111/j.1095-8649.1999.tb00722.x

Mendonça, F. Z., Volpato, G. L., Costa-Ferreira, R. S., & Gonçalves-De-Freitas, E. (2010). Substratum choice for nesting in male Nile tilapia *Oreochromis niloticus. Journal of Fish Biology, 77*(6), 1439–1445. https://doi.org/10.1111/j.1095-8649.2010.02754.x

Miyai, C. A., Carretero Sanches, F. H., Costa, T. M., Colpo, K. D., Volpato, G. L., & Barreto, R. E. (2011). The correlation between subordinate fish eye colour and received attacks: A negative social feedback mechanism for the reduction of aggression during the formation of dominance hierarchies. *Zoology, 114*(6), 335–339. https://doi.org/10.1016/j.zool.2011.07.001

Moreira, P. S. A., & Volpato, G. L. (2004). Conditioning of stress in Nile tilapia. *Journal of Fish Biology, 64*(4), 961–969. https://doi.org/10.1111/j.1095-8649.2004.00362.x

Mota, V. C., Martins, C. I. M., Eding, E. H., Canário, A. V. M., & Verreth, J. A. J. (2014). Steroids accumulate in the rearing water of commercial recirculating aquaculture systems. *Aquacultural Engineering, 62*, 9–16. https://doi.org/10.1016/j.aquaeng.2014.07.004

Nehemia, A., Maganira, J. D., & Rumisha, C. (2012). Length-Weight relationship and condition factor of tilapia species grown in marine and fresh water ponds. *Agriculture and Biology Journal of North America, 3*(3), 117–124. https://doi.org/10.5251/abjna.2012.3.3.117.124

Nelson, J. S. (2006). *Fishes of the world*. John Wiley & Sons.

Nepomnyashchikh, V. A., & Gremyachikh, V. A. (1993). Relationship between the trajectory structure and asymmetry of choice of the movement direction in tilapia *Oreochromis mossambicus* Peters (Cichlidae). *Zhurnal Obshchei Biologii, 54*, 619–626.

Nepomnyashchikh, V. A., & Izvekov, E. I. (2006). Variability of the behavioral laterality in Teleostei (Pisces). *Journal of Ichthyology, 46*(S2), S235–S242. https://doi.org/10.1134/S0032945206110142

Ng, W. K., & Romano, N. (2013). A review of the nutrition and feeding management of farmed tilapia throughout the culture cycle. *Reviews in Aquaculture, 5*(4), 220–254. https://doi.org/10.1111/raq.12014

Nivelle, R., Gennotte, V., Kalala, E. J. K., Ngoc, N. B., Muller, M., Mélard, C., & Rougeot, C. (2019). Temperature preference of Nile tilapia (*Oreochromis niloticus*) juveniles induces spontaneous sex reversal. *PLOS ONE, 14*(2), e0212504. https://doi.org/10.1371/journal.pone.0212504

Noble, C., Kadri, S., Mitchell, D. F., & Huntingford, F. A. (2007). Influence of feeding regime on intraspecific competition, fin damage and growth in 1+ Atlantic salmon parr (*Salmo salar* L.) held in freshwater production cages. *Aquaculture Research, 38*(11), 1137–1143. https://doi.org/10.1111/j.1365-2109.2007.01777.x

Noble, C., Jones, H. A. C., Damsgård, B., Flood, M. J., Midling, K. Ø., Roque, A., Sæther, B. S., & Cottee, S. Y. (2012). Injuries and deformities in fish: Their potential impacts upon aquacultural production and welfare. *Fish Physiology and Biochemistry, 38*(1), 61–83. https://doi.org/10.1007/s10695-011-9557-1

Noble, C., Gismervik, K., Iversen, M. H., Kolarevic, J., Nilsson, J., Stien, L. H., & Turnbull, J. F. (Eds.). (2018). *Welfare Indicators for farmed Atlantic salmon: Tools for assessing fish welfare*. Nofima.

Noble, C., Iversen, M. H., Lein, I., Kolarevic, J., Johansen, L. H., Burgerhout, E., Puvanendran, V., Kousoulaki, K., Aas, G. H., Stene, A., & Espmark, Å. M. (2019). *An introduction to Operational and Laboratory based Welfare Indicators for ballan wrasse (Labrus bergylta)*, FHF

Noble, C., Gismervik, K., Iversen, M. H., Kolarevic, J., Nilsson, J., Stien, L. H., & Turnbull, J. F. (Eds.). (2020). *Welfare Indicators for farmed rainbow trout: Tools for assessing fish welfare: Nofima.*

North, B. P., Ellis, T., Knowles, T., Bron, J. E., & Turnbull, J. F. (2008). The use of stakeholder focus groups to identify indicators for the on-farm assessment of trout welfare. *Fish welfare*. Blackwell Publishing.

Obirikorang, K. A., Agbo, N. W., Obirikorang, C., Adjei-Boateng, D., Ahiave, S. E., & Skov, P. V. (2019). Effects of water flow rates on growth and welfare of Nile tilapia (*Oreochromis niloticus*) reared in a recirculating aquaculture system. *Aquaculture International, 27*(2), 449–462. https://doi.org/10.1007/s10499-019-00342-0

Obirikorang, K. A., Opoku, E. N., & Gyampoh, B. A. (2022). Feed digestion, growth and disease prevalence in Nile tilapia (*Oreochromis niloticus*) cultured at different water exchange rates in a Recirculating Aquaculture System. *Aquaculture Studies, 22*(3). https://doi.org/10.4194/AQUAST565

Paracueles, C. N. L. (2015). Toxicity of ammonia on *Oreochromis niloticus* fingerlings. *Journal of Environment and Aquatic Resources, 3*, 96–100. https://doi.org/10.48031/msunjear.2015.03.09

Paredes-Trujillo, A., Velázquez-Abunader, I., Papiol, V., Del Rio-Rodríguez, R. E., & Vidal-Martínez, V. M. (2021). Negative effect of ectoparasite burdens on the condition factor from farmed tilapia *Oreochromis niloticus* in the Yucatan, Mexico. Veterinary Parasitology, 292, 109393. https://doi.org/10.1016/j.vetpar.2021.109393.

Park, K. H., & Choi, S. H. (2012). The effect of mistletoe, *Viscum album coloratum*, extract on innate immune response of Nile tilapia (*Oreochromis niloticus*). *Fish and Shellfish Immunology, 32*(6), 1016–1021. https://doi.org/10.1016/j.fsi.2012.02.023

Park, K. H., & Choi, S. (2014). Effects of *Prunella vulgaris Labiatae* extract on specific and non-specific immune responses in tilapia (*Oreochromis niloticus*). *Journal of Animal Science and Technology, 56*, 3. https://doi.org/10.1186/2055-0391-56-3

Pavlidis, M., & Samaras, A. (2020). Mediterranean fish welfare: Guide to good practices and assessment indicators. *Hellenic Aquaculture* Producers Association (HAPO).

Pedrazzani, A. S., Quintiliano, M. H., Bolfe, F., Sans, E. C. O., & Molento, C. F. M. (2020). Tilapia on-farm welfare assessment protocol for semi-intensive production systems. *Frontiers in Veterinary Science, 7*, 606388. https://doi.org/10.3389/fvets.2020.606388

Pepels, P. P. L. M., & Balm, P. H. M. (2004). Ontogeny of corticotropin-releasing factor and of hypothalamic–pituitary–interrenal axis responsiveness to stress in tilapia (*Oreochromis mossambicus*, Teleostei). *General and Comparative Endocrinology, 139*(3), 251–265. https://doi.org/10.1016/j.ygcen.2004.09.013

Planellas, R., S., & Garcia de Leaniz, C. (2024). Fishes. In R. C. Hubrecht & J. Kirkwood (Eds.), *The UFAW handbook on the care and management of laboratory and other research animals*. John Wiley & Sons.

Popma, T., & Masser, M. (1999). *Tilapia life history and biology*.

Popper, A. N. (2003). Effects of anthropogenic sounds on fishes. *Fisheries, 28*(10), 24–31. https://doi.org/10.1577/1548-8446(2003)28[24:EOASOF]2.0.CO;2

Pullin, R. S. V. (1991). Cichlids in aquaculture. In M. A. Keenleyside (Ed.), *Cichlid fishes: Behaviour, ecology and evolution*. Chapman & Hall.

Quagrainie, K., Cai, J., & Hishamunda, N. (2017). *Social and economic performance of tilapia farming in Africa.* Food and Agriculture Organization of the United Nations.

Rey, S., Treasurer, J., Pattillo, C., & Mcadam, B. J. (2021). Using model selection to choose a size-based condition index that is consistent with operational welfare indicators. *Journal of Fish Biology*, 99(3), 782–795. https://doi.org/10.1111/jfb.14761

Rodriguez-Barreto, D., Rey, O., Uren-Webster, T. M., Castaldo, G., Consuegra, S., & Garcia de Leaniz, C. (2019). Transcriptomic response to aquaculture intensification in Nile tilapia. *Evolutionary Applications, 12*(9), 1757–1771. https://doi.org/10.1111/eva.12830

Ross, L. G. (2000). Environmental physiology and energetics. In M. C. Beveridge & B. J. Mcandrew (Eds.), *Tilapias: Biology and exploitation.* Springer.

Royal Society for the Prevention of Cruelty to Animals. (2018). *RSPCA standards for farmed Atlantic salmon.* Royal Society for the Prevention of Cruelty to Animals.

Royal Society for the Prevention of Cruelty to Animals. (2020). *RSPCA standards for farmed rainbow trout.* Royal Society for the Prevention of Cruelty to Animals.

Royal Society for the Prevention of Cruelty to Animals. (2021). RSPCA welfare standards for Farmed Atlantic salmon, *Southwater, Horsham, West Sussex RH13 9RS.* Royal Society for the Prevention of Cruelty to Animals.

Sadoul, B., & Geffroy, B. (2019). Measuring cortisol, the major stress hormone in fishes. *Journal of Fish Biology, 94*(4), 540–555. https://doi.org/10.1111/jfb.13904

Sakyi, M. E., Cai, J., Tang, J., Xia, L., Li, P., Abarike, E. D., Kuebutornye, F. K. A., & Jian, J. (2020). Short term starvation and re-feeding in Nile tilapia (*Oreochromis niloticus*, Linnaeus 1758): Growth measurements, and immune responses. *Aquaculture Reports, 16*, 100261. https://doi.org/10.1016/j.aqrep.2019.100261

Salzburger, W. (2018). Understanding explosive diversification through cichlid fish genomics. *Nature Reviews. Genetics, 19*(11), 705–717. https://doi.org/10.1038/s41576-018-0043-9

Segner, H., Sundh, H., Buchmann, K., Douxfils, J., Sundell, K. S., Mathieu, C., Ruane, N., Jutfelt, F., Toften, H., & Vaughan, L. (2012). Health of farmed fish: Its relation to fish welfare and its utility as welfare indicator. *Fish Physiology and Biochemistry, 38*(1), 85–105. https://doi.org/10.1007/s10695-011-9517-9

Segner, H., Reiser, S., Ruane, N., Rösch, R., Steinhagen, D., & Vehanen, T. (2019). *Welfare of fishes in aquaculture.* Food and Agriculture Organization.

Seibel, H., Baßmann, B., & Rebl, A. (2021). Blood will tell: What hematological analyses can reveal about fish welfare. *Frontiers in Veterinary Science,* 8, 616955. https://doi.org/10.3389/fvets.2021.616955

Sevier, A., Smith, R., Benfey, T., Danzmann, R., Bernier, N., & Moccia, R. (2019). Effects of biodensity on the growth, stress physiology, and welfare of Arctic charr (*Salvelinus alpinus*) in freshwater. *Comparative Biochemistry and Physiology. Part A, Molecular and Integrative Physiology, 231,* 91–103. https://doi.org/10.1016/j.cbpa.2019.01.021

Shahkar, E., Park, G., Lee, D.-M., Kwak, S., & Bai, S. C. (2015). Effects of dietary Macsumsuk® supplementation on growth performance, haematological parameters, disease resistance and body composition of juvenile Nile tilapia, *Oreochromis niloticus* L. *Journal of Applied Animal Research, 43*(2), 125–130. https://doi.org/10.1080/09712119.2014.928626

Shalaby, A., Khattab, Y., & Abdel Rahman, A. (2006). Effects of garlic (*Allium sativum*) and chloramphenicol on growth performance, physiological parameters and survival of Nile tilapia (*Oreochromis niloticus*). *Journal of Venomous Animals and Toxins Including Tropical Diseases, 12,* 172–201.

Soares, S., Green, D. M., Turnbull, J. F., Crumlish, M., & Murray, A. G. (2011). A baseline method for benchmarking mortality losses in Atlantic salmon (*Salmo salar*) production. *Aquaculture, 314*(1–4), 7–12. https://doi.org/10.1016/j.aquaculture.2011.01.029

Souza, R. L. D., Lima, E. C. R. D., Melo, F. Pd., Ferreira, M. G. P., & Correia, E. D. S. (2019). The culture of Nile tilapia at different salinities using a biofloc system. *Revista Ciência Agronômica, 50*(2), 267–275. https://doi.org/10.5935/1806-6690.20190031

Stickney, R. R. (1993). Tilapia. In R. R. Stickney (Ed.), *Culture of Nonsalmonid freshwater fishes.* CRC Press.

Stickney, R. R. (2017). Tilapia feeding habits and environmental tolerances. In P. W. Perschbacher & R. R. Stickney (Eds.), *Tilapia in intensive co-culture.*

Stien, L. H., Bracke, M. B. M., Folkedal, O., Nilsson, J., Oppedal, F., Torgersen, T., Kittilsen, S., Midtlyng, P. J., Vindas, M. A., Øverli, Ø., & Kristiansen, T. S. (2013). Salmon Welfare Index Model (SWIM 1.0): A semantic model for overall welfare assessment of caged Atlantic salmon: Review of the selected welfare indicators and model presentation. *Reviews in Aquaculture, 5*(1), 33–57. https://doi.org/10.1111/j.1753-5131.2012.01083.x

Stien, L. H., Bracke, M., Noble, C., & Kristiansen, T. S. (2020). Assessing fish welfare in aquaculture. In *The welfare of fish.*, 303–321. Springer. https://doi.org/10.1007/978-3-030-41675-1_13

Stubbe Solgaard, H. S., & Yang, Y. (2011). Consumers' perception of farmed fish and willingness to pay for fish welfare. *British Food Journal, 113*(8), 997–1010. https://doi.org/10.1108/00070701111153751

Suresh, A. V., & Lin, C. K. (1992). Tilapia culture in saline waters: A review. *Aquaculture, 106*(3–4), 201–226. https://doi.org/10.1016/0044-8486(92)90253-H

Tacon, A. G. J., & Metian, M. (2008). Global overview on the use of fish meal and fish oil in industrially compounded aquafeeds: Trends and future prospects. *Aquaculture, 285*(1–4), 146–158. https://doi.org/10.1016/j.aquaculture.2008.08.015

Tatemoto, P., & Serra, M. (2021). Low luminosity reduces agonistic behaviour of Nile tilapia. *Journal of Applied Animal Welfare Science, 24*(4), 434–441. https://doi.org/10.1080/10888705.2021.1888293

Tatemoto, P., Valença-Silva, G., Queiroz, M. R., & Broom, D. M. (2021). Living with low environmental complexity increases fear indicators in Nile tilapia. *Animal Behaviour, 174*, 169–174. https://doi.org/10.1016/j.anbehav.2021.02.006

Teletchea, F., & Fontaine, P. (2014). Levels of domestication in fish: Implications for the sustainable future of aquaculture. *Fish and Fisheries, 15*(2), 181–195. https://doi.org/10.1111/faf.12006

Teodózio de Araújo, F. C., Ribeiro, R. P., Campos, E. C., Todesco, H., Tsujii, K. M., Mantovani, L. S. C., Ribeiro, R. F., Carvalho, J. C., Casetta, J., & Lopera-Barrero, N. M. (2022). Could serum glucose be a selection criterion in Nile tilapia breeding programs? *Aquaculture, 548*, 737573.

Torrezani, C. S., Pinho-Neto, C. F., Miyai, C. A., Sanches, F. H. C., & Barreto, R. E. (2013). Structural enrichment reduces aggression in Tilapia rendalli. *Marine and Freshwater Behaviour and Physiology, 46*(3), 183–190. https://doi.org/10.1080/10236244.2013.805053

Trevi, S., Uren Webster, T.M., Consuegra, S., & Garcia de Leaniz, C. (2023). The benefits of microalgae replacement in fish nutrition: a meta-analysis. *Scientific Reports, 13*. https://doi.org/10.1038/s41598-023-29183-x

Trevi, S., Uren Webster, T.M., Consuegra, S., & Garcia de Leaniz, C. (2024). Effects of micro-algae dietary oil replacement on growth, omega-3 deposition and gut microbiome composition of Nile tilapia (*Oreochromis niloticus*). *Aquaculture, Fish & Fisheries, e164*, 1–15. https://doi.org/10.1002/aff2.164

Turner, G. F., & Robinson, R. L. (2000). Reproductive biology, mating systems and parental care. *Tilapias*. Springer.

Ufodikel, E., & Wada, R. (1991). Feeding habits oftilapia, Sarotherodon nilotica (Perciformis: Cichlidae) fry in Jos,. Nigeria. *Revista de Biología Tropical*, 189–192.

Uren Webster, T. M., Rodriguez-Barreto, D., Consuegra, S., & Garcia de Leaniz, C. (2020). Cortisol-related signatures of stress in the fish microbiome. *Frontiers in Microbiology, 11*, 1621. https://doi.org/10.3389/fmicb.2020.01621

Vanderzwalmen, M., Mcneill, J., Delieuvin, D., Senes, S., Sanchez-Lacalle, D., Mullen, C., Mclellan, I., Carey, P., Snellgrove, D., Foggo, A., Alexander, M. E., Henriquez, F. L., & Sloman, K. A. (2021). Monitoring water quality changes and ornamental fish behaviour during commercial transport. *Aquaculture, 531*, 735860. https://doi.org/10.1016/j.aquaculture.2020.735860

Van De Vis, J. W., Poelman, M., Lambooij, E., Bégout, M. L., & Pilarczyk, M. (2012). Fish welfare assurance system: Initial steps to set up an effective tool to safeguard and monitor farmed fish welfare at a company level. *Fish Physiology and Biochemistry, 38*(1), 243–257. https://doi.org/10.1007/s10695-011-9596-7

Vera Cruz, E. M., & Brown, C. L. (2007). The influence of social status on the rate of growth, eye color pattern and insulin-like growth factor-I gene expression in Nile tilapia, *Oreochromis niloticus. Hormones and Behavior, 51*(5), 611–619. https://doi.org/10.1016/j.yhbeh.2007.02.010

Vera Cruz, E. M., & Tauli, M. P. (2015). Eye color pattern during isolation indicates stress-coping style in Nile tilapia *Oreochromis niloticus* L. *International Journal of Scientific Research in Knowledge, 3*(7), 181–186. https://doi.org/10.12983/ijsrk-2015-p0181-0186

Vieira, B. R. M., Guermandi, I. I., Bellot, M. S., Camargo-Dos-Santos, B., Favero-Neto, J., & Giaquinto, P. C. (2021). The effects of tryptophan supplementation on stress and aggression in Nile tilapia. *Journal of Applied Ichthyology, 37*(4), 578–584. https://doi.org/10.1111/jai.14186

Vijayan, M. M., Pereira, C., Grau, E. G., & Iwama, G. K. (1997). Metabolic responses associated with confinement stress in tilapia: The role of cortisol. *Comparative Biochemistry and Physiology Part C, 116*(1), 89–95. https://doi.org/10.1016/S0742-8413(96)00124-7

Villarroel, M., Miranda-De La Lama, G. C., Escobar-Álvarez, R., & Moratiel, R. (2022). Fish welfare in urban aquaponics: Effects of fertilizer for lettuce (*Lactuca sativa* L.) on some physiological stress indicators in Nile tilapia (*Oreochromis niloticus* L.). *Water, 14*(6), 935. https://doi.org/10.3390/w14060935

Volpato, G. L., Luchiari, A. C., Duarte, C. R., Barreto, R. E., & Ramanzini, G. C. (2003). Eye color as an indicator of social rank in the fish Nile tilapia. *Brazilian Journal of Medical and Biological Research, 36*(12), 1659–1663. https://doi.org/10.1590/s0100-879x2003001200007

Volpato, G. L., Gonçalves-De-Freitas, E., & Fernandes-De-Castilho, M. (2007). Insights into the concept of fish welfare. *Diseases of Aquatic Organisms, 75*(2), 165–171. https://doi.org/10.3354/dao075165

Wang, L. H., & Tsai, C. L. (2000). Effects of temperature on the deformity and sex differentiation of tilapia, *Oreochromis mossambicus*. *Journal of Experimental Zoology, 286*(5), 534–537. https://doi.org/10.1002/(SICI)1097-010X(20000401)286:5<534::AID-JEZ11>3.0.CO;2-2

Wang, Z.-C., Zhao, M.-M., Fang, Z., & Chen, D.-W. (2017). Nitric oxide euthanasia: A potential procedure for improving animal welfare and fillet color of tilapia (*Oreochromis niloticus*). *Aquaculture International, 25*(5), 1845–1856. https://doi.org/10.1007/s10499-017-0162-x

Watanabe, W. O., Fitzsimmons, K., & Yi, Y. (2006). Farming tilapia in saline waters. In C. E. Lim & C. D. Webster (Eds.), *Tilapia: Biology, culture, and nutrition*. CRC Press.

Weirup, L., Schulz, C., & Seibel, H. (2022). Fish welfare evaluation index (fWEI) based on external morphological damage for rainbow trout (*Oncorhynchus mykiss*) in flow through systems. *Aquaculture, 556*, 738270. https://doi.org/10.1016/j.aquaculture.2022.738270

Winberg, S., Øverli, Ø., & Lepage, O. (2001). Suppression of aggression in rainbow trout (*Oncorhynchus mykiss*) by dietary L-tryptophan. *Journal of Experimental Biology, 204*(22), 3867–3876. https://doi.org/10.1242/jeb.204.22.3867

Wu, H., Shinoda, R., Murata, M., Matsumoto, H., Ohnuki, H., & Endo, H. (2019). Real-time fish stress visualization came true: A novel multi-stage color-switching wireless biosensor system. *Biosensors and Bioelectronics, 130*, 360–366. https://doi.org/10.1016/j.bios.2018.09.042

Xu, J., Liu, Y., Cui, S., & Miao, X. (2006). Behavioral responses of tilapia (*Oreochromis niloticus*) to acute fluctuations in dissolved oxygen levels as monitored by computer vision. *Aquacultural Engineering, 35*(3), 207–217. https://doi.org/10.1016/j.aquaeng.2006.02.004

Yilmaz, S., Ergün, S., Kaya, H., & Gürkan, M. (2014). Influence of *Tribulus terrestris* extract on the survival and histopathology of *Oreochromis mossambicus* (Peters, 1852) fry before and after *Streptococcus iniae* infection. *Journal of Applied Ichthyology, 30*(5), 994–1000. https://doi.org/10.1111/jai.12458

Yılmaz, S., Acar, ÜKesbiç, O. S., Gültepe, N., & Ergün, S. (2015). Effects of dietary allspice, *Pimenta dioica* powder on physiological responses of *Oreochromis mossambicus* under low pH stress. *SpringerPlus, 4*, 719. https://doi.org/10.1186/s40064-015-1520-7

Yilmaz, S. (2019a). Effects of dietary blackberry syrup supplement on growth performance, antioxidant, and immunological responses, and resistance of Nile tilapia, *Oreochromis niloticus* to *Plesiomonas shigelloides. Fish and Shellfish Immunology, 84*, 1125–1133. https://doi.org/10.1016/j.fsi.2018.11.012

Yilmaz, S. (2019b). Effects of dietary caffeic acid supplement on antioxidant, immunological and liver gene expression responses, and resistance of Nile tilapia, *Oreochromis niloticus* to *Aeromonas veronii. Fish and Shellfish Immunology, 86*, 384–392. https://doi.org/10.1016/j.fsi.2018.11.068

Yin, G., Ardo, L., Jeney, Z., Xu, P., & Jeney, G. (2008). Chinese herbs (*Lonicera japonica* and *Ganoderma lucidum*) enhance non-specific immune response of tilapia, *Oreochromis niloticus*, and protection against *Aeromonas hydrophila. Diseases in Asian aquaculture VI.* Fish Health Section, Asian Fisheries Society, 269–282.

Zaki, M. A. A., Alabssawy, A. N., Nour, A. E.-A. M., El Basuini, M. F., Dawood, M. A. O., Alkahtani, S., & Abdel-Daim, M. M. (2020). The impact of stocking density and dietary carbon sources on the growth, oxidative status and stress markers of Nile tilapia (*Oreochromis niloticus*) reared under biofloc conditions. *Aquaculture Reports, 16*, 100282. https://doi.org/10.1016/j.aqrep.2020.100282

Zanolo, R., Leonhardt, J. H., Silva e Souza, A. T., & Yamamura, M. H. (2009). The influence of branchial parasitism by monogenoid trematodes on the development of Nile tilapia (*Oreochromis niloticus*) Linnaeus, 1757 bred in net-pond systems in Capivara Dam, PR. *Revista Brasileira de Parasitologia Veterinaria*, 18(1), 47–52. https://doi.org/10.4322/rbpv.01801009.

Zheng, H., Liu, R., Zhang, R., & Hu, Y. (2014). A method for real-time measurement of respiratory rhythms in medaka (*Oryzias latipes*) using computer vision for water quality monitoring. *Ecotoxicology and Environmental Safety, 100*, 76–86. https://doi.org/10.1016/j.ecoenv.2013.11.016

Chapter 9

Tilapia bacterial diseases

Esteban Soto, Taylor I. Heckman, Gustavo Ramirez Paredes, Divya Rose, Matt J. Griffin, Thomas P. Loch, Benjamin R. LaFrentz and Alvin C. Camus

Introduction

Intensified management practices in tilapia aquaculture, involving both outdoor bodies of water and indoor water reuse systems with concomitant reduction in water quality and elevated levels of stress, have resulted in an increased incidence and severity of disease. Among the most encountered virulent agents responsible for mortalities in tilapia aquaculture are the bacterial pathogens *Streptococcus* spp., *Francisella orientalis*, columnaris-causing bacteria (*Flavobacterium* spp.), *Edwardsiella* spp., *Aeromonas* spp. and *Vibrio* spp. Other bacteria, including rickettsia-like organisms, *Shewanella* spp. and *Pseudomonas* spp., have been reported as opportunistic pathogens in tilapia and will not be discussed in this chapter.

Streptococcosis

Piscine streptococcosis broadly refers to fish disease caused by lactic acid bacteria in the order Lactobacillales. *Streptococcus iniae*, *S. agalactiae*, *S. dysgalactiae*, *S. parauberis* and *S. ictaluri*, as well as several species outside of the *Streptococcus* genus such as *Lactococcus* spp., *Enterococcus faecalis* and *Vagococcus salmoniarum* (Vendrell *et al.*, 2006; Jantrakajorn *et al.*, 2014), are included. *Streptococcus agalactiae* is the most prevalent species in cultured tilapia, accounting for

Kim Thompson, Craig Shoemaker and David C. Little (eds) **Tilapia**
DOI: 10.1079/9781800629455.0009,

80–90% of cases in Asia and South America, followed by *S. iniae* (Chen *et al.*, 2012; Jantrakajorn *et al.*, 2014; Ortega Asencios *et al.*, 2016). *Lactococcus garvieae*, *L. petauri*, *S. dysgalactiae* and *E. faecalis* are emerging pathogens of tilapia (Abu-Elala *et al.*, 2020; Catão Egger *et al.*, 2023). The aetiologic agents of tilapia streptococcosis share common morphological and phenotypic characteristics. They are Gram-positive, non-motile, non-spore-forming, ovoid or spherical cocci 0.5–2.0 μm in diameter that are catalase and typically oxidase negative. These fastidious, facultatively anaerobic bacteria form pairs or chains in liquid media and translucent or white colonies on agar media (Amal and Zamri-Saad, 2011; Abu-Elala *et al.*, 2019).

Geographical distribution, host range and transmission

Streptococcal species have a worldwide distribution and cause disease in most regions where tilapia are cultured (Shoemaker *et al.*, 2017a). *Streptococcus agalactiae* and *S. iniae* infections in both wild and farmed tilapia have been reported broadly in the Americas, the Middle East, Africa and Asia (Amal and Zamri-Saad, 2011; Baums *et al.*, 2013; Bwalya *et al.*, 2020a). Outbreaks of lactococcosis have occurred in Taiwan, Indonesia, Brazil, Zambia and Egypt (Evans *et al.*, 2009; Tsai *et al.*, 2013; Anshary *et al.*, 2014; Osman *et al.*, 2017; Bwalya *et al.*, 2020b; Catão Egger *et al.*, 2023). *Streptococcus dysgalactiae* has been reported in cultured tilapia in Brazil, Indonesia, Taiwan, Egypt and the United States (Netto *et al.*, 2011; Abdelsalam *et al.*, 2013; Osman *et al.*, 2017; Liao *et al.*, 2020; Shoemaker *et al.*, 2022a), while *E. faecalis* has been implicated in wild and farmed tilapia mortalities in Egypt, Thailand and Bangladesh (Abu-Elala *et al.*, 2020; Akter *et al.*, 2021). Because *Lactococcus* spp., *S. dysgalactiae* and *E. faecalis* are emerging pathogens in tilapia, it is likely that they will expand into additional regions. Transmission of streptococcal species in tilapia aquaculture is largely horizontal. Streptococci may be introduced to production systems through several sources, complicating prevention efforts. Reports of the year-round presence of *Streptococcus* spp. in water and sediments near fish farms (Kitao *et al.*, 1979; Nguyen *et al.*, 2002) and the ability of streptococcal and lactococcal agents to form biofilms

(Chideroli *et al.*, 2017; Akoğlu, 2020; Heckman and Soto, 2021; Shahin *et al.*, 2022) suggests the possibility of environmental reservoirs. Carrier fish harbouring the bacteria have also been implicated in transmission, as *S. iniae* and *S. agalactiae* have been isolated from the skin and tissues of apparently healthy farmed and wild fish (Zlotkin *et al.*, 1998; Bromage *et al.*, 1999; Zamri-Saad *et al.*, 2014; Sun *et al.*, 2016). Healthy carrier broodstock may also pass these pathogens to progeny, indicating that seed stock could spread streptococcosis (Anshary *et al.*, 2014; Pradeep *et al.*, 2016). Contaminated feed is an additional potential source of infection (Kim *et al.*, 2007; Anshary *et al.*, 2014), and vectors such as wild ichthyophagous birds or external parasites have been proposed to circulate the bacteria between farms and animals, respectively (Vendrell *et al.*, 2006; Xu *et al.*, 2007; Chideroli *et al.*, 2017). Sources of *Enterococcus* spp. are more apparent, reported to be a result of water pollution with human waste (Zahran *et al.*, 2019; Akter *et al.*, 2021). Ingestion is considered the primary route of streptococcal infection. Orally infected fish may then shed sufficient concentrations of bacteria into the water for transmission to cohabitants through the skin, gills or nares (Bromage and Owens, 2002; Vendrell *et al.*, 2006; Baums *et al.*, 2013; Zamri-Saad *et al.*, 2014; Shoemaker *et al.*, 2017a). Cannibalism of dead fish is commonly observed, perpetuating outbreaks and disease spread (Shoemaker *et al.*, 2000; Heckman *et al.*, 2020).

Diagnosis

Observation of compatible clinical signs, followed by isolation of Gram-positive, catalase-negative cocci, constitutes a presumptive diagnosis of streptococcosis but specific identification requires further biochemical and molecular analysis. Positive samples are most consistently isolated from the brain, but bacteria can also be recovered from the kidney, spleen and liver (Ortega Asencios *et al.*, 2016; Chideroli *et al.*, 2017; Shoemaker *et al.*, 2017). These bacteria require nutrient-rich media for growth, and agar supplemented with 5% sheep's blood allows for observation of haemolytic activity. Selective agars containing colistin, thallium acetate and/or oxolinic acid to suppress Gram-negative bacterial growth have also been developed (Abdelsalam *et al.*, 2013; Shoemaker *et al.*, 2017). *Enterococcus* spp. are the only agents of piscine streptococcosis that grow on MacConkey agar (Abu-Elala *et al.*, 2019).

Biochemical profiling using the API· 20 Strep, API· Rapid ID 32 Strep and Vitek· two systems (bioMérieux, France) may aid in diagnosis, but the combined effects of interspecies similarity and intraspecies phenotypic heterogeneity may lead to misidentification or inconclusive results (Evans *et al.*, 2006; Anshary *et al.*, 2014; Osman *et al.*, 2017; Abu-Elala *et al.*, 2019; Abu-Elala *et al.*, 2020; Koh *et al.*, 2020). Molecular methods are suggested for definitive identification. Traditional PCR using the 16S rRNA gene, 16S–23S intergenic regions or species-specific primer pairs generally allow characterization to the species level for all but *Lactococcus* spp. (Meyburgh *et al.*, 2017; Mishra *et al.*, 2018; Maekawa *et al.*, 2020; Heckman *et al.*, 2024). Quantitative PCR (qPCR) is more sensitive than conventional methods and valuable for detection in tissue and environmental samples (Han *et al.*, 2011; Tsai *et al.*, 2013; Su *et al.*, 2016; Torres-Corral and Santos, 2021). Colorimetric loop-mediated isothermal amplification (LAMP) assays are similarly sensitive and are a lower-tech option for diagnosis of streptococcal species in the field (Jung *et al.*, 2010; Han *et al.*, 2011; Cai *et al.*, 2012). Further strain identification and discrimination requires multigene analysis. Multiplex PCR can be used for capsular typing of *S. agalactiae* (Imperi *et al.*, 2010) and virulence gene profiling in *S. iniae* (Baums *et al.*, 2013). Pulsed field gel electrophoresis (PFGE) has high discriminatory power for typing *S. iniae*, *S. agalactiae*, and *S. dysgalactiae* fish isolates (Weinstein *et al.*, 1997; Tsai *et al.*, 2012; Costa *et al.*, 2014; Sun *et al.*, 2016). Multilocus sequence analysis or typing (MLSA/T), considered the 'gold standard' for bacterial typing, is available as well (Jones *et al.*, 2003; Ruiz-Garbajosa *et al.*, 2006; Ferrario *et al.*, 2013; Mishra *et al.*, 2018; Heckman *et al.*, 2020). Compared with PFGE, MLSA/T is easier to replicate and can be combined with whole-genome analysis to provide improved depth of information on pathogen evolution and epidemiology (Martens *et al.*, 2008; Li *et al.*, 2009).

Pathology

Streptococcosis is a multifactorial disease and infections may progress differently depending on the bacterial species and strain, and environmental conditions. Mortality rates are generally 30–50% but can reach upwards of 75% in acute outbreaks (Perera *et al.*, 1994; Eldar *et al.*, 1995; Bromage and Owens, 2002; Abu-Elala *et al.*, 2020). Alternatively, chronic streptococcosis can cause persistent, daily losses of fewer fish over longer periods (Li *et al.*, 2014; Chideroli *et al.*, 2017). These bacterial species share a common pathogenesis strategy of adhesion and invasion of mucosal epithelial cells followed by dissemination in the bloodstream to the brain, heart and other target organs (Pradeep *et al.*, 2016; Abu-Elala *et al.*, 2020).

Clinical signs and gross lesions

Central nervous system involvement is reflected in the common clinical signs of streptococcosis. Afflicted fish often show behavioural changes such as erratic swimming and loss of orientation in the water column. Fish become lethargic, unresponsive to food and may remain grouped at the bottom of the tank. Ocular changes, including uni- or bilateral exophthalmia, corneal opacity, hyphema and hypopyon (Fig. 9.1A), are common. Darkened skin is typical, accompanied by areas of haemorrhage, and commonly abscesses and ulcers (Netto *et al.*, 2011; Ortega Asencios *et al.*, 2016; Abu-Elala *et al.*, 2019; Shoemaker *et al.*, 2020). Pale gills, stiff or eroded fins, body curvature, abdominal distension, scale detachment and excess mucus production have also been reported (Vendrell *et al.*, 2006; Abdelsalam *et al.*, 2013; Shoemaker *et al.*, 2017a; Zahran *et al.*, 2019; Akter *et al.*, 2021). In some cases, death is rapid and not preceded by readily observable clinical signs (Bromage and Owens, 2002; Heckman *et al.*, 2020). Internally, the body cavity may contain bloody or purulent fluid and, typically, organ and tissue adhesions (Fig. 9.1B). The spleen is generally enlarged and congested. The liver is also commonly enlarged and either pale or congested (Vendrell *et al.*, 2006; Shoemaker *et al.*, 2017a; Abu-Elala *et al.*, 2020). Haemorrhaging of internal organs may be present,

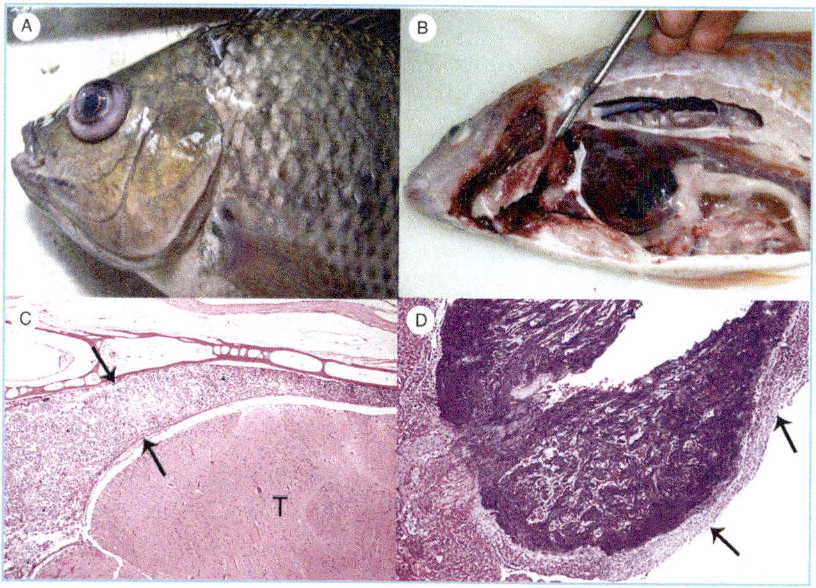

Figure 9.1 Gross and microscopic lesions in Nile tilapia fingerlings challenged with *Streptococcus agalactiae* by intracoelomic injection. (A), exophthalmia and periocular oedema. (B), marked hepatic congestion and splenomegaly. (C), severe meningitis characterized by intense granulomatous inflammatory cell infiltrates (arrows) that fill the cranial cavity and envelop the telencephalon (T). (D), epicarditis with expansion of the bulbus arteriosus by granulomatous inflammatory infiltrates (arrows). Fig. 9.4A and B were supplied by Dr Juan A. Morales, Universidad Nacional, Costa Rica. Fig. 9.1C and D were sourced from Soto *et al.* (2016b).

particularly in cases of *Lactococcus* infection (Meyburgh *et al.*, 2017; Abu-Elala *et al.*, 2020; Shahin *et al.*, 2022). The brain may be soft and oedematous, potentially with areas of congestion (Zamri-Saad *et al.*, 2014; Chideroli *et al.*, 2017; Abu-Elala *et al.*, 2020). These clinical signs, while characteristic of streptococcosis, are not pathognomonic and diagnosis must be coupled with bacterial identification.

Histopathology

Histopathological assessment of tilapia streptococcosis typically reveals multi-organ inflammation, particularly of the brain, heart and eyes (Fig. 9.1C,D). Tilapia can display meningitis, encephalitis, myocarditis, pericarditis, epicarditis, endocarditis and endo- or panopthalmitis (Chang and Plumb, 1996; Perera *et al.*, 1998; Chen *et al.*, 2007; Baums *et al.*, 2013; Abu-Elala *et al.*, 2020). Perisplenitis, peritonitis, oophoritis, orchitis and necrosis or degeneration of regions of spleen, liver and kidney have also been reported (Miyazaki *et al.*, 1984; Abuseliana *et al.*, 2011). HyperplAsia and fusion of gill lamella may be observed (Abu-Elala *et al.*, 2020). Inflammation can be either suppurative or granulomatous (Baums *et al.*, 2013; Ortega Asencios *et al.*, 2016; Martínez-Lara *et al.*, 2021). Inflammatory infiltrates are dominated by macrophages and accompanied by Gram-positive cocci free in tissues and within macrophage cytoplasm (Baums *et al.*, 2013; Soto *et al.*, 2015a; Ortega Asencios *et al.*, 2016; Abu-Elala *et al.*, 2020). Intracellular survival in macrophages and other host cells has been demonstrated by piscine streptococcal pathogens and appears integral for their dissemination (Zlotkin *et al.*, 2003; Avci *et al.*, 2014; Soto *et al.*, 2015a; Iregui *et al.*, 2016; Zahran *et al.*, 2019). Bacterial distribution, localization and severity of lesions may differ among different bacterial species, although the causative agents cannot be reliably diagnosed histologically. For example, the presence of cocci in both circulation and tissue is reported to be more prominent with *S. agalactiae* than *S. iniae* (Shoemaker *et al.*, 2017a), and Abu-Elala *et al.* (2020) reported that lesions caused by *S. agalactiae* and *L. garvieae* are more severe and widely distributed than those induced by *E. faecalis*.

Protection and control strategies

Prevention and control of streptococcosis begins with informed management and husbandry practices. The risk of introducing bacterial species or new strains can be reduced by purchasing pathogen-free stock, quarantining new fish, using uncontaminated feeds and, where possible, limiting interactions between cultured fish and wild animals. Keeping equipment and facilities clean and properly disinfected can help prevent spread and curtail outbreaks. Environmental parameters such as low dissolved oxygen, high pH (>8.0), high ammonia and high (>30°C) or fluctuating water temperatures can predispose tilapia to streptococcosis (Shoemaker *et al.*, 2017a). High stocking densities and external parasites can stress fish and damage skin, respectively, promoting opportunistic bacterial invasion and increased mortality (Xu *et al.*, 2007; Xu *et al.*, 2009; Zamri-Saad *et al.*, 2014). Therefore, maintaining water quality, appropriate stocking densities and limiting parasite numbers may aid in the prevention of epizootics. Proper nutrition is also important, although evidence is mixed regarding the potential benefits of probiotics and immunostimulants (Shoemaker *et al.*, 2017a; Foysal *et al.*, 2019). Additional measures such as reduced feeding during outbreaks and the prompt removal of moribund and dead fish can reduce bacterial spread and pathogen loads.

Antimicrobials are often employed in attempts to limit fish losses. Florfenicol, oxytetracycline and erythromycin are the most common treatments for streptococcosis in tilapia, although antimicrobial availability differs by country (Vendrell *et al.*, 2006; Zamri-Saad *et al.*, 2014; Chideroli *et al.*, 2017). AQUAFLOR· (florfenicol, Merck Animal Health) is the only approved antibiotic for streptococcosis in the United States and requires a veterinary feed directive. However, the efficacy of medicated feeds is limited when fish are anorexic, and by the bacterium's ability to survive intracellularly, necessitating multiple treatments, and favouring the development of carrier fish and resistance to treatments (Ghittino *et al.*, 2003; Agnew and Barnes, 2007; Chideroli *et al.*, 2017). Multi-drug resistance among the various agents of streptococcosis is a growing issue for animal and human health, including numerous reports involving fish isolates (Russo *et al.*, 2006; Chideroli *et al.*, 2017; Meyburgh *et al.*, 2017; Mishra *et al.*, 2018; Abu-Elala *et al.*, 2020; Lin *et al.*, 2020). The prophylactic use of antimicrobials is strongly discouraged for this reason and is prohibited in many regions (Shoemaker *et al.*, 2017a).

Vaccination programmes can significantly reduce antimicrobial usage and the incidence of disease (Soto *et al.*, 2015b). Several commercial vaccines for use in tilapia aquaculture have been developed for use in certain countries and offer different ranges of protection. AQUAVAC Strep Sa and Sa1 are injectable vaccines from Merck Animal Health that protect against *S. agalactiae* biotypes 2 and 1, respectively, while AQUAVAC Strep Si targets *S. iniae* and is delivered by injection or immersion. ALPHA JECT· Micro 1 Tila is an injectable vaccine for *S. agalactiae* serotype 1b (biotype 2) from Pharmaq. MEVAC Aqua Strept, available in Egypt, is the only polyvalent commercial vaccine to offer protection against *S. agalactiae*, S. *iniae*, *L. garvieae* and *E. faecalis*. It uses a nanoparticulate adjuvant Montanide™ IMS 1312 VG (Seppic, France) to improve uptake at mucosal sites and is administered by injection or immersion (Abu-Elala *et al.*, 2019). There is no commercial vaccine for *S. dysgalactiae* in tilapia.

In countries like the United States where no commercial vaccines for streptococcosis are available, autogenous vaccines prepared from circulating strains can be used to limit future outbreaks (Meyburgh *et al.*, 2017; Bwalya *et al.*, 2020b). Traditional, inactivated whole-cell preparations available at the commercial scale can provide high levels of protection against serotypically homogenous strains of a specific species but generally require labour-intensive injection administration and the use of adjuvants or booster doses (Soto *et al.*, 2015b). In addition, bacterial strain diversity within species and serotypes limits the extent of their protection, leading to vaccine escape and reinfection of vaccinated stocks with new variants (Millard *et al.*, 2012; Chideroli *et al.*, 2017; Pumchan *et al.*, 2020). Substantial efforts to improve the efficacy, ease of delivery and protective range of streptococcal vaccines have been made, including bacterins, live attenuated vaccines, recombinant subunit vaccines, DNA vaccines and bacterial ghosts for *S. iniae* and *S. agalactiae* (Ma *et al.*, 2020; Pumchan *et al.*, 2020; Miccoli *et al.*, 2021; Wang *et al.*, 2021; Heckman *et al.*, 2022). While every vaccine type offers distinct advantages and disadvantages, relatively high levels of protection have been achieved with each formulation in laboratory settings. The advent of whole-genome sequencing and bioinformatic analysis will inform vaccine design, especially in advancing protection against genetically and serotypically diverse strains.

Recently, several programmes have been initiated for selective breeding of tilapia for resistance to streptococcal pathogens (Kayansamruaj *et al.*, 2023). Research has demonstrated substantial additive genetic variation in resistance to *S. iniae* (LaFrentz *et al.*, 2016; Shoemaker *et al.*, 2017b; LaFrentz *et al.*, 2020) and

S. agalactiae (Shoemaker *et al.*, 2017b; Suebsong *et al.*, 2019; Sukhavachana *et al.*, 2019; LaFrentz *et al.*, 2020; Joshi *et al.*, 2021a), demonstrating that improvement in resistance to these pathogens is possible. In addition, genome-wide association studies have been conducted to enhance and improve selective breeding. Sukhavachana *et al.* (2020) identified a panel of single-nucleotide polymorphism (SNP) markers that explained ~10% of the genetic variation in resistance to *S. agalactiae*. Vela-Avitúa *et al.* (2023) identified quantitative trait loci (QTLs) that explained 12–16% of the genetic variance in resistance to *S. iniae*, and these QTL's were demonstrated to be effective in marker-assisted selection (MAS) for resistance to *S. iniae*. Through these selective breeding programmes, tilapia with enhanced disease resistance are currently available to tilapia producers.

Francisellosis

The family Francisellaceae (Sjöstedt, 2005) order Thiotrichales, class Gammaprotobacteria currently comprises four validly described genera: *Allofrancisella* (Qu *et al.*, 2016), *Cysteiniphilum* (Liu *et al.*, 2017), *Francisella* Dorofe'ev, 1947 in Approved Lists 1980 (Skerman *et al.*, 1980) and *Pseudofrancisella* (Zheng *et al.*, 2019). Despite the wide range of aquatic environments in which these bacteria can be found, only species of the genus *Francisella* have been identified as pathogens of aquatic organisms.

There are, at present, four *Francisella* species relevant to aquaculture: *F. halioticida* (Brevik *et al.*, 2011), '*Francisella marina*' *aka* '*Francisella salimarina*' (Soto *et al.*, 2018; Li *et al.* 2020), *F. orientalis* and *F. noatunensis* (Ramirez-Paredes *et al.*, 2020), all of which are facultative intracellular pathogens that can evade the immune response of their host by entering and replicating in host phagocytic cells, resulting in severe granulomatous disease. *Francisella orientalis* was previously misplaced as '*F. noatunensis* subsp. *orientalis*' and mainly affects warmwater fish species such as the three-line grunt (Kamaishi *et al.*, 2005), hybrid striped bass (Ostland *et al.*, 2006), tilapia (Soto *et al.*, 2009; Ramírez-Paredes *et al.*, 2017) and ornamental species like French grunt, Caesar grunt (Soto *et al.*, 2014), Indo-Pacific reef fish (Camus *et al.*, 2013) and ornamental cichlids (Lewisch *et al.*, 2014; López-Crespo *et al.*, 2019; Chang *et al.*, 2021). *Francisella orientalis* (Ramírez-Paredes, 2015; Ramirez-Paredes *et al.*, 2020) is the only species reported to date as the cause of francisellosis in tilapia, and is therefore the main focus of this summary.

Geographical distribution, host range and transmission routes

Francisella orientalis affects a wide range of marine and freshwater fish species and has worldwide distribution. In the tilapia industry it has been described causing outbreaks in species of different genera, including *Oreochromis* spp., *Sarotherodon* spp., *Tilapia* spp. and various hybrids of these species (Birkbeck *et al.*, 2011; Colquhoun and Duodu, 2011; Ramírez-Paredes *et al.*, 2017; Pulpipat *et al.*, 2019).

Horizontal transmission is well known for *F. orientalis* (Kamaishi *et al.*, 2005; Colquhoun and Duodu, 2011), while vertical transmission has also been suggested in tilapia (Pradeep *et al.*, 2017). As such, movement, transportation and importation of affected commercial tilapia can act as a mechanism for bacterial spread (Birkbeck *et al.*, 2011). While little is known concerning potential vectors capable of maintaining the bacterium in the absence of a suitable primary fish host, it has been suggested that farmed and wild fish, including ornamentals, could serve as asymptomatic carriers and also play a role in disease spread (Camus *et al.*, 2013).

Francisella orientalis occurs in brackish, fresh- and seawater environments. The optimum temperature range for outbreaks on tilapia farms is 20–25°C. Similarly, an association between temperature and mortality was observed in captive marine grunts (Soto *et al.*, 2014). In farmed tilapia the disease is typically controlled by reducing stocking densities and post-handling stress, as well as using water temperatures > 28°C where possible. Although *F. orientalis* is highly infectious, survival outside of fish hosts is not well documented. It has been suggested that it may enter a viable but non-culturable (VBNC) state after a number of weeks in either fresh- or salt water

(Soto and Revan, 2012). Bacteria in a VBNC state fail to propagate on media on which they normally grow but remain alive and are capable of metabolic activity. In addition, *F. orientalis* is capable of producing biofilms that play a role in environmental persistence. Biofilms can form within 24 hours and persist for more than 5 days in fresh-, brackish and marine water at temperatures < 30°C. Biofilms also render the bacteria more resistant to certain disinfectants (Soto *et al.*, 2015c).

Diagnosis

Preliminary diagnosis of francisellosis is based on the history, mortality pattern and gross and histopathologic findings. Differential diagnoses in tilapia include other granulomatous diseases caused by intracellular bacteria including mycobacteriosis, nocardiosis and edwardsiellosis. Definitive diagnosis should be based on the presence of two or more of the following: positive bacterial isolation, molecular detection or compatible histologic lesions (Ramírez-Paredes *et al.*, 2017).

Francisella orientalis is fastidious and requires enriched selective media for successful isolation. The bacterial cells are small coccobacilli (0.2–0.4 μm wide × 0.4–1.9 μm long), Gram-negative, oxidase-negative, catalase-positive, non-motile, non-sporulating, aerobic, pleomorphic and biochemically unreactive in most standard bacteriological tests. On agar, colonies appear visible in 3–5 days at 25–30°C (Fig. 9.2A). Isolation of *F. orientalis* is achieved by streaking homogenized spleen or kidney from diseased fish, ideally with visible nodular lesions, onto cystine heart agar supplemented 1:1 (v/v) with 2% bovine haemoglobin (CHAH) (BD, Oxford, UK) or animal (horse, rabbit, fish, sheep) blood (Kamaishi *et al.*, 2005; Olsen *et al.*, 2006; Birkbeck *et al.*, 2007; Soto *et al.*, 2009; Lewisch *et al.*, 2014; Ruane *et al.*, 2015; Lin *et al.*, 2016; Ramírez-Paredes *et al.*, 2017). Other commercially available media, such as modified Thayer Martin, provide selective isolation of *F. orientalis*. Shoemaker *et al.* (2022b) utilized this medium to recover *F. orientalis* from laboratory-infected fish following culture of brain tissues in freshly dead fish.

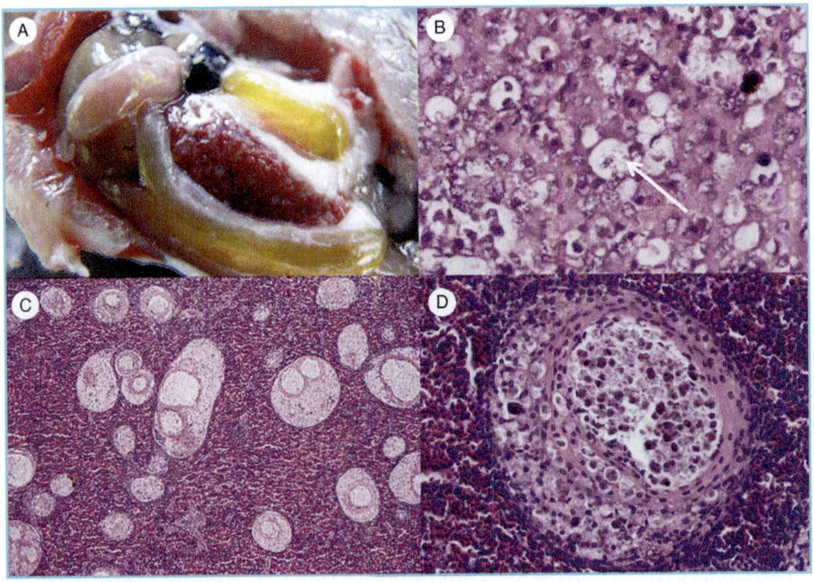

Figure 9.2 Gross and microscopic lesions of francisellosis in a natural outbreak of disease in Nile tilapia (*O. niloticus*) fingerlings caused by *F. orientalis*. (A), splenomegaly with multifocal white nodules corresponding to widespread granuloma formation. (B), macrophage with cytoplasmic vacuole containing small intracellular coccobacilli (arrow) consistent with *F. orientalis*. (C), chronic francisellosis with widespread granuloma formation in the spleen. (D), higher magnification of splenic granuloma with central region of pigment-laden macrophages surrounded by broad mantles of flattened and epithelioid macrophages.

Growth in liquid media is possible using Modified Mueller-Hinton II cation-adjusted broth, supplemented with 2% IsoVitaleX (BD, Oxford, UK) and 0.1% D-(+)-glucose ACS reagent (Sigma-Aldrich, Dorset, UK) (MMHB; Baker *et al.*, 1985). Other liquid media that support growth of *F. orientalis* are Bacto heart infusion broth (BD) supplemented with 10 g/l glucose and 1 g/l L-cystine, and Growth medium B1817 (Ramirez-Paredes *et al.*, 2020).

A range of methods are available for molecular detection of *F. orientalis*. Traditionally these methods support the definitive diagnosis of piscine francisellosis in the laboratory and/or on the farm. The genus-specific conventional PCR developed by Forsman *et al.* (1994) is a useful tool for the detection in fish tissues of members of the genus *Francisella* (including *F. orientalis*) when signs of disease are evident and for the identity confirmation of suspected colonies without DNA sequencing. Highly sensitive quantitative PCR protocols specific for *F. orientalis* (and *F. noatunensis*) have been published by Duodu *et al.* (2012) and Soto *et al.* (2010). In addition, *F. orientalis in situ* hybridization and immunohistochemical assays for use in formalin-fixed, paraffin-embedded tissues (FFPE) have been described (Hsieh *et al.*, 2007; Soto *et al.*, 2012c; Dong *et al.*, 2016). On-farm molecular tests include a LAMP assay (Pradeep *et al.*, 2017) and a recombinase polymerase amplification (RPA) assay (Shahin *et al.*, 2018). The latter is a rapid, simple, sensitive, specific and cost-effective method for the detection of francisellosis in the field.

Clinical signs and gross lesions

Clinical signs of francisellosis in tilapia are non-specific and vary in severity during the acute and chronic stages of the disease (Birkbeck *et al.*, 2011; Colquhoun and Duodu, 2011). Mortality rates on farms can vary from 10 to 90% depending on fish size and age, water temperature and quality and extent of handling (Soto *et al.*, 2009a, b, 2012b, 2013b). Anorexia and abnormal swimming are frequently observed during outbreaks, and fish may display coelomic distention due to coelomitis, ascites and organomegaly (Fig. 9.2B). Externally, fish may appear pale or darkened, with petechial to ecchymotic haemorrhages, ulcers, frayed fins and exophthalmia. Gills are typically pale and may contain white nodules in severe cases. Internally the most significant gross change is the presence of widespread, pale, tan- or cream-coloured nodules dispersed predominantly in the anterior kidney, posterior kidney and spleen, usually accompanied by marked renomegaly and splenomegaly (Fig. 9.2C, D). Additional lesions can occur in the gills, liver, eyes and, sporadically, the gastrointestinal tract and mesenteric fat.

Histopathology

Necrotizing, granulomatous lesions and discrete granuloma formation characterize the histopathologic features of francisellosis in tilapia (Birkbeck *et al.*, 2011; Colquhoun and Duodu, 2011; Camus *et al.*, 2013). The hallmark of infection is the presence of granulomas dominated by macrophages containing the small coccoid bacteria within a single, large, clear cytoplasmic vacuole. The bacteria are best visualized with Giemsa stain rather than haematoxylin and eosin or tissue Gram stains. Bacterial numbers are highest early in the course of infection and decrease over time. In chronic infections, granulomas can develop central regions of coagulative necrosis with or without fibrous encapsulation. Lesions can affect virtually any tissue but are most severe in the spleen, anterior kidney and haematopoietic areas of the posterior kidney, often resulting in extensive parenchymal loss. Less commonly affected are the gills, heart, central nervous system, liver and gastrointestinal submucosa (Hsieh *et al.*, 2006; Mauel *et al.*, 2007; Soto *et al.*, 2011a, b, 2013). A more detailed description of lesion development following immersion challenge is found in Soto *et al.* (2013).

Protective and control strategies

Francisella orientalis can cause disease alone or in co-infections with other bacteria and viruses (Assis *et al.*, 2017; Huang *et al.*, 2020; Nguyen *et al.*, 2020). Numerous studies have demonstrated the potential of vaccination for the prevention and control of piscine francisellosis (Krossøy *et al.*, 2007; Schrøder *et al.*, 2009; Soto *et al.*, 2009b; Müller *et al.*, 2011; Soto *et al.*, 2011b, 2014b; Brudal *et al.*, 2014; Ramírez-Paredes, 2015; Lagos *et al.*, 2017; Lampe *et al.*, 2017a, b; Shahin *et al.*, 2019; Pulpipat *et al.*, 2020; Oliveira *et al.*, 2022). Currently no licensed vaccines are available, although the use of autogenous vaccines based on whole inactivated cells has proven successful reducing mortalities during the winter in countries where these are accessible, such as Brazil.

Programmes have been initiated to improve the resistance of tilapia to francisellosis through selective breeding, and improved tilapia are available for producers. Moderate to high heritabilities were demonstrated for resistance to *F. orientalis* demonstrating the feasibility of improving disease resistance (Joshi *et al.*, 2021b; Shoemaker *et al.*, 2022b). Importantly, both research programmes have shown that breeding for resistance to *F. orientalis* will not have a negative impact on growth, harvest weight or resistance to *S. agalactiae*.

While salinity has no effect on disease outcomes resulting from francisellosis, increasing water temperature from 25 to 30°C prevents the development of clinical signs and mortality in *F. orientalis*-challenged fish (Soto *et al.*, 2012b). Under experimental conditions, gradually increasing the temperature to 30°C over a 2-week period and maintaining it there for 3 months, while concurrently providing oxytetracycline-medicated feed to supply 3 gm of antibiotic per 100 pounds of fish daily for 10 days, resulted in a significant decrease in mortalities. As a result, the combination of temperature manipulation and antimicrobial therapy is recommended to reduce mortality (Soto *et al.*, 2014a, b).

Antimicrobial susceptibility testing for *F. orientalis*, based on methods developed for *F. tularensis*, indicates low minimal inhibitory concentration (MIC) for florfenicol and oxytetracycline (Baker *et al.*, 1985; Soto *et al.*, 2016a, b; Ramírez-Paredes *et al.*, 2017), and in tilapia naturally infected or laboratory challenged with *F. orientalis*, treatment with both antibiotics reduced mortalities (Mauel *et al.*, 2003; Soto *et al.*, 2010b). Fortuitously, both florfenicol and oxytetracycline are approved for aquacultural use in most countries.

Should depopulation and disinfection of affected facilities become necessary, Virkon® S (1% or more for 10 min), bleach (10% or ~6000 ppm available chlorine for 10 min) and hydrogen peroxide (3% for 10 min) all provide effective biocidal activity against *F. orientalis* in both the water column and biofilms (Soto *et al.*, 2015a, b).

Columnaris disease

Columnaris disease outbreaks in fish were first reported over a century ago (Davis, 1922) and, collectively, continue to be amongst the most insidious and destructive diseases of farmed fish worldwide. Although three species within the genus *Flavobacterium* (*F. columnare*,

F. psychrophilum and *F. branchiophilum*) are the most widely recognized aetiological agents of flavobacterial fish diseases (Davis, 1922; Borg, 1948; Holt, 1988; Wakabayashi *et al.*, 1989), taxa belonging to multiple genera (*Flavobacterium*, *Chryseobacterium*, *Tenacibaculum* and others) within the Family *Flavobacteriaceae* and *Weeksellaceae* (Phylum Bacterioidetes; Bernardet and Nakagawa, 2006; García-López *et al.*, 2019; i.e. flavobacteria) have been linked with systemic infections, disease outbreaks and mortality in fish (reviewed in Loch and Faisal, 2015).

In this context, a range of flavobacteria have been isolated from healthy and diseased tilapia. For example, Mwega *et al.* (2019) explored the prevalence of flavobacterial infections in apparently healthy wild and farmed Nile tilapia in Tanzania. The authors identified 67 isolates belonging to the genera *Flavobacterium* and *Chryseobacterium*, and included species such as *F. indicum*, *F. aquatile*, *F. oryzae*, *C. indologenes*, *C. joostei* and unidentified *Chryseobacterium* spp. Ezzeldeen *et al.* (2015) also recovered *C. indologenes* from healthy and diseased Nile tilapia in Egypt. Flemming *et al.* (2007) recovered 32 *Flavobacterium* spp. isolates from different fish species in South Africa and tested the virulence of five isolates in Mozambique tilapia, revealing their capacity to induce mortality and gross disease signs consistent with other flavobacterial infections (e.g. gill necrosis, skin ulceration and fin rot). Sebastião *et al.* (2019) isolated *F. inkyongense* from diseased chocolate cichlids and demonstrated a lack of pathogenicity to Nile tilapia. Similarly, a range of flavobacteria have been recovered from pond water used to rear tilapia (e.g. *F. tilapiae*; Chen *et al.*, 2013), although their pathogenicity remains unclear.

Although a range of flavobacteria appear to be facultative tilapia pathogens, those species that cause columnaris disease are a significant impediment to tilapia aquaculture and will be the focus of the remainder of this section.

Columnaris disease was first described in 1917 at the US Fisheries Biological Station in Fairport, Iowa, USA (Davis, 1922) and was associated with grossly appreciable signs that included skin ulcers, frayed fins and severe gill damage. Davis (1922) performed a thorough documentation of the disease, including occurrence, cause, pathogenesis, methods of infection and treatment and control strategies. Although unable to culture the aetiological bacterium, Davis proposed the name *Bacillus columnaris* based on cell morphology and its tendency to form columns or haystacks on affected tissues when observed via light microscopy. In 1944 the bacterium was successfully isolated and cultured in low-nutrient media (Ordal and Rucker, 1944) and was reclassified as *Chondrococcus columnaris.* The bacterium has since been reclassified as *Cytophaga columnaris* (Garnjobst, 1945), *Flexibacter columnaris* (Bernardet and Grimont, 1989), *Flavobacterium columnare* (Bernardet *et al.*, 1996) and, recently, into four species of *Flavobacterium* (LaFrentz *et al.*, 2022; see below).

Research has demonstrated that *F. columnare* exhibits substantial genetic variation that has important implications for virulence and host species preference. A phylogenomic study by LaFrentz *et al.* (2018) revealed four distinct genetic groups within the species with host preferences, and it was suggested that these may represent unique bacterial species (Kayansamruaj *et al.*, 2017; LaFrentz *et al.*, 2018). For example, there is an association between genetic group one isolates and salmonids, as well as an association between genetic group four isolates and tilapia (LaFrentz *et al.*, 2018). LaFrentz *et al.* (2022) further characterized the four genetic groups using a polyphasic approach based on phenotypic and genomic analyses. The results afforded differentiation between the four genetic groups, and it was proposed that genetic group one isolates remain recognized as *F. columnare* whereas isolates of genetic groups 2, 3 and 4 be recognized as novel *Flavobacterium* species for which the names *F. covae* sp. nov., *F. davisii* sp. nov. and *F. oreochromis* sp. nov. were, respectively, proposed and subsequently validated (Oren and Garrity, 2022). Collectively these four species are referred to as 'columnaris-causing bacteria' (CCB). Although laboratory studies have demonstrated that *F. columnare, F. covae* and *F. davisii* can cause columnaris disease in tilapia (Shoemaker and LaFrentz, 2015), *F. oreochromis* is commonly recovered from cases of columnaris disease in the tilapia industry (LaFrentz *et al.*, 2018) and reports of natural *F. davisii* infections in tilapia also exist (Guan *et al.*, 2022; Shoemaker *et al.*, 2022a).

Geographical distribution, host range and transmission routes

CCB are presumed to have a worldwide distribution and have been recovered from infected fish in Africa, Asia, Europe and Australia, as well as from North and South America. *Flavobacterium columnare*, *F. covae* and *F. davisii* appear to be more widely distributed than *F. oreochromis*, with the latter being found primarily in South America, Central America and Asia (LaFrentz *et al.*, 2018). This distribution may be due to the association of *F. oreochromis* and tilapia, with these regions represented by substantial tilapia industries.

It has been suggested that nearly all freshwater (as well as some brackish) fish are susceptible to CCB infections (Davis, 1922; Starliper and Schill, 2011; Declercq *et al.*, 2013). Columnaris disease outbreaks regularly occur in farmed fish such as carp, salmon, trout, tilapia and catfish; however, mortality events in wild and feral fish populations also occur (Loch *et al.*, 2013; Scott and Bollinger, 2014).

CCB are readily transmitted fish to fish, whereby those that survive an initial infection can become carriers that cyclically shed the bacterium (Fujihara and Nakatani, 1971). Moreover, CCB released from dead tilapia are highly virulent and can induce rapid disease among healthy cohabitants (Shoemaker and LaFrentz, 2015), stressing the importance of rapidly removing mortalities. The bacteria can survive for long periods in water (Arias *et al.*, 2012), which increases the risk of columnaris disease outbreaks in farmed fish reared in open-water sources potentially harbouring the agent (Kunttu *et al.*, 2012). Of interest is the potential for CCB to form biofilms on inert surfaces, where it may serve as a reservoir for infection (Welker *et al.*, 2005; Cai *et al.*, 2013).

Diagnosis

Diagnosis of columnaris disease is based on characteristic clinical signs and detection and identification of CCB. A presumptive diagnosis can be made when (1) clinical signs of columnaris disease are observed in moribund fish; (2) long, slender Gram-negative rods are observed in wet mounts from lesions; and (3) putative CCB colonies are recovered *in vitro*. The clinical signs of columnaris disease are described below.

Wet mounts of tissues collected from the gills or skin lesions of moribund fish will reveal long, slender, rod-shaped bacteria that often aggregate into columns or haystacks (Fig. 9.3) and may exhibit gliding motility. Bacterial numbers are often highest on the leading margins of lesions. Cells of CCB are typically 0.3–0.5 μm wide and 3–10 μm long (Fig. 9.4A); however, cells up to 20 μm in length may occur.

Primary isolation of CCB should be from the gills, external lesions, and/or internal organs. To support bacterial growth, low-nutrient microbiological media such as Shieh, modified Shieh, tryptone yeast extract salts (TYES), Hsu-Shotts or cytophaga agar are required (reviewed in Loch and Faisal, 2017), as well as R2A and nutrient agar. The incorporation of antibiotics also aids in media selectivity. Samples should be streaked for isolation onto a suitable

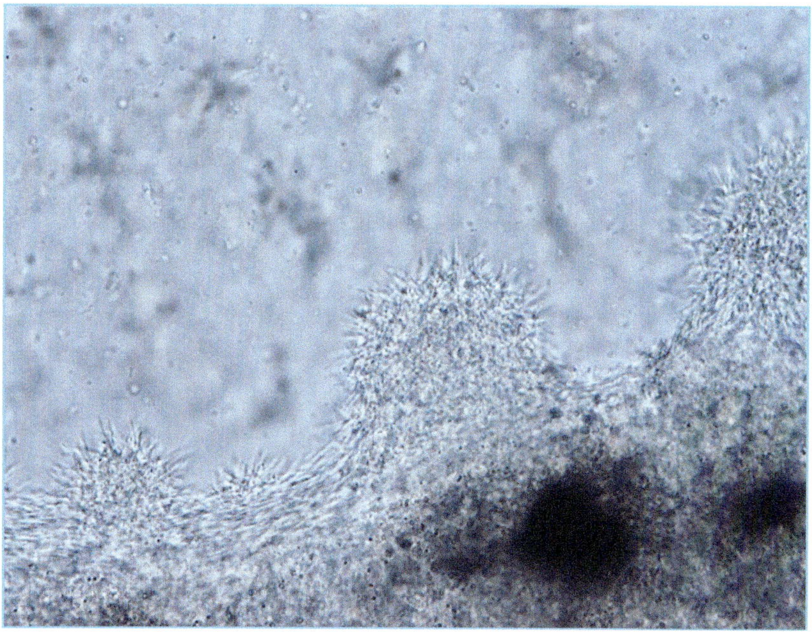

Figure 9.3 Wet-mount skin scrape preparation of ulcerated skin from a moribund tilapia displaying columns or haystacks of aggregated, long, slender bacterial rods characteristic of bacteria that cause columnaris disease.

Figure 9.4 Cell morphology, colony morphology and gross lesions of columnaris disease in tilapia (*Oreochromis* sp.). (A), Gram stain of columnaris disease-causing bacteria demonstrating the typical long, slender morphology of the Gram-negative rods. (B), colonies of columnaris disease-causing bacteria exhibiting characteristic flat, rhizoid morphology with irregular margins and yellow coloration. (C), patchy yellow–tan areas of necrosis in the gills of a tilapia. (D), hybrid tilapia exhibiting depigmented lesions on the caudal peduncle corresponding to rapidly spreading areas of acute erosion and ulceration. Fig. 9.4C was supplied by Dr Juan A. Morales, Universidad Nacional, Costa Rica.

medium and incubated at 20–30°C for 2–3 days. Most CCB produce yellow rhizoid colonies with irregular margins that adhere tightly to the agar (Fig. 9.4B); however other colony morphologies also occur (Kunttu *et al.*, 2009).

Confirmation of bacterial identity may be accomplished via several serological and molecular based assays (reviewed in Loch and Faisal, 2017), the most common of which is PCR analysis on putative CCB recovered from moribund fish or directly on affected tissues. Mabrok *et al.* (2020) developed PCR primers based on the chondroitin AC lyase gene, which are specific for all CCB. A positive result is indicated by the presence of the specific amplicon at 287 base pairs (BP). LaFrentz *et al.* (2019) developed a multiplex PCR that can assign an unknown CCB to one of the four *Flavobacterium* species. A positive result is indicated by the presence of a specific amplicon at 415 bp for *F. columnare*, 320 bp for *F. covae*, 894 bp for *F. davisii* or 659 bp for *F. oreochromis*.

Clinical signs and gross lesions

All life stages of tilapia are susceptible to columnaris disease. The disease was recently reported in tilapia eggs and larvae (Guan *et al.*, 2022), but is more common in fry and fingerlings. Behavioural signs may include lethargy, loss of appetite and hanging at the surface of the water with rapid opercular movements (Amin *et al.*, 1988). Gross external lesions are easily recognized and include frayed fins and fin rot, skin depigmentation that can progress to ulceration and necrotic gill lesions. The gills of infected fish may show mild to severe pale areas of necrosis (Fig. 9.4C) and gill filaments may be covered with a thick film of mucus (Amin *et al.*, 1988). Skin lesions can form around the dorsal fin, resulting in the so-called symmetrical saddleback lesion and on other parts of the body. Lesions can appear depigmented and often assume a yellowish discoloration from bacterial flexirubin-type pigment production (Fig. 9.4D).

Histopathology

Few studies have described the histopathology of columnaris disease specifically in tilapia. Amin *et al.* (1988) examined the gills of numerous diseased tilapia, describing dilation of blood vessels in the gill arches, primary gill filaments and telangiectasis of secondary gill lamellae. In addition, hyperplasia of gill lamellae and necrosis of gill filaments were observed.

Histopathology of skin lesions in other fish species demonstrate epidermal cell necrosis with sloughing of the epidermis (Morrison *et al.*, 1981; Bullard *et al.*, 2011; Bullard *et al.*, 2013), usually accompanied by cellular debris and numerous CCB cells replacing the epidermis. Bacteria can extend into the dermis, and inflammatory changes are usually relatively mild with small numbers of mononuclear cells. Internally, few pathological changes have been observed.

Protective and control strategies

The first line of defence against columnaris disease is a strong biosecurity and fish health management plan (Faisal *et al.*, 2017) that emphasizes good management practices to provide proper environmental conditions and reduce risk factors (e.g. stress, handling, rapid changes in temperature, poor water quality and abrasions of the skin surface). However, even with these in place, epizootics may occur and require treatment. Treatment for columnaris disease generally includes external bath treatments and/or antibiotics. External treatments include bath administration of salt, copper sulfate, potassium permanganate, hydrogen peroxide, chloramine-T and quaternary ammonium compounds (Declercq *et al.*, 2013). Antibiotics such as florfenicol and oxytetracycline are commonly used to control columnaris disease in tilapia and other fish species. The approach taken for treatment is highly dependent upon the nature of the epizootic and the type of system in which the fish are cultured.

Wonmongkol *et al.* (2018) suggested the potential for selective breeding of Nile tilapia for resistance to columnaris disease. Low to moderate heritability estimates were reported. Given the impact of columnaris disease on tilapia production, it is likely that other programmes will incorporate breeding for resistance to columnaris disease.

Currently there are no approved vaccines for the prevention of columnaris disease in tilapia; however, substantial research to this end is being conducted. Vaccine research has included testing bacterins (i.e. killed whole-cell vaccines) delivered via different routes (Grabowski *et al.*, 2004; Leal *et al.*, 2010). Although these have provided minimal efficacy, recent research indicates that whole-cell vaccines coated with chitosan and nanovaccines improve vaccine uptake and efficacy (Kitiyodom *et al.*, 2019a, b). Live attenuated vaccines delivered through the water have resulted in moderate levels of protection (Mohammed *et al.*, 2013). A review of efforts to develop CCB vaccines for tilapia and other fish species can be found in Kayansamruaj *et al.* (2020). Other potential prevention methods under investigation include phage therapy (Laanto *et al.*, 2015), selective breeding (Wonmongkol *et al.*, 2018) and probiotics. As research progresses on pathogenesis, virulence factors, genetic diversity and immune responses, it is anticipated that improved methods to prevent and control columnaris disease in tilapia will follow.

Edwardsiellosis

The genus *Edwardsiella* are Gram-negative, enteric bacterial pathogens initially reported from a range of host species including eels, seals, snakes and humans (Hoshina, 1962; King and Adler, 1964; Ewing *et al.*, 1965; Sakazaki, 1965; Sakazaki and Tamura, 1975). The type species, *E. tarda*, represents organisms that were concurrently described from multiple laboratories, including the 'Bartholomew' group from King and Adler (1964), the 'Asakusa' group from Sakazaki (1965) and 'Bacterium 1483–59' from Ewing *et al.* (1965). Originally, *Edwardsiella* encompassed enteric bacteria with unique phenotypic characters that did not fit classification schemes for genera within the family Enterobacteriaciae. Comparative phylogenomic analysis has restructured the order Enterobacteriales, and *Edwardsiella* are now classified within the family Hafniaceae along with the genera *Hafnia* and *Obesumbacterium* (Adeolu *et al.*, 2016).

The first reports of *Edwardsiella* infections in fish date from the early 1970s, when *E. tarda* was deemed the cause of emphysematous putrefactive disease in farm-raised channel catfish (*Ictalurus punctatus*) in Arkansas, USA (Meyer and Bullock, 1973). Since these initial reports, *E. tarda* has become a globally recognized fish pathogen. In 1981, Hawke *et al.* described *Edwardsiella ictaluri* as the aetiologic agent of enteric septicaemia of catfish (ESC), an important disease of catfish aquaculture in the south-eastern United States (Hawke and Khoo, 2004; Wise *et al.*, 2004). Initially thought specific to channel catfish, subsequent reports of *E. ictaluri* causing infections in non-ictalurid fish have included tilapia, zebrafish and other commercially important species (Kasornchandra *et al.*, 1987; Baxa *et al.*, 1990; Tucker and Robinson, 1990; Crumlish *et al.*, 2002; Keskin *et al.*,

2004; Iwanowicz *et al.*, 2006; Mainous *et al.*, 2010; Soto *et al.*, 2012a, b; Hawke *et al.*, 2013). The historical literature of *Edwardsiella* was further complicated by the recent segregation of *E. tarda* into three discrete taxa, *E. tarda*, *E. piscicida* and *E. anguillarum* (Abayneh *et al.*, 2013; Shao *et al.*, 2015). While bona fide *E. tarda* remains a pathogen of concern for some fish species, *E. piscicida* and *E. anguillarum* are now more widely recognized as significant fish pathogens (Griffin *et al.*, 2017, 2020)

Edwardsiella spp. in tilapia were first reported in the early 1980s, when *E. tarda* was isolated from gut contents of *Tilapia nilotica* during forensic investigations of sporadic cases of diarrhoea involving humans in the tropics (Vandepitte *et al.*, 1980). Plumb and Sanchez (1983) showed that tilapia are susceptible to experimental infection with *E. ictaluri*, although *E. tarda* has been the most-reported *Edwardsiella* sp. associated with disease in tilapia. The first documentation of natural *Edwardsiella* spp. infections in tilapia (*Sarotherodon niloticus*) was in the mid-1980s, when Miyashita (1984) isolated *E. tarda* during mortality events on aquaculture farms in Japan. Since then, numerous reports of *E. tarda* in tilapia have emerged. However, in the absence of accompanying molecular data, it is difficult to discern whether these early reports involve bona fide *E. tarda*, *E. piscicida* or *E. anguillarum*. Regardless, prevention and/or treatment of *E. tarda* infection in tilapia has received significant research investment (Park and Jeong, 1996; Igarashi *et al.*, 2002; Kwon *et al.*, 2006; Pirarat *et al.*, 2006; Pongsak and Phumkhachorn, 2010; Baba *et al.*, 2016; Zeng *et al.*, 2017; Sherif *et al.*, 2021).

While Plumb and Sanchez (1983) described experimental infection of tilapia with *E. ictaluri*, the first natural outbreak in pond culture was reported in 2012 (Soto *et al.*, 2012a, b). Natural outbreaks of *E. ictaluri* in juvenile red tilapia hybrids were also reported from open-cage culture in South-east Asia, with mortality rates exceeding 50%. Affected fish presented with lesions similar to those induced by *F. orientalis*, suggesting that *E. ictaluri* may be under-reported in this region (Dong *et al.*, 2019). Following the initial reports by Dong *et al.* (2019), 26 additional outbreaks occurred in Vietnam. Recurrence of infection in domestic stocks indicates that *E. ictaluri* has become established within farmed tilapia cultured in this region and should be considered as a differential diagnosis and pathogen of concern for tilapia (Nhinh *et al.*, 2023).

Perhaps the most important *Edwardsiella* spp. associated with tilapia is *E. anguillarum*. First reported as a pathogen of eels in the mid-2010s (Shao *et al.*, 2015), numerous reports have since identified *E. anguillarum* as a threat to tilapia production (Armwood *et al.*, 2019; Oh *et al.*, 2020; da Costa *et al.*, 2022; López-Porras *et al.*, 2022). While the host range for *E. anguillarum* is expanding, the historical significance of *E. tarda* vs *E. anguillarum* in tilapia remains convoluted. With the recognition of *E. anguillarum* as congruous with isolates formerly categorized as 'atypical, fish-pathogenic *E. tarda*', it is plausible that *E. anguillarum* has impacted tilapia culture since before Shao *et al.* (2015) recognized it as a discrete taxon. This is supported by molecular surveys, which revealed that several *Edwardsiella* spp. isolates from disease outbreaks in Brazil were misclassified, which has probably obscured the presence of *E. anguillarum* in South America (da Costa *et al.*, 2022). Similarly, Oh *et al.* (2020) documented the first case of *E. anguillarum* from a tilapia farm in Korea, while recognizing that previous outbreaks in the region may have been inaccurately attributed to *E. tarda*. Likewise, *E. anguillarum* was implicated in chronic mortality events in tilapia fry and fingerlings from a culture facility in Costa Rica, again calling into question whether *E. anguillarum* is under-reported due to the inability to phenotypically distinguish *E. anguillarum* from *E. tarda*, as well as its non-specific clinical signs and lesions that are similar to those induced by *F. orientalis* and *E. ictaluri* (Soto *et al.*, 2012a, c; López-Porras *et al.*, 2022).

Geographical distribution, host range and transmission

Edwardsiella spp. have been reported from an assortment of host species across the globe, of which *E. tarda* is the most widely reported, with accounts from North America, South America, Europe, Asia, Australia and Africa (Meyer and Bullock, 1973; Eaves *et al.*, 1990; Clavijo *et al.*, 2002; Alcaide *et al.*, 2006; Davies *et al.*, 2018; Wamala *et al.*, 2018). Nevertheless, with the reorganization of the genus in the mid-2010s and segregation of

E. tarda into three discrete taxa, these reports could be erroneous, particularly if they occurred prior to the recognition of *E. piscicida* or *E. anguillarum*. Initially, *E. ictaluri* was limited to North America and thought to be specific to channel catfish. However, the geographic and host range for *E. ictaluri* has expanded to include a diverse range of temperate and tropical non-ictalurid freshwater fish species including *Danio* spp., knifefish (order Gymnotiformes), *Oreochromis* spp., *Clarias* spp., *Pangasius* sp., ayu (*Plecoglossus altivelis*), Chinook salmon, etc. (Kasornchandra *et al.*, 1987; Baxa *et al.*, 1990; Tucker and Robinson, 1990; Crumlish *et al.*, 2002; Keskin *et al.*, 2004; Iwanowicz *et al.*, 2006; Mainous *et al.*, 2010; Soto *et al.*, 2012a, b; Hawke *et al.*, 2013; Phillips *et al.*, 2017).

As enteric pathogens, *Edwardsiella* infections occur primarily by the faecal–oral route, with reports of increased disease incidence in ponds with actively feeding fish (Wise *et al.*, 2004). In tilapia challenged experimentally with *E. ictaluri*, the bacterium was detected in the epidermal mucus layer, stomach and proximal intestine, alluding to cutaneous and oral transmission routes (Soto *et al.*, 2013a). Similarly, experimental evidence using green fluorescent protein methodologies and a blue gourami infection model indicate an oral portal of entry for *E. tarda* (Ling *et al.*, 2001). While transmission routes for *E. piscicida* have not been confirmed, histological analyses of diagnostic submissions reveal marked inflammation in the gills of infected channel and hybrid catfish, suggesting a possible infection route (Lester Khoo, personal communication). Comparably, the portal of entry for *E. anguillarum* has not been determined but is thought to follow similar routes of infection. As is typical with other bacterial infections, horizontal transmission among cohorts probably occurs because dead and infected fish seed the environment and pass bacteria on to their cohabitants.

Diagnosis

Diagnosis of *Edwardsiella* spp. infection is based on host species, disease signs and lesions, water temperatures conducive to outbreaks, followed by bacterial isolation and identification. Bacterial isolation is achieved by inoculation of brain, kidney or splenic tissue onto general nutrient media, such as brain heart infusion, Mueller Hinton or tryptic soy agar plates, and subsequent incubation at 28–37°C for 24–48 hours. Media can be supplemented with 5% defibrinated sheep blood to visualize haemolytic activity. β-haemolysis by *E. ictaluri* is minimal, but more pronounced with *E. tarda*, *E. piscicida* and *E. anguillarum*. *Edwardsiella ictaluri* does not grow well >30°C and grows much slower than other taxa, often requiring 48 hours at 28°C to produce observable (1–2 mm) colonies compared with 24 hours for *E. tarda*, *E. piscicida* and *E. anguillarum*. All produce small, round, slightly raised, greyish-white colonies. Bacterial identification to the genus level is based on biochemical characterization of isolates, while species identification may require molecular testing. *Edwardsiella ictaluri* is indole and oxidase negative, while *E. tarda*, *E. piscicida* and *E. anguillarum* are indole positive and oxidase negative. As such, the indole and oxidase tests coupled with growth temperature and incubation period differentiate *E. ictaluri* from other species. However, differentiation of *E. tarda*, *E. piscicida* and *E. anguillarum* requires molecular confirmation using species-specific PCR primers and/or molecular probes (Griffin *et al.*, 2014; Reichley *et al.*, 2017; da Costa *et al.*, 2022) or sequencing of resolute molecular markers such as *gyrB* or *sodB* (Yamada and Wakabayashi, 1999; Griffin *et al.*, 2014; Reichley *et al.*, 2017).

Clinical signs and gross lesions

Clinical signs and gross lesions of edwardsiellosis, regardless of species, are generally non-pathognomonic and can include lethargy, erratic swimming, discoloration, exophthalmia, corneal opacity, reddening of the anal papilla, pale gills, congestion, ascites, renomegaly, splenomegaly and pale necrotic foci in the spleen, kidney and liver (Miyazaki and Kaige, 1985; Park *et al.*, 2012; Soto *et al.*, 2012a, Soto *et al.*, 2013a; Iregui *et al.*, 2016; Armwood *et al.*, 2019; Oh *et al.*, 2020). Diseased tilapia infected with *E. ictaluri* display white necrotic spots on the liver, spleen and head kidney, with limited notable gross external changes (Dong *et al.*, 2019; Nhinh *et al.*, 2023). Renomegaly and splenomegaly, accompanied by similar multifocal nodules in the internal organs, were reported in Nile tilapia fingerlings naturally infected with *E. ictaluri* and were reproduced in laboratory challenges. These gross changes are like those observed in tilapia francisellosis, mycobacteriosis and nocardiosis. Externally, affected *E. ictaluri*-infected tilapia exhibit variable darkening or pallor and pale gills resulting from bacterial-induced anaemia (Soto *et al.*, 2012a; Dong *et al.*, 2019; Nhinh *et al.*, 2023). Nile tilapia naturally infected with *E. anguillarum* exhibit comparable gross internal changes (Armwood *et al.*, 2019).

Histology

Tilapia infected with *E. ictaluri* exhibit severe inflammation of the spleen, head kidney and liver, with tissue parenchyma being filled with macrophage aggregates (Fig. 9.5). Chronic inflammation can be observed in the gills, with substantial hypertrophy and lamellar fusion. Limited inflammation can also be observed in the submucosa, heart and gastrointestinal tract of infected fish (Soto *et al.*, 2012a; 2013). Granulomas in the spleen containing clumps of rod-shaped bacterial aggregates can be viewed under high magnification (Fig. 9.5C). Pyknotic splenic cells and melano-macrophage centres were also observed in tilapia infected with *E. ictaluri* (Dong *et al.*, 2019), with congestion in the brain and hepatic tissues and lipidosis in the liver (Nhinh *et al.*, 2023). Similarly, tilapia (*Oreochromis mossambicus*) experimentally injected with an isolate identified as *E. tarda* resulted in erosions and muscular necrosis, with abundant bacteria and inflammatory cells at the injection site. Focal necrosis with bacteria-laden inflammatory cells were reported from the liver, spleen, kidneys and heart (Kaige *et al.*, 1986). As the infection progressed, affected tissue was walled off by layered epithelioid cells leading to granuloma formation in the liver, spleen and kidneys, with the liver, followed by the spleen, being most severely affected. The gills were also necrotic, with abundant inflammatory infiltrates and hyperplasia (Miyazaki and Kaige, 1985). However, the work by Kaige and others was performed prior to the recognition of *E. piscicida* or *E. anguillarum* and must be considered within that context.

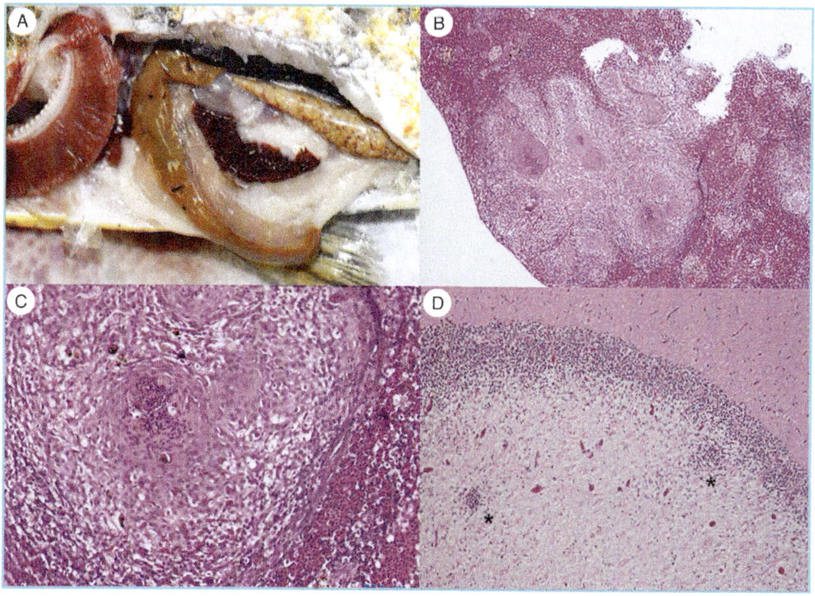

Figure 9.5 Edwardsiellosis in a laboratory-infected tilapia hybrid (*Oreochromis* spp.) fingerling caused by *Edwardsiella ictaluri*. (A), splenomegaly with multifocal white nodules represented histologically by granuloma formation. (B), chronic infection with widespread, multifocal, coalescing granulomas in the spleen. (C), higher magnification of a developing splenic granuloma with central necrosis surrounded by an outer rim of macrophages and lymphocytes. (D), encephalitis with small foci of gliosis (*) within the white matter of the optic tectum. Fig. 9.5A, D from Soto *et al.* (2013). Figure 9.5B,C from Soto *et al.* (2012a).

In a spontaneous *E. anguillarum* outbreak in Nile tilapia, acute lesions were necrotizing with large numbers of free and phagocytized Gram-negative bacterial rods. As lesions became chronic, bacterial numbers declined when peripheral zones of epithelioid macrophages developed into discrete granulomas with large central regions of necrotic debris (Fig. 9.6). Lesions were most common in the spleen and haematopoietic areas of the kidneys but also involved the liver, heart, gills, pseudobranch, ocular choroid rete, swim bladder rete, gastrointestinal wall, skeletal muscle and central nervous system (Armwood *et al.*, 2019). Inflammation surrounding the olfactory tract suggested ascending infection to the brain through the olfactory epithelium, similar to the route taken by *E. ictaluri* in channel

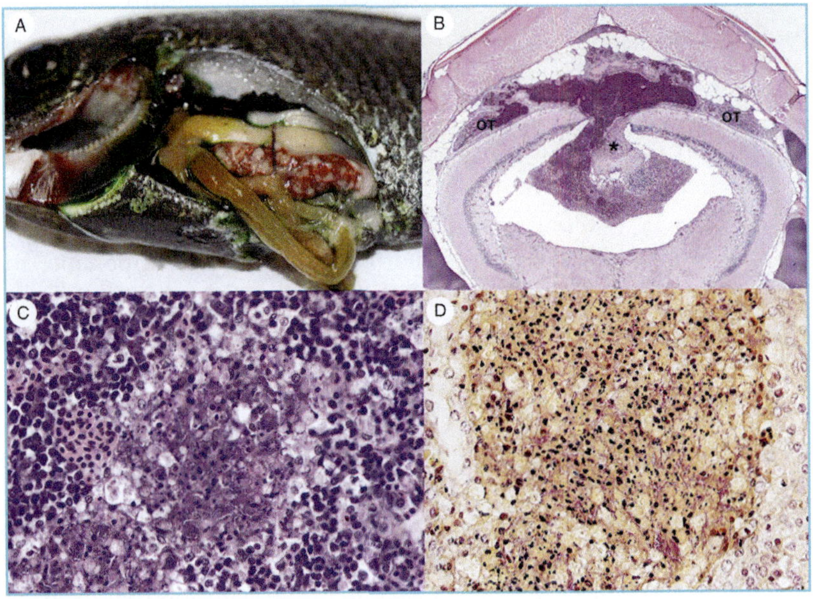

Figure 9.6 Edwardsiellosis in tilapia (*Oreochromis* spp.) fingerlings caused by *E. anguillarum*. (A), splenomegaly with multifocal white nodules corresponding to widespread granuloma formation in a laboratory-infected tilapia hybrid. (B), brain. Inflammatory infiltrates, dominated by macrophages, mixed with necrotic cellular debris and bacteria, partially fill the third ventricle, efface the torus longitudinalis (asterisk) and extend along the meningeal surface of the optic tectum (OT), partially filling the cranial cavity. (C), early splenic granuloma formation with a central area of necrosis surrounded by inflammatory cells composed largely of macrophages and lymphocytes. (D), Gram staining of splenic granuloma with large numbers of Gram-negative rods (*E. anguillarum*). Fig. 9.6B–D sourced from Armwood *et al.* (2019).

catfish (Morrison and Plumb, 1994; Armwood *et al.*, 2019). Similar changes were reported in tank-reared tilapia in South Korea (Oh *et al.*, 2020). Despite severe meningitis, tilapia lack the cranial fontanelle necessary for development of the so-called 'hole in the head' lesion described with *E. piscicida* and *E. ictaluri* infections in ictalurids (Hawke and Khoo, 2004; Griffin *et al.*, 2019). Infection by *E. ictaluri* in Nile tilapia and *E. piscicida* in largemouth bass also induced multi-organ necrosis and granulomas with intracellular bacterial rods (Soto *et al.*, 2012a; 2013; Camus *et al.*, 2019; Dong *et al.*, 2019; Nhinh *et al.*, 2023).

Control and management strategies

Antibiotics are commonly used to treat bacterial infections in aquacultural operations, and *Edwardsiella* spp. are naturally sensitive to tetracyclines, aminoglycosides, most beta-lactams, quinolones, antifolates, chloramphenicol, nitrofurantoin and fosfomycin, while less responsive to macrolides, lincosamides, streptogramins, glycopeptides, rifampin, fusidic acid and oxacillin (Stock and Wiedemann, 2001). Reichley *et al.* (2017) determined MICs for 39 antimicrobial compounds against representatives for each *Edwardsiella* sp. from a range of hosts, demonstrating a high degree of variability among isolates. While the majority were sensitive to quinolones, carbapenems and cephalosporins, there was marked variability in the response to macrolides, aminoglycosides, penicillins and tetracyclines, particularly erythromycin, amoxicillin, oxytetracycline and spectinomycin.

While many compounds have demonstrated efficacy against *Edwardsiella*, approval for the use of antimicrobials varies globally. In the United States, Romet-30® (5:1 sulfadimethoxine and ormetoprim), Terramycin® (Oxytetracycline) and Aquaflor® (Florfenicol) are the only approved antibiotics for use in food fish and require veterinary prescription. 'Off-label' use is also possible (in some cases) when a valid veterinary–client–patient relationship is in place, and when *in vitro* evidence of susceptibility is available. Concerns over microbial resistance resulting from repeated or prolonged antibiotic use have led to the use of VFD or prescriptions (Usui *et al.*, 2014; World Health Organization, 2016; Qiao *et al.*, 2018). Evidence suggests that some isolates of *E. ictaluri* and *E. piscicida* share multidrug-resistance plasmids, indicating horizontal transfer among congeners in aquaculture systems (Divya Divya, 2021). Multi-drug-resistant isolates of bona fide *E. tarda* have also been identified in *O. niloticus* in Egypt and hybrid red tilapia in Thailand (Niu *et al.*, 2019; Algammal *et al.*, 2022), and multi-drug resistance is widespread among *E. ictaluri* populations in Vietnamese tilapia culture, including that to amoxicillin, sulfamethoxazole/trimethoprim and erythromycin (Nhinh *et al.*, 2023). These reports highlight the need for more judicious antibiotic use and the development of effective alternatives to antibiotics in tilapia aquaculture (Niu *et al.*, 2019).

Considering the increase in antibiotic resistance, preventative strategies such as probiotic dietary supplements, particularly lactic acid bacteria, have garnered significant research investment. However, protection of fish from bacterial infections has produced mixed results, indicating that different *Lactobacillus* spp. provide varying levels of effectiveness against *Edwardsiella* spp. Pimentel and Katagiri (2008) observed enhanced serum killing of *E. tarda* in fish fed diets supplemented with *L. salivarius* and *L. rhamnosus*, and fish fed *L. brevis*, *L. plantarum*, *L. rhamnosus* and *L. salivarius* had lower mortality following *E. tarda* challenge than fish fed a basal diet. The greatest protection was provided by *L. rhamnosus*. Similarly, Pirarat *et al.* (2006) reported reduced mortality in *O. niloticus* fed diets supplemented with *L. rhamnosus* then challenged with *E. piscicida* isolate E381 (reported as *E. tarda*). Sherif *et al.* (2021) also saw significant improvement in the survival of *O. niloticus* fed *L. plantarum*-supplemented diets and subsequently challenged with *E. piscicida* (reported as *E. tarda*; GenBank MN871777). These promising findings demonstrate that probiotics have practical application in combatting bacterial infections in aquaculture and are worthy of further investigation.

As a means of prevention, substantial evidence indicates that a live-attenuated vaccine is efficacious against *Edwardsiella* spp. in farm-raised catfish in the United States (Wise *et al.*, 2015; Nho *et al.*, 2017; Griffin *et al.*, 2020; Wise *et al.*, 2020; Sayed *et al.*, 2021; López-Porras *et al.*, 2022). However, serological differences between isolates of *E. ictaluri* derived from catfish and other fish species, including tilapia, raise concerns that efficacious vaccines against one bacterial strain in one culture species may not be effective in others (Hawke *et al.*, 2013; Griffin *et al.*, 2016). As such, it is recommended that vaccines targeting *Edwardsiella* spp. in tilapia be derived from isolates cultured from diseased tilapia. Alternative vaccine strategies against *Edwardsiella* spp. in tilapia, including bacterial ghosts, recombinants and bacterins, have met with varying degrees of success (Kwon *et al.*, 2006; Trung Cao *et al.*, 2014). As with other culture species, the development of *Edwardsiella* vaccines in tilapia has probably been hindered by taxonomic ambiguity (Dubey *et al.*, 2019; da Costa *et al.*, 2022). As was the case in US catfish aquaculture, the adoption of vaccine strategies in tilapia appears to be impeded more by a lack of logistically straightforward delivery protocols than a scarcity of effective vaccines (Griffin *et al.*, 2017).

Motile aeromonad septicaemia and vibriosis

Aeromonas spp. are Gram-negative, catalase- and oxidase-positive, glucose fermenters, nitrate degrading and 2,4-diamino-6,7-di-iso-propylpteridine phosphate (vibriostatic factor, O/129)-resistant rods belonging to the family Aeromonadaceae, order Aeromonadales, class Gammaproteobacteria (Janda and Abbott, 2010; Hossain and Heo, 2021). The genus *Aeromonas* (from the Greek *aer* meaning air or gas and *mona* meaning units) was first proposed in 1936 to describe gas-producing bacteria; however, the first isolates may have been described in 1890 (Gonçalves Pessoa *et al.*, 2019). Over 35 *Aeromonas* spp. have been described, although *A. hydrophila*, *A. sobria*, *A. veronii*, *A. jandaei*, *A. schubertii*, *A. dhakensis* and *A. cavieae* are recognized aetiologic agents of motile aeromonad septicemia (MAS) in tilapia (Li and Cai, 2011; Dong *et al.*, 2017; Liu *et al.*, 2018; Soto-Rodriguez *et al.*, 2018).

Vibriosis in tilapia is caused by several members of the family Vibrionaceae, including *Vibrio anguillarum*, *V. parahaemolyticus*, *V. vulnificus*, *V. alginolyticus*, *V. harveyi*, *V. mimicus* and *V. cholerae*. However, there are over 143 species within the family Vibrionaceae, and more can cause disease in fish (Sanches-Fernandes *et al.*, 2022). Most vibrio are Gram-negative, halophilic, oxidase-positive, curved or straight bacterial rods, motile due to polar flagella.

Geographical distribution, host range and transmission routes

The agents of MAS and vibriosis in tilapia are ubiquitous in marine and brackish environments although some, such as *V. cholerae*, can be found in freshwater (Ina-Salwany *et al.*, 2019), particularly when temperatures are >15°C. *Aeromonas* spp. are predominantly found in fresh and brackish waters with high organic loads (Janda and Abbott, 2010), while *Vibrio* spp. are primarily found in brackish and marine environments. However, *V. vulnificus*, *V. mimicus* and *V. cholerae* have been reported in freshwater. Transmission typically occurs horizontally, and both taxa are capable of persisting in water and on contaminated equipment for long periods without the need of a fish host (De Silva and Heo, 2022; Noorian *et al.*, 2023). *Aeromonas* spp. and *Vibrio* spp. are effective biofilm formers and viable, non-culturable forms have been described in both genera (Maalej *et al.*, 2004; Wagley, 2023). Transmission is particularly effective when fish with external lesions, such as skin ulcers, are exposed to the bacteria. The oral route is also an efficient means of infection and is associated with the common cannibalistic behaviour of tilapia (Fouz *et al.*, 2002).

All tilapia species and their hybrids should be regarded as susceptible to infection by these bacteria, although environmental stress and opportunistic infection play an important role in disease development. Stressful conditions conducive to immunosuppression can include overstocking, poor water quality, extreme temperatures (particularly >30°C) and poor nutrition (Abdelsalam *et al.*, 2021; Zhao *et al.*, 2022). However, highly virulent strains acting as primary pathogens are known to exist in fish species such as the channel catfish (Baumgartner *et al.*, 2017). In addition, co-infections with other pathogenic agents including viruses (tilapia lake virus, infectious spleen and kidney necrosis virus), parasites (*Ichthyophthirius multifiliis*) and other bacteria (*F. orientalis*, *S. agalactiae* and *F. columnare*) are common (Abdelsalam *et al.*, 2021; Assane *et al.*, 2022; Puneeth *et al.*, 2022).

Equally important, some *Aeromonas* spp. and *Vibrio* spp. are capable of infecting and causing disease in other animal taxa including mammals, and human infections following the handling of contaminated tilapia are not uncommon. Zoonotic infections range from local cellulitis to potentially life-threating conditions (Ziarati *et al.*, 2022).

Diagnosis

Vibrio spp. and motile *Aeromonas* spp. are non-fastidious organisms that grow on most common laboratory media and are frequently isolated from aquatic environments. Careful interpretation is needed when providing diagnosis because both are part of the normal flora of mucosal sites in tilapia and are aggressive colonizers, particularly of open wounds and the internal organs of fresh dead fish (Sreedharan *et al.*, 2012; Elgendy *et al.*, 2016). Rapid growth of these bacteria from external tissues, including the gills and skin, or those exposed to the aquatic milieu, such as ulcerated surfaces, occurs readily and can often overgrow other bacterial or fungal pathogens, masking their presence. Selective and differential media such as thiosulfate-citrate-bile salts-sucrose (TCBS) agar or cellobiose-polymyxin B-colistin (CPC) agar are particularly useful in suspected cases of vibriosis where mixed bacterial contamination is likely (Kobayashi *et al.*, 1963; Høi *et al.*, 1998). The high pH and presence of bile salts provides selective advantages for *Vibrio* spp. growth while suppressing other Gram-negative and -positive bacteria. Similarly, semi-selective media such as Rimler Shotts agar (Shotts and Rimler, 1973) or ampicillin dextrin agar plates (Hardy Diagnostics, Santa Maria, CA) can be used for isolation of motile *Aeromonas* spp. (Hanson *et al.*, 2014).

Isolation of *Vibrio* spp. and *Aeromonas* spp. from internal organs including kidney, spleen or brain is highly suggestive of septicaemia. Overnight growth is typically appreciated in nutrient media such as tryptic soy agar and brain heart infusion with or without blood supplementation. Most isolates recovered from tilapia are motile and alpha- or beta-haemolytic. Differentiation between motile *Aeromonas* spp. and *Vibrio* spp. can be carried out easily with the use of the vibriostatic agent O/129. Further identification to genus and species level can be achieved with the use of other biochemical tests, MALDI-TOFF or molecular methods. The latter are preferred when definitive diagnosis is needed, because both genera include species with high degrees of diversity.

Clinical signs and gross lesions

Clinical manifestations of MAS and vibriosis in tilapia are similar in many respects. Outbreaks can be peracute, with high mortality and few observable changes, to chronic with ongoing low-level mortalities that accumulate over time and produce more observable lesions (Plumb and Hanson, 2011). Clinical signs and gross changes in affected fish are non-specific and include anorexia, lethargy, erratic swimming, pallor or melanosis, coelomic distention, lepidorthosis, oedema, unilateral and bilateral exophthalmia, corneal opacity, petechial to ecchymotic skin haemorrhages, haemorrhagic skin ulcers, anal prolapse and fin erosion (Fig. 9.7A, B). Internally, organomegaly, bloody ascites and adhesions between coelomic organs, as well as congestion and haemorrhages involving the coelomic organs and brain are seen (Li and Cai, 2011; Dong *et al.*, 2017; Liu *et al.*, 2018; Soto-Rodriguez *et al.*, 2018; Elgendy *et al.*, 2022).

Histology

Histological features of MAS and vibriosis in tilapia are typical of heamorrhagic septicaemias induced by multiple Gram-negative bacterial species (Fig. 9.7C, D). While generally described as haemorrhagic and necrotizing, detailed descriptions of histopathological changes are limited. The most prominent lesions associated with MAS in tilapia involve the liver, spleen and kidney, with additional changes in the brain and intestine described by some authors.

Severely congested livers with degenerative hepatocellular changes and pigment deposition, presumed hemosiderin, probably result from erythrocyte destruction and are consistent with gross observations of anaemia (Dong *et al.*, 2017; Soto-Rodriguez *et al.*, 2018; Ren *et al.*, 2019; Ren *et al.*, 2020). Liu *et al.* (2018) describe white foci of necrosis up to 2 mm affecting the livers, kidneys and spleens of Nile tilapia infected with *A. schubertii*. Microscopically, lesions contained large numbers of short bacterial rods surrounded by necrotic tissue and inflammatory cell infiltrates. Changes induced by *A. schubertii* in the spleen include increased prominence of ellipsoids, perivascular oedema, depletion of haematopoietic cells and

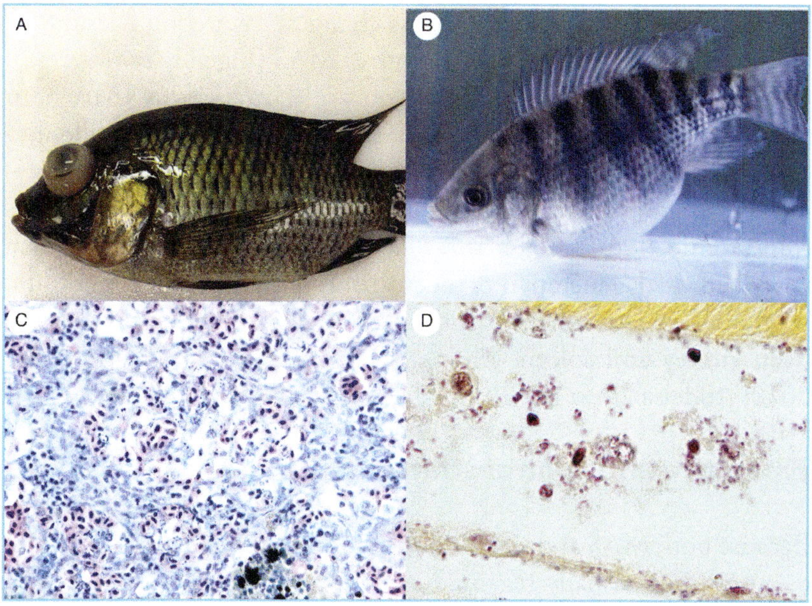

Figure 9.7 Systemic vibriosis and aeromoniasis in naturally infected tilapia (*Oreochromis* spp.) (A), severe exophthalmia and periocular oedema in a tilapia hybrid infected by *Vibrio vulnificus*. (B), severe ascites resulting in coelomic distension due to *Aeromonas* spp. infection. (C), Giemsa-stained section of head kidney with *V. vulnificus* bacterial rods free within vascular spaces. (D), Brown and Hopps Gram-stained histological section demonstrating Gram-negative *V. vulnificus* bacteria free within the periocular space.

thromboses (Ren *et al.*, 2019, 2020). In a report of hybrid tilapia experimentally infected with *A. dhakensis*, necrosis in the same organs with lymphocytic inflammation was accompanied by intravascular bacterial rods and widespread pyknotic and karyorrhectic erythrocytes. Intraperitoneal injection of an extracellular product (ECP) culture filtrate reproduced the lesions and caused mortalities, indicating a role for liberated bacterial toxins as virulence factors in the disease pathogenesis (Soto-Rodriguez *et al.*, 2018). Additional lesions associated with *Aeromonas* spp. infections include haemorrhages in the renal interstitium and spleen, congestion of capillaries in the renal glomerula and brain and intestinal mucosal degeneration (Dong *et al.*, 2017; Liu *et al.*, 2018; Ren *et al.*, 2020).

As with clinical signs and gross changes, microscopic lesions of vibriosis in tilapia are similar to those of MAS (Elgendy *et al.*, 2022). Co-infections between *Aeromonas* spp. and *V. alginolyticus* have been reported, highlighting the need for appropriate methods to identify bacteria, determine antibiotic sensitivities and investigate underlying environmental stressors conducive to outbreaks of mixed opportunistic infection (Abdelsalam *et al.*, 2021). Microscopic lesions are described as degenerative or necrotizing with variable congestion, haemorrhage and inflammatory infiltration primarily affecting the liver, kidney and spleen (El-Sharaby *et al.*, 2017; Sumithra *et al.*, 2019; Abdelsalam *et al.*, 2021; Elgendy *et al.*, 2022)

Protective and control strategies

Because both MAS and vibriosis are typically associated with stressful events and are usually regarded as opportunistic diseases in tilapia, ensuring good husbandry practices is the first, and probably the most important, preventative action against them. The association of poor water quality, co-infections and suboptimal management is frequently associated with the diseases and should be corrected prior to initiation of any other prophylactic or therapeutic practice. Vaccines against MAS and vibriosis have been developed and are available in most countries where tilapia are cultured. Comprehensive reviews of vaccines against both are available in the literature (Ina-Salwany *et al.*, 2019; Ji *et al.*, 2020; Shirajum Monir *et al.*, 2020). In general, due to the great diversity that exists within the respective genera, most vaccines are serotype and biotype specific (Izabela *et al.* 2015). Autogenous vaccines are often effective, while immunization against heterologous isolates typically fails. Therefore, complete characterization of the suspected offending isolate is needed for effective prophylaxis. Inactivated whole-cell killed (WCK) vaccines are the most common in use against MAS and vibriosis in tilapia. WCK vaccines have been reported to provide some degree of protection against MAS (Ruangpan *et al.*, 1986; Aly *et al.*, 2015), as well as *V. vulnificus* biotype I (Shoemaker *et al.*, 2011), *V. anguillarum* (Gallage *et al.*, 2016) and *V. harveyi* (Abu Nor *et al.*, 2020). Injectable vaccines are generally more effective (Shirajum Monir *et al.*, 2020), and most WCK vaccines are administered by intracoelomic injection. However, oral and immersion routes are attractive for cultured tilapia due to their ease of administration and reduced stress on fish.

Although less available, other types of vaccines including recombinant, DNA and live-attenuated formulations have been developed against MAS and vibriosis (Anuradha *et al.*, 2010; Pridgeon and Klesius, 2011; Pridgeon *et al.*, 2013; Fernandez *et al.*, 2014; de Santana Lacerda *et al.*, 2015). However, concerns over reversion of the vaccine strain to virulence and the use of genetically modified organisms in aquaculture have limited the acceptance of these formulations.

Future perspectives

Bacterial diseases remain a pervasive problem in tilapia aquaculture worldwide. The warm temperatures and intensive rearing conditions typically employed in the sector present optimal conditions for bacterial growth and infection, increasing outbreak prevalence and severity. Disease risk for common tilapiine bacterial pathogens can be mitigated by the same management interventions: decreased stocking densities, appropriate feed rates, water quality monitoring, strict biosecurity practices and prompt removal of mortalities from culture systems. In addition, as our understanding of the taxonomy, transmission and pathogenesis of emerging and established diseases improves, so too have targeted methods to combat them. Protective measures through selective breeding programmes, nutritional regimes and commercial or autogenous vaccines may be regionally available for the various aetiologic agents discussed. Cross-protective mucosal vaccines, however, are lacking, and are a significant area of ongoing research. Stringent antimicrobial use may also help limit animal losses, although antimicrobial resistance is a pressing concern. As such, development of safe, practical and widely effective strategies for disease management continues to be a priority in supporting sustainable development of the tilapia culture industry.

References

Abayneh, T., Colquhoun, D.J. and Sørum, H. (2013) *Edwardsiella piscicida* sp. nov., a novel species pathogenic to fish. *Journal of Applied Microbiology* 114(3), 644–654. DOI: 10.1111/jam.12080.

Abdelsalam, M., Asheg, A. and Eissa, A.E. (2013) *Streptococcus dysgalactiae*: An emerging pathogen of fishes and mammals. *International Journal of Veterinary Science and Medicine* 1(1), 1–6. DOI: 10.1016/j.ijvsm.2013.04.002.

Abdelsalam, M., Ewiss, M.A.Z., Khalefa, H.S., Mahmoud, M.A., Elgendy, M.Y. *et al.* (2021) Coinfections of *Aeromonas* spp., *Enterococcus faecalis*, and *Vibrio alginolyticus* isolated from farmed Nile tilapia and African catfish in Egypt, with an emphasis on poor water quality. *Microbial Pathogenesis* 160, 105213. DOI: 10.1016/j.micpath.2021.105213.

Abu-Elala, N.M., Samir, A., Wasfy, M. and Elsayed, M. (2019) Efficacy of injectable and immersion polyvalent vaccine against streptococcal infections in broodstock and offspring of Nile tilapia (*Oreochromis niloticus*). *Fish and Shellfish Immunology* 88, 293–300. DOI: 10.1016/j.fsi.2019.02.042.

Abu-Elala, N.M., Abd-Elsalam, R.M. and Younis, N.A. (2020) Streptococcosis, Lactococcosis and Enterococcosis are potential threats facing cultured Nile tilapia (*Streptococcus agalactiae*) production. *Aquaculture Research* 51(10), 4183–4195. DOI: 10.1111/are.14760.

Abu Nor, N., Zamri-Saad, M., Md Yasin, I.S., Salleh, A., Mustaffa-Kamal, F. *et al.* (2020) Efficacy of whole cell inactivated *Vibrio harveyi* vaccine against vibriosis in a marine red hybrid tilapia (*Oreochromis niloticus* × *O. mossambicus)* model. *Vaccines* 8(4), 734. DOI: 10.3390/vaccines8040734.

Abuseliana, A.F., Mohd Daud, H.H., Abdul Aziz, S., Bejo, S.K. and Alsaid, M. (2011) Pathogenicity of *Streptococcus agalactiae* isolated from a fish farm in Selangor to juvenile red tilapia (*Oreochromis* sp.). *Journal of Animal and Veterinary Advances* 10(7), 914–919. DOI: 10.3923/javaa.2011.914.919.

Adeolu, M., Alnajar, S., Naushad, S. and Gupta, R.S. (2016) Genome-based phylogeny and taxonomy of the "Enterobacteriales": proposal for enterobacteriales order nov. divided into the families *Enterobacteriaceae*, *Erwiniaceae* fam. nov., *Pectobacteriaceae* fam. nov., *Yersiniaceae* fam. nov., *Hafniaceae* fam. nov., *Morganellaceae* fam. nov., and *Budviciaceae* fam. nov. *International.*

Agnew, W. and Barnes, A.C. (2007) *Streptococcus iniae*: An aquatic pathogen of global veterinary significance and a challenging candidate for reliable vaccination. *Veterinary Microbiology* 122(1–2), 1–15. DOI: 10.1016/j.vetmic.2007.03.002.

Akoğlu, A. (2020) The effect of some environmental conditions on planktonic growth and biofilm formation by some lactic acid bacteria isolated from a local cheese in Turkey. *Biotechnology Letters* 42(3), 481–492. DOI: 10.1007/s10529-020-02794-4.

Akter, T., Foysal, M.J., Alam, M., Ehsan, R., Paul, S.I. *et al.* (2021) Involvement of *Enterococcus* species in streptococcosis of Nile tilapia in Bangladesh. *Aquaculture* 531, 735790. DOI: 10.1016/j.aquaculture.2020.735790.

Alcaide, E., Herraiz, S. and Esteve, C. (2006) Occurrence of *Edwardsiella tarda* in wild European eels *Anguilla anguilla* from Mediterranean Spain. *Diseases of Aquatic Organisms* 73(1), 77–81. DOI: 10.3354/dao073077.

Algammal, A.M., Abo Hashem, M.E., Alfifi, K.J., Al-Otaibi, A.S., Alatawy, M. *et al.* (2022) Sequence analysis, antibiogram profile, virulence and antibiotic resistance genes of XDR and MDR *Gallibacterium anatis* isolated from layer chickens in Egypt. *Infection and Drug Resistance* 15, 4321–4334. DOI: 10.2147/IDR.S377797.

Aly, S.M., Albutti, A.S., Rahmani, A.H., Nashwa, M. and Atti, N.M.A. (2015) The response of new-season Nile tilapia to *Aeromonas hydrophila* vaccine. *International Journal of Clinical and Experimental Medicine* 8(3), 4508–4514.

Amal, M.N. and Zamri-Saad, M. (2011) Streptococcosis in tilapia (*Oreochromis niloticus*): A review. *Pertanika Journal of Tropical Agricultural Science* 34, 195–206.

Amin, N.E., Abdallah, I.S., Faisal, M., Easa, M.E.S., Alaway, T. *et al.* (1988) Columnaris infection among cultured Nile tilapia *Oreochromis niloticus*. *Antonie van Leeuwenhoek* 54(6), 509–520. DOI: 10.1007/BF00588387.

Anshary, H., Kurniawan, R.A., Sriwulan, S., Ramli, R. and Baxa, D.V. (2014) Isolation and molecular identification of the etiological agents of streptococcosis in Nile tilapia (*Oreochromis niloticus*) cultured in net cages in Lake sentani, Papua, Indonesia. *SpringerPlus* 3, 627. DOI: 10.1186/2193-1801-3-627.

Anuradha, K., Foo, H.L., Mariana, N.S., Loh, T.C., Yusoff, K.M.D. *et al.* (2010) Live recombinant *Lactococcus lactis* vaccine expressing aerolysin genes D1 and D4 for protection against *Aeromonas hydrophila* in tilapia (*Oreochromis niloticus*). *Journal of Applied Microbiology* 109(5), 1632–1642. DOI: 10.1111/j.1365-2672.2010.04789.x.

Arias, C.R., LaFrentz, S., Cai, W. and Olivares-Fuster, O. (2012) Adaptive response to starvation in the fish pathogen *Flavobacterium columnare*: Cell viability and ultrastructural changes. *BMC Microbiology* 12, 266–266. DOI: 10.1186/1471-2180-12-266.

Armwood, A.R., Camus, A.C., López-Porras, A., Ware, C., Griffin, M.J. *et al.* (2019) Pathologic changes in cultured Nile tilapia (*Oreochromis niloticus*) associated with an outbreak of *Edwardsiella anguillarum*. *Journal of Fish Diseases* 42(10), 1463–1469. DOI: 10.1111/jfd.13058.

Assane, I.M., Prada-Mejia, K.D., Gallani, S.U., Weiser, N.F., Valladão, G.M.R. *et al.* (2022) *Enterogyrus* spp. (Monogenea: Ancyrocephalinae) and *Aeromonas jandaei* co-infection associated with high mortality following transport stress in cultured Nile tilapia. Epub September 7, 2021. PubMed: 34406699, *Transboundary and Emerging Diseases* 69(4), e276–e287. DOI: 10.1111/tbed.14295.

Assis, G.B., Tavares, G.C., Pereira, F.L., Figueiredo, H.C. and Leal, C.A. (2017) Natural coinfection by *Streptococcus agalactiae* and *Francisella noatunensis* subsp. *orientalis* in farmed Nile tilapia (*Oreochromis niloticus* L.). *Journal of Fish Diseases* 40(1), 51–63. DOI: 10.1111/jfd.12493.

Avci, H., Birincioglu, S.S., Tanrikul, T.T., Epikmen, E.T., Metin, N. *et al.* (2014) Experimental *Lactococcus garvieae* infection in rainbow trout, *Oncorhynchus mykiss,* Walbaum 1792: A comparative histopathological and immunohistochemical study. *Journal of Fish Diseases* 37(5), 481–495. DOI: 10.1111/jfd.12132.

Baba, E., Acar, Ü., Öntaş, C., Kesbiç, O.S. and Yılmaz, S. (2016) Evaluation of citrus limon peels essential oil on growth performance, immune response of Mozambique tilapia *Oreochromis mossambicus* challenged with *Edwardsiella tarda*. *Aquaculture* 465, 13–18.

Baker, C.N., Hollis, D.G. and Thornsberry, C. (1985) Antimicrobial susceptibility testing of *Francisella tularensis* with a modified mueller-hinton broth. *Journal of Clinical Microbiology* 22(2), 212–215. DOI: 10.1128/jcm.22.2.212-215.1985.

Baumgartner, W.A., Ford, L. and Hanson, L. (2017) Lesions caused by virulent *Aeromonas hydrophila* in farmed catfish (*Ictalurus punctatus* and *I. punctatus* × *I. furcatus*) in Mississippi. *Journal of Veterinary Diagnosis Investigation* 29(5), 747–751. DOI: 10.1177/1040638717708584.

Baums, C.G., Hermeyer, K., Leimbach, S., Adamek, M., Czerny, C.P. *et al.* (2013) Establishment of a model of *Streptococcus iniae* meningoencephalitis in Nile tilapia (*Oreochromis niloticus*). *Journal of Comparative Pathology* 149(1), 94–102. DOI: 10.1016/j.jcpa.2012.10.003.

Baxa, D.V., Groff, J.M., Wishkovsky, A. and Hedrick, R.P. (1990) Susceptibility of non-ictalurid fishes to experimental infection with *Edwardsiella ictaluri*. *Diseases of Aquatic Organisms* 8, 113–117. DOI: 10.3354/dao008113.

Bernardet, J.-F. and Grimont, P.A.D. (1989) Deoxyribonucleic acid relatedness and phenotypic characterization of *Flexibacter columnaris* sp. nov., nom. rev., *Flexibacter psychrophilus* sp. nov., nom. rev., and *Flexibacter maritimus* Wakabayashi, Hikida, and Masumura 1986. *International Journal of Systematic Bacteriology* 39(3), 346–354. DOI: 10.1099/00207713-39-3-346.

Bernardet, J.-F., Segers, P., Vancanneyt, M., Kersters, K. and Vandamme, P. (1996) Cutting a gordian knot: emended classification and description of the genus *Flavobacterium*, emended description of the family *Flavobacteriaceae*, and proposal of *Flavobacterium hydatis* nom. nov. (basonym, *Cytophaga aquatilis* Strohl and Tait 1978). *International Journal of Systematic Bacteriology* 46(1), 128–148. DOI: 10.1099/00207713-46-1-128.

Bernardet, J.-F. and Nakagawa, Y. (2006) An introduction to the family flavobacteriaceae. In: Dworkin, M., Falkow, S., Rosenberg, E., Schleifer, K. and Stackebrandt, E. (eds) *The Prokaryotes: A Handbook on the Biology of Bacteria*. Springer-Verlag, pp. 455–480.

Birkbeck, T.H., Bordevik, M., Frøystad, M.K. and Baklien, A. (2007) Identification of francisella sp. from Atlantic salmon, *Salmo salar* L., in Chile. *Journal of Fish Diseases* 30(8), 505–507. DOI: 10.1111/j.1365-2761.2007.00837.x.

Birkbeck, T.H., Feist, S.W. and Verner-Jeffreys, D.W. (2011) *Francisella* infections in fish and shellfish. *Journal of Fish Diseases* 34(3), 173–187. DOI: 10.1111/j.1365-2761.2010.01226.x.

Borg, A.F. (1948) *Studies on myxobacteria associated with diseases in salmonid fishes*. In University of Washington.

Brevik, O.J., Ottem, K.F., Kamaishi, T., Watanabe, K. and Nylund, A. (2011) *Francisella halioticida* sp. nov., a pathogen of farmed giant abalone (*Haliotis gigantea*) in Japan. *Journal of Applied Microbiology* 111(5), 1044–1056. DOI: 10.1111/j.1365-2672.2011.05133.x.

Bromage, E.S. and Owens, L. (2002) Infection of barramundi *Lates calcarifer* with *Streptococcus iniae*: Effects of different routes of exposure.

Diseases of Aquatic Organisms 52(3), 199–205. DOI: 10.3354/dao052199.

Bromage, E.S., Thomas, A. and Owens, L. (1999) *Streptococcus iniae*, a bacterial infection in barramundi *Lates calcarifer*. *Diseases of Aquatic Organisms* 36(3), 177–181. DOI: 10.3354/dao036177.

Brudal, E., Ulanova, L.S., O Lampe, E., Rishovd, A.L., Griffiths, G. *et al.* (2014) Establishment of three *Francisella* infections in zebrafish embryos at different temperatures. *Infection and Immunity* 82(6), 2180–2194. DOI: 10.1128/IAI.00077-14.

Bullard, S.A., McElwain, A. and Arias, C.R. (2011) Scanning electron microscopy of "saddleback" lesions associated with experimental infections of *Flavobacterium columnare* in channel catfish, *Ictalurus punctatus* (Siluriformes: Ictaluridae), and zebrafish, *Danio rerio* (Cypriniformes: Cyprinidae). *Journal of the World Aquaculture Society* 42(6), 906–913. DOI: 10.1111/j.1749-7345.2011.00527.x.

Bullard, S.A., Mohammed, H. and Arias, C.R. (2013) First record of the fish pathogen *Flavobacterium columnare genomovar* II from bluegill, *Lepomis macrochirus* (Rafinesque), with observations on associated lesions. *Journal of Fish Diseases* 36(4), 447–451. DOI: 10.1111/jfd.12005.

Bwalya, P., Simukoko, C., Hang'ombe, B.M., Støre, S.C., Støre, P. *et al.* (2020a) Characterization of *Streptococcus*-like bacteria from diseased *Oreochromis niloticus* farmed on Lake Kariba in Zambia. *Aquaculture* 523, 735185. DOI: 10.1016/j.aquaculture.2020.735185.

Bwalya, P., Hang'ombe, B.M., Gamil, A.A., Munang'andu, H.M., Evensen, Ø. *et al.* (2020b) A whole-cell *Lactococcus garvieae* autovaccine protects Nile tilapia against infection. *PLOS ONE* 15(3), e0230739. DOI: 10.1371/journal.pone.0230739.

Cai, S.H., Wang, B., Lu, Y.S., Jian, J.C. and Wu, Z.H. (2012) Development of loop-mediated isothermal amplification method for rapid detection of *Streptococcus iniae*, the causative agent of streptococcosis in fish. *Journal of Basic Microbiology* 52(2), 116–122. DOI: 10.1002/jobm.201100082.

Cai, W., De La Fuente, L. and Arias, C.R. (2013) Biofilm formation by the fish pathogen *Flavobacterium columnare*: Development and parameters affecting surface attachment. *Applied and Environmental Microbiology* 79(18), 5633–5642. DOI: 10.1128/AEM.01192-13.

Camus, A., Griffin, M., Armwood, A. and Soto, E. (2019) A spontaneous outbreak of systemic *Edwardsiella piscicida* infection in largemouth bass, *Micropterus salmoides* in California, USA. *Journal of Fish Diseases* 42(5), 759–763. DOI: 10.1111/jfd.12961.

Camus, A.C., Dill, J.A., McDermott, A.J., Clauss, T.M., Berliner, A.L. *et al.* (2013) *Francisella noatunensis* subsp. *orientalis* infection in indo-pacific reef fish entering the United States through the ornamental fish trade. *Journal of Fish Diseases* 36(7), 681–684. DOI: 10.1111/jfd.12071.

Catão Egger, R.C., Rosa, J.C.C., Resende, L.F.L., Pádua, S.B., Oliveira Barbosa, F. *et al.* (2023) Emerging fish pathogens *Lactococcus petauri* and *L. garvieae* in Nile tilapia (*Oreochromis niloticus*) farmed in Brazil. *Aquaculture* 565, 739093. DOI: 10.1016/j.aquaculture.2022.739093.

Chang, C.H., Poudyal, S., Pulpipat, T., Wang, P.C. and Chen, S.C. (2021) Pathological manifestations of *Francisella orientalis* in the green Texas cichlid (*Herichthys cyanoguttatus*). *Animals: An Open Access Journal from MDPI* 11(8), 2284. DOI: 10.3390/ani11082284.

Chang, P.H. and Plumb, J.A. (1996) Histopathology of experimental *Streptococcus* sp. infection in tilapia, *Oreochromis niloticus* (L.), and channel catfish, *Ictalurus punctatus* (Ratinesque). *Journal of Fish Diseases* 19(3), 235–241. DOI: 10.1111/j.1365-2761.1996.tb00130.x.

Chen, C.Y., Chao, C.-B. and Bowser, P. (2007) Comparative histopathology of *Streptococcus iniae* and *Streptococcus agalactiae*-infected tilapia. *Bulletin of the European Association of Fish Pathologists* 27.

Chen, M., Li, L.P., Wang, R., Liang, W.W., Huang, Y. *et al.* (2012) PCR detection and PFGE genotype analyses of streptococcal clinical isolates from tilapia in China. *Veterinary Microbiology* 159(3–4), 526–530. DOI: 10.1016/j.vetmic.2012.04.035.

Chen, W.M., Huang, W.C., Young, C.C. and Sheu, S.Y. (2013) *Flavobacterium tilapiae* sp. nov., isolated from a freshwater pond, and emended descriptions of *Flavobacterium defluvii* and *Flavobacterium johnsoniae*. *International Journal of Systematic and Evolutionary Microbiology* 63(3), 827–834. DOI: 10.1099/ijs.0.041178-0.

Chideroli, R.T., Amoroso, N., Mainardi, R.M., Suphoronski, S.A., Padua, S.B. *et al.* (2017) Emergence of a new multidrug-resistant and highly virulent serotype of *Streptococcus agalactiae* in fish farms from Brazil. *Aquaculture* 479, 45–51. DOI: 10.1016/j.aquaculture.2017.05.013.

Clavijo, A., Conroy, G., Conroy, D., Santander, J. and Aponte, F.E. (2002) First report of *Edwardsiella tarda* from tilapias in Venezuela. *Bulletin of the European Association of Fish Pathologists* 22, 280–282.

Colquhoun, D.J. and Duodu, S. (2011) *Francisella* infections in farmed and wild aquatic organisms. *Veterinary Research* 42(1), 47. DOI: 10.1186/1297-9716-42-47.

Costa, F.A., Leal, C.A., Leite, R.C. and Figueiredo, H.C. (2014) Genotyping of *Streptococcus dysgalactiae* strains isolated from Nile tilapia, *Oreochromis niloticus* (L.). *Journal of Fish Diseases* 37(5), 463–469. DOI: 10.1111/jfd.12125.

Crumlish, M., Dung, T.T., Turnbull, J.F., Ngoc, N.T.N. and Ferguson, H.W. (2002) Identification of *Edwardsiella ictaluri* from diseased freshwater catfish, *Pangasius hypophthalmus* (sauvage), cultured in the Mekong Delta. Vietnam. *Journal of Fish Diseases* 25(12), 733–736. DOI: 10.1046/j.1365-2761.2002.00412.x.

da Costa, A.R., Chideroli, R.T., Lanes, G.C., Ferrari, N.A., Chicoski, L.M. *et al.* (2022) Multiplex PCR assay for correct identification of the fish pathogenic species of *Edwardsiella* genus reveals the presence of *E. anguillarum* in South America in strains previously characterized as *E. tarda*. *Journal of Applied Microbiology* 132(6), 4225–4235. DOI: 10.1111/jam.15538.

Davies, Y.M., Xavier, M.G., Cunha, M.P.V., Franco, L.S., Santos, S.L.P. *et al.* (2018) *Edwardsiella tarda* outbreak affecting fishes and aquatic birds in Brazil. *Veterinary Quarterly* 95, 105.

Davis, H.S. (1922) A new bacterial disease of freshwater fishes. *Bulletin of the United States Bureau of Fisheries* 38, 261–280.

de Alexandre Sebastião, F., LaFrentz, B.R., Shelley, J.P., Stevens, B., Marancik, D. *et al.* (2019) *Flavobacterium inkyongense* isolated from ornamental cichlids. *Journal of Fish Diseases* 42(9), 1309–1313. DOI: 10.1111/jfd.13043.

Declercq, A.M., Haesebrouck, F., Den Broeck, W., Bossier, P. and Decostere, A. (2013) Columnaris disease in fish: A review with emphasis on bacterium-host interactions. *Veterinary Research* 44(1), 27. DOI: 10.1186/1297-9716-44-27.

de Santana Lacerda, I.P., Goncalves, Y.S., Oliveira, S.T.L., Demarqui, F.N., Krewer, C. da C. *et al.* (2015) Efficacy of *Aeromonas hydrophila* S-layer bacterins with different protein profiles as a vaccine in Nile tilapia (*Oreochromis niloticus*). *African Journal of Microbiology Research* 9(29), 1770–1777.

De Silva, L.A.D.S. and Heo, G.J. (2022, December 24) Biofilm formation of pathogenic bacteria isolated from aquatic animals. *Archives of Microbiology* 205(1), 36. DOI: 10.1007/s00203-022-03332-8.

Divya, D. (2021) Phenotypic and genotypic characterization and comparison of *Edwardsiella ictaluri* isolates derived from catfish and ornamental fish species. MSc Thesis, Mississippi State University.

Dong, H.T., Gangnonngiw, W., Phiwsaiya, K., Charoensapsri, W., Nguyen, V.V. *et al.* (2016) Duplex PCR assay and in situ hybridization for detection of *Francisella* spp. and *Francisella noatunensis* subsp. orientalis in red tilapia. *Diseases of Aquatic Organisms* 120(1), 39–47. DOI: 10.3354/dao03021.

Dong, H.T., Techatanakitarnan, C., Jindakittikul, P., Thaiprayoon, A., Taengphu, S. *et al.* (2017) *Aeromonas jandae* and *Aeromonas veronii* caused disease and mortality in Nile tilapia, *Oreochromis niloticus* (L.). *Journal of Fish Diseases* 40(10), 1395–1403. DOI: 10.1111/jfd.12617.

Dong, H.T., Senapin, S., Jeamkunakorn, C., Nguyen, V.V., Nguyen, N.T. *et al.* (2019) Natural occurrence of Edwardsiellosis caused by *Edwardsiella ictaluri* in farmed hybrid red tilapia (*Oreochromis* sp.) in Southeast Asia. *Aquaculture* 499, 17–23. DOI: 10.1016/j.aquaculture.2018.09.007.

Dorofe'ev, K.A. (1947) Classification of the causative agent of tularemia. *Symp. Res. Works Inst. Epidemiol. Mikrobiol. (Chita) (Russ.), 1* 170–180.

Dubey, S., Maiti, B., Kim, S.H., Sivadasan, S.M., Kannimuthu, D. *et al.* (2019) Genotypic and phenotypic characterization of *Edwardsiella* isolates from different fish species and geographical areas in Asia: Implications for vaccine development. *Journal of Fish Diseases* 42(6), 835–850. DOI: 10.1111/jfd.12984.

Duodu, S., Larsson, P., Sjödin, A., Soto, E., Forsman, M. *et al.* (2012) Real-time PCR assays targeting unique DNA sequences of fish-pathogenic *Francisella noatunensis* subspecies *noatunensis* and *orientalis. Diseases of Aquatic Organisms* 101(3), 225–234. DOI: 10.3354/dao02514.

Eaves, P., Ketterer, P.J., Anderson, I.G., Beumer, J.P. and Beumer, J.P. (1990) The isolation and identification of *Edwardsiella tarda* from a diseased native Australian eel (*Anguillare inhardtii*). *Australian Veterinary Journal* 67(9), 336–337. DOI: 10.1111/j.1751-0813.1990.tb07819.x.

Eldar, A., Bejerano, Y., Livoff, A., Horovitcz, A. and Bercovier, H. (1995) Experimental streptococcal meningo-encephalitis in cultured fish. *Veterinary Microbiology* 43(1), 33–40. DOI: 10.1016/0378-1135(94)00052-x.

Elgendy, M.Y., Kenawy, A.M. and Noor El-Deen, A.E. (2016) *Gyrodactylus anguillae* and *Vibrio vulnificus* infections affecting cultured eel, *Anguilla anguilla. Comunicata Scientiae* 7(1), 1–11. DOI: 10.14295/cs.v7i1.1248.

Elgendy, M.Y., Abdelsalam, M., Kenawy, A.M. and Ali, S.E. (2022) Vibriosis outbreaks in farmed Nile tilapia (*Oreochromis niloticus*) caused by *Vibrio mimicus* and *V. cholerae*. *Aquaculture International* 30(5), 2661–2677. DOI: 10.1007/s10499-022-00921-8.

El-Sharaby, S.M.A., Abd-Elgaber, M., Tarabees, R., Khalil, R.H., Ali, M.N. *et al.* (2017) Bacteriological and histopathological studies on vibrio species isolated from naturally infected freshwater fish in delta region, Egypt. *Advances in Animal Sciences* 6, 17–26.

Evans, J.J., Pasnik, D.J., Klesius, P.H. and Al-Ablani, S. (2006) First report of *Streptococcus agalactiae* and *Lactococcus garvieae* from a wild bottlenose dolphin (*Tursiops truncatus*). *Journal of Wildlife Diseases* 42(3), 561–569. DOI: 10.7589/0090-3558-42.3.561.

Evans, J.J., Klesius, P.H. and Shoemaker, C.A. (2009) First isolation and characterization of *Lactococcus garvie*ae from brazilian Nile tilapia, *Oreochromis niloticus* (L.), and pintado, *Pseudoplathystoma corruscans* (Spix and Agassiz). *Journal of Fish Diseases* 32(11), 943–951. DOI: 10.1111/j.1365-2761.2009.01075.x.

Ewing, W.H., McWhorter, A.C., Escobar, M.R. and Lubin, A.H. (1965) *Edwardsiella*, a new genus of Enterobacteriaceae based on a new species *E. Tarda*. *International Bulletin of Bacteriological Nomenclature and Taxonomy* 15(1), 33–38. DOI: 10.1099/00207713-15-1-33.

Ezzeldeen, N.A., Alshehri, A.N.Z., Abdelmoneum, M.A., Khaled, F.M., Eissa, A.E. *et al.* (2015) Phenotypic and genotypic characterization of *Flavobacterium* spp. isolated from wild and cultured fish. *Advances in Environmental Biology* 9, 160–168.

Faisal, M., Samaha, H. and Loch, T.P. (2017) Planning a fish health program. In: Jeney, G. (ed.) *Fish Diseases: Prevention and Control Strategies*. Academic Press, pp. 221–248.

Fernandez, J.B., Yambot, A.V. and Almeria, O. (2014) Vaccination of Nile tilapia (*Oreochromis niloticus*) using lipopolysaccharide (LPS) prepared from *Aeromonas hydrophila*. *International Journal of Fauna and Biological Studies* 1, 1–3.

Ferrario, C., Ricci, G., Milani, C., Lugli, G.A., Ventura, M. *et al.* (2013) *Lactococcus garvieae*: Where is it from? A first approach to explore the evolutionary history of this emerging pathogen. *PLOS ONE* 8(12), e84796. DOI: 10.1371/journal.pone.0084796.

Flemming, L., Rawlings, D. and Chenia, H. (2007) Phenotypic and molecular characterisation of fish-borne *Flavobacterium johnsoniae*-like isolates from aquaculture systems in South Africa. *Research in Microbiology* 158(1), 18–30. DOI: 10.1016/j.resmic.2006.09.001.

Forsman, M., Sandström, G. and Sjöstedt, A. (1994) Analysis of 16S ribosomal DNA sequences of *Francisella* strains and utilization for determination of the phylogeny of the genus and for identification of strains by PCR. *International Journal of Systematic Bacteriology* 44(1), 38–46. DOI: 10.1099/00207713-44-1-38.

Fouz, B., Alcaide, E., Barrera, R. and Amaro, C. (2002) Susceptibility of Nile tilapia (*Oreochromis niloticus*) to *Vibriosis* due to Vibrio vulnificus biotype 2 (serovar E). *Aquaculture* 212(1–4), 21–30. DOI: 10.1016/S0044-8486(02)00002-9.

Foysal, M.J., Alam, M., Momtaz, F., Chaklader, M.R., Siddik, M.A.B. *et al.* (2019) Dietary supplementation of garlic (*Allium sativum*) modulates gut microbiota and health status of tilapia (*Oreochromis niloticus*) against *Streptococcus iniae* infection. *Aquaculture Research* 50(8), 2107–2116. DOI: 10.1111/are.14088.

Fujihara, M.P. and Nakatani, R.E. (1971) Antibody production and immune responses of rainbow trout and coho salmon to *Chondrococcus columnaris*. *Journal of the Fisheries Research Board of Canada* 28(9), 1253–1258. DOI: 10.1139/f71-191.

Gallage, S., Katagiri, T., Endo, M., Futami, K., Endo, M. *et al.* (2016) Influence of moderate hypoxia on vaccine efficacy against *Vibrio anguillarum* in *Oreochromis niloticus* (Nile tilapia). *Fish and Shellfish Immunology* 51, 271–281. DOI: 10.1016/j.fsi.2016.02.024.

García-López, M., Meier-Kolthoff, J.P., Tindall, B.J., Gronow, S., Woyke, T. *et al.* (2019) Analysis of 1,000 type-strain genomes improves taxonomic classification of bacteroidetes. *Frontiers in Microbiology* 10, 2083. DOI: 10.3389/fmicb.2019.02083.

Garnjobst, L. (1945) *Cytophaga columnaris* (Davis) in pure culture: A myxobacterium pathogenic to fish. *Journal of Bacteriology* 49(2), 113–128. DOI: 10.1128/jb.49.2.113-128.1945.

Ghittino, C., Latini, M., Agnetti, F., Panzieri, C., Lauro, L. *et al.* (2003) Emerging pathologies in aquaculture: Effects on production and food safety. *Veterinary Research Communications* 27(1 (Suppl 1)), 471–479. DOI: 10.1023/B:VERC.0000014204.37722.b6.

Gonçalves Pessoa, R.B.de OliveiraW.FMarques, D.S.C., Dos Santos Correia, M.T., Carvalho, E.V.M.M. *et al.* (2019) The genus *Aeromonas*: A general approach. *Microbiology and Pathogology* 130, 81–94. DOI: 10.1016/j.micpath.2019.02.036.

Grabowski, L.D., LaPatra, S.E. and Cain, K.D. (2004) Systemic and mucosal antibody response in tilapia, *Oreochromis niloticus* (L.), following immunization with *Flavobacterium columnare. Journal of Fish Diseases* 27(10), 573–581. DOI: 10.1111/j.1365-2761.2004.00576.x.

Griffin, M.J., Ware, C., Quiniou, S.M., Steadman, J.M., Gaunt, P.S. *et al.* (2014) *Edwardsiella piscicida* identified in the Southeastern USA by *gyrb* sequence, species-specific and repetitive sequence-mediated PCR. *Diseases of Aquatic Organisms* 108(1), 23–35. DOI: 10.3354/dao02687.

Griffin, M.J., Reichley, S.R., Greenway, T.E., Quiniou, S.M., Ware, C. *et al.* (2016) Comparison of *Edwardsiella ictaluri* isolates from different hosts and geographic origins. *Journal of Fish Diseases* 39(8), 947–969. DOI: 10.1111/jfd.12431.

Griffin, M.J., Greenway, T.E. and Wise, D.J. (2017) *Edwardsiella* spp. In: Woo, P. and Cipriano, R. (eds) *Fish Viruses and Bacteria.* Pathobiology and Protection, CAB International, pp. 190–210.

Griffin, M.J., Reichley, S.R., Baumgartner, W.A., Aarattuthodiyil, S., Ware, C. *et al.* (2019) Emergence of *Edwardsiella piscicida* in farmed channel ♀, *Ictalurus punctatus* × blue ♂, *Ictalurus furcatus*, hybrid catfish cultured in Mississippi. *Journal of the World Aquaculture Society* 50(2), 420–432. DOI: 10.1111/jwas.12533.

Griffin, M.J., Soto, E. and Wise, D.J. (2020) *Edwardsiella* spp. In: Woo, P.T.K., Leong, J. and Buchmann, K. (eds) *Climate Change and Infectious Fish Diseases.* CABI Publishing.

Guan, B., Cai, Y., Zhou, Y., Zhao, Z., Wang, Y. *et al.* (2022) Pathogen identification, risk factor and preventive measure of a columnaris disease outbreak in tilapia (*Oreochromis niloticus*) eggs and larvae from a tilapia hatchery. *Aquaculture* 561, 738718. DOI: 10.1016/j.aquaculture.2022.738718.

Han, H.J., Jung, S.J., Oh, M.J. and Kim, D.H. (2011) Rapid and sensitive detection of *Streptococcus iniae* by loop-mediated isothermal amplification (LAMP). *Journal of Fish Diseases* 34(5), 395–398. DOI: 10.1111/j.1365-2761.2011.01242.x.

Hanson, L.A., Liles, M.R., Hossain, M.J., Griffin, M.J. and Hemstreet, W.G. (2014) Motile Aeromonas septicemia. American Fisheries Society fish health section blue book. Available at: https://units.fisheries.org/fhs/wp-content/uploads/sites/30/2017/08/1.2.9-Motile-Aeromonas-Septicemia-2014.pdf

Hawke, J.P. and Khoo, L.H. (2004) Infectious disease. In: Tucker, C.S. and Hargreaves, J.A. (eds) *Biology and Culture of the Channel Catfish*, Vol. 1. Amsterdam, Netherlands, pp. 387–443.

Hawke, J.P., Kent, M., Rogge, M., Baumgartner, W., Wiles, J. *et al.* (2013) Edwardsiellosis caused by *Edwardsiella ictaluri* in laboratory populations of zebrafish *Danio rerio*. *Journal of Aquatic Animal Health* 25(3), 171–183. DOI: 10.1080/08997659.2013.782226.

Heckman, T.I. and Soto, E. (2021) *Streptococcus iniae* biofilm formation enhances environmental persistence and resistance to antimicrobials and disinfectants. *Aquaculture* 540, 736739. DOI: 10.1016/j.aquaculture.2021.736739.

Heckman, T.I., Griffin, M.J., Camus, A.C., LaFrentz, B.R., Morick, D. *et al.* (2020) Multilocus sequence analysis of diverse *Streptococcus iniae* isolates indicates an underlying genetic basis for phenotypic heterogeneity. *Diseases of Aquatic Organisms* 141, 53–69. DOI: 10.3354/dao03521.

Heckman, T.I., Shahin, K., Henderson, E.E., Griffin, M.J. and Soto, E. (2022) Development and efficacy of *Streptococcus iniae* live-attenuated vaccines in Nile tilapia, *Oreochromis niloticus*. *Fish and Shellfish Immunology* 121, 152–162. DOI: 10.1016/j.fsi.2021.12.043.

Høi, L., Dalsgaard, I. and Dalsgaard, A. (1998) Improved isolation of *Vibrio vulnificus* from seawater and sediment with cellobiose-colistin agar. *Applied and Environmental Microbiology* 64(5), 1721–1724. DOI: 10.1128/AEM.64.5.1721-1724.1998.

Holt, R.A. (1988) *Cytophaga psychrophila,* the causative agent of bacterial cold-water disease in salmonid fish. In Oregon State University.

Hoshina, T. (1962) On a new bacterium, *Paracolobactrum anguillimortiferum* n. sp. *Bulletin of the Japanese Society for the Science of Fish* 28(2), 162–164. DOI: 10.2331/suisan.28.162.

Hossain, S. and Heo, G.J. (2021) Ornamental fish: A potential source of pathogenic and multidrug-resistant motile *Aeromonas* spp. *Letters in Applied Microbiology* 72(1), 2–12. DOI: 10.1111/lam.13373.

Hsieh, C.Y., Tung, M.C., Tu, C., Chang, C.D. and Tsai, S.S. (2006) Enzootics of visceral granulomas associated with *Francisella*-like organism infection in tilapia (*Oreochromis* spp.). *Aquaculture* 254(1–4), 129–138. DOI: 10.1016/j.aquaculture.2006.03.044.

Hsieh, C.Y., Wu, Z.B., Tung, M.C. and Tsai, S.S. (2007) PCR and in situ hybridization for the detection and localization of a new pathogen

Francisella-like bacterium (FLB) in ornamental cichlids. *Diseases of Aquatic Organisms* 75(1), 29–36. DOI: 10.3354/dao075029.

Huang, Y., Cai, S., Jian, J., Liu, G. and Xu, L. (2020) Co-infection of infectious spleen and kidney necrosis virus and *Francisella* sp. in farmed pearl gentian grouper (♀ *Epinephelus fuscoguttatus* × ♂ *E. lanceolatus*) in China-A case report. *Aquaculture* 526, 735409. DOI: 10.1016/j.aquaculture.2020.735409.

Igarashi, A.I.T., Iida, T. and Crosa, J.H. (2002) Iron-acquisition ability of *Edwardsiella tarda* with involvement in its virulence. *Fish Pathology* 37(2), 53–57. DOI: 10.3147/jsfp.37.53.

Imperi, M., Pataracchia, M., Alfarone, G., Baldassarri, L., Orefici, G. *et al.* (2010) A multiplex PCR assay for the direct identification of the capsular type (ia to IX) of *Streptococcus agalactiae*. *Journal of Microbiological Methods* 80(2), 212–214. DOI: 10.1016/j.mimet.2009.11.010.

Ina-Salwany, M.Y., Al-Saari, N., Mohamad, A., Mursidi, F.A., Mohd-Aris, A. *et al.* (2019) Vibriosis in fish: A review on disease development and prevention. *Journal of Aquatic Animal Health* 31(1), 3–22. DOI: 10.1002/aah.10045.

Iregui, C.A., Comas, J., Vásquez, G.M. and Verján, N. (2016) Experimental early pathogenesis of *Streptococcus agalactiae* infection in red tilapia *Oreochromis* spp. *Journal of Fish Diseases* 39(2), 205–215. DOI: 10.1111/jfd.12347.

Iwanowicz, L.R., Griffin, A.R., Cartwright, D.D. and Blazer, V.S. (2006) Mortality and pathology in brown bullheads *Ameiurus nebulosus* associated with a spontaneous *Edwardsiella ictaluri* outbreak under tank culture conditions. *Diseases of Aquatic Organisms* 70(3), 219–225. DOI: 10.3354/dao070219.

Izabela, Pd.S.L., Yasmin, S.G., Samira, S.T.L., Fabio, N.D., Cristina, C.C. *et al.* (2015) Efficacy of *Aeromonas hydrophila* S-layer bacterins with different protein profiles as a vaccine in Nile tilapia (*Oreochromis niloticus*). *African Journal of Microbiology Research* 9(29), 1770–1777. DOI: 10.5897/AJMR2015.7586.

Janda, J.M. and Abbott, S.L. (2010) The genus *Aeromonas*: Taxonomy, pathogenicity, and infection. *Clinical Microbiology Reviews* 23(1), 35–73. DOI: 10.1128/CMR.00039-09.

Jantrakajorn, S., Maisak, H. and Wongtavatchai, J. (2014) Comprehensive investigation of streptococcosis outbreaks in cultured Nile tilapia, *Oreochromis niloticus*, and red tilapia, *Oreochromis* sp., of Thailand. *Journal of the World Aquaculture Society* 45(4), 392–402. DOI: 10.1111/jwas.12131.

Ji, Q., Wang, S., Ma, J. and Liu, Q. (2020) A review: Progress in the development of fish Vibrio spp. vaccines. *Immunology Letters* 226, 46–54. DOI: 10.1016/j.imlet.2020.07.002.

Jones, N., Bohnsack, J.F., Takahashi, S., Oliver, K.A., Chan, M.S. *et al.* (2003) Multilocus sequence typing system for group B *Streptococcus*. *Journal of Clinical Microbiology* 41(6), 2530–2536. DOI: 10.1128/JCM.41.6.2530-2536.2003.

Joshi, R., Skaaurd, A. and Tola Alvarez, A. (2021a) Experimental validation of genetic selection for resistance against *Streptococcus agalactiae* via different routes of infection in the commercial Nile tilapia breeding programme. *Journal of Animal Breeding and Genetics* 138(3), 338–348. DOI: 10.1111/jbg.12516.

Joshi, R., Almeida, D.B., Costa, A.R., Skaarud, A., Pádua Pereira, U. *et al.* (2021b) Genomic selection for resistance to Francisellosis in commercial Nile tilapia population: Genetic and genomic parameters, correlation with growth rate and predictive ability. *Aquaculture* 537, 736515. DOI: 10.1016/j.aquaculture.2021.736515.

Jung, M.Y., Chang, Y.H. and Kim, W. (2010) A real-time PCR assay for detection and quantification of *Lactococcus garvieae*. *Journal of Applied Microbiology* 108(5), 1694–1701. DOI: 10.1111/j.1365-2672.2009.04568.x.

Kaige, N., Miyazaki, T. and Kubota, S.S. (1986) A histopathological study of Edwardsiellosis in tilapia-experimental infection. *Fish Pathology* 21(2), 95–99.

Kamaishi, T., Fukuda, Y., Nishiyama, M., Kawakami, H., Matsuyama, T. *et al.* (2005) Identification and pathogenicity of intracellular *Francisella* bacterium in three-line grunt *Parapristipoma trilineatum*. *Fish Pathology* 40(2), 67–71. DOI: 10.3147/jsfp.40.67.

Kasornchandra, J., Rogers, W.A. and Plumb, J.A. (1987) *Edwardsiella ictaluri* from walking catfish, *Clarias Batrachus* L., in Thailand. *Journal of Fish Diseases* 10(2), 137–138. DOI: 10.1111/j.1365-2761.1987.tb00729.x.

Kayansamruaj, P., Dong, H.T., Hirono, I., Kondo, H., Senapin, S. *et al.* (2017) Comparative genome analysis of fish pathogen *Flavobacterium columnare* reveals extensive sequence diversity within the species. *Infection, Genetics and Evolution* 54, 7–17. DOI: 10.1016/j.meegid.2017.06.012.

Kayansamruaj, P., Areechon, N. and Unajak, S. (2020) Development of fish vaccine in Southeast Asia: A challenge for the sustainability of SE

Asia aquaculture. *Fish and Shellfish Immunology* 103, 73–87. DOI: 10.1016/j.fsi.2020.04.031.

Kayansamruaj, P., Dinh-Hung, N., Srisapoome, P., Na-Nakorn, U. and Chatchaiphan, S. (2023) Genomics-driven prophylactic measures to increase streptococcosis resistance in tilapia. *Journal of Fish Diseases* 46(6), 597–610. DOI: 10.1111/jfd.13763.

Keskin, O., Secre, S., Izgur, M., Turkymaz, S. and Mkaksya, R.S. (2004) *Edwardsiella ictaluri* infection in *Oncorynchus mykiss* (rainbow trout). *Turkish Journal of Veterinary and Animal Sciences* 28, 649–653.

Kim, J.H., Gomez, D.K., Choresca, C.H. and Park, S.C. (2007) Detection of major bacterial and viral pathogens in trash fish used to feed cultured flounder in Korea. *Aquaculture* 272(1–4), 105–110. DOI: 10.1016/j.aquaculture.2007.09.008.

King, B.M. and Adler, D.L. (1964) A previously undescribed group of Enterobacteriaceae. *American Journal of Clinical Pathology* 41(2), 230–232.

Kitao, T., Aoki, T. and Iwata, K. (1979) Epidemiological study on streptococcicosis of cultured yellowtail (*Seriola quinqueradiata*)-I. *Nippon Suisan Gakkaishi* 45(5), 567–572. DOI: 10.2331/suisan.45.567.

Kitiyodom, S., Kaewmalun, S., Nittayasut, N., Suktham, K., Surassmo, S. *et al.* (2019a) The potential of mucoadhesive polymer in enhancing efficacy of direct immersion vaccination against *Flavobacterium columnare* infection in tilapia. *Fish and Shellfish Immunology* 86, 635–640. DOI: 10.1016/j.fsi.2018.12.005.

Kitiyodom, S., Yata, T., Yostawornkul, J., Kaewmalun, S., Nittayasut, N. *et al.* (2019b) Enhanced efficacy of immersion vaccination in tilapia against columnaris disease by chitosan-coated "pathogen-like" mucoadhesive nanovaccines. *Fish and Shellfish Immunology* 95, 213–219. DOI: 10.1016/j.fsi.2019.09.064.

Kobayashi, T., Enomoto, S., Sakazaki, R. and Kuwahara, S. (1963) A new selective isolation medium for the *Vibrio* group; on a modified Nakanishi's medium. *Nihon Saikingaku Zasshi. Japanese Journal of Bacteriology* 18, 387–392. DOI: 10.3412/jsb.18.387.

Koh, T.H., Binte Abdul Rahman, N. and Sessions, O.M. (2020) Comparative genomic analysis of *Streptococcus dysgalactiae* subspecies *dysgalactiae*, an occasional cause of zoonotic infection. *Pathology* 52(2), 262–266. DOI: 10.1016/j.pathol.2019.09.016.

Krossøy, B., Maaseide, N.P., Ottem, K.F., Nylund, A. and Knappskog, D.H. (2007) Grado, Italy. book of abstracts p330 Development of a bacterin vaccine against *francisella* sp.; a pathogen causing serious problems for the growing cod (*Gadus morhua*) farming industry in Norway. In: *The 13th International Conference of Fish and Shellfish Diseases*, pp. 17–21. Available at: http://eafp.org/wp-content/uploads/2016/11/grado-abstractbookfinal.pdf

Kunttu, H.M.T., Suomalainen, L.R., Jokinen, E.I. and Valtonen, E.T. (2009) *Flavobacterium columnare* colony types: Connection to adhesion and virulence?. *Microbial Pathogenesis* 46(1), 21–27. DOI: 10.1016/j.micpath.2008.10.001.

Kunttu, H.M.T., Sundberg, L.R., Pulkkinen, K. and Valtonen, E.T. (2012) Environment may be the source of *Flavobacterium columnare* outbreaks at fish farms. *Environmental Microbiology Reports* 4(4), 398–402. DOI: 10.1111/j.1758-2229.2012.00342.x.

Kwon, S.R., Nam, Y.K., Kim, S.K. and Kim, K.H. (2006) Protection of tilapia (*Oreochromis mosambicus*) from edwardsiellosis by vaccination with *Edwardsiella tarda* ghosts. *Fish & Shellfish Immunology* 20(4), 621–626. DOI: 10.1016/j.fsi.2005.08.005.

Laanto, E., Bamford, J.K.H., Ravantti, J.J. and Sundberg, L.R. (2015) The use of phage FCL-2 as an alternative to chemotherapy against columnaris disease in aquaculture. *Frontiers in Microbiology* 6, 829. DOI: 10.3389/fmicb.2015.00829.

LaFrentz, B. R., Lozano, C. A., Shoemaker, C. A., García, J. C., Xu, D.-H., Løvoll, M., & Rye, M. (2016). Controlled challenge experiment demonstrates substantial additive genetic variation in resistance of Nile tilapia (*Oreochromis niloticus*) to *Streptococcus iniae*. Aquaculture, 458, 134–139. DOI: 10.1016/j.aquaculture.2016.02.034.

LaFrentz, B.R., García, J.C., Waldbieser, G.C., Evenhuis, J.P., Loch, T.P. *et al.* (2018) Identification of four distinct phylogenetic groups in *Flavobacterium columnare* with fish host associations. *Frontiers in Microbiology* 9, 452. DOI: 10.3389/fmicb.2018.00452.

LaFrentz, B.R., García, J.C. and Shelley, J.P. (2019) Multiplex PCR for genotyping *Flavobacterium columnare. Journal of Fish Diseases* 42(11), 1531–1542. DOI: 10.1111/jfd.13068.

LaFrentz, B.R., Lozano, C.A., Shoemaker, C.A., García, J.C., Ospina-Arango, J.F. *et al.* (2020) Genetic (co)variation between harvest weight and resistance to both *Streptococcus iniae* and *S. agalactiae* capsular type ib in Nile tilapia (*Oreochromis niloticus*). *Aquaculture* 529, 735726. DOI: 10.1016/j.aquaculture.2020.735726.

LaFrentz, B.R., Králová, S., Burbick, C.R., Alexander, T.L., Phillips, C.W, *et al.* (2022) The fish pathogen *Flavobacterium columnare* represents four distinct species: *Flavobacterium columnare, Flavobacterium covae* sp. nov., *Flavobacterium davisii* sp. nov. and *Flavobacterium oreochromis* sp. nov., and emended description of *Flavobacterium columnare. Systematic and Applied Microbiology* 45(2), 126293. DOI: 10.1016/j.syapm.2021.126293.

Lagos, L., Tandberg, J.I., Repnik, U., Boysen, P., Ropstad, E. *et al.* (2017) Characterization and vaccine potential of membrane vesicles produced by *Francisella noatunensis* subsp. orientalis in an adult zebrafish model. *Clinical and Vaccine Immunology* 24(5), e00557–16. DOI: 10.1128/CVI.00557-16.

Lampe, E.O., Tandberg, J.I., Rishovd, A.L. and Winther-Larsen, H.C. (2017a) *Francisella noatunensis* ssp. *noatunensis* iglC deletion mutant protects adult zebrafish challenged with acute mortality dose of wild-type strain. *Diseases of Aquatic Organisms* 123(2), 123–140. DOI: 10.3354/dao03087.

Lampe, E.O., Zingmark, C., Tandberg, J.I., Thrane, I.M.P., Brudal, E. *et al.* (2017b) *Francisella noatunensis* subspecies noatunensis CLpB deletion mutant impairs development of francisellosis in a zebrafish model. *Vaccine* 35(52), 7264–7272. DOI: 10.1016/j.vaccine.2017.11.009.

Leal, C.A.G., Carvalho-Castro, G.A., Sacchetin, P.S.C., Lopes, C.O., Moraes, A.M. *et al.* (2010) Oral and parenteral vaccines against *Flavobacterium columnare*: Evaluation of humoral immune response by ELISA and *in vivo* efficiency in Nile tilapia (*Oreochromis niloticus*). *Aquaculture International* 18(4), 657–666. DOI: 10.1007/s10499-009-9287-x.

Lewisch, E., Dressler, A., Menanteau-Ledouble, S., Saleh, M. and El-Matbouli, M. (2014) Francisellosis in ornamental African cichlids in Austria. *Bulletin of the European Association of Fish Pathologists* 34(2), 63–70. Available at: https://eafp.org/download/2014-volume34/issue_2/34-2-063-Lewisch.pdf

Li, L.H., Luo, H.M., Feng, J.H., Ming, Y.Z., Zheng, M.L. *et al.* (2020) *Francisella salimarina* sp. nov., isolated from coastal seawater. *International Journal of Systematic and Evolutionary Microbiology* 70(5), 3264–3272. DOI: 10.1099/ijsem.0.004164.

Li, W., Raoult, D. and Fournier, P.E. (2009) Bacterial strain typing in the genomic era. *FEMS Microbiology Reviews* 33(5), 892–916. DOI: 10.1111/j.1574-6976.2009.00182.x.

Li, Y. and Cai, S.H. (2011) Identification and pathogenicity of *Aeromonas sobria* on tail-rot disease in juvenile tilapia *Oreochromis niloticus. Current Microbiology* 62(2), 623–627. DOI: 10.1007/s00284-010-9753-8.

Li, Y.W., Liu, L., Huang, P.R., Fang, W., Luo, Z.P. *et al.* (2014) Chronic streptococcosis in Nile tilapia, *Oreochromis niloticus* (L.), caused by *Streptococcus agalactiae. Journal of Fish Diseases* 37(8), 757–763. DOI: 10.1111/jfd.12146.

Liao, P.-C., Tsai, Y.-L., Chen, Y.-C., Wang, P.-C., Liu, S.-C. *et al.* (2020) Analysis of streptococcal infection and correlation with climatic factors in cultured tilapia *Oreochromis* spp. in Taiwan. *Applied Sciences* 10(11), 4018. DOI: 10.3390/app10114018.

Lin, Q., Li, N., Fu, X., Hu, Q., Chang, O. *et al.* (2016) An outbreak of granulomatous inflammation associated with *Francisella noatunensis* subsp. orientalis in farmed tilapia (*Oreochromis nolitcus* x *O. aureus*) in China. *Chinese Journal of Oceanology and Limnology* 34(3), 460–466. Available at: http://doi.org/10.1007/s00343-016-4311-2

Lin, Y.S., Kweh, K.H., Koh, T.H., Lau, Q.C. and Abdul Rahman, N.B. (2020) Genomic analysis of *Lactococcus garvieae* isolates. *Pathology* 52(6), 700–707. DOI: 10.1016/j.pathol.2020.06.009.

Ling, S.H., Wang, X.H., Lim, T.M. and Leung, K.Y. (2001) Green fluorescent protein-tagged *Edwardsiella tarda* reveals portal of entry in fish. *FEMS Microbiology Letters* 194(2), 239–243. DOI: 10.1111/j.1574-6968.2001.tb09476.x.

Liu, C., Chang, O.Q., Zhang, D.F., Li, K.B., Wang, F. *et al.* (2018) *Aeromonas schubertii* as a cause of multi-organ necrosis in internal organs of Nile tilapia, *Oreochromis niloticus. Journal of Fish Diseases* 41(10), 1529–1538. DOI: 10.1111/jfd.12848.

Liu, L., Salam, N., Jiao, J.Y., E, S.M., Chen, C. *et al.* (2017) *Cysteiniphilum litorale* gen. nov., sp. nov., isolated from coastal seawater. *International Journal of Systematic and Evolutionary Microbiology* 67(7), 2178–2183. DOI: 10.1099/ijsem.0.001917.

Loch, T.P. and Faisal, M. (2015) Emerging flavobacterial infections in fish: A review. *Journal of Advanced Research* 6(3), 283–300. DOI: 10.1016/j.jare.2014.10.009.

Loch, T.P. and Faisal, M. (2017) *Flavobacterium* spp. In: Woo, P.T.K. and Cipriano, R. (eds) *Fish Viruses and Bacteria: Pathobiology and Protection.* CA B International, pp. 211–232.

Loch, T.P., Fujimoto, M., Woodiga, S.A., Walker, E.D., Marsh, T.L. *et al.* (2013) Diversity of fish-associated flavobacteria of

Michigan. *Journal of Aquatic Animal Health* 25(3), 149–164. DOI: 10.1080/08997659.2012.758189.

López-Crespo, R.A., Martínez-Chavarría, L.C., Lugo-García, A.T., Romero-Romero, L.P., García-Márquez, L.J. *et al.* (2019) Outbreak of francisellosis (*Francisella noatunensis* subsp. *orientalis*) in cultured neon jewel cichlids *Hemichromis bimaculatus* from Morelos, Mexico. *Diseases of Aquatic Organisms* 137(2), 125–130. DOI: 10.3354/dao03429.

López-Porras, A., Griffin, M.J., Ware, C., Richardson, B.M., Greenway, T.E. *et al.* (2022) Cross-protective efficacy of a live-attenuated *Edwardsiella ictaluri* vaccine against heterologous *Edwardsiella piscicida* isolates in channel and channel × blue catfish hybrids. *Journal of Fish Diseases* 45(7), 1001–1010. DOI: 10.1111/jfd.13623.

Ma, Y., Hao, L., Liang, Z., Ma, J., Ke, H. *et al.* (2020) Characterization of novel antigenic vaccine candidates for Nile tilapia (*Oreochromis niloticus*) against *Streptococcus agalactiae* infection. *Fish and Shellfish Immunology* 105, 405–414. DOI: 10.1016/j.fsi.2020.07.024.

Maalej, S., Gdoura, R., Dukan, S., Hammami, A. and Bouain, A. (2004) Maintenance of pathogenicity during entry into and resuscitation from viable but nonculturable state in *Aeromonas hydrophila* exposed to natural seawater at low temperature. *Journal of Applied Microbiology* 97(3), 557–565. DOI: 10.1111/j.1365-2672.2004.02336.x.

Mabrok, M., Chokmangmeepisarn, P., LaFrentz, B.R., Kayansamruaj, P., Dong, H.T. *et al.* (2020) Development of a species-specific polymerase chain reaction for highly sensitive detection of *Flavobacterium columnare* targeting chondroitin AC lyase gene. *Aquaculture* 521, 734597. DOI: 10.1016/j.aquaculture.2019.734597.

Maekawa, S., Wang, Y.T., Yoshida, T., Wang, P.C. and Chen, S.C. (2020) Group C *Streptococcus dysgalactiae* infection in fish. *Journal of Fish Diseases* 43(9), 963–970. DOI: 10.1111/jfd.13211.

Mainous, M.E., Smith, S.A. and Kuhn, D.D. (2010) Effect of common aquaculture chemicals against *Edwardsiella ictaluri* and *Edwardsiella tarda*. *Journal of Aquatic Animal Health* 22(4), 224–228. DOI: 10.1577/H10-020.1.

Martens, M., Dawyndt, P., Coopman, R., Gillis, M., Vos, P. *et al.* (2008) Advantages of multilocus sequence analysis for taxonomic studies: A case study using 10 housekeeping genes in the genus *Ensifer* (including former *Sinorhizobium*). *International Journal of Systematic and Evolutionary Microbiology* 58(1), 200–214. DOI: 10.1099/ijs.0.65392-0.

Martínez-Lara, P., Martínez-Porchas, M., Gollas-Galván, T., Hernández-López, J. and Robles-Porchas, G.R. (2021) Granulomatosis in fish aquaculture: A mini review. *Reviews in Aquaculture* 13(1), 259–268. DOI: 10.1111/raq.12472.

Mauel, M.J., Miller, D.L., Frazier, K., Liggett, A.D., Styer, L. *et al.* (2003) Characterization of a piscirickettsiosis-like disease in Hawaiian tilapia. *Diseases of Aquatic Organisms* 53(3), 249–255. DOI: 10.3354/dao053249.

Mauel, M.J., Soto, E., Moralis, J.A. and Hawke, J. (2007) A piscirickettsiosis-like syndrome in cultured Nile tilapia in Latin america with *Francisella* spp. as the pathogenic agent. *Journal of Aquatic Animal Health* 19(1), 27–34. DOI: 10.1577/H06-025.1.

Meyburgh, C.M., Bragg, R.R. and Boucher, C.E. (2017) *Lactococcus garvieae*: An emerging bacterial pathogen of fish. *Diseases of Aquatic Organisms* 123(1), 67–79. DOI: 10.3354/dao03083.

Meyer, F.P. and Bullock, G.L. (1973) *Edwardsiella tarda*, a new pathogen of channel catfish (*Ictalurus punctatus*). *Applied Microbiology* 25(1), 155–156. DOI: 10.1128/am.25.1.155-156.1973.

Miccoli, A., Manni, M., Picchietti, S. and Scapigliati, G. (2021) State-of-the-art vaccine research for aquaculture use: The case of three economically relevant fish species. *Vaccines* 9(2). DOI: 10.3390/vaccines9020140.

Millard, C.M., Baiano, J.C.F., Chan, C., Yuen, B., Aviles, F. *et al.* (2012) Evolution of the capsular operon of *Streptococcus iniae* in response to vaccination. *Applied and Environmental Microbiology* 78(23), 8219–8226. DOI: 10.1128/AEM.02216-12.

Mishra, A., Nam, G.H., Gim, J.A., Lee, H.E., Jo, A. *et al.* (2018) Current challenges of *Streptococcus* infection and effective molecular, cellular, and environmental control methods in aquaculture. *Molecules and Cells* 41(6), 495–505. DOI: 10.14348/molcells.2018.2154.

Miyashita, T. (1984) *Pseudomonas fluorescens* and *Edwardsiella tarda* isolated from diseased tilapia. *Fish Pathology* 19(1), 45–50. DOI: 10.3147/jsfp.19.45.

Miyazaki, T. and Kaige, N. (1985) Comparative histopathology of edwardsiellosis in fishes. *Fish Pathology* 20(2/3), 219–227. DOI: 10.3147/jsfp.20.219.

Miyazaki, T., Kubota, S.S., Kaige, N. and Miyashita, T. (1984) A histopathological study of streptococcal disease in tilapia. *Fish Pathology* 19(3), 167–172. DOI: 10.3147/jsfp.19.167.

Mohammed, H., Olivares-Fuster, O., LaFrentz, S. and Arias, C.R. (2013) New attenuated vaccine against columnaris disease in fish: Choosing the right parental strain is critical for vaccine efficacy. *Vaccine* 31(45), 5276–5280. DOI: 10.1016/j.vaccine.2013.08.052.

Morrison, C., Cornick, J., Shum, G. and Zwicker, B. (1981) Microbiology and histopathology of "saddleback" disease of underyearling Atlantic salmon, *Salmo salar* L. *Journal of Fish Diseases* 4(3), 243–258. DOI: 10.1111/j.1365-2761.1981.tb01131.x.

Morrison, E.E. and Plumb, J.A. (1994) Olfactory organ of channel catfish as a site of experimental *Edwardsiella ictaluri* infection. *Journal of Aquatic Animal Health* 6(2), 101–109. DOI: 10.1577/1548-8667(1994)0062.3.CO;2.

Müller, A., Booman, M., Hori, T., Kahlke, T., Seppola, M. *et al.* (2011) Mangalore India. Book of abstracts p215. The immune response against bacterial antigens in Atlantic cod (*Gadus morhua* L.) using microarray analysis. In: *The 8th Symposium on Diseases in Asian Aquaculture (DAA8)*, pp. 21–25. Available at: http://www.fhs-afs.net/pdf/DAA8%20Book%20of%20Abstracts.pdf

Mwega, E., Colquhoun, D.J., Tuntufye, H., Mdegela, R., Mutoloki, S. *et al.* (2019) Isolation and characterization of Flavobacteriaceae from farmed and wild Nile tilapia in Tanzania. *Journal of Aquatic Animal Health* 31(1), 23–30. DOI: 10.1002/aah.10048.

Netto, L.N., Leal, C.A. and Figueiredo, H.C. (2011) *Streptococcus dysgalactiae* as an agent of septicaemia in Nile tilapia, *Oreochromis niloticus* (L.). *Journal of Fish Diseases* 34(3), 251–254. DOI: 10.1111/j.1365-2761.2010.01220.x.

Nguyen, H.T., Kanai, K. and Yoshikoshi, K. (2002) Ecological investigation of *Streptococcus iniae* in cultured Japanese flounder (*Paralichthys olivaceus*) using selective isolation procedures. *Aquaculture* 205(1–2), 7–17. DOI: 10.1016/S0044-8486(01)00667-6.

Nguyen, V.V., Dong, H.T., Senapin, S., Kayansamruaj, P., Pirarat, N. *et al.* (2020) Synergistic infection of *Ichthyophthirius multifiliis* and *Francisella noatunensis* subsp. *orientalis* in hybrid red tilapia (oreochromis sp.). *Microbial Pathogenesis* 147, 104369. DOI: 10.1016/j.micpath.2020.104369.

Nhinh, D.T., Hoa, D.T., Giang, N.T.H., Van, K., Dang, L.T. *et al.* (2023) Synergistic infection of *Edwardsiella ictaluri* and *Flavobacterium oreochromis* in cage cultured tilapia (*Oreochromis* sp.). *Journal of Fish Diseases* 46(10), 1125–1136. DOI: 10.1111/jfd.13832.

Nho, S.W., Abdelhamed, H., Karsi, A. and Lawrence, M.L. (2017) Improving the safety of a live attenuated *Edwardsiella ictaluri* vaccine against enteric septicemia of catfish and evaluation of efficacy. *Veterinary Microbiology* 210, 83–90. DOI: 10.1016/j.vetmic.2017.09.004.

Niu, G., Wongsathein, D., Boonyayatra, S. and Khattiya, R. (2019) Occurrence of multiple antibiotic resistance and genotypic characterization in *Edwardsiella tarda* isolated from cage-cultured hybrid red tilapia (*Oreochromis* sp.) in the Ping River, Northern Thailand. *Aquaculture Research* 50, 3643–3652. DOI: 10.1111/are.14322.

Noorian, P., Hoque, M.M., Espinoza-Vergara, G. and McDougald, D. (2023) Environmental reservoirs of pathogenic *Vibrio* spp. and their role in disease: The list keeps expanding. *Advances in Experimental Medicine and Biology* 1404, 99–126. DOI: 10.1007/978-3-031-22997-8_6.

Oh, W. T., Jun, J. W., Kim, H. J., Giri, S. S., Yun, S., Kim, S. G., Kim, S. W., Kang, J. W., Han, S. J., Kwon, J. and Park, S. C. (2020). Characterization and pathological analysis of a virulent Edwardsiella anguillarum strain isolated from Nile tilapia (*Oreochromis niloticus*) in Korea. *Frontiers in veterinary science*, 7, 14. DOI: 10.3389/fvets.2020.00014

Oliveira, T.F., Quieróz, G.A., Leibowitz, M.P. and Gomes Leal, C.A.G. (2022) Development of an inactivated whole cell vaccine through immersion for immunoprophylaxis of *francisella orientalis* infections in Nile tilapia (*Oreochromis niloticus* L.) fingerlings and juveniles. *Fish and Shellfish Immunology* 127, 405–411. DOI: 10.1016/j.fsi.2022.06.040.

Olsen, A.B., Mikalsen, J., Rode, M., Alfjorden, A., Hoel, E. *et al.* (2006) A novel systemic granulomatous inflammatory disease in farmed Atlantic cod, *Gadus morhua* L., associated with a bacterium belonging to the genus *Francisella. Journal of Fish Diseases* 29(5), 307–311. DOI: 10.1111/j.1365-2761.2006.00714.x.

Ordal, E.J. and Rucker, R.R. (1944) Pathogenic myxobacteria. *Experimental Biology and Medicine* 56(1), 15–18. DOI: 10.3181/00379727-56-14572.

Oren, A. and Garrity, G. (2022) Valid publication of new names and new combinations effectively published outside the IJSEM. *International Journal of Systematic and Evolutionary Microbiology* 72(5). DOI: 10.1099/ijsem.0.005331.

Ortega Asencios, Y., Barreiro Sánchez, F., Bueno Mendizábal, H., Huancaré Pusari, K., Ostos Alfonso, H. *et al.* (2016) First report of *Streptococcus agalactiae* isolated from *Oreochromis niloticus* in Piura, Peru: Molecular identification and histopathological lesions. *Aquaculture Reports* 4, 74–79. DOI: 10.1016/j.aqrep.2016.06.002.

Osman, K.M., Al-Maary, K.S., Mubarak, A.S., Dawoud, T.M., Moussa, I.M.I. *et al.* (2017) Characterization and susceptibility of streptococci and enterococci isolated from Nile tilapia (*Oreochromis niloticus*) showing septicaemia in aquaculture and wild sites in Egypt. *BMC Veterinary Research* 13(1), 357. DOI: 10.1186/s12917-017-1289-8.

Ostland, V.E., Stannard, J.A., Creek, J.J., Hedrick, R.P., Ferguson, H.W. *et al.* (2006) Aquatic *Francisella*-like bacterium associated with mortality of intensively cultured hybrid striped bass. *Morone chrysops* × *M. saxatilis*. *Diseases of Aquatic Organisms* 72(2), 135–145. DOI: 10.3354/dao072135.

Park, K.H. and Jeong, H.D. (1996) Enhanced resistance against *Edwardsiella tarda* infection in tilapia (*Oreochromis niloticus*) by administration of protein-bound polysaccharide. *Aquaculture* 143(2), 135–143. DOI: 10.1016/0044-8486(95)01224-9.

Park, S.B., Aoki, T. and Jung, T.S. (2012) Pathogenesis of and strategies for preventing *Edwardsiella tarda* infection in fish. *Veterinary Research* 43(1), 67. DOI: 10.1186/1297-9716-43-67.

Perera, R.P., Johnson, S.K., Collins, M.D. and Lewis, D.H. (1994) *Streptococcus iniae* associated with mortality of *Tilapia nilotica* × *T. aurea* hybrids. *Journal of Aquatic Animal Health* 6(4), 335–340. DOI: 10.1577/1548-8667(1994)0062.3.CO;2.

Perera, R.P., Fiske, R.A. and Johnson, S.K. (1998) Histopathology of hybrid tilapias infected witha biotype of *Streptococcus iniae*. *Journal of Aquatic Animal Health* 10(3), 294–299. DOI: 10.1577/1548-8667(1998)0102.0.CO;2.

Phillips, A.C.N., Reichley, S.R., Ware, C. and Griffin, M.J. (2017) *Edwardsiella ictaluri* infection in *Pangasius* catfish imported from West Bengal into the Southern Caribbean. *Journal of Fish Diseases* 40(6), 743–756. DOI: 10.1111/jfd.12552.

Pimentel, S.S. and Katagiri, T. (2008) Differences of probiotic effects on *Edwardsiella tarda* challenged Nile tilapia (*Oreochromis niloticus*) fed with four *Lactobacillus* species. *Aquaculture Science* 56(3), 401–408.

Pirarat, N., Kobayashi, T., Katagiri, T., Maita, M. and Endo, M. (2006) Protective effects and mechanisms of a probiotic bacterium *Lactobacillus rhamnosus* against experimental *Edwardsiella tarda* infection in tilapia (*Oreochromis niloticus*). *Veterinary Immunology and Immunopathology* 113(3–4), 339–347. DOI: 10.1016/j.vetimm.2006.06.003.

Plumb, J.A. and Hanson, L.A. (2011) Bacterial diseases of tilapia. In: *Health Maintenance and Principal Microbial Diseases of Cultured Fishes*. John Wiley & Sons, pp. 445–463.

Plumb, J.A. and Sanchez, D.J. (1983) Susceptibility of five species of fish to *Edwardsiella ictaluri*. *Journal of Fish Diseases* 6(3), 261–266. DOI: 10.1111/j.1365-2761.1983.tb00075.x.

Pongsak, R. and Phumkhachorn, P. (2010) Assessment of synergistic efficacy of carvacrol and cymene against *Edwardsiella tarda in vitro* and in tilapia (*Oreochromis niloticus*). *African Journal of Microbiology Research* 4, 420–425. Available at: https://api.semanticscholar.org/CorpusID:85731444

Pradeep, P.J., Suebsing, R., Sirthammajak, S., Kampeera, J., Jitrakorn, S. *et al.* (2016) Evidence of vertical transmission and tissue tropism of streptococcosis from naturally infected red tilapia (*Oreochromis* spp.). *Aquaculture Reports* 3, 58–66. DOI: 10.1016/j.aqrep.2015.12.002.

Pradeep, P.J., Suebsing, R., Sirithammajak, S., Kampeera, J., Turner, W. *et al.* (2017) Vertical transmission and concurrent infection of multiple bacterial pathogens in naturally infected red tilapia (*Oreochromis* spp.). *Aquaculture Research* 48(6), 2706–2717. DOI: 10.1111/are.13102.

Pridgeon, J.W. and Klesius, P.H. (2011) Development and efficacy of novobiocin and rifampicin-resistant *Aeromonas hydrophila* as novel vaccines in channel catfish and Nile tilapia. *Vaccine* 29(45), 7896–7904. DOI: 10.1016/j.vaccine.2011.08.082.

Pridgeon, J.W., Klesius, P.H. and Yildirim-Aksoy, M. (2013) Attempt to develop live attenuated bacterial vaccines by selecting resistance to gossypol, proflavine hemisulfate, novobiocin, or ciprofloxacin. *Vaccine* 31(18), 2222–2230. DOI: 10.1016/j.vaccine.2013.03.004.

Pulpipat, T., Lin, K.H., Chen, Y.H., Wang, P.C. and Chen, S.C. (2019) Molecular characterization and pathogenicity of *Francisella noatunensis* subsp. *orientalis* isolated from cultured tilapia (*Oreochromis* sp.) in Taiwan. *Journal of Fish Diseases* 42(5), 643–655. DOI: 10.1111/jfd.12964.

Pulpipat, T., Maekawa, S., Wang, P.C. and Chen, S.C. (2020) Immune responses and protective efficacy of a formalin-killed *Francisella noatunensis* subsp. *orientalis* vaccine evaluated through intraperitoneal and immersion challenge methods in *Oreochromis niloticus*. *Vaccines* 8(2), 163. DOI: 10.3390/vaccines8020163.

Pumchan, A., Krobthong, S., Roytrakul, S., Sawatdichaikul, O., Kondo, H. *et al.* (2020) Novel chimeric multiepitope vaccine for streptococcosis disease in Nile tilapia (*Oreochromis niloticus* linn.). *Scientific Reports* 10(1), 603. DOI: 10.1038/s41598-019-57283-0.

Puneeth, T.G., Pallavi, B., Vilasini, U., Kushala, K.B., Nithin, M.S. *et al.* (2022) Large scale mortality in cultured Nile tilapia (*Oreochromis niloticus*): Natural co-infection with *Aeromonas hydrophila* and *Streptococcus iniae. Iranian Journal of Veterinary Research* 23(3), 219–228. DOI: 10.22099/IJVR.2022.41796.6084.

Qiao, M., Ying, G.G., Singer, A.C. and Zhu, Y.G. (2018) Review of antibiotic resistance in China and its environment. *Environment International* 110, 160–172. DOI: 10.1016/j.envint.2017.10.016.

Qu, P.H., Li, Y., Salam, N., Chen, S.Y., Liu, L. *et al.* (2016) *Allofrancisella inopinata* gen. nov., sp. nov. and *Allofrancisella frigidaquae* sp. nov., isolated from water-cooling systems, and transfer of *Francisella guangzhouensis* Qu *et al.* 2013 to the new genus as *Allofrancisella guangzhouensis* comb. nov. *International Journal of Systematic and Evolutionary Microbiology* 66(11), 4832–4838. DOI: 10.1099/ijsem.0.001437.

Ramírez-Paredes, J.G. (2015) The fish pathogen *Francisella orientalis*: Characterisation and vaccine development. Available at: http://hdl.handle.net/1893/21822

Ramírez-Paredes, J.G., Thompson, K.D., Metselaar, M., Shahin, K., Soto, E. *et al.* (2017) A polyphasic approach for phenotypic and genetic characterization of the fastidious aquatic pathogen *Francisella noatunensis* subsp. *orientalis. Frontiers in Microbiology* 8, 2324. DOI: 10.3389/fmicb.2017.02324.

Ramírez-Paredes, J.G., Mendoza-Roldan, M.A., Lopez-Jimena, B., Shahin, K., Metselaar, M. *et al.* (2019) Whole cell inactivated autogenous vaccine effectively protects red Nile tilapia (*Oreochromis niloticus*) against francisellosis via intraperitoneal injection. *Journal of Fish Diseases* 42(8), 1191–1200. DOI: 10.1111/jfd.13041.

Ramirez-Paredes, J.G., Larsson, P., Thompson, K.D., Penman, D.J., Busse, H.J. *et al.* (2020) Reclassification of *Francisella noatunensis* subsp. *orientalis* Ottem *et al.* 2009 as *Francisella orientalis* sp. nov., *Francisella noatunensis* subsp. *chilensis* subsp. nov. and emended description of *Francisella noatunensis.* 70(3), 2034–2048. DOI: 10.1099/ijsem.0.004009.

Reichley, S.R., Ware, C., Steadman, J., Gaunt, P.S., García, J.C. *et al.* (2017) Comparative phenotypic and genotypic analysis of *Edwardsiella* isolates from different hosts and geographic origins, with emphasis on isolates formerly classified as *E. tarda*, and evaluation of diagnostic methods. *Journal of Clinical Microbiology* 55(12), 3466–3491. DOI: 10.1128/JCM.00970-17.

Ren, Z., Cai, Y., Wang, S., Liu, S., Li, A. *et al.* (2019) First case of *Aeromonas schubertii* infection in brackish water wild Nile tilapia, *Oreochromis niloticus*, in China. *Aquaculture* 501, 247–254. DOI: 10.1016/j.aquaculture.2018.11.036.

Ren, Z., Wang, S., Cai, Y., Wu, Y., Tian, L. *et al.* (2020) Antioxidant capacity, non-specific immunity, histopathological analysis and immune-related gene expression in Nile tilapia *Oreochromis niloticus* infected with *Aeromonas schubertii. Aquaculture* 529, 735642. DOI: 10.1016/j.aquaculture.2020.735642.

Ruane, N.M., Bolton-Warberg, M., Rodger, H.D., Colquhoun, D.J., Geary, M. *et al.* (2015) An outbreak of francisellosis in wild-caught Celtic sea Atlantic cod, *Gadus morhua* L., juveniles reared in captivity. *Journal of Fish Diseases* 38(1), 97–102. DOI: 10.1111/jfd.12210.

Ruangpan, L., Kitao, T. and Yoshida, T. (1986) Protective efficacy of *Aeromonas hydrophila* vaccines in Nile tilapia. *Veterinary Immunology and Immunopathology* 12(1–4), 345–350. DOI: 10.1016/0165-2427(86)90139-x.

Ruiz-Garbajosa, P., Bonten, M.J.M., Robinson, D.A., Top, J., Nallapareddy, S.R. *et al.* (2006) Multilocus sequence typing scheme for *Enterococcus faecalis* reveals hospital-adapted genetic complexes in a background of high rates of recombination. *Journal of Clinical Microbiology* 44(6), 2220–2228. DOI: 10.1128/JCM.02596-05.

Russo, R., Mitchell, H. and Yanong, R.P.E. (2006) Characterization of *Streptococcus iniae* isolated from ornamental cyprinid fishes and development of challenge models. *Aquaculture* 256(1–4), 105–110. DOI: 10.1016/j.aquaculture.2006.02.046.

Sakazaki, R. (1965) A proposed group of the family Enterobacteriaceae, The Asakusa group. *International Bulletin of Bacteriological Nomenclature and Taxonomy* 15(1), 45–47. DOI: 10.1099/00207713-15-1-45.

Sakazaki, R. and Tamura, K. (1975) Priority of the specific epithet *Anguillimortiferum* over the specific epithet *tarda* in the name of the organism presently known as *Edwardsiella tarda. International Journal of Systematic Bacteriology* 25(2), 219–220. DOI: 10.1099/00207713-25-2-219.

Sanches-Fernandes, G.M.M., Sá-Correia, I. and Costa, R. (2022) Vibriosis outbreaks in aquaculture: Addressing environmental and public health concerns and preventive therapies using gilthead seabream farming as a model system. *Frontiers in Microbiology* 13, 904815. DOI: 10.3389/fmicb.2022.904815.

Sayed, M., Ozdemir, O., Essa, M., Olivier, A., Karsi, A. *et al.* (2021) Virulence and live vaccine potential of *Edwardsiella piscicida* phoP and phoQ mutants in catfish against *Edwardsiellosis*. *Journal of Fish Diseases* 44(9), 1463–1474. DOI: 10.1111/jfd.13453.

Schrøder, M.B., Ellingsen, T., Mikkelsen, H., Norderhus, E.A. and Lund, V. (2009) Comparison of antibody responses in Atlantic cod (*Gadus morhua* L.) to *Vibrio anguillarum*, *Aeromonas salmonicida* and *Francisella* sp. *Fish and Shellfish Immunology* 27(2), 112–119. DOI: 10.1016/j.fsi.2008.11.016.

Scott, S.J. and Bollinger, T.K. (2014) *Flavobacterium columnare*: An important contributing factor to fish die-offs in southern lakes of Saskatchewan, Canada. *Journal of Veterinary Diagnostic Investigation* 26(6), 832–836. DOI: 10.1177/1040638714553591.

Shahin, K., Gustavo Ramirez-Paredes, J., Harold, G., Lopez-Jimena, B., Adams, A. *et al.* (2018) Development of a recombinase polymerase amplification assay for rapid detection of *Francisella noatunensis* subsp. *orientalis*. *PLOS ONE* 13(2), e0192979. DOI: 10.1371/journal.pone.0192979.

Shahin, K., Shinn, A.P., Metselaar, M., Ramirez-Paredes, J.G., Monaghan, S.J. *et al.* (2019) Efficacy of an inactivated whole-cell injection vaccine for Nile tilapia, *Oreochromis niloticus* (L), against multiple isolates of *Francisella noatunensis* subsp. *orientalis* from diverse geographical regions. *Fish and Shellfish Immunology* 89, 217–227. DOI: 10.1016/j.fsi.2019.03.071.

Shahin, K., Veek, T., Heckman, T.I., Littman, E., Mukkatira, K. *et al.* (2022) Isolation and characterization of *Lactococcus garvieae* from rainbow trout, *Onchorhyncus mykiss*, from California, USA. *Transboundary and Emerging Diseases* 69(4), 2326–2343. DOI: 10.1111/tbed.14250.

Shao, S., Lai, Q., Liu, Q., Wu, H., Xiao, J. *et al.* (2015) Phylogenomics characterization of a highly virulent *Edwardsiella* strain ET080813T encoding two distinct T3SS and three T6SS gene clusters: Propose a novel species as *Edwardsiella anguillarum* sp. nov. *Systematic and Applied Microbiology* 38(1), 36–47. DOI: 10.1016/j.syapm.2014.10.008.

Sherif, A.H., Gouda, M.Y., Al-Sokary, E.T. and Elseify, M.M. (2021) *Lactobacillus plantarum* enhances immunity of Nile tilapia *Oreochromis niloticus* challenged with *Edwardsiella tarda*. *Aquaculture Research* 52(3), 1001–1012. DOI: 10.1111/are.14955.

Shirajum Monir, M., Yusoff, S.M., Mohamad, A. and Ina-Salwany, M.Y. (2020, June) Vaccination of tilapia against motile *Aeromonas septicemia*: A review. *Journal of Aquatic Animal Health* 32(2), 65–76. DOI: 10.1002/aah.10099.

Shoemaker, C.A. and LaFrentz, B.R. (2015) Lack of association between *Flavobacterium columnare* genomovar and virulence in hybrid tilapia *Oreochromis niloticus* (L.) × *Oreochromis aureus* (Steindachner). *Journal of Fish Diseases* 38(5), 491–498. DOI: 10.1111/jfd.12262.

Shoemaker, C.A., Evans, J.J. and Klesius, P.H. (2000) Density and dose: Factors affecting mortality of *Streptococcus iniae* infected tilapia (*Oreochromis niloticus*). *Aquaculture* 188(3–4), 229–235. DOI: 10.1016/S0044-8486(00)00346-X.

Shoemaker, C.A., Lafrentz, B.R. and Klesius, P.H. (2011, November) Vaccination of sex reversed hybrid tilapia (*Oreochromis niloticus* × *O. aureus*) with an inactivated *Vibrio vulnificus* vaccine. *Biologicals* 39(6), 424–429. DOI: 10.1016/j.biologicals.2011.08.004.

Shoemaker, C.A., Xu, D. and Soto, E. (2017a) *Streptococcus iniae* and *S. agalactiae*. In: Woo, P.T.K. and Cipriano, R.C. (eds) *Fish Viruses and Bacteria: Pathobiology and Protection*. CABI Publishing, pp. 298–313. DOI: 10.1079/9781780647784.0298.

Shoemaker, C.A., Lozano, C.A., LaFrentz, B.R., García, J.C., Soto, E. *et al.* (2017b) Additive genetic variation in resistance of Nile tilapia (*Oreochromis niloticus*) to *Streptococcus iniae* and *S. agalactiae* capsular type Ib: Is genetic resistance correlated?. *Aquaculture* 468, 193–198. DOI: 10.1016/j.aquaculture.2016.10.022.

Shoemaker, C.A., LaFrentz, B.R., García, J.C., Xu, D. and Beck, B.H. (2020) Streptococcosis in fish. In: *In FHS Blue Book: Suggested Procedures for the Detection and Identification of Certain Finfish and Shellfish Pathogens*, 2020 ed. Available at: https://units.fisheries.org/fhs/fish-health-sectionblue-book-2020

Shoemaker, C.A., Heckman, T.I., Soto, E., LaFrentz, B.R., Beck, B.H. *et al.* (2022a) Coinfection by group C *Streptococcus dysgalactiae* subsp. *equisimilis* and *Flavobacterium davisii* in Nile tilapia from the United States. *Bulletin of the European Association of Fish Pathologists,* October (2). DOI: 10.48045/001c.38695.

Shoemaker, C.A., Lozano, C.A., LaFrentz, B.R., Mumma, P., Vela-Avitúa, S. *et al.* (2022b) Additive genetic variation in resistance of Nile tilapia (*Oreochromis niloticus*) to *Francisella orientalis* and its genetic (co) variation to both harvest weight and resistance to *Streptococcus agalactiae* Ib. *Aquaculture* 561, 738736. DOI: 10.1016/j.aquaculture.2022.738736.

Shotts, E.B.J. and Rimler, R. (1973) Medium for the isolation of *Aeromonas hydrophila*. *Applied Microbiology* 26(4), 550–553. DOI: 10.1128/am.26.4.550-553.1973.

Sjöstedt, A.B. (2005) *Francisella*. In: Brenner, D.J., Krieg, N.P., Staley, J.T. and Garrity, G.M. (eds) *The Proteobacteria, Part B, Bergey's Manual of Systematic Bacteriology*, 2nd ed. Springer, N.Y, pp. 200–210. DOI: 10.1002/9781118960608.gbm01215.

Skerman, V.B.D., McGowan, V. and Sneath, P.H.A. (1980) Approved lists of bacterial names. *International Journal of Systematic Bacteriology* 30(1), 225–420. Available at: http://eprints.icrisat.ac.in/id/eprint/11965

Soto, E. and Revan, F. (2012) Culturability and persistence of *Francisella noatunensis* subsp. *orientalis* (syn. *Francisella asiatica*) in sea- and freshwater microcosms. *In Microbial Ecology* 63(2), 398–404. DOI: 10.1007/s00248-011-9932-6.

Soto, E., Griffin, M., Arauz, M., Riofrio, A., Martinez, A., & Cabrejos, M. E. (2012a). Edwardsiella ictaluri as the causative agent of mortality in cultured Nile tilapia. *Journal of Aquatic Animal Health*, 24(2), 81-90. DOI: 10.1080/08997659.2012.675931

Soto, E., Hawke, J.P., Fernandez, D. and Morales, J.A. (2009a) *Francisella* sp., an emerging pathogen of tilapia, *Oreochromis niloticus* (L.), in Costa Rica. *Journal of Fish Diseases* 32(8), 713–722. DOI: 10.1111/j.1365-2761.2009.01070.x.

Soto, E., Fernandez, D. and Hawke, J.P. (2009b) Attenuation of the fish pathogen *Francisella* sp. by mutation of the iglC* gene. *Journal of Aquatic Animal Health* 21(3), 140–149. DOI: 10.1577/H08-056.1.

Soto, E., Bowles, K., Fernandez, D. and Hawke, J.P. (2010a) Development of a real-time PCR assay for identification and quantification of the fish pathogen *Francisella noatunensis* subsp. *orientalis*. *Diseases of Aquatic Organisms* 89(3), 199–207. DOI: 10.3354/dao02204.

Soto, E., Endris, R.G. and Hawke, J.P. (2010b) *In vitro* and *in vivo* efficacy of florfenicol for treatment of *Francisella asiatica* infection in tilapia. *Antimicrobial Agents and Chemotherapy* 54(11), 4664–4670. DOI: 10.1128/AAC.00206-10.

Soto, E., Baumgartner, W., Wiles, J. and Hawke, J.P. (2011a) *Francisella asiatica* as the causative agent of piscine francisellosis in cultured tilapia (*Oreochromis* sp.) in the United States. *Journal of Veterinary Diagnostic Investigation* 23(4), 821–825. DOI: 10.1177/1040638711407058.

Soto, E., Wiles, J., Elzer, P., Macaluso, K. and Hawke, J.P. (2011b) Attenuated *Francisella asiatica* iglC mutant induces protective immunity to francisellosis in tilapia. *Vaccine* 29(3), 593–598. DOI: 10.1016/j.vaccine.2010.06.040.

Soto, E., Abrams, S.B. and Revan, F. (2012b) Effects of temperature and salt concentration on *Francisella noatunensis* subsp. *orientalis* infections in Nile tilapia *Oreochromis niloticus*. *Diseases of Aquatic Organisms* 101(3), 217–223. DOI: 10.3354/dao02533.

Soto, E., Illanes, O., Hilchie, D., Morales, J.A., Sunyakumthorn, P. *et al.* (2012c) Molecular and immunohistochemical diagnosis of *Francisella noatunensis* subsp. *orientalis* from formalin-fixed, paraffin-embedded tissues. *Journal of Veterinary Diagnostic Investigation* 24(5), 840–845. DOI: 10.1177/1040638712452108.

Soto, E., Illanes, O., Revan, F., Griffin, M., & Riofrio, A. (2013a). Bacterial distribution and tissue targets following experimental Edwardsiella ictaluri infection in Nile tilapia *Oreochromis niloticus*. *Diseases of Aquatic Organisms*, 104(2), 105-112. DOI: 10.3354/dao02593

Soto, E., Kidd, S., Mendez, S., Marancik, D., Revan, F. *et al.* (2013b) *Francisella noatunensis* subsp. *orientalis* pathogenesis analyzed by experimental immersion challenge in Nile tilapia, *Oreochromis niloticus* (L.). *Veterinary Microbiology* 164(1–2), 77–84. DOI: 10.1016/j.vetmic.2013.01.024.

Soto, E., Primus, A.E., Pouder, D.B., George, R.H., Gerlach, T.J. *et al.* (2014a) Identification of *Francisella noatunensis* in novel host species french grunt (*Haemulon flavolineatum*) and Caesar grunt (*Haemulon carbonarium*). *Journal of Zoo and Wildlife Medicine* 45(3), 727–731. DOI: 10.1638/2014-0059R.1.

Soto, E., Brown, N., Gardenfors, Z.O., Yount, S., Revan, F. *et al.* (2014b) Effect of size and temperature at vaccination on immunization and protection conferred by a live attenuated *Francisella noatunensis* immersion vaccine in red hybrid tilapia. *Fish and Shellfish Immunology* 41(2), 593–599. DOI: 10.1016/j.fsi.2014.10.009.

Soto, E., Wang, R., Wiles, J., Baumgartner, W., Green, C. *et al.* (2015a) Characterization of isolates of *Streptococcus agalactiae* from diseased farmed and wild marine fish from the U.S. Gulf Coast, Latin America, and Thailand. *Journal of Aquatic Animal Health* 27(2), 123–134. DOI: 10.1080/08997659.2015.1032439.

Soto, E., Griffin, M.J. and Tobar, J.A. (2015b) Mucosal vaccines. In: *Mucosal Health in Aquaculture*, 1st edn. In Academic Press, pp. 300–323. Available at: https://www.elsevier.com/books/mucosal-health-in-aquaculture/beck/978-0-12-417186-2

Soto, E., Halliday-Simmonds, I., Francis, S., Kearney, M.T. and Hansen, J.D. (2015c) Biofilm formation of *Francisella noatunensis* subsp. *orientalis*. *Veterinary Microbiology* 181(3–4), 313–317. DOI: 10.1016/j.vetmic.2015.10.007.

Soto, E., Halliday-Simmonds, I., Francis, S., Fraites, T., Martínez-López, B. *et al.* (2016a) Improved broth microdilution method for antimicrobial susceptibility testing of *Francisella noatunensis orientalis*. *Journal of Aquatic Animal Health* 28(3), 199–207. DOI: 10.1080/08997659.2016.1185051.

Soto, E., Zayas, M., Tobar, J., Illanes, O., Yount, S. *et al.* (2016b) Laboratory-controlled challenges of Nile tilapia (*Oreochromis niloticus*) with *Streptococcus agalactiae*: Comparisons between immersion, oral, intracoelomic and intramuscular routes of infection. *Journal of Comparative Pathology* 155(4), 339–345. DOI: 10.1016/j.jcpa.2016.09.003.

Soto, E., Griffin, M.J., Morales, J.A., Calvo, E.B., Alexandre Sebastião, F. *et al.* (2018) *Francisella marina* sp. nov., etiologic agent of systemic disease in cultured spotted rose snapper (*Lutjanus guttatus*) in Central America. *Applied and Environmental Microbiology* 84(16), e00144–18. DOI: 10.1128/AEM.00144-18.

Soto-Rodriguez, S.A., Lozano-Olvera, R., Garcia-Gasca, M.T., Abad-Rosales, S.M., Gomez-Gil, B. *et al.* (2018) Virulence of the fish pathogen *Aeromonas dhakensis*: Genes involved, characterization and histopathology of experimentally infected hybrid tilapia. *Diseases of Aquatic Organisms* 129(2), 107–116. DOI: 10.3354/dao03247.

Sreedharan, K., Philip, R. and Singh, I.S.B. (2012) Virulence potential and antibiotic susceptibility pattern of motile aeromonads associated with freshwater ornamental fish culture systems: A possible threat to public health. *Brazilian Journal of Microbiology* 43(2), 754–765. DOI: 10.1590/S1517-83822012000200040.

Starliper, C.E. and Schill, W. (2011) Flavobacterial diseases: Columnaris disease, coldwater disease and bacterial gill disease. In: Woo, P.T.K. and Bruno, D.B. (eds) *Fish Diseases and Disorders*. CA B International, pp. 606–631.

Stock, I. and Wiedemann, B. (2001) Natural antibiotic susceptibilities of *Edwardsiella tarda*, *E. ictaluri*, and *E. hoshinae*. *Antimicrobial Agents and Chemotherapy* 45(8), 2245–2255. DOI: 10.1128/AAC.45.8.2245-2255.2001.

Su, Y.L., Feng, J., Li, Y.W., Bai, J.S. and Li, A.X. (2016) Development of a quantitative PCR assay for monitoring *Streptococcus agalactiae* colonization and tissue tropism in experimentally infected tilapia. *Journal of Fish Diseases* 39(2), 229–238. DOI: 10.1111/jfd.12358.

Suebsong, W., Poompuang, S., Srisapoome, P., Koonawootrittriron, S., Luengnaruemitchai, A. *et al.* (2019) Selection response for *Streptococcus agalactiae* resistance in Nile tilapia *Oreochromis niloticus*. *Journal of Fish Diseases* 42(11), 1553–1562. DOI: 10.1111/jfd.13074.

Sukhavachana, S., Poompuang, S., Onming, S. and Luengnaruemitchai, A. (2019) Heritability estimates and selection response for resistance to *Streptococcus agalactiae* in red tilapia *Oreochromis* spp. *Aquaculture* 502, 384–390. DOI: 10.1016/j.aquaculture.2018.12.075.

Sukhavachana, S., Tongyoo, P., Massault, C., McMillan, N., Leungnaruemitchai, A. *et al.* (2020) Genome-wide association study and genomic prediction for resistance against *Streptococcus agalactiae* in hybrid red tilapia (*Oreochromis* spp.). *Aquaculture* 525, 735297. DOI: 10.1016/j.aquaculture.2020.735297.

Sumithra, T.G., Reshma, K.J., Anusree, V.N., Sayooj, P., Sharma, S.R.K. *et al.* (2019) Pathological investigations into *Vibrio vulnificus* infection in genetically improved farmed tilapia (*Oreochromis niloticus* L.) cultured at a floating cage farm of India. *Aquaculture* 511, 734217. DOI: 10.1016/j.aquaculture.2019.734217.

Sun, J., Fang, W., Ke, B., He, D., Liang, Y. *et al.* (2016) Inapparent *Streptococcus agalactiae* infection in adult/commercial tilapia. *Scientific Reports* 6(1), 26319. DOI: 10.1038/srep26319.

Torres-Corral, Y. and Santos, Y. (2021) Development of a real-time PCR assay for detection and quantification of *Streptococcus iniae* using the lactate permease gene. *Journal of Fish Diseases* 44(1), 53–61. DOI: 10.1111/jfd.13267.

Trung Cao, T., Tsai, M.A., Yang, C.D., Wang, P.C., Kuo, T.Y. *et al.* (2014) Vaccine efficacy of glyceraldehyde-3-phosphate dehydrogenase (GAPDH) from *Edwardsiella ictaluri* against *E. tarda* in tilapia. *Journal of General and Applied Microbiology* 60(6), 241–250. DOI: 10.2323/jgam.60.241.

Tsai, M.A., Wang, P.C., Liaw, L.L., Yoshida, T. and Chen, S.C. (2012) Comparison of genetic characteristics and pathogenicity of *Lactococcus garvieae* isolated from aquatic animals in Taiwan. *Diseases of Aquatic Organisms* 102(1), 43–51. DOI: 10.3354/dao02516.

Tsai, M.A., Wang, P.C., Cao, T.T., Liao, P.C., Liaw, L.L. *et al.* (2013) Immunoprotection of glyceraldehyde-3-phosphate dehydrogenase

(GAPDH) from *Lactococcus garvieae* against Lactococcosis in tilapia. *Journal of General and Applied Microbiology* 59(6), 437–449. DOI: 10.2323/jgam.59.437.

Tucker, C.S. and Robinson, E.H. (1990) *Channel catfish farming handbook.* Van Nostrand Reinhold.

Usui, M., Ozawa, S., Onozato, H., Kuge, R., Obata, Y. *et al.* (2014) Antimicrobial susceptibility of indicator bacteria isolated from chickens in Southeast Asian countries (Vietnam, Indonesia and Thailand). *Journal of Veterinary Medical Science* 76(5), 685–692.

Vandepitte, J., Damme, L., Fofana, Y. and Desmyter, J. (1980) *Edwarsiella tarda* et *Plesiomonas shigelloides.* leur rôle comme agents de diarrhées et leur épidémiologie. *Bulletin de La Société de Pathologie Exotique et de Ses Filiales* 73(2), 139–149.

Vela-Avitúa, S., LaFrentz, B.R., Lozano, C.A., Shoemaker, C.A., Ospina-Arango, J.F. *et al.* (2023) Genome-wide association study for *Streptococcus iniae* in Nile tilapia (*Oreochromis niloticus*) identifies a significant QTL for disease resistance. *Frontiers in Genetics* 14, 1078381. DOI: 10.3389/fgene.2023.1078381.

Vendrell, D., Balcázar, J.L., Ruiz-Zarzuela, I., Blas, I., Gironés, O. *et al.* (2006) *Lactococcus garvieae* in fish: A review. *Comparative Immunology, Microbiology and Infectious Diseases* 29(4), 177–198. DOI: 10.1016/j.cimid.2006.06.003.

Wagley, S. (2023) The viable but non-culturable (VBNC) state *in vibrio* species: Why studying the VBNC state now is more exciting than ever. *Advances in Experimental Medicine and Biology* 1404, 253–268. DOI: 10.1007/978-3-031-22997-8_13.

Wakabayashi, H., Huh, G.J. and Kimura, N. (1989) *Flavobacterium branchiophila* sp. nov., a causative agent of bacterial gill disease of freshwater fishes. *International Journal of Systematic Bacteriology* 39(3), 213–216. DOI: 10.1099/00207713-39-3-213.

Wamala, S.P., Mugimba, K.K., Mutoloki, S., Evensen, Ø., Mdegela, R. *et al.* (2018) Occurrence and antibiotic susceptibility of fish bacteria isolated from *Oreochromis niloticus* (Nile tilapia) and *Clarias gariepinus* (African catfish) in Uganda. *Fisheries and Aquatic Sciences* 21(1), 6. DOI: 10.1186/s41240-017-0080-x.

Wang, Q., Zhang, C., Xu, L., Chen, J. and Wang, X. (2021) Characterization of *Streptococcus iniae* ghost vaccine and its immunization in Nile tilapia (*Oreochromis niloticus*). *Aquaculture Research* 52(4), 1359–1368. DOI: 10.1111/are.14990.

Weinstein, M.R., Litt, M., Kertesz, D.A., Wyper, P., Rose, D. *et al.* (1997) Invasive infections due to a fish pathogen, *Streptococcus iniae. S. Iniae* Study Group. *New England Journal of Medicine* 337(9), 589–594. DOI: 10.1056/NEJM199708283370902.

Welker, T.L., Shoemaker, C.A., Arias, C.R. and Klesius, P.H. (2005) Transmission and detection of *Flavobacterium columnare* in channel catfish *Ictalurus punctatus*. *Diseases of Aquatic Organisms* 63(2–3), 129–138. DOI: 10.3354/dao063129.

Wise, D.J., Camus, A.C., Schwedler, T.E. and Terhune, J.S. (2004) 15 health management. *Developments in Aquaculture and Fisheries Science* 34, 444–503. DOI: 10.1016/S0167-9309(04)80017-6.

Wise, D.J., Greenway, T.E., Byars, T.S., Griffin, M.J. and Khoo, L.H. (2015) Oral vaccination of channel catfish against enteric septicemia of catfish using a live attenuated *Edwardsiella ictaluri* isolate. *Journal of Aquatic Animal Health* 27(2), 135–143. DOI: 10.1080/08997659.2015.1032440.

Wise, D.J., Greenway, T.E., Byars, T.S., Kumar, G., Griffin, M.J. *et al.* (2020) Validation of *Edwardsiella ictaluri* oral vaccination platform in experimental pond trials. *Journal of the World Aquaculture Society* 51(2), 346–363. DOI: 10.1111/jwas.12675.

Wonmongkol, P., Sukhavachana, S., Ampolsak, K., Srisapoome, P., Suwanasopee, T. *et al.* (2018) Genetic parameters for resistance against *Flavobacterium columnare* in Nile tilapia *Oreochromis niloticus* (Linnaeus, 1758). *Journal of Fish Diseases* 41(2), 321–328. DOI: 10.1111/jfd.12728.

World Health Organization (2016) World Health Organization. Global action plan on antimicrobial resistance. Available at: https://www.who.int/publications/i/item/9789241509763

Xu, D.H., Shoemaker, C.A. and Klesius, P.H. (2007) Evaluation of the link between gyrodactylosis and Streptococcosis of Nile tilapia, *Oreochromis niloticus* (L.). *Journal of Fish Diseases* 30(4), 233–238. DOI: 10.1111/j.1365-2761.2007.00806.x.

Xu, D.H., Shoemaker, C.A. and Klesius, P.H. (2009) Enhanced mortality in Nile tilapia *Oreochromis niloticus* following coinfections with ichthyophthiriasis and streptococcosis. *Diseases of Aquatic Organisms* 85(3), 187–192. DOI: 10.3354/dao02073.

Yamada, Y. and Wakabayashi, H. (1999) Identification of fish-pathogenic strains belonging to the genus *Edwardsiella* by sequence analysis of *sodB*. *Fish Pathology* 34(3), 145–150. DOI: 10.3147/jsfp.34.145.

Zahran, E., Mahgoub, H.A., Abdelhamid, F., Sadeyen, J.-R. and Risha, E. (2019) Experimental pathogenesis and host immune responses of *Enterococcus faecalis* infection in Nile tilapia (*Oreochromis niloticus*). *Aquaculture* 512, 734319. DOI: 10.1016/j.aquaculture.2019.734319.

Zamri-Saad, M., Azmai, M.N.A., Abd, S.Z. and Rashid, Z. (2014) Control and prevention of streptococcosis in cultured tilapia in Malaysia: A review. *Pertanika Journal of Tropical Agricultural Science* 37(4), 389–410.

Zeng, Z.H., Du, C.C., Liu, S.R., Li, H., Peng, X.X. *et al.* (2017) Glucose enhances tilapia against *Edwardsiella tarda* infection through metabolome reprogramming. *Fish and Shellfish Immunology* 61, 34–43. DOI: 10.1016/j.fsi.2016.12.010.

Zhao, Y., Wang, Y., Song, L., Li, S., Chen, X. *et al.* (2022) Effects of chronic alkalinity stress on immune response in Nile tilapia (*Oreochromis niloticus*) infected with *Aeromonas hydrophila. Aquaculture* 561. DOI: 10.1016/j.aquaculture.2022.738599.

Zheng, M.L., Jiao, J.Y., Dong, L., Han, M.X., Li, L.H. *et al.* (2019) *Pseudofrancisella aestuarii* gen. nov., sp. nov., a novel member of the family Francisellaceae isolated from estuarine seawater. *Antonie van Leeuwenhoek* 112(6), 877–886. DOI: 10.1007/s10482-018-01220-0.

Ziarati, M., Zorriehzahra, M.J., Hassantabar, F., Mehrabi, Z., Dhawan, M. *et al.* (2022) Zoonotic diseases of fish and their prevention and control. *Veterinary Quarterly* 42(1), 95–118. DOI: 10.1080/01652176.2022.2080298.

Zlotkin, A., Hershko, H. and Eldar, A. (1998) Possible transmission of *Streptococcus iniae* from wild fish to cultured marine fish. *Applied and Environmental Microbiology* 64(10), 4065–4067. DOI: 10.1128/AEM.64.10.4065-4067.1998.

Zlotkin, A., Chilmonczyk, S., Eyngor, M., Hurvitz, A., Ghittino, C. *et al.* (2003) Trojan horse effect: Phagocyte-mediated *Streptococcus iniae* infection of fish. *Infection and Immunity* 71(5), 2318–2325. DOI: 10.1128/IAI.71.5.2318-2325.2003.

Chapter 10

Tilapia viral diseases

Saengchan Senapin, Janina Costa, Kim Thompson and Ha Thanh Dong

Introduction

This chapter summarizes all viral infections that have been reported in tilapia. Eleven tilapia viral pathogens and their taxonomic classification are listed in Table 10.1. This chapter addresses (1) tilapia viral pathogens, (2) their geographical distribution, host range and transmission routes, (3) pathology and diagnosis and (4) protective and control strategies based on current scientific knowledge. Due to limited information, infections of koi herpes virus (KHV) and frog virus 3 (FV3) in tilapia are not included in this chapter.

Iridovirus infection

The Iridoviridae family is one of the most important families of viruses in finfish. These viruses infect a wide range of fish, including many freshwater and marine species. Viruses in the family Iridoviridae have been classified into two subfamilies (Alphairidovirinae and Betairidovirinae) comprising seven genera: *Ranavirus*, *Lymphocystivirus*, *Megalocytivirus*, *Iridovirus*, *Chloriridovirus*, *Decapodiridovirus* and *Daphniairidovirus* (ICTV, 2022a). The first three of these genera, belonging to the Alphairidovirinae subfamily, have been found to infect bony fish, reptiles and amphibians while the last four, belonging to the Betairidovirinae subfamily, have been linked to diseases in invertebrates, primarily insects and

Kim Thompson, Craig Shoemaker and David C. Little (eds) **Tilapia**
DOI: 10.1079/9781800629455.0010,

Table 10.1 List of tilapia viral pathogens.

Family	Genus	Species	Virus name
Amnoonviridae	*Tilapinevirus*	*Tilapinevirus tilapiae*	Tilapia lake virus (TiLV)
Parvoviridae	*Chaphamaparvovirus*	n/a	Tilapia parvovirus (TiPV)
Birnaviridae	*Aquabirnavirus*	*Infectious pancreatic necrosis virus*	Infectious pancreatic necrosis virus (IPNV)
Herpesviridae	*Herpes-like virus*[a]		Tilapia larvae encephalitis virus (TLEV)
Alloherpesviridae	*Cyprinivirus*	*Cyprinid herpesvirus 3*	Koi herpes virus (KHV)[b]
Nodaviridae	*Betanodavirus*	*Red-spotted grouper nervous necrosis virus*	Nervous necrosis virus (NNV)
Iridoviridae	*Megalocytivirus*	*Infectious spleen and kidney necrosis virus*	Infectious spleen and kidney necrosis virus (ISKNV)
		n/a	Tilapia Megalocytivirus
	Ranavirus	n/a	Bohle virus
		Frog virus 3	Frog virus 3[b]
	Lymphocystivirus	n/a	Lymphocystis disease virus (LCDV)

[a]named in Shlapobersky *et al.* (2010)
[b]not discussed in this chapter. Available information in Chengula *et al.* (2017) and Wahidi *et al.* (2019a, b)
n/a, not assigned by ICTV as of February 2022

crustaceans (Kurita and Nakajima, 2012; Chinchar *et al.*, 2017a). Note that all members of the Iridoviridae family are referred to as 'iridovirids' (rather than iridoviruses) to avoid confusion with the genus *Iridovirus* (Chinchar *et al.*, 2017b). In tilapia, four iridovirids have been reported: infectious spleen and kidney necrosis virus (ISKNV; *Megalocytivirus*), tilapia *Megalocytivirus*, Bohle iridovirus (BIV; *Ranavirus*) and lymphocystis disease virus (LCDV; *Lymphocystivirus*) (Machimbirike *et al.*, 2019; Ramírez-Paredes *et al.*, 2021; Shahin *et al.*, 2021; Figueiredo *et al.*, 2022).

ISKNV

Infectious agent

Megalocytivirus ISKNV is a large, enveloped icosahedral virus (120–300 nm in diameter) with a linear, double-stranded DNA genome of 111–112 kb (NCBI) and replicates in both the nucleus and cytoplasm of the cell (Murti *et al.*, 1985; Chinchar *et al.*, 2009). ISKNV virions from infected tilapia have a diameter of 105–175 nm (Smith *et al.*, 1997; Subramaniam *et al.*, 2016; Ramírez-Paredes *et al.*, 2021; Figueiredo *et al.*, 2022). Although partial genome sequences of ISKNV from tilapia have been published (Dong *et al.*, 2015; Ramírez-Paredes *et al.*, 2021; Figueiredo *et al.*, 2022), no complete genome sequences have yet been reported.

Geographical distribution, host range and transmission routes

Geographical distribution: chronologically, the first report of ISKNV infection in tilapia was from Thailand in 2015 and 2016 (Dong *et al.*, 2015; Suebsing *et al.*, 2016), and subsequently in the United States in 2016 (Subramaniam *et al.*, 2016), then Ghana (Ramírez-Paredes *et al.*, 2021) and Brazil (Figueiredo *et al.*, 2022) in 2021. However, disease outbreaks associated with irido-like viruses were previously reported in tilapia in 1997 and 1998, with the former case relating to fingerlings imported from Florida to Idaho while the latter were imported from Florida to Canada (Smith *et al.*, 1997; McGrogan *et al.*, 1998). ISKNV are genetically classified into genotypes I and II based on major capsid protein (*MCP*) gene sequences. The ISKNV strain found in Brazil belonged to genotype I (Figueiredo *et al.*, 2022) while the genotype of other tilapia ISKNV strains has not been confirmed.

Host range: ISKNV has an extensive host range. In addition to infecting Nile tilapia (*Oreochromis niloticus*) and red hybrid tilapia (*O. niloticus* ×*O. mossambicus*) (Dong *et al.*, 2015; Subramaniam *et al.*, 2016; Suebsing *et al.*, 2016; Ramírez-Paredes *et al.*, 2021; Figueiredo *et al.*, 2022), ISKNV has been identified in >150 fish species worldwide, including farmed fish and ornamental fish from both marine and freshwater environments (Jeong *et al.*, 2008; Wang *et al.*, 2011; Kurita & Nakajima, 2012; Sriwanayos *et al.*, 2013; Subramaniam *et al.*, 2014; Dong *et al.*, 2015; Sah Putra *et al.*, 2020). Economically important aquaculture fish species susceptible to the virus include grouper (*Epinephelus* spp.), mandarin fish (*Siniperca chuatsi*), giant gourami (*Osphronemus goramy*) and Asian sea bass (*Lates calcarifer*), while examples of susceptible ornamental fish include orbiculate batfish (*Platax orbicularis*), Banggai cardinalfish (*Pterapogon kauderni*) and marble goby (*Oxyeleotris marmorata*) (He *et al.*, 2001; Wang *et al.*, 2011; Kurita & Nakajima, 2012; Sriwanayos *et al.*, 2013; Dong *et al.*, 2017a; Cartner *et al.*, 2019; Johan & Zainathan, 2020; Swaminathan *et al.*, 2021). In addition, freshwater species commonly used as research models, including zebrafish (*Danio rerio*), southern platy (*Xiphophorus maculatus*) and guppy (*Poecilia reticulata*), have also been listed as naturally susceptible to ISKNV (Rimmer *et al.*, 2015; Bermúdez *et al.*, 2018; Cartner *et al.*, 2019).

Transmission routes: although there is no information available on the routes of transmission of ISKNV in tilapia, experimental transmission studies in other fish hosts have shown that ISKNV is likely to be transferred horizontally by cohabitation, contaminated water, cannibalism or ingestion of infected fish tissues (Go & Whittington, 2006; Australian Department of Agriculture, 2014). According to Suebsing *et al.* (2016), ISKNV-positive gonad tissue in infected red hybrid tilapia, detected by loop-mediated isothermal amplification (LAMP), indicates the potential for vertical transmission of ISKNV in tilapia.

Pathology and diagnosis

Clinical signs and gross lesions

High levels of mortality (≥50%) have been associated with natural ISKNV infections in tilapia fry, juveniles, grow-out and broodstock (Smith *et al.*, 1997; Dong *et al.*, 2015; Subramaniam *et al.*, 2016; Ramírez-Paredes *et al.*, 2021; Figueiredo *et al.*, 2022). Diseased fish appeared lethargic, displayed gill pallor and produced ascites (Smith *et al.*, 1997; McGrogan *et al.*, 1998; Subramaniam *et al.*, 2016; Ramírez-Paredes *et al.*, 2021; Figueiredo *et al.*, 2022). Other clinical signs included darkening of the skin, erratic swimming, skin haemorrhages, skin nodules, anorexia, rayed fins, exophthalmia, ocular opacity and loss, loss of scales and excessive mucus secretion (Smith *et al.*, 1997; McGrogan *et al.*, 1998; Ramírez-Paredes *et al.*, 2021; Figueiredo *et al.*, 2022). At necropsy, fish were found to have severe ascites and swollen and haemorrhagic organs, particularly the spleen, liver and kidney (Smith *et al.*, 1997; McGrogan *et al.*, 1998; Ramírez-Paredes *et al.*, 2021; Figueiredo *et al.*, 2022).

Histopathology

Large numbers of basophilic hypertrophied cells, known as megalocytes, were seen in several tissues, most notably the kidney, spleen and liver, with smaller numbers found in the gills, heart, eyes and intestinal submucosa (McGrogan *et al.*, 1998; Ramírez-Paredes *et al.*, 2021; Figueiredo *et al.*, 2022). There were also necrotic lesions associated with the presence of megalocytes (Smith *et al.*, 1997; McGrogan *et al.*, 1998; Ramírez-Paredes *et al.*, 2021).

Molecular diagnosis

Several diagnostic techniques are available for the detection of ISKNV in tilapia. A World Organization for Animal Health (WOAH; formally known as OIE)-recommended PCR technique for red seabream iridovirus (RSIV)/ISKNV (Kurita *et al.*, 1998; WOAH, 2021) was used to screen for ISKNV infection in tilapia from the Ghanaian and Brazilian outbreaks (Ramírez-Paredes *et al.*, 2021; Figueiredo *et al.*, 2022). Because this approach amplifies both RSIV and ISKNV, confirmation of positive amplified products by DNA sequencing is required. Semi- and nested-PCR approaches have been used to increase the sensitivity of ISKNV detection in tilapia (Rimmer *et al.*, 2012; Dong *et al.*, 2015; Ramírez-Paredes *et al.*, 2021). A commercially available kit, namely Path-ISKNV-EASY (Primerdesign, Southampton, UK), was applied to detect and quantify ISKNV in the disease outbreaks reported in Ghana (Ramírez-Paredes *et al.*, 2021). Although other publicly available quantitative PCR (qPCR) approaches effectively detect ISKNV in other fish species, they have yet to be tested in tilapia. LAMP, an isothermal amplification technique, has also been used to detect ISKNV in tilapia samples (Suebsing *et al.*, 2016).

Cell culture

Bluegill fry (BF-2) cells were found to be the optimal host cells for isolation of ISKNV amongst the various cell lines (BF-2, E-11, GF and EPC) tested (Smith *et al.*, 1997; Ramírez-Paredes *et al.*, 2021). Following 5 days of post-viral inoculation, cytopathic effects (CPEs) were observed as enlarged, rounded, phase-bright cells (Ramírez-Paredes *et al.*, 2021).

Protective and control strategies

There is currently no treatment for ISKNV infections. Prevention of transmission and spread of the virus is, therefore, the most effective strategy available for controlling the disease. Fish movement and importation should be conducted in strict accordance with biosecurity control procedures. Handling stress can act as a predisposing factor for ISKNV outbreaks (Figueiredo *et al.*, 2022). Other risk factors contributing to ISKNV outbreaks should be determined, and appropriate measures taken to deal with these as part of normal tilapia-farming activities. Effective diagnostics for ISKNV detection, particularly in asymptomatic fish, are critical and urgently needed (e.g. for fish trade). Although effective vaccines have been developed against ISKNV for other fish species (Fu *et al.*, 2012; Huang *et al.*, 2012; Dong *et al.*, 2013; Fu *et al.*, 2015; Li *et al.*, 2015), no ISKNV vaccines have been reported for tilapia. A vaccine derived from RSIV provided only partial protection in Asian sea bass naturally infected with ISKNV (Dong *et al.*, 2015). Autogenous vaccines, which use the same strain as found in the epidemic region, may provide better levels of protection (Barnes *et al.*, 2022). Furthermore, rapid development and timely regulatory approval for autogenous vaccines would greatly benefit the tilapia aquaculture industry. Additional considerations include the development of multivalent vaccinations for ISKNV and other pathogenic bacteria commonly co-infecting farmed tilapia (McGrogan *et al.*, 1998; Dong *et al.*, 2015; Ramírez-Paredes *et al.*, 2021). Future selective breeding technologies to develop ISKNV-resistant populations of tilapia will allow the industry to recover from, and reduce, the significant losses caused by ISKNV infections.

A novel tilapia megalocytivirus

The only incidence of a novel *Megalocytivirus* infection in tilapia (*Oreochromis* spp.) to date has been reported in an aquaculture farm in Southern California, USA, in the spring of 2019 (Shahin *et al.*, 2021). The clinical case involved 4.5-month-old fingerlings displaying signs of lethargy, anorexia, abnormal swimming and low mortality levels. Non-specific gross abnormalities included fin erosion, cutaneous melanosis, gill pallor and coelomic distension. Internal abnormalities detected at necropsy included ascites, hepatomegaly, renomegaly, splenomegaly and multifocal yellow–white lesions in the spleen and kidney (Shahin *et al.*, 2021) (Fig. 10.1). Several approaches were used to identify and characterize this tilapia novel *Megalocytivirus*, including molecular biology, histopathology, cell culture, experimental challenge and electron microscopy (Shahin *et al.*, 2021). Phylogenetic analysis of the concatenated sequences of putative amino acid sequences of the *MCP* gene and DNA polymerase (*DPol*) gene sequences revealed that this novel tilapia *Megalocytivirus* was most closely related to scale drop disease

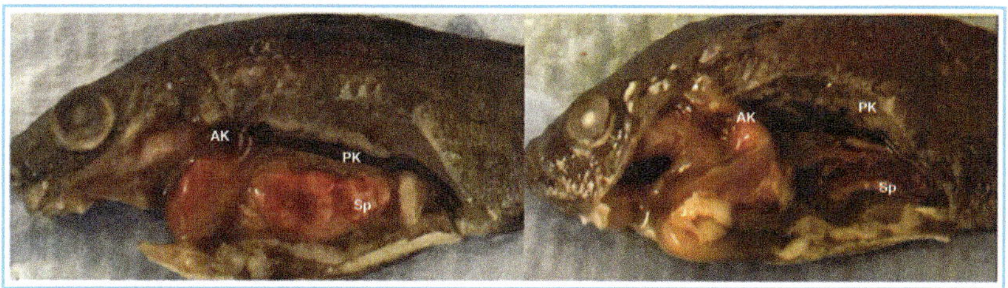

Figure 10.1 Gross findings in tilapia (*Oreochromis* spp.) naturally infected with a novel *Megalocytivirus* and *Francisella orientalis*. Fish showed classic gross lesions of piscine francisellosis, including splenomegaly and renomegaly accompanied by multifocal yellow–white nodules, ascites and cutaneous melanosis. AK, anterior kidney; PK, posterior kidney; Sp, spleen. Photographs and caption with permission from Shahin *et al.* (2021).

virus (SDDV) and European chub iridovirus (ECIV) (Shahin *et al.*, 2021). The virus showed 92.26% identity to the SDDV sequence based on a partial *DPol* sequence. A probe-based qPCR targeting the *MCP* gene has been developed for detection of the tilapia *Megalocytivirus*. The virus can infect Mozambique tilapia brain cells (OmB), causing a cytopathic effect in the cells such as cellular vacuolation, cell rounding and detachment 6–10 days post inoculation. Icosahedral viral particles with an average diameter of 115.4 ± 5.8 nm and containing dense central cores could be seen in the cytoplasm of the infected OmB cells by transmission electron microscopy (TEM) (Fig. 10.2). An experimental infection trial using virus prepared from the transfected cell line was performed in Nile tilapia fingerlings, which resulted in a cumulative mortality rate of 39% 16 days following intracoelomic injection. Histopathological examination of the challenged fish revealed necrotic foci in haematopoietic tissues (kidney and spleen). There was no evidence of histopathological lesions associated with the scale loss seen in the SDDV infections described by Gibson-Kueh *et al.* (2012) or Senapin *et al.* (2019) in Asian sea bass. It was noted that the bacterium *Francisella orientalis* was also recovered from the naturally *Megalocytivirus*-affected

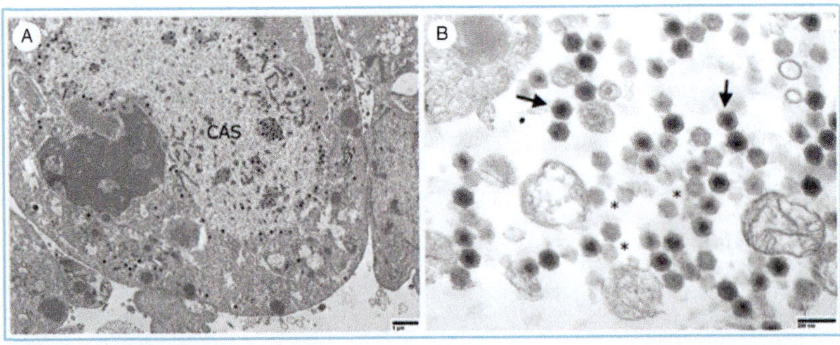

Figure 10.2 TEMs of *Megalocytivirus*-infected OmB cells. (A) virus particles were observed in the cell cytoplasm and within a large cytoplasmic assembly site (CAS). Particles were individualized and formed small paracrystalline arrays (scale bar, 1 µm). (B) higher-magnification image (scale bar, 200 nm) of virions with icosahedral nucleocapsids and electron-dense nucleic acid cores surrounded by a translucent zone and outer nucleocapsid layer (arrows). Empty nucleocapsids were scattered throughout (asterisks). Photograph and caption with permission from Shahin *et al.* (2021).

tilapia (Shahin *et al.*, 2021). However, its impact on the fish and its interaction with the tilapia *Megalocytivirus* have yet to be explored. Pathogen surveillance and its significance to tilapia aquaculture warrant further investigation to ensure appropriate control measures.

BIV

BIV was first isolated from the ornate borrowing frog *Lymnodynastes ornatus* in an area called Bohle in Queensland, Australia (Speare & Smith, 1992). Despite the rarity of BIV-associated natural disease, several species of amphibians, reptiles and fish have been shown to be susceptible to BIV through experimental infection (Moody & Owens, 1994; Cullen *et al.*, 1995; Ariel & Owens, 1997; Ariel *et al.*, 2015; Maclaine *et al.*, 2019). Tilapia is the only natural fish host for BIV recorded to date, with only one disease case reported. Massive mortalities reaching 100%, caused by BIV, occurred in a fry population of tilapia *Oreochromis mossambicus* housed in an aquatic disease laboratory in North Queensland (Ariel & Owens, 1997). The fry were from broodfish that had survived a die-off event (Ariel & Owens, 1997), suggesting that the virus can infect both adult fish and fry. The name 'spinning tilapia' (ST) syndrome was given to the clinically diseased fry due to their erratic spinning behaviour during swimming (Ariel & Owens, 1997). Dark coloration, fin clamping, unresponsiveness, sinking motionlessly to the bottom of the tank, lethargy and anorexia were also seen in the affected fish. Infection trials were performed with naïve tilapia fry and barramundi fingerlings (*Lates calcarifer*) through cannibalism of infected fish. Consequently, the ST syndrome and mortalities resulted in both fish species (Ariel and Owens, 1997). Both infected tilapia and barramundi showed similar histological characteristics. Tilapia kidney showed shrinkage of renal tubules, haemorrhaging and infiltration of eosinophilic granular cells whereas tilapia muscles showed focal myolysis. Focal necrosis of the spleen and kidney was observed in infected barramundi. TEM of the diseased barramundi specimens showed the presence of viral particles in the cytoplasm of cells, with an icosahedral shape (~140 nm) and an electron-dense core (~110 nm), indicative of *Iridovirus* morphology (Ariel & Owens,

1997). However, according to the case report, attempts to isolate the virus from these fish on bluegill fry (BF2) cells were unsuccessful (Ariel & Owens, 1997). The genomic sequence of BIV from infected tilapia has yet to be established. As a result, a PCR design for detection of BIV in tilapia has not been possible although it is not known whether PCR detection methods validated for BIV in frogs (Pallister *et al.*, 2007; Jaramillo *et al.*, 2012; Stilwell *et al.*, 2018) could be applied to BIV infecting tilapia. Consequently, the spread and effect of BIV infection in tilapia farming are unknown and no control measures have been addressed.

LCDV

Lymphocystis disease was one of the first viral diseases to be described in fish in the 19th century due to the massive size of infected cells observed in the skin of affected fish (Wolf, 1988; Leong, 2008). LCDV particles were detected by electron microscopy in 1962 (Walker, 1962). The virus is widespread, from the Arctic to tropical regions, marine and freshwater, and has been identified in >140 fish species (Anders, 1989). Species of LCDV-susceptible fish belong to genera of higher-evolutionary bony fish, and the virus has not been reported in less-evolved fish such as salmonids and cyprinids (Borrego *et al.*, 2017).

Lymphocystis virus has double-stranded DNA, with a genome of 102–208.5 kbp. The particle size of the virus is between 198 and 240 nm depending on the host species (Essbauer and Ahne, 2001; Jancovich *et al.*, 2012; ICTV, 2022b) and is icosahedral, with two outer protein layers and filaments. Four species of LCDV have been described: LCDV-1 and LCDV-2 (LCDV-C or China) isolated from flounder (*Paralichthys olivaceus*), while LCDV-3 (LCDV-Sa) and LCDV-4 were isolated from gilthead seabream (*Sparus aurata*) and whitemouth croaker (*Micropogonias furnieri*), respectively (ICTV, 2022b).

LCVD has been described in *Tilapia* and *Oreochromis* spp. from Lakes Victoria, Kitangiri and George in East Africa (Paperna, 1973). This author described the presence of clusters of hypertrophy cells (up to 5 mm) under the skin epithelium, which were composed of up to 200 cells each 260–330 μm in size and enclosed within an eosinophilic capsule. However, the hypertrophied cells described in tilapia differ from those described in other fish species, with many large vacuoles present in the cell cytoplasm, while large distinct basophilic bodies were absent. The cytoplasm of some vacuolated cells had diffuse basophilic material, and cell size was much smaller than the usual 700 μm observed for other infected fish species. In non-vacuolated cells, round inclusion bodies of reticular texture were observed. This was the first and only report of LCDV in tilapia and the only existing information on the histological description of LCDV in tilapia.

LCDV thrives in the fish farm environment because it is transmitted horizontally through live feed, with rotifers and artemia acting as possible reservoirs or vertically through eggs (Cano *et al.*, 2009, 2013). It does not cause massive mortalities but is still regarded as an important disease for aquaculture worldwide because the typical skin lesions decrease the economic value of the fish, increase fish susceptibility to other infectious diseases and reduce growth (Barja, 2004; Golomazou & Panagiotaki, 2019).

The real impact of this disease on tilapia farming is unknown. However, LCDV has caused 10–15% mortality in carp in India, 20–40% mortality in gilthead seabream in Tunisia and is recognized as a factor affecting the development of mariculture in China (Xing *et al.*, 2006; Haddad-Boubaker *et al.*, 2013; Shahi *et al.*, 2020). With the rapid development of tilapia culture globally, it is expected that LCDV outbreaks might eventually be more frequent, with the potential to damage global tilapia production. Several molecular (Cano *et al.*, 2007; Li *et al.*, 2010; Valverde *et al.*, 2016) and serological methods (García-Rosado *et al.*, 2002; Cano *et al.*, 2006; Sheng *et al.*, 2013) have been developed for LCDV diagnostics that could be used for LCDV outbreaks in tilapia. The knowledge acquired in relation to LCDV infection in other fish species, including vaccine development and the use of food additives, could be applicable to tilapia production (Zheng *et al.*, 2006; Tian *et al.*, 2008; Carballo *et al.*, 2020).

Tilapia lake virus disease

Infectious agent

Tilapia lake virus disease (TiLVD), also known as syncytial hepatitis of tilapia (SHT), is caused by *Tilapinevirus tilapiae* or tilapia lake virus (TiLV), a member of the family Amnoonviridae. TiLV is an enveloped RNA virus with a diameter of approximately 55–100 nm. The TiLV genome is composed of ten negative-sense, single-stranded RNA segments ranging in length from 465 to 1641 base pairs (total genome of 10.323 kb) (Eyngor *et al.*, 2014; Ferguson *et al.*, 2014; Bacharach *et al.*, 2016; Adams *et al.*, 2017). Although the viral genome is predicted to encode 14 putative proteins, their functions are unknown except for segment 1, which contains an open reading frame (ORF) with weak amino acid homology (17% identity, 37% segment coverage) to the influenza C virus PB1 sub unit (Bacharach *et al.*, 2016; Acharya *et al.*, 2019). Phylogenetic analysis based on either segment 1 or the concatenate genome suggested the presence of two genetic clades, namely Thai and Israeli clades (Pulido *et al.*, 2019; Debnath *et al.*, 2020; Taengphu et al., 2020). Like other segmented RNA viruses, genomic reassortments have been reported in several TiLV isolates (Chaput *et al.*, 2020; Thawornwattana *et al.*, 2021; Verma *et al.*, 2022). Natural disease outbreaks have reportedly resulted in a mortality rate of 20–90%, leading to significant economic loss for tilapia farmers (Eyngor *et al.*, 2014; Ferguson *et al.*, 2014; Tang *et al.*, 2021).

Geographical distribution, host range and transmission routes

Geographical distribution: TiLV infection has been detected in at least 18 tilapia-farming countries or regions worldwide including Israel, Ecuador, Colombia, Egypt, Thailand, India, Indonesia, Malaysia, the Philippines, Bangladesh, the Taiwan Province of China, Peru, the United Republic of Tanzania, Uganda, Mexico, the United States of America, China and Vietnam (Jansen *et al.*, 2019; Surachetpong *et al.*, 2020; Tang *et al.*, 2021; Tran *et al.*, 2022; He *et al.*, 2023). The disease and its causative agent were initially discovered from outbreaks which occurred during 2011–2012 in Ecuador and Israel (Eyngor *et al.*, 2014; Ferguson *et al.*, 2014). However, it is assumed that the virus is associated with usual die-offs seen in farmed fish in these countries since 2009 (Eyngor *et al.*, 2014; Bacharach *et al.*, 2016). In Thailand, although TiLV was first reported in early 2017 (Dong *et al.*, 2017b; Surachetpong *et al.*, 2017), retrospective diagnostics of archived samples indicate that the virus had been present in the kingdom since at least 2012 (Dong *et al.*, 2017c). Genomic epidemiology studies performed by Thawornwattana *et al.*, 2021 suggest that the presence of virus originated from some time between 2003 and 2009, in agreement with the fact that suspected TiLVD-associated mortality was first observed in 2009 (Eyngor *et al.*, 2014; Bacharach *et al.*, 2016). Active movement of live fish for aquaculture may have contributed to the virus's transboundary spread (Dong *et al.*, 2017c).

Host range: TiLV is reported to infect tilapia species in the families Cichlidae, Osphronemidae and Danionidae (Table 10.2), which includes farmed species (Nile tilapia, red hybrid tilapia, Mozambique tilapia and giant gourami), wild tilapia, *Sarotherodon galilaeus* and zebrafish, a laboratory model fish species (Table 10.2). Current scientific knowledge suggests that TiLV infects all life stages of susceptible tilapia, although young fish (fingerlings and juveniles) are more vulnerable than larger fish.

Transmission: TiLV is known to be transmitted both vertically and horizontally. Different developmental stages of tilapia collected from affected hatcheries tested positive for TiLV, including fertilized eggs and fry, suggesting possible vertical transmission (Dong *et al.*, 2017c). Subsequently, an experimental infection of Nile tilapia broodstock with TiLV demonstrated that the virus caused systemic infection and that it can enter the reproductive organs (ovary and testis) via the recirculation system, and fertilized eggs produced from infected broodstock also tested positive for TiLV (Dong *et al.*, 2020). TiLV was found in the ovary and testis of naturally and experimentally infected tilapia broodstock, as well as in 2-day-old

Table 10.2 Host range of TiLV.

Scientific name	Common name	Infection type (N/E)	References
Family Cichlidae			
Oreochromis niloticus	Nile tilapia	N, E	Ferguson *et al.*, 2014; Eyngor *et al.*, 2014
O. niloticus × *O. mossambicus* (or *Oreochromis* sp.)	Hybrid red tilapia (or red tilapia)	N, E	Dong *et al.*, 2017b; Surachetpong *et al.*, 2017
O. niloticus × *O. aureus*	Grey tilapia	N, E	Eyngor *et al.*, 2014; Kembou Tsofack *et al.*, 2017; Mugimba *et al.*, 2018
O. mossambicus	Mozambique tilapia	E	Waiyamitra *et al.*, 2021
Sarotherodon galilaeus	Mango tilapia	N	Eyngor *et al.*, 2014
Aulonocara spp.	Malawi cichlid	E	Yamkasem *et al.*, 2021a
Family Osphronemidae			
Osphronemus goramy	Giant gourami	N, E	Jaemwimol *et al.*, 2018; Chiamkunakorn *et al.*, 2019
Family Danionidae			
Danio rerio	Zebrafish	E	Rakus *et al.*, 2020

N, natural infection; E, experimental infection

fry (Yamkasem *et al.*, 2019). Cohabitation of TiLV-infected tilapia with naïve fish successfully induced disease in the latter, which resulted in 56–80% mortality within 2 weeks of starting the trial (Eyngor *et al.*, 2014; Liamnimitr *et al.*, 2018). Horizontal transmission of TiLV is likely to occur through either cannibalism of infected sick/dead fish or shedding of virus from infected fish to water.

Pathology and diagnosis

Clinical signs and gross lesions

Fish with TiLVD display a variety of abnormal behaviours including lethargy, loss of appetite, swimming at the water's surface, erratic

Figure 10.3 Nile tilapia naturally infected with TiLV showing overall darkening, skin colour discoloration, gill pallor and abdominal enlargement. Internally the liver was discoloured and watery, with yellow liquid and gas accumulating in the intestine. The spleen was swollen and dark. Image by H. T. Dong.

swimming, loss of balance or cessation of schooling. Externally, signs of skin erosion, scale protrusion, gill pallor, abdominal distension, anaemia and exophthalmia have also been reported (Ferguson *et al.*, 2014; Bacharach *et al.*, 2016; Dong *et al.*, 2017b; Fathi *et al.*, 2017; Surachetpong *et al.*, 2017; Tattiyapong *et al.*, 2017) (Fig. 10.3). Pale body and reddish opercula usually occur in diseased hybrid red tilapia, while a darkened body colour is usually observed in diseased Nile tilapia. Internally, diseased fish exhibit a variety of abnormalities, including pale, necrotic and watery liver. In many cases, liver tissue develops an unusual green or dark colour. In addition, infected fish had an enlarged spleen and gall bladder, as well as fluid accumulation in the intestine and abdomen cavity (Dinh-Hung *et al.*, 2021; Tang *et al.*, 2021).

Histopathology

Syncytial hepatitis (the presence of giant cells containing multiple nuclei in the liver) is considered the pathognomonic lesion caused by TiLV infection (Ferguson *et al.*, 2014; del-Pozo *et al.*, 2017; Dong *et al.*, 2017b; Behera *et al.*, 2018; Tang *et al.*, 2021). In addition, several stereotypical histopathological patterns have been described in the liver, including cell dissociation, focal necrosis, loss of stored fats and foamy cytoplasm of hepatocytes, which contains eosinophilic inclusions or lipoprotein-like droplets. Degeneration of pancreatic tissue with infiltration of inflammatory cells, and hepatic vacuolization, have also been reported (Ferguson *et al.*, 2014; Dong *et al.*, 2017b; Tattiyapong *et al.*, 2017; Jansen *et al.*, 2019).

Histopathological changes have also been described in other organs of infected fish, including spleen, kidney, brain, gills, the gastrointestinal tract and reproductive organs. The major changes observed include tissue degeneration, increasing melano-macrophage centres (MMCs) in lymphoid tissues, increasing cellular debris-laden macrophages in the spleen, gastric mucosal epithelial necrosis, meningitis and inflammation with infiltration of leukocytes in multiple organs (Eyngor *et al.*, 2014; Ferguson *et al.*, 2014; Dong *et al.*, 2017b; Surachetpong *et al.*, 2017; Behera *et al.*, 2018; Tattiyapong *et al.*, 2018; Senapin *et al.*, 2019; Dong *et al.*, 2020). Most recently, Dinh-Hung *et al.* (2021) published a comprehensive report on the localization of TiLV and histopathological changes in various parts of the brain that shed light on neurological disorders in diseased fish.

Molecular diagnostic methods

Numerous molecular diagnostic methods have been developed for the detection and quantification of TiLV, including conventional PCR with reverse transcription (RT-PCR) (Eyngor *et al.*, 2014), nested RT-PCR (Kembou Tsofack *et al.*, 2017), semi-nested RT-PCR (Dong *et al.*, 2017b; Taengphu *et al.*, 2020), SYBR Green RT-qPCR (Tattiyapong *et al.*, 2018) and TaqMan probe-based RT-qPCR (Waiyamitra *et al.*, 2018; Taengphu *et al.*, 2022).

Even though their sensitivity varies, these methods are capable of detecting TiLV from clinically sick fish and some have been reported to detect TiLV in subclinically infected tilapia (Senapin *et al.*, 2019; Taengphu *et al.*, 2020, 2022). A recently developed RT-qPCR method targeting TiLV genome segment 9 has been fully validated and is recommended for detection and quantification of TiLV not only in fish samples, but also in environmental water samples (Taengphu *et al.*, 2022). In addition, some low-cost RT-LAMP methods have been recommended for pond-side diagnosis (Phusantisampan *et al.*, 2019; Yin *et al.*, 2019; Kampeera *et al.*, 2021). Delamare-Deboutteville *et al.* (2021) employed a specific PCR for TiLV diagnosis in combination with rapid amplicon sequencing using Oxford Nanopore Technologies (ONT) for pathogen genotyping. This promising platform is particularly useful for rapid identification and typing of emerging infectious pathogens in new geographical locations. Please refer to the original publications for details on each detection protocol.

In situ hybridization (ISH) methods using TiLV-specific probes (Bacharach *et al.*, 2016; Dong *et al.*, 2017b; Dinh-Hung *et al.*, 2021) and immunohistochemistry (IHC) using a polyclonal rabbit immunoglobulin G (IgG) against TiLV (Piewbang *et al.*, 2021) have been employed for the detection and localization of TiLV in different organs of infected fish. These methods are useful for studying host–pathogen interaction and disease pathogenesis.

Cell culture

Viable TiLV can be isolated from infected fish using several permissive cell lines, including E-11 derived from snakehead fish (*Channa striata*) and tilapia primary brain cells (Eyngor *et al.*, 2014), OmB and TmB derived from *O. mossambicus* (Kembou Tsofack *et al.*, 2017), CFF derived from *Pristolepis fasciatus* (Behera *et al.*, 2018), OnlB and OnlL derived from Nile tilapia liver and brain (Thangaraj *et al.*, 2018), TiB derived from tilapia brain (Wang *et al.*, 2018) or CAMB derived from hybrid snakehead brain (Wang *et al.*, 2020). Both fresh and frozen specimens (better preserved at −80°C) collected from diseased fish (e.g. liver, brain, kidney, spleen, reproductive organs, serum and mucus) are useful for virus detection (Tattiyapong *et al.*, 2017; Liamnimitr *et al.*, 2018; Dong *et al.*, 2020). Cytopathic effects usually appear after 4–12 days of inoculation with filtered virion-containing homogenate using cells cultured at 25°C. The cell culture supernatant containing TiLV can be preserved at −80°C for further investigation.

Protective and control strategies

There is currently no cure for TiLV infections. Implementation of suitable biosecurity practices is recommended to prevent the introduction and transboundary spread of TiLV. Diagnostic screening for selection of TiLV-negative fish prior to fish movements or stocking may help reduce the risk of disease outbreaks (Jansen *et al.*, 2019; Tang *et al.*, 2021). Several studies have demonstrated vertical transmission of TiLV and encouraged the use of TiLV-free broodstock to produce TiLV-free seed (Yamkasem *et al.*, 2019; Dong *et al.*, 2020). Farm disease records suggest that stress from transportation and stocking of fingerlings and juveniles may play a significant role in the development of disease outbreaks. Thus, farmers should mitigate against stress from transportation and handling and enhance fish immunity by feeding immunostimulants during periods of high risk (Jansen *et al.*, 2019; Tang *et al.*, 2021).

Disinfection of equipment, culture water and the pond bottom between production cycles may also help prevent viral transmission. Numerous disinfectants have been shown to be effective at inhibiting TiLV replication, including iodine (2.5 ppm), sodium hypochlorite (NaOCl, 10 ppm), hydrogen peroxide (H_2O_2, 300 ppm), formalin (80 ppm) and Virkon™ (0.5%) (Jaemwimol *et al.*, 2019). Soto *et al.* (2019) identified disinfectants capable of inactivating TiLV following 30 min of exposure, including buffered povidone iodine (PVP iodine, 50 ppm) complex and household bleach (free chlorine, 20 ppm).

A combination of biosecurity measures, improved genetic resistance through selective breeding programmes and vaccination have been effective in reducing the number of viral disease outbreaks in aquaculture. Although commercial TiLV vaccines are not yet available, several laboratories have demonstrated vaccination as a promising approach for prevention of TiLVD. The first TiLV vaccine was developed in Israel using TiLV strains that had been attenuated through subsequent cell-culture passages. The prototype for these vaccines demonstrated a relative percentage survival (RPS) of 56–58% (Bacharach & Eldar, 2016). Tattiyapong *et al.* (2020) reported that tilapia can produce anti-TiLV antibodies within 7–10 days post vaccination and, when fish were re-exposed to TiLV by artificial infection, an increase in antibody levels occurred within 2 weeks post-infection, suggesting that the fish had developed humoral memory to the virus. Subsequently Zeng *et al.* developed a β-propiolactone-inactivated vaccine combined with the adjuvant Montanide IMS 1312, which resulted in 85.7% RPS (Zeng *et al.*, 2021). This vaccine induced increased expression of several immune-related genes (*TNFα*, *IL1β*, *IFNγ*, *CD4*, *MHC-Ia* and *MHC-II*), elicited specific IgM and neutralizing antibodies at 3 weeks post immunization and increased immunity significantly following a booster vaccination. Similarly, Mai *et al.* (2021) reported that injectable heat- and formalin-killed vaccines were both effective at preventing TiLV infection in Nile tilapia following a booster dose of the vaccine, with RPS of 71.3 and 79.6%, respectively. These vaccines stimulated both adaptive and innate immunity, induced the expression of three immunoglobulin classes (IgM, IgD and IgT) and induced not only systemic but also mucosal IgM responses in the vaccinated fish. When tested in tilapia broodstock, these vaccines elicited anti-TiLV IgM in both male and female broodstock, and these maternal antibodies were transferrable to both fertilized eggs and 1–3 day old larvae. Interestingly, these maternal antibodies conferred 80–95% protection against TiLV challenge in naïve juvenile tilapia via passive immunization. This study also suggests that vaccination of tilapia broodstock with TiLV HKV may be a strategy for prevention of TiLV infection in tilapia fertilized eggs and larvae (Mai *et al.*, 2022). TiLV vaccines based on DNA and recombinant proteins have also been tested in laboratory trials. Primary vaccination with a segment 8 (VP20)-based DNA vaccine, followed by a booster dose with a recombinant VP20 (rVP20) in adjuvant M402, resulted in 72.5% protection in vaccinated fish, which was significantly higher than the level obtained with DNA or rVP20 vaccines alone (Zeng *et al.*, 2021). Although several promising vaccines have been developed and tested in the laboratory, scaling up these vaccines at an affordable price for the tilapia industry in low- and middle-income countries remains a significant challenge.

Tilapia parvovirus infection

Infectious agent

Tilapia parvovirus (TiPV) is a novel parvovirus first discovered from faecal samples of crocodiles fed with tilapia in China by next-generation sequencing (NGS). High levels of TiPV were detected in crocodile faeces and tilapia intestines (Du *et al.*, 2019). In 2020, Liu *et al.* reported mass mortality of adult tilapia (500–600 g) associated with this virus in another province in China (Liu *et al.*, 2020). The latter study isolated and characterized the virus, and experimental infection fulfilled Koch's postulates.

TiPV is a linear, single-stranded DNA (ssDNA), non-enveloped, spherical-shaped virus approximately 30 nm in diameter. An almost complete genome (4,269 bp) of TiPV has been identified that contains a 208 bp 5' untranslated region (UTR), a 396 bp (ORF1), 1,875 bp of complete non-structural protein 1 (NS1), 504 bp of NS2, 216 bp of ORF2, 1665 bp of complete capsid protein (VP1) and a partial (46 bp) 3' UTR (Liu *et al.*, 2020). Genome organization of TiPV is shown in Fig. 10.4. The genome contained two major ORFs encoding NS1 (618 amino

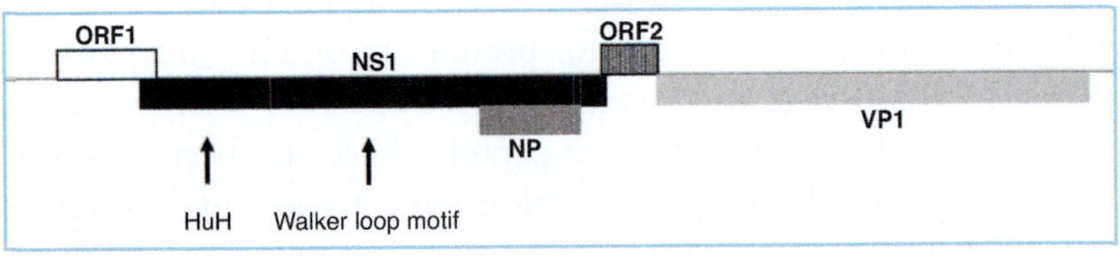

Figure 10.4 Genome organization of TiPV (adapted from Liu *et al.*, 2020)

acids (aa)) and VP1 (554 aa), respectively. The HuH motif (u denotes a hydrophobic residue) and a Walker loop motif were identified in the NS1 protein, between amino acids 78–80 and 264–360, respectively. Currently, TiPV is proposed as a novel member of the genus *Chapparvovirus* of the family Parvoviridae (Du *et al.*, 2019; Liu *et al.*, 2020).

Geographical distribution, host range and transmission routes

Geographical distribution: TiPV has been reported in multiple geographical locations in China (Du *et al.*, 2019; Liu *et al.*, 2020) and Thailand (Yamkasem *et al.*, 2021b; Piewbang *et al.*, 2022). In all reported natural disease events in both countries, tilapia with TiPV were found to be co-infected with other pathogens such as *Streptococcus* sp. (Liu *et al.*, 2020) and TiLV (Yamkasem *et al.*, 2021b; Piewbang *et al.*, 2022). A relatively high prevalence (23.1−64.6%) of TiPV was detected from internal organs (kidney and spleen) of juvenile tilapia (10–30 g) collected from different provinces in China between 2015 and 2019 (Liu *et al.*, 2020).

Host range: TiPV infection has been reported in Nile tilapia (*O. niloticus*) in China and hybrid red tilapia (*Oreochromis* sp.) in Thailand.

Transmission routes: unavailable.

Pathology and diagnosis

Clinical signs and gross lesions

The reported disease events in China have been associated with cage-cultured adult Nile tilapia, with cumulative mortality reaching 60–70%. Clinically sick fish showed lethargy, anorexia and abnormal swimming behaviour (darting or corkscrew movements), skin haemorrhages and occasional exophthalmia and pronounced ocular lesions (Liu *et al.*, 2020). Experimental infection using purified virus derived from cell culture or tissue homogenates obtained from severely diseased fish resulted in 90% mortality at 11 days post infection (dpi). Diseased fish exhibited clinical signs of haemorrhages and ocular lesions (Liu *et al.*, 2020). The co-infection cases in Thailand occurred in hybrid red tilapia, mainly juveniles (10–30 g), with fewer cases in larger fish (300–800 g) and reported cumulative mortality ranging from 50 to 75%. Diseased fish exhibited signs of lethargy, haemorrhaging and focal ulceration on their body surface, exophthalmia and abnormal swimming (Yamkasem *et al.*, 2021b; Piewbang *et al.*, 2022).

Histopathology

Histopathological changes were described in multiple organs (spleen, kidney, liver, brain, heart and gills) in TiPV-infected fish from China. In moderately and severely diseased fish, necrotic splenocytes exhibited condensed and marginated nuclei and virus inclusion bodies. Other generalized changes included tissue degeneration with infiltration of lymphocytes and macrophages. In the spleen and kidney there were increasing numbers of melano-macrophage centres. The brain was described as having enlarged blood vessels, severe cerebral cortex oedema and vacuolated and marginated nuclear chromatin in neurons. Other changes included oedema of the renal glomerulus, enlarged hepatic sinusoids, disorder and disruption of the heart muscle trabeculae and swollen and congested gill lamellae (Liu *et al.*, 2020). With respect to disease cases in Thailand, general histopathological lesions were described in fish with TiLV–TiPV co-infection (Piewbang *et al.*, 2022); however, the pathognomonic lesion caused by TiPV remains unknown.

Molecular diagnosis

Liu *et al.* published several sets of primers for both conventional PCR and qPCR to detect TiPV (Liu *et al.*, 2020). A SYBR Green qPCR assay targeting the NS1 gene was then employed to investigate tissue tropism and for quantification of viral loads. The highest viral load was detected in the kidney of diseased fish ($3.5 \times 10^{7.32 \pm 0.21}$/mg), followed by spleen ($4.2 \times 10^{7.12 \pm 0.35}$/mg), intestine ($4.3 \times 10^{6.37 \pm 0.36}$/mg), heart ($5.1 \times 10^{6.25 \pm 0.27}$/mg) and brain ($1.8 \times 10^{6.09 \pm 0.24}$/mg).

A significantly lower viral load was detected in the gills, liver and eyes ($2.8 \times 10^{3.78 \pm 0.32}$ to $1.8 \times 10^{4.32 \pm 0.17}$/mg) and was undetected in muscle samples. This assay was also used for surveillance of TiPV infection in tilapia in several provinces in China (Liu *et al.*, 2020). Unfortunately, optimization of the method, including analytical and diagnostic specificity/sensitivity, was not described. Most recently, Yamkasem *et al.* published a TaqMan probe-based qPCR for detection of this virus (Yamkasem *et al.*, 2022). The detection limit of the assay was ten copies/µl, with no cross-amplification evident when tested with other viruses or bacteria. The TaqMan qPCR and SYBR Green qPCR assays had comparable levels of sensitivity when tested against field samples.

Cell culture

The tilapia brain cell line (TiB) established from Nile tilapia was the first reported as a permissive cell line for propagation of TiPV (Liu *et al.*, 2020). Following inoculation of TiB with tissue homogenates from naturally infected fish, changes in cell morphology were observed by three dpi and CPE (cell shrinkage, rounding and cell fusion with cytoplasmic vacuolization) was evident at six dpi. The cell monolayer was completely detached by eight dpi. After three passages, virus titres reached $10^{4.2 \pm 0.38}$ $TCID_{50}$ (50% tissue culture infectious dose)/ml (Liu *et al.*, 2020). Piewbang *et al.* reported that both TiLV and TiPV from homogenates of fish co-infected with both viruses could replicate in tilapia brain cells and E-11 cells; however, it was not possible to separate the viruses from the mixed culture (Piewbang *et al.*, 2022).

Protective and control strategies

Although TiPV has been reported in China and Thailand (two major tilapia-producing countries), the reported disease events involved the presence of other pathogenic agents (*Streptococcus* sp. and TiLV). On the other hand, there was a relatively high prevalence of TiPV detected from randomly collected tilapia in multiple locations in China, suggesting that further investigation is required to assess the impact of TiPV as a primary or opportunistic agent. Nevertheless, increased awareness and implementation of biosecurity practices are needed to avoid transboundary spread of this virus.

Viral encephalopathy and retinopathy

Infectious agent

Viral encephalopathy and retinopathy (VER) is WOAH's official name for the viral disease caused by nervous necrosis virus (NNV) (WOAH, 2019). The disease is also referred to as sea bass viral encephalitis (SVE), fish viral encephalitis (FVE) or viral encephalitis and retinitis. However, it is most frequently referred to as viral nervous necrosis (VNN) (Bellance & Gallet de Saint-Aurin, 1988; Yoshikoshi and Inoue, 1990; Comps *et al.*, 1994; Tanaka *et al.*, 2004). The economic importance of VER has increased over the years, with up to 100% mortalities seen in post-hatch larvae, fingerlings and juvenile fish (Arimoto *et al.*, 1993a; Johansen *et al.*, 2004a). Reports of VER in Nile tilapia were recorded in France (Bigarré *et al.*, 2009), Thailand (Keawcharoen *et al.*, 2015), Indonesia (Prihartini *et al.*, 2015, Brazil (de Padua, 2017) and Egypt (Taha *et al.*, 2020), with mortality levels ranging between 70 and 100% (Keawcharoen *et al.*, 2015; Taha *et al.*, 2020).

The causative agent of the disease was first described as a picorna-like virus (Glazebrook *et al.*, 1990; Mori *et al.*, 1992) and subsequently placed in the Nodaviridae family and genus *Betanodavirus* (Comps *et al.*, 1994; Curtis *et al.*, 2001), although the virus is often referred to as NNV. *Betanodavirus* particles are icosahedral, with a diameter of 23–25 nm, are non-enveloped, lack a poly(A) sequence at the 3' terminus and their nucleic acids are composed of two single-stranded, positive-sense RNA molecules (ssRNA), RNA1 and RNA2, with a molecular weight of 1.01×10^6 and 0.49×10^6 Da, respectively (Glazebrook *et al.*, 1990; Comps *et al.*, 1994). The two ssRNA have three ORFs, RNA1, RNA2 and a subgenomic RNA3 that is collinear with the 3' terminus of RNA1 (Nagai and Nishizawa,

1999). Although NNV isolates from Nile tilapia have not been fully characterized, the viral particles observed in the Egypt outbreak ranged from 38.2 to 53.5 nm in diameter (Taha *et al.*, 2020), which is larger than NNV particles detected from other fish species.

The size of RNA1 varies from 981 to 983 aa (Nagai and Nishizawa, 1999; Tan *et al.*, 2001; Sommerset and Nerland, 2004) and it encodes a 100 kDa, RNA-dependent RNA polymerase (RdRp) (Nagai and Nishizawa, 1999), which is host species dependent and is involved in the virus' temperature dependency (Nishizawa *et al.*, 1995a; Hata *et al.*, 2010).

The size of RNA2 varies according to viral strain, ranging from 1367 nt in guppy (*Poecilia reticulate*) NNV to 1433 nt in greasy grouper (*Epinephelus tauvina*) NNV. However, the coat protein (CP) encoded by RNA2 has the same molecular weight for all host species (37 kDa) (Nishizawa *et al.*, 1995a; Delsert *et al.*, 1997; Tan *et al.*, 2001; Hegde *et al.*, 2003). The subgenomic RNA3, 371 nt in size, encodes for two small proteins, B1 and B2. B1 is translated on the same reading frame as RdRp and has been suggested to be an anti-necrotic factor (Chen *et al.*, 2009). B2 overlaps with the C terminus of CP and is translated in more than one reading frame. It has been suggested that B2 also acts as a necrotic death factor, either impairing ATP synthesis, acting as an inhibitor of short interfering RNA (siRNA) or inhibiting fish interferon production (Fenner *et al.*, 2006a; Chen *et al.*, 2007; Su & Hong, 2010; Su *et al.*, 2014).

Phylogenetic analysis based on the T4 region of RNA2 has been used to group NNV isolates into five distinct genotypes: BFNNV (barfin flounder nervous necrosis virus), RGNNV (red-spotted grouper nervous necrosis virus), SJNNV (striped jack nervous necrosis virus), TPNNV (tiger puffer nervous necrosis virus) and TNV (turbot necrosis virus) (Nishizawa *et al.*, 1997; Johansen *et al.*, 2004b). These genotypes reflect the virus' adaptation to distinct hosts and temperature ranges, with each genotype having a specific host temperature range (BFNNV, 15–20°C; RGNNV, 25–30°C; SJNNV, 20–25°C; and TPNNV, 20°C) (Iwamoto *et al.*, 2000). All *Betanodavirus* isolates detected in Nile tilapia belong to the RGNNV genotype (Bigarré *et al.*, 2009; Yanuhar *et al.*, 2018; Taha *et al.*, 2020).

The occurrence of reassortants between RGNNV and SJNNV has been observed only in the Iberian Peninsula/Southern Europe and the Adriatic Sea, with distinct RNA rearrangements taking place. While the Adriatic Sea (Croatia and Italy) reassortment is composed of an RNA1-SJNNV genotype and an RNA2-RGNNV genotype (SJNNV/RGNNV) (Toffolo *et al.*, 2007), that of the Iberian Peninsula is composed of an RNA1-RGNNV genotype and an RNA2-SJNNV genotype (RGNNV/SJNNV) (Olveira *et al.*, 2009; Panzarin *et al.*, 2012). The presence of a co-infection with both reassortments is also possible (Lopez-Jimena *et al.*, 2010), as well as infections between species produced in the same facility (Volpe *et al.*, 2020; Vázquez-Salgado *et al.*, 2022).

The four genotypes cluster into three distinct serotype groups, with group A neutralizing SJNNV, group B neutralizing TPNNV and group C neutralizing RGNNV and BFNNV (Morit *et al.*, 2003). Monoclonal and polyclonal antibodies raised against RGNNV cross-reacted and neutralized BFNNV, but not SJNNV (Costa, 2005).

Geographical distribution, host range and transmission routes

Geographical distribution: VER was first observed in 1985 in European sea bass (*Dicentrarchus labrax* L.) cultured in Martinique Bellance and Gallet de Saint-Aurin in 1988, but soon after was found in Australia, northern Europe, North America, the Mediterranean and South-east Asia (Glazebrook *et al.*, 1990; Breuil *et al.*, 1991; Mori *et al.*, 1991; Renault *et al.*, 1991; Grotmol *et al.*, 1995; Chi *et al.*, 1997; Curtis *et al.*, 2001; Lin *et al.*, 2001; Starkey *et al.*, 2001; Johnson *et al.*, 2002; Oh *et al.*, 2002; Athanassopoulou *et al.*, 2003; Ucko *et al.*, 2004; Ransangan and Manin, 2010; Kara *et al.*, 2014).

Host range: *Betanodavirus* has been reported in >120 wild and farmed fish species (Costa and Thompson, 2016). It was believed to be restricted to marine species until it was detected in freshwater species in 2000 (Athanassopoulou *et al.*, 2003; Hegde *et al.*, 2003). *Betanodavirus* was first reported in farmed Nile tilapia in 2009

(Bigarré *et al.*, 2009). Previously, Skliris and Richards (1999a) had experimentally infected *O. mossambicus* (Peters) with a European sea bass NNV isolate, but no mortalities were observed.

Transmission: Transmission of NNV can occur vertically, with the virus detected in fertilized eggs, sperm, ovarian fluid and hatched eggs (Arimoto *et al.*, 1992; Breuil *et al.*, 2002; Nallala *et al.*, 2021). The virus has also been seen in reproductive organs, the digestive tract, liver and kidney. The stress of reproduction and spawning appears to promote viral replication (Valero *et al.*, 2015).

Cohabitation challenges have demonstrated that the virus is also transmitted horizontally (Arimoto *et al.*, 1993b; Skliris & Richards, 1999b; Hick *et al.*, 2011), which is a concern in regard to cage-to-cage transmission of the disease within aquaculture systems. Several fish and invertebrate species have been identified as potential reservoirs for the virus (Skliris and Richards, 1998; Castri *et al.*, 2001; Gomez *et al.*, 2008a; b). The origin of tilapia VER outbreaks remains undetermined.

Pathology and diagnosis

Clinical signs and gross lesions

Fish with VER exhibit lethargy and changes in coloration (pale or dark), anorexia, sinking and resting on the bottom of the tanks and floating upside down at the water surface. Typical behaviours associated with VER are darting or swimming in a corkscrew fashion, rotating, spinning, horizontal looping and abrupt whirling (Glazebrook *et al.*, 1990; Yoshikoshi & Inoue, *et al.*, 1990; Bloch *et al.*, 1991; Mori *et al.*, 1991; Nguyen *et al.*, 1994; Chua *et al.*, 1995; Fukuda *et al.*, 1996; Grotmol *et al.*, 1997; Munday and Nakai, 1997; Zafran *et al.*, 2000; Munday *et al.*, 2002). Nile tilapia confirmed as NNV positive also displayed these neurological disturbances (Keawcharoen *et al.*, 2015; Taha *et al.*, 2020), together with other signs of VER including ocular opacity, lethargy, anorexia and dark coloration (Bigarré *et al.*, 2009; Prihartini *et al.*, 2015; Taha *et al.*, 2020).

Histopathology

The histopathological lesions observed in *Betanodavirus*-infected fish are characteristic of this disease. *Betanodavirus* is a neurotropic virus that causes pathology only in neuronal tissues (Yoshikoshi and Inoue, 1990; Mori *et al.*, 1991), although virus-like particles have been observed in gill pillar cells and tissue lesions in Atlantic halibut (*Hippoglossus hippoglossus*; Grotmol *et al.*, 1997). Pathognomonic lesions associated with NNV infection include extensive lesions associated with the central nervous system (CNS), with vacuolation seen in the retina and brain tissue and the mid- and hind-brain showing neuronal degeneration (Glazebrook *et al.*, 1990; Nishizawa *et al.*, 1995a).

Vascular degeneration or neuronal necrosis was observed in the spinal cord, olfactory bulb, cerebrum, optic lobe of the forebrain and retinal tissue of infected Nile tilapia (Bigarré *et al.*, 2009; Keawcharoen *et al.*, 2015). Cytoplasmatic vacuolation was observed in these tissues, together with sporadic nuclear pyknosis and karyolysis (Bigarré *et al.*, 2009). Vacuolation and neuronal necrosis in the eye were observed mainly in the inner nuclear layer, although there were lesions present in the outer nuclear and ganglion cell layers. The spinal cord had severe vacuolation, and while Bigarré *et al.* (2009) observed vacuolation associated with neuronal degeneration and swelling, Keawcharoen *et al.* (2015) reported that such vacuolation was in the dendrites and cytoplasm of megalocells.

Molecular diagnostic methods

Numerous molecular methods have been developed for the detection and quantification of NNV over the years. The first RT-PCR was developed by Nishizawa *et al.* (1994), and has since become the most-used RT-PCR method for detection of the virus. It targets the RNA2 region T4 (592–1017 nt), producing a 426 bp fragment with forward primer F2 (5'-CGTGTCAGTCATGTGTCGCT-3') and reverse primer R3 (3'-AGAAGTGGGCACAACTGAGC5'), with a detection limit of 0.1 fg. A nested RT-PCR has proven useful for detection of the virus in asymptomatic carriers (Thiery *et al.*, 1999; Gomez *et al.*, 2004). Several real-time methods have been developed, included a nucleic acid sequence-based amplification assay (Starkey *et al.*, 2004), and SYBR Green and TaqMan technologies have been used to develope RT-qPCR and qPCR for both RNA 1 and RNA 2 (Ciulli *et al.*, 2006). A rapid-diagnosis protocol has been developed for the detection of NNV by laboratories with restricted facilities and, as pond-side tests, LAMP and real-time-LAMP (Sung & Lu, 2009; Mekata *et al.*, 2015; Hwang *et al.*, 2016; Yong *et al.*, 2017). Positive staining of the virus with ISH, *in situ* PCR or *in situ* RT-PCR has demonstrated its presence in testis, fin, tectum opticum, torus semicircularis, cerebellum, thalamus, hypothalamus and medulla (Kuo *et al.*, 2011; Valero *et al.*, 2015; Kim *et al.*, 2018). All of these molecular techniques have been developed with brain tissue from infected fish, and the use of such a biological matrix requires that the fish be sacrificed. Ferreira *et al.* (2019) described the use of non-lethal samples such as blood, caudal fin and gill for the detection of NNV in infected fish.

NNV was first detected in tilapia using primer set F2 and R3 (Nishizawa *et al.*, 1994); however, once both RNA strands were sequenced, detection was performed by substituting primer F2 with PVP75 (5'-CGTGTCGGTGCTGTGT-3'), because tilapia NNV has three nucleotide substitutions in the F2 region (Bigarré *et al.*, 2009). The approach used to identify NNV in tilapia farmed in Thailand used an RT-PCR primer set developed against NNV isolated from distinct marine species from Thailand (Keawcharoen *et al.*, 2015), while a nested PCR was used to identify NNV in Tilapia seed in Indonesia (Prihartini *et al.*, (2015).

Cell culture

The propagation of NNV has been achieved with the use of several continuous fish cell lines. The list is extended and includes SNN-1 and its clone E-11, derived from the embryo of snakehead fish (*Channa striata*) (Frerichs *et al.*, 1996; Iwamoto *et al.*, 2001), GF-1 derived from the fin of grouper (*Epinephelus coioides*) (Chi *et al.*, 1999), GB derived from grouper brain (yellow grouper, *Epinephelus awoara*) (Lai *et al.*, 2001a), SAF-1 derived from the fin of gilthead seabream (*Sparus auratus* L.) (Bandín *et al.*, 2006), EAGB derived from the brain of grouper (*Epinephelus akaara*) (Huang *et al.*, 2011) and GF derived from grunt (Jitrakorn *et al.*, 2020). However, the cell lines most used for NNV propagation and quantification of cytopathic effects are still SSN-1 and E-11. The propagation of *Betanodavirus* in cell culture is achieved by homogenizing brain tissue in Hanks' balanced salt solution (HBSS), supplemented with 2% FBS and antibiotics, filtration through a sterile 0.22 μm membrane and inoculation into the selected cell line (WOAH, 2019) using pre-adsorption or simultaneous inoculation methodologies (Costa, 2005). Cells are incubated at a temperature as close as possible to that of the host, and CPE is usually observed at 4–7 dpi. The propagation of *Betanodavirus* from Nile tilapia infected tissue is restricted to the study of Bigarré *et al.* (2009), who had limited success because no CPE was observed at 6 dpi.

Serological techniques

Serological techniques, such as enzyme-linked immunosorbent assay (ELISA), have been widely used to study the antibody response of fish following vaccination in both wild and farmed fish to assess whether they have been exposed to *Betanodavirus*. ELISA can be used both as a non-lethal method for screening broodstock and as a diagnostic tool for NNV detection (Arimoto *et al.*, 1992; Mushiake *et al.*, 1992; Nishizawa *et al.*, 1995b; Breuil *et al.*, 2000; Liu *et al.*, 2006; Venkata Satyanarayana *et al.*, 2021). These assays have been developed as antigen capture, indirect, competitive or sandwich ELISAs, using monoclonal and polyclonal antibodies to either fish IgM or NNV particles, recombinant proteins, purified viral particles or virus in cell culture supernatant (Arimoto *et al.*, 1993b; Watanabe *et al.*, 2000; Shieh & Chi, 2005; Fenner *et al.*, 2006b; Jaramillo *et al.*, 2016; Gye & Nishizawa, 2020; Venkata Satyanarayana *et al.*, 2021; Makesh *et al.*, 2022). Components in cell culture medium have been shown to inhibit the binding of antigens (Choi *et al.*, 2014), however, in an ELISA developed to detect antibodies against NNV (Gye & Nishizawa, 2018). One application of serology recently reported is a rapid diagnostic test using lateral flow technology. This was developed to detect RGNNV particles in brain tissue, with a specificity of 100%, sensitivity of 94.92%, diagnostic effectiveness of 96.81% and no cross-reaction with other known fish viruses (Shyam *et al.*, 2020). Antibodies developed against NNV have also been used in IHC to facilitate localization of the virus in lesions within infected tissues (Breuil *et al.*, 2001; Lai *et al.*, 2001b).

No serology studies have been reported for tilapia *Betanodavirus*. *Betanodavirus* isolates obtained from other fish species can be detected using the same antibodies if these have been produced against the same genotype as the virus isolate. Because all tilapia *Betanodavirus* isolates detected to date belong to genotype RGNNV, serology technologies should be transferable to detection of tilapia NNV or antibodies produced by tilapia against the virus using commercially available anti-tilapia IgM antibodies.

Protective and control strategies

Betanodavirus is one of the most stable viruses to affect fish (Frerichs *et al.*, 1996). The most efficient method of controlling the virus is through strict biosecurity measures to prevent the virus from entering the farming system. The virus can be completely inactivated by heat treatment, ultraviolet light (UV), ozone, alkaline conditions (pH 12) and with chemicals such as sodium hypochlorite, calcium hypochlorite and benzalkonium chloride (Arimoto *et al.*, 1996; Frerichs *et al.*, 2000).

The susceptibility of fish to *Betanodavirus* infection depends on the age of the fish, with the hatchery stage of the life cycle particularly susceptible (Toffan *et al.*, 2021). To help prevent infection in the hatchery, measures such as screening of broodstock for the presence of antibodies against NNV and screening of ovarian fluid, sperm and eggs for the presence of the virus should be performed. Disinfection of fertilized eggs, screening of hatched larvae by RT-PCR and the introduction of a policy of limiting the number of spawnings have also helped reduce NNV-related mortalities (Mushiake *et al.*, 1994; Mori *et al.*, 1998; Watanabe *et al.*, 1998; Breuil *et al.*, 2000; Dalla Valle *et al.*, 2000; Grotmol and Totland, 2000; Watanabe *et al.*, 2000; Buchan *et al.*, 2006; Panzarin *et al.*, 2010; Nishioka *et al.*, 2011; Watanabe *et al.*, 2013). Breeding for resistance is an option with high heritability in species including Atlantic cod, European sea bass and Asian sea bass, in which it has been shown that resistance to NNV can be increased by genetic improvement (Bangera *et al.*, 2011; Wang *et al.*, 2017; Palaiokostas *et al.*, 2018; Griot *et al.*, 2021). Vaccination of broodstock is another promising method for stopping the spread of the virus from broodstock to eggs (Kai *et al.*, 2010). While commercial vaccines based on inactivated virus are available for other fish species, no commercial vaccines have been licensed for use in tilapia and there are also no published studies relating to vaccination of tilapia against *Betanodavirus*.

Infectious pancreatic necrosis virus

There have only been two reports of infectious pancreatic necrosis virus (IPNV) being detected in farmed tilapia. The first was related to *Tilapia mossambicus* in the Lukang region of Taiwan by Hedrick *et al.* in 1983. It was not until 2018 that IPNV was detected for the second time in tilapia, this time in Nile tilapia, farmed in Kenya (Mulei *et al.*, 2018). In each case, no clinical signs of disease were associated with the virus. The viruses in both reports were genetically similar and identical to the European IPNV AB strain (Machimbirike *et al.*, 2019).

Infectious agent

The virus is a member of the genus *Aquabirnavirus* in the family Birnaviridae. It is bi-segmented, with dsRNA composed of two linear segments, A and B. Segment A encodes the major structural proteins VP2, VP3, the viral protease VP4 and the small non-structural protein, VP5da and is 3092 bp in length, while segment B encodes for the polymerase VP1 and is 2748 bp in length (Santi *et al.*, 2005; Munang'andu *et al.*, 2016). The virus is a non-enveloped, icosahedral virion with a diameter of around 60 nm (McAllister, 1983;

Rodriguez Saint-Jean *et al.*, 2003). Genotyping is based on the VP2 sequence, with six different genogroups identified, these correlating with the serotyping system for the virus (Blake *et al.*, 2001; Cutrín *et al.*, 2004). The amino acids in the VP2 capsid on the outer surface of the virus are highly variable and are responsible for the antigenic diversity between different aquabirnaviruses, and are used to type the virus (Munang'andu *et al.*, 2016).

Geographical distribution, host range and transmission routes

IPNV has a diverse geographical range, with the virus having been detected in Australia, Canada, Chile, China, the European mainland, Ireland, Japan, Korea, Laos, New Zealand, Norway, Scotland, Thailand and the USA (Chong, 2022). Infectious pancreatic necrosis (IPN), the disease caused by the virus, is associated with high mortality levels in young salmonid fish. In contrast, fish older than 6 months tend to suffer subclinical infection with lower levels of mortality (McAllister, 1983). The virus is predominantly found in marine environments but has been isolated from both marine and freshwater fish (salmonids and non-salmonids) and other aquatic animals (Machimbirike *et al.*, 2019).

Both horizontal and vertical transmission has been reported for IPNV (McAllister, 1983). Survivors of an IPNV outbreak can become carriers, shedding the virus into the water and transmitting the virus horizontally. In contrast, subclinically infected fish can transmit the virus vertically to fertilized eggs (Wolf, 1988). The IPNV in tilapia from Taiwan, identified as an isolate closely related to the European isolate AB strain, implies the possible intercontinental spread of the virus (Machimbirike *et al.*, 2019). In Kenya, two affected tilapia farms were supplied with water from water bodies stocked with rainbow trout infected with the virus; the tilapia appears to have been horizontally infected with the virus through contaminated water from the trout farms. The third tilapia farm had no shared stocking activities or water sources with these farms, and virus transmission may have occurred through sharing of harvesting nets (Mulei *et al.*, 2018).

Pathology and diagnosis

Clinical signs and gross lesions

No clinical signs or gross lesions were associated with IPNV in the tilapia from Taiwan or Kenya. Clinical symptoms typical of IPN include darkening of body coloration, exophthalmia, abdominal distension and faecal mucoid casts. The pathognomonic pathology of IPN in salmonids is acute catarrhal enteritis (McAllister, 1983; Rodriguez Saint-Jean *et al.*, 2003). Affected fish can have petechial haemorrhaging on their pectoral and pelvic fins and exhibit bursts of corkscrew swimming or whirling behaviour (McAllister, 1983). Internally the liver and spleen appear pale and enlarged (McAllister, 1983; Rodriguez Saint-Jean *et al.*, 2003), and petechial haemorrhaging covers the visceral organs, including the pyloric caeca and pancreas (McAllister, 1983; Rodriguez Saint-Jean *et al.*, 2003).

Histopathology

There are no reports of pathology in the tissues of the IPNV-positive tilapia from Taiwan or Kenya (see Chong (2022) for an overview of the histopathological changes expected from clinical IPN). This includes severe pancreatic acinar cell necrosis, intracytoplasmic inclusions and sloughing of the intestinal epithelium of the pylorus, pyloric caeca and anterior intestine.

Molecular and serological diagnosis

Virus identification in the infected tilapia in Taiwan was based on virus isolation in CHSE-214 cells, neutralization assays and comparison of virion polypeptides and RNA with different IPNV serogroups (Hedrick *et al.*, 1983), while PCR and amplicon sequencing identified the IPNV from tilapia infected in Kenya (Mulei *et al.*, 2018). Both were identified as resembling the European isolate AB strain of IPNV (Machimbirike *et al.*, 2019). For further information, an overview of isolation, identification, serology and molecular diagnostics for IPNV is provided by Chong (2022).

Protective and control strategies

IPNV is challenging to inactivate in the aquatic environment (Ruane *et al.*, 2007). The virus remains infective for >7 months at 10°C in municipal tap water, but its infectivity decreases in untreated freshwater or seawater between 14 and 17 days (Rodriguez Saint-Jean *et al.*, 2003). A variety of disinfectants are effective against the virus, including Virkon (1–2% w/v); solutions with pH >12; UV >1200 J/m^2; temperatures > 60°C; and ozone at 0.1–2.0 mg/L (Ruane *et al.*, 2007) or chlorine at 200 mg/L for 1 hour (McAllister, 1983). Husbandry practices that reduce stress or the use of IPNV-free eggs (i.e. by screening broodstock and fish eggs) can help to reduce IPN outbreaks (McAllister, 1983). Surface disinfection of eggs is not regarded as an effective method to prevent vertical transmission of the virus (Bullock *et al.*, 1976).

Vaccination is another option to protect tilapia from IPNV if disease outbreaks impact production. Because IPNV strains from tilapia are genetically similar and identical to the European IPNV strain, current IPNV isolates produced for salmonids may cross-protect against tilapia strains. The inclusion of an rVP2 antigen in commercial vaccines has been shown to offer enhanced protection (Rimstad, 2014). To warn of any upcoming issues with IPNV in tilapia farming, Machimbirike *et al.* (2019) suggested that IPNV should be considered a pathogen for active surveillance in tilapia-farming countries.

Tilapia larvae encephalitis virus

Tilapia larvae encephalitis virus (TLEV) is a herpesvirus that infects tilapia larvae. It was first detected in laboratory-reared gynogenetic blue tilapia (*O. aureus*) larvae in Israel in 2010. Its name originates from the whirling behaviour displayed by affected larvae (Shlapobersky *et al.*, 2010). High levels of mortality have been associated with the virus. The full impact of this virus on tilapia aquaculture remains to be determined, as no other cases have been reported since those documented by Shlapobersky *et al.* (2010).

Infectious agent

Virions were observed in the cytoplasm of brain tissue from diseased tilapia larvae. These were icosahedral hexagonal in shape, 100 nm in diameter and had an electron-dense core. Molecular characterization of the virus was performed using pairwise alignment of a TLEV-partial sequence. Phylogenetic analysis placed the virus within the family Herpesviridae and was unrelated to Alloherpesviridae and Iridoviridae. Compared with other herpesviruses, it had the highest similarity (76%) and identity (64%) to turtle herpesvirus GTHV-Ha.

Geographical distribution, host range and transmission routes

To date, TLEV has been reported only in Israel, and there are no reports of species other than tilapia being affected. Shlapobersky *et al.* (2010) showed horizontal transmission between different tilapia species using a cohabitation challenge model in which morbid blue larvae were used to infect naïve red tilapia (a hybrid of different tilapia species originating from the Philipinnes (Galman and Avtalion, 1983), Nile tilapia (originating from Ghana) and *Sarotherodon galilaeus* (from the Sea of Galilee). A single peak of mortality was noted in the naïve fish after cohabiting with the blue larvae, and a two-peak mortality pattern was evidenced in the offspring of surviving females for all three tilapia species. The authors also showed vertical transmission from these females, with sexually mature males and females being potential carriers and vertical spreaders of the virus (Shlapobersky *et al.*, 2010).

Pathology and diagnosis

Clinical signs and gross lesions

Symptoms of TELV infection included whirling swimming behaviour and dark pigmentation of the skin and fins. Infected larvae had herpes virions within their brain tissue and, together with the observed whirling swimming behaviour, this implied damage to the central nervous system of the fish. High levels of mortality were also associated with the infection (92–98%).

In a subsequent study, Sinyakov *et al.* (2011) evaluated the phagocyte-mediated reactive oxygen species (ROS) response of tilapia larvae to TELV as an indicator of innate immunity impairment by the virus. The virus impaired the ability of ROS-producing cells to produce ROS in both vertically and horizontally infected fish, indicating severe TLEV-imposed phagocyte dysfunction in affected larvae.

Histopathology

No histopathological lesions were described by Shlapobersky *et al.* (2010).

Molecular diagnosis

A PCR-based assay has been developed to detect TELV, which is available to screen fish and farming systems for the virus (Shlapobersky *et al.*, 2010).

Protective and control strategies

The high levels of mortality associated with the virus suggest that TELV poses a threat to tilapia farming. Fish from two unrelated external hatcheries in Israel, kibbutz Hamaapil and Gan Shmuel, displayed similar symptoms to TELV-infected fish (Shlapobersky *et al.*, 2010). The authors suggest that other farmers might have overlooked the disease. Further studies are needed to establish the status of TLEV infections in other tilapia and non-tilapia species. This would inform on the need for, and design of, protective and control strategies. The authors suggest that systematic PCR screening of broodstocks should be considered to identify TLEV carriers.

Conclusion

The viruses outlined in this chapter have the potential to severely impact tilapia aquaculture globally. For tilapia farmers, TiLV and ISKNV are considered the most significant viruses carrying a high risk of transboundary spread. There has been an increase in the number of studies conducted on TiLV. Viral infections can cause high levels of mortality in tilapia stock, resulting in lost productivity and revenue for farmers and affecting food supplies, the economy and employment in tilapia-producing countries. The risk is that these viruses will be introduced to unaffected regions through the movement of subclinically infected fish or infected fish products without appropriate control measures being in place. More research is clearly needed to establish the impact of these viruses on tilapia aquaculture, and to determine their geographical distribution. Information is needed on the presence of these viruses within farming systems, their persistence in the environment and further virus characterization: e.g. genetic types and virulence. Risk factors that predispose fish to disease by these viruses also need to be determined, including the fish's response to the virus. Understanding the potential issues these viruses can have on tilapia farming will inform on the need for control strategies and what form these should take. This will allow action to be taken before they have the opportunity to become a significant concern for the tilapia industry.

References

Acharya, V., Chakraborty, H.J., Rout, A.K., Balabantaray, S., Behera, B.K. *et al.* (2019) Structural characterization of open reading frame-encoded functional genes from Tilapia Lake Virus (TiLV). *Molecular*

Biotechnology 61(12), 945–957. Available at: https://doi.org/10.1007/s12033-019-00217-y

Adams, M.J., Lefkowitz, E.J., King, A.M.Q., Harrach, B., Harrison, R.L. *et al.* (2017) Changes to taxonomy and the international code of virus classification and nomenclature ratified by the international committee on taxonomy of viruses. *Archives of Virology* 162(8), 2505–2538. Available at: https://doi.org/10.1007/s00705-017-3358-5

Anders, K. (1989) Lymphocysytis diseases of fishes. In: Ahne, W. and Kurstak, E. (eds) *Viruses of Lower Vertebrates*. Springer, pp. 141–160.

Ariel, E. and Owens, L. (1997) Epizootic mortalities in tilapia *Oreochromis mossambicus*. *Diseases of Aquatic Organisms* 29, 1–6. Available at: https://doi.org/10.3354/dao029001

Ariel, E., Wirth, W., Burgess, G., Scott, J. and Owens, L. (2015) Pathogenicity in six Australian reptile species following experimental inoculation with Bohle iridovirus. *Diseases of Aquatic Organisms* 115(3), 203–212. Available at: https://doi.org/10.3354/dao02889

Arimoto, M., Mushiake, K., Mizuta, Y., Nakai, T., Muroge, K. *et al.* (1992) Detection of Striped Jack Nervous Necrosis Virus(SJNNV) by Enzyme-Linked Immunosorbent Assay(ELISA). *Fish Pathology* 27(4), 191–195. Available at: https://doi.org/10.3147/jsfp.27.191

Arimoto, M., Maruyama, K. and Furusawa, I. (1993a) Epizootiology of viral vervous vecrosis (VNN) in striped jack. *Fish Pathology* 29, 19–24.

Arimoto, M., Mori, K., Nakai, T., Muroga, K. and Furusawa, I. (1993b) Pathogenicity of the causative agent of viral nervous necrosis disease in Striped Jack, pseudocaranx dentex (Bloch and Schneider). *Journal of Fish Diseases* 16(5), 461–469. Available at: https://doi.org/10.1111/j.1365-2761.1993.tb00879.x

Arimoto, M., Sato, J., Maruyama, K., Mimura, G. and Furusawa, I. (1996) Effect of chemical and physical treatments on the inactivation of Striped Jack Nervous Necrosis Virus (SJNNV). *Aquaculture* 143(1), 15–22. Available at: https://doi.org/10.1016/0044-8486(96)01261-6

Athanassopoulou, F., Billinis, C., Psychas, V. and Karipoglou, K. (2003) Viral encephalopathy and retinopathy of *Dicentrarchus labrax* (L.) farmed in fresh water in Greece. *Journal of Fish Diseases* 26(6), 361–365. Available at: https://doi.org/10.1046/j.1365-2761.2003.00458.x

Australian Department of Agriculture (2014) Importation of freshwater ornamental fish: A review of biosecurity risks associated with gourami Iridovirus and related viruses. Final import risk analysis report. Retrieved February 19, 2022. Department of Agriculture. Available at: https://www.agriculture.gov.au/sites/default/files/style%20library/images/daff/__data/assets/pdffile/0004/2404309/gourami-ira.pdf

Bacharach, E. and Eldar, A. (2016) Tilapia lake virus vaccines. Google. Retrieved February 19, 2022. Available at: https://patents.google.com/patent/US20160354458A1/en

Bacharach, E., Mishra, N., Briese, T., Zody, M.C., Kembou Tsofack, J.E. *et al.* (2016) Characterization of a novel orthomyxo-like virus causing mass Die-Offs of tilapia. *mBio* 7(2), e00431–e00416. Available at: https://doi.org/10.1128/mBio.00431-16

Bandín, I., Olveira, J.G., Borrego, J.J., Dopazo, C.P. and Barja, J.L. (2006) Susceptibility of the fish cell line SAF-1 to Betanodavirus. *Journal of Fish Diseases* 29(10), 633–636. Available at: https://doi.org/10.1111/j.1365-2761.2006.00757.x

Bangera, R., Ødegård, J., Præbel, A.K., Mortensen, A. and Nielsen, H.M. (2011) Genetic correlations between growth rate and resistance to vibriosis and viral nervous necrosis in Atlantic cod (*Gadus morhua* L.). *Aquaculture* 317(1–4), 67–73. Available at: https://doi.org/10.1016/j.aquaculture.2011.04.018

Barja, J.L. (2004) Report about fish viral diseases. In: Alvarez-Pellitero, P., Barja, J.L., Basurco, B., Berthe, F. and Toranzo, A.E. (eds) *Mediterranean Aquaculture Diagnostic Laboratories.* CIHEAM, pp. 91–102.

Barnes, A.C., Rudenko, O., Landos, M., Dong, H.T., Lusiastuti, A. *et al.* (2022) Autogenous vaccination in aquaculture: A locally enabled solution towards reduction of the global antimicrobial resistance problem. *Reviews in Aquaculture* 14(2), 907–918. Available at: https://doi.org/10.1111/raq.12633

Behera, B.K., Pradhan, P.K., Swaminathan, T.R., Sood, N., Paria, P. *et al.* (2018) Emergence of tilapia lake virus associated with mortalities of farmed Nile tilapia *Oreochromis niloticus* (Linnaeus 1758) in India. *Aquaculture* 484, 168–174. Available at: https://doi.org/10.1016/j.aquaculture.2017.11.025

Bellance, R. and Gallet de Saint-Aurin, D. (1988) L'encephalite virale du loup de mer. *Caraïbes Medical* 2, 105–144.

Bermúdez, R., Losada, A.P., Azevedo, A.M., Guerra-Varela, J., Pérez-Fernández, D. *et al.* (2018) First description of a natural infection with spleen and kidney necrosis virus in zebrafish. *Journal of Fish Diseases* 41, 1283–1294. Available at: https://doi.org/10.1111/jfd.12822

Bigarré, L., Cabon, J., Baud, M., Heimann, M., Body, A. *et al.* (2009) Outbreak of betanodavirus infection in tilapia, *Oreochromis niloticus*

(L.), in fresh water. *Journal of Fish Diseases* 32(8), 667–673. Available at: https://doi.org/10.1111/j.1365-2761.2009.01037.x

Blake, S., Ma, J.Y., Caporale, D.A., Jairath, S. and Nicholson, B.L. (2001) Phylogenetic relationships of aquatic birnaviruses based on deduced amino acid sequences of genome segment A cDNA. *Diseases of Aquatic Organisms* 45(2), 89–102. Available at: https://doi.org/10.3354/dao045089

Bloch, B., Gravningen, K. and Larsen, J.L. (1991) Encephalomyelitis among turbot associated with a picornavirus-like agent. *Diseases of Aquatic Organisms* 10, 65–70. Available at: https://doi.org/10.3354/dao010065

Borrego, J.J., Valverde, E.J., Labella, A.M. and Castro, D. (2017) Lymphocystis disease virus: Its importance in aquaculture. *Reviews in Aquaculture* 9(2), 179–193. Available at: https://doi.org/10.1111/raq.12131

Breuil, G., Bonami, J.R., Pepin, J.F. and Pichot, Y. (1991) Viral infection(picorna-like virus) associated with mass mortalities in hatchery-reared sea bass (*Dicentrarchus labrax*) larvae and juveniles. *Aquaculture* 97(2–3), 109–116. Available at: https://doi.org/10.1016/0044-8486(91)90258-9

Breuil, G., Pépin, J., Castric, J., Fauvel, C. and Thiéry, R. (2000) Detection of serum antibodies against nodavirus in wild and farmed adult sea bass: Application to the screening of broodstock in sea bass hatcheries. *Bulletin of the European Association of Fish Pathologists* 20, 95–100.

Breuil, G., Mouchel, O., Fauvel, C. and Pépin, J.F. (2001) Sea bass *Dicentrarchus labrax* nervous necrosis virus isolates with distinct pathogenicity to sea bass larvae. *Diseases of Aquatic Organisms* 45(1), 25–31. Available at: https://doi.org/10.3354/dao045025

Breuil, G., Pépin, J.F.P., Boscher, S. and Thiéry, R. (2002) Experimental vertical transmission of nodavirus from broodfish to eggs and larvae of the sea bass, *Dicentrarchus labrax* (L.). *Journal of Fish Diseases* 25(12), 697–702. Available at: https://doi.org/10.1046/j.1365-2761.2002.00406.x

Buchan, K.A.H., Martin-Robichaud, D.J., Benfey, T.J., MacKinnon, A. and Boston, L. (2006) The efficacy of ozonated seawater for surface disinfection of haddock (*Melanogrammus aeglefinus*) eggs against piscine nodavirus. *Aquacultural Engineering* 35(1), 102–107. Available at: https://doi.org/10.1016/j.aquaeng.2005.10.001

Bullock, G.L., Rucker, R.R., Amend, D., Wolf, K. and Stuckey, H.M. (1976) Infectious pancreatic necrosis: Transmission with iodine-treated and nontreated eggs of brook trout (*Salvelinus fontinalis*). *Journal of the Fisheries Research Board of Canada* 33(5), 1197–1198. Available at: https://doi.org/10.1139/f76-152

Cano, I., Alonso, M.C., Garcia-Rosado, E., Saint-Jean, S.R., Castro, D, *et al.* (2006) Detection of lymphocystis disease virus (LCDV) in asymptomatic cultured gilt-head seabream (*Sparus aurata*, L.) using an immunoblot technique. SR. Jeonju University, *Veterinary Microbiology* 113(1–2), 137–141. Available at: https://doi.org/10.1016/j.vetmic.2005.10.038

Cano, I., Ferro, P., Alonso, M.C., Bergmann, S.M., Römer-Oberdörfer, A. *et al.* (2007) Development of molecular techniques for detection of lymphocystis disease virus in different marine fish species. *Journal of Applied Microbiology* 102(1), 32–40. Available at: https://doi.org/10.1111/j.1365-2672.2006.03066.x

Cano, I., Lopez-Jimena, B., Garcia-Rosado, E., Ortiz-Delgado, J.B., Alonso, M.C. *et al.* (2009) Detection and persistence of lymphocystis disease virus (LCDV) in *Artemia* sp. *Aquaculture* 291(3–4), 230–236. Available at: https://doi.org/10.1016/j.aquaculture.2009.03.018

Cano, I., Valverde, E.J., Garcia-Rosado, E., Alonso, M.C., Lopez-Jimena, B. *et al.* (2013) Transmission of lymphocystis disease virus to cultured gilthead seabream, *Sparus aurata* L., larvae. *Journal of Fish Diseases* 36(6), 569–576. Available at: https://doi.org/10.1111/jfd.12011

Carballo, C., Mateus, A.P., Maya, C., Mantecón, L., Power, D.M. *et al.* (2020) Microalgal extracts induce larval programming and modify growth and the immune response to bioactive treatments and LCDV in Senegalese sole post-larvae. *Fish and Shellfish Immunology* 106, 263–272. Available at: https://doi.org/10.1016/j.fsi.2020.07.020

Cartner, S.C., Eisen, J.S., Farmer, S.C., Guillemin, K.J., Kent, M.L. *et al.* (2019) *The zebrafish in biomedical research: Biology, husbandry, diseases, and research applications*. Academic Press.

Castri, J., Thiéry, R., Jeffroy, J., Kinkelin, P. and Raymond, J.C. (2001) Sea bream *Sparus aurata*, an asymptomatic contagious fish host for nodavirus. *Diseases of Aquatic Organisms* 47(1), 33–38. Available at: https://doi.org/10.3354/dao047033

Chaput, D.L., Bass, D., Alam, M.M., Hasan, N.A., Stentiford, G.D. *et al.* (2020) The segment matters: Probable reassortment of tilapia lake virus (TiLV) complicates phylogenetic analysis and inference of geographical origin of new isolate from Bangladesh. *Viruses* 12(3), 258. Available at: https://doi.org/10.3390/v12030258

Chen, L.J., Su, Y.C. and Hong, J.R. (2009) Betanodavirus non-structural protein B1: A novel anti-necrotic death factor that modulates cell death in early replication cycle in fish cells. *Virology* 385(2), 444–454. Available at: https://doi.org/10.1016/j.virol.2008.11.048

Chen, S.P., Wu, J.L., Su, Y.C. and Hong, J.R. (2007) Anti-bcl-2 family members, zfBcl-x(L) and zfMcl-1a, prevent cytochrome c release from cells undergoing Betanodavirus-induced secondary necrotic cell death. *Apoptosis* 12(6), 1043–1060. Available at: https://doi.org/10.1007/s10495-006-0032-x

Chengula, A.A., Kasanga, C., Mutoloki, S., Mdegela, R. and Evensen, Ø. (2017) Molecular characterization of ranaviruses detected in Nile tilapia (*Oreochromis niloticus*). In: *4th International Symposium on Ranaviruses, Budapest, Hungary, 70.*

Chi, S.C., Lo, C.F., Kou, G.H., Chang, P.S., Peng, S.E. *et al.* (1997) Mass mortalities associated with viral nervous necrosis (VNN) disease in two species of hatchery-reared grouper *Epinephelus fuscoguttatus* and *Epinephelus akaara* (Temminck and Schlegel). *Journal of Fish Diseases* 20(3), 185–193. Available at: https://doi.org/10.1046/j.1365-2761.1997.00291.x

Chi, S.C., Hu, W.W. and Lo, B.J. (1999) Establishment and characterization of a continuous cell line (GF-1) derived from grouper, *Epinephelus coioides* (Hamilton): A cell line susceptible to grouper nervous necrosis virus (GNNV). *Journal of Fish Diseases* 22(3), 173–182. Available at: https://doi.org/10.1046/j.1365-2761.1999.00152.x

Chiamkunakorn, C., Machimbirike, V.I., Senapin, S., Khunrae, P., Dong, H.T. *et al.* (2019) Blood and liver biopsy for the non-destructive screening of tilapia lake virus. *Journal of Fish Diseases* 42(11), 1629–1636. Available at: https://doi.org/10.1111/jfd.13076

Chinchar, V.G., Hyatt, A., Miyazaki, T. and Williams, T. (2009) Family iridoviridae: poor viral relations no longer. In: *Lesser Known Large dsDNA Viruses.* pp. 123–170. Available at: https://doi.org/10.1007/978-3-540-68618-7_4

Chinchar, V.G., Hick, P., Ince, I.A., Jancovich, J.K., Marschang, R. *et al.* (2017a) ICTV virus taxonomy profile: *Iridoviridae*. *Journal of General Virology* 98(5), 890–891. Available at: https://doi.org/10.1099/jgv.0.000818

Chinchar, V.G., Waltzek, T.B. and Subramaniam, K. (2017b) Ranaviruses and other members of the family *Iridoviridae*: Their place in the virosphere. *Virology* 511, 259–271. Available at: https://doi.org/10.1016/j.virol.2017.06.007

Choi, B., Gye, H.J., Oh, M.J. and Nishizawa, T. (2014) Cell culture medium inhibits antigen binding used in an ELISA for detection of antibodies against nervous necrosis virus. *Journal of Aquatic Animal Health* 26(3), 168–172. Available at: https://doi.org/10.1080/08997659.2014.922516

Chong, R.S.-M. (2022) Infectious pancreatic necrosis. In: Kibenge, F.S.B., Baldisserotto, B. and Chong, R.S.-M. (eds) *Aquaculture Pathophysiology Volume I. Finfish Diseases*, 1st edn. Acedemic Press, pp. 165–175.

Chua, F.H.C., Loo, J.J. and Wee, J.Y. (1995) Mass mortality in juvenile greasy grouper, *epinephelus tauvina*, associated with vacuolating encephalopathy and retinopathy. In: Shariff, M., Arthur, J.R. and Subasinghe, R.P. (eds) *Diseases in Asian Aquaculture II*. Fish Health Section, Asian Fisheries Society, pp. 235–241.

Ciulli, S., Galletti, E., Gallina, L., Vaccari, F. and Prosperi, S. (2006) Detection and quantification of betanodavirus by real time PCR. *Veterinary Research Communications* 30(S1), 235–238. DOI: 10.1007/s11259-006-0049-1.

Comps, M., Pépin, J.F. and Bonami, J.R. (1994) Purification and characterization of two fish encephalitis viruses (FEV) infecting *Lates calcarifer* and *Dicentrarchus labrax*. *Aquaculture* 123(1–2), 1–10. DOI: 10.1016/0044-8486(94)90114-7.

Costa, J.Z. (2005) B-cell epitopes in fish nodavirus. (Doctor in Philosophy Doctor in Philosophy). University of Stirling.

Costa, J.Z. and Thompson, K.D. (2016) Understanding the interaction between betanodavirus and its host for the development of prophylactic measures for viral encephalopathy and retinopathy. *Fish and Shellfish Immunology* 53, 35–49. DOI: 10.1016/j.fsi.2016.03.033.

Cullen, B.R., Owens, L. and Whittington, R.J. (1995) Experimental infection of Australian anurans (*Limnodynastes terraereginae* and *Litoria latopalmata*) with Bohle iridovirus. *Diseases of Aquatic Organisms* 23, 83–92. DOI: 10.3354/dao023083.

Curtis, P.A., Drawbridge, M., Iwamoto, T., Nakai, T., Hedrick, R.P. *et al.* (2001) Nodavirus infection of juvenile white sea bass, *Atractoscion nobilis*, cultured in Southern California: First record of viral nervous necrosis (VNN) in North America. *Journal of Fish Diseases* 24(5), 263–271. DOI: 10.1046/j.1365-2761.2001.00292.x.

Cutrín, J.M., Barja, J.L., Nicholson, B.L., Bandín, I., Blake, S. *et al.* (2004) Restriction fragment length polymorphisms and sequence analysis: An approach for genotyping infectious pancreatic necrosis virus reference strains and other aquabirnaviruses isolated from Northwestern Spain. *Applied and Environmental Microbiology* 70(2), 1059–1067. Available at: DOI: https://doi.org/10.1128/AEM.70.2.1059-1067.2004.

Dalla Valle, L., Zanella, L., Patarnello, P., Paolucci, L., Belvedere, P. *et al.* (2000) Development of a sensitive diagnostic assay for fish nervous necrosis virus based on RT-PCR plus nested PCR. *Journal of Fish Diseases* 23(5), 321–327. DOI: 10.1046/j.1365-2761.2000.00255.x.

Debnath, P.P., Delamare-Deboutteville, J., Jansen, M.D., Phiwsaiya, K., Dalia, A. *et al.* (2020) Two-year surveillance of tilapia lake virus (TiLV) reveals its wide circulation in tilapia farms and hatcheries from multiple districts of Bangladesh. *Journal of Fish Diseases* 43(11), 1381–1389. DOI: 10.1111/jfd.13235.

Delamare-Deboutteville, J., Taengphu, S., Gan, H.M., Kayansamruaj, P., Debnath, P.P. *et al.* (2021) Rapid genotyping of tilapia lake virus (TiLV) using nanopore sequencing. *Journal of Fish Diseases* 44(10), 1491–1502. DOI: 10.1111/jfd.13467.

del-Pozo, J., Mishra, N., Kabuusu, R., Cheetham, S., Eldar, A. *et al.* (2017) Syncytial hepatitis of tilapia (*Oreochromis niloticus* L.) is associated with orthomyxovirus-like virions in hepatocytes. *Veterinary Pathology* 54(1), 164–170. DOI: 10.1177/0300985816658100.

Delsert, C., Morin, N. and Comps, M. (1997) A fish encephalitis virus that differs from other nodaviruses by its capsid protein processing. *Archives of Virology* 142(12), 2359–2371. DOI: 10.1007/s007050050248.

de Padua, B. (2017) Doenças virais na tilapicultura Brasileira: O que teremos pela frente?. Retrieved February 19, 2022. Available at: https://www.aquaculturebrasil.com/coluna/133/doencas-virais-na-tilapicultura-brasileira:-o-que-teremos-pela-frente

Dinh-Hung, N., Sangpo, P., Kruangkum, T., Kayansamruaj, P., Rung-Ruangkijkrai, T. *et al.* (2021) Dissecting the localization of tilapia tilapinevirus in the brain of the experimentally infected Nile tilapia, *Oreochromis niloticus* (L.). *Journal of Fish Diseases* 44(8), 1053–1064. DOI: 10.1111/jfd.13367.

Dong, C., Xiong, X., Luo, Y., Weng, S., Wang, Q. *et al.* (2013) Efficacy of a formalin-killed cell vaccine against infectious spleen and kidney necrosis virus (ISKNV) and immunoproteomic analysis of its major immunogenic proteins. *Veterinary Microbiology* 162(2–4), 419–428. DOI: 10.1016/j.vetmic.2012.10.026.

Dong, H.T., Nguyen, V.V., Le, H.D., Sangsuriya, P., Jitrakorn, S. *et al.* (2015) Naturally concurrent infections of bacterial and viral pathogens in disease outbreaks in cultured Nile tilapia (*Oreochromis niloticus*)

farms. *Aquaculture* 448, 427–435. DOI: 10.1016/j.aquaculture.2015.06.027.

Dong, H.T., Jitrakorn, S., Kayansamruaj, P., Pirarat, N., Rodkhum, C. *et al.* (2017a) Infectious spleen and kidney necrosis disease (ISKND) outbreaks in farmed barramundi (*Lates calcarifer*) in Vietnam. *Fish and Shellfish Immunology* 68, 65–73. DOI: 10.1016/j.fsi.2017.06.054.

Dong, H.T., Siriroob, S., Meemetta, W., Santimanawong, W., Gangnonngiw, W. *et al.* (2017b) Emergence of tilapia lake virus in Thailand and an alternative semi-nested RT-PCR for detection. *Aquaculture* 476, 111–118. DOI: 10.1016/j.aquaculture.2017.04.019.

Dong, H.T., Ataguba, G.A., Khunrae, P., Rattanarojpong, T. and Senapin, S. (2017c) Evidence of TiLV infection in tilapia hatcheries from 2012 to 2017 reveals probable global spread of the disease. *Aquaculture* 479, 579–583. DOI: 10.1016/j.aquaculture.2017.06.035.

Dong, H.T., Senapin, S., Gangnonngiw, W., Nguyen, V.V., Rodkhum, C. *et al.* (2020) Experimental infection reveals transmission of tilapia lake virus (TiLV) from tilapia broodstock to their reproductive organs and fertilized eggs. *Aquaculture* 515, 734541. DOI: 10.1016/j.aquaculture.2019.734541.

Du, J., Wang, W., Chan, J.F., Wang, G., Huang, Y. *et al.* (2019) Identification of a novel Ichthyic parvovirus in Marine species in Hainan island, China. *Frontiers in Microbiology* 10, 2815. DOI: 10.3389/fmicb.2019.02815.

Essbauer, S. and Ahne, W. (2001) Viruses of lower vertebrates. *Journal of Veterinary Medicine. B, Infectious Diseases and Veterinary Public Health* 48(6), 403–475. DOI: 10.1046/j.1439-0450.2001.00473.x.

Eyngor, M., Zamostiano, R., Kembou Tsofack, J.E., Berkowitz, A., Bercovier, H. *et al.* (2014) Identification of a novel RNA virus lethal to tilapia. *Journal of Clinical Microbiology* 52(12), 4137–4146. DOI: 10.1128/JCM.00827-14.

Fathi, M., Dickson, C., Dickson, M., Leschen, W., Baily, J. *et al.* (2017) Identification of tilapia lake virus in Egypt in Nile tilapia affected by "summer mortality"syndrome. *Aquaculture* 473, 430–432. DOI: 10.1016/j.aquaculture.2017.03.014.

Fenner, B.J., Du, Q., Goh, W., Thiagarajan, R., Chua, H.K. *et al.* (2006a) Detection of betanodavirus in *Juvenile barramundi, Lates calcarifer* (Bloch), by antigen capture ELISA. *Journal of Fish Diseases* 29(7), 423–432. DOI: 10.1111/j.1365-2761.2006.00736.x.

Fenner, B.J., Thiagarajan, R., Chua, H.K. and Kwang, J. (2006b) Betanodavirus B2 is an RNA interference antagonist that facilitates intracellular viral RNA accumulation. *Journal of Virology* 80(1), 85–94. DOI: 10.1128/JVI.80.1.85-94.2006.

Ferguson, H.W., Kabuusu, R., Beltran, S., Reyes, E., Lince, J.A. *et al.* (2014) Syncytial hepatitis of farmed tilapia, *Oreochromis niloticus* (L.): A case report. *Journal of Fish Diseases* 37(6), 583–589. DOI: 10.1111/jfd.12142.

Ferreira, I.A., Costa, J.Z., Macchia, V., Dawn Thompson, K. and Baptista, T. (2019) Detection of betanodavirus in experimentally infected European seabass (*Dicentrarchus labrax*, Linnaeus 1758) using non-lethal sampling methods. *Journal of Fish Diseases* 42(8), 1097–1105. DOI: 10.1111/jfd.13015.

Figueiredo, H.C.P., Tavares, G.C., Dorella, F.A., Rosa, J.C.C., Marcelino, S.A.C. *et al.* (2022) First report of infectious spleen and kidney necrosis virus in Nile tilapia in Brazil. *Transboundary and Emerging Diseases* 69(5), 3008–3015. DOI: 10.1111/tbed.14217.

Frerichs, G.N., Rodger, H.D. and Peric, Z. (1996) Cell culture isolation of piscine neuropathy nodavirus from juvenile sea bass, *Dicentrarchus labrax*. *Journal of General Virology* 77(9), 2067–2071. DOI: 10.1099/0022-1317-77-9-2067.

Frerichs, G.N., Tweedie, A., Starkey, W.G. and Richards, R.H. (2000) Temperature, ph and electrolyte sensitivity and heat, UV and disisnfectant inactivation of sea bass (*Dicentrarchus labrax*) neuropathy nodavirus. *Aquaculture* 185(1–2), 13–24. DOI: 10.1016/S0044-8486(99)00337-3.

Fu, X., Li, N., Lai, Y., Liu, L., Lin, Q. *et al.* (2012) Protective immunity against iridovirus disease in mandarin fish, induced by recombinant major capsid protein of infectious spleen and kidney necrosis virus. *Fish and Shellfish Immunology* 33(4), 880–885. DOI: 10.1016/j.fsi.2012.07.012.

Fu, X., Li, N., Lin, Q., Guo, H., Liu, L. *et al.* (2015) Early protein ORF086 is an effective vaccine candidate for infectious spleen and kidney necrosis virus in mandarin fish *Siniperca chuatsi*. *Fish and Shellfish Immunology* 46(2), 200–205. DOI: 10.1016/j.fsi.2015.05.052.

Fukuda, Y., Nguyen, H.D., Furuhashi, M. and Nakai, T. (1996) Mass mortality of cultured sevenband grouper, *Epinephelus septemfascinatus*, associated with viral nervous Necrosis. *Fish Pathology* 31(3), 165–170. DOI: 10.3147/jsfp.31.165.

Galman, O. and Avtalion, R.R. (1983) A preliminary investigation of the characteristics of red tilapias from the Philippines and Taiwan. In: Fishelson, L. and Yaron, Z. (eds) *Proceedings of the International Symposium on Tilapia in Aquaculture*, Tel Aviv University Press, pp. 291–301.

García-Rosado, E., Castro, D., Cano, I., Pérez-Prieto, S.I. and Borrego, J.J. (2002) Serological techniques for detection of lymphocystis virus in fish. *Aquatic Living Resources* 15(3), 179–185. DOI: 10.1016/S0990-7440(02)01174-9.

Gibson-Kueh, S., Chee, D., Chen, J., Wang, Y.H., Tay, S. *et al.* (2012) The pathology of "scale drop syndrome"in Asian sea bass, *Lates calcarifer* Bloch, a first description. *Journal of Fish Diseases* 35(1), 19–27. DOI: 10.1111/j.1365-2761.2011.01319.x.

Glazebrook, J.S., Heasman, M.P. and de Beer, S.W. (1990) Picorna-like viral particles associated with mass mortalities in larval barramundi, *Lates calcarifer* Bloch. *Journal of Fish Diseases* 13(3), 245–249. DOI: 10.1111/j.1365-2761.1990.tb00780.x.

Go, J. and Whittington, R. (2006) Experimental transmission and virulence of a megalocytivirus (family *Iridoviridae*) of dwarf gourami (*Colisa lalia*) from Asia in murray cod (*Maccullochella peelii peelii*) in Australia. *Aquaculture* 258(1–4), 140–149. DOI: 10.1016/j.aquaculture.2006.04.033.

Golomazou, E. and Panagiotaki, P. (2019) Lymphocystis disease virus (LCDV) in aquatic environment. In: Nriagu, J. (ed.) *Encyclopedia of Environmental Health*, 2nd ed. Elsevier, pp. 158–162.

Gomez, D.K., Sato, J., Mushiake, K., Isshiki, T., Okinaka, Y. *et al.* (2004) PCR-based detection of betanodaviruses from cultured and wild marine fish with no clinical signs. *Journal of Fish Diseases* 27(10), 603–608. DOI: 10.1111/j.1365-2761.2004.00577.x.

Gomez, D.K., Baeck, G.W., Kim, J.H., Choresca, C.H.J. and Park, S.C. (2008a) Genetic analysis of betanodaviruses in subclinically infected aquarium fish and invertebrates. *Current Microbiology* 56(5), 499–504. DOI: 10.1007/s00284-008-9116-x.

Gomez, D.K., Baeck, G.W., Kim, J.H., Choresca, C.H.J. and Park, S.C. (2008b) Molecular detection of betanodavirus in wild marine fish populations in Korea. *Journal of Veterinary Diagnostic Investigation* 20(1), 38–44. DOI: 10.1177/104063870802000107.

Griot, R., Allal, F., Phocas, F., Brard-Fudulea, S., Morvezen, R. *et al.* (2021) Genome-wide association studies for resistance to viral nervous necrosis in three populations of European sea bass (*Dicentrarchus labrax*) using a novel 57k SNP array DlabChip. *Aquaculture* 530, 735930. DOI: 10.1016/j.aquaculture.2020.735930.

Grotmol, S. and Totland, G.K. (2000) Surface disinfection of Atlantic halibut *Hippoglossus hippoglossus* eggs with ozonated sea-water inactivates nodavirus and increases survival of the larvae. *Diseases of Aquatic Organisms* 39(2), 89–96. DOI: 10.3354/dao039089.

Grotmol, S., Totland, G.K., Kvellestad, A., Fjell, K. and Olsen, A.B. (1995) Mass mortality of larval and juvenile hatchery-reared halibut (*Hippoglossus hippoglossus* L.) associated with the presence of virus-like particles in vacuolated lesions in the central nervous system and retina. *Bulletin of the European Association of Fish Pathologists* 15, 176–180.

Grotmol, S., Totland, G.K., Thorud, K. and Hjeltnes, B.K. (1997) Vacuolating encephalopathy and retinopathy associated with a nodavirus-like agent: A probable cause of mass mortality of cultured larval and juvenile atlantic halibut *Hippoglossus hippoglossus*. *Diseases of Aquatic Organisms* 29, 85–97. DOI: 10.3354/dao029085.

Gye, H.J. and Nishizawa, T. (2018) Reducing background optical density in enzyme-linked immunosorbent assay for detecting nervous necrosis virus (NNV)-specific igm by immobilizing fish sera. *Aquaculture* 485, 93–100. DOI: 10.1016/j.aquaculture.2017.11.039.

Gye, H.J. and Nishizawa, T. (2020) Reproducibility of antigen-immobilized enzyme-linked immunosorbent assay (ELISA) and sandwich ELISA for quantitative detection of NNV particles. *Journal of Virological Methods* 275, 113754. DOI: 10.1016/j.jviromet.2019.113754.

Haddad-Boubaker, S., Bouzgarou, N., Fakhfakh, E., Khayech, M., Mohamed, S.B. *et al.* (2013) Detection and genetic characterisation lymphocystis disease virus (LCDV) isolated during disease outbreaks in cultured gilthead sea bream *Sparus aurata* in Tunisia. *Fish Pathology* 48(3), 101–104. DOI: 10.3147/jsfp.48.101.

Hata, N., Okinaka, Y., Iwamoto, T., Kawato, Y., Mori, K. *et al.* (2010) Identification of RNA regions that determine temperature sensitivities in betanodaviruses. *Archives of Virology* 155(10), 1597–1606. DOI: 10.1007/s00705-010-0736-7.

He, J.G., Deng, M., Weng, S.P., Li, Z., Zhou, S.Y. *et al.* (2001) Complete genome analysis of the mandarin fish infectious spleen and kidney necrosis iridovirus. *Virology* 291(1), 126–139. DOI: 10.1006/viro.2001.1208.

He, T., Zhang, Y., Gao, L., Miao, B., Zheng, J. *et al.* (2023) Identification and pathogenetic study of tilapia lake virus (TiLV) isolated from naturally diseased tilapia. *Aquaculture* 565, 739166. DOI: 10.1016/j.aquaculture.2022.739166.

Hedrick, R.P., Fryer, J.L., Chen, S.N. and Kou, G.H. (1983) Characteristics of four birnaviruses isolated from fish in Taiwan. *Fish Pathology* 18(2), 91–97. DOI: 10.3147/jsfp.18.91.

Hegde, A., Teh, H.C., Lam, T.J. and Sin, Y.M. (2003) Nodavirus infection in freshwater ornamental fish, guppy, *Poicelia* reticulata-comparative characterization and pathogenicity studies. *Archives of Virology* 148(3), 575–586. DOI: 10.1007/s00705-002-0936-x.

Hick, P., Schipp, G., Bosmans, J., Humphrey, J. and Whittington, R. (2011) Recurrent outbreaks of viral nervous necrosis in intensively cultured barramundi (*Lates calcarifer*) due to horizontal transmission of betanodavirus and recommendations for disease control. *Aquaculture* 319(1–2), 41–52. DOI: 10.1016/j.aquaculture.2011.06.036.

Huang, X., Huang, Y., Ouyang, Z. and Qin, Q. (2011) Establishment of a cell line from the brain of grouper (*Epinephelus akaara*) for cytotoxicity testing and virus pathogenesis. *Aquaculture* 311(1–4), 65–73. DOI: 10.1016/j.aquaculture.2010.11.037.

Huang, Z., Tang, J., Li, M., Fu, Y., Dong, C. *et al.* (2012) Immunological evaluation of *Vibrio alginolyticus*, *Vibrio harveyi*, *Vibrio vulnificus* and infectious spleen and kidney necrosis virus (ISKNV) combined-vaccine efficacy in *Epinephelus coioides*. *Veterinary Immunology and Immunopathology* 150(1–2), 61–68. DOI: 10.1016/j.vetimm.2012.08.008.

Hwang, J., Suh, S.S., Park, M., Oh, M.J., Kim, J.O. *et al.* (2016) Detection of coat protein gene of nervous necrosis virus using loop-mediated isothermal amplification. *Asian Pacific Journal of Tropical Medicine* 9(3), 235–240. DOI: 10.1016/j.apjtm.2016.01.035.

ICTV (2022a) Family. *Iridoviridae*. Available at: https://ictv.global/taxonomy/taxondetails?taxnode_id=202103725

ICTV (2022b) Genus. *Lymphocystivirus*. Available at: https://ictv.global/taxonomy/taxondetails?taxnode_id=202103727 Retrieved February 22, 2022

Iwamoto, T., Nakai, T., Mori, K., Arimoto, M. and Furusawa, I. (2000) Cloning of fish cell line SSN-1 for piscine nodavirus. *Diseases of Aquatic Organisms* 43(2), 81–89. DOI: 10.3354/dao043081.

Iwamoto, T., Mori, K., Arimoto, M. and Nakai, T. (2001) A combined cell-culture and RT-PCR method for rapid detection of piscine nodaviruses. *Journal of Fish Diseases* 24(4), 231–236. DOI: 10.1046/j.1365-2761.2001.00291.x.

Jaemwimol, P., Rawiwan, P., Tattiyapong, P., Saengnual, P., Kamlangdee, A. *et al.* (2018) Susceptibility of important warm water fish species to tilapia lake virus (TiLV) infection. *Aquaculture* 497, 462–468. DOI: 10.1016/j.aquaculture.2018.08.028.

Jaemwimol, P., Sirikanchana, K., Tattiyapong, P., Mongkolsuk, S. and Surachetpong, W. (2019) Virucidal effects of common disinfectants against tilapia lake virus. *Journal of Fish Diseases* 42(10), 1383–1389. DOI: 10.1111/jfd.13060.

Jancovich, J.K., Chinchar, V.G., Hyatt, A., Miyazaki, T., Williams, T. *et al.* (2012) Family *Iridoviridae*. In: King, A.M.Q., Adams, M.J., Carstens, E.B. and Lefkowitz, E.J. (eds) *Virus Taxonomy: Ninth Report of the International Committee on Taxonomy of Viruses*. Elsevier Academic Press, pp. 193–210.

Jansen, M.D., Dong, H.T. and Mohan, C.V. (2019) Tilapia lake virus: A threat to the global tilapia industry?. *Reviews in Aquaculture* 11(3), 725–739. DOI: 10.1111/raq.12254.

Jaramillo, D., Tweedie, A., Becker, J.A., Hyatt, A., Crameri, S. *et al.* (2012) A validated quantitative polymerase chain reaction assay for the detection of ranaviruses (Family *Iridoviridae*) in fish tissue and cell cultures, using EHNV as A model. *Aquaculture* 356–357, 186–192. DOI: 10.1016/j.aquaculture.2012.05.017.

Jaramillo, D., Hick, P., Deece, K., Tweedie, A., Kirkland, P. *et al.* (2016) Comparison of ELISA formats for detection of antibodies specific for nervous necrosis virus (betanodavirus) in the serum of immunized barramundi *Lates calcarifer* and Australian bass *Macquaria novemaculeata*. *Aquaculture* 451, 33–38. DOI: 10.1016/j.aquaculture.2015.08.015.

Jeong, J.B., Kim, H.Y., Jun, L.J., Lyu, J.H., Park, N.G. *et al.* (2008) Outbreaks and risks of infectious spleen and kidney necrosis virus disease in freshwater ornamental fishes. *Diseases of Aquatic Organisms* 78(3), 209–215. DOI: 10.3354/dao01879.

Jitrakorn, S., Gangnonngiw, W., Bunnontae, M., Manajit, O., Rattanarojpong, T. *et al.* (2020) Infectious cell culture system for concurrent propagation and purification of megalocytivirus ISKNV and nervous necrosis virus from Asian sea bass (*Lates calcarifer*). *Aquaculture* 520, 734931. DOI: 10.1016/j.aquaculture.2020.734931.

Johan, C.A.C. and Zainathan, S.C. (2020) Megalocytiviruses in ornamental fish: A review. *Veterinary World* 13(11), 2565–2577. DOI: 10.14202/vetworld.2020.2565-2577.

Johansen, R., Grove, S., Svendsen, A.K., Modahl, I. and Dannevig, B.H. (2004a) A sequential study of pathological findings in Atlantic halibut, *Hippoglossus hippoglossus* (L.), throughout one year after an acute outbreak of viral encephalopathy and retinopathy. *Journal of Fish Diseases* 27(6), 327–341. DOI: 10.1111/j.1365-2761.2004.00548.x.

Johansen, R., Sommerset, I., Tørud, B., Korsnes, K., Hjortaas, M.J. *et al.* (2004b) Characterization of nodavirus and viral encephalopathy and retinopathy in farmed turbot, *Scophthalmus maximus* (L.). *Journal of Fish Diseases* 27(10), 591–601. DOI: 10.1111/j.1365-2761.2004.00581.x.

Johnson, S.C., Sperker, S.A., Leggiadro, C.T., Groman, D.B., Griffiths, S.G. *et al.* (2002) Identification and characterization of a piscine neuropathy and nodavirus from juvenile Atlantic cod from the Atlantic coast of North America. *Journal of Aquatic Animal Health* 14(2), 124–133. DOI: 10.1577/1548-8667(2002)0142.0.CO;2.

Kai, Y.H., Su, H.M., Tai, K.T. and Chi, S.C. (2010) Vaccination of grouper broodfish (*Epinephelus tukula*) reduces the risk of vertical transmission by nervous necrosis virus. *Vaccine* 28(4), 996–1001. DOI: 10.1016/j.vaccine.2009.10.132.

Kampeera, J., Dangtip, S., Suvannakad, R., Khumwan, P., Senapin, S. *et al.* (2021) Reverse transcription loop-mediated isothermal amplification (RT-LAMP) combined with colorimetric gold nanoparticle (AuNP) probe assay for visual detection of tilapia lake virus (TiLV) in Nile and red hybrid tilapia. *Journal of Fish Diseases* 44(10), 1595–1607. DOI: 10.1111/jfd.13482.

Kara, H.M., Chaoui, L., Derbal, F., Zaidi, R., Boisséson, C. *et al.* (2014) Betanodavirus-associated mortalities of adult wild groupers *Epinephelus marginatus* (Lowe) and *Epinephelus costae* (Steindachner) in Algeria. *Journal of Fish Diseases* 37(3), 273–278. DOI: 10.1111/jfd.12020.

Keawcharoen, J., Techangamsuwan, S., Ponpornpisit, A., Lombardini, E.D., Patchimasiri, T. *et al.* (2015) Genetic characterization of a betanodavirus isolated from a clinical disease outbreak in farm-raised tilapia *Oreochromis niloticus* (L.) in Thailand. *Journal of Fish Diseases* 38(1), 49–54. DOI: 10.1111/jfd.12200.

Kembou Tsofack, J.E., Zamostiano, R., Watted, S., Berkowitz, A., Rosenbluth, E. *et al.* (2017) Detection of tilapia lake virus in clinical samples by culturing and nested reverse transcription-PCR. *Journal of Clinical Microbiology* 55(3), 759–767. DOI: 10.1128/JCM.01808-16.

Kim, J.O., Kim, J.O., Kim, S.J., Kim, W.S. and Oh, M.J. (2018) Development of double labeling in situ hybridization using RNA probes for genome detection of nervous necrosis virus (NNV). *Molecular and Cellular Probes* 42, 18–24. DOI: 10.1016/j.mcp.2018.10.003.

Kuo, H.C., Wang, T.Y., Chen, P.P., Chen, Y.M., Chuang, H.C. *et al.* (2011) Real-time quantitative PCR assay for monitoring of nervous necrosis virus infection in grouper aquaculture. *Journal of Clinical Microbiology* 49(3), 1090–1096. DOI: 10.1128/JCM.01016-10.

Kurita, J. and Nakajima, K. (2012) Megalocytiviruses. *Viruses* 4(4), 521–538. DOI: 10.3390/v4040521.

Kurita, J., Nakajima, K., Hirono, I. and Aoki, T. (1998) Polymerase chain reaction (PCR) amplification of DNA of red sea bream iridovirus (RSIV). *Fish Pathology* 33(1), 17–23. DOI: 10.3147/jsfp.33.17.

Lai, Y., Chiu, H.C., Murali, S., Guo, I., Chen, S.-C. *et al.* (2001a) *In vitro* neutralization by monoclonal antibodies against yellow grouper nervous necrosis (YGNNV) and immunolocalization of virus infection in yellow grouper, *Epinephelus awoara* (Temminck and Schlegel). *Journal of Fish Diseases* 24(4), 237–244. DOI: 10.1046/j.1365-2761.2001.00293.x.

Lai, Y., Murali, S., Chiu, H.-C., Ju, H.-Y., Lin, Y.-S. *et al.* (2001b) Propagation of yellow grouper nervous necrosis virus (YGNNV) in a new nodavirus-susceptible cell line from yellow grouper, *Epinephelus awoara* (Temminck and schlegel), brain tissue. *Journal of Fish Diseases* 24(5), 299–309. DOI: 10.1046/j.1365-2761.2001.00303.x.

Leong, J.C. (2008) Fish viruses. *Encyclopedia of Virology* 227–234.

Li, N., Fu, X., Guo, H., Lin, Q., Liu, L. *et al.* (2015) Protein encoded by ORF093 is an effective vaccine candidate for infectious spleen and kidney necrosis virus in Chinese perch *Siniperca chuatsi*. *Fish and Shellfish Immunology* 42(1), 88–90. DOI: 10.1016/j.fsi.2014.10.008.

Li, Q., Yue, Z., Liu, H., Liang, C., Zheng, X. *et al.* (2010) Development and evaluation of a loop-mediated isothermal amplification assay for rapid detection of lymphocystis disease virus. *Journal of Virological Methods* 163(2), 378–384. DOI: 10.1016/j.jviromet.2009.10.028.

Liamnimitr, P., Thammatorn, W., U-thoomporn, S., Tattiyapong, P. and Surachetpong, W. (2018) Non-lethal sampling for tilapia lake virus detection by RT-qPCR and cell culture. *Aquaculture* 486, 75–80. DOI: 10.1016/j.aquaculture.2017.12.015.

Lin, L., He, J., Mori, K.-I., Nishioka, T., Wu, J.L. *et al.* (2001) Mass mortalities associated with viral nervous necrosis in hatchery-reared groupers in the People's Republic of China. *Fish Pathology* 36(3), 186–188. DOI: 10.3147/jsfp.36.186.

Liu, W., Hsu, C.H., Chang, C.Y., Chen, H.H. and Lin, C.S. (2006) Immune response against grouper nervous necrosis virus by vaccination of virus-like particles. *Vaccine* 24(37–39), 6282–6287. DOI: 10.1016/j.vaccine.2006.05.073.

Liu, W., Zhang, Y., Ma, J., Jiang, N., Fan, Y. *et al.* (2020) Determination of a novel parvovirus pathogen associated with massive mortality in adult tilapia. *PLOS Pathogens* 16(9), e1008765. DOI: 10.1371/journal.ppat.1008765.

Lopez-Jimena, B., Cherif, N., Garcia-Rosado, E., Infante, C., Cano, I. *et al.* (2010) A combined RT-PCR and dot-blot hybridization method reveals the coexistence of SJNNV and RGNNV betanodavirus genotypes in wild meagre (*Argyrosomus regius*). *Journal of Applied Microbiology* 109(4), 1361–1369. DOI: 10.1111/j.1365-2672.2010.04759.x.

Machimbirike, V.I., Jansen, M.D., Senapin, S., Khunrae, P., Rattanarojpong, T. *et al.* (2019) Viral infections in tilapines: More than just tilapia lake virus. *Aquaculture* 503, 508–518. DOI: 10.1016/j.aquaculture.2019.01.036.

Maclaine, A., Forzán, M.J., Mashkour, N., Scott, J. and Ariel, E. (2019) Pathogenesis of bohle iridovirus (Genus *Ranavirus*) in experimentally infected juvenile eastern water dragons (*Intellagama lesueurii lesueurii*). *Veterinary Pathology* 56(3), 465–475. DOI: 10.1177/0300985818823666.

Mai, T.T., Kayansamruaj, P., Taengphu, S., Senapin, S., Costa, J.Z. *et al.* (2021) Efficacy of heat-killed and formalin-killed vaccines against tilapia tilapinevirus in juvenile Nile tilapia (*Oreochromis niloticus*). *Journal of Fish Diseases* 44(12), 2097–2109. DOI: 10.1111/jfd.13523.

Mai, T.T., Kayansamruaj, P., Soontara, C., Kerddee, P., Nguyen, D.-H., Senapin, S. *et al.* (2022) Immunization of Nile tilapia (*Oreochromis niloticus*) broodstock with tilapia lake virus (TiLV) inactivated vaccines elicits protective antibody and passive maternal antibody transfer. *Vaccines* 10, 167.

Makesh, M., Venkata Satyanarayana, N., Selvamano, S., Jayaprakash, N.S., Bera, A. *et al.* (2022) Nervous necrosis virus titration and antigen quantitation by indirect sandwich enzyme linked immunosorbent assay. *Journal of Virological Methods* 305, 114536. DOI: 10.1016/j.jviromet.2022.114536.

McAllister, P.E. (1983) Infectious pancreatic necrosis (IPN) of salmonid fishes. Retrieved December 27, 2022. USFW Publications p. 141. US Fish and Wildlife. Available at: https://digitalcommons.unl.edu/cgi/viewcontent.cgi?article=1140&context=usfwspubs

McGrogan, D.G., Ostland, V.E., Byrne, P.J. and Ferguson, H.W. (1998) Systemic disease involving an iridovirus-like agent in cultured tilapia,

Oreochromis niloticus L. a case report. *Journal of Fish Diseases* 21(2), 149–152. DOI: 10.1046/j.1365-2761.1998.00082.x.

Mekata, T., Satoh, J., Inada, M., Dinesh, S., Harsha, P. *et al.* (2015) Development of simple, rapid and sensitive detection assay for grouper nervous necrosis virus using real-time loop-mediated isothermal amplification. *Journal of Fish Diseases* 38(10), 873–879. DOI: 10.1111/jfd.12297.

Moody, N.J.G. and Owens, L. (1994) Experimental demonstration of the pathogenicity of a frog virus, Bohle iridovirus, for a fish species, barramundi *Lates calcarifer. Diseases of Aquatic Organisms* 18, 95–102. DOI: 10.3354/dao018095.

Mori, K., Nakai, T., Nagahara, M., Muroga, K., Mekuchi, T. *et al.* (1991) Viral disease in hatchery-reared larvae and juveniles of redspotted grouper. *Gyobyo Kenkyo* 26, 209–210.

Mori, K., Nakai, T., Muroga, K., Arimoto, M., Mushiake, K. *et al.* (1992) Properties of a new virus belonging to nodaviridae found in larval striped jack (*Pseudocaranx dentex*) with nervous necrosis. *Virology* 187(1), 368–371. DOI: 10.1016/0042-6822(92)90329-n.

Mori, K., Mushiake, K. and Arimoto, M. (1998) Control measures for viral nervous necrosis in striped Jack. *Fish Pathology* 33(4), 443–444. DOI: 10.3147/jsfp.33.443.

Morit, K., Mangyoku, T., Iwamoto, T., Arimoto, M., Tanaka, S. *et al.* (2003) Serological relationships among genotypic variants of betanodavirus. *Diseases of Aquatic Organisms* 57(1–2), 19–26. DOI: 10.3354/dao057019.

Mugimba, K.K., Chengula, A.A., Wamala, S., Mwega, E.D., Kasanga, C.J. *et al.* (2018) Detection of tilapia lake virus (ti LV) infection by PCR in farmed and wild Nile tilapia (*Oreochromis niloticus*) from Lake Victoria. *Journal of Fish Diseases* 41(8), 1181–1189. DOI: 10.1111/jfd.12790.

Mulei, I.R., Nyaga, P.N., Mbuthia, P.G., Waruiru, R.M., Njagi, L.W. *et al.* (2018) Infectious pancreatic necrosis virus isolated from farmed rainbow trout and tilapia in Kenya is identical to European isolates. *Journal of Fish Diseases* 41, 1191–1200. DOI: 10.1111/jfd.12807.

Munang'andu, H.M., Mutoloki, S. and Evensen, Ø. (2016) Birnaviruses of aquatic animals. In: Kibenge, F.S.B. and Godoy, M.G. (eds) *Aquaculture Virology*. Elsevier, Inc., Academic Press, pp. 237–250.

Munday, B.L. and Nakai, T. (1997) Special topic review: Nodaviruses as pathogens in larval and juvenile marine finfish. *World Journal of Microbiology and Biotechnology* 13(4), 375–381. DOI: 10.1023/A:1018516014782.

Munday, B.L., Kwang, J.and Moody.(2002) Betanodavirus infections of teleost fish: a review. *Journal of Fish Diseases* 25(J), 127–142.

Murti, K.G., Goorha, R. and Granoff, A. (1985) An unusual replication strategy of an animal iridovirus. *Advances in Virus Research* 30, 1–19. DOI: 10.1016/s0065-3527(08)60446-7.

Mushiake, K., Arimoto, M., Furusawa, T., Furusawa, I., Nakai, T. *et al.* (1992) Detection of antibodies against striped jack nervous necrosis virus (SJNNV) from brood stocks of Striped Jack. *Nippon Suisan Gakkaishi* 58(12), 2351–2356. DOI: 10.2331/suisan.58.2351.

Mushiake, K., Nishizawa, T., Nakai, T., Furusawa, I. and Muroga, K. (1994) Control of VNN in striped jack: selection of spawners based on the detection of SJNNV gene by polymerase chain reaction (PCR). *Fish Pathology* 29(3), 177–182. DOI: 10.3147/jsfp.29.177.

Nagai, T. and Nishizawa, T. (1999) Sequence on the nonstructural protein gene encoded by RNA 1 of striped jack nervous necrosis virus. *Journal of General Virology* 80(11), 3019–3022. DOI: 10.1099/0022-1317-80-11-3019.

Nallala, V.S., Makesh, M., Radhika, K., Sathish Kumar, T., Raja, P. *et al.* (2021) Characterization of red-spotted grouper nervous necrosis virus isolated from ovarian fluids of asymptomatic wild Asian seabass, *Lates calcarifer*. *Aquaculture* 542, 736846. DOI: 10.1016/j.aquaculture.2021.736846.

Nguyen, H.D., Mekuchi, T., Imura, K., Nakai, T., Nishizawa, T. *et al.* (1994) Occurrence of viral nervous necrosis (VNN) in hatchery-reared juvenile Japanese flounder *Paralichthys olivaceus*. *Fisheries Science* 60(5), 551–554. DOI: 10.2331/fishsci.60.551.

Nishioka, T., Mori, K.-I., Sugaya, T., Takeuchi, H., Tsuzaki, T. *et al.* (2011) Viral nervous necrosis (VNN) in seed production of red tilefish branchiostegus japonicus and its control. *Aquaculture Science* 59, 275–282.

Nishizawa, T., Mori, K., Nakai, T., Furusawa, I., Muroga, K. *et al.* (1994) Polymerase chain reaction (PCR) amplification of RNA of striped jack nervous necrosis virus (SJNNV). *Diseases of Aquatic Organisms* 18, 103–107. DOI: 10.3354/dao018103.

Nishizawa, T., Kise, M., Nakai, T. and Muroga, K. (1995a) Neutralizing monoclonal antibodies to striped jack nervous necrosis virus (SJNNV). *Fish Pathology* 30(2), 111–114. DOI: 10.3147/jsfp.30.111.

Nishizawa, T., Mori, K., Furuhashi, M., Nakai, T., Furusawa, I. *et al.* (1995b) Comparison of the coat protein genes of five fish nodaviruses, the causative agent of viral nervous necrosis in marine fish. *Journal of General Virology* 76(7), 1563–1569. DOI: 10.1099/0022-1317-76-7-1563.

Nishizawa, T., Furuhashi, M., Nagai, T., Nakai, T. and Muroga, K. (1997) Genomic classification of fish nodaviruses by molecular phylogenetic analysis of the coat protein gene. *Applied and Environmental Microbiology* 63(4), 1633–1636. DOI: 10.1128/aem.63.4.1633-1636.1997.

Oh, M.-J., Jung, S.-J., Kim, S.-R., Rajendran, K.V., Kim, Y.-J. *et al.* (2002) A fish nodavirus associated with mass mortality in hatchery-reared red drum, *Sciaenops ocellatus*. *Aquaculture* 211(1–4), 1–7. DOI: 10.1016/S0044-8486(01)00877-8.

Olveira, J.G., Souto, S., Dopazo, C.P., Thiéry, R., Barja, J.L. *et al.* (2009) Comparative analysis of both genomic segments of betanodaviruses isolated from epizootic outbreaks in farmed fish species provides evidence for genetic reassortment. *Journal of General Virology* 90(12), 2940–2951. DOI: 10.1099/vir.0.013912-0.

Palaiokostas, C., Cariou, S., Bestin, A., Bruant, J.S., Haffray, P. *et al.* (2018) Genome-wide association and genomic prediction of resistance to viral nervous necrosis in European sea bass (*Dicentrarchus labrax*) using RAD sequencing. *Genetics, Selection, Evolution* 50(1), 30. DOI: 10.1186/s12711-018-0401-2.

Pallister, J., Gould, A., Harrison, D., Hyatt, A., Jancovich, J. *et al.* (2007) Development of real-time PCR assays for the detection and differentiation of Australian and European ranaviruses. *Journal of Fish Diseases* 30(7), 427–438. DOI: 10.1111/j.1365-2761.2007.00828.x.

Panzarin, V., Patarnello, P., Mori, A., Rampazzo, E., Cappellozza, E. *et al.* (2010) Development and validation of a real-time TaqMan PCR assay for the detection of betanodavirus in clinical specimens. *Archives of Virology* 155(8), 1193–1203. DOI: 10.1007/s00705-010-0701-5.

Panzarin, V., Fusaro, A., Monne, I., Cappellozza, E., Patarnello, P. *et al.* (2012) Molecular epidemiology and evolutionary dynamics of betanodavirus in Southern Europe. *Infection, Genetics and Evolution* 12(1), 63–70. DOI: 10.1016/j.meegid.2011.10.007.

Paperna, I. (1973) Lymphocystis in fish from east African lakes. *Journal of Wildlife Diseases* 9(4), 331–335. DOI: 10.7589/0090-3558-9.4.331.

Phusantisampan, T., Tattiyapong, P., Mutrakulcharoen, P., Sriariyanun, M. and Surachetpong, W. (2019) Rapid detection of tilapia lake virus using a one-step reverse transcription loop-mediated isothermal amplification assay. *Aquaculture* 507, 35–39. DOI: 10.1016/j.aquaculture.2019.04.015.

Piewbang, C., Tattiyapong, P., Techangamsuwan, S. and Surachetpong, W. (2021) Tilapia lake virus immunoglobulin G (TiLV IgG) antibody: Immunohistochemistry application reveals cellular tropism of TiLV infection. *Fish and Shellfish Immunology* 116, 115–123. DOI: 10.1016/j.fsi.2021.06.017.

Piewbang, C., Tattiyapong, P., Khemthong, M., Lachroje, S., Boonrungsiman, S. *et al.* (2022) Dual infections of tilapia parvovirus (TiPV) and tilapia lake virus (TiLV) in multiple tilapia farms: Their impacts, genetic diversity, viral tropism, and pathological effects. *Aquaculture* 550, 737887. DOI: 10.1016/j.aquaculture.2022.737887.

Prihartini, N.C., Yanuhar, U.and Maftuch(2015) Betanodavirus infections in tilapia seed (*Oreochromis* sp.). In: *Indonesia Journal of Life Science and Biomedicine*, Vol. 5. pp. 104–106.

Pulido, L.L.H., Mora, C.M., Hung, A.L., Dong, H.T. and Senapin, S. (2019) Tilapia lake virus (TiLV) from Peru is genetically close to the Israeli isolates. *Aquaculture* 510, 61–65. DOI: 10.1016/j.aquaculture.2019.04.058.

Rakus, K., Mojzesz, M., Widziolek, M., Pooranachandran, N., Teitge, F. *et al.* (2020) Antiviral response of adult zebrafish (*Danio rerio*) during tilapia lake virus (TiLV) infection. *Fish and Shellfish Immunology* 101, 1–8. DOI: 10.1016/j.fsi.2020.03.040.

Ramírez-Paredes, J.G., Paley, R.K., Hunt, W., Feist, S.W., Stone, D.M. *et al.* (2021) First detection of infectious spleen and kidney necrosis virus (ISKNV) associated with massive mortalities in farmed tilapia in Africa. *Transboundary and Emerging Diseases* 68(3), 1550–1563. DOI: 10.1111/tbed.13825.

Ransangan, J. and Manin, B.O. (2010) Mass mortality of hatchery-produced larvae of Asian seabass, *Lates calcarifer* (bloch), associated with viral nervous necrosis in Sabah, Malaysia. *Veterinary Microbiology* 145(1–2), 153–157. DOI: 10.1016/j.vetmic.2010.03.016.

Renault, T., Haffner, P., Baudin-Laurencin, F., Breuil, G. and Bonami, J.R. (1991) Mass mortalities in hatchery-reared sea bass (*Lates calcarifer*) larvae associated with the presence in the brain and retina of virus-like particles. *Bulletin of the European Association of Fish Pathologists* 11, 68–73.

Rianto Wah, B.R., Yanuhar, U., Fadjar, M. and Andayani, S. (2019a) Clinical and molecular study of Koi herpesvirus (KHV) emerged in *Oreochromis niloticus* from Indonesia. *Asian Journal of Scientific Research* 12(3), 316–322. DOI: 10.3923/ajsr.2019.316.322.

Rimmer, A.E., Becker, J.A., Tweedie, A. and Whittington, R.J. (2012) Development of a quantitative Polymerase Chain Reaction (qPCR) assay for the detection of Dwarf Gourami Iridovirus (DGIV) and other megalocytiviruses and comparison with the Office International des Epizooties (OIE) reference PCR protocol. *Aquaculture* 358–359, 155–163. DOI: 10.1016/j.aquaculture.2012.06.034.

Rimmer, A.E., Becker, J.A., Tweedie, A., Lintermans, M., Landos, M. *et al.* (2015) Detection of dwarf gourami iridovirus (infectious spleen and kidney necrosis virus) in populations of ornamental fish prior to and after importation into Australia, with the first evidence of infection in domestically farmed platy (*Xiphophorus maculatus*). *Preventive Veterinary Medicine* 122(1–2), 181–194. DOI: 10.1016/j.prevetmed.2015.09.008.

Rimstad, E. (2014) Vaccination against infectious pancreatic necrosis. In: Gudding, R., Lillehaug, A. and Evensen, Ø. (eds) *Fish Vaccination.* John Wiley & Sons, Ltd, pp. 303–312.

Rodriguez Saint-Jean, S., Borrego, J.J. and Perez-Prieto, S.I. (2003) Infectious pancreatic necrosis virus: Biology, pathogenesis, and diagnostic methods. *Advances in Virus Research* 62, 113–165. DOI: 10.1016/s0065-3527(03)62003-8.

Ruane, N., Geoghegan, F.O. and Cinneide, M. (2007) Infectious pancreatic necrosis virus and its impact on the Irish salmon aquaculture and wild fish sectors. *Marine Environment and Health Series.* Marine Institute 30.

Sah Putra, B., Hick, P.M., Hall, E., Whittington, R.J., Khairul, R. *et al.* (2020) Prevalence of infectious spleen and kidney necrosis virus (ISKNV), nervous necrosis virus (NNV) and ectoparasites in juvenile Epinephelus spp. farmed in Aceh, Indonesia. *Pathogens* 9(7), 578. DOI: 10.3390/pathogens9070578.

Santi, N., Song, H., Vakharia, V.N. and Evensen, Ø. (2005) Infectious pancreatic necrosis virus VP5 is dispensable for virulence and persistence. *Journal of Virology* 79(14), 9206–9216. DOI: 10.1128/JVI.79.14.9206-9216.2005.

Senapin, S., Dong, H.T., Meemetta, W., Gangnonngiw, W., Sangsuriya, P. *et al.* (2019) Mortality from scale drop disease in farmed *Lates calcarifer* in Southeast Asia. *Journal of Fish Diseases* 42(1), 119–127. DOI: 10.1111/jfd.12915.

Shahi, N., Mallik, S.K., Kala, K., Joshi, N., Patiyal, R.S. *et al.* (2020) Seasonal emergence of benign epidermal tumor in farm-reared adult grass carp (*Ctenopharyngodon idella*) caused by lymphocystis disease virus at Uttarakhand, India. *Aquaculture* 526, 735408. DOI: 10.1016/j.aquaculture.2020.735408.

Shahin, K., Subramaniam, K., Camus, A.C., Yazdi, Z., Yun, S. *et al.* (2021) Isolation, identification and characterization of a novel megalocytivirus from cultured tilapia (*Oreochromis* spp.) from Southern California, USA. *Animals: An Open Access Journal from MDPI* 11(12), 3524. DOI: 10.3390/ani11123524.

Sheng, X., Xu, X. and Zhan, W. (2013) Development and application of antibody microarray for lymphocystis disease virus detection in fish. *Journal of Virological Methods* 189(2), 243–249. DOI: 10.1016/j.jviromet.2013.02.015.

Shieh, J.R. and Chi, S.C. (2005) Production of monoclonal antibodies against grouper nervous necrosis virus (GNNV) and development of an antigen capture ELISA. *Diseases of Aquatic Organisms* 63(1), 53–60. DOI: 10.3354/dao063053.

Shlapobersky, M., Sinyakov, M.S., Katzenellenbogen, M., Sarid, R., Don, J. *et al.* (2010) Viral encephalitis of tilapia larvae: Primary characterization of a novel herpes-like virus. *Virology* 399(2), 239–247. DOI: 10.1016/j.virol.2010.01.001.

Shyam, K.U., Jeong, H.-N., Oh, M.-J., Kim, C.-S. and Kim, W.-S. (2020) Development of a lateral flow immuno-chromatic strip assay for the detection of nervous necrosis virus (NNV, RGNNV genotype). *Aquaculture* 520, 734944. DOI: 10.1016/j.aquaculture.2020.734944.

Sinyakov, M.S., Belotsky, S., Shlapobersky, M. and Avtalion, R.R. (2011) Vertical and horizontal transmission of tilapia larvae encephalitis virus: The bad and the ugly. *Virology* 410(1), 228–233. DOI: 10.1016/j.virol.2010.11.004.

Skliris, G.P. and Richards, R.H. (1998) Assessment of the susceptibility of the brine shrimp *Artemia salina* and the rotifer *Brachionus plicatilis* to experimental nodavirus infections. *Aquaculture* 169(1–2), 133–141. DOI: 10.1016/S0044-8486(98)00330-5.

Skliris, G.P. and Richards, R.H. (1999a) Induction of nodavirus disease in seabass, *Dicentrarchus labrax*, using different infection models. *Virus Research* 63(1–2), 85–93. DOI: 10.1016/s0168-1702(99)00061-1.

Skliris, G.P. and Richards, R.H. (1999b) Nodavirus isolated from experimentally infected tilapia *Oreochromis mossambicus* (Peters). *Journal of Fish Diseases* 22(4), 315–318. DOI: 10.1046/j.1365-2761.1999.00171.x.

Smith, C.E., Ramsey, D., Speer, C.A. and Blixt, A. (1997) *Histopathology associated with an irido-like virus infection of tilapia*, Vol. 25. Filson Historical Society/AFS Newsletter, pp. 6–9.

Sommerset, I. and Nerland, A.H. (2004) Complete sequence of RNA1 and subgenomic RNA3 of Atlantic halibut nodavirus (AHNV). *Diseases of Aquatic Organisms* 58(2–3), 117–125. DOI: 10.3354/dao058117.

Soto, E., Yun, S. and Surachetpong, W. (2019) Susceptibility of tilapia lake virus to buffered povidone-iodine complex and chlorine. *Aquaculture* 512, 734342. DOI: 10.1016/j.aquaculture.2019.734342.

Soto, E., Illanes, O., Hilchie, D., Morales, J. A., Sunyakumthorn, P., Hawke, J. P., Goodwin, A. E., Riggs, A., Yanong, R. P., Pouder, D. B., Francis-Floyd, R., Arauz, M., Bogdanovic, L., & Castillo-Alcala, F. (2012). Molecular and immunohistochemical diagnosis of Francisella noatunensis subsp. orientalis from formalin-fixed, paraffin-embedded tissues [Article]. *Journal of Veterinary Diagnostic Investigation*, 24(5), 840-845. https://doi.org/10.1177/1040638712452108

Speare, R. and Smith, J.R. (1992) An iridovirus-like agent isolated from the ornate burrowing frog *Limnodynastes ornatus* in northern Australia. *Diseases of Aquatic Organisms* 14, 51–57. DOI: 10.3354/dao014051.

Sriwanayos, P., Francis-Floyd, R., Stidworthy, M.F., Petty, B.D., Kelley, K. *et al.* (2013) Megalocytivirus infection in orbiculate batfish *Platax orbicularis*. *Diseases of Aquatic Organisms* 105(1), 1–8. DOI: 10.3354/dao02594.

Starkey, W.G., Ireland, J.H., Muir, K.F., Jenkins, M.E., Roy, W.J. *et al.* (2001) Nodavirus infection in Atlantic cod and Dover sole in the UK. *Veterinary Record* 149(6), 179–181. DOI: 10.1136/vr.149.6.179.

Starkey, W.G., Millar, R.M., Jenkins, M.E., Ireland, J.H., Muir, K.F. *et al.* (2004) Detection of piscine nodaviruses by real-time nucleic acid sequence based amplification (NASBA). *Diseases of Aquatic Organisms* 59(2), 93–100. DOI: 10.3354/dao059093.

Stilwell, N.K., Whittington, R.J., Hick, P.M., Becker, J.A., Ariel, E. *et al.* (2018) Partial validation of a taqman real-time quantitative PCR for the detection of ranaviruses. *Diseases of Aquatic Organisms* 128(2), 105–116. DOI: 10.3354/dao03214.

Su, Y.C. and Hong, J.R. (2010) Betanodavirus B2 causes ATP depletion-induced cell death via mitochondrial targeting and complex II inhibition *in vitro* and *in vivo*. *Journal of Biological Chemistry* 285(51), 39801–39810. DOI: 10.1074/jbc.M110.164988.

Su, Y.C., Chiu, H.W., Hung, J.C. and Hong, J.R. (2014) Betanodavirus B2 protein induces hydrogen peroxide production, leading to Drp1-recruited mitochondrial fragmentation and cell death via mitochondrial targeting. *Apoptosis* 19(10), 1457–1470. DOI: 10.1007/s10495-014-1016-x.

Subramaniam, K., Shariff, M., Omar, A.R., Hair-Bejo, M. and Ong, B.L. (2014) Detection and molecular characterization of infectious spleen and kidney necrosis virus from major ornamental fish breeding states in Peninsular Malaysia. *Journal of Fish Diseases* 37(7), 609–618. DOI: 10.1111/jfd.12152.

Subramaniam, K., Gotesman, M., Smith, C.E., Steckler, N.K., Kelley, K.L. *et al.* (2016) Megalocytivirus infection in cultured Nile tilapia *Oreochromis niloticus*. *Diseases of Aquatic Organisms* 119(3), 253–258. DOI: 10.3354/dao02985.

Suebsing, R., Pradeep, P.J., Jitrakorn, S., Sirithammajak, S., Kampeera, J. *et al.* (2016) Detection of natural infection of infectious spleen and kidney necrosis virus in farmed tilapia by hydroxynapthol blue-loop mediated isothermal amplification assay. *Journal of Applied Microbiology* 121(1), 55–67. DOI: 10.1111/jam.13165.

Sung, C.H. and Lu, J.K. (2009) Reverse transcription loop-mediated isothermal amplification for rapid and sensitive detection of nervous necrosis virus in groupers. *Journal of Virological Methods* 159(2), 206–210. DOI: 10.1016/j.jviromet.2009.03.022.

Surachetpong, W., Janetanakit, T., Nonthabenjawan, N., Tattiyapong, P., Sirikanchana, K. *et al.* (2017) Outbreaks of tilapia lake virus infection, Thailand, 2015–2016. *Emerging Infectious Diseases* 23(6), 1031–1033. DOI: 10.3201/eid2306.161278.

Surachetpong, W., Roy, S.R.K. and Nicholson, P. (2020) Tilapia lake virus: The story so far. *Journal of Fish Diseases* 43(10), 1115–1132. DOI: 10.1111/jfd.13237.

Swaminathan, T.R., Raj, N.S., Preena, P.G., Pradhan, P.K., Sood, N. *et al.* (2021) Infectious spleen and kidney necrosis virus-associated large-scale mortality in farmed giant gourami *Osphronemus goramy*, in India. *Journal of Fish Diseases* 44(12), 2043–2053. DOI: 10.1111/jfd.13519.

Taengphu, S., Sangsuriya, P., Phiwsaiya, K., Debnath, P.P., Delamare-Deboutteville, J. *et al.* (2020) Genetic diversity of tilapia lake virus genome segment 1 from 2011 to 2019 and a newly validated semi-nested RT-qPCR method. *Aquaculture* 526, 735423. DOI: 10.1016/j.aquaculture.2020.735423.

Taengphu, S., Kayansamruaj, P., Kawato, Y., Delamare-Deboutteville, J., Mohan, C.V. *et al.* (2022) Concentration and quantification of

Tilapia tilapinevirus from water using a simple iron flocculation coupled with probe-based RT-qPCR. *PeerJ* 10, e13157. DOI: 10.7717/peerj.13157.

Taha, E., Shawky, M., Ahmed, B., Moustafa, M., Yousif, A. *et al.* (2020) Emergence of viral nervous necrosis is associated with mass mortality in hatchery-reared tilapia (*Oreochromis niloticus*) in Egypt. *Aquaculture International* 28(5), 1811–1823. DOI: 10.1007/s10499-020-00559-4.

Tan, C., Huang, B., Chang, S.F., Ngoh, G.H., Munday, B. *et al.* (2001) Determination of the complete nucleotide sequences of RNA1 and RNA2 from greasy grouper (*Epinephelus tauvina*) nervous necrosis virus, Singapore strain. *Journal of General Virology* 82(3), 647–653. DOI: 10.1099/0022-1317-82-3-647.

Tanaka, S., Takagi, M. and Miyazaki, T. (2004) Histopathological studies on viral nervous necrosis of sevenband grouper, *Epinephelus septemfasciatus* thunberg, at the grow-out stage. *Journal of Fish Diseases* 27(7), 385–399. DOI: 10.1111/j.1365-2761.2004.00559.x.

Tang, K.F.J., Bondad-Reantaso, M.G., Surachetpong, W., Dong, H.T., Fejzic, N, *et al.* (2021) Tilapia lake virus disease strategy manual. *FAO fisheries and aquaculture circular.* Retrieved February 19, 2022. Food and Agriculture Organization, 1220. Available at: https://www.fao.org/3/cb7293en/cb7293en.pdf

Tattiyapong, P., Dachavichitlead, W. and Surachetpong, W. (2017) Experimental infection of tilapia lake virus (tilv) in Nile tilapia (*Oreochromis niloticus*) and red tilapia (*Oreochromis* spp.). *Veterinary Microbiology* 207, 170–177. DOI: 10.1016/j.vetmic.2017.06.014.

Tattiyapong, P., Sirikanchana, K. and Surachetpong, W. (2018) Development and validation of a reverse transcription quantitative polymerase chain reaction for tilapia lake virus detection in clinical samples and experimentally challenged fish. *Journal of Fish Diseases* 41(2), 255–261. DOI: 10.1111/jfd.12708.

Tattiyapong, P., Dechavichitlead, W., Waltzek, T.B. and Surachetpong, W. (2020) Tilapia develop protective immunity including a humoral response following exposure to tilapia lake virus. *Fish and Shellfish Immunology* 106, 666–674. Available at: https://doi.org/10.1016/j.fsi.2020.08.031

Thangaraj, R.S., Ravi, C., Kumar, R.G., Dharmaratnam, A., Valaparambil Saidmuhammed, B.V. *et al.* (2018) Derivation of two tilapia (*Oreochromis niloticus*) cell lines for efficient propagation of tilapia lake virus (TiLV). *Aquaculture* 492, 206–214. DOI: 10.1016/j.aquaculture.2018.04.012.

Thawornwattana, Y., Dong, H.T., Phiwsaiya, K., Sangsuriya, P., Senapin, S. *et al.* (2021) Tilapia lake virus (TiLV): Genomic epidemiology and its early origin. *Transboundary and Emerging Diseases* 68(2), 435–444. DOI: 10.1111/tbed.13693.

Thiery, R., Raymond, J.C. and Castric, J. (1999) Natural outbreak of viral encephalopathy and retinopathy in juvenile sea bass, *Dicentrarchus labrax*: Study by nested reverse transcriptase–polymerase chain reaction. *Virus Research* 63(1–2), 11–17. DOI: 10.1016/s0168-1702(99)00053-2.

Tian, J.Y., Sun, X.Q. and Chen, X.G. (2008) Formation and oral administration of alginate microspheres loaded with pDNA coding for lymphocystis disease virus (LCDV) to Japanese flounder. *Fish and Shellfish Immunology* 24(5), 592–599. DOI: 10.1016/j.fsi.2008.01.009.

Toffan, A., Biasini, L., Pretto, T., Abbadi, M., Buratin, A. *et al.* (2021) Age dependency of RGNNV/SJNNV viral encephalo-retinopathy in gilthead sea bream (*Sparus aurata*). *Aquaculture* 539, 736605. DOI: 10.1016/j.aquaculture.2021.736605.

Toffolo, V., Negrisolo, E., Maltese, C., Bovo, G., Belvedere, P. *et al.* (2007) Phylogeny of betanodaviruses and molecular evolution of their RNA polymerase and coat proteins. *Molecular Phylogenetics and Evolution* 43, 298–308.

Tran, T.H., Nguyen, V.T.H., Bui, H.C.N., Tran, Y.B.T., Tran, H.T.T, *et al.* (2022) Tilapia lake virus (TiLV) from vietnam is genetically distantly related to tilv strains from other countries. *Journal of Fish Diseases* 45(9), 1389–1401. DOI: 10.1111/jfd.13669.

Ucko, M., Colorni, A. and Diamant, A. (2004) Nodavirus infections in israeli mariculture. *Journal of Fish Diseases* 27(8), 459–469. DOI: 10.1111/j.1365-2761.2004.00565.x.

Valero, Y., Arizcun, M., Esteban, M.Á., Bandín, I., Olveira, J.G. *et al.* (2015) Nodavirus colonizes and replicates in the testis of gilthead seabream and European sea bass modulating its immune and reproductive functions. *PLOS ONE* 10(12), e0145131. DOI: 10.1371/journal.pone.0145131.

Valverde, E.J., Cano, I., Labella, A., Borrego, J.J. and Castro, D. (2016) Application of a new real-time polymerase chain reaction assay for surveillance studies of lymphocystis disease virus in farmed gilthead seabream. *BMC Veterinary Research* 12, 71. DOI: 10.1186/s12917-016-0696-6.

Vázquez-Salgado, L., Olveira, J.G., Dopazo, C.P. and Bandín, I. (2022) Interspecies transmission between *Solea senegalensis* and *Sparus aurata* of reassortant nervous necrosis virus (NNV) strains and effect of stress on the outcome of the infection. *Aquaculture* 547, 737519. DOI: 10.1016/j.aquaculture.2021.737519.

Venkata Satyanarayana, N., Makesh, M., Sain, A., Jayaprakash, N.S., Kailasam, M. *et al.* (2021) Non-lethal screening of Asian seabass (*Lates calcarifer*) by monoclonal antibody based indirect enzyme linked immunosorbent assay for viral nervous necrosis. *Fish and Shellfish Immunology Reports* 2, 100011. DOI: 10.1016/j.fsirep.2021.100011.

Verma, D.K., Sood, N., Paria, A., Swaminathan, T.R., Mohan, C.V. *et al.* (2022) Reassortment and evolutionary dynamics of tilapia lake virus genomic segments. *Virus Research* 308, 198625. DOI: 10.1016/j.virusres.2021.198625.

Volpe, E., Gustinelli, A., Caffara, M., Errani, F., Quaglio, F. *et al.* (2020) Viral nervous necrosis outbreaks caused by the RGNNV/SJNNV reassortant betanodavirus in gilthead sea bream (*Sparus aurata*) and European sea bass (*Dicentrarchus labrax*). *Aquaculture* 523, 735155. DOI: 10.1016/j.aquaculture.2020.735155.

Wahidi, B. R., Yanuhar, U., Fadjar, M. and Andayani, S. (2019a) Clinical and molecular study of Koi Herpesvirus (KHV) emerged in *Oreochromis niloticus* from Indonesia. *Asian Journal of Scientific Research 12*, 316-322. DOI: 10.3923/ajsr.2019

Wahidi, B.R., Yanuhar, U., Mohamad Fadjar, M.M. and Andayani, S. (2019b) Pathognomonic features and ultrastructural of Koi herpesvirus infected *Oreochromis niloticus. Biodiversitas Journal of Biological Diversity* 20(2), 497–503. DOI: 10.13057/biodiv/d200228.

Waiyamitra, P., Tattiyapong, P., Sirikanchana, K., Mongkolsuk, S., Nicholson, P. *et al.* (2018) A taqman RT-qpcr assay for tilapia lake virus (TiLV) detection in tilapia. *Aquaculture* 497, 184–188. DOI: 10.1016/j.aquaculture.2018.07.060.

Waiyamitra, P., Piewbang, C., Techangamsuwan, S., Liew, W.C. and Surachetpong, W. (2021) Infection of tilapia tilapinevirus in Mozambique tilapia (*Oreochromis mossambicus*), a globally vulnerable fish species. *Viruses* 13(6), 1104. DOI: 10.3390/v13061104.

Walker, R. (1962) Fine structure of lymphocystis virus of fish. *Virology* 18, 503–505. DOI: 10.1016/0042-6822(62)90047-8.

Wang, L., Liu, P., Huang, S., Ye, B., Chua, E. *et al.* (2017) Genome-wide association study identifies loci associated with resistance to viral nervous necrosis disease in Asian seabass. *Marine Biotechnology* 19(3), 255–265. DOI: 10.1007/s10126-017-9747-7.

Wang, Q., Zeng, W.W., Li, K.B., Chang, O.Q., Liu, C. *et al.* (2011) Outbreaks of an iridovirus in marbled sleepy goby, *Oxyeleotris marmoratus* (Bleeker), cultured in southern China. *Journal of Fish Diseases* 34(5), 399–402. DOI: 10.1111/j.1365-2761.2011.01244.x.

Wang, Y., Wang, Q., Zeng, W., Yin, J., Li, Y. *et al.* (2018) Establishment and characterization of a cell line from tilapia brain for detection of tilapia lake virus. *Journal of Fish Diseases* 41(12), 1803–1809. DOI: 10.1111/jfd.12889.

Wang, Y., Li, Z., Wang, Q., Zeng, W., Li, Y. *et al.* (2020) Establishment of a brain cell line obtained from hybrids of *Channa argus* × *Channa maculata* for the detection of tilapia lake virus. *Microbial Pathogenesis* 138, 103810. DOI: 10.1016/j.micpath.2019.103810.

Watanabe, K., Suzuki, S., Nishizawa, T., Suzuki, K., Yoshimizu, M. *et al.* (1998) Control strategy for viral nervous necrosis of barfin flounder. *Fish Pathology* 33(4), 445–446. Available at: https://doi.org/10.3147/jsfp.33.445

Watanabe, K.I., Nishizawa, T. and Yoshimizu, M. (2000) Selection of brood stock candidates of barfin flounder using an ELISA sustem with recombinant protein of barfin flounder nervous necrosis virus. *Diseases of Aquatic Organisms* 41(3), 219–223. DOI: 10.3354/dao041219.

Watanabe, K., Ide, K., Iwasaki, T., Satoh, J., Mori, K. *et al.* (2013) Disinfection of fertilized grouper eggs with electrolyzed seawater for prevention of viral nervous necrosis. *Fish Pathology* 48(1), 5–8. DOI: 10.3147/jsfp.48.5.

WOAH (2019) Chapter 2.3.12. viral encephalopathy and retinopathy. In: *Manual of Diagnostic Tests for Aquatic Animals.* Available at: https://www.woah.org/fileadmin/Home/eng/Health_standards/aahm/current/chapitre_viral_encephalopathy_retinopathy.pdf Retrieved February 15, 2023

WOAH (2021) Chapter 2.3.8. red sea bream iridoviral disease. In: *Manual of Diagnostic Tests for Aquatic Animals.* Available at: https://www.woah.org/fileadmin/Home/eng/Health_standards/aahm/current/chapitre_rsbid.pdf Retrieved February 15, 2023

Wolf, K. (1988) *Fish viruses and fish viral diseases.* Cornell University Press.

Xing, J., Sheng, X. and Zhan, W. (2006) Lymphocystis disease and diagnostic methods in China. *Aquaculture Asia Magazine* XI, 30–33.

Yamkasem, J., Tattiyapong, P., Kamlangdee, A. and Surachetpong, W. (2019) Evidence of potential vertical transmission of tilapia lake virus. *Journal of Fish Diseases* 42(9), 1293–1300. DOI: 10.1111/jfd.13050.

Yamkasem, J., Piewbang, C., Techangamsuwan, S., Pierezan, F., Soto, E. *et al.* (2021a) Susceptibility of ornamental african cichlids *Aulonocara* spp. to experimental infection with tilapia lake virus. *Aquaculture* 542, 736920. Available at: https://doi.org/10.1016/j.aquaculture.2021.736920

Yamkasem, J., Tattiyapong, P., Gorgoglione, B. and Surachetpong, W. (2021b) Uncovering the first occurrence of *Tilapia parvovirus* in Thailand in tilapia during co-infection with *Tilapia tilapinevirus*. *Transboundary and Emerging Diseases* 68(6), 3136–3144. DOI: 10.1111/tbed.14143.

Yamkasem, J., Tattiyapong, P. and Surachetpong, W. (2022) Development and application of TaqMan probe-based quantitative PCR assays for the detection of tilapia parvovirus. *Journal of Fish Diseases* 45(3), 379–386. DOI: 10.1111/jfd.13565.

Yanuhar, U., Christi, N. and Arfiati, D. (2018) Genetic characterization of viral nervous necrosis infects tilapia (*Oreochromis* sp) in indonesia 1st international conference in one health (ICOH 2017), Malang, Indonesia. In: pp. 174–179. DOI: 10.2991/icoh-17.2018.34.

Yin, J., Wang, Q., Wang, Y., Li, Y., Zeng, W. *et al.* (2019) Development of a simple and rapid reverse transcription–loopmediated isothermal amplification (RT-LAMP) assay for sensitive detection of tilapia lake virus. *Journal of Fish Diseases* 42(6), 817–824. DOI: 10.1111/jfd.12983.

Yong, C.Y., Yeap, S.K., Omar, A.R. and Tan, W.S. (2017) Advances in the study of nodavirus. *PeerJ* 5, e3841. DOI: 10.7717/peerj.3841.

Yoshikoshi, K. and Inoue, K. (1990) Viral nervous necrosis in hatchery-reared larvae and juveniles of Japanese parrot fish, *Oplenathus fasciatus* (Temminck and Schlegel). *Journal of Fish Diseases* 13(1), 69–77. DOI: 10.1111/j.1365-2761.1990.tb00758.x.

Zafran, I., Koesharyani, I., Johnny, F., Yuasa, K., Harada, T. *et al.* (2000) Nervous necrosis in humpback grouper cromileptes altivelis larvae and juveniles in Indonesia. *Fish Pathology* 35(2), 95–96. DOI: 10.3147/jsfp.35.95.

Zeng, W., Wang, Y., Chen, X., Wang, Q., Bergmann, S.M. *et al.* (2021) Potency and efficacy of VP20-based vaccine against tilapia lake virus using different prime-boost vaccination regimens in tilapia. *Aquaculture* 539, 736654. DOI: 10.1016/j.aquaculture.2021.736654.

Zheng, F.-R., Sun, X.-Q., Liu, H.-Z. and Zhang, J.-X. (2006) Study on the distribution and expression of a DNA vaccine against lymphocystis disease virus in Japanese flounder (*Paralichthys olivaceus*). *Aquaculture* 261(4), 1128–1134. DOI: 10.1016/j.aquaculture.2006.05.020.

Chapter 11

Disease diagnosis for tilapia pathogens

Partho Pratim Debnath, Nopadon Pirarat, Kim D. Thompson and Channarong Rodkhum

Introduction

Diseases in tilapia

Tilapia, which comprises over 100 species, is the world's second-largest farmed fish species after carp and is cultivated in over 135 countries worldwide, with production ranging from extensive household ponds to large commercial operations (Ng and Romano, 2013; Amal *et al.*, 2018; Jansen *et al.*, 2019). They are affected by various viral, bacterial, fungal or parasitic pathogens. Therefore, understanding the diseases affecting tilapia is required before an effective diagnosis can be performed.

Tilapia harbour three different forms of parasite protozoa: ciliates, flagellates and sporozoan (Saptiani *et al.*, 2017). Ectoparasitic infections in tilapia include ciliated protozoans, such as *Trichodina* sp. and *Ichthyophthirius multifiliis* (Shoemaker *et al.*, 2000; Pantoja, 2012). The most common protozoan diseases affecting tilapia are trichodiniosis, ichthyophthiriasis/white spot disease, chilodonellosis, ichthyobodoosis and myxosporidia (Abdel-Baki *et al.*, 2015; Valladão *et al.*, 2016; Taghreed Ibrahim, 2020). Monogenetic trematodes such as Dactylogyridae, Cichlidogyrides and Gyrodactylidae cause monogenesis in tilapia, while digenetic trematodes such as *Clinostomum tilapiae* and *C. complanatum* can be a severe issue in tilapia aquaculture, resulting in mass mortalities in fingerling

Kim Thompson, Craig Shoemaker and David C. Little (eds) **Tilapia**
DOI: 10.1079/9781800629455.0011, © CAB International, 2025

and juvenile fish (Roux and Avenant-Oldewage, 2010; El-Sayed, 2020; Taghreed Ibrahim, 2020). Various nematodes, cestodes and acanthocephalans have been found in both wild and cultured tilapia (Ramadan, 1991; Aloo *et al.*, 1995; Omoregie *et al.*, 1995). Parasitic crustaceans that infest wild and farmed tilapia include copepods such as *Ergasilus* spp., *Lernaea* spp., *Caligus* spp. and *Lamperoglena* spp., branchiurans such as *Argulus* spp. and *Dolops* spp. and isopods such as *Alitropus typus* (Douëllou & Erlwanger, 1994).

Several viral infections have been reported in tilapia production in recent years. Nine viral diseases have been reported, including six DNA and three RNA viruses. Although tilapia parvovirus (TiPV) has just been reported (Liu *et al.*, 2020; Yamkasem *et al.*, 2021), tilapia lake virus (TiLV) is the most widespread viral disease in tilapia aquaculture to date (Eyngor *et al.*, 2014; Debnath *et al.*, 2020; Surachetpong *et al.*, 2020). Other recognized viral infections include infectious spleen and kidney necrosis virus (ISKNV), infectious pancreatic necrosis virus (IPNV), nervous necrosis virus (NNV), tilapia larvae encephalitis virus (TLEV), Bohle virus (Ranavirus), iridovirus-like agent and lymphocystis disease virus (LCDV) (McGrogan *et al.*, 1998; Shlapobersky *et al.*, 2010; Keawcharoen *et al.*, 2015; Chinchar *et al.*, 2017; Mulei *et al.*, 2018; Taha *et al.*, 2020).

Tilapia are susceptible to a variety of bacterial diseases, including streptococcosis (Anshary *et al.*, 2014; Costa *et al.*, 2014; Kayansamruaj *et al.*, 2014a, b); francisellosis (Soto *et al.*, 2009; Nguyen *et al.*, 2015; Lin *et al.*, 2016); edwardsiellosis (Clavijo *et al.*, 2002; Soto et al., 2012); and bacterial haemorrhagic septicaemia (Soto-Rodriguez *et al.*, 2013; Dong *et al.*, 2015; Eissa *et al.*, 2015; Austin & Austin, 2016). In addition, hahellosis, a novel bacterial disease caused by *Hahella chejuensis*, has been shown to induce red egg syndrome (Senapin *et al.*, 2016).

Fungal infections are 'secondary diseases' because they proliferate in necrotic tissues associated with injuries, bacterial or parasitic lesions and dead and rotting eggs, and under poor culture conditions. A variety of fungal diseases have been identified in wild and farmed tilapia, such as saprolegniasis, caused by the water mould oomycete *Saprolegnia ubiquitous* (Goodwin, 2012; Zahran *et al.*, 2017), branchiomycosis (gill rot), which is caused by two oomycetes – *Branchiomyces sanguinis* and *B. demigrans* (Goodwin, 2012) – and ichthyophoniasis, which is one of the most economically and environmentally devastating diseases that affects farmed Nile tilapia (El-Ghany & El-Ashram, 2008; Yokota *et al.*, 2008).

Importance of disease diagnosis

The impact of infectious disease can be reduced through prompt diagnosis and treatment. Diagnosis can increase treatment efficacy and reduce long-term consequences. Undetected infected fish can spread disease to non-infected fish on the same or neighbouring farms and cause transboundary transmission of the disease. Early detection can help prevent or control disease outbreaks.

Similarly, accurate disease diagnosis can aid in determining the exact nature of a disease, allowing the implementation of an effective treatment strategy to avoid excessive therapy, abuse of unwanted chemicals and misuse of antibiotics. Furthermore, reducing antibiotic usage will assist in protecting the environment from the spread of antimicrobial-resistant bacteria. Although disease diagnosis appears to be an additional cost, it can be more cost effective by avoiding high treatment costs, minimizing disease occurrence, preventing production losses and conserving the environment. Preceding the implementation of any mitigation measures on farms, it is vital to have a thorough understanding of the infectious agents affecting tilapia, including the clinical symptoms, pathology and epidemiology of infections, as well as understanding diagnostic tests and pathogen characteristics.

Conducting pathogen/disease diagnosis in tilapia

Investigation of diseases is a systematic and meticulous process that must be strictly adhered to. Fundamental factors for proper and comprehensive disease investigation include:

- laboratory cleanliness to prevent carryover and cross-contamination
- effective documentation to track farm and disease information
- instructing farmers on submitting samples and detailing the number, type and preservation of samples
- arranging all necropsy materials and sampling materials, e.g. fixatives and culture media, prior to receiving the samples
- sampling of moribund or live fish, because dead fish cannot be used for disease diagnosis
- documenting all clinical observations (external, internal and behavioural) of diseased fish
- labelling and storing samples appropriately, including duplicate samples, for additional testing
- following standard diagnosis methodology to ensure the accuracy of results
- maintaining systematic records of diagnosis reports
- complying with national and international standards for biological and laboratory waste disposal
- using standard disinfection protocols for all reusable laboratory items.

Disease information, biological samples and necropsy for pathogen diagnosis

Disease information

Disease investigation and diagnosis require basic information on the disease occurrence, which should be as comprehensive as possible. Basic information includes location, farm, farmer, farming system, biosecurity, stocking information, disease and mortality trends, stressors and other factors (Table 11.1), collected using either online tools or paper-based survey questionnaires.

Biological sample collection

The primary objective of biological sample collection is to perform laboratory testing of fish to investigate unexpected death, disease

Table 11.1 Description of the relevant information detailing disease incidence.

Topics	Details
Enumerator information	Name, email and mobile number
Farmer consent	Farmer's consent to provide information and permission for sample analysis
Farmer contact	Farm name and contact details, including address, mobile/telephone number and email
Farm location	Information about regions, districts, provinces and country, as well as geographic coordinates
Farmer information	Gender, age, education, occupation, role on farm, aquaculture experiences
Farming information	Farming system (commercial/homestead/both, case/earthen pond/cement tank), farm size, farm type (seasonal, perennial), water source, culture type (monoculture or polyculture, extensive/semi-intensive/intensive/super-intensive), polyculture species and shared waterbody
Stock details	Tilapia source, size, sex, stocking density, for polyculture information about co-cultivated species (species name, source, size and stocking density, etc.)
Biosecurity measures	Fallowing duration, biosecurity measures during pre-stocking, post-stocking, between production cycles, disease incidence and post disease incidence
Disease information	Clinical signs, mortality rate, duration and pattern, disease control measures, e.g. medication, chemical treatment, etc., treatment efficacy and relevant factors
Water quality parameter	Information about water quality parameters including pH, dissolved oxygen, salinity, total dissolved solids, ammonia, nitrate, nitrite, etc., before and during disease incidence
Stress factors	Information about any stress factors such as weather, water, handling or chemicals

symptoms, increasing mortality, disease surveillance and disease investigation. The diagnosis accuracy depends on the quality of the samples submitted. Therefore, collecting appropriate samples is one of the most crucial aspects of disease diagnosis. Diagnostic results could be inaccurate without proper samples. Consequently, standardized sample collection procedures should be followed, detailing sample type and size, target tissue, fixation and storage.

Sample types

When studying disease epidemics, fish samples (known as purposive samples) likely to be positive for the infectious pathogen(s) of interest are collected (Tavornapanich *et al.*, 2020). The fish may be recently deceased, moribund, exhibiting disease symptoms or behaving differently from the rest of the stock. In addition, establishing the age range of susceptible fish and environmental conditions (e.g. water temperature) that facilitate pathogen infection is relevant for sample collection. Fish that have already died tend to be unsuitable candidates for evaluation because they decay rapidly and become hosts to numerous bacteria and parasites, degrading them further. Similarly, fish with no signs of disease may be healthy and sampling these will not assist in determining the disease problem.

Sample size

Investigation of a clinical outbreak requires a logical and systematic approach, informed by clinical experience, to allow appropriate individuals and diagnostic tests to be selected. The minimal number of samples required for analysis depends on various factors, including the certainty of results, the sensitivity and specificity of the test and the availability of resources (Tavornapanich *et al.*, 2020). Several online resources, such as WinEpi, FreeCalc – EpiTools, are available for estimation of sample size for sampling purposes.

Target tissue

Disease diagnosis accuracy relies on the appropriate tissues being taken during sampling. Inappropriate tissue selection for disease diagnosis can lead to erroneous identification. Internal tissues are preferable for bacterial culture and isolation because these are less likely to harbour contaminants, whereas external tissues like ulcers and gills do. Consequently, the kidney is regarded as one of the best reservoirs of bacterial infection (Austin, 2019). In addition, the liver, spleen and brain have also been used from sick tilapia for bacterial pathogen isolation and identification. Internal organs, including liver, spleen, kidney and brain, are typically utilized to diagnose viral diseases in tilapia; for example, the optimal organ to sample for TiLV or TiPV detection is the liver; however, these viruses can also be detected in the spleen, kidney and brain of sick fish (Debnath *et al.*, 2020; Dong *et al.*, 2022). Likewise, the kidney and spleen are the optimal tissues for diagnosing ISKNV in tilapia (Ramírez-Paredes *et al.*, 2021).

Sample fixation and storage

Histopathology: preparation of tissue sections for histopathology requires a certain level of expertise. The thickness of the dissected tissue for fixation should typically not exceed 5 mm. Fixation of tissue should take place as soon as possible after dissection because tissues from tilapia, as with other aquatic animals, deteriorate rapidly after death due to autolysis. Neutral buffered formalin (NBF, 10%) is recommended for fixation of tilapia tissues for histopathology (Martins *et al.*, 2018; Miki *et al.*, 2018). Individual tissue must be kept in 10% NBF for 24–48 hours and then replaced in 70% ethanol to prevent over-hardening and ensure long-term preservation. To adjust 10% NBF to a neutral pH, quantities of a buffer – commonly sodium phosphate – are added. The following formula is recommended: 100 ml of formalin (37–40% stock solution), 900 ml of distilled water, 4 g/L monobasic NaH_2PO_2 and 6.5 g/L dibasic/anhydrous Na_2HPO_4. A 10% formalin solution is 10% of the 37–40% stock solution; the actual amount of dissolved formaldehyde is only 3.7–4.0% (formula available at https://www2.nau.edu/fpm/documents/PBSandFormalin.pdf). During tissue fixation the tissue:fixative ratio should be 1:10.

Transmission electron microscopy (TEM): fixation is the initial step for preserving cells or tissue in a state as similar to its living state as possible (Park *et al.*, 2016). Chemical fixation is the most common method for sample fixation for TEM, with glutaraldehyde most frequently used. Tissue blocks (1 mm^3 cubes or smaller) are typically fixed in 2.5% glutaraldehyde from 4 hours to overnight at 4–8°C (Ivanchenko *et al.*, 2021). The protocol for preparation of 2.5% glutaraldehyde is as follows:

To prepare 100 ml of 2.5% glutaraldehyde in 0.1 M sodium cacodylate buffer (pH 7.2–7.4):

0.2 M sodium cacodylate buffer	50 ml
25% EM grade glutaraldehyde	10 ml
Distilled water	40 ml

The fixative is stable for about 3 weeks at 4–8°C (formula available at https://emunit.hku.hk/documents/SamplePreparationTechnique.pdf).

Molecular diagnostics (PCR/sequencing): freezing is one of the methods most commonly used to preserve samples for DNA or RNA analysis (Straube & Juen, 2013). Long-term storage often involves freezing at 80°C or in liquid nitrogen (196°C) (Rissanen *et al.*, 2010; Michaud and Foran, 2011); for short-term storage, 20–28°C is recommended (Seutin *et al.*, 1991; Matsuo *et al.*, 1999). However, it can be challenging to freeze samples collected in the field immediately, and there is always the possibility that they will thaw during transport. In this instance, ethanol (75–96%) can be effectively utilized to preserve samples for PCR, sequence analysis or microsatellite amplification (Straube and Juen, 2013). However, there is some controversy about using this method for RNA viruses, which can rapidly degrade in ethanol-preserved samples. Therefore, RNALater is the preferred method for preserving biological samples containing RNA viruses (Kruse *et al.*, 2017).

Necropsy

Necropsy area or room: necropsy of moribund or deceased tilapia is essential in diagnosing the cause of the disease. A designated area should be allocated at the farm or the laboratory to perform the necropsy, where the risk of contamination can be minimized while ensuring compliance with hygiene and safety standards. The designated space must have appropriate lighting and ventilation, water and electricity facilities and a surface that is easily disinfected.

Equipment: depending on the size of the tilapia to be sampled, a set of necropsy instruments that is required includes:

- instruments to measure and weigh fish (scale ruler and balance) (Fig. 11.1)
- dissection items including necropsy tray, dissecting box, scissors, forceps and scalpel blade with blade holder (Fig. 11.1)
- items to disinfect and sterilize surfaces and tools (e.g. ethanol spray bottles, spirit lamps or gas burners, tissue paper, etc. (Fig. 11.2))
- items for sample storage (ice, tubes of different sizes and tube holders) (Fig. 11.2).

Post-mortem of tilapia: the aim of dissection is to expose the body cavity (Fig. 11.3). The heart is located between and posterior to the gills. The spleen, a triangular-shaped organ with a flat surface and dark red colour, is found on the left dorsal side of the abdomen,

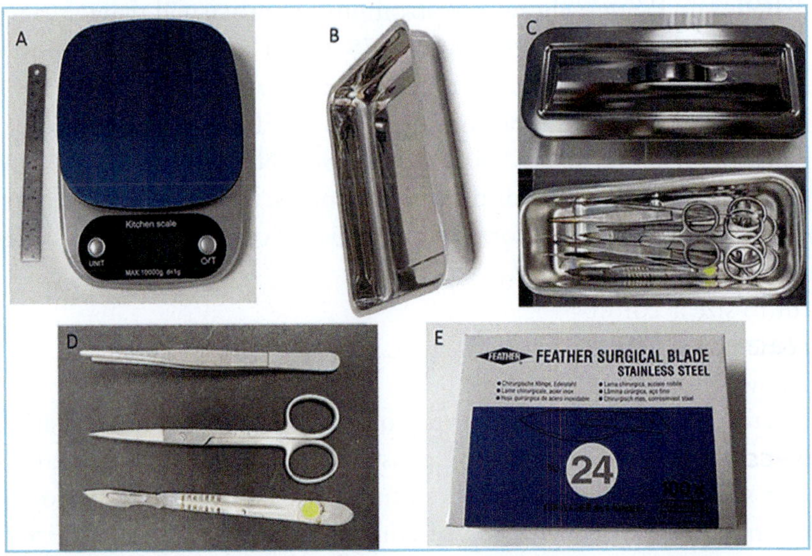

Figure 11.1 Necropsy materials necessary for measurement and dissection of fish. (A) measuring ruler and scale; (B) sampling tray; (C) dissection box; (D) scissors, forceps and scalpel blade holder; (E) scalpel blades

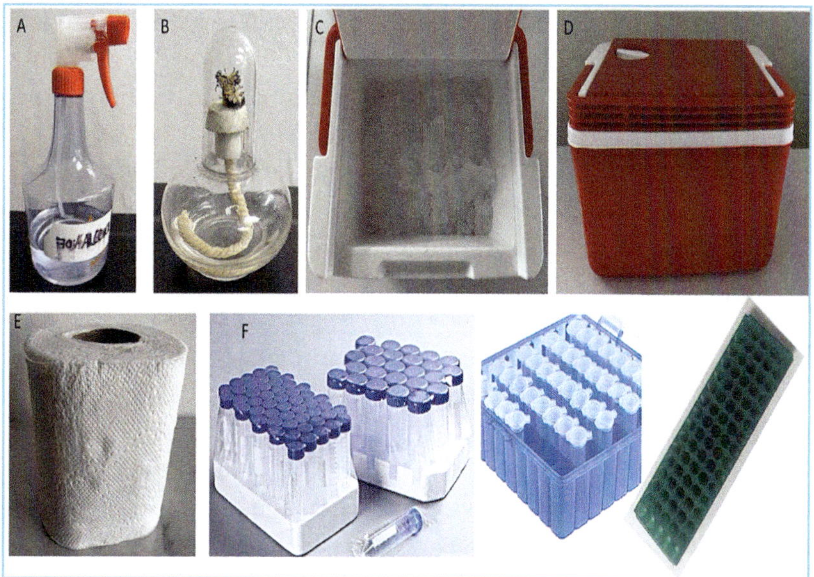

Figure 11.2 Items required for disinfection and sample storage. (A) ethanol spray bottle; (B) spirit lamp; (C) ice; (D) ice box; (E) tissue paper; (F) sample-storage tubes and racks

across from the stomach. The tilapia kidney is retroperitoneal, located near the spinal column, dark reddish-brown and may be noticeably lobed or closely embedded along the vertebrae. The head kidney is generally separated from the trunk kidney or merged. The liver is often a brownish-red colour. The brain resides in the skull, caudal and dorsal to the eyes. When sampling, the target organ must be cut to size according to the instructions given below. Organs can be pooled or stored separately providing the correct sampling regime is followed, i.e. correct preservatives and sample:size ratio (Fig. 11.3).

Recommendations

- Avoid sampling and dissecting dead tilapia due to post-mortem alterations and undesirable microbes responsible for after-death cell decomposition.
- Complete dissection as soon as possible to avoid cell autolysis owing to delayed dissection.

- During dissecting, avoid sample-to-sample cross-contamination.
- It is essential to prevent physical damage to tissues during sampling for histopathology and TEM.
- Tissue samples should not be too large to allow effective permeation of the fixative.
- The fixative volume should be sufficient to ensure that sample quality is preserved.

Physical observation and microscopy

Physical observation

The evaluation of gross physical clinical signs is the first step in disease diagnosis. This assesses changes in behaviour and external and internal clinical symptoms. Some tilapia diseases have a limited number of distinct clinical signs while others have none.

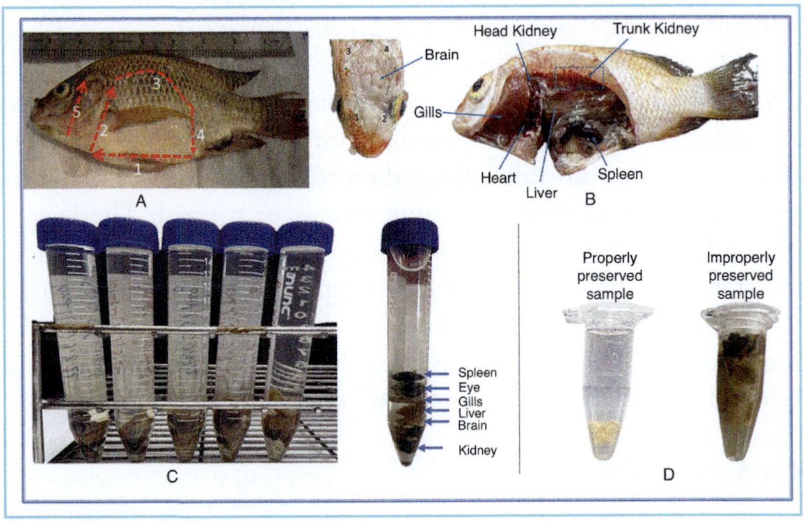

Figure 11.3 Necropsy and sample preservation method. (A) direction of tilapia dissection for open body cavity; (B) Internal organs of tilapia; (C) demonstration of sample preservation for histopathology purposes in buffered formalin; (D) demonstration of appropriate and inappropriate sample preservation in 95% ethanol for PCR analysis, with a sample preservative ratio of 1:10

Behavioural changes

Changes in tilapia behaviour can commence from the onset of the disease. Generally, poor appetite, gasping, lethargy, cessation of schooling and irregular swimming or swirling can be observed (Fig. 11.4). Major bacterial diseases in tilapia, such as motile *Aeromonas* septicaemia (MAS), exhibit significant alterations in behaviour, including sluggish movement, surface swimming, erratic swimming and a diminished appetite (Wang & Xu, 1985; Liu *et al.*, 1993; Li & Cai, 2011; Dong *et al.*, 2017a). Furthermore, streptococcosis in tilapia includes irregular swimming, sideways swimming, spinning and lethargy (Amal & Zamri-Saad, 2011; Zhang, 2021) while fatigue and impaired swimming are characteristic of francisellosis (Jantrakajorn & Wongtavatchai, 2016). Similarly, behavioural changes due to viral diseases have been reported in tilapia, including loss of balance and erratic swimming for viral nervous necrosis (VNN); lethargy, slow swimming and spinning for BIV; whirling for birnavirus and TLEV; lethargy for ISKNV; and lethargy, swimming at the water surface, cessation of schooling, swirling and loss of appetite for TiLV (McGrogan *et al.*, 1998; Shlapobersky *et al.*, 2010; Keawcharoen *et al.*, 2015; Subramaniam *et al.*, 2016; Surachetpong *et al.*, 2020). Common behavioural indicators can be present in various tilapia diseases, and it is not always straightforward to determine which diseases are responsible for the symptoms.

External physical clinical symptoms

Any anomalies in the external organs of tilapia are categorized as external clinical signs. Various clinical signs can be an indicator of a particular disease, but identical clinical signs are also found between diseases and therefore there are no clinical signs that can be used to specifically ascertain the cause of an infection in tilapia. Symptoms include exophthalmia, eye shrinkage and opacity, ulcers or lesions on the body surface, haemorrhages, tail and fin rot, skin discoloration, scale protrusion, abdominal distension etc. (Fig. 11.4). External clinical signs recorded for a variety of bacterial diseases in tilapia include fin rot, lesions on the body

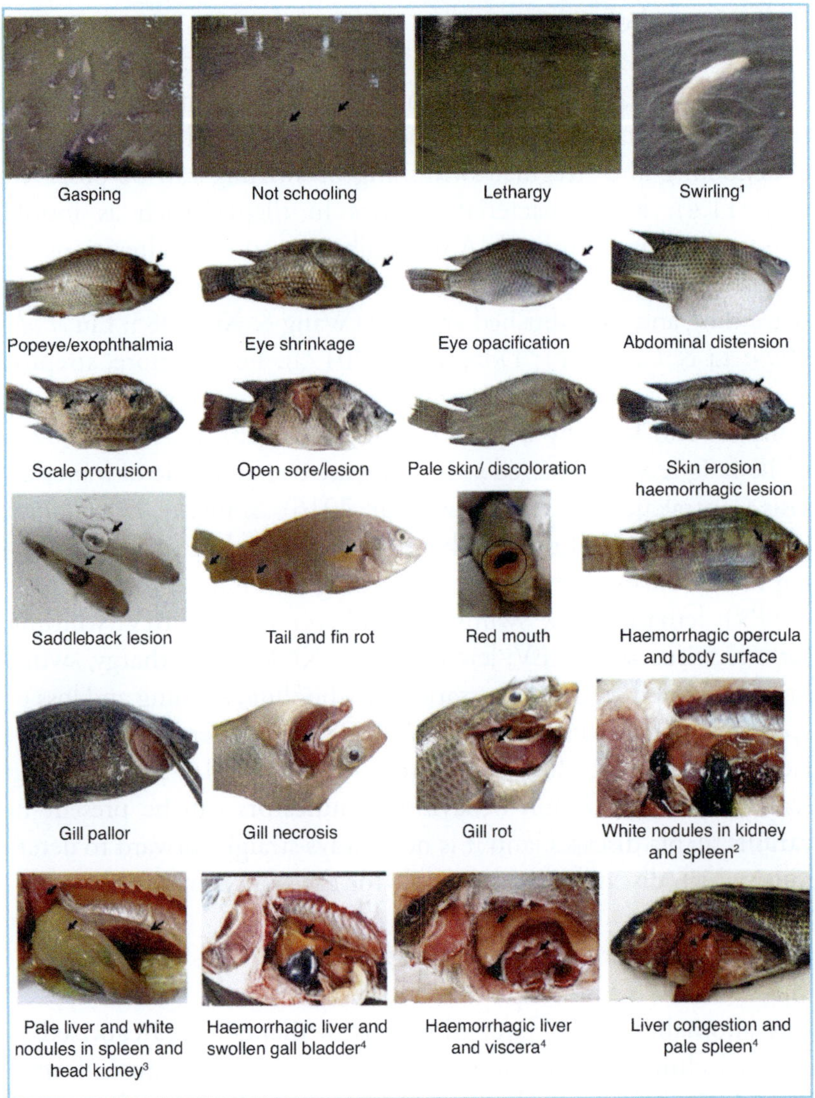

Figure 11.4 Clinical manifestations of diverse tilapia diseases[1–4]

surface, abdominal distension, skin haemorrhages, exophthalmia and blackened skin for MAS (Li & Cai, 2011; Shirajum Monir *et al.*, 2020); exophthalmia, melanosis of the skin, haemorrhagic lesions on the body and fins, corneal opacity and dark coloration for streptococcosis (Tung *et al.*, 1987; Darwish & Griffin, 2002;

Oviedo-Bolaños *et al.*, 2021); mild exophthalmia, moderate abdominal distension and scale protrusion for francisellosis (Ortega *et al.*, 2016); discoloration of the body, ulcers in muscles and skin, profuse mucus secretion, saddleback, fin rot and eroded tails (Roberts & Sommerville, 1982; Elgendy *et al.*, 2022a, b); and exophthalmia and corneal opacity, as well as haemorrhages, erosions and ulcers on the external body surface for edwardsiellosis (Elgendy *et al.*, 2022c). Likewise, various visible signs of viral infection in tilapia include exophthalmia, scale protrusion, stomach distension, ulcerated body surface and haemorrhages caused by TiLV (Debnath *et al.*, 2020; Surachetpong *et al.*, 2020); body pallor, exophthalmia, scale protrusion, skin haemorrhages and dorsal muscle haemorrhages caused by TiPV (Yamkasem *et al.*, 2021); and endophthalmitis, skin purulent sores, skin haemorrhages, skin ulcers, loss of scales, excess mucus and haemorrhages induced by ISKNV (Ramírez-Paredes *et al.*, 2021).

Internal clinical symptoms

Observations of anomalies in the liver, spleen, kidney, heart and brain of tilapia are considered internal clinical signs of disease. These include liver, spleen and kidney pallor; haemorrhages and blood congestion in the liver; white nodules on the liver, kidney and/or spleen; and a bloated gall bladder, amongst others (Fig. 11.4). White nodules in the liver, kidney and spleen have been documented as a unique clinical symptom of francisellosis and edwardsiellosis in tilapia (Soliman *et al.*, 2009; Nguyen *et al.*, 2015). Streptococcosis has been associated with the deposition of ascitic fluid in the abdominal cavity, granulomas in the brain and a fibrous layer covering internal organs such as the heart (Al-Harbi, 1994; Oviedo-Bolaños *et al.*, 2021). Ascites, enlarged and haemorrhagic internal organs, including the spleen, heart, brain, liver and kidney, have been attributed to ISKNV (Ramírez-Paredes *et al.*, 2021), while pallor of the gills and visceral organs, particularly the liver, severe abdominal ascites and liver haemorrhages have been ascribed to BIV (Smith *et al.*, 1998). As with behavioural and external symptoms, assigning specific internal symptoms to a particular disease is difficult.

General smear microscopy

Smear microscopy, also known as wet mounts, is a simple technique widely used in clinical laboratories. It is considered a level two method for identification of diseases, and plays an essential role in diagnosing and managing tilapia disease. Wet mounts are used to observe motile organisms such as parasite infestation and some bacterial infections, with mucus (skin and gill scrapings), gill clips, faecal samples, fluid samples from the peritoneal cavity and blood samples used as samples. During sample collection, fish should be handled promptly and carefully. Moreover, the mucus layer on the skin provides protection against pathogen entry so it is important to disrupt this as little as possible. A general anaesthetic is recommended to prevent the fish from experiencing excessive stress and trauma.

As with other fish species, skin is the first line of defence against infectious pathogens in tilapia, and cutaneous lesions are prevalent in diseased fish. Mucus smears are frequently used to evaluate suspected cutaneous bacterial or fungal infections and ectoparasite infestations. The surface of the skin and gills is scraped to collect a small quantity of mucus. It is possible to scrape with the blunt edge of a scalpel blade, a spatula or the edge of a microscope slide. After placing the sampled mucus on a microscope slide, a drop of distilled water is applied before placing a coverslip onto the slide. The specimen is examined under a microscope as a wet-mount preparation.

Fin biopsies are conducted when lesions are observed on the fins. The fin should be spread, and a small triangular wedge cut from the fin with fine-tipped scissors. The wedge should be placed on a microscope slide, a drop of distilled water added and examined under a microscope.

When respiratory diseases in tilapia are suspected or necrosis appears on the gills, a gill biopsy is an effective diagnostic technique. Simply cut and remove a few primary lamellae with the tips of the scissors, place them on a microscope slide with a drop of distilled water and study them under a microscope.

To observe pathogens in internal organs, e.g. liver, kidney and spleen, a sterile loop inserted into the organ can be used to collect a small amount of sample, which is then placed on a microscope slide with a drop of distilled water and observed under the microscope. Similarly, blood or peritoneal fluid can be collected using a sterile syringe and placed on a microscope slide; the blood or fluid is then smeared with a cover slide and covered with a coverslip for observation under the microscope.

Gram staining of smears can help inform on the nature of the bacterial infection. Gram-positive cocci were observed on the spleen smear of diseased tilapia (Fig. 11.5) (Liao *et al.*, 2020). Streptococcus sp. was reported by employing Gram staining of smeared kidney, liver and spleen tissue (Fig. 11.5) (Dong, 2018). Similarly, Dong (2018) showed the presence of *Flavobacterium columnare* through wet mount of saddleback lesions in tilapia (Fig. 11.5).

Histopathology

Specific histopathological changes are used as diagnostic indicators for bacterial and viral diseases in tilapia. Histopathology is the microscopic analysis of ultra-thin tissue sections to assess alterations in tissues caused by infection (Gurcan *et al.*, 2009; Johny *et al.*, 2022). Gills, brain, heart, liver, spleen, kidney and other infected tissue are sampled for histopathology based on the presence of lesions or clinical indicators, or in the absence of apparent clinical signs. Slide preparation includes tissue processing, embedding, sectioning and staining. Briefly, the tissue is placed in cassettes, rinsed for 1 hour under running tap water to eliminate buffer formalin solution and then transferred to a tissue-processing machine. Tissue processing involves passing the tissue through a series of water/alcohol combinations, followed by a Clearent, xylene (HistoClear) or equivalent and finally through melted paraffin, to produce a matrix to support the tissues during sectioning. Tissues embedded in paraffin are sectioned then stained, typically with haematoxylin and eosin (H&E). Tissue slides are deparaffinized by immersion for 5 min in three changes of xylene (xylene I, II and III). Thereafter the slides are rehydrated for 5 min with an ethanol series (100, 100, 95, 80 and 70%). After soaking for 5 min in distilled water the sides are submerged in haematoxylin solution for 10 min followed by immersion

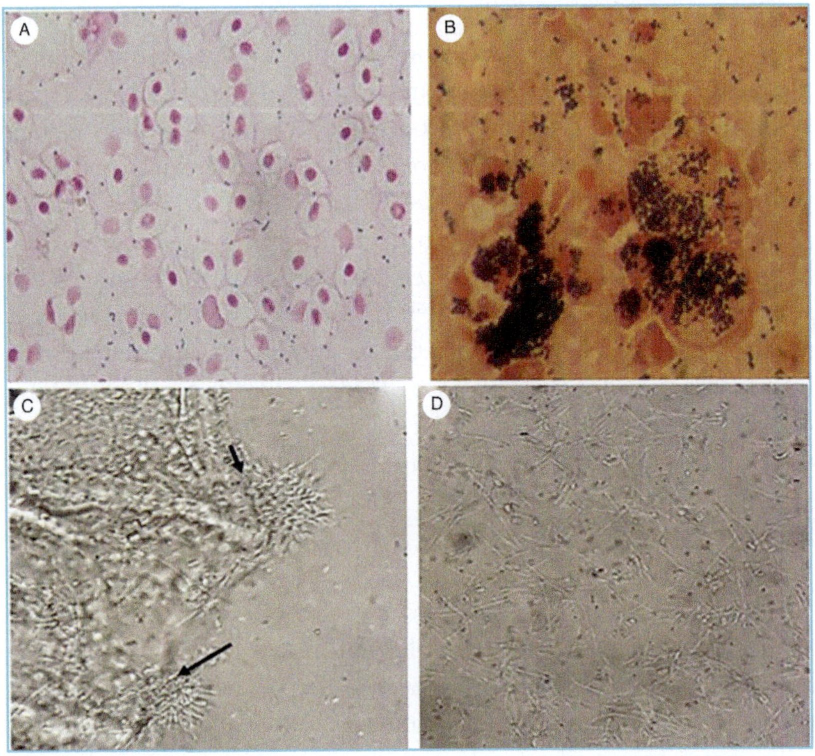

Figure 11.5 Identification of bacterial agents in diseased tilapia using wet mounts. (A) cocci bacteria on spleen smear with Gram staining (Liao *et al.*, 2020); (B) *Streptococcus* sp. on kidney smear with Gram staining (Dong, 2018); (C,D) Wet mount of saddleback lesion showing clumps of long, rod-shaped *Flavobacterium columnare* (Dong, 2018)

for 10 min in flowing water before transfer to an eosin solution for 5 min. Tissue sections are dehydrated through an ethanol series (70, 80, 95, 95 and 100%), each for 5 min. The slides are air-dried for at least 30 min at room temperature then re-immersed for 5 min in xylene I, II and III before mounting tissue sections with a Paramount solution and glass coverslip. The prepared slide is examined under a light microscope to detect pathological alterations (Mumford *et al.*, 2007; CarloAlberto, 2010; Hewitson *et al.*, 2010).

Pathological changes in the liver of tilapia infected with *Streptococcus agalactiae* include hyperactivation of melanomacrophage centres with overcrowded melanophores in the pancreas,

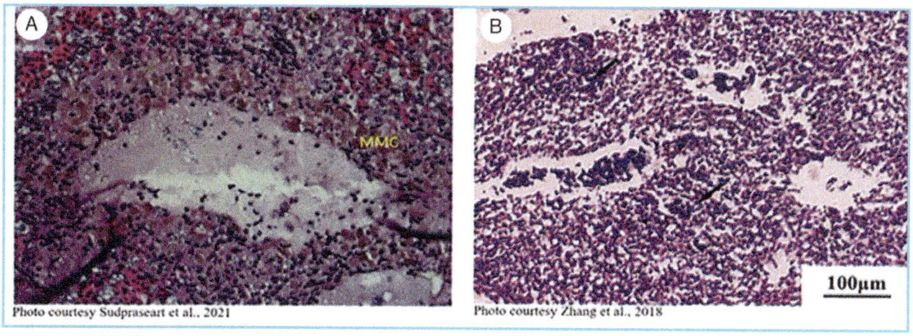

Figure 11.6 Histopathological changes in the liver (A) and head kidney (B) of tilapia infected with *S. agalactiae* exhibited melanomacrophage centres (MMC) and macrophage infiltration

severe hepatocyte degeneration and deposition of melanophores, oedema and septicaemia caused by clumping of bacteria within the hepatic vessel, with and without haemolysis, while head kidney revealed evident leukomonocyte infiltration (Fig. 11.6) (Zhang *et al.*, 2018; Sudpraseart et al., 2021).

The typical pathological lesion of tilapia with francisellosis is multifocal granulomas in multiple internal organs (Fig. 11.7) (Dong *et al.*, 2016; Poudyal *et al.*, 2020; Chang *et al.*, 2021).

Edwardsiellosis of tilapia notably causes histological alterations in several internal organs, including severe multifocal necrosis, pyogranulomas, pyknosis and karyorrhexis in the spleen, along with the presence of melanomacrophages and heterophils around the bile duct of the liver (Fig. 11.8) (Dong *et al.*, 2019; Oh *et al.*, 2020; Machimbirike *et al.*, 2022).

The typical pathological lesions reported in tilapia with MAS are the degradation of hepatocytes and hemosiderin deposition around the blood arteries and liver cells (Fig. 11.8). In addition, the brain and liver demonstrate severe blood congestion whereas the posterior kidney exhibits renal tubule shrinkage, degenerative alterations in both tubules and glomeruli, interstitial oedema and leukocytic infiltration (Dong *et al.*, 2017a; Abdel-Latif *et al.*, 2022). Viral infections also lead to various pathological lesions in numerous internal organs. Infection with TiPV revealed Cowdry type A inclusion bodies (CAIBs) in pancreatic acinar cells in the

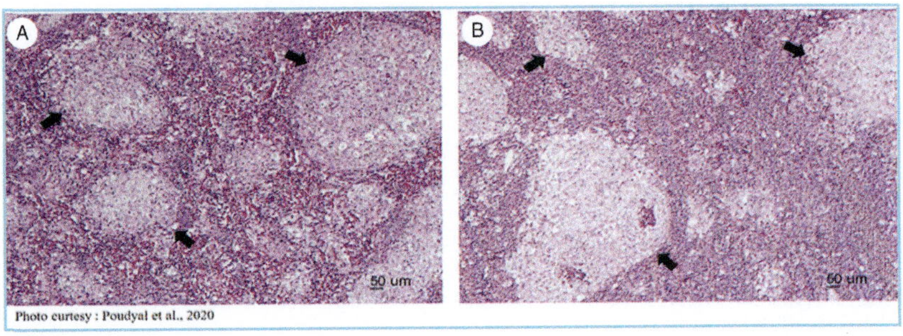

Figure 11.7 Multiple granulomas are evident in the spleen (A) and head kidney (B)

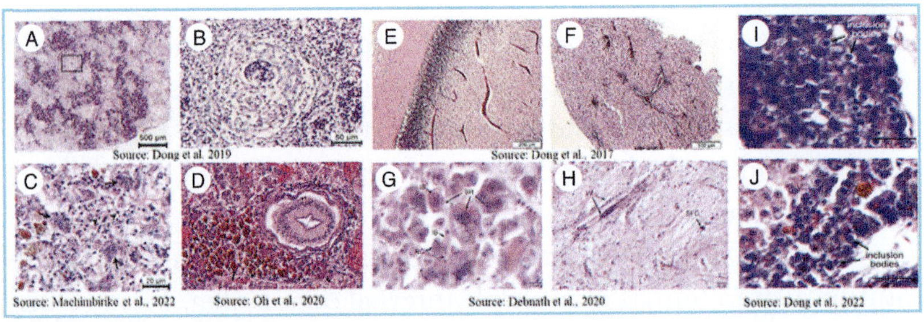

Figure 11.8 Histopathological changes in internal organs of tilapia. Pathological lesions due to *edwardsiellosis* (A) multifocal necrosis and pyogranulomas, (B) high magnification of a splenic granuloma, and (C) cell pyknosis (arrow head) and karyorrhexis (arrow) in the spleen; (D) Melanomacrophage and heterophils (arrow) on the liver; (E &F) severe blood congestion in brain and liver due to MAS; (G, H) The liver (G) and brain (H) sections of TiLV positive Tilapia. IB stands for intracytoplasmic inclusion body; Inf, infiltration of lymphocytic inflammatory cells; NC, normal cell; SFC, syncytium formation cell; SH, syncytial hepatocytes; (I &J) pathological lesion due to TiPV; (I) typical CAIB in pancreatic tissue next to the intestine and (J) within the hepatopancreas (arrow).

hepatopancreas, which has been recognized as the distinctive lesion for TiPV (Fig. 11.8); these are believed to be intranuclear inclusions, characterized by chromatin margination and a distinctive halo encircling a central basophilic inclusion body (IB) (Dong *et al.*, 2022). Similarly, syncytial hepatitis, also known as multinucleated giant cells, has been revealed as a pathognomonic lesion of TiLV (Ferguson *et al.*, 2014; Dong *et al.*, 2017b; Debnath *et al.*, 2020). In addition, intracytoplasmic IBs and foamy cytoplasm in the liver, infiltration of lymphocytic inflammatory cells and syncytium formation cells in the brain have all been attributed to TiLV infection (Dong *et al.*, 2017b; Debnath *et al.*, 2020) (Fig. 11.8).

Megalocytes are abundant in the lamina propria and pancreatic acinar tissue of tilapia infected with ISKNV. In addition, the spleen displays significant eosinophilia and vacuolated macrophage aggregates (Fig. 11.8) (Ramírez-Paredes *et al.*, 2021). Furthermore, VNN disorders in tilapia caused by *Betanodavirus* are detectable through histological changes in the brain and eyes triggered by the formation of vacuoles, with IBs as the primary pathognomic characteristic (Fig. 11.8) (Prihartini *et al.*, 2015). In addition, neuronal necrosis was observed in forebrain retina, spinal cord, olfactory bulb and optic lobe (Keawcharoen *et al.*, 2015).

Culture-based identification of potential tilapia pathogens

Overview

Culture-based disease diagnosis involves cultivating the microorganism under laboratory conditions (Koo, 2022); it has been possible to isolate a variety of bacteria, fungi, parasites and viruses using such methods. Growth of microorganisms on either solid or liquid nutritive media provides sufficient quantities of the organism to facilitate their identification (Vazquez-Pertejo, 2020) and enable antimicrobial susceptibility testing. Three main types of culture are used: solid, liquid and cell cultures. Solid- and liquid-based cultures are generally used for diagnosing bacterial infections in tilapia while cell cultures are predominantly used to identify viral diseases.

Solid culture

For solid cultures, agar, a long, unbranched polysaccharide chain, is added to the medium at a concentration of 1.5–2.0% where it solidifies at 37°C (Acharya, 2021). Solid media are used to cultivate pure cultures of bacteria or to isolate bacteria for assessment of colony attributes (Acharya, 2021). Bacterial growth on media can appear as either mucoid, spherical, smooth, granular, fibrous, asymmetrical or punctate on a solid substrate (Aiman, 2022). Commonly used culture media include, among others, MacConkey agar, nutrient agar and blood agar.

Liquid culture

A nutrient-rich 'medium' or 'broth' is used to grow a liquid culture of bacteria (Koo, 2022), containing appropriate levels of nutrients for bacterial growth but no gelling agents such as gelatine or agar (Acharya, 2021). Liquid cultures are used to cultivate bacteria for motility and antibiogram testing, to recover bacteria from lyophilized or stock cultures and to assess bacterial metabolism, toxin and enzyme production (Acharya, 2021). Nutrient broth, glucose broth, brain–heart infusion (BHI) broth, alkaline peptone water (APW) and tryptic soy broth (TSB) are often used as liquid media. Since liquid cultures frequently contain many microbial species, these are typically less effective in the diagnosis of bacteria than solid cultures. (Kragh *et al.*, 2018).

Cell culture

Viruses cannot be grown on solid or liquid media because, as intracellular parasites, they can replicate only inside living host cells (Wang *et al.*, 2018). By infecting animal cells with viruses in culture, it is possible to study the effect of the virus on the cell. Many viruses, for instance, have a 'deleterious' or 'cytopathic' impact on cells that can be observed (Koo, 2022). Cell cultures are considered the 'gold standard' diagnosis approach in virology.

Culture-dependent bacterial diseases diagnosis

Multiple diagnostic methods, including conventional diagnostics, serology and molecular techniques, have been applied to identify bacteria associated with disease in tilapia aquaculture (Austin, 2019). The crucial point is how to select the most appropriate method to

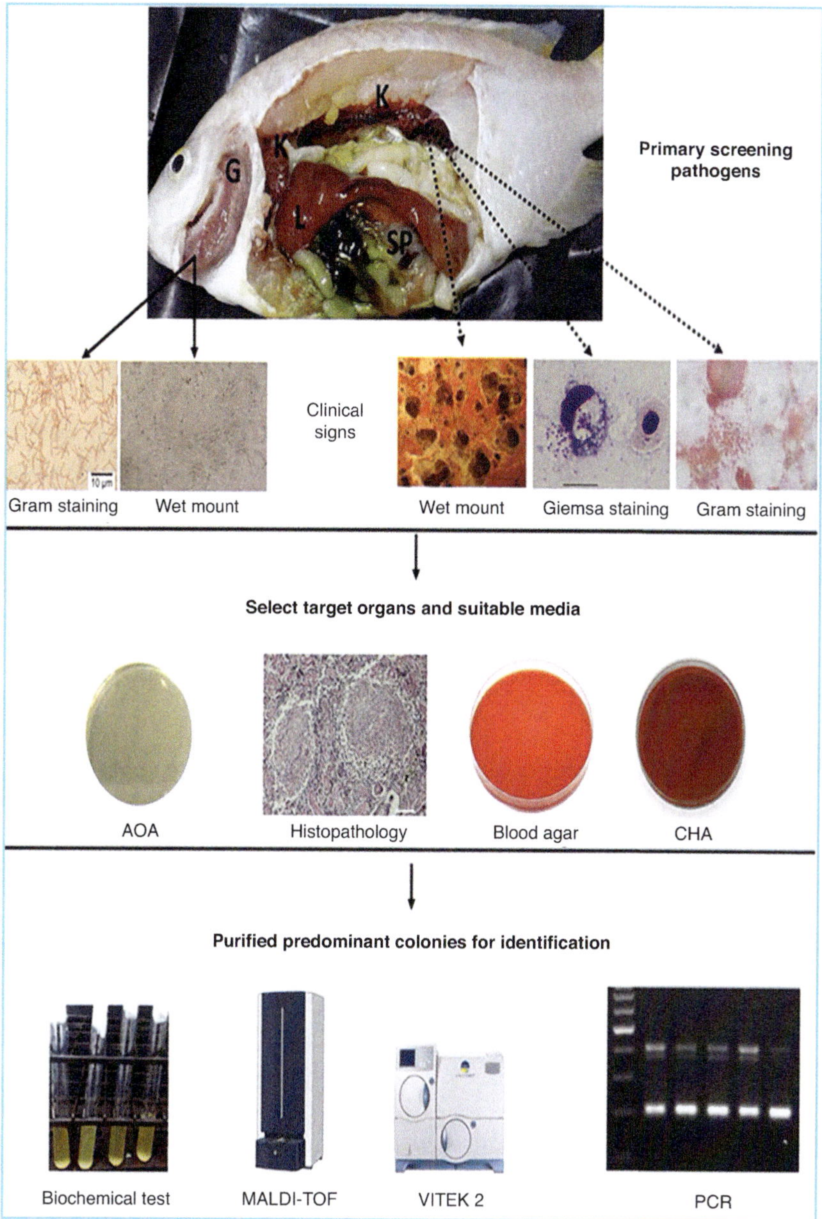

Figure 11.9 Flow diagram for identification of bacteria associated with disease in tilapia.

obtain reliable results. The prognosis begins by observing clinical symptoms and then performing wet-mount and Gram staining to choose the appropriate culture medium (Fig. 11.9). This serves as a source of nutrients to support the growth of microorganisms *in vitro* (Aiman, 2022). Culture media are divided into seven categories according to their practical applicability and chemical composition (Aiman, 2022): basal, enriched, selective, enrichment, indicator or differential, transport and assay media.

Basal media

Basal media, also termed general-purpose media, are simple media that promote the growth of the majority of non-selective bacteria (Acharya, 2021). These are common, non-selective media containing sources of carbon and nitrogen. Typically these media are employed for the initial isolation of a microorganism. Peptone water, nutrient broth and nutrient agar are regarded as basal media.

Enriched media

Supplementary nutrients, such as blood, serum and egg yolk, etc., are incorporated into the basal medium to develop an enriched medium (Acharya, 2021). Enriched media permit the growth of customized microorganisms but limit the growth of fastidious microorganisms that require vitamins and growth-promoting ingredients (Aiman, 2022). Blood agar is an example of an enriched medium, prepared by the addition of 5–10% (by volume) blood.

Selective media

Selective media restrict the growth of certain bacteria while allowing others to grow. These are agar-based (solid) media that enable the isolation of individual colonies. The selective growth of microorganisms is controlled by the inclusion of chemicals such as antibiotics, dyes and bile salts or by modification of pH (Singh, 2022). For example, TCBS agar contains bile salts to limit the growth of undesirable bacteria and is selective for *Vibrio cholera*. By fermenting sucrose, *V. cholera* produces an acid that functions as an indicator called bromothymol blue, resulting in the appearance of yellow colonies (Aiman, 2022). Middlebrook 7H10 Agar Base is another example of a selective medium, used for *Mycobacterium* isolation, cultivation and sensitivity testing; it contains malachite green, which inhibits the growth of other bacterial species (Wormser and Hanna, 2004).

Enrichment media

Enrichment media are liquid media used prior to plating on a solid selective medium, to increase the relative concentration of specific microorganisms in the culture (Aiman, 2022). The selective medium provides an environment and conditions that inhibit the growth of unwanted bacteria (Acharya, 2021). APW can recover pathogens from faecal samples, for example.

Indicator, or differential, media

Due to the presence of specific markers, such as dyes or metabolic substrates added to the medium, colonies of other microbial species exhibit different colours. These media separate bacteria based on the colour of colonies growing on the same plate; biochemical features such as neutral red, phenol red and methylene blue indicate the organism's growth (Singh, 2022). Blood agar, for example, is used to distinguish between haemolytic and non-haemolytic bacteria. Three forms of blood cell lysis are detected on blood agar: alpha, beta and gamma haemolysis (Acharya, 2021). Although these media facilitate the growth of various microorganisms, their ability to lyse blood cells varies, which helps in differentiation of bacterial colonies. TCBS media contains sucrose, consumed by fermented microorganisms, helping to distinguish these from non-fermented microorganisms. Based on this, bacterial colonies with a distinct colour develop on the media, allowing bacterial identification and differentiation (Acharya, 2021).

Table 11.2 Media commonly used for culture of tilapia bacterial pathogens.

Name, type and composition of medium	Preparation method	Culture conditions	Uses of medium	References
Tryptic soy agar (TSA) General-purpose medium Agar, tryptone, soya peptone, sodium chloride	Mix 15g of agar, 15g of tryptone, 5g of soy peptone and 5g of sodium chloride with 1,000 ml of distilled water. Dissolve and boil the mixture. Autoclave the agar for 15–20 min at 121°C and 15 pounds per square inch of pressure. Pour roughly 20 ml of agar per sterilized Petri plate and wait until the agar solidifies. Store at 4°C.	28–30°C for 24–48 hours	Primarily used as an initial growth medium for assessing colony morphology, maintaining a pure culture and generating adequate growth for subsequent biochemical testing and storage. TSA slants are utilized to store and transport bacterial cultures.	(Gerhardt *et al.*, 1994; Standard, 2010)
Blood agar: TSA with 5% sheep blood Enriched medium, also used as differential medium in allowing the detection of haemolysis TSA (as mentioned above) + 5% sheep blood, pH should be 7.2–7.6 (7.4).	Prepare TSA as instructed above and cool to 55 ± 5°C. Add 5% sheep blood and stir thoroughly. Pour the agar on to sterile Petri dishes, approximately 20 ml, on each plate, and let it settle. Preserve at 4°C.	28–30°C for 24–48 hours	Blood agar is a general-purpose, enriched medium used to culture a vast array of pathogens, including extremely fastidious species. Additionally, it is used to identify and distinguish haemolytic bacteria, particularly *Streptococcus* species. Furthermore, it is applied as a differential medium for detecting haemolysis (the destruction of red blood cells) caused by cytolytic toxins emitted by particular strains of *Bacillus*, *Streptococcus*, *Enterococcus*, *Staphylococcus* and *Aerococcus*.	(Standard, 2010; Aryal, 2022b)

Nutrient agar General-purpose medium 0.5% Peptone, 0.3% beef extract/yeast extract, 1.5% agar and 0.5% NaCl	Suspend 28g of nutritious agar powder in 1l of distilled water then boil while stirring to dissolve all components. Autoclave the dissolved mixture for 15 min at 12°C. Allow the agar to cool but not solidify after autoclaving. Pour agar into each plate and allow it to solidify at room temperature. Keep the plates in the refrigerator.	28–30°C for 24–48 hours	Nutrient agar is a nutrient medium utilized for the culture of microorganisms that facilitates the growth of a wide variety of non-selective bacteria.	(Aryal, 2022a)
Cysteine heart agar (CHA) Selective medium 10% Sheep blood, 100 units/ml polymixin B	Suspend 51g of CHA agar in 1000 ml of distilled water. Heat the medium to boiling to dissolve it entirely. Autoclave at 15 pounds of pressure (121°C) for 15 min. Cool to 50–60°C, add 10% sheep blood aseptically, stir firmly and pour into sterile Petri dishes.	28–30°C for 2–4 days	Cystine heart agar supplemented with 10% sheep blood is suggested as a selective medium for recovering *Francisella* sp., whereas polymyxin B inhibits the majority of Enterobacteriaceae.	(Soto *et al.*, 2009)

Continued

Table 11.2 (continued)

Name, type and composition of medium	Preparation method	Culture conditions	Uses of medium	References
Anacker and Ordal agar (AOA) Selective medium Tryptone, yeast extract, sodium acetate, beef extract, agar	Take 17.5g of AO agar with 1l of distilled water. Dissolve by bringing to the boil while frequently stirring, then blend and distribute into final containers. To sterilize, autoclave at 121°C (pressure 15 lb) for 15 min. Cool to 50°C and add neomycin at a concentration of 0.5 mg/ml and polymyxin B at 200 IU/ml. Pour the medium into sterile Petri dishes for solidification and preservation in the refrigerator.	30°C for 24–48 hours	AOA medium is regarded as selective for the primary isolation of *Flavobacterium* sp. from diseased tilapia.	(Anacker and Ordal, 1955; Daskalov *et al.*, 1999; Dong *et al.*, 2015)

Middlebrook 7H10 medium +Middlebrook OADC growth supplement Selective and enriched medium **Middlebrook 7H10 agar composition:** Ammonium sulfate, L-glutamic acid, monopotassium phosphate, disodium phosphate, sodium citrate, ferric ammonium citrate, magnesium sulfate, calcium chloride, zinc sulfate, copper sulfate, pyridoxine hydrochloride, biotin, malachite green **Middlebrook OADC growth supplement composition:** Bovine albumin fraction V, dextrose, catalase, oleic acid, sodium chloride	Prepare 19.47g of Middlebrook 7H10 agar base with 900 ml of distilled water. Add 5 ml of glycerol and boil to dissolve the medium completely. Autoclave the media at 15 pounds of pressure (121°C) for 10 mins. Cool to 45°C and add 20 ml of Middlebrook OADC growth supplement aseptically. Mix thoroughly and pour into screw-capped tubes or a Petri dish.	23–30°C for 2–3 months.	Middlebrook 7H10 agar enriched with Middlebrook OADC growth supplement is a selective medium for isolating and identifying *Mycobacterium* sp. from diseased tilapia and other fish.	(Rhodes *et al.*, 2005; Delghandi *et al.*, 2020)

Continued

Table 11.2 (continued)

Name, type and composition of medium	Preparation method	Culture conditions	Uses of medium	References
Thiosulfate–citrate–bile salts–sucrose (TCBS) agar Selective medium Sucrose, dipeptone, sodium citrate, sodium thiosulfate, sodium chloride, yeast extract, ox bile (oxgall), sodium cholate, ferric citrate, bromothymol blue, thymol blue, agar	Suspend 88.1g of dehydrated medium in 1l of distilled water while constantly swirling until the medium is completely dissolved. There is no obligation to autoclave; allow the solution to cool to 50°C and pour it into clean Petri plates.	28°C for 24–48 hours	TCBS agar is a selective medium for identifying *Vibrio cholerae* and other enteropathogenic *Vibrio* (especially *V. parahaemolyticus*).	(Aryal, 2022c; Mamdouh Y. Elgendy *et al.*, 2022a)
Mueller–Hinton agar (MHA) Assay medium Beef extract, starch, acid hydrolysate of casein, agar	Disperse 38g of the medium in 1000 ml of distilled water. The medium must be heated with regular stirring and brought to a boil for 1 min to completely dissolve. Autoclave for 15 min at 121°C. Bring to ambient temperature. Pour cooled Mueller–Hinton Agar into sterile Petri dishes. Allow the medium to become solidified. Store the plates at 2–8°C.	28°C for 24 hours	MHA is a standard solid medium recommended for assessing the susceptibility of bacteria to antimicrobial drugs using the Kirby–Bauer method of diffusion or dilution in agar.	(Samir *et al.*, 2021; Alazab *et al.*, 2022)

Transport media

Transport media are buffered solutions containing carbohydrates, peptones and other nutrients (except growth factors) that help maintain bacterial survival during transport without permitting their proliferation (Rijal, 2022). The incorporation of charcoal neutralizes inhibitory factors. These media prevent specimen desiccation, maintain the pathogen:commensal ratio and suppress the overgrowth of undesirable microorganisms (Acharya, 2021). *Vibrio* spp. can be transported using Cary Blair transfer medium and APW. Pike's medium is used to transfer streptococci.

Assay media

These media are used to assess vitamins, amino acids and antibiotics. For instance, when evaluating whether bacteria are resistant to antibiotics, Mueller–Hinton agar (1.7% agar) facilitates the diffusion of medicines throughout the medium. In addition it contains starch, which can absorb toxins produced by bacteria. On this medium, antibiotics are surrounded by a zone of inhibition (Aiman, 2022).

Key media used for bacterial culture in tilapia

Bacterial culture entails the inoculation of a clinical sample on solid or liquid medium and incubation under suitable conditions for the pathogen to grow. Bacteriologists use a variety of culture media, including TSA, blood agar, CHAB, AOA, TCBS and MHA, depending on the presumed nature of the disease and the aim of bacterial isolation. Table 11.2 lists the type, composition and examples of culture media most commonly used to diagnose bacterial diseases in tilapia.

Practical guidelines for bacterial isolation and biochemical characterization

Bacterial characterization starts with the isolation, purification and identification of bacteria. It is recommended to use both general-purpose and selective media to reduce the likelihood of missing key microorganisms in the specimen, considering the clinical symptoms and suspected diagnosis. For instance, TSA medium supplemented with 5% sheep's blood can grow a large variety of fastidious bacteria. Selective media such as AOA can be used to recover *Flavobacterium* spp. if fish are exhibiting gill necrosis; CHAB medium is advised for growing *Francisella* spp. if there is evidence of white nodules in the internal organs. Internal organs such as the brain, kidney, spleen and liver are the most reliable source for isolation of bacteria because these tissues typically harbour disease-causing bacteria during an infection. By contrast, infected tissues exposed to the environment, such as lesions on the body's surface or gills, carry an inherent risk of contamination, making it difficult to determine whether the resulting bacterial growth is the pathogen causing the disease, a secondary invader or an accidental contaminant (Austin, 2019). Internal organs sampled from moribund fish or fish that have just died are less likely to be contaminated with bacteria compared with fish that had been dead for several hours.

Isolation of bacterial colonies using the streak-plate procedure

The purpose of the streak-plate method is to obtain pure cultures of bacteria, or colonies, from mixed populations by employing simple mechanical separation (Sanders, 2012). A cluster of genetically similar cells is a colony. Unlike a single cell, it is visible to the naked eye and consists of millions of cells originating from a single bacterium initially put on the plate. In the streak-plate technique, a mixture of cells is dispersed across the surface of a semi-solid medium in a Petri dish, where fewer bacterial cells are deposited at widely separated sites on the medium's surface and, following incubation, form colonies. Using the quadrant-streak approach, which permits sequential dilution of the original microbial material, it is possible to extract well-isolated discrete colonies. As the original sample is diluted by streaking it over successive quadrants, the number of organisms declines. By the third, fourth or fifth quadrant a few organisms have been transferred on the inoculating loop, producing

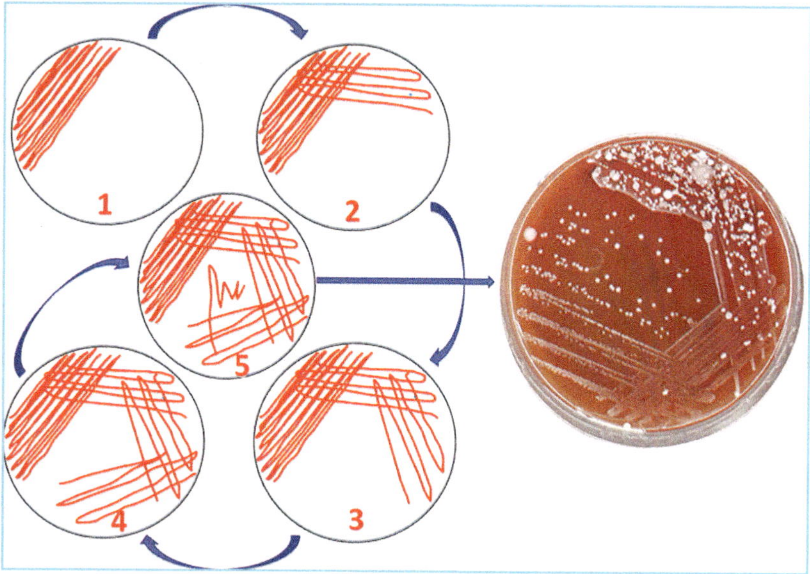

Figure 11.10 Method for bacterial cultivation on solid media using quadrant-streaking

several isolated colonies. Using an inoculating loop, tissue samples obtained from tilapia (kidney, spleen) are streaked onto the first quadrant of an agar plate using a back-and-forth motion. As shown in Fig. 11.10, these streaks are extended into the plate's second, third and fourth quadrants. Before spreading the streak from one quarter to another, the loop is flamed and allowed to cool. The plate is inverted to prevent condensation forming on the lid and leaking onto the colonies. Plates are incubated following the culture conditions outlined in Table 11.2.

Pure bacterial culture

The primary culture plate is inspected for bacterial colonies following incubation. Dense growth from infected tissues is usually regarded as an indicator of the causative pathogen's recovery. Although the objective is to recover pure cultures, the criteria for selection of colonies are subjective and necessitate the expertise of the microbiologist to ensure that meaningful colonies are chosen for further investigation (Austin, 2019). In most instances it is advisable to select representatives of the dominating colony types for pure culture. Colonies are selected based on their size, shape and colour, then gently touched with a sterile loop and streaked onto a plate using the technique described above. Following incubation, colonies on the agar plate are observed and the purity of the culture assessed. Pure cultures should exhibit identical colony traits and not overlap.

Primary biochemical test for bacterial characterization

Gram staining is the first biochemical test performed on bacteria recovered from a pure culture; this is followed by acid-fast, catalase, oxidase, motility and oxidation fermentation assays. The primary advantage of these tests is that most bacteria may be assigned to a genus or limited group of genera using the findings of a small number of well-selected tests.

Gram staining. Gram staining is key to the phenotypic classification of bacteria, which classifies bacteria based on whether they retain crystal violet stain (Gram-positive, blue to purple) or not (Gram-negative, red to pink) (Smith and Hussey, 2005). This also highlights cell morphology (e.g. bacilli, cocci) and cell arrangement (e.g. clumps, chains, diploids). Gram staining starts with heat-fixing the specimen material to a slide, which requires four basic staining steps: applying a primary stain (crystal violet), adding a mordant (Gram's Iodine), rapid decolorization with decolorizer and counterstaining with safranin (Davies *et al.*, 1983; McClelland, 2001). Gram-positive bacteria include *Staphylococcus* spp. and *Streptococcus* spp., amongst others, while Gram-negative bacteria include *Escherichia coli* and *Salmonella* spp.

Acid-fast staining. Acid-fast staining is a differential technique initially devised by Ziehl and then improved by Neelsen, and therefore this technique is known as Ziehl–Neelsen staining. Acid fastness is demonstrated when an organism resists decolorization by strong acids or mixes of ethanol and mineral acid; this is typical of a small number of bacteria and, when positive, is indicative of mycobacteria

(Ziehl, 1882; Neelsen, 1883). *Nocardia* is occasionally acid-resistant but rarely withstands the intense decolorization that mycobacteria may sustain (Barrow & Feltham, 1993).

Motility. Motility, the ability of an organism to move independently through propeller-like flagella specific to bacteria, has long been recognized as a useful taxonomic tool and defining biological characteristic of microorganisms (Jordan *et al.*, 1934; Leifson, 1951; Kelly & Fulton, 1953). Motility is highly significant regarding identification in Enterobacteriaceae, because flagella are seen typically in bacilli but only in a few flagellated cocci (Jordan *et al.*, 1934; Jarrell & McBride, 2008). Although it is feasible to study motility using a hanging-drop method, the preferred method is semi-solid agar. Motile bacteria 'swarm' and form a visible, scattered growth on semi-solid agar media. Sulphide indole motility (SIM) medium, a mixed differential medium, is most frequently employed for this purpose (Sulkin & Willett, 1940; Greene *et al.*, 1951). The formula for this medium is the combination of 30g of peptone, 3g of beef extract, 0.2g of ferrous ammonium sulfate ($(NH_4)_2Fe(SO_4)_2(H_2O)_6$), 0.2g of sodium thiosulfate ($Na_2S_2O_3(H_2O)_5$) and 3g of agar with 1l of distilled water, which is then stirred and brought to the boil to dissolve the agar. Next, distribute 5 ml aliquots into screw-capped test tubes and autoclave at 121°C for 15 min at 15 pounds per square inch. To test for motility, select a properly isolated young colony (18–24 hours) with a sterile needle and make a single stab down the centre of the tube, approximately halfway into the medium (Patricia & Laura, 2011). Noteworthy is the incubation temperature; most motile organisms are motile at lower temperatures (e.g. 15–25°C) but may not be motile at the optimal growth temperature (37°C) (Austin and Austin, 2016). After overnight incubation, dispersed growth away from the inoculation line indicates a positive motility test.

Catalase test. The catalase test identifies the enzyme catalase, which catalyses the release of oxygen from hydrogen peroxide (H_2O_2). It is used to distinguish between bacteria that produce the enzyme catalase, such as staphylococci, and those that do not, such as streptococci. The catalase test is advantageous for determining aerobic and obligatory anaerobic bacteria, because anaerobes typically lack the enzyme (McLeod & Gordon, 1923; Mahon *et al.*, 2011). Generally 3% H_2O_2 is used for culture (Clarke and Cowan, 1952; MacFaddin, 2000) whereas 15% H_2O_2 is employed for catalase detection in anaerobes (Bartelt, 2000). The standard approach for catalase testing begins with transferring a well-isolated colony to a clean slide. Add one drop of 3% H_2O_2 then observe for the instant production of bubbles; positive reactions are immediately identifiable by the development of bubbles. Observation of bubble formation under a magnifying glass or microscope can assess weakly positive responses. A reaction is catalase-negative if it does not generate bubbles (no catalase available to hydrolyse the H_2O_2).

Oxidase test. The oxidase test detects the enzyme cytochrome oxidase, indophenol oxidase (Gaby & Free, 1958; Oser, 1965; MacFaddin, 2000). In the electron transport cycle, cytochrome oxidase transports electrons from a donor molecule to oxygen. The oxidase reagent includes a chromogenic reducing agent, a chemical that alters colour when oxidized (Gaby & Free, 1958; Gerhardt *et al.*, 1981; Liu, 1986). In addition to the Kovács oxidase reagent, the Gordon and McLeod reagent and the Gaby and Hadley oxidase test reagent, several oxidase test reagents are commercially available (Gordon & McLeod, 1928; Kovacs, 1956; Gaby & Hadley, 1957). The oxidase assay can be executed, among other ways, on filter paper, in a Petri dish, with a swab, on an impregnated oxidase test strip or in a test tube, with the filter paper test the method most prominently employed. A few drops of oxidase reagent are added to a Whatman filter paper. Using a loop, choose a well-isolated colony from a plate of freshly cultured bacteria (18–24 hours old) and rub it onto the prepared filter paper. Bacteria are oxidase-positive if the colour changes to dark purple in 5–10 s, while they are delayed oxidase-positive if the colour changes colour in 60–90 s. By contrast, oxidase-negative bacteria are those whose colour does not change or whose reaction time exceeds 2 min (Gerhardt *et al.*, 1981). The test is sensitive and valuable for classifying and identifying bacteria (Steel,

1961). It can also be used to differentiate between pseudomonads and enteric bacteria.

Conformity tests for characterization of bacteria

Primary biochemical tests usually characterize bacteria up to the genus level. Routine biochemical tests, or matrix-assisted laser desorption ionization time-of-flight mass spectrometry (MALDI-TOF MS), are often used to characterize bacteria to the species level. This preliminary diagnosis can be validated using molecular diagnostic techniques, as explained later in the chapter.

Routine biochemical tests. Most bacteria are classified based on their response to route biochemical tests. Various biochemical kits are commercially available which, although not explicitly designed for aquatic pathogens, have been found by several researchers to be effective after making simple modifications (Popovic, 2008; Crumlish, 2017). The most commonly used kits are the API 20E, API ZYM, API 20NE, API 50CH, API 50L, Vitek system, Biolog-GN, Biolog-GP, Enterotubes and RapidID 32 (Popovic, 2008; Crumlish, 2017; Austin, 2019). These tests are conducted in the laboratory under controlled conditions using standard operating protocols. Before their implementation the tests should be evaluated for sensitivity and specificity, and this should be revisited frequently to ensure that the most appropriate methods are implemented.

MALDI-TOF MS. This is a high-throughput method that compares the protein fingerprint generated by microbial cells, which is then compared to a database of reference spectra using algorithms incorporated into the system. This method has recently been introduced into clinical microbiology laboratories for the identification of microorganisms (Seng *et al.*, 2009; Rodríguez-Sánchez *et al.*, 2014; Singhal *et al.*, 2015; Zhou *et al.*, 2017) and is being used more frequently for identification and classification of microorganisms (Fenselau and Demirev, 2001; Welker and Moore, 2011; Lavigne *et al.*, 2013). An overview of the method used has been described by Patel (2015). A single colony from a pure bacterial culture is picked with a wooden toothpick, smeared onto a MALDI-TOF MS target plate, overlaid with 1 μl of formic acid and allowed to dry for 5 min before adding 1–2 ml of matrix solution containing cyano-4-hydroxycinnamic acid (CHCA) dissolved in 50% acetonitrile, 2.5% trifluoroacetic acid and 70% formic acid. The target plate is dried for 5 min before placing it in the MALDI-TOF MS chamber. MALDI-TOF MS uses pulsed UV radiation aided by matrix crystallization to evaporate the sample's biopolymer and accelerate the particles within the vacuum chamber. The biopolymer is detected and given as mass spectra by the detector (Patel, 2015). The manufacturer-recommended, real-time (RT) identification score criteria are then applied: a score of ≥2000 suggests a species-level identification, a score between ≥1700 and <2000 indicates a genus-level identification and a score <1700 indicates no reliable identification. MALDI-TOF MS is an ongoing revolution in the clinical microbiology laboratory due to its ability to identify pathogens at the serotype or strain level, and to profile drug resistance within minutes (Romero-Gómez *et al.*, 2012; Østergaard *et al.*, 2015).

Cell culture techniques for virus isolation

Viruses are obligate intracellular parasites, meaning that they can multiply only within living host cells. Cell cultures are recognized as the 'gold standard' in virology due to their numerous implications for virus detection, identification, propagation, isolation, confirmation and characterization (Leland & French, 1988; Wang *et al.*, 2018). Cell culture refers to the extraction of cells from an animal or plant and their subsequent growth in a suitable artificial environment. They may be extracted directly from the tissue and disaggregated using mechanical or enzymatic methods to prepare primary cultures, or acquired as a developed cell line.

Primary cell culture and subculture

To prepare primary cell lines from tilapia, choose a healthy individual and disinfect the external surface with 75% ethanol. Target tissue should be dissected under aseptic conditions and dipped three times in sterile phosphate buffered saline (PBS; pH 7.4) containing antibiotics and antimycotics at a final concentration of 1000 IU/ml penicillin, 1000 µg/ml streptomycin and 2.5 µg/ml amphotericin B. Tissue is transferred to a fresh plate, thoroughly chopped then centrifuged for 10 min at 200 g at room temperature. The pellet is then resuspended in L-15 medium supplemented with 20% foetal bovine serum (FBS), 200 IU/ml penicillin, 200 µg/ml streptomycin and 0.5 µg/ml amphotericin before transfer to a tissue culture flask with incubation at 27°C. Confluent monolayers of the primary cell cultures should be separated using 0.25% trypsin solution-EDTA, subcultured at a ratio of 1:2–1:5 and then maintained in L-15 medium supplemented with 20% FBS, 200 IU/ml penicillin, 200 µg/ml streptomycin and 0.5 µg/ml amphotericin B. After 10–20 subcultures, the concentration of FBS can be lowered to 10% without antibiotics being added to the culture medium (Dong *et al.*, 2008; Wang *et al.*, 2018).

Cell culture using developed cell lines

Many established cell lines have been used to isolate viruses from tilapia (see Table 11.3). These can be purchased from various cell line collections.

Virus isolation

Virus culture is carried out in the laboratory, where virus samples are inoculated into different cell lines. The virus's potential to infect is assessed based on changes in cell morphology, known as a cytopathic effect (CPE). Samples of organs including the brain, kidney, liver, spleen and heart are aseptically collected from diseased tilapia. The tissues are homogenized in nine volumes of Hanks' buffered saline solution (HBSS) or PBS, centrifuged at 3000*g* for 10 min and the supernatant filtered through a 0.22-µm filter. To infect cells with virus from tilapia, cell monolayers (about 90% confluence) are washed twice with HBSS/PBS and incubated for 1 hour at 25°C with 500 µl of the virus filtrate. The cells are then washed with HBSS/PBS and L-15 media with 5% FBS is added to the infected

Table 11.3 Cell lines employed in the recovery of various tilapia viruses.

SL no.	Cell line name	Tissue of origin	Fish species
1	CIK	*Ctenopharyngodon idella* kidney	*C. idella*
2	CCB	Common carp brain	*Cyprinus carpio*
3	CrCB	*Carassius auratus* brain	*C. auratus*
4	FHM	Fathead minnow	*Pimephales promelas*
5	CFF	Caudal fin of *Puntius fasciatus*	*Puntius fasciatus*
6	CAMB	Brain of the hybrid snakehead	*Channa argus* ♀ × *C. maculata* ♂
7	CAMK	Hybrid snakehead kidney	*Channa argus* ♀ × *C. maculata* ♂
8	SS	*Siniperca chuatsi* brain	*S. chuatsi*
9	GrE	*Gobiocypris rarus* epithelioid	*G. rarus*
10	ZF4	Zebrafish fry	*Danio rerio*
11	MSF	*Micropterus salmoides* fin	*M. salmoides*
12	CHSE-214	Chinook salmon embryo	*Oncorhynchus tshawytscha*
13	BF2	Caudal trunk of 1-year-old bluegill	*Lepomis macrochirus*
14	E11 (SSN1)	Snakehead fish, whole fry tissue	*Ophicephalus striatus*
15	TiB	Tilapia brain	*Oreochromis niloticus*
16	OmB	Mozambique tilapia brain	*O. mossambicus*
17	TmB	Mozambique tilapia bulbus arteriosus	*O. mossambicus*
18	On1B	Brain of Nile tilapia	*Oreochromis niloticus*
19	On1L	Liver of Nile tilapia	*Oreochromis niloticus*

cells. The cells are then incubated at 25°C and monitored for 21 days for the development of a CPE.

CPEs of tilapia viruses

CPEs are the morphological alterations that virus infections can elicit in cell culture. The degree of apparent cell alteration generated by viral infection varies depending on the type of virus, host cell type, multiplicity of infection (MOI) and other aspects. Rounding, detachment, syncytia or fusion, shrinkage, increased refractivity, aggregation, loss of adhesion, cell lysis or death are the morphological changes most frequently noted following viral infection.

Cell death is a common virus-induced CPE. Many viruses induce apoptosis or lysis to kill cells, resulting in the most severe kind of CPE. Rapid cell shrinkage, pyknosis and detachment from the glass occur over the entire monolayer. For example, the CPE in CHSE-214 cells caused by IPNV is shown in Fig. 11.11. Syncytia formation is a distinct kind of CPE. Viral proteins used by viruses to induce fusion with the host cell are synthesized on the plasma membrane of infected cells by enveloped viruses. Cell-to-cell fusion, which results in the development of syncytia (multinucleated giant cell), occurs when these viral proteins attach to receptors on the surface of adjacent cells. CPE formation in cells infected with TiLV is an example of syncytia formation (Fig. 11.11). In addition, rounding, detachment and/or clumping of adhering cells are morphological alterations associated with CPE. In cells infected with the ISKNV, CPEs appear as enlarged, rounded, phase-bright cells (Fig. 11.11). Furthermore, a typical CPE induced by TiPV infection features cell rounding, fusion and shrinkage along with cytoplasmic vacuolization (Fig. 11.11).

Importance of cell culture in the diagnosis of tilapia disease

Early disease detection and proactive management strategies are essential for effective control of emerging viral diseases in tilapia aquaculture, especially since limited treatment options are available

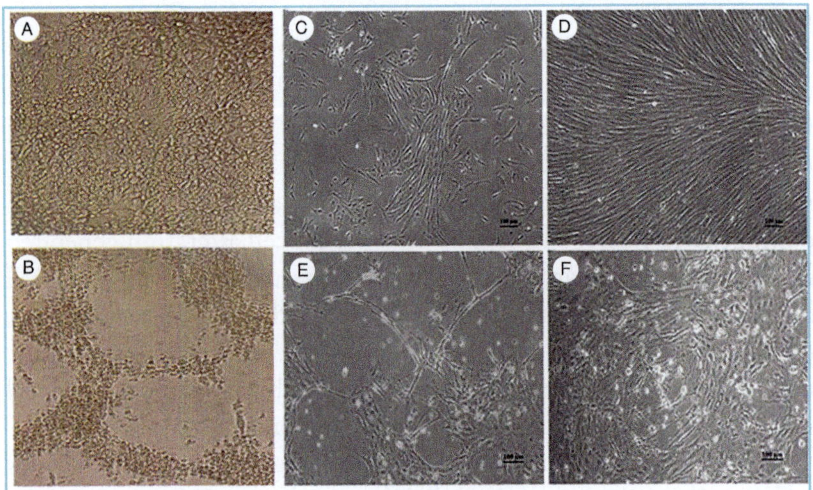

Source (A, B): Authors of this chapter Source (C,D,E,F): Liu et al., 2020

Figure 11.11 CHSE-214 cells infected with infectious pancreatic necrosis virus. (A) normal E11 cells; (B) CPE with cell shrinkage and syncytial development following TiLV infection; cytopathic effect in Tilapia brain cells (TiB) by TiPV; (C) TiB cells at passage 1, 10 days; (D) TiB cells at passage 2, 3 days; (E) TiB cells infected with TiPV at passage 3 at 5 days post-infection; (F) TiB cells infected with TiPV at passage 6 at 3 days post-infection (scale bar, 100 μm) (Liu, *et al.*, 2020).

for these. Knowledge is needed relating to the infectious cycle of the virus, its mode of infection, its pathogenicity, possible host range and methods of inhibiting viral replication to develop management strategies. These can be investigated *in vitro* using fish cell lines.

Molecular-based identification of pathogens in tilapia aquaculture

Overview

Molecular techniques are faster and more accurate than traditional diagnostic approaches, with greater sensitivity and specificity for pathogen identification. Over the past 15 years or so, significant progress has been made in the use of molecular tools to detect fish pathogens. These are now commonly used to diagnose fish diseases, as well as to understand the epidemiology of the diseases. Molecular techniques generally used include polymerase chain reaction (PCR), restriction enzyme digestion, probe hybridization, *in situ* hybridization (ISH) and microarray. Because of the improved sensitivity of molecular methods compared with conventional detection methods, pathogens can be identified in asymptomatic fish, thus potentially helping to prevent disease outbreaks. Overall, molecular techniques have reduced the time needed for diagnosis and offer higher specificity, sensitivity and quantification of the infectious agent. The various molecular-based technologies used to diagnose tilapia diseases are shown in Table 11.4.

PCR

The PCR process involves amplifying a specific segment of DNA using short oligonucleotide primers that anneal to either end of the target segment, which amplifies both strands of DNA and commences DNA synthesis using a thermostable DNA polymerase (Cunningham, 2002). Many copies of the target sequence are generated during its exponential amplification, permitting the identification of a known pathogen sequence. PCR-based technology has numerous advantages in disease diagnostics. It is precise, highly sensitive and is a more rapid and effective detection method than traditional methods, particularly for detection of low-quantity, slow-growing, difficult-to-cultivate or uncultivable pathogens. Furthermore, due to the high sensitivity of the detection limit, this technique is frequently utilized to screen subclinical samples for pathogens. In contrast, the principal drawbacks of PCR-based approaches include false-positive results due to cross- or carry-over contamination, the detection sensitivity limit of the designed method and the limited detection spectrum of the assay for concurrent identification of numerous species, virulence factors or drug resistance (Fernández *et al.*, 2008). Several types of PCR have been used to diagnose tilapia diseases, including conventional, nested, semi-nested, multiplex, reverse transcriptase and quantitative PCR.

Table 11.4 Molecular diagnostic techniques used to diagnose major tilapia diseases.

Name of disease/ aetiological agent	Type of pathogen	Reported tilapia species	Molecular diagnostic technique used	References
Tilapia *Megalocytivirus*	Icosahedral, double-stranded DNA virus	*Oreochromis* spp.	qPCR	(Shahin *et al.*, 2021)
Tilapia parvovirus (TiPV)	Non-enveloped, spherical, single-stranded DNA virus	Nile tilapia (*O. niloticus*) Red hybrid tilapia (*Oreochromis* spp.)	qPCR, *in situ* hybridination	(Liu, *et al.*, 2020)
Tilapia lake virus (TiLV) diseases caused by *Tilapia tilapinevirus*	Enveloped, negative-sense, single-stranded RNA virus	Nile tilapia (*O. niloticus*) Grey tilapia (*O. niloticus* × *O. aureus*) Red tilapia (*Oreochromis* spp.) Red hybrid tilapia (*O. niloticus* x *O. mossambicus*) Mozambique tilapia (*O. mossambicus*) Mango tilapia (*Sarotherodon galilaeus*) Malawi cichlid (*Aulonocara* spp.)	Semi-nested RT-PCR, Nested RT-PCR, RT-qPCR, *in situ* hybridization	(Eyngor *et al.*, 2014; Dong *et al.*, 2017b; Tattiyapong *et al.*, 2018; Taengphu *et al.*, 2020; Piewbang *et al.*, 2021)
Infectious spleen and kidney necrosis virus (ISKNV) diseases caused by *Megalocytivirus*	Large icosahedral, circularly permutated, terminally redundant, double-stranded DNA virus	Nile tilapia (*O. niloticus*) Red hybrid tilapia (*O. niloticus* × *O. mossambicus*) Flag cichlid (*Laetacara curviceps*)	Single-step conventional PCR, qPCR, loop-mediated isothermal amplification assay (LAMP), *in situ* hybridization	(Kurita *et al.*, 1998; Ramírez-Paredes *et al.*, 2021; OIE, 2012; Wang *et al.*, 2009; Suebsing *et al.* (2016))

Infectious pancreatic necrosis virus (IPNV) diseases by viruses belonging to the family Birnaviridae, genus *Aquabirnavirus*	Naked, icosahedral, double-stranded RNA virus	Nile tilapia (*O. niloticus*) Mozambique tilapia (*O. mossambicus*)	Single-step and nested RT-PCR, multiplex PCR, LAMP,	(Yoshimizu and Kimura, 1985; Mortensen *et al.*, 1993; Saint-Jean *et al.*, 2003; Soliman *et al.*, 2009; Suebsing *et al.*, 2011; Mulei *et al.*, 2018)
Nervous necrosis virus (NNV) disease caused by the virus under the family Nodaviridae and genus *Betanodavirus*	Non-enveloped, positive-sense, single-stranded RNA virus	Nile tilapia (*O. niloticus*)	RT-PCR, nested RT-PCR, RPA, *in situ* hybridization	(Do *et al.*, 2013; Keawcharoen *et al.*, 2015; Pirarat *et al.*, 2011; Prihartini *et al.*, 2015; Taha *et al.*, 2020)
Tilapia larvae encephalitis virus (TLEV) disease caused by the virus under the family Herpesviridae	Encapsulated, positive-sense, homologue RNA virus	Nile tilapia (*O. niloticus*) Blue tilapia larvae (*Oreochromis aureus*) Mango tilapia (*S. galilaeus*) Red hybrid tilapia (*O. niloticus* × *O. mossambicus*)	RT-PCR	(Shlapobersky *et al.*, 2010)
Bohle virus infection (*Ranavirus*)	Linear, non-enveloped, double-stranded DNA virus	Nile tilapia (*O. niloticus*) Mozambique tilapia (*O. mossambicus*)	Conventional PCR	(Harp and Petranka, 2006)

Continued

Table 11.4 (continued)

Name of disease/ aetiological agent	Type of pathogen	Reported tilapia species	Molecular diagnostic technique used	References
Streptococcosis disease Key species reported causing diseases in tilapia: *Streptococcus agalactiae*, *S. iniae* To date, only four serotypes in aquatic animals and tilapia have been identified: Ia, Ib, II and III. Serotype IX is a novel serotype for tilapia.	Spherical or ovoid in shape, non-motile, non-spore forming, Gram-positive bacteria	Nile tilapia (*O. niloticus*) Red tilapia (*Oreochromis* sp.) Red hybrid tilapia (*O. niloticus* × *O. mossambicus*) Blue tilapia (*O. aureus*) Longfin tilapia (*O. macrochir*) Three-spot tilapia (*O. andersonii*) Redbreast tilapia (*Tilapia rendalli*) Tanganyika tilapia (*O. tanganicae*)	PCR, multiplex PCR, real-time PCR, DNA microarray, MALDI-TOF-MS	(Berridge *et al.*, 1998; Teng *et al.*, 2002; Klesius *et al.*, 2006; Shi *et al.*, 2012; Rodkhum *et al.*, 2012; Assis *et al.*, 2017; Mishra *et al.*, 2018)
Columnaris disease *Flavobacterium columnare*. Six genomovars of *F. columnare* isolates: I, II, II-A, II-B, I/II and III	Long, slender, Gram-negative rods, non-flagellated, bacillary bacteria that are motile on agar medium and prevalent in freshwater environments.	Nile tilapia (*O. niloticus*) Red tilapia (*Oreochromis* sp.)	Nested PCR, real-time PCR, multiplex PCR, RPA, DNA microarray	(Amin *et al.*, 1988; Chang *et al.*, 2012; Kannika *et al.*, 2017; Kayansamruaj *et al.*, 2017; Sahoo *et al.*, 2010; Lievens *et al.*, 2011; LaFrentz *et al.*, 2018; Mabrok *et al.*, 2021)
Francisellosis Caused by *Francisella orientalis*	Gram-negative, pleomorphic, intracellular, facultative bacterial pathogen	Nile tilapia (*O. niloticus*) Red tilapia (*Oreochromis* sp.) Hybrid red tilapia (*O. niloticus* ×*O. mossambicus*)	PCR, real-time PCR, RPA, *in silico* DNA–DNA hybridization (DDH)	(Soto *et al.*, 2013; Ortega *et al.*, 2016; Shahin *et al.*, 2018; Soto *et al.*, 2019; Ramirez-Paredes *et al.*, 2020)

Motile Aeromonas septicaemia (MAS) Species identified for this infection: *A. hydrophila* *A. schubertii* *A. sobria* *A. salmonicida* *A. dhakensis* *A. veronii* *A. jandaei*	Motile, Gram-negative, straight rods (bacillus to coccibacillus shape) bacterium	Nile tilapia (*O. niloticus*) Mozambique tilapia (*O. mossambicus*) Blue tilapia (*O. aureus*) Hybrid tilapia (*O. mossambicus* x *O. niloticus*)	Nested PCR, real-time PCR, multiplex PCR, multiplex nested PCR, LAMP, RPA, DNA microarray, MALDI-TOF-MS	(Soler *et al.*, 2003; Szczuka and Kaznowski, 2004; Chang *et al.*, 2009; Yogananth *et al.*, 2009; Uma *et al.*, 2010; Subashkumar *et al.*, 2014; Bartkova *et al.*, 2017; Chapela *et al.*, 2018; Piamsomboon *et al.*, 2020; Qu *et al.*, 2021; Zhang *et al.*, 2014b; Zhang *et al.*, 2014a; Zhou *et al.*, 2014)
Edwardsiellosis Key species reported as a causative agent: *E. ictaluri* *E. tarda*	Gram-negative, short, rod-shaped, facultative anaerobic bacterium	Nile tilapia (*O. niloticus*) Hybrid red tilapia (*Oreochromis* sp.)	Real-time PCR, multiplex PCR, LAMP, DNA microarray, MALDI-TOF-MS.	(Yeh *et al.*, 2005; Chang *et al.*, 2009; Sakai *et al.*, 2009; Chang *et al.*, 2012; Reichley *et al.*, 2015; Miniero Davies *et al.*, 2018; Piamsomboon *et al.*, 2020; Zhang *et al.*, 2014a)

Conventional PCR involves a pair of primers complementary to the sequence of interest (Pestana *et al.*, 2010). Improving sensitivity and specificity depends on the primers' design. Primers should be sufficiently lengthy to provide a high annealing temperature and limit the likelihood of non-specific primer annealing. However, excessively long primers may encourage non-specific annealing to sections of DNA that are not precisely complementary to the primer sequence. To enhance diagnostic sensitivity, semi-nested and nested PCR techniques have been implemented. In semi-nested PCR, primers used in the first round of amplification are either substituted with forward or reverse primers for the second and subsequent cycles, i.e., amplification is performed twice using either two forward and one reverse or one forward and two reverse primer sets. Nested PCR is a technique that minimizes non-specific DNA template amplification. In this PCR-based approach, two sets of primers conduct two consecutive PCRs. Primers covering the target sequence and additional sequences bordering both ends of the target sequence are used in the first reaction. On the product of the first PCR, a second reaction is executed with primers that bind to the target sequence and are within the amplified region of the first PCR (Dwivedi *et al.*, 2017). Second-round PCR diminishes non-specific binding since most amplicons from the initial reaction include only the target sequence and its flanking sequences. Due to the second round of amplification, nested PCR requires a lesser amount of target DNA and, as a result, it is more sensitive. The main disadvantages of this PCR-based assay include the increased risk of carryover of amplified DNA, which results in cross-contamination, and the increase in specificity may not be sufficient to justify the use of a more sophisticated and time-consuming procedure. Nested PCR would be advised only when the detection levels reached with single-round PCR are insufficient to detect the pathogen in terms of sensitivity or specificity (Altinok & Kurt, 2003).

Multiplex PCR is analogous to conducting multiple PCR reactions concurrently in a single PCR tube. Multiplex PCR employs distinct primer pairs in a single reaction to amplify several pathogens simultaneously. This approach contains two or more primer pairs in a single reaction tube and simultaneously targets two or more DNA templates. It is a reasonably simple molecular technique used to detect many pathogens in a single PCR run (Wilton & Cousins, 1992; Williams *et al.*, 1999; González *et al.*, 2004). It is essential to use primer pairs appropriate to the target gene, and the PCR products (amplicons) should have sizes specific to diverse DNA sequences. In addition, each primer set's annealing temperature must be tuned for optimal performance within a single reaction and amplicon size. Targeting many pathogens simultaneously saves time and reagents; hence, additional data can be gleaned from a single test run (Dwivedi *et al.*, 2017). Reverse transcriptase PCR (RT-PCR) synthesizes complementary DNA (cDNA) from RNA by reverse transcription (RT) followed by PCR amplification of a specific cDNA (Dwivedi *et al.*, 2017). The desired RNA sequence can be transcribed using either random hexanucleotides or sequence-specific primers (Fernández *et al.*, 2008). The reverse transcriptase PCR technique is predominantly employed to detect RNA targets, such as RNA viruses. In addition, RT-PCR detects specific mRNA and measures gene expression levels (Altinok and Kurt, 2003).

Real-time PCR, commonly referred to as quantitative PCR (qPCR), is a method used for detecting and quantifying sequence-specific PCR products as they develop in 'real time' during the amplification process. This method identifies and measures fluorescent signal intensities that increase directly with respect to the amount of PCR product in a reaction. Pathogen load is measured in cycle threshold (Ct values), which is inversely proportional to the amount of target nucleic acid present in a sample. There are two significant fluorescence monitoring systems for DNA amplification: probe based and non-probe based. The former technique employs amplicon-specific fluorescent probes that emit fluorescent signals only upon hybridization with their corresponding target sequence. For example, probe types available include TaqMan, fluorescence resonance energy transfer (FRET), scorpion and molecular beacon. In contrast, the non-probe-based method utilizes SYBR Green I, which fluoresces upon binding to double-stranded DNA (dsDNA). Real-time PCR is sensitive to non-specific amplification since it employs specific probes, has a minimal risk of contamination because no post-PCR is necessary and is more specific, sensitive and reliable. On the other hand, this approach is unsuitable for amplifying multiple target sequences in a single reaction using multiple PCR primer pairs. In addition, its set-up needs a high level of expertise and higher expenditures for equipment and reagents (Altinok & Kurt, 2003; Fernández *et al.*, 2008).

Loop-mediated isothermal amplification (LAMP) amplifies DNA under isothermal conditions with excellent specificity, efficiency and speed. This approach utilizes a DNA polymerase and a set of four uniquely designed primers (pairs of inner and outer primers) that identify six different sequences on the target DNA (Notomi *et al.*, 2000; Notomi *et al.*, 2015). There has been a significant increase in the use of this method due to its ability to synthesise up to 10^9 copies in <1 hour using basic equipment such as a water bath or heating block under isothermal conditions (65°C). LAMP products can be detected by either measuring the turbidity generated by the precipitation of magnesium pyrophosphate or agarose gel electrophoresis (Mori *et al.*, 2001). Nevertheless, stem-loop DNA structures will generate ladder-like patterns during gel electrophoresis. Consequently, intercalating dyes such as calcein, SYBR green I and HNB detect DNA products by either colour change or fluorescence (Wong *et al.*, 2018).

Recombinase polymerase amplification (RPA) is a highly sensitive and selective isothermal amplification technology that operates at 37–42°C, requires minimal sample preparation and may amplify as few as one to ten DNA target copies in <20 min (Lobato & O'Sullivan, 2018). This technique is used for rapid detection of pathogens in the field with higher amplification sensitivity than the LAMP method. The RPA process utilizes enzymes known as recombinases that bind oligonucleotide primers with their homologous sequences in duplex DNA to generate a recombinase–primer complex. The DNA strand displaced by the recombinase is stabilized by single-stranded binding (SSB) protein. Primer elongation begins by the strand displacing DNA polymerase if the target sequence is present. This procedure is repeated and leads to exponential duplex

DNA amplification (Lobato & O'Sullivan, 2018; Li *et al.*, 2020). RPA products are detected using gel electrophoresis. In contrast, fluorogenic probes such as TwistAmpTM Exo or TwistAmpTM fpg probes (TwistDx, Cambridge, UK) or oligochromatographic lateral flow (LF) strips can also be employed to detect and analyse the RPA product (Wu *et al.*, 2019).

Hybridization-based methods

ISH is a crucial molecular laboratory technique for localizing a sequence of DNA or RNA in a biological sample, either in a preserved tissue section or whole tissue, by hybridizing the complementary strand of a nucleotide probe to a particular sequence (Jensen, 2014; Cui *et al.*, 2016; Prudent & Raoult, 2019). A 'probe', consisting of a small quantity of single-stranded DNA (ssDNA) labelled with a chemical or fluorescent dye, is injected into a biological sample and put on a glass plate. The tagged probe explores the biological material for its complementary sequence and binds to it. The place where the probe has been attached may then be identified using a microscope. This approach has been used to identify and localize various tilapia pathogens, as shown in Table 11.4.

DNA microarray (DNA chip or biochip) is another hybridization method employed to diagnose tilapia diseases. Microarray technology is a laboratory technique involving the attachment of thousands to millions of known nucleic acid fragments to a solid surface, sometimes referred to as a 'chip'. The unknown DNA molecules are fragmented using restriction endonucleases, and fluorescent dyes are added to the fragments; these are then exposed to DNA chip probes. The target DNA fragments, and their corresponding sequences, subsequently attach to these DNA probes and the unattached DNA fragments are removed by washing. The target DNA fragments may be identified by their fluorescence emission when a laser beam is passed across them. A computer records the fluorescence emission pattern and DNA recognition (Govindarajan *et al.*, 2012; Bumgarner, 2013;). DNA microarray analysis is extremely rapid, sensitive and specific in concurrent identification of various DNA fragments.

Next-generation sequencing technology for pathogen identification

Traditional methods of identifying tilapia pathogens, such as culture-based techniques and microscopy, can have significant limitations. They are time-consuming, lack specificity and often fail to detect a broad spectrum of pathogens. In contrast, next-generation sequencing (NGS) offers a comprehensive, swift and precise alternative for pathogen detection and characterization (Dong *et al.*, 2023). NGS represents an array of sequencing technologies capable of concurrently sequencing millions of DNA or RNA fragments using platforms such as Illumina MiSeq. These versatile methods enable a wide range of applications, from targeted gene sequencing and gene expression analysis to epigenomics. NGS is particularly useful for focused studies where specific genomic regions or types of genetic variation are of interest.

NGS encompasses a variety of techniques, each tailored to genomic research needs, notably differentiated by long- and short-read sequencing methods (Heather & Chain, 2016). Technologies such as Pacific Biosciences (PacBio) and Oxford Nanopore excel in long-read sequencing and can process DNA fragments that span hundreds of kilobases. This capability makes them highly effective in mapping complex genomic regions, streamlining genome assembly and identifying gene isoforms. However, these methods are associated with higher error rates and increased financial outlay compared with short-read sequencing. In contrast, short-read sequencing, performed using platforms such as Illumina, produces DNA fragment sequences ranging from 100 to 500 bases in length. This approach provides greater sequence accuracy and is particularly useful in identifying single-nucleotide variants and small insertions or deletions (indels). It is more cost effective for large-scale studies due to its scalability and high throughput. The data analysis for short-read sequencing is well established, with robust bioinformatics tools available. Nevertheless, short reads have limitations in resolving complex genomic regions and are less effective for novel genome assembly (Amarasinghe *et al.*, 2020).

Whole-genome sequencing (WGS), a specific application of NGS, entails the sequencing of an organism's entire genomic DNA. While it is more straightforward for bacterial genomes, advancements in sequencing technologies, particularly those capable of long-read sequencing like the Oxford Nanopore MinionTM, show promise in generating extensive genomic datasets from numerous samples and more complex pathogens. WGS provides a comprehensive view of a pathogen's genome, capturing all genetic variations including minor mutations, structural variations and chromosomal rearrangements. It is invaluable in research necessitating a complete genetic overview, such as in detailed genetic mapping, novel genetic variant identification and comprehensive microbial genome analysis for epidemiological studies (Bayliss *et al.*, 2017).

The emergence and spread of significant bacterial pathogens in aquaculture have been closely monitored using molecular typing methods, as detailed by Bayliss *et al.* (2017). The introduction of multi-locus sequence typing (MLST) in the late 1990s marked a pivotal advancement in molecular typing. The accessibility of MLST data online has facilitated molecular epidemiological monitoring (Urwin & Maiden, 2003). Initially, MLST was limited by its use of a small set of genes, constraining its precision in distinguishing between localized outbreaks and endemic cases or in tracking the spread of specific clones at a global or national scale (Pérez-Losada *et al.*, 2013; Habib *et al.*, 2014). However, the integration of WGS has considerably enhanced MLST databases, shifting the focus from a mere seven genes to potentially thousands. This expansion has yielded extensive, clinically pertinent information ranging from antimicrobial resistance to precise strain identification, pivotal for epidemiological analysis and associated disease management (Dong *et al.*, 2023). The ability to identify sub-variants within bacterial strains has played a crucial role in understanding the spread of pathogens at various levels within the fish production system, often revealing distinct geographic distributions (Bayliss *et al.*, 2017).

The application of NGS in tilapia pathogen identification presents several advantages. It can detect numerous pathogens in a single run and provides quicker and more accurate results than traditional methods. Its use is especially beneficial for early pathogen detection, critical in averting large-scale outbreaks in tilapia farming. NGS also offers insights into pathogen genetic diversity, aiding in understanding disease dynamics and resistance mechanisms (Dong *et al.*, 2023). However, NGS as a diagnostic tool is not without challenges; the costs of sequencing and the need for specialized bioinformatics skills are significant barriers. Furthermore, interpreting NGS data, particularly in distinguishing between pathogenic and non-pathogenic organisms, demands careful analysis.

A substantial number of complete reference genomes are already available for various pathogens associated with aquaculture, including some that can affect tilapia (Bayliss *et al.*, 2017). NGS has been employed in disease diagnosis for tilapia, enabling rapid and precise identification of pathogens associated with streptococcosis, TiLV and other bacterial pathogens of tilapia (Dong *et al.*, 2023). It has the potential to play a key role in monitoring and surveillance of tilapia populations, facilitating early pathogen detection. In addition, NGS supports the research and development of targeted treatments and vaccines by improving our understanding of disease mechanisms. As NGS technology evolves and becomes more cost effective, its application in tilapia health management is anticipated to grow, potentially revolutionizing disease management strategies for tilapia.

Concluding remarks

Effective health management in aquaculture relies on the prompt detection and diagnosis of pathogens in affected fish, both clinically and subclinically, and in environmental samples. The use of conventional diagnostic procedures has notably aided the identification of infectious agents in cultured tilapia; however, these methods are slow and less sensitive. It is frequently impossible to make an accurate diagnosis based on the clinical symptoms of infectious diseases in fish. Nevertheless, the speed and sensitivity of conventional diagnostic approaches, such as the observation of CPEs on cell cultures in the case of viruses and the isolation and growth in specialized culture conditions, particular biochemical and serological assays or *in situ* microscopic examination, are often insufficient when dealing with specific microorganisms.

The application of molecular techniques to diagnosing tilapia disease and other fish diseases is advancing. In addition to implementing more accurate, sensitive and specific disease diagnosis techniques, molecular techniques are now being used as on-site diagnostic tools, which are crucial for the tilapia culture industry in terms of point-of-care and resource-limited settings. Once molecular approaches began to be utilized in the study of tilapia infections – for example, to investigate pathogenicity mechanisms, antibiotic resistance and so on – it became evident that there is potential for their application in the clinical sector. However, the greatest obstacle in tilapia disease diagnostics is the development of lower-priced, field-applicable diagnostic tests that unskilled individuals can use. Ultimately, the development of molecular diagnostic technologies will be crucial for advancing new, more effective treatment programmes and preventative strategies, such as early identification of carriers or environmental samples.

Notes

1 https://www.fao.org/fileadmin/user_upload/faoweb/FI/news/VirtualCourseTiLV2021/Day5_1a.pdf;

2 https://www.researchgate.net/publication/277559558_Francisella_noatunensis_subsp_orientalis_an_emerging_bacterial_pathogen_affecting_cultured_red_tilapia_Oreochromis_sp_in_Thailand/figures?lo=1

3 https://aquahealth.wordpress.com/2018/10/11/natural-occurrence-of-edwardsiellosis-caused-by-edwardsiella-ictaluri-in-farmed-hybrid-red-tilapia-oreochromis-sp-in-southeast-asia/

4 https://www.researchgate.net/publication/282816011_Genomic_comparison_between_pathogenic_Streptococcus_agalactiae_isolated_from_Nile_tilapia_in_Thailand_and_fish-derived_ST7_strains/figures?lo=1

References

Abdel-Baki, A.A., Zayed, E., Sakran, T. and Al-Quraishy, S. (2015) A new record of *Myxobolus brachysporus* and *M. israelensis* in the tilapia (*Oreochromis niloticus*) collected from the Nile river, Egypt. *Saudi Journal of Biological Sciences* 22(5), 539–542. DOI: 10.1016/j.sjbs.2015.01.003.

Abdel-Latif, H.M.R., Shukry, M. and Abd-Elaziz, R.A. (2022) Clinico-pathological findings and expression of inflammatory cytokines, apoptosis, and oxidative stress-related genes draw mechanistic insights in Nile tilapia reared under ammonia-N exposure and *Aeromonas hydrophila* challenge. *Fish and Shellfish Immunology* 127, 1–12. DOI: 10.1016/j.fsi.2022.06.001.

Acharya, T. (2021). Bacterial culture media: Classification, types, uses. https://microbeonline.com/types-of-bacteriological-culture-medium/

Aiman, F. (2022). Microbial culture media- definition, types, examples, uses. https://microbenotes.com/types-of-culture-media/#application-of-culture-media

Alazab, A., Sadat, A. and Younis, G. (2022) Prevalence, antimicrobial susceptibility, and genotyping of *Streptococcus agalactiae* in tilapia fish (*Oreochromis niloticus*) in Egypt. *Journal of Advanced Veterinary and Animal Research* 9(1), 95–103. DOI: 10.5455/javar.2022.i573.

Al-Harbi, A.H. (1994) First isolation of streptococcus sp. from hybrid tilapia (*Oreochromis niloticus* × *O. aureus*) in Saudi Arabia. *Aquaculture* 128(3–4), 195–201. DOI: 10.1016/0044-8486(94)90308-5.

Aloo, P., Okelo, R.O. and Ntiba, M.J. (1995) Helminth parasites of *Oreochromis leucostictus* (Trewavas) in Lake Naivasha, Kenya. In: *Pan African Fisheries Congress on Sustainable Development of Fisheries in Africa, Food Standards Agency.*

Altinok, İ. and Kurt, İ. (2003) Molecular diagnosis of fish diseases: A review. *Turkish Journal of Fisheries and Aquatic Sciences* 3(2).

Amal, M.N.A. and Zamri-Saad, M. (2011) Streptococcosis in tilapia (*Oreochromis niloticus*): A review. Review, *Pertanika Journal of Tropical Agricultural Science* 34(2), 195–206.

Amal, M.N.A., Koh, C.B., Nurliyana, M., Suhaiba, M., Nor-Amalina, Z, *et al.* (2018) A case of natural co-infection of tilapia lake virus and *Aeromonas veronii* in a Malaysian red hybrid tilapia (*Oreochromis niloticus* × *O. mossambicus*) farm experiencing high mortality. *Aquaculture* 485, 12–16. DOI: 10.1016/j.aquaculture.2017.11.019.

Amarasinghe, S.L., Su, S., Dong, X., Zappia, L., Ritchie, M.E, *et al.* (2020) Opportunities and challenges in long-read sequencing data analysis. *Genome Biology* 21(1), 30. DOI: 10.1186/s13059-020-1935-5.

Amin, N.E., Abdallah, I.S., Faisal, M., Easa, M.E.S., Alaway, T, *et al.* (1988) Columnaris infection among cultured Nile tilapia *Oreochromis niloticus. Antonie van Leeuwenhoek* 54(6), 509–520. DOI: 10.1007/BF00588387.

Anacker, R.L. and Ordal, E.J. (1955) Study of a bacteriophage infecting the myxobacterium *Chondrococcus columnaris. Journal of Bacteriology* 70(6), 738–741. DOI: 10.1128/jb.70.6.738-741.1955.

Anshary, H., Kurniawan, R.A., Sriwulan, S., Ramli, R. and Baxa, D.V. (2014) Isolation and molecular identification of the etiological agents of streptococcosis in Nile tilapia (*Oreochromis niloticus*) cultured in net cages in lake sentani, papua, indonesia. *SpringerPlus* 3(1), article 627. DOI: 10.1186/2193-1801-3-627.

Aryal, S. (2022a) Nutrient agar: Composition, preparation and uses. Available at: https://microbiologyinfo.com/nutrient-agar-composition-preparation-and-uses/

Aryal, S. (2022b) Blood agar- composition, preparation, uses and pictures. Available at: https://microbiologyinfo.com/blood-agar-composition-preparation-uses-and-pictures/

Aryal, S. (2022c) Thiosulfate-citrate-bile salts-sucrose (TCBS) agar-composition, principle, uses, preparation and colony morphology. Available at: https://microbiologyinfo.com/thiosulfate-citrate-bile-salts-sucrose-tcbs-agar-composition-principle-uses-preparation-and-colony-morphology/

Assis, G.B.N., Pereira, F.L., Zegarra, A.U., Tavares, G.C., Leal, C.A, *et al.* (2017) Use of MALDI-TOF mass spectrometry for the fast identification of gram-positive fish pathogens. *Frontiers in Microbiology* 8, 1492. DOI: 10.3389/fmicb.2017.01492.

Austin, B. (2019) Methods for the diagnosis of bacterial fish diseases. *Marine Life Science and Technology* 1(1), 41–49. DOI: 10.1007/s42995-019-00002-5.

Austin, B. and Austin, D.A. (2016) *Bacterial fish pathogens: Disease of farmed and wild fish (6th ed)*. Springer. DOI: 10.1007/978-3-319-32674-0.

Barrow, G.I. and Feltham, R.K.A. (1993) *Manual for the identification of medical bacteria.* Cambridge University Press.

Bartelt, M. (2000) *Diagnostic bacteriology, a study guide.* F. A. Davis Company.

Bartkova, S., Kokotovic, B., Skall, H.F., Lorenzen, N. and Dalsgaard, I. (2017) Detection and quantification of *Aeromonas salmonicida* in fish tissue by real-time PCR. *Journal of Fish Diseases* 40(2), 231–242. DOI: 10.1111/jfd.12505.

Bayliss, S.C., Verner-Jeffreys, D.W., Bartie, K.L., Aanensen, D.M., Sheppard, S.K, *et al.* (2017) The promise of whole genome pathogen sequencing for the molecular epidemiology of emerging aquaculture pathogens. *Frontiers in Microbiology* 8, 121. DOI: 10.3389/fmicb.2017.00121.

Berridge, B.R., Fuller, J.D., De Azavedo, J., Low, D.E., Bercovier, H. *et al.* (1998) Development of specific nested oligonucleotide PCR primers for the *Streptococcus iniae* 16S-23S ribosomal DNA intergenic spacer. *Journal of Clinical Microbiology* 36(9), 2778–2781. DOI: 10.1128/JCM.36.9.2778-2781.1998.

Bumgarner, R. (2013) Overview of DNA microarrays: Types, applications, and their future. Chapter 22, Unit–22.1, *Current Protocols in Molecular Biology*. DOI: 10.1002/0471142727.mb2201s101.

CarloAlberto, R. (2010) Histology protocols. *European Journal of Histochemistry* 54. DOI: 10.4081/ejh.2010.e27.

Chang, C.H., Poudyal, S., Pulpipat, T., Wang, P.C. and Chen, S.C. (2021) Pathological manifestations of *Francisella orientalis* in the green texas cichlid (*Herichthys cyanoguttatus*). Article, *Animals* 11(8), Article 2284. DOI: 10.3390/ani11082284.

Chang, C.I., Wu, C.C., Cheng, T.C., Tsai, J.M. and Lin, K.J. (2009) Multiplex nested-polymerase chain reaction for the simultaneous detection of *Aeromonas hydrophila, Edwardsiella tarda, Photobacterium damselae* and *Streptococcus iniae,* four important fish pathogens in subtropical Asia. *Aquaculture Research* 40(10), 1182–1190. DOI: 10.1111/j.1365-2109.2009.02214.x.

Chang, C.I., Hung, P.H., Wu, C.C., Cheng, T.C., Tsai, J.M. *et al.* (2012) Simultaneous detection of multiple fish pathogens using a naked-eye readable DNA microarray. *Sensors* 12(3), 2710–2728. DOI: 10.3390/s120302710.

Chapela, M.J., Ferreira, M., Ruiz-Cruz, A., Martin-Varela, I., Fernández-Casal, J. *et al.* (2018) Application of real-time PCR for early diagnosis of diseases caused by *Aeromonas salmonicida, Vibrio anguillarum,* and *Tenacibaculum maritimum* in turbot: A field study. Note, *Journal of Applied Aquaculture* 30(1), 76–89. DOI: 10.1080/10454438.2017.1406419.

Chinchar, V.G., Hick, P., Ince, I.A., Jancovich, J.K., Marschang, R, *et al.* (2017) ICTV virus taxonomy profile: Iridoviridae. *Journal of General Virology* 98(5), 890–891. DOI: 10.1099/jgv.0.000818.

Clarke, P.H. and Cowan, S.T. (1952) Biochemical methods for bacteriology. *Journal of General Microbiology* 6(1–2), 187–197. DOI: 10.1099/00221287-6-1-2-187.

Clavijo, A.M., Conroy, G., Conroy, D.A., Santander, J. and Aponte, F. (2002) First report of *Edwardsiella tarda* from tilapias in Venezuela. *Bulletin of the European Association of Fish Pathologists* 22(4), 280–282.

Costa, F.A.A., Leal, C.A.G., Leite, R.C. and Figueiredo, H.C.P. (2014) Genotyping of streptococcus dysgalactiae strains isolated from Nile tilapia, *Oreochromis niloticus* (L.). *Journal of Fish Diseases* 37(5), 463–469. DOI: 10.1111/jfd.12125.

Crumlish, M. (2017) Bacterial diagnosis and control in fish and shellfish. *Diagnosis and Control. of Diseases of Fish and Shellfish* 5–14.

Cui, C., Shu, W. and Li, P. (2016) Fluorescence *in situ* hybridization: Cell-based genetic diagnostic and research applications. *Frontiers in Cell and Developmental Biology* 4, 89. DOI: 10.3389/fcell.2016.00089.

Cunningham, C.O. (2002) Molecular diagnosis of fish and shellfish diseases: Present status and potential use in disease control. *Aquaculture* 206(1–2), 19–55. DOI: 10.1016/S0044-8486(01)00864-X.

Darwish, A.M. and Griffin, B.R. (2002) Study shows oxyetracycline controls streptococcus in tilapia. *Global Aquaculture Advocate* 5(6), 34–35.

Daskalov, H., Austin, D.A. and Austin, B. (1999) An improved growth medium for *Flavobacterium psychrophilum. Letters in Applied Microbiology* 28(4), 297–299. DOI: 10.1046/j.1365-2672.1999.00522.x.

Davies, J.A., Anderson, G.K., Beveridge, T.J. and Clark, H.C. (1983) Chemical mechanism of the gram stain and synthesis of a new electron-opaque marker for electron microscopy which replaces the iodine mordant of the stain. *Journal of Bacteriology* 156(2), 837–845. DOI: 10.1128/jb.156.2.837-845.1983.

Debnath, P.P., Delamare-Deboutteville, J., Jansen, M.D., Phiwsaiya, K., Dalia, A, *et al.* (2020) Two-year surveillance of tilapia lake virus (tilv) reveals its wide circulation in tilapia farms and hatcheries from multiple districts of Bangladesh. *Journal of Fish Diseases* 43(11), 1381–1389. DOI: 10.1111/jfd.13235.

Delghandi, M.R., El-Matbouli, M. and Menanteau-Ledouble, S. (2020) Mycobacteriosis and infections with non-tuberculous mycobacteria in aquatic organisms: A review. *Microorganisms* 8(9), 1368. DOI: https://www.mdpi.com/2076–2607/8/9/1368.%2010.3390/microorganisms8091368.

Do, J.W., Lee, N.S., Jung, S.H., Kim, K.K., Choi, H.S, *et al.* (2013) Histopathologic characterization of viral pathogens in cultured olive flounder, *Paralichthys olivaceus,* using *in-situ* hybridization methods. *Journal of Fish Pathology* 26(3), 163–171. DOI: 10.7847/jfp.2013.26.3.163.

Dong, C., Weng, S., Shi, X., Xu, X., Shi, N, *et al.* (2008) Development of a mandarin fish siniperca chuatsi fry cell line suitable for the study of infectious spleen and kidney necrosis virus (ISKNV). *Virus Research* 135(2), 273–281. DOI: 10.1016/j.virusres.2008.04.004.

Dong, H.T. (2018) Bacterial diseases of tilapia. FAO meeting documents. Available at: https://www.fao.org/fi/static-media/MeetingDocuments/TiLV/dec2018/p15.pdf

Dong, H.T., LaFrentz, B., Pirarat, N. and Rodkhum, C. (2015) Phenotypic characterization and genetic diversity of *Flavobacterium columnare* isolated from red tilapia. *Oreochromis Sp., in Thailand. Journal of Fish Diseases* 38(10), 901–913. DOI: 10.1111/jfd.12304

Dong, H.T., Nguyen, V.V., Kayansamruaj, P., Gangnonngiw, W., Senapin, S, *et al.* (2016) *Francisella noatunensis* subsp. orientalis infects striped catfish (*Pangasianodon hypophthalmus*) and common carp (*Cyprinus carpio*) but does not kill the hosts. *Aquaculture* 464, 190–195. DOI: 10.1016/j.aquaculture.2016.06.033.

Dong, H.T., Techatanakitarnan, C., Jindakittikul, P., Thaiprayoon, A., Taengphu, S, *et al.* (2017a) *Aeromonas jandaei* and *Aeromonas veronii* caused disease and mortality in Nile tilapia, *Oreochromis niloticus* (L.). *Journal of Fish Diseases* 40(10), 1395–1403. DOI: 10.1111/jfd.12617.

Dong, H.T., Siriroob, S., Meemetta, W., Santimanawong, W., Gangnonngiw, W, *et al.* (2017b) Emergence of tilapia lake virus in thailand and an alternative semi-nested RT-PCR for detection. *Aquaculture* 476, 111–118. DOI: 10.1016/j.aquaculture.2017.04.019.

Dong, H.T., Senapin, S., Jeamkunakorn, C., Nguyen, V.V., Nguyen, N.T, *et al.* (2019) Natural occurrence of *Edwardsiellosis* caused by *Edwardsiella ictaluri* in farmed hybrid red tilapia (*Oreochromis* sp.) in Southeast Asia. *Aquaculture* 499, 17–23. DOI: 10.1016/j.aquaculture.2018.09.007.

Dong, H.T., Sangpo, P., Dien, L.T., Mai, T.T., Linh, N.V, *et al.* (2022) Usefulness of the pancreas as a prime target for histopathological diagnosis of tilapia parvovirus (tipv) infection in Nile tilapia, *Oreochromis niloticus*. *Journal of Fish Diseases* 45(9), 1323–1331. DOI: 10.1111/jfd.13663.

Dong, H.T., Chaijarasphong, T., Barnes, A.C., Delamare-Deboutteville, J., Lee, P.A, *et al.* (2023) From the basics to emerging diagnostic technologies: what is on the horizon for tilapia disease diagnostics. *Reviews in Aquaculture* 15(S1), 186–212. DOI: 10.1111/raq.12734.

Douëllou, L. and Erlwanger, K.H. (1994) Crustacean parasites of fishes in lake kariba, zimbabwe, preliminary results. *Hydrobiologia* 287(3), 233–242. DOI: 10.1007/BF00006372.

Dwivedi, S., Purohit, P., Misra, R., Pareek, P., Goel, A, *et al.* (2017) Diseases and molecular diagnostics: A step closer to precision medicine. *Indian Journal of Clinical Biochemistry* 32(4), 374–398. DOI: 10.1007/s12291-017-0688-8.

Eissa, I.A.M., Maather, E., Mona, S., Desuky, E., Mona, Z. *et al.* (2015) *Aeromonas veronii* sobria a causative agent of mass mortalities in cultured Nile tilapia in-Sharkia governorate. *Egypt. Life Science Journal* 12(5), 90–97.

Elgendy, M.Y., Abdelsalam, M., Kenawy, A.M. and Ali, S.E. (2022a) Vibriosis outbreaks in farmed Nile tilapia (*Oreochromis niloticus*) caused by vibrio mimicus and *V. cholerae*. *Aquaculture International* 30(5), 2661–2677. DOI: 10.1007/s10499-022-00921-8.

Elgendy, M.Y., Abdelsalam, M., Mohamed, S.A. and Ali, S.E. (2022b) Molecular characterization, virulence profiling, antibiotic susceptibility, and scanning electron microscopy of *Flavobacterium columnare* isolates retrieved from Nile tilapia (*Oreochromis niloticus*). *Aquaculture International* 30(2), 845–862. DOI: 10.1007/s10499-021-00819-x.

Elgendy, M.Y., Sherif, A.H., Kenawy, A.M. and Abdelsalam, M. (2022c) Phenotypic and molecular characterization of the causative agents of *Edwardsiellosis* causing Nile tilapia (*Oreochromis niloticus*) summer mortalities. *Microbial Pathogenesis* 169, article 105620. DOI: 10.1016/j.micpath.2022.105620.

El-Ghany, N.A.A. and El-Ashram, A.M.M.M. (2008) Diagnosis of ichthyophoniasis in *Oreochromis niloticus* in Egypt by polymerase chain reaction (pcr). In: *International Symposium on Tilapia in Aquaculture*, pp. 1307–1328.

El-Sayed, A.-F.M. (2020) Stress and diseases. In: El-Sayed, A.-F.M. (ed.) *Tilapia Culture (2nd Ed)*. Academic Press, pp. 205–243. DOI: 10.1016/B978–0-12–816509–6.00009–4.

Eyngor, M., Zamostiano, R., Kembou Tsofack, J.E., Berkowitz, A., Bercovier, H, *et al.* (2014) Identification of a novel RNA virus lethal to tilapia. *Journal of Clinical Microbiology* 52(12), 4137–4146. DOI: 10.1128/JCM.00827-14.

Fenselau, C. and Demirev, P.A. (2001) Characterization of intact microorganisms by MALDI mass spectrometry. *Mass Spectrometry Reviews* 20(4), 157–171. DOI: 10.1002/mas.10004.

Ferguson, H.W., Kabuusu, R., Beltran, S., Reyes, E., Lince, J.A, *et al.* (2014) Syncytial hepatitis of farmed tilapia, *Oreochromis niloticus* (L.): A case report. *Journal of Fish Diseases* 37(6), 583–589. DOI: 10.1111/jfd.12142.

Fernández, L., Álvarez, B., Menéndez, A., Méndez, J. and Guijarro, J.A. (2008) Molecular tools for monitoring infectious diseases in aquaculture species. *Dyn Biochem. Process Biotechnol. Mol. Biol* 2, 233–240.

Gaby, W.L. and Free, E. (1958) Differential diagnosis of pseudomonas-like microorganisms in the clinical laboratory. *Journal of Bacteriology* 76(4), 442–444. DOI: 10.1128/jb.76.4.442-444.1958.

Gaby, W. L., & Hadley, C. (1957). Practical laboratory test for the identification of *Pseudomonas aeruginosa*. Journal of Bacteriology, 74(3), 356–358. DOI: https://doi.org/10.1128/jb.74.3.356-358.1957.

Gerhardt, P., Murray, R.G.E., Costilow, R.N., Nester, E.W., Wood, W.A. *et al.* (1981) *Manual and methods for general bacteriology*. ASM Press.

Gerhardt, P., Murray, R. G. E., Wood, W. A., & Krieg, N. R. (1994). Methods for general and molecular bacteriology. http://lib.ugent.be/catalog/rug01:000309577. American Society for Microbiology.

González, S.F., Krug, M.J., Nielsen, M.E., Santos, Y. and Call, D.R. (2004) Simultaneous detection of marine fish pathogens by using multiplex PCR and a DNA microarray. *Journal of Clinical Microbiology* 42(4), 1414–1419. DOI: 10.1128/JCM.42.4.1414-1419.2004.

Goodwin, A.E. (2012) 4.2.1. branchiomycosis. In: *Handbook of Diagnostic Techniques*. pp. 1–8.

Gordon, J. and McLeod, J.W. (1928) The practical application of the direct oxidase reaction hi bacteriology. *Journal of Pathology and Bacteriology* 31(2), 185–190.

Govindarajan, R., Duraiyan, J., Kaliyappan, K. and Palanisamy, M. (2012) Microarray and its applications. *Journal of Pharmacy & Bioallied Sciences* 4(Suppl 2), S310–S312. DOI: 10.4103/0975-7406.100283.

Greene, R.A., Blum, E.F., DeCORO, C.T., Fairchild, R.B., Kaplan, M.T, *et al.* (1951) Rapid methods for the detection of motility. *Journal of Bacteriology* 62(3), 347. DOI: 10.1128/jb.62.3.347-347.1951.

Gurcan, M.N., Boucheron, L.E., Can, A., Madabhushi, A., Rajpoot, N.M, *et al.* (2009) Histopathological image analysis: A review. *IEEE Reviews in Biomedical Engineering* 2, 147–171. DOI: 10.1109/RBME.2009.2034865.

Habib, C., Houel, A., Lunazzi, A., Bernardet, J.F., Olsen, A.B, *et al.* (2014) Multilocus sequence analysis of the marine bacterial genus tenacibaculum suggests parallel evolution of fish pathogenicity and endemic colonization of aquaculture systems. *Applied and Environmental Microbiology* 80(17), 5503–5514. DOI: 10.1128/AEM.01177-14.

Harp, E.M. and Petranka, J.W. (2006) Ranavirus in wood frogs (*Rana sylvatica*): Potential sources of transmission within and between ponds. *Journal of Wildlife Diseases* 42(2), 307–318. DOI: 10.7589/0090-3558-42.2.307.

Heather, J.M. and Chain, B. (2016) The sequence of sequencers: the history of sequencing DNA. *Genomics* 107(1), 1–8. DOI: 10.1016/j.ygeno.2015.11.003.

Hewitson, T.D., Wigg, B. and Becker, G.J. (2010) Tissue preparation for histochemistry: fixation, embedding, and antigen retrieval for light microscopy. In: Hewitson, T.D. and Darby, I.A. (eds) *Histology Protocols*. Humana Press, pp. 3–18. DOI: 10.1007/978-1-60327-345-9_1.

Ivanchenko, M.V., Indzhykulian, A.A. and Corey, D.P. (2021) Electron microscopy techniques for investigating structure and composition of hair-cell stereociliary bundles. *Frontiers in Cell and Developmental Biology* 9, 744248. DOI: 10.3389/fcell.2021.744248.

Jansen, M.D., Dong, H.T. and Mohan, C.V. (2019) Tilapia lake virus: A threat to the global tilapia industry. *Reviews in Aquaculture* 11(3), 725–739. DOI: 10.1111/raq.12254.

Jantrakajorn, S. and Wongtavatchai, J. (2016) Francisella infection in cultured tilapia in Thailand and the inflammatory cytokine response. *Journal of Aquatic Animal Health* 28(2), 97–106. DOI: 10.1080/08997659.2015.1135198.

Jarrell, K.F. and McBride, M.J. (2008) The surprisingly diverse ways that prokaryotes move. *Nature Reviews. Microbiology* 6(6), 466–476. DOI: 10.1038/nrmicro1900.

Jensen, E. (2014) Technical review: *In situ* hybridization. *The Anatomical Record* 297(8), 1349–1353. DOI: 10.1002/ar.22944.

Johny, T.K., Swaminathan, T.R., Sood, N., Pradhan, P.K. and Lal, K.K. (2022) A panoptic review of techniques for finfish disease diagnosis: the status quo and future perspectives. *Journal of Microbiological Methods* 196, 106477. DOI: 10.1016/j.mimet.2022.106477.

Jordan, E.O., Caldwell, M.E. and Reiter, D. (1934) Bacterial motility. *Journal of Bacteriology* 27(2), 165–174. DOI: 10.1128/jb.27.2.165-174.1934.

Kannika, K., Pisuttharachai, D., Srisapoome, P., Wongtavatchai, J., Kondo, H, *et al.* (2017) Molecular serotyping, virulence gene profiling and pathogenicity of *Streptococcus agalactiae* isolated from tilapia farms in thailand by multiplex PCR. *Journal of Applied Microbiology* 122(6), 1497–1507. DOI: 10.1111/jam.13447.

Kayansamruaj, P., Pirarat, N., Hirono, I. and Rodkhum, C. (2014a) Increasing of temperature induces pathogenicity of *Streptococcus agalactiae* and the up-regulation of inflammatory related genes in infected Nile tilapia (*Oreochromis niloticus*). *Veterinary Microbiology* 172(1–2), 265–271. DOI: 10.1016/j.vetmic.2014.04.013.

Kayansamruaj, P., Pirarat, N., Katagiri, T., Hirono, I. and Rodkhum, C. (2014b) Molecular characterization and virulence gene profiling of pathogenic *Streptococcus agalactiae* populations from tilapia (*Oreochromis* sp.) farms in Thailand. *Journal of Veterinary Diagnostic Investigation* 26(4), 488–495. DOI: 10.1177/1040638714534237.

Kayansamruaj, P., Dong, H.T., Hirono, I., Kondo, H., Senapin, S. and Rodkhum, C. (2017). Comparative genome analysis of fish pathogen *Flavobacterium columnare* reveals extensive sequence diversity within the species [Article]. *Infection, Genetics and Evolution*, 54, 7–17. https://doi.org/10.1016/j.meegid.2017.06.012

Keawcharoen, J., Techangamsuwan, S., Ponpornpisit, A., Lombardini, E.D., Patchimasiri, T, *et al.* (2015) Genetic characterization of a betanodavirus isolated from a clinical disease outbreak in farm-raised tilapia *Oreochromis niloticus* (L.) in Thailand. *Journal of Fish Diseases* 38(1), 49–54. DOI: 10.1111/jfd.12200.

Kelly, A.T. and Fulton, M. (1953) Use of triphenyl tetrazolium in motility test medium. *American Journal of Clinical Pathology* 23(5), 512. DOI: 10.1093/ajcp/23.5_ts.512.

Klesius, P., Evans, J., Shoemaker, C., Yeh, H., Goodwin, A.E, *et al.* (2006) Rapid detection and identification of *Streptococcus iniae* using a monoclonal antibody-based indirect fluorescent antibody technique. *Aquaculture* 258(1–4), 180–186. DOI: 10.1016/j.aquaculture.2005.06.040.

Koo, I. (2022) Using cultures for infectious disease diagnosis. Very well Health. Available at: https://www.verywellhealth.com/diagnosis-of-an-infection-microbial-cultures-1958751

Kovacs, N. (1956) Identification of pseudomonas pyocyanea by the oxidase reaction. *Nature* 178(4535), 703–703. DOI: 10.1038/178703a0.

Kragh, K.N., Alhede, M., Rybtke, M., Stavnsberg, C., Jensen, P.Ø, *et al.* (2018) The inoculation method could impact the outcome of microbiological experiments. *Applied and Environmental Microbiology* 84(5). DOI: 10.1128/AEM.02264-17.

Kruse, C.P.S., Basu, P., Luesse, D.R. and Wyatt, S.E. (2017) Transcriptome and proteome responses in rnalater preserved tissue of *Arabidopsis thaliana. PLOS ONE* 12(4), e0175943. DOI: 10.1371/journal.pone.0175943.

Kurita, J., Nakajima, K., Hirono, I. and Aoki, T. (1998) Polymerase chain reaction (PCR) amplification of DNA of red sea bream iridovirus (RSIV). *Fish Pathology* 33(1), 17–23. DOI: 10.3147/jsfp.33.17.

LaFrentz, B.R., García, J.C., Waldbieser, G.C., Evenhuis, J.P., Loch, T.P, *et al.* (2018) Identification of four distinct phylogenetic groups in flavobacterium columnare with fish host associations. *Frontiers in Microbiology* 9, article 452. DOI: 10.3389/fmicb.2018.00452.

Lavigne, J.P., Espinal, P., Dunyach-Remy, C., Messad, N., Pantel, A, *et al.* (2013) Mass spectrometry: A revolution in clinical microbiology. *Clinical Chemistry and Laboratory Medicine* 51(2), 257–270. DOI: 10.1515/cclm-2012-0291.

Leifson, E. (1951) Staining, shape and arrangement of bacterial flagella. *Journal of Bacteriology* 62(4), 377–389. DOI: 10.1128/jb.62.4.377-389.1951.

Leland, D.S. and French, M.L.V. (1988) Virus isolation and identification. In: Lennette, E.H., Halonen, P., Murphy, F.A., Balows, A. and Hausler, W.J. (eds) *Laboratory Diagnosis of Infectious Diseases Principles and Practice, II Viral, Rickettsial, and Chlamydial Diseases.* Springer, pp. 39–59. DOI: 10.1007/978–1-4612–3900–0_3.

Leland, D.S. and Ginocchio, C.C. (2007) Role of cell culture for virus detection in the age of technology. *Clinical Microbiology Reviews* 20(1), 49–78. DOI: 10.1128/CMR.00002-06.

Li, H., Yuan, G., Luo, Y., Yu, Y., Ai, T, *et al.* (2020) Rapid and sensitive detection of infectious spleen and kidney necrosis virus by recombinase polymerase amplification combined with lateral flow dipsticks. *Aquaculture* 519, 734926. DOI: 10.1016/j.aquaculture.2020.734926.

Li, Y. and Cai, S.H. (2011) Identification and pathogenicity of *Aeromonas sobria* on tail-rot disease in juvenile tilapia *Oreochromis niloticus. Current Microbiology* 62(2), 623–627. DOI: 10.1007/s00284-010-9753-8.

Liao, P.-C., Tsai, Y.-L., Chen, Y.-C., Wang, P.-C., Liu, S.-C, *et al.* (2020) Analysis of streptococcal infection and correlation with climatic factors in cultured tilapia *Oreochromis* spp. in Taiwan. *Applied Sciences* 10(11), 4018. DOI: https://www.mdpi.com/2076-3417/10/11/4018.

Lievens, B., Frans, I., Heusdens, C., Justé, A., Jonstrup, S.P, *et al.* (2011) Rapid detection and identification of viral and bacterial fish pathogens using a DNA array-based multiplex assay. *Journal of Fish Diseases* 34(11), 861–875. DOI: 10.1111/j.1365-2761.2011.01304.x.

Lin, Q., Li, N., Fu, X., Hu, Q., Chang, O, *et al.* (2016) An outbreak of granulomatous inflammation associated with *Francisella noatunensis* subsp. *orientalis* in farmed tilapia (*Oreochromis niloticus* × *O. aureus*) in China. *Chinese Journal of Oceanology and Limnology* 34(3), 460–466. DOI: 10.1007/s00343-016-4311-2.

Liu, J.-K. (1986) N,N,N'-N'-tetramethyl-p-phenylenediamine-dependent cytochrome oxidase analyses of bacillus species. *International Journal of Systematic Bacteriology* 36(1), 38–46. DOI: 10.1099/00207713-36-1-38.

Liu, W., Zhang, Y., Ma, J., Jiang, N., Fan, Y, *et al.* (2020) Determination of a novel parvovirus pathogen associated with massive mortality in adult tilapia. *PLOS Pathogens* 16(9), article e1008765. DOI: 10.1371/journal.ppat.1008765.

Liu, X., Zhou, X. and Mo, X. (1993) Characteristics on pathogen of caudal fin rot disease of tilapia (*Oreochromis niloticus*) in mariculture. *Tropical Oceanography/Redai Haiyang* 12, 100–103.

Lobato, I.M. and O'Sullivan, C.K. (2018) Recombinase polymerase amplification: Basics, applications and recent advances. *Trends in Analytical Chemistry* 98, 19–35. DOI: 10.1016/j.trac.2017.10.015.

Mabrok, M., Elayaraja, S., Chokmangmeepisarn, P., Jaroenram, W., Arunrut, N, *et al.* (2021) Rapid visualization in the specific detection of *Flavobacterium columnare*, a causative agent of freshwater columnaris using a novel recombinase polymerase amplification (RPA) combined with lateral flow dipstick (LFD) assay. *Aquaculture* 531, article 735780. DOI: 10.1016/j.aquaculture.2020.735780.

MacFaddin, J.F. (2000) *Biochemical tests for identification of medical bacteria*, 3rd ed ed. Lippincott Williams & Wilkins.

Machimbirike, V.I., Crumlish, M., Dong, H.T., Santander, J., Khunrae, P, *et al.* (2022) *Edwardsiella ictaluri*: A systemic review and future perspectives on disease management. Review, *Reviews in Aquaculture* 14(3), 1613–1636. DOI: 10.1111/raq.12665.

Mahon, C.R., Lehman, D.C. and Manuselis, G. (2011) *Textbook of diagnostic microbiology*, 4th ed ed. W. B. Saunders Company.

Martins, M., Cardoso, L., Furtado, W., Tancredo, K. and Lehmann, N. (2018) *Histopathology guide for freshwater fish*. UFSC.

Matsuo, S., Sugiyama, T., Okuyama, T., Yoshikawa, K., Honda, K, *et al.* (1999) Preservation of pathological tissue specimens by freeze-drying for immunohistochemical staining and various molecular biological analyses. *Pathology International* 49(5), 383–390. DOI: 10.1046/j.1440-1827.1999.00887.x.

McClelland, R. (2001) Gram's stain: the key to microbiology. *MLO: Medical Laboratory Observer* 33(4), 20–22, .

McGrogan, D.G., Ostland, V.E., Byrne, P.J. and Ferguson, H.W. (1998) Systemic disease involving an iridovirus-like agent in cultured tilapia, *Oreochromis niloticus* L. – A case report. *Journal of Fish Diseases* 21(2), 149–152. DOI: 10.1046/j.1365-2761.1998.00082.x.

McLeod, J.W. and Gordon, J. (1923) Catalase production and sensitiveness to hydrogen peroxide amongst bacteria: with a scheme of classification based on these properties. *Journal of Pathology and Bacteriology* 26(3), 326–331. DOI: 10.1002/path.1700260304.

Michaud, C.L. and Foran, D.R. (2011) Simplified field preservation of tissues for subsequent DNA analyses. *Journal of Forensic Sciences* 56(4), 846–852. DOI: 10.1111/j.1556-4029.2011.01771.x.

Miki, M., Ohishi, N., Nakamura, E., Furumi, A. and Mizuhashi, F. (2018) Improved fixation of the whole bodies of fish by a double-fixation method with formalin solution and Bouin's fluid or Davidson's fluid. *Journal of Toxicologic Pathology* 31(3), 201–206. DOI:10.1293/tox.2018-0001

Miniero Davies, Y., Oliveira, M.G., Paulo Vieira Cunha, M., Soares Franco, L., Pulecio Santos, S.L, *et al.* (2018) *Edwardsiella tarda* outbreak affecting fishes and aquatic birds in brazil. *Veterinary Quarterly* 38(1), 99–105. DOI: 10.1080/01652176.2018.1540070.

Mishra, A., Nam, G.H., Gim, J.A., Lee, H.E., Jo, A. *et al.* (2018) Current challenges of streptococcus infection and effective molecular, cellular, and environmental control methods in aquaculture. *Molecules and Cells* 41(6), 495–505. Available at: https://www.scopus.com/inward/record.uri?eid=2-s2.0–85056508470&partnerID=40&md5=6f714b14ce14cbf593b115ebf4e57700. https://doi.org/10.14348/molcells.2018.2154

Mori, Y., Nagamine, K., Tomita, N. and Notomi, T. (2001) Detection of loop-mediated isothermal amplification reaction by turbidity derived from magnesium pyrophosphate formation. *Biochemical and Biophysical Research Communications* 289(1), 150–154. DOI: 10.1006/bbrc.2001.5921.

Mortensen, S.H., Evensen, Ø., Rødseth, O.M. and Hjeltnes, B.K. (1993) The relevance of infectious pancreatic necrosis virus (IPNV) in farmed norwegian turbot (*Scophthalmus maximus*). *Aquaculture* 115(3–4), 243–252. DOI: 10.1016/0044-8486(93)90140-T.

Mulei, I.R., Nyaga, P.N., Mbuthia, P.G., Waruiru, R.M., Njagi, L.W, *et al.* (2018) Infectious pancreatic necrosis virus isolated from farmed rainbow trout and tilapia in Kenya is identical to European isolates. *Journal of Fish Diseases* 41(8), 1191–1200. DOI: 10.1111/jfd.12807.

Mumford, S., Heidel, J., Smith, C., Morrison, J., Macconnell, B. *et al.* (2007) Fish histology and histopathology 4th ed. United States Fish and Wildlife Service.

Neelsen, F. (1883). Ein casuistischer Beitrag zur Lehre von der Tuberkulose. Centrabl med Wissenschaften, 28, 497–501.

Ng, W.K. and Romano, N. (2013) A review of the nutrition and feeding management of farmed tilapia throughout the culture cycle. *Reviews in Aquaculture* 5(4), 220–254. DOI: 10.1111/raq.12014.

Nguyen, V. V., Dong, H. T., Senapin, S., Pirarat, N., & Rodkhum, C. (2015). *Francisella noatunensis* subsp. *orientalis*, an emerging bacterial pathogen affecting cultured red tilapia (*Oreochromis* sp.) in Thailand. Autonomous Robots, 1–6.

Notomi, T., Okayama, H., Masubuchi, H., Yonekawa, T., Watanabe, K, *et al.* (2000) Loop-mediated isothermal amplification of DNA. *Nucleic Acids Research* 28(12), E63. DOI: 10.1093/nar/28.12.e63.

Notomi, T., Mori, Y., Tomita, N. and Kanda, H. (2015) Loop-mediated isothermal amplification (LAMP): principle, features, and future prospects. *Journal of Microbiology* 53(1), 1–5. DOI: 10.1007/s12275-015-4656-9.

Oh, W.T., Jun, J.W., Kim, H.J., Giri, S.S., Yun, S, *et al.* (2020) Characterization and pathological analysis of a virulent *Edwardseillla anguillarum* strain isolated from Nile tilapia (*Oreochromis niloticus*) in korea. *Frontiers in Veterinary Science* 7, article. DOI: 10.3389/fvets.2020.00014.

Omoregie, E., Ufodike, E.B.C. and Onwuliri, C.O.E. (1995) A comparative survey of helminth parasites of the Nile tilapia, *Oreochromis niloticus* from a fish farm and petroleum polluted freshwater. *Aquatic Sciences* 10, 7–14.

Ortega, C., Mancera, G., Enríquez, R., Vargas, A., Martínez, S, *et al.* (2016) First identification of *Francisella noatunensis* subsp. *orientalis* causing mortality in mexican tilapia *Oreochromis* spp. *Diseases of Aquatic Organisms* 120(3), 205–215. DOI: 10.3354/dao02999.

Oser, B. (1965) In: Hawk, L. (ed.) *Physiological Chemistry. Enzymes and Their Action: Cell Respiration*, 14th ed. McGraw-Hill.

Østergaard, C., Hansen, S. G., & Møller, J. K. (2015). Rapid first-line discrimination of methicillin resistant *Staphylococcus aureus* strains using MALDI-TOF MS. International Journal of Medical Microbiology, 305(8), 838–847. DOI: https://doi.org/10.1016/j.ijmm.2015.08.002.

Oviedo-Bolaños, K., Rodríguez-Rodríguez, J.A., Sancho-Blanco, C., Barquero-Chanto, J.E., Peña-Navarro, N, *et al.* (2021) Molecular identification of *Streptococcus* sp. and antibiotic resistance genes present in tilapia farms (*Oreochromis niloticus*\) from the northern pacific region, costa rica. *Aquaculture International* 29(5), 2337–2355. DOI: 10.1007/s10499-021-00751-0.

Pantoja, M.F.W. (2012) Protozoan and metazoan parasites of Nile tilapia *Oreochromis niloticus* cultured in Brazil. Fishing engineer, L.N.R., fishing engineer, M.D.R., zoo technician, D.M., and Tavares-Dias M, *Revista MVZ Córdoba* 17(1), 2812–2819.

Park, C.-H., Kim, H.-W., Rhyu, I.J. and Uhm, C.-S. (2016) How to get well-preserved samples for transmission electron microscopy. *Applied Microscopy* 46(4), 188–192. DOI: 10.9729/AM.2016.46.4.188.

Patel, R. (2015). MALDI-TOF MS for the diagnosis of infectious diseases. Clinical Chemistry, 61(1), 100–111. https://doi.org/10.1373/clinchem.2014.221770

Patricia, S. and Laura, C. (2011) Motility test medium protocol. American Society for Microbiology. Available at: https://asm.org/ASM/media/Protocol-Images/Motility-Test-Medium-Protocol.pdf?ext=.pdf

Pérez-Losada, M., Cabezas, P., Castro-Nallar, E. and Crandall, K.A. (2013) Pathogen typing in the genomic sera: MLST and the future of molecular epidemiology. *Infection, Genetics and Evolution* 16, 38–53. DOI: 10.1016/j.meegid.2013.01.009.

Pestana, E.A., Belak, S., Diallo, A., Crowther, J.R. and Viljoen, G.J. (2010) Real-time PCR – the basic principles. In: Pestana, E., Belak, S., Diallo, A., Crowther, J.R. and Viljoen, G.J. (eds) *Early, Rapid and Sensitive Veterinary Molecular Diagnostics – Real Time PCR Applications.* Springer Netherlands, pp. 27–46. DOI: 10.1007/978-90-481-3132-7_3.

Piamsomboon, P., Jaresitthikunchai, J., Hung, T.Q., Roytrakul, S. and Wongtavatchai, J. (2020) Identification of bacterial pathogens in cultured fish with a custom peptide database constructed by matrix-assisted laser desorption/ionization time-of-flight mass spectrometry (MALDI-TOF MS). *BMC Veterinary Research* 16(1), article 52. DOI: 10.1186/s12917-020-2274-1.

Piewbang, C., Tattiyapong, P., Techangamsuwan, S. and Surachetpong, W. (2021) Tilapia lake virus immunoglobulin G (TiLV IgG) antibody: immunohistochemistry application reveals cellular tropism of TiLV infection. *Fish and Shellfish Immunology* 116, 115–123. DOI: 10.1016/j.fsi.2021.06.017.

Pirarat, N., Pinpimai, K., Endo, M., Katagiri, T., Ponpornpisit, A, *et al.* (2011) Modulation of intestinal morphology and immunity in Nile tilapia (*Oreochromis niloticus*) by lactobacillus rhamnosus GG. *Research in Veterinary Science* 91(3), e92–e97. DOI: 10.1016/j.rvsc.2011.02.014.

Popovic, N.T. (2008) Commercial phenotypic tests (API 20E) in diagnosis of fish bacteria: A review. *Veterinární Medicína* 52(2), 49–53. DOI: 10.17221/2058-vetmed.

Poudyal, S., Pulpipat, T., Wang, P.C. and Chen, S.C. (2020) Comparison of the pathogenicity of *Francisella orientalis* in Nile tilapia (*Oreochromis niloticus*), Asian seabass (lates calcarifer) and largemouth bass (*Micropterus salmoides*) through experimental intraperitoneal infection. *J. Fish Dis* 43(9), 1097–1106. DOI: 10.1111/jfd.13217.

Prihartini, N.C., Yanuhar, U. and Maftuch, (2015) Betanodavirus infections in tilapia seed (*Oreochromis* sp.), in Indonesia. *Journal of Life Sciences and Biomedicine* 5, 106–109.

Prudent, E. and Raoult, D. (2019) Fluorescence in situ hybridization, a complementary molecular tool for the clinical diagnosis of infectious diseases by intracellular and fastidious bacteria. *FEMS Microbiology Reviews* 43(1), 88–107. DOI: 10.1093/femsre/fuy040.

Qu, Y., Wang, Q., Li, Y., Wang, Y., Yin, J, *et al.* (2021) Development of a real-time recombinase polymerase amplification assay for rapid detection of *Aeromonas hydrophila*. *Journal of Fish Diseases* 44(4), 469–477. DOI: 10.1111/jfd.13291.

Ramadan, H.H. (1991) Effect of host species, sex, length, diet and different seasons on the parasitic infection of tilapia fish in Lake Manzala. *Journal of King Abdulaziz University-Marine Sciences* 2(1), 81–91. DOI: 10.4197/mar.2-1.6.

Ramirez-Paredes, J.G., Larsson, P., Thompson, K.D., Penman, D.J. and Busse, H.J, *et al.* (2020) Reclassification of *Francisella noatunensis* subsp. *orientalis* Ottem et al. 2009 as *Francisella orientalis* sp. nov., *Francisella noatunensis* subsp. *chilensis* subsp. nov. and emended description of *Francisella noatunensis*. International journal of systematic and evolutionary microbiology, 70(3), 2034–2048. DOI: 10.1099/ijsem.0.004009.

Ramírez-Paredes, J.G., Paley, R.K., Hunt, W., Feist, S.W., Stone, D.M, *et al.* (2021) First detection of infectious spleen and kidney necrosis virus (ISKNV) associated with massive mortalities in farmed tilapia in Africa. *Transboundary and Emerging Diseases* 68(3), 1550–1563. DOI: 10.1111/tbed.13825.

OIE (2012) Chapter 2.3.7. Red sea bream iridoviral disease In: Manual of Diagnostic Tests for Aquatic Animals 7th ed., World Organisation for Animal Health, Paris, pp. 345–356.

Reichley, S.R., Ware, C., Greenway, T.E., Wise, D.J. and Griffin, M.J. (2015) Real-time polymerase chain reaction assays for the detection and quantification of *Edwardsiella tarda*, *Edwardsiella piscicida*, and *Edwardsiella piscicida*–like species in catfish tissues and pond water. *Journal of Veterinary Diagnostic Investigation* 27(2), 130–139. DOI: 10.1177/1040638714566672.

Redi, C. (2010) Histology protocols. *European Journal of Histochemistry* 54(2). DOI: 10.4081/ejh.2010.e27.

Rhodes, M.W., Kator, H., McNabb, A., Deshayes, C., Reyrat, J.M, *et al.* (2005) *Mycobacterium pseudoshottsii* sp. nov., a slowly growing chromogenic species isolated from chesapeake bay striped bass (*Morone saxatilis*). *International Journal of Systematic and Evolutionary Microbiology* 55(3), 1139–1147. DOI: 10.1099/ijs.0.63343-0.

Rijal, N. (2022) Transport media used in microbiology LabNisha Rijal. Available at: https://microbeonline.com/transport-medium-bacterial-viral-sample-transport-used-microbiology-laboratory/

Rissanen, A.J., Kurhela, E., Aho, T., Oittinen, T. and Tiirola, M. (2010) Storage of environmental samples for guaranteeing nucleic acid yields for molecular microbiological studies. *Applied Microbiology and Biotechnology* 88(4), 977–984. DOI: https://doi.org/10.1007/s00253-010-2838-2.

Roberts, R.J. and Sommerville, C. (1982) Diseases of tilapia. *Biology and Culture of Tilapias* 247–263.

Rodkhum, C., Kayansamruaj, P., Pirarat, N. and Wongtawatchai, J. (2012) Duplex PCR for simultaneous and unambiguous detection of *Streptococcus iniae* and *Streptococcus agalactiae* associated with Streptococcosis of cultured tilapia in Thailand. *Thai Journal of Veterinary Medicine* 42(2), 153–158. DOI: 10.56808/2985-1130.2376.

Rodríguez-Sánchez, B., Sánchez-Carrillo, C., Ruiz, A., Marín, M., Cercenado, E, *et al.* (2014) Direct identification of pathogens from positive blood cultures using matrix-assisted laser desorption-ionization time-of-flight mass spectrometry. *Clinical Microbiology and Infection* 20(7), 421–O427. DOI: 10.1111/1469-0691.12455.

Romero-Gómez, M.P., Gómez-Gil, R., Paño-Pardo, J.R. and Mingorance, J. (2012) Identification and susceptibility testing of microorganism by direct inoculation from positive blood culture bottles by combining MALDI-TOF and vitek-2 compact is rapid and effective. *Journal of Infection* 65(6), 513–520. DOI: 10.1016/j.jinf.2012.08.013.

Roux, L.E. and Avenant-Oldewage, A. (2010) Checklist of the fish parasitic genus *Cichlidogyrus* (Monogenea), including its cosmopolitan distribution and host species. *African Journal of Aquatic Science* 35(1), 21–36. DOI: 10.2989/16085914.2010.466632.

Sahoo, L., Das, B.K., Parhi, J. and Mukherjee, S.C. (2010) DNA fingerprinting of *Flavobacterium columnare* using RAPD-PCR. *Indian Journal of Microbiology* 50(1), 10–16. DOI: 10.1007/s12088-010-0070-5.

Saint-Jean, S.R., Borrego, J.J. and Perez-Prieto, S.I., SR (2003) Infectious pancreatic necrosis virus: biology, pathogenesis, and diagnostic methods. In: *In Advances in Virus Research*. Jeonju University. Academic Press, p. 62. : DOI: https://doi.org/10.1016/S0065-3527(03)62003-8.

Sakai, T., Yuasa, K., Sano, M. and Iida, T. (2009) Identification of *Edwardsiella ictaluri* and *E. tarda* by species-specific polymerase chain reaction targeted to the upstream region of the fimbrial gene. *Journal of Aquatic Animal Health* 21(2), 124–132. DOI: 10.1577/H08-061.1.

Samir, S., Awad, A. and Younis, G. (2021) Prevalence, virulence determinants and antimicrobial-resistant profile of *Edwardsiella tarda* isolated from Nile tilapia (*Oreochromis niloticus*) in Egypt. *Advances in Animal and Veterinary Sciences* 10(5), 1031–1038. DOI: 10.17582/JOURNAL.AAVS/2022/10.5.1031.1038.

Sanders, E.R. (2012) Aseptic laboratory techniques: Plating methods. *Journal of Visualized Experiments: JoVE* 63(63), e3064. DOI: 10.3791/3064.

Saptiani, G., Pebrianto, C.A.A., Hardi, E.H. and Ardhani, F. (2017) Diversity and prevalence of ectoparasites associated with cultured fish from coal ponds in east. Indonesia.

Senapin, S., Dong, H.T., Meemetta, W., Siriphongphaew, A., Charoensapsri, W, *et al.* (2016) *Hahella chejuensis* is the etiological agent of a novel red egg disease in tilapia (*Oreochromis* spp.) hatcheries in Thailand. *Aquaculture* 454, 1–7. DOI: 10.1016/j.aquaculture.2015.12.013.

Seng, P., Drancourt, M., Gouriet, F., Scola, B., Fournier, P.E, *et al.* (2009) Ongoing revolution in bacteriology: Routine identification of bacteria by matrix-assisted laser desorption ionization time-of-flight mass spectrometry. *Clinical Infectious Diseases* 49(4), 543–551. DOI: 10.1086/600885.

Seutin, G., White, B.N. and Boag, P.T. (1991) Preservation of avian blood and tissue samples for DNA analyses. *Canadian Journal of Zoology* 69(1), 82–90. DOI: 10.1139/z91-013.

Shahin, K., Gustavo Ramirez-Paredes, J.G., Harold, G., Lopez-Jimena, B., Adams, A, *et al.* (2018) Development of a recombinase polymerase amplification assay for rapid detection of *Francisella noatunensis* subsp. *orientalis*. *PLOS ONE* 13(2), article e0192979. DOI: 10.1371/journal.pone.0192979.

Shahin, K., Subramaniam, K., Camus, A.C., Yazdi, Z., Yun, S, *et al.* (2021) Isolation, identification and characterization of a novel megalocytivirus from cultured tilapia (*Oreochromis* spp.) from Southern California, USA. *Animals (Basel)* 11(12). DOI: 10.3390/ani11123524.

Shi, Y.-H., Chen, J., Li, C.-H., Lu, X.-J., Zhang, D.-M, *et al.* (2012) Detection of bacterial pathogens in aquaculture samples by DNA microarray analysis. *Aquaculture* 338–341, 29–35. DOI: 10.1016/j.aquaculture.2012.01.009.

Shirajum Monir, M., Yusoff, S.M., Mohamad, A. and Ina-Salwany, M.Y. (2020) Vaccination of tilapia against motile *Aeromonas septicemia*: A review. *Journal of Aquatic Animal Health* 32(2), 65–76. DOI: 10.1002/aah.10099.

Shlapobersky, M., Sinyakov, M.S., Katzenellenbogen, M., Sarid, R., Don, J, *et al.* (2010) Viral encephalitis of tilapia larvae: Primary characterization of a novel herpes-like virus. *Virology* 399(2), 239–247. DOI: 10.1016/j.virol.2010.01.001.

Shoemaker, C.A., Evans, J.J. and Klesius, P.H. (2000) Density and dose: Factors affecting mortality to *Streptococcus iniae*-infected tilapia (*Oreochromis niloticus*). *Aquaculture* 188(3–4), 229–235. DOI: 10.1016/S0044-8486(00)00346-X.

Singh, A. (2022) Culture media: Classification, types, and relevance. Available at: https://conductscience.com/culture-media/

Singhal, N., Kumar, M., Kanaujia, P.K. and Virdi, J.S. (2015) MALDI-TOF mass spectrometry: An emerging technology for microbial identification and diagnosis. *Frontiers in Microbiology* 6, 791. : DOI: https://doi.org/10.3389/fmicb.2015.00791.

Smith, A.C. and Hussey, M.A. (2005). American Society for Microbiology. Available at: https://asm.org/getattachment/5c95a063–326b-4b2f-98ce-001de9a5ece3/gram-stain-protocol-2886.pdf

Smith, C.E., Ramsey, D. and Speer, C.A. (1998) Histopathology associated with an irido-like virus infection of tilapia, *Tilapia nilotica*[Article]. Aquaculture'98 – Book of Abstracts.

Soler, L., Figueras, M.J., Chacón, M.R., Guarro, J. and Martinez-Murcia, A.J. (2003) Comparison of three molecular methods for typing aeromonas popoffii isolates. *Antonie Van Leeuwenhoek* 83(4), 341–349. DOI: 10.1023/A:1023312415276.

Soliman, H., Midtlyng, P.J. and El-Matbouli, M. (2009) Sensitive and rapid detection of infectious pancreatic necrosis virus by reverse transcription loop mediated isothermal amplification. *Journal of Virological Methods* 158(1–2), 77–83. DOI: 10.1080/08997659.2012.675931.

Soto, E., Hawke, J.P., Fernandez, D. and Morales, J.A. (2009) *Francisella* sp., an emerging pathogen of tilapia, *Oreochromis niloticus* (L.), in costa rica. *Journal of Fish Diseases* 32(8), 713–722. DOI: 10.1111/j.1365-2761.2009.01070.x.

Soto, E., Illanes, O., Hilchie, D., Morales, J. A., Sunyakumthorn, P., Hawke, J.P., Goodwin, A.E., Riggs, A., Yanong, R.P., Pouder, D.B., Francis-Floyd, R., Arauz, M., Bogdanovic, L., & Castillo-Alcala, F. (2012). Molecular and immunohistochemical diagnosis of *Francisella noatunensis* subsp. orientalis from formalin-fixed, paraffin-embedded tissues [Article]. *Journal of Veterinary Diagnostic Investigation, 24*(5), 840–845. https://doi.org/10.1177/1040638712452108

Soto, E., Kidd, S., Mendez, S., Marancik, D., Revan, F, *et al.* (2013) *Francisella noatunensis* subsp. *orientalis* pathogenesis analyzed by experimental immersion challenge in Nile tilapia, *Oreochromis niloticus* (L.). *Veterinary Microbiology* 164(1–2), 77–84. DOI: 10.1016/j.vetmic.2013.01.024.

Soto, E., Shahin, K., Talhami, J.J., Griffin, M.J., Adams, A, *et al.* (2019) Characterization of *Francisella noatunensis* subsp. *orientalis* isolated from Nile tilapia *Oreochromis niloticus* farmed in lake Yojoa, Honduras. *Diseases of Aquatic Organisms* 133(2), 141–145. DOI: 10.3354/dao03333.

Soto-Rodriguez, S.A., Cabanillas-Ramos, J., Alcaraz, U., Gomez-Gil, B. and Romalde, J.L. (2013) Identification and virulence of *Aeromonas dhakensis, Pseudomonas mosselii* and microbacterium paraoxydans isolated from Nile tilapia, *Oreochromis niloticus,* cultivated in Mexico. *Journal of Applied Microbiology* 115(3), 654–662. DOI: 10.1111/jam.12280.

Standard, T.A. (2010) Diagnosis of streptococcosis in tilapia. Available at: https://www.acfs.go.th/standard/download/eng/STD_TILAPIA.pdf

Steel, K.J. (1961) The oxidase reaction as a taxonomic tool. *Journal of General Microbiology* 25(2), 297–306. DOI: 10.1099/00221287-25-2-297.

Straube, D. and Juen, A. (2013) Storage and shipping of tissue samples for DNA analyses: A case study on earthworms. *European Journal of Soil Biology* 57, 13–18. DOI: 10.1016/j.ejsobi.2013.04.001.

Subashkumar, R., Thayumanavan, T. and Lakshmanaperumalsamy, P. (2014) RAPD and ERIC – PCR typing of virulent *Aeromonas hydrophila* isolated from children with acute diarrhoea. *Journal of Cell and Molecular Biology* 12(1–2), 47–53.

Sudpraseart, C., Wang, P.C., & Chen, S.C. (2021). Phenotype, genotype and pathogenicity of *Streptococcus agalactiae* isolated from cultured tilapia (*Oreochromis spp.*) in Taiwan [Article]. *Journal of Fish Diseases,* 44(6), 747–756. https://doi.org/10.1111/jfd.13296

Subramaniam, K., Gotesman, M., Smith, C.E., Steckler, N.K., Kelley, K.L, *et al.* (2016) Megalocytivirus infection in cultured Nile tilapia *Oreochromis niloticus. Diseases of Aquatic Organisms* 119(3), 253–258. DOI: 10.3354/dao02985.

Suebsing, R., Oh, M.J. and Kim, J.H. (2011) Evaluation of rapid and sensitive reverse transcription loop-mediated isothermal amplification method for detecting infectious pancreatic necrosis virus in chum salmon (*Oncorhynchus keta*). *Journal of Veterinary Diagnostic Investigation* 23(4), 704–709. DOI: 10.1177/1040638711407897.

Suebsing, R., Pradeep, P.J., Jitrakorn, S., Sirithammajak, S., Kampeera, J, *et al.* (2016) Detection of natural infection of infectious spleen and kidney necrosis virus in farmed tilapia by hydroxynapthol blue-loop-mediated isothermal amplification assay. *Journal of Applied Microbiology* 121(1), 55–67. DOI: 10.1111/jam.13165.

Sulkin, S.E. and Willett, J.C. (1940) A triple sugar-ferrous sulfate medium for use in identification of enteric organisms. *Journal of Laboratory and Clinical Medicine* 25(6), 649–653.

Surachetpong, W., Roy, S.R.K. and Nicholson, P. (2020) Tilapia lake virus: The story so far. *Journal of Fish Diseases* 43(10), 1115–1132. DOI: 10.1111/jfd.13237.

Szczuka, E. and Kaznowski, A. (2004) Typing of clinical and environmental *Aeromonas* sp. strains by random amplified polymorphic DNA PCR, repetitive extragenic palindromic PCR, and enterobacterial repetitive intergenic consensus sequence PCR. *Journal of Clinical Microbiology* 42(1), 220–228. DOI: 10.1128/JCM.42.1.220-228.2004.

Taengphu, S., Sangsuriya, P., Phiwsaiya, K., Debnath, P.P., Delamare-Deboutteville, J, *et al.* (2020) Genetic diversity of tilapia lake virus genome segment 1 from 2011 to 2019 and a newly validated semi-nested RT-PCR method. *Aquaculture* 526, 735423. DOI: 10.1016/j.aquaculture.2020.735423.

Taghreed, I. (2020) Diseases of Nile tilapia with special emphasis on water pollution. *Journal of Environmental Science and Technology* 13(1), 29–56. DOI: 10.3923/jest.2020.29.56.

Taha, E., Shawky, M., Ahmed, B., Moustafa, M., Yousif, A, *et al.* (2020) Emergence of viral nervous necrosis is associated with mass mortality in hatchery-reared tilapia (*Oreochromis niloticus*) in Egypt. *Aquaculture International* 28(5), 1811–1823. DOI: 10.1007/s10499-020-00559-4.

Tattiyapong, P., Sirikanchana, K. and Surachetpong, W. (2018) Development and validation of a reverse transcription quantitative polymerase chain reaction for tilapia lake virus detection in clinical samples and experimentally challenged fish. *Journal of Fish Diseases* 41(2), 255–261. DOI: 10.1111/jfd.12708.

Tavornapanich, S., Brun, E. and Dverdal Jansen, M. (2020) 1. guidelines for on-farm sampling for targeted surveillance to certify disease freedom, diagnosis in case of mortalities and for analysis of mortalities caused by unknown aetiology. CIHEAM, *Diagnostic Manual for the Main Pathogens in European Sea Bass and Gilthead Sea Bream Aquaculture* 75. Available at: http://om.ciheam.org/om/pdf/b75/00007934.pdf

Teng, L.J., Hsueh, P.R., Tsai, J.C., Chen, P.W., Hsu, J.C, *et al.* (2002) GroESL sequence determination, phylogenetic analysis, and species differentiation for viridans group streptococci. *Journal of Clinical Microbiology* 40(9), 3172–3178. DOI: 10.1128/JCM.40.9.3172-3178.2002.

Tung, M.C., Chen, S.C. and Tsai, S.S. (1987) General septicemia of streptococcal infection in cage-culture tilapia, *Tilapia mossambica*, in Southern Taiwan. *COA Fisheries Series* 10, 187–197.

Uma, A., Rebecca, G., Meena, S. and Saravanabava, K. (2010) PCR detection of putative aerolysin and hemolysin genes in an *Aeromonas hydrophila* isolate from infected koi carp (*Cyprinus carpio*). *Tamilnadu Journal of Veterinary and Animal Sciences* 6(1), 31–33.

Urwin, R. and Maiden, M.C.J. (2003) Multi-locus sequence typing: A tool for global epidemiology. *Trends in Microbiology* 11(10), 479. DOI: 10.1016/j.tim.2003.08.006.

Valladão, G.M.R., Alves, L.O. and Pilarski, F. (2016) Trichodiniasis in Nile tilapia hatcheries: Diagnosis, parasite:host-stage relationship and treatment. *Aquaculture* 451, 444–450. DOI: 10.1016/j.aquaculture.2015.09.030.

Vazquez-Pertejo, M.T. (2020) Culture. MSD Manuals. Available at: https://www.msdmanuals.com/professional/infectious-diseases/laboratory-diagnosis-of-infectious-disease/immunologic-tests-for-infectious-disease

Wang, C.S., Chao, S.Y., Ku, C.C., Wen, C.M. and Shih, H.H. (2009) PCR amplification and sequence analysis of the major capsid protein gene of megalocytiviruses isolated in taiwan. *Journal of Fish Diseases* 32(6), 543–550. DOI: 10.1111/j.1365-2761.2009.01043.x.

Wang, Y., Wang, Q., Zeng, W., Yin, J., Li, Y, *et al.* (2018) Establishment and characterization of a cell line from tilapia brain for detection of tilapia lake virus. *Journal of Fish Diseases* 41(12), 1803–1809. DOI: 10.1111/jfd.12889.

Wang, Z.G. and Xu, B.H. (1985) Studies on the pathogenic bacteria of the "rotten-skin" diseases of the Nile tilapia (*Tilapia nilotica*). *Journal of Fisheries of China/Shuichan Xuebao* 9, 217–221.

Welker, M. and Moore, E.R. (2011) Applications of whole-cell matrix-assisted laser-desorption/ionization time-of-flight mass spectrometry in systematic microbiology. *Systematic and Applied Microbiology* 34(1), 2–11. DOI: 10.1016/j.syapm.2010.11.013.

Williams, K., Blake, S., Sweeney, A., Singer, J.T. and Nicholson, B.L. (1999) Multiplex reverse transcriptase PCR assay for simultaneous detection of three fish viruses. *Journal of Clinical Microbiology* 37(12), 4139–4141. DOI: 10.1128/JCM.37.12.4139-4141.1999.

Wilton, S. and Cousins, D. (1992) Detection and identification of multiple mycobacterial pathogens by DNA amplification in a single tube. *PCR Methods and Applications* 1(4), 269–273. DOI: 10.1101/gr.1.4.269.

Wong, Y.P., Othman, S., Lau, Y.L., Radu, S. and Chee, H.Y. (2018) Loop-mediated isothermal amplification (LAMP): A versatile technique for detection of micro-organisms. *Journal of Applied Microbiology* 124(3), 626–643. DOI: 10.1111/jam.13647.

Wormser, G.P. and Hanna, B.A. (2004) Manual of clinical microbiology. $189.95 (cloth). clinical infectious diseases. *Yolken Washington* 38(8), 1199–1200. DOI: 10.1086/383067.

Wu, L., Ye, L., Wang, Z., Cui, Y. and Wang, J. (2019) Utilization of recombinase polymerase amplification combined with a lateral flow strip for detection of *Perkinsus beihaiensis* in the oyster *Crassostrea hongkongensis. Parasites and Vectors* 12(1), 360. DOI: 10.1186/s13071-019-3624-3.

Yamkasem, J., Tattiyapong, P., Gorgoglione, B. and Surachetpong, W. (2021) Uncovering the first occurrence of tilapia parvovirus in thailand in tilapia during co-infection with tilapia tilapinevirus. *Transboundary and Emerging Diseases* 68(6), 3136–3144. DOI: 10.1111/tbed.14143.

Yeh, H.Y., Shoemaker, C.A. and Klesius, P.H. (2005) Evaluation of a loop-mediated isothermal amplification method for rapid detection of channel *Ictalurus punctatus* punctatus important bacterial pathogen *Edwardsiella ictaluri. Journal of Microbiological Methods* 63(1), 36–44. DOI: 10.1016/j.mimet.2005.02.015.

Yogananth, N., Bhakyaraj, R., Chanthuru, A., Anbalagan, T. and Mullai Nila, K. (2009) Detection of virulence gene in *Aeromonas hydrophila* isolated from fish samples using PCR technique. *Global Journal of Biotechnology and Biochemistry* 4(1), 51–53.

Yokota, M., Watanabe, S., Hatai, K., Kurata, O., Furihata, M, *et al.* (2008) Transmission of the parasite ichthyophonus hoferi in cultured rainbow trout and comparison of epidemic models. *Journal of Aquatic Animal Health* 20(4), 207–214. DOI: 10.1577/H06-043.1.

Yoshimizu, M. and Kimura, T. (1985) A coagglutination test with antibody-sensitized staphylococci for rapid and simple diagnosis of bacterial and viral diseases of fish. *Fish Pathology* 20(2/3), 243–261. DOI: 10.3147/jsfp.20.243.

Zahran, E., Hafez, E.E., Mohd Altaf Hossain, F., Elhadidy, M. and Shaheen, A.A. (2017) Saprolegniosis in Nile tilapia: Identification, molecular characterization, and phylogenetic analysis of two novel pathogenic saprolegnia strains. *Journal of Aquatic Animal Health* 29(1), 43–49. DOI: 10.1080/08997659.2016.1259691.

Zhang, D.F., Zhang, Q.Q. and Li, A.H. (2014a) Development of a multiplex PCR assay for rapid and simultaneous detection of four genera of fish pathogenic bacteria. *Letters in Applied Microbiology* 59(5), 471–478. DOI: 10.1111/lam.12303.

Zhang, X., Yang, W., Wu, H., Gong, X. and Li, A. (2014b) Multilocus sequence typing revealed a clonal lineage of *Aeromonas hydrophila* caused motile aeromonas septicemia outbreaks in pond-cultured cyprinid fish in an epidemic area in central China. *Aquaculture* 432, 1–6. DOI: 10.1016/j.aquaculture.2014.04.017.

Zhang, Z. (2021) Research advances on tilapia streptococcosis. *Pathogens* 10(5), 558. DOI: 10.3390/pathogens10050558.

Zhang, Z., Lan, J., Li, Y., Hu, M., Yu, A, *et al.* (2018) The pathogenic and antimicrobial characteristics of an emerging *Streptococcus agalactiae* serotype IX in tilapia. *Microbial Pathogenesis* 122, 39–45. DOI: 10.1016/j.micpath.2018.05.053.

Zhou, M., Yang, Q., Kudinha, T., Sun, L., Zhang, R, *et al.* (2017) An improved in-house MALDI-TOF MS protocol for direct cost-effective identification of pathogens from blood cultures. *Frontiers in Microbiology* 8, 1824. DOI: 10.3389/fmicb.2017.01824.

Zhou, Q.J., Wang, L., Chen, J., Wang, R.N., Shi, Y.H, *et al.* (2014) Development and evaluation of a real-time fluorogenic loop-mediated isothermal amplification assay integrated on a microfluidic disc chip (on-chip LAMP) for rapid and simultaneous detection of ten pathogenic bacteria in aquatic animals. *Journal of Microbiological Methods* 104, 26–35. DOI: 10.1016/j.mimet.2014.06.008.

Ziehl, F. (1882) Zur Fårbung des tuberkelbacillus. *DMW – Deutsche Medizinische Wochenschrift* 8(33), 451. DOI: 10.1055/s-0029-1196721.

Chapter 12

Concurrent infections of tilapia

De-Hai Xu and Craig A. Shoemaker

Introduction

Tilapia are among the most important farm-raised fish in the world, accounting for ~10% of finfish production in 2016 (Food and Agriculture Organization, 2018). Production of tilapia is 50% in semi-intensive or intensive ponds, 25% in cages and 10% in intensive recirculating systems (Fitzsimmons, 2006). Tilapia are considered more resistant to pathogens than many other cultured species. As the intensity of culture increases in ponds, cages and recirculating systems to achieve high fish production, stocking density, feed rates and other aquaculture procedures increase and result in increasing stress on fish and deteriorating water quality, leading to high incidence and severity of diseases in tilapia (Shoemaker *et al.*, 2006).

Tilapia are affected by bacterial, fungal, parasitic and viral pathogens. Disease resulting from infection is responsible for major losses in tilapia aquaculture. Bacterial diseases are the most common in intensive fish-rearing facilities. Mortalities in tilapia have been reported to be caused by bacterial pathogens *Aeromonas* sp., *Edwardsiella tarda*, *Enterococcus* sp., *Francisella orientalis*, *Flavobacterium columnare*, *Streptococcus* sp., *Staphylococcus* sp. and *Pseudomonas* sp. (Shoemaker *et al.*, 2006; Hamouda *et al.*, 2019; Ramirez-Paredes *et al.*, 2020). Parasites in tilapia are mainly ectoparasites, including *Trichodina* sp., *Trichodinella* sp., *Ichthyophthirius multifiliis*, *Ichthyobodo* sp., *Dactylogyrus* sp., *Gyrodactylus* sp. and *Oodinium* sp. (Shoemaker *et al.*, 2006; Indahsari *et al.*, 2019). El-Sharouny and Badran (1995)

Kim Thompson, Craig Shoemaker and David C. Little (eds) **Tilapia**
DOI: 10.1079/9781800629455.0012, © CAB International, 2025

recovered 17 fungal species from tilapia among which, *Saprolegnia parasitica* and *S. ferax* resulted in mortalities in experimental fish. Fungal problems of tilapia may be significant in hatcheries producing fry. If *Saprolegnia* or other water molds become established in incubating eggs, all eggs may be lost (Plumb, 1997). Recent outbreaks of viral disease in tilapia caused by tilapia lake virus (TiLV) were reported in 15 countries located in Asia, Africa and South America (Jansen *et al.*, 2018; Machimbirike *et al.*, 2019; Nicholson *et al.*, 2019). Other reported viral infections include infectious pancreatic necrosis virus (IPNV), nervous necrosis virus (NNV), tilapia larvae encephalitis virus (TLEV), Bohle iridovirus (BIV), iridovirus-like agent, infectious spleen and kidney necrosis virus (ISKNV) and lymphocystis disease virus (LCDV). Except for TiLV, other viruses are less common and only seven cases of disease have been reported in 11 papers since 1975 (Machimbirike *et al.*, 2019).

Fish pathogens are normally present in natural waters but typically do not cause disease until environmental factors affect either fish, pathogen(s) or both (Raman *et al.*, 2013). Diverse groups of fish pathogens cohabiting in aquaculture systems may show similar preferences for environmental conditions or fish hosts, so coinfections occur. In an aquatic environment, fish may be coinfected by the same group of pathogens (homologous), such as coinfection by two or more bacterial species. In other cases, fish may be coinfected by pathogens from different groups (heterologous), such as coinfection by parasitic and bacterial species. The interactions between co-occurring pathogens can be either synergistic or antagonistic (Bradley & Jackson, 2008; Abdel-Latif *et al.*, 2020). Synergistic effects occur when one pathogen induces immune suppression and hinders fish immune response against subsequent infections by another pathogen, leading to an increase in infection severity and mortality (Bradley & Jackson, 2008; Kotob *et al.*, 2016). Antagonistic effects, however, result from the competition of pathogens for nutrients or microenvironments thus limiting the population size of the infectious pathogen(s) (Kotob *et al.*, 2016).

Most studies of tilapia diseases have been conducted for a single pathogen (a parasite, a bacterium, a fungus or a virus). The study design allows control of a single variable (i.e. one pathogen) but does not fully reflect conditions in aquaculture, whereby tilapia are likely exposed to multiple pathogens. Concurrent infection is defined as the simultaneous infection of a host by two or more pathogens (Cox, 2001; Kotob *et al.*, 2016). Other terms used to describe concurrent infections include coinfections, dual infections, concomitant infections, multiple infections, mixed infections, polymicrobial infections and secondary infections (Cox, 2001; Kotob *et al.*, 2016; Abdel-Latif *et al.*, 2020). Increasing evidence suggests that coinfections contribute to the severity of infectious diseases in tilapia, especially bacterial diseases (Xu *et al.*, 2007; Nicholson *et al.*, 2019). In this chapter we review studies on coinfections of tilapia by various pathogens, evaluate the impact of coinfections on fish disease susceptibility, explore the effects of coinfections on fish immune responses and discuss possible management strategies to limit coinfections in tilapia.

Natural concurrent infection

Concurrent infection in ponds

Parasites, bacteria and viruses are common inhabitants in water of fish farms. Recent field studies showed that tilapia were naturally coinfected by homologous or heterologous pathogens. Between 2013 and 2015 an increasing number of tilapia farms in Egypt experienced high summer morbidity. In September 2015, Nicholson *et al.* (2017) visited eight commercial tilapia farms located in different regions of the Nile delta and observed morbidities ranging from 43 to 100%. They randomly sampled 13–40 fish per farm to identify pathogens responsible for the summer mortalities of tilapia. They identified *Aeromonas* sp. (*A. hydrophilia*, *A. veronii*, *A. ichthiosmia*, *A. jandaei* or *A. enteropelogenes*) on every farm and TiLV in 50% of the farms investigated. Coinfections of TiLV and *A. veronii* or TiLV and *A. enteropelogenes* were discovered in 50% of tilapia farms, as well as coinfections among *Aeromonas* sp.

To investigate mass mortalities of tilapia in Thailand during May to December 2016, Nile tilapia and red hybrid tilapia were collected from 52 outbreaks across 18 farms spanning nine different geographical areas of Thailand (Nicholson *et al.*, 2019). Three fish displaying clinical signs and three healthy fish were collected from each site for disease screening. Different fish pathogens were identified from those samples including *Aeromonas* sp., *Streptococcus* sp. and TiLV. The results showed that, among those 52 outbreaks, 31% were TiLV coinfections with bacteria, mainly TiLV/*Aeromonas* sp. The other 15% of cases were attributed to a single TiLV infection and 14% were positive for bacterial infection without TiLV (Nicholson *et al.*, 2019).

During the summer in 2016, heavy mortalities were reported among Nile tilapia reared in earthen ponds in Behera Province, Egypt, with mortalities occurring for at least 1 week (Abdel-Latif & Khafaga, 2020). A total of 40 affected fish (moribund or fresh dead) were sampled for parasites and bacteria. *Gyrodactylus cichlidarum* was observed on the skin and gills of affected fish. *Aeromonas hydrophila* was isolated and identified from fish using biochemical tests and PCR. Most fish showed concurrent infection of *G. cichlidarum* and *A. hydrophila*. Investigation results showed that poor water quality associated with high temperature, together with natural coinfection of *G. cichlidarum* and *A. hydrophila*, induced high mortality of cultured tilapia during that summer (Abdel-Latif & Khafaga, 2020).

Concurrent infection by *A. hydrophila* and *E. tarda* was reported from red hybrid tilapia from four commercial fish farms located in Kelantan, Malaysia (Lee & Wendy, 2017). The diseased fish showed hypopigmentation, exophthalmia, ocular opacity and abdominal distension. The cumulative mortality of coinfected tilapia approached 10% (Lee & Wendy, 2017).

Concurrent infection in floating cages

In floating cage systems, fish are cultured at high density and exposed to an environment harboring a variety of different pathogens making fish susceptible to multiple infections (Dong *et al.*, 2015). From August 2015 to October 2016 a study was carried out at the Três Marias reservoir in the south-east of Brazil to determine the seasonal dynamics of bacterial pathogens (Delphino *et al.*, 2019). A total of 2233 fish were sampled from six farms rearing Nile tilapia in commercial floating cages. Several bacterial species were isolated, including *A. hydrophila*, *A. jandaei*, *A. veronii*, *A. caviae*, *E. tarda*, *F. orientalis* (Fo), *Streptococcus agalactiae*, and *S. iniae*. The main bacteria detected were *S. agalactiae*, infecting mostly adult tilapia throughout the culture period, and Fo, infecting mainly younger tilapia during the cooler months. A total of 33 cases of coinfection were detected, including coinfection by *S. iniae*/Fo, *S. agalactiae*/*E. tarda*, *S. agalactiae*/*A. hydrophila* and *S. iniae*-Fo/*A. hydrophila* in farm-raised Nile tilapia.

Disease outbreaks occurred in floating cages and resulted in high mortalities in two tilapia farms along the Mekong River in Thailand in 2014. To investigate the causes of these outbreaks, diseased Nile tilapia were collected from two farms to isolate and identify pathogens (Dong *et al*., 2015). Five bacterial species and one virus were concurrently found in the diseased fish. The most predominant bacteria were *Flavobacterium columnare* and *A. veronii*. Other pathogens included *S. agalactiae*, *Plesiomonas shigeloides*, *Vibrio cholerae* and *Iridovirus*. Among those pathogens, *F. columnare* and *A. veronii* co-occurred in 100% of the diseased fish from both farms, but most diseased fish were infected by three pathogens comprising the following combinations: *F. columnare*/*A. veronii*/*S. agalactiae*, *F. columnare*/*A. veronii*/*V. cholerae* and *F. columnare*/*A. veronii*/Iridovirus (Dong *et al*., 2015).

Concurrent infection with homologous pathogens

Bacterial concurrent infection

Natural concurrent infections were reported in farm-raised tilapia by *Streptococcus* sp., *F. orientalis* (Fo), *E. tarda* and *A. hydrophila* in Brazil (Delphino *et al*., 2019) and by *F. columnare*, *A. veronii*, *S. agalactiae* and *V. cholerae* in Thailand (Dong *et al*., 2015). In the autumn of 2015, Nile tilapia mortalities were observed from a commercial farm (Assis *et al*., 2017). Both *S. agalactiae* and Fo were isolated and identified from the coinfected fish. A laboratory challenge trial was conducted to evaluate the pathogenicity of *S. agalactiae* and Fo strains isolated from these coinfected fish. Four experimental groups were used in the challenge trial: group one was injected intraperitoneally (IP) with *S. agalactiae* (10^5 colony-forming units (CFU) per fish); group two with Fo (10^5 CFU per fish); group three with *S. agalactiae* plus Fo (10^5 CFU per fish); and group 4 with 0.1 ml of sterile PBS as control. Challenged fish were monitored over 15 days for mortality and the presence of bacteria. The results showed mortality of 88% in group 1 (*S. agalactiae* only), 38% in group 2 (Fo only), 100% in group 3 (*S. agalactiae* and Fo) and no mortalities in the control group, thus demonstrating the potential synergistic effects of bacterial concurrent infection.

A coinfection study of Fo and *S. agalactiae* was conducted in a laboratory trial using hybrid red tilapia (*O. niloticus* × *O. mossambicus*) to evaluate survival and immune response (Sirimanapong *et al.*, 2018). Fish were divided into four groups with the following treatments: group 1, no Fo/no *S. agalactiae* control; group 2, Fo infection only; group 3, *S. agalactiae* only; group 4, Fo/*S. agalactiae* coinfection. The experimental results showed 100, 100, 46.6 and 3.3% survival in groups 1–4, respectively. A synergistic effect of Fo and *S. agalactiae* coinfection was thus demonstrated. Fish coinfected with Fo and *S. agalactiae* died more rapidly and suffered higher mortality than those infected by Fo or *S. agalactiae* separately. Vertical transmission and concurrent infection of multiple bacterial pathogens (Fo and *Shewanella putrefaciens*) were also reported in naturally infected red tilapia (Pradeep *et al.*, 2017).

Parasitic concurrent infection

Species of the ciliated protozoan *Trichodina* and monogenoidean platyhelminths are common parasites of cultured fish (Martins *et al.*, 2011). In a study evaluating the effect of parasitism in Nile tilapia, Martins *et al.* (2011) found that Nile tilapia from a pond were naturally coinfected with *Trichodina heterodentata* and *Gyrodactylus cichlidarum* with prevalence of 100 and 70%, respectively.

To study the concurrent infection of parasites, *Tilapia guinensis* were collected from Opi lakes in Enugu State, Nigeria, monthly from November 2007 to October 2008 (Echi *et al.*, 2009). The fish were coinfected with three clinostomatid metacercariae, *Clinostomum tilapiae*, *C. complanatum* and *Euclinostomum heterostomum*. These digenetic metacercariae were encysted in tilapia tissues including the submucosa of the oral cavity, gill chamber, operculum, muscle, abdominal cavity, mesentery and viscera (Echi *et al.*, 2009). Infection by metacercariae affects the growth and survival of fish or causes disfiguration, so their market value as a food fish is diminished (Echi *et al.*, 2009). *Tilapia zilii* collected from the same lake from November 2007 to October 2008 were also found to be concurrently infected with metacercariae from the same three species reported above (Echi *et al.*, 2012).

Concurrent infection with heterologous pathogens

Parasitic and bacterial concurrent infection

Parasites and bacteria are common pathogens of cultured tilapia, and the concurrent infection of parasites and bacteria results in synergistic effects in tilapia (Xu *et al.*, 2007, 2009; Eissa *et al.*, 2010; Martins *et al.*, 2011). A study was conducted to evaluate whether Nile tilapia infected with *Gyrodactylus* were more susceptible to *S. iniae* (Xu *et al.*, 2007). Fish were divided into four groups and received the following treatments: (1) *G. niloticus*/*S. iniae* coinfection, (2) solo bacterial infection, (3) solo parasitic infection and (4) no infection. After fish were challenged by *S. iniae* immersion for 1 hour, mortality was monitored for 2 weeks. The study results demonstrated that tilapia with a solo bacterial infection or solo parasitic infection had very low mortality (7 and 0%, respectively) but concurrent infection resulted in a significantly higher mortality of 42% (Xu *et al.*, 2007).

In a study to evaluate effect of concurrent infection of parasite and *S. iniae* on fish (Martins *et al.*, 2011), Nile tilapia naturally coinfected with *T. heterodentata* and *G. cichlidarum* were divided into three groups. One group maintained the coinfection with both parasites; one was exposed to additional 40,000 *I. multifiliis* theronts per fish, so that fish were coinfected with three parasites; the third group was treated twice with potassium permanganate and once with formalin on three consecutive days. Following treatment, no parasites were recovered from the treated fish. Fish with different parasite infection(s) were then challenged by IP injection of *S. iniae* at 1.55×10^6 CFU per fish. Mortality was recorded for 18 days post challenge. The results demonstrated a synergistic effect of coinfection of parasites and *S. iniae* on tilapia. Fish free from parasitism showed higher survival (48%) than those coinfected with *Trichodina* and *Gyrodactylus* (38%). Fish parasitized by three parasites showed the lowest survival rate (28%) following challenge (Martins *et al.*, 2011).

Mass mortalities occurred in earthen ponds polycultured with Nile tilapia and African sharptooth catfish (*Clarias gariepinus*) located at an aquaculture station in Sharkiya, Egypt in the early summer of 2009. Mortalities among the two farmed species were close to 75% of the entire earthen pond population (Eissa *et al.*, 2010). Clinical examination revealed that >75% (38 of 50 fish) of the fish examined were concurrently infected with *Myxobolus tilapiae* and *F. columnare*. Many fish exhibited typical signs of hole-in-the-head-like lesions caused by the myxosporean spores. The study concluded that the two pathogens, together with poor water quality, led to mass fish mortality (Eissa *et al.*, 2010).

Effects of parasitic load on concurrent infection

A study was conducted to determine the effects of parasite load on Nile tilapia mortality following *S. iniae* coinfection (Xu *et al.*, 2009). Fish were distributed into three groups, with 25 fish per tank and four tanks per group. Fish in each group were exposed to one of three theront loads of *I. multifiliis* (0, 10,000 or 20,000 theronts per fish). Fish in two tanks of each group were challenged with *S. iniae*, with the other two tanks serving as non-*S. iniae* challenge controls. The results demonstrated that mortality increased in fish with a high parasitic load (20,000 theronts per fish) when compared with those having a low parasitic load (10,000 theronts per fish) following exposure to an equal number of *S. iniae* and subjected to the same conditions (Xu *et al.*, 2009). Mortality was 0, 0 and

4%, respectively, for fish exposed to parasite only at 0, 10,000 and 20,000 theronts per fish. Tilapia infected with only *S. iniae* showed a 20% cumulative mortality. Mortalities increased from 38% in fish exposed to *I. multifiliis* at 10,000 theronts per fish to 88% in fish at 20,000 theronts per fish following *S. iniae* coinfection. Exposure to high theront numbers significantly increased fish mortality following bacterial coinfection (Xu *et al.*, 2009).

Effects of parasite developmental size on concurrent infection

As well as parasite load, parasite developmental size may influence fish mortality. A trial was conducted to determine the effects of parasite size on fish mortality following *S. iniae* coinfection (Xu *et al.*, 2009). Nile tilapia split into six groups received one of the following treatments: (1) no pathogen infection, (2) infection by *S. iniae* only, (3) infection by *I. multifiliis* only, (4) challenged with *S. iniae* at hour 4 post *I. multifiliis* infection, (5) challenged with *S. iniae* on day 2 post *I. multifiliis* infection and (6) challenged with *S. iniae* on day 4 post *I. multifiliis* infection. The life stages of the parasite *I. multifiliis* include an infective theront, a parasitic trophont and a reproductive tomont. Following fish infection, trophonts expanded significantly in size, their volume in fish increasing significantly from 1.3×10^4 μm^3 at hour 4 to 2×10^7 μm^3 at day 4 post parasitism (Xu *et al.*, 2009). Mortality increased from 10.0 to 47.5% when fish were challenged with *S. iniae* from hour 4 to day 4 post parasitism. Fish parasitized with well-developed trophonts (day 4) suffered higher mortality than those infected by young trophonts (hour 4) following *S. iniae* challenge. Continuous rotation of trophonts damaged fish skin and gills (Xu *et al.*, 2000), and the increase in trophont volume resulted in greater tissue damage. Damaged skin, lesions and ulcers are putative routes for bacterial invasion and subsequent manifestation of disease (Xu *et al.*, 2007). The results from this study (Xu *et al.*, 2009) demonstrated that parasite size directly enhanced *S. iniae* invasion, and that coinfection by *I. multifiliis* and *S. iniae* greatly increased tilapia mortality.

Bacterial and viral concurrent infection

Concurrent infections of bacteria and viruses were reported for tilapia in Thailand (Dong *et al.*, 2015; Surachetpong *et al.*, 2017; Nicholson *et al.*, 2019) and in Egypt (Nicholson *et al.*, 2017). A case of natural coinfection of TilV and *A. veronii* was reported in a Malaysian red hybrid tilapia farm experiencing high mortality (Amal *et al.*, 2018). In hybrid tilapia reared in earthen ponds, 300–1800 dead fish were recorded daily for about 3 weeks. A total of 20 affected morbid tilapia were collected for investigation of outbreak pathogens, with *A. veronii* isolated from 50% and TiLV from 20% of sampled tilapia. All fish positive for TiLV were coinfected with *A. veronii* (Amal *et al.*, 2018). The interaction of *A. veronii* and TiLV appeared synergistic and increased the severity of the disease (Amal *et al.*, 2018).

In late 2018, high mortality in tilapia (>50% of production) was reported in intensive cage culture across Lake Volta in Ghana (Ramírez-Paredes *et al.*, 2019). Diseased fish showed darkening, erratic swimming and abdominal distension with associated ascites. ISKNV was detected in diseased fish cells with electron microscopy and confirmed by quantitative real-time polymerase chain reaction (qPCR). The results suggest that ISKNV was the cause of the disease on the investigated farms and likely played a key role in mortality. One common observation of coinfections with *S. agalactiae* and other tilapia bacterial pathogens (*Aeromonas* spp., *E. tarda, Plesiomonas shigelloides*) further suggests that these may interact to cause severe disease and high mortality (Ramírez-Paredes *et al.*, 2019).

Fungal and bacterial concurrent infection

Fungal infections in fish are generally considered as being secondary to other pathogens such as bacterial or parasitic infections. Cutuli *et al.* (2015) described a case of skin and subcutaneous mycosis in tilapia caused by *Fusarium oxysporum* coinfection with *A. hydrophila*. A group of 2000 tilapia were bought from a local fish farm and held in 400-l tanks in a recirculating system. One week later, 50%

of fish had died. Examination of moribund and dead fish revealed the presence of fungal lesions on the skin of the head, and bacteria were isolated from internal organs. The etiological agent was identified as *F. oxysporum* by morphological and molecular analyses. The bacterial isolate was identified and confirmed as *A. hydrophila* by culture, DNA sequencing and histopathological findings. *Fusarium oxysporum* likely acted as the primary invader, promoting bacterial coinfection by tissue damage and resulting in fish death (Cutuli *et al.*, 2015).

A similar study investigated concurrent infection of Nile tilapia with the bacterium *S. iniae* and fungus *Candida albicans*. Four groups of tilapia were experimentally infected: (1) with *S. iniae* only, (2) with *C. albicans* only, (3) with both *S. iniae* and *C. albicans* and (4) non-infected control (Oda *et al.*, 2016). Following infection, fish were observed for 4 weeks for clinical abnormalities, mortality and biochemical and histopathological changes. Fish in group 3 showed higher mortality (76%) than those in groups 1 (62%), 2 (57%) and 4 (0%). Fish infected with both pathogens showed more severe clinical signs and post-mortem change (Oda *et al.*, 2016).

In January 2010, mass mortalities among Nile tilapia and African sharptoothed catfish occurred at the Mariotteya Stream, Egypt. The magnitude of mass kills approached several thousands of dead and dying fish, with clinical signs including multiple skin ulcers and respiratory and osmoregulatory failure. The cause of mortality was identified as a concurrent infection by *Pseudomonas fluorescens* and *C. albicans*, exacerbated by poor water quality (Eissa *et al.*, 2013).

Multiple coinfections

Multiple coinfections are infections by two or more genetically different pathogens where each pathogen causes harm to the host. Multiple coinfections usually occur in high-density culture systems, such as ponds or cages. During 2015 and 2016, 32 outbreaks occurred at fish farms in Thailand with mortality rates ranging from 20 to 90% noted among tilapia farms (Surachetpong *et al.*, 2017). To investigate the suspected pathogens, samples of between ten and 30 moribund fish and between five and ten apparently healthy fish were collected from each outbreak location. Results showed that fish were infected by monogenean parasites (*Gyrodactylus* and *Dactylogyrus*), ciliate protozoa (*Trichodina*), bacteria (*Flavobacterium* sp., *Aeromonas* sp. and *Streptococcus* sp.), and fish collected from 22 of 32 locations were positive for TiLV (Surachetpong *et al.*, 2017). Except for concurrent infection by two pathogens (Table 12.2), multiple coinfections were noted in many outbreaks, including external parasites, bacteria and TiLV (Surachetpong *et al.*, 2017).

Impact of concurrent infection

Increased fish mortality

Many studies have found that concurrent infection greatly increased tilapia mortality, including concurrent infection by either homologous pathogens (Table 12.1) or heterologous pathogens (Table 12.2). Results from the majority of laboratory trials demonstrated synergistic interaction between two or multiple pathogens on fish mortality (Tables 12.1 and 12.2). Laboratory trial designs included several groups of fish subject to solo pathogen infection or concurrent infection by two or multiple pathogens. The trial designs make it possible to compare the effect of infection by a singular pathogen to two or multiple pathogens.

In a coinfection study of homologous pathogens, Sirimanapong *et al.* (2018) reported that hybrid tilapia coinfected with bacteria Fo and *S. agalactiae* died more rapidly and suffered much higher mortality (97%) than those infected by Fo only (0%) or *S. agalactiae* only (53%). High fish mortality has also been reported in tilapia concurrently infected by heterologous pathogens (Table 12.2).

Enhanced bacterial entry and increased bacterial load in fish

In concurrent infection, synergistic effects occur when the first pathogen enhances infection by the second pathogen, leading to

Table 12.1 Concurrent infections by different homologous pathogens in tilapia.

Fish species	Study	First pathogen	Second pathogen	Interaction	References
Bacterial co-infection					
Nile tilapia	Field	*F. columnare–A. veronii*	*S. agalactiae*	ND	Dong *et al.*, 2015
Nile tilapia	Field	*F. columnare–A. veronii*	*V. cholerae*	ND	
Red tilapia	Field	*A. hydrophila*	*E. tarda*	ND	Lee and Wendy, 2017
Nile tilapia	Laboratory	*F. orientalis*	*S. agalactiae*	Synergistic	Assis *et al.*, 2017
Red tilapia	Laboratory	*F. orientalis*	*Shewanella putrefaciens*	Synergistic	Pradeep *et al.*, 2017
Red tilapia	Laboratory	*F. orientalis*	*S. agalactiae*	Synergistic	Sirimanapond *et al.*, 2018
Nile tilapia	Field	*S. agalactiae*	*S. iniae*	ND	
Nile tilapia	Field	*S. iniae*	*A. hydrophila*	ND	
Nile tilapia	Field	*S. dysgalactiae* subsp. *equisimilis*	*F. davisii*	ND	Shoemaker *et al.* (2022)
Parasitic co-infection					
T. guinensis	Field	*Clinostomum tilapiae, C. complanatum, Euclinostomum heterostomum*			Echi *et al.*, 2009
Tilapia zilii	Field	*C. tilapiae, C. complanatum, E. heterostomum*			Echi *et al.*, 2012
Nile tilapia	Field	*Trichodina heterodentata*	*Gyrodactylus cichlidarum*	ND	Martins *et al.*, 2011
Nile tilapia	Laboratory	*G. cichlidarum*	*Cichlidogyrus sclerosus*	Synergistic	Zhi *et al.*, 2018

Table 12.2 Concurrent infections by different heterologous pathogens in tilapia.

Fish species	Study	First pathogen	Second pathogen	Interaction	References
Parasitic–bacterial co-infection					
Nile tilapia	Laboratory	*Gyrodactylus niloticus*	S. iniae	Synergistic	Xu *et al.*, 2007
Nile tilapia	Laboratory	*I. multifiliis*	*S. iniae*	Synergistic	Xu *et al.*, 2009
Nile tilapia	Field	*Myxobolus tilapiae*	*F. columnare*	Synergistic	Eissa *et al.*, 2010
Nile tilapia	Laboratory	*Trichodina–Gyrodactylus*	*S. iniae*	Synergistic	Martins *et al.*, 2011
Nile tilapia	Laboratory	*Trichodina–Gyrodactylus*–Ich	*S. iniae*	Synergistic	
Red tilapia	Laboratory	*I. multifiliis*	*F. columnare*	Synergistic	Xu *et al.*, 2014
Hybrid tilapia	Laboratory	*Trichodina* sp.	*F. columnare*	Synergistic	Xu *et al.*, 2015
Nile tilapia	Field	*G. cichlidarum*	*A. hydrophila*	Synergistic	Abdel-Latif & Khafaga, 2020
Nile tilapia	Field	*Centrocestus–M. tilapiae*	S. agalactiae	Synergistic	Eissa *et al.*, 2021, Eissa *et al.*, 2021
Parasitic–viral co-infection					
Tilapia sp.	Field	Ectoparasites[a]	TiLV	ND	Surachetpong *et al.*, 2017
Bacterial–viral co-infection					
Nile tilapia	Field	*F. columnare–A. veronii*	Iridovirus	ND	Dong *et al.*, 2015
Tilapia sp.	Field	TiLV	*Flavobacterium* sp.	ND	Surachetpong *et al.*, 2017
Tilapia sp.	Field	TiLV	*Aeromonas* sp.	ND	
Nile tilapia	Field	TiLV	*Aeromonas* sp.	Synergistic	Nicholson *et al.*, 2017
Red tilapia	Field	*A. veronii*	TiLV	Synergistic	Amal *et al.*, 2018

Continued

Table 12.2 (continued)

Fish species	Study	First pathogen	Second pathogen	Interaction	References
Nile tilapia	Field	TiLV	*Aeromonas* sp.	Synergistic	Nicholson *et al.*, 2019
Red tilapia	Field	TiLV	*Aeromonas* sp.	Synergistic	
Red tilapia	Field	TiLV	*Streptococcus* sp.	Synergistic	
Red tilapia	Laboratory	TiLV	*A. hydrophila*	Synergistic	
Mossambic	Field	TiLV	*A. veronii*	ND	Suresh *et al.*, 2023,
Nile tilapia	Field	Infectious spleen and kidney necrosis virus (ISKNV)	*Streptococcus* sp.	ND	Ramírez-Paredes *et al.*, 2019
Fungal–bacterial co-infection					
Nile tilapia	field	*Pseudomonas fluorescens*	*Candida albicans*	ND	Eissa *et al.*, 2013
Nile tilapia	Laboratory	*Fusarium oxysporum*	*A. hydrophila*	Synergistic	Cutuli *et al.*, 2015
Nile tilapia	Laboratory	*S. iniae*	*C. albicans*	Synergistic	Oda *et al.*, 2016
Multiple co-infection					
Tilapia sp.	Field	Ectoparasites	Bacteria & TiLV[a]	ND	Surachetpong *et al.*, 2017
Red tilapia	Field	Ectoparasites	Bacteria & TiLV[a]	ND	

[a]Ectoparasites *Gyrodactylus*, *Dactylogyrus and Trichodina* sp.; *bacteria Flavobacterium*, *Aeromonas* and *Streptococcus*; virus TiLV

an increase in the severity of the infection(s) and fish mortality. This synergistic effect is commonly noted in parasitic/bacterial or fungal/bacterial concurrent infection. Many external parasites cause mechanical damage to the skin or gills. For example, *Gyrodactylus* sp. attach to fish gills, fins and skin by a posterior attachment haptor with a large anchor and small marginal hooklets (Hoffman, 1985). *Trichodina* sp., commonly seen in cultured tilapia, have a sclerotized ring of interlocking denticles (Shoemaker *et al.*, 2006). The invasion and movement of these parasites in fish, from one site to another, caused mechanical damage to the epithelium. These lesions are potential portals of entry for bacterial pathogens, thus rendering parasitized fish more susceptible to bacterial invasion (Xu *et al.*, 2007).

A study was conducted to evaluate the susceptibility of hybrid tilapia to *F. columnare* following parasitism by *I. multifiliis*, including mortality and bacterial load (Xu *et al.*, 2014). Fish received one of the following treatments: (1) non-infected control, (2) infected by *I. multifiliis* alone, (3) infected by *F. columnare* alone or (4) coinfected by *I. multifiliis* and *F. columnare*. The bacterial load in fish tissues were quantified by qPCR. The results demonstrated that parasitized tilapia showed significantly higher mortality (60%) when exposed to *F. columnare* than non-parasitized fish (29%). The bacterial load of *F. columnare* in the kidney of fish infected by *I. multifiliis* was 8180 GE/mg (genome equivalents per mg of tissue), which was 17-fold higher than that of non-parasitized fish (471 GE/mg). *I. multifiliis* parasitism of tilapia enhanced *F. columnare* invasion and resulted in a higher bacterial load in tissue (Xu *et al.*, 2014).

Parasites serve as vectors for bacteria

Studies have been conducted to demonstrate that parasites can act as vectors in the transmission of bacteria to fish. Xu *et al.* (2012) used *I. multifiliis* and fluorescent *E. ictaluri* as a model to study the interactions among parasite, bacterium and fish. Following exposure of theronts to fluorescent *E. ictaluri* for 4 hours, these were then fixed in formalin, washed with sterile water, harvested by centrifugation and suspended in sterile water. The number of theronts carrying fluorescent *E. ictaluri* was counted using a flow cytometer. The results showed that >60% of theronts carried *E. ictaluri*. Xu *et al.* (2012) further evaluated whether tomonts could pass fluorescent bacteria to theronts and infect fish. They found that all tomonts (100%) carried the fluorescent bacterium following exposure to *E. ictaluri*. The bacterium survived and replicated during tomont division. Among theronts released from tomonts, 66% were observed with attached fluorescent *E. ictaluri*. When catfish were exposed to these theronts, 60% of fish were positive for *E. ictaluri* as determined by fluorescent microscopy. Fluorescent *E. ictaluri* were observed on skin and gill wet-mounts from fish. The trial demonstrated that tomonts exposed to *E. ictaluri* could pass *E. ictaluri* to infective theronts, with these then transmitting the bacterium to the fish. This study provided evidence that *I. multifiliis* can vector *E. ictaluri* to fish.

A study was conducted to determine whether the ectoparasite *Gyrodactylus niloticus* could carry the bacterial pathogen *S. iniae* as a mechanical vector in tilapia (Xu *et al.*, 2007). Tilapia infected by *G. niloticus* were exposed to *S. iniae* at 7.5×10^9 CFU/L in water for 1 hour, or by IP injection at 10^8 CFU per fish. *Gyrodactylus* were collected from live or moribund fish, treated with 300 IU penicillin/ml and 300 μg streptomycin/ml for 15 min and washed thoroughly with sterile water. Washed *G. niloticus* samples were then plated on blood agar to culture and identify *S. iniae*. *Streptococcus iniae* was isolated from 60 and 40% of *G. niloticus* collected from fish infected by IP injection and immersion, respectively, at 24 hours post challenge (Xu *et al.*, 2007). These results suggested that *G. niloticus* can serve as vector to carry the bacterial pathogen *S. iniae*.

Immunosuppression in coinfection

Zhi *et al.* (2018) conducted a study to evaluate the expression of immune-related genes and demonstrate immunosupression in Nile tilapia coinfected by the parasites *G. cichlidarum* and *Cichlidogyrus sclerosus*. *G. cichlidarum* is viviparous and mainly parasitizes the fish body surface while *C. sclerosus* is oviparous and inhabits the gills (Zhi *et al.*, 2018). Nile tilapia were divided into four groups and received one of the following treatments: (1) *G. cichlidarum* (GC) infection only, (2) *C. sclerosus* (CS) infection only, (3) GC and CS concurrent infection or (4) no infection. Expression patterns of immune-related genes were compared among fish free from parasites, those infected by a singular parasite and those with two-parasite coinfection. The results demonstrated that concurrent infection resulted in significantly lower expressions of transforming growth factor beta (TGF-β) in both skin and gills at day 8 post infection. Similar low-level expression was observed in interleukin-1β and tumor necrosis factor (TNF)-α in the gills. *Gyrodactulus cichlidarum* infection on the body surface significantly downregulated the expression of TNF-α, TGF-β and COX-2 in the gills (Zhi *et al.*, 2018). More work is needed in this area to understand the impact of multiple pathogens on the fish immune response.

Health management

Prevention is an important factor for fish health management in aquaculture. To prevent the introduction and spread of diseases onto a fish farm, fish from sources outside the farm need to be quarantined and/or examined for pathogens. Fish should be treated prior to stocking if parasites are detected (Xu *et al.*, 2015). Close monitoring of fish for disease signs during quarantine is another important step in fish health management. Monitoring can detect sick fish and identify the cause of disease at an early stage so that appropriate treatment may be initiated. When parasites are detected in ponds, treatment may be considered if these are susceptible and treatments economically feasible (Xu *et al.*, 2015). When evaluating losses caused by parasitic disease, indirect losses caused by bacterial pathogens should also be considered.

Anti-parasite treatment reduces bacterial load and improves fish survival

The mucus and epidermis are the first line of defense in fish and act as protective shields against invasive microorganisms. Parasite infections can disrupt the first line of defense in the skin and gills, thereby creating putative routes for bacterial invasion and increased susceptibility to bacterial pathogens (Lom, 1995). A study was conducted to evaluate whether treatment of hybrid tilapia parasitized by *Trichodina* sp. would improve fish survival and reduce *F. columnare* infection following exposure of fish to *F. columnare* (Xu *et al.*, 2015). Hybrid tilapia parasitized by *Trichodina* sp. were divided into three groups. The first group received no anti-parasite treatment, the second were bath-treated with 150 mg/l formalin for 1 hour while the third were treated twice with 150 mg/l formalin, for 1 hour each, at 2-day intervals. In this study, a single treatment of parasitized fish markedly reduced parasite prevalence and intensity. Almost no *Trichodina* sp. was detected in those parasitized fish treated twice. All fish were then exposed to *F. columnare* by immersion challenge. Fish not treated anti-parasitically showed significantly higher mortality (38%) than those treated once (17%) following exposure to *F. columnare*. Fish treated twice showed lower mortality (6%) than those treated only once (17%) (Xu *et al.*, 2015).

Xu *et al.* (2015) also sampled fish randomly from each group 1 and 3 days post *F. columnare* challenge. Gill, liver and kidney tissues were used to quantify *F. columnare* via qPCR. The non-treated fish showed a significantly higher load of *F. columnare* in gill, kidney and liver compared with those treated with formalin. The bacterial load of non-treated fish was 27,075 GE/mg, 12-fold higher than that of fish treated once (2250 GE/mg) and 39-fold higher than that of fish treated twice (699 GE/mg). Formalin treatment of parasitized fish reduced both bacterial invasion and load in fish tissues and ultimately improved fish survival (Xu *et al.*, 2015). This work suggests that prevention and treatment of parasites in fish are an important part of fish health management that will not only reduce the direct damage caused by the parasite but will also reduce fish mortality due to bacterial infection.

Impact of concurrent parasitism on vaccine efficacy

Martins *et al.* (2011) conducted a study to determine whether parasitism of fish affects vaccine efficacy. Three groups of Nile tilapia were vaccinated with a killed *S. iniae* bacterin by IP injection:

group 1, free from parasites; group 2, coinfected with two parasites (*T. heterodentata* and *G. cichlidarum*); and group 3, coinfected with three parasites (*T. heterodentata*, *G. cichlidarum* and *I. multifiliis*). Twenty-one days post vaccination, fish were challenged by *S. iniae* and survival of the three groups was compared. A reduction in vaccine performance in group 3 was noted, with a relative percentage survival of 72% compared with 95 and 92%, respectively, in groups 1 and 2 (Martins *et al.*, 2011). The authors suggested that the stress induced by multiple infections and/or the impact of Ich infection on overall immune response likely influenced vaccine efficacy.

Conclusion and future perspectives

This chapter provides a summary of studies, conducted between 2000 and 2023, on concurrent infection of tilapia by various pathogens. Concurrent infections naturally occur in different species of tilapia worldwide. Two or multiple homologous or heterologous pathogens can act synergistically to enhance disease development, increase fish mortality and result in heavy economic losses in farmed tilapia. Studies on concurrent infections by parasites and bacteria have demonstrated that parasitism serves as a portal of entry to enhance bacterial invasion, and that parasites may vector bacterial fish pathogens. More research is needed to better understand the impacts of concurrent infection on physiological and immune responses in tilapia. Greater knowledge in these areas may lead to improved health management strategies aimed at limiting the impact of coinfection on tilapia health.

References

Abdel-Latif, H.M.R. and Khafaga, A.F. (2020) Natural coinfection of cultured Nile tilapia *Oreochromis niloticus* with *Aeromonas hydrophila* and *Gyrodactylus cichlidarum* experiencing high mortality during summer. *Aquaculture Research* 00, 1–13. DOI: 10.1111/are.14538.

Abdel-Latif, H.M.R., Dawood, M.A.O., Menanteau-Ledouble, S. and El-Matbouli, M. (2020) The nature and consequences of coinfections in tilapia: A review. *Journal of Fish Diseases* 00, 1–14. DOI: 10.1111/jfd.13164.

Amal, M.N.A., Koh, C.B., Nurliyana, M., Suhaiba, M., Nor-Amalina, Z. *et al.* (2018) A case of natural coinfection of tilapia lake virus and *Aeromonas veronii* in A malaysian red hybrid tilapia (*Oreochromis niloticus* × *O. mossambicus*) farm experiencing high mortality. *Aquaculture* 485, 12–16. DOI: 10.1016/j.aquaculture.2017.11.019.

Assis, G.B., Tavares, G.C., Pereira, F.L., Figueiredo, H.C. and Leal, C.A. (2017) Natural coinfection by *Streptococcus agalactiae* and *Francisella noatunensis* subsp. *Orientalis* in farmed Nile tilapia (*Oreochromis niloticus* L). *Journal of Fish Diseases* 40(1), 51–63. DOI: 10.1111/jfd.12493.

Bradley, J.E. and Jackson, J.A. (2008) Measuring immune system variation to help understand host–pathogen community dynamics. *Parasitology* 135(7), 807–823. DOI: 10.1017/S0031182008000322.

Cox, F.E.G. (2001) Concomitant infections, parasites and immune responses. *Parasitology* S23–S38. DOI: 10.1017/s003118200001698x.

Cutuli, M.T., Gibello, A., Rodriguez-Bertos, A., Blanco, M.M., Villarroel, M. GiraldoA.. (2015) Skin and subcutaneous mycoses in tilapia (*Oreochromis niloticus*) caused by *Fusarium oxysporum* in coinfection with *Aeromonas hydrophila*. *Medical Mycology Case Reports* 9, 7–11. DOI: 10.1016/j.mmcr.2015.06.002.

Delphino, M.K.V.C., Leal, C.A.G., Gardner, I.A., Assis, G.B.N., Roriz, G.D. *et al.* (2019) Seasonal dynamics of bacterial pathogens of Nile tilapia farmed in a brazilian reservoir. *Aquaculture* 498, 100–108. DOI: 10.1016/j.aquaculture.2018.08.023.

Dong, H.T., Nguyen, V.V., Le, H.D., Sangsuriya, P., Jitrakorn, S. *et al.* (2015) Naturally concurrent infections of bacterial and viral pathogens in disease outbreaks in cultured Nile tilapia (*Oreochromis niloticus*) farms. *Aquaculture* 448, 427–435. DOI: 10.1016/j.aquaculture.2015.06.027.

Echi, P., Okafor, F. and Eyo, J. (2009) Coinfection and morphometrics of three clinostomatids (digenea: Clinostomatidae) in *Tilapia guinensis* bleeker, 1862 from opi lake. *Nigeria. Biological Research* 7(1), 432–436.

Echi, P., Eyo, J., Okafor, F., Onyishi, G. and Ivoke, N. (2012) First record of co–infection of three clinostomatid parasites in cichlids (osteichthyes: cichlidae) in a tropical freshwater lake. *Iranian Journal of Public Health* 41(7), 86–90.

Eissa, A.E., Zaki, M.M. and Aziz, A.A. (2010) *Flavobacterium columnare/Myxobolus tilapiae* concurrent infection in the earthen pond reared Nile tilapia (*Oreochromis niloticus*) during the early summer. *Interdisciplinary Bio Central* 2(2), 5. DOI: 10.4051/ibc.2010.2.2.0005.

Eissa, A.E., Tharwat, N.A. and Zaki, M.M. (2013) Field assessment of the mid winter mass kills of trophic fishes at mariotteya stream, egypt: Chemical and biological pollution synergistic model. *Chemosphere* 90(3), 1061–1068. DOI: 10.1016/j.chemosphere.2012.09.010.

Eissa, A.E., Attia, M.M., Elgendy, M.Y., Ismail, G.A., Sabry, N.M, *et al.* (2021) *Streptococcus, Centrocestus formosanus* and *Myxobolus tilapiae* concurrent infections in farmed Nile tilapia (*Oreochromis niloticus*). *Microbial Pathogenesis* 158, 105084. DOI: 10.1016/j.micpath.2021.105084.

El-Sharouny, H.M. and Badran, R.A.M. (1995) Experimental transmission and pathogenicity of some zoosporic fungi to tilapia fish. *Mycopathologia* 132(2), 95–103. DOI: 10.1007/BF01103781.

Fitzsimmons, K. (2006) Prospect and potential for global production. In: Lim, C. and Webster, C.D. (eds) *Tilapia Biology, Culture and Nutrition.* The Haworth Press, pp. 51–72.

Food and Agriculture Organization (2018) The state of world fisheries and aquaculture. meeting the sustainable development goals. Meeting the sustainable development goals. Food and Agriculture Organization. Available at: http://www.fao.org/3/i9540en/i9540en.pdf

Hamouda, A.H., Moustafa, E.M. and Zayed, M.M. (2019) Overview on the most prevailing bacterial diseases infecting *Oreochromis niloticus* at aswan fish hatchery egypt. *Advances in Animal and Veterinary Sciences* 7(11), 950–961.

Hoffman, G.L. (1985) Worm parasites (helminthes and leeches). In: Plumb, J.A. (ed.) *Principal Diseases of Farm Raised Catfish. Southern Cooperative Series* Bulletin 225. Auburn University. Auburn, pp. 33–48.

Indahsari, M., Kismiyati, Ulkhaq, M.F. (2019) Prevalence and intensity of ectoparasites of tilapia (*Oreochromis niloticus*) in ponds with low, medium and high stocking density. *IOP Conference Series: Earth and Environmental Science* 236(12108), 1–6. DOI: 10.1088/1755-1315/236/1/012108.

Jansen, M.D., Dong, H.T. and Mohan, C.V. (2018) Tilapia lake virus: A threat to the global tilapia industry? *Reviews in Aquaculture* 11(3), 725–739. DOI: 10.1111/raq.12254.

Kotob, M.H., Menanteau-Ledouble, S., Kumar, G., Abdelzaher, M. and El-Matbouli, M. (2016) The impact of coinfections on fish: A review. *Veterinary Research* 47(1), 98. DOI: 10.1186/s13567-016-0383-4.

Lee, S.W. and Wendy, W. (2017) Antibiotic and heavy metal resistance of *Aeromonas hydrophila* and *Edwardsiella tarda* isolated from red hybrid tilapia (*Oreochromis* spp.) coinfected with motile aeromonas septicemia and edwardsiellosis. *Veterinary World* 10(7), 803–807. DOI: 10.14202/vetworld.2017.803-807.

Lom, J. (1995) Protozoan and metazoan infections. In: Woo, P.T.K. (ed.) *Fish Diseases and Disorders*. CABI Publishing.

Machimbirike, V.I., Jansen, M.D., Senapin, S., Khunrae, P., Rattanarojpong, T. *et al.* (2019) Viral infections in tilapines: more than just tilapia lake virus. Review, *Aquaculture* 503, 508–518. Available at: DOI: https://doi.org/10.1016/j.aquaculture.2019.01.036.

Martins, M.L., Shoemaker, C.A., Xu, D.H. and Klesius, P.H. (2011) Effect of parasitism on vaccine efficacy against *Streptococcus iniae* in Nile tilapia. *Aquaculture* 314(1–4), 18–23. DOI: 10.1016/j.aquaculture.2011.01.022.

Nicholson, P., Fathi, M.A., Fischer, A., Mohan, C., Schieck, E, *et al.* (2017) Detection of tilapia lake virus in Egyptian fish farms experiencing high mortalities in 2015. *Journal of Fish Diseases* 40(12), 1925–1928. DOI: 10.1111/jfd.12650.

Nicholson, P., Mon-on, N., Jaemwimol, P., Tattiyapong, P. and Surachetpong, W. (2019) Coinfection of tilapia lake virus and Aeromonas hydrophila synergistically increased mortality and worsened the disease severity in tilapia (*Oreochromis* spp.). *Aquaculture (in Press).*

Oda, S.S., Tohamy, H.G. and Massoud, R.G. (2016) Pathological alterations in *Nile tilapia* experimentally infected with Streptococcus iniae and Candida albicans. *Turkish Journal of Fisheries and Aquatic Sciences* 16(4), 779–788.

Plumb, J.A. (1997) Infectious diseases of tilapia. In: *In Tilapia Aquaculture in the Americas*, Vol. 1. World Aquaculture Society, pp. 212–228.

Pradeep, P.J., Suebsing, R., Sirithammajak, S., Kampeera, J., Turner, W. *et al.* (2017) Vertical transmission and concurrent infection of multiple bacterial pathogens in naturally infected red tilapia (*Oreochromis* spp). *Aquaculture Research* 48(6), 2706–2717. DOI: 10.1111/are.13102.

Raman, R.P., Prakash, C., Makesh, M. and Pawar, N.A. (2013) Environmental stress mediated diseases of fish: An overview. *Advances in Fish Research* 5, 141–158.

Ramírez-Paredes, J.G., Paley, R.K., Hunt, W., Feist, S.W., Stone, D.M. *et al.* (2019) First detection of Infectious Spleen and Kidney Necrosis Virus (ISKNV) associated with massive mortalities in farmed tilapia in Africa. *bioRxiv* 680538. DOI: 10.1101/680538.

Ramirez-Paredes, J.G., Larsson, P., Thompson, K.D., Penman, D.J. and Busse, H.J, *et al.* (2020) Reclassification of *Francisella noatunensis* subsp.

orientalis Ottem et al. 2009 as *Francisella orientalis* sp. nov., *Francisella noatunensis* subsp. *chilensis* subsp. nov. and emended description of *Francisella noatunensis. International Journal of Systematic and Evolutionary Microbiology* 70(3), 2034–2048. DOI: 10.1099/ijsem.0.004009.

Shoemaker, C.A., Xu, D.H., Joyce, J.E. and Klesius, P.H. (2006) Parasites and diseases. In: Lim, C. and Webster, C.D. (eds) *Tilapias: Biology, Culture and Nutrition*. The Haworth Press, Inc, pp. 561–582.

Shoemaker, C.A., Heckman, T.I., Soto, E., LaFrentz, B.R., Beck, B.H. *et al.* (2022) Coinfection by group C streptococcus dysgalactiae subsp. equisimilis and Flavobacterium davisii in Nile tilapia from the united states. *Bulletin of the European Association of Fish Pathologists* 42(2), 22–33. DOI: 10.48045/001c.38695.

Sirimanapong, W., Thompson, K.D., Shinn, A.P., Adams, A. and Withyachumnarnkul, B. (2018) *Streptococcus agalactiae* infection kills red tilapia with chronic *Francisella noatunensis* infection more rapidly than the fish without the infection. *Fish and Shellfish Immunology* 81, 221–232. DOI: 10.1016/j.fsi.2018.07.022.

Surachetpong, W., Janetanakit, T., Nonthabenjawan, N., Tattiyapong, P., Sirikanchana, K. *et al.* (2017) Outbreaks of tilapia lake virus infection, thailand, 2015–2016. *Emerging Infectious Diseases* 23(6), 1031–1033. DOI: 10.3201/eid2306.161278.

Suresh, T., Nithin, M.S., Kushala, K.B., Girisha, S.K., Shivakumar, V.B. *et al.* (2023) Largescale mortality of *Oreochromis mossambicus* in lakes and reservoirs of Karnataka due to coinfection of Tilapia Lake Virus (TiLV) and multidrug-resistant *Aeromonas veronii*: An emerging fish disease in India. *Aquaculture* 565. DOI: 10.1016/j.aquaculture.2022.739077.

Xu, D.H., Klesius, P.H., Shoemaker, C.A. and Evans, J.J. (2000) The early development of *Ichthyophthirius multifiliis* in channel catfish in vitro. *Journal of Aquatic Animal Health* 12(4), 290–296. DOI: 10.1577/1548-8667(2000)0122.0.CO;2.

Xu, D.H., Shoemaker, C.A. and Klesius, P.H. (2007) Evaluation of the link between gyrodactylosis and streptococcosis of Nile tilapia, *Oreochromis niloticus. Journal of Fish Diseases* 30(4), 233–238. DOI: 10.1111/j.1365-2761.2007.00806.x.

Xu, D.H., Shoemaker, C.A. and Klesius, P.H. (2009) Enhanced mortality in Nile tilapia, *Oreochromis niloticus* (L.) following coinfections with *Ichthyophthiriasis* and *Streptococcosis. Diseases of Aquatic Organisms* 85(3), 187–192. DOI: 10.3354/dao02073.

Xu, D.H., Shoemaker, C.A. and Klesius, P.H. (2012) Ichthyophthirius multifiliis as a potential vector of *Edwardsiella ictaluri* in channel catfish. *FEMS Microbiology Letters* 329(2), 160–167. DOI: 10.1111/j.1574-6968.2012.02518.x.

Xu, D.H., Shoemaker, C.A. and LaFrentz, B.R. (2014) Enhanced susceptibility of hybrid tilapia to *Flavobacterium columnare* after parasitism by *Ichthyophthirius multifiliis*. *Aquaculture* 430, 44–49. DOI: 10.1016/j.aquaculture.2014.03.041.

Xu, D.H., Shoemaker, C.A. and Zhang, D. (2015) Treatment of Trichodina sp. reduced load of F*lavobacterium columnare* and improved survival of hybrid tilapia. *Aquaculture Reports* 2, 126–131. DOI: 10.1016/j.aqrep.2015.09.007.

Zhi, T., Xu, X., Chen, J., Zheng, Y., Zhang, S. *et al.* (2018) Expression of immune-related genes of Nile tilapia *Oreochromis niloticus* after *Gyrodactylus cichlidarum* and *Cichlidogyrus sclerosus* infections demonstrating immunosupression in coinfection. *Fish and Shellfish Immunology* 80, 397–404. DOI: 10.1016/j.fsi.2018.05.060.

Chapter 13

Tilapia-borne zoonoses

Ruth Zadoks, Joke van der Giessen and Olga Haenen

Introduction

Zoonoses can be described as diseases caused by infectious agents that are transmissible between animals and humans (bidirectional), or transmissible from animals to humans (unidirectional), with terms such as reverse zoonosis or anthroponosis used to describe transmission in the opposite direction, i.e. from humans to animals. In this chapter we will cover organisms communicable between tilapia and humans, regardless of the direction of transmission, for organisms that are known to cause disease in people. The direction of transmission may not be known, or there may be spillover and spillback from one host to another, so that 'who gave it to whom' changes over time. Many zoonotic organisms cause disease in some host species or individuals, but not in others. Indeed, absence of recognizable signs or symptoms of disease is a major contributor to transmission of zoonotic organisms, especially foodborne ones, because apparently healthy people or fish harbouring pathogens are difficult to identify and segregate from food supply or food preparation activities. Some communicable entities, such as antimicrobial resistance genes, are not even organisms, but segments of genetic code that may be present as part of a communicable organism or in the environment. Physical or chemical contaminants that may be shared between tilapia and people, e.g. microplastics or anthropogenic chemicals, will not be considered in this chapter.

In addition to variable interpretations of the concept of zoonoses, there are variable levels of evidence to support public health risks

Kim Thompson, Craig Shoemaker and David C. Little (eds) **Tilapia**
DOI: 10.1079/9781800629455.0013, © CAB International, 2025

associated with communicable organisms from tilapia. The mere presence of a macro- or microparasite that can cause disease does not in and of itself constitute evidence of a public health risk but merely a hazard, i.e. a potential risk. Authors of scientific publications are prone to suggest public health risks, partly to prevent hazards translating into actual disease, but possibly also to enhance the visibility or impact of their work. In this chapter we consider pathogens from tilapia that are proven to cause disease in humans and those that currently have an unproven potential to do so, at least to our knowledge, which is based on publications from many researchers, years and countries, but in a limited number of languages. This raises another important point, notably the mismatch between the geographic distribution of tilapia and the geographic distribution of wealth and scientific endeavour. In other words, tilapia pathogens and tilapia zoonoses, like those of other aquatic species, are under-represented in the English-language scientific literature, which is dominated by zoonoses from mammalian and avian species, despite the fact that tens of millions of people in Asia are infected with foodborne trematodes of aquatic origin (including from tilapia). In this chapter we use the name 'tilapia' primarily to refer to Nile tilapia, *Oreochromis niloticus* (Linnaeus, 1758) or hybrid tilapia, *Oreochromis* spp., unless stated otherwise. We hope that this chapter will help to raise awareness of the distribution and potential impact of zoonoses from tilapia and stimulate research and control programmes that benefit the health of tilapia, humans and their shared environment.

Parasitic zoonoses

Trematodes or flatworms are parasites with a complex life cycle involving one or two intermediate hosts and diverse definitive hosts, which may include fish and people. Foodborne zoonotic trematodes (FZT) can cause infection in humans via the consumption of contaminated food, including raw fish, resulting in severe, often chronic, liver, lung or intestinal disease. Major aquatic FZT include the liver flukes *Fasciola*, *Chlonorchis sinensis* and *Opisthorchis viverrini* and the lung fluke *Paragonimus*, which are estimated to affect tens of millions of people worldwide (FAO/WHO, 2014). In tilapia, however, only intestinal flukes are reported. Fishborne intestinal flukes, with heterophyids and echinostomes as the most important groups, have a multi-host life cycle, including snails as first intermediate host, fish (including tilapia) as second intermediate host and fish-eating birds or mammals (including humans) as definitive host. Tilapia or other fish contain the larval encysted metacercariae whereas definitive hosts harbour the adult flukes that may cause clinical signs and symptoms (Chi *et al.*, 2008; Lobna *et al.*, 2010).

Intestinal FZT species reported in tilapia include the major heterophyids *Stellantchasmus falcatus*, *Haplorchis pumilio* and *Centrocestus formosans* and the minor heterophyids *Procerovum varium* in Asia (e.g. China, Thailand, Vietnam) (Nguyen *et al.*, 2007; Li *et al.*, 2013; Wiriya *et al.*, 2013; Chai & Jung, 2020) and *C. formosans* in South America (Brazil, Costa Rica) (Arguedas Cortés *et al.*, 2010; Pinto *et al.*, 2014). Non-zoonotic parasitic trematodes of tilapia are not considered here. The prevalence of metacercariae of FZT in tilapia aquaculture is generally low, with reports ranging from 0 to ca. 10% depending on study, country and production system. The comparatively low prevalence of FZT in tilapia, which is accompanied by low infection intensity, may be related to their anatomy and physiology, or their behaviour, notably the depth of the water column they occupy, which may not be ideal habitat for intermediate host snails (Nguyen *et al.*, 2007; Chi *et al.*, 2008). Considering the size of tilapia stocks in aquaculture systems, low prevalence still represents a massive number of positive fish. To date there is no English-language literature on human infections with FZT from tilapia. This could mean that such infections are uncommon, of limited clinical significance, difficult to diagnose (due to infrastructural, economic or biological reasons), not routinely reported (e.g. non-notifiable) and/or poorly covered in Western scientific literature even if they are diagnosed. Indeed, symptoms of heterophyid infection in humans are generally mild, although abdominal pain and diarrhoea may occur, and heterophyid eggs may cause eosinophilic granulomas in important organs such as the brain, heart and spinal cord (Elsheikha & Elshazly, 2008; FAO/WHO, 2014).

The reported prevalence of FZT metacercariae varies by production system. For example, 0–3% prevalence was reported for tilapia raised in urban wastewater ponds and lakes, and 10% for fish from rural garden–cage–pond systems in Vietnam (Nguyen *et al.*, 2007; Van De *et al.*, 2012). In Thailand the prevalence was ca. 2% for caged tilapia and up to 10% in ponds (Wiriya *et al.*, 2013). Differences may also exist between production stages, with 2.4 and 8.9% prevalence reported for nursery and grow-out ponds, respectively, in Vietnam (Chi *et al.*, 2008). Conversely, the prevalence was higher in nursery ponds (1.1–6.2%) than in grow-out ponds (0–0.5%) in China, for both monoculture and polyculture or integrated systems (Li *et al.*, 2013). Wild-caught tilapia may harbour FZT, but data are very scarce; zero out of six and five out of ten wild-caught tilapia from China and Thailand, respectively, were reported as FZT positive (Van De *et al.*, 2012; Li *et al.*, 2013). Thus, inferences on the risk posed by wild-caught fish to farmed tilapia, through either direct introduction or the consumption of wild-caught fish by intermediate hosts such as dogs and cats, are largely speculative. All in all, the importance of FZT in tilapia in Asia or South America is minor compared with that in other fish species or compared with the importance of bacterial zoonoses.

The situation is different in Egypt where fish-borne FZT, notably heterophyiasis, is endemic in the Nile Delta and where encysted metacercariae have been detected in both Nile tilapia and red-bellied tilapia, or St Peter's fish (*Coptodon zillii,* formerly known as *Tilapia zillii* (Gervais, 1848)). Unhygienic conditions, such as disposal of human waste directly into surface waters where fish are produced, combined with a culture of consuming salted but unheated fish, contributes to perpetuation of the life cycle (Lobna *et al.*, 2010). The major heterophyids *Heterophyes heterophyes*, *Haplorchis yokogawai* and *Pygidiopsis genata* were detected in both tilapia species from brackish water. Only the latter two parasites were detected in freshwater tilapia (both fish species), whereby the difference may be mediated through the habitat of intermediate host snails (Elsheikha and Elshazly, 2008; Lobna *et al.*, 2010). The prevalence of metacercariae varies with season, fish size and water type, and ranges from 15.4 to 46.4% in wild-caught fish (Lobna *et al.*, 2010). Thus, tilapia production in aquaculture systems could potentially reduce the risk of fish-borne FZT and its associated public health burden, whilst also to a large extent contributing to meeting a growing consumer demand.

Bacterial zoonoses

Streptococcus iniae

Streptococcus iniae, a gram-positive coccus, is one of the main aetiological agents of warm-water streptococcosis in fish, including tilapia. It was first reported in dolphins in the 1970s, and subsequently emerged as an important cause of disease in both freshwater and marine aquaculture species in the 1980s (Agnew & Barnes, 2007). Reports of human cases followed in the 1990s, initially from North America rather than from those countries where *S. iniae* affected aquaculture. The first known case of human *S. iniae* infection occurred in Texas in 1991, with a second case in Ottawa, Canada, in 1994, but it wasn't until a cluster of cases occurred in Greater Toronto, Canada, in 1995–1996 that the connection with aquaculture could be demonstrated (Centers for Disease Control and Prevention (CDC), 1996). Over the course of a year, nine cases of invasive *S. iniae* infection in people were reported, mostly manifesting as cellulitis of the hand. Almost all cases occurred in people who had handled live or freshly killed raw fish, mostly tilapia, and who had sustained percutaneous injuries whilst doing so (Weinstein *et al.*, 1997). Thus, human *S. iniae* infection is primarily a contact zoonosis rather than a foodborne zoonosis.

Since then, *S. iniae* has been reported in tilapia from Israel, Japan, Taiwan and the USA (Agnew & Barnes, 2007), and in people from Vancouver, Canada (Goh *et al.*, 1998), Hong Kong (Lau *et al.*, 2006), Singapore (Koh *et al.*, 2004) and the USA (Facklam *et al.*, 2005), possibly reflecting the quality of the healthcare and diagnostics systems as much as the distribution of the disease (Lau *et al.*, 2003). Atypical manifestation of the bacteria may also hamper their recognition (Sun *et al.*, 2007). In addition to bacteraemic cellulitis, clinical manifestations in sporadic human cases of *S. iniae* infection include severe and potentially life-threatening diseases such as endocarditis (Weinstein *et al.*, 1997), osteomyelitis (Lau *et al.*, 2003; Koh *et al.*, 2004) and septic arthritis (Lau *et al.*, 2006). Risk factors for human *S. iniae* infection include Asian (predominantly Chinese) ethnic origin, old age, pre-existing conditions and preparation of fresh fish, again implying contact exposure or topical acquisition rather than foodborne transmission (Weinstein *et al.*, 1997; Lau *et al.*, 2003; Facklam *et al.*, 2005; Sun *et al.*, 2007; Koh *et al.*, 2009). Work on fish farms has not been linked to *S. iniae* infection in people (Koh *et al.*, 2004). Apart from the outbreak in Toronto and one case from Singapore, *S. iniae* infections have not been specifically linked to tilapia, which probably reflects the lack of epidemiological data as well as the wide fish host range of the organism (Agnew & Barnes, 2007).

Streptococcus agalactiae

Of such concern is the emergence of *S. agalactiae* or group B *Streptococcus* (GBS) as a foodborne pathogen in Asia that the Food and Agriculture Organization of the United Nations devoted a Risk Profile to it in 2021 (FAO, 2021), even though recent reviews of zoonotic infections of aquatic origin barely mentioned the organism (Haenen *et al.*, 2013; Gauthier, 2015). There are numerous similarities and differences between *S. agalactiae* and *S. iniae*. Like *S. iniae*, *S. agalactiae* is a gram-positive coccus, a cause of summer streptococcosis in tilapia and other fishes and a pathogen of humans (Barkham *et al.*, 2019). As for *S. iniae*, emergence of *S. agalactiae* in aquaculture is temporally and possibly causally linked to the intensification of aquaculture (Barkham *et al.*, 2019), and transmission of the pathogen from fish to people is associated with food preferences, with comparatively high incidence reported among ethnic Chinese communities but not among fish farmers or fishmongers in Singapore (Tan *et al.*, 2016; Kalimuddin *et al.*, 2017). Both pathogens cause severe invasive disease, with manifestations of *S. agalactiae* disease including bacteraemia, endocarditis, meningitis and bone and joint infections (Tan *et al*., 2016; Kalimuddin *et al*., 2017).

There are numerous differences too. Unlike *S. iniae*, *S. agalactiae* is commonly carried by humans and is well known as a cause of disease and mortality in pregnant women, babies and infants (Seale *et al*., 2017). Of the many types of *S. agalactiae* known to occur in humans or other mammalian hosts, only three clades are associated with fishes: clonal complex (CC) 7, CC283 and CC552, with tilapia constituting a major host species for all three clades (Delannoy *et al*., 2013) (Fig. 13.1). Of those clades, CC552 is non-pathogenic to humans due to its reduced genome size and its preference for growth at low temperature, which is compatible with poikilothermic fishes and frogs but not with warm-blooded humans (Delannoy *et al*., 2013; Rosinski-Chupin *et al*., 2013). CC7 has been reported in humans, notably neonates in Japan, and in fishes, but its zoonotic potential is hypothetical because there is no evidence of zoonotic transmission (Evans *et al*., 2008). Only one clade, CC283, has been proven to be

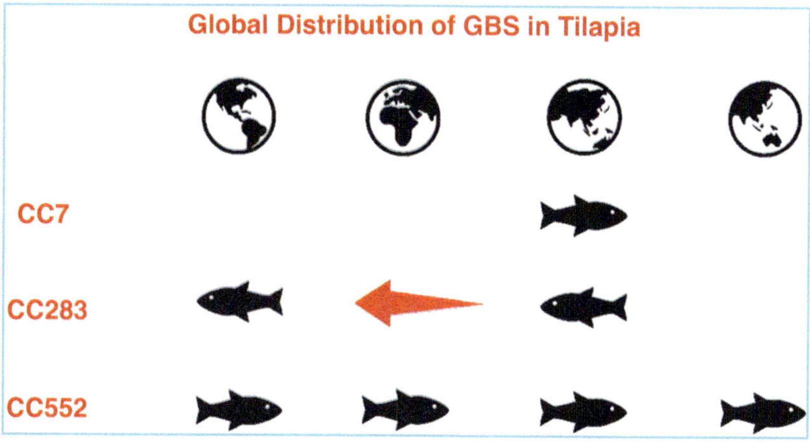

Figure 13.1 Globally, three clades or clonal complexes (CC) of *S. agalactiae*, also known as Group B *Streptococcus* (GBS), are found in tilapia. Occurrence of CC7 in fishes is largely limited to Asia, CC552 is the only clade to occur across all continents with tilapia aquaculture and CC283 is the only clade with proven zoonotic transmission via consumption of raw fish. Although CC283 is primarily reported from Asia, where it has occurred for several decades, its introduction into South America was recently reported.

zoonotic and, in contrast to *S. iniae*, its transmission is foodborne via raw fish consumption, e.g. yu-sheng, a Chinese-style raw fish dish popular in Singapore, rather than based on contact (Rajendram *et al.*, 2016; Tan *et al.*, 2016; Kalimuddin *et al.*, 2017). Tilapia is reportedly rarely used in this dish, but there is some uncertainty around the accuracy of the identification of fish species used for food preparation (Barkham *et al.*, 2019).

It is conceivable that ST283 originated in people, spilled over into tilapia via human effluent, adapted through acquisition of fish-associated mobile genetic elements (Delannoy *et al.*, 2016), amplified in aquaculture systems and then spilled back into humans via raw fish consumption. Surprisingly, infections with CC283, mostly represented by its namesake sequence type (ST) 283, predominantly occur in people without underlying comorbidities, i.e. they pose a threat to otherwise healthy people (Kalimuddin *et al.*, 2017; Barkham *et al.*, 2019). Where only tens of cases of human *S. iniae* infection have been reported (Haenen *et al.*, 2013), hundreds of cases of *S. agalactiae* ST283 infection have been reported, all suspected to be fish-borne (Barkham *et al.*, 2019). For *S. agalactiae*, as for *S. iniae*, this is likely to be an underestimate, affected, among other things, by both non-notifiable status and limitations to diagnostic and health care infrastructure in many parts of Asia (Barkham *et al.*, 2019).

Hypervirulent pathogenic *S. agalactiae* ST283 has been reported from people in Hong Kong, Lao PDR, Singapore, Thailand and Vietnam, and from tilapia in Malaysia, Thailand, Vietnam and Singapore (Ip *et al.*, 2016; Barkham *et al.*, 2019; FAO, 2021). Although the link with raw fish consumption has been established based on epidemiological and genomic data and the impact of interventions in Singapore (Chau *et al.*, 2017; Kalimuddin *et al.*, 2017; Barkham *et al.*, 2019), the link with tilapia is more tenuous. Black tilapia is listed among the fish species testing positive for *S. agalactiae* at retail, but so were many other species, including those that are not known to be naturally infected with the pathogen (Barkham *et al.*, 2019). Outbreaks of *S. agalactiae* in tilapia aquaculture can have major impact in terms of morbidity, mortality and economic loss but the organism has also been detected in apparently healthy tilapia, which could result in exposure of consumers when eating the fish raw (Li *et al.*, 2014; Barkham *et al.*, 2019; FAO, 2021). In recent years hypervirulent ST283 has spread to tilapia aquaculture in Brazil, probably via international movement of fish stocks or fry, a mechanism also described for non-zoonotic CC552 (Kawasaki *et al.*, 2018; Leal *et al.*, 2019). At the time of writing, no human cases of foodborne ST283 infection had been reported from Brazil. Considering the appetite for tilapia ceviche in South America, and for tilapia sashimi in other parts of the world (see section below on enteric bacteria), it will be important to monitor how the emergence and expansion of zoonotic *S. agalactiae* ST283 in tilapia aquaculture and humans unfolds.

Vibrio spp.

From a public health perspective, the most important *Vibrio* species is the cholera toxin-producing *V. cholerae* serogroup O1, which causes the life-threatening diarrhoeal disease cholera (Son *et al.*, 2011). *Vibrio cholerae*, a gram-negative bacillus, is associated with contaminated aquatic environments and has been found in seawater and freshwater ponds used for aquaculture, including freshwater tilapia production ponds (Tey *et al.*, 2015). It was detected in the intestines of tilapia after experimental exposure, and in unspecified tilapia samples (Tey *et al.*, 2015; Hounmanou *et al.*, 2019). Tilapia may be a putative reservoir host for survival and transmission of *V. cholerae* O1 Biotype El Tor, given that experimental exposure to high doses resulted in tilapia gut colonization and fish-to-fish transmission. Whether these processes occur under natural conditions is yet to be determined (Hounmanou *et al.*, 2019). Foodborne transmission of *V. cholerae* from tilapia has been documented in hand-reared water birds, in particular great cormorants (*Phalacrocorax carbo*) (Laviad-Shitrit *et al.*, 2019), and bird migration may contribute to the geographical spread of cholera. There is no evidence for *V. cholerae* as cause of disease in tilapia and it is not a proven fish-borne zoonosis (Gauthier, 2015). Similarly, *V. parahaemolyticus* is a major foodborne pathogen but with scant support for finfish as source, although *V. parahaemolyticus* carrying *trh*, an indicator of human pathogenicity, has been detected in the gut, gills and flesh of tilapia from Bangladesh (Siddique *et al.*, 2021). In other studies, *V. parahaemolyticus* was found in saltwater but not in freshwater ponds used for tilapia aquaculture (Tey *et al.*, 2015).

The most important *Vibrio* species, from the perspective of tilapia health and in the context of zoonotic disease, is *V. vulnificus*, a halophilic (salt-loving), gram-negative bacillus. Within the species *V. vulnificus* biotypes 1, 2 and 3 were recognized, with further subdivision into serovars, e.g. serovars A and E within biotype 2, all of which have been reported as cause of disease in tilapia, either naturally or experimentally (Fouz *et al.*, 2010; Gauthier, 2015; Cho *et al.*, 2018; Choi *et al.*, 2021). Biotypes do not align with the lineage nomenclature that was subsequently proposed based on core genome analysis of *V. vulnificus* (Roig *et al.*, 2017). For example, biotype 2 is polyphyletic, probably due to acquisition by multiple lineages of a transferable virulence plasmid that encodes resistance to innate immunity of eels and other teleosts (Amaro *et al.*, 2015). Biotype 2 serovar E is a zoonotic clade with a worldwide distribution that is linked to aquaculture (Amaro *et al.*, 2015; Roig *et al.*, 2017). In Israel biotype 3 has been reported repeatedly from tilapia, and as a zoonotic agent in people with vibriosis resulting from exposure during tilapia farming or handling. The name *V. vulnificus* pathovar *piscis* has been proposed for the pathovar grouping all fish-virulent isolates, regardless of biotype (Roig *et al.*, 2017).

In Bangladesh *V. vulnificus* was found in the water and in gills, muscle and intestines of tilapia from coastal (saltwater) aquaculture but not in freshwater tilapia or ponds (Mahmud *et al.*, 2010). Humans can acquire *V. vulnificus* through ingestion, which may lead to septicaemia and mortality as high as 50% (Choi *et al.*, 2021). Alternatively, infection can be acquired through exposure of open wounds by contaminated seawater or sea food, or following skin puncture by spines from infected fish (Mahmud *et al.*, 2010; Haenen *et al.*, 2013; Gauthier, 2015). This may lead to progressive tissue and life-threatening necrosis, so-called necrotizing fasciitis, occasionally requiring amputation, or to secondary bacterial sepsis, with mortality up to 25% (Said *et al.*, 1998; Chan *et al.*,

1999; Choi *et al.*, 2021). Infection due to *V. vulnificus* biotype 3, sometimes called the Israeli biotype, caused by direct injury while purchasing, cleaning or handling live fish, was first described in Israel in 1996–1997. Follow-up studies from 1998 to 2005 identified 132 additional cases, almost all (95%) associated with percutaneous fish exposure, mostly (83%) to tilapia (Zaidenstein *et al.*, 2008). Recently the zoonotic strain was isolated from diseased farmed tilapia from India (Sumithra *et al.*, 2019) and the eastern Mediterranean (Carmona-Salido *et al.*, 2021a, b), which is a basis for further concern.

Although most *V. vulnificus* infections affect the hands, the feet and legs may also be involved following penetrating skin wounds caused by St Peter's fish, with the youngest known patient a 14-year-old boy who stepped on a St Peter's fish and the oldest a 81-year-old woman who had been carrying a shopping bag from which the fish's spine protruded (Calif *et al.*, 2003). Fatal *V. vulnificus* infection has been reported in Hong Kong following a prick from the dorsal fin of a tilapia (Chan *et al.*, 1999) and in Canada after handling and ingestion of tilapia (Vinh *et al.*, 2006). Both cases were characterized by underlying health issues and presented with bullous cutaneous lesions and sepsis. A third patient, also from Canada but without underlying health issues, made a full recovery following *V. vulnificus* infection acquired when descaling a tilapia that she had bought live at a local market (Majere and Cortina, 2017). Factors that may have contributed to the emergence of *V. vulnificus* include (1) the intensification of aquaculture; (2) increased recognition of infections; (3) changes in marketing strategies, whereby the sale of live tilapia not kept on ice was seen as a driver of increased case numbers in Israel (Bisharat, 2002); and (4) rising levels of *Vibrio* with increase in water temperatures and low-to-moderate salinity (Calif *et al.*, 2003). It should be borne in mind, however, that the vast majority of *V. vulnificus* infections are derived from shellfish and eel (Haenen *et al.*, 2014) and that the comparatively high contribution of tilapia as a source of infection in Israel may be associated with religiously motivated consumption practices (Bisharat, 2002; Calif *et al.*, 2003).

Potential zoonoses

Aeromonas spp.

Aeromonas spp. are emerging human pathogens that can cause gastrointestinal and extra-intestinal infections, including skin and soft tissue infections and systemic infections (Haenen *et al.*, 2013; Chenia & Duma, 2017). The major human pathogenic *Aeromonas* sp., *A. hydrophila*, is also the main species causing atypical motile aeromonad septicaemia (MAS) in tilapia (Shirajum Monir *et al.*, 2020). Other motile *Aeromonas* spp. that may cause significant disease outbreaks in tilapia, often as secondary infection, include *A. sobria*, *A. veronii* and A. *caviae*, which are comparatively rare in humans (Gauthier, 2015; Shirajum Monir *et al.*, 2020). Human *Aeromonas* infections have been associated with aquatic environments and handling of shellfish or ornamental fish, but there is no evidence of zoonotic transmission between tilapia and people (Haenen *et al.*, 2013; Gauthier, 2015). With the current state of knowledge, *Aeromonas* from tilapia constitutes a public health hazard rather than a proven public health risk.

Edwardsiella spp.

As with *Aeromonas,* several members of the genus *Edwardsiella* have been detected as a cause of disease in both tilapia and humans (Haenen *et al.*, 2013; Gauthier, 2015), including *E. tarda*, *E. piscicida* and *E. anguillarium*. Zoonotic risk is sometimes mentioned in publications on *Edwardsiella* infection in tilapia (Miniero Davies *et al.*, 2018; Oh *et al.*, 2020; Sun *et al.*, 2020), but there is no direct evidence for transmission between tilapia and humans, whether from case studies, risk factor studies or molecular studies. Although host adaptation markers have been reported within *E. tarda* and *E. ictaluri*, both species have high genome plasticity (Wang *et al.*, 2009; Suezawa *et al.*, 2016; Machimbirike *et al.*, 2021) and it will be important to monitor the evolution of *Edwardsiella* spp. lest zoonotic variants emerge or amplify, possibly in association with the intensification of tilapia aquaculture, as observed for *S. agalactiae* ST283.

Mycobacterium marinum

Although there are several studies on experimental challenge of tilapia with *Mycobacterium marinum* (Wolf & Smith, 1999; Manrique *et al.*, 2019), descriptions of naturally occurring infection or disease in tilapia are very rare (Lara-Flores *et al.*, 2014). Thus, whilst *M. marinum* is undisputedly a zoonotic pathogen causing swimmer granuloma and more severe skin infections in humans, acquired from handling fish (Lahey, 2003; Haenen *et al.*, 2014), there is scant evidence for tilapia as source of infection. A rare exception is a risk factor study of *M. marinum* infections in response to a small outbreak (four cases) in New York City, USA. Risk factors for infection included skin injury to the finger or hand and purchasing tilapia or whiting (Yacisin *et al.*, 2017). A definitive food or environmental source for the outbreak could not be established, leaving the role of tilapia as a source of human *M. marinum* infections largely speculative and rare at best.

Recycling, waste and zoonoses

It is estimated that 80% of global sewage production is returned to the environment without (adequate) treatment (Fig. 13.2). Sewage can be used as a source of nutrients in agriculture and aquaculture, or it can pollute agriculture and aquaculture environments unintentionally (Mhongole *et al.*, 2017; Burket *et al.*, 2018). Together with desirable nutrients, undesirable anthropogenic chemicals, including antimicrobial residues, and microorganisms, including human and animal pathogens, are recycled. As discussed in the section above on *S. agalactiae*, this may have contributed to the emergence of the tilapia- and human-pathogenic strain ST283. Other human pathogens may also be recycled via sewage, water and aquatic species, including tilapia. In this section we introduce human pathogens that have been detected in tilapia but where evidence of disease causation is lacking.

Helicobacter pylori

Helicobacter pylori is primarily known as a cause of gastritis and gastric and duodenal ulcers in humans, with occasional detection in other mammalian species. Its main reservoir is the human stomach, allowing for faecal–oral transmission and transmission through contaminated food and water, among other routes (Abdel-Moein *et al.*, 2015; Cortés-Sánchez, 2021). In a study of 315 farmed, wild-caught, freshwater, brackish and marine fishes, only tilapia samples tested positive for *H. pylori* by both antigen testing and PCR. This included farmed and wild-caught tilapia from freshwater streams, the river Nile and a saltwater lake. Among catfish from the same water bodies, *H. pylori* was not detected but, considering the number of samples and low prevalence, this may not constitute a statistically or biologically significant difference. The role of *H. pylori* from tilapia

Figure 13.2 Floating fish farm with toilet draining into the Mekong River. Globally, it is estimated that 80% of sewage is released into the environment without adequate treatment, whether intentionally, to recycle nutrients from human, pig or poultry waste, or unintentionally. Photo, R. N. Zadoks.

as a foodborne pathogen is yet to be proven (Abdel-Moein *et al.*, 2015).

Enteric bacteria

Major enteric pathogens of humans include *Campylobacter* spp., verotoxin-producing *Escherichia coli* (VTEC), non-typhoidal *Salmonella* (NTS) and *Listeria monocytogenes*. Reports of the first two in tilapia are exceedingly rare and are limited to post-harvest or post-import detection, suggesting contamination by handlers or the handling environment (Wang *et al.*, 2011; Onjong *et al.*, 2018). Detection of NTS, including a variety of serovars, has been reported from live and processed tilapia. In live tilapia, NTS is found on external surfaces and in the intestine, with intestinal prevalence <4% in two studies from Africa (Youssef *et al.*, 1992; Khalefa *et al.*, 2021) and 40% in a study from Asia (Li *et al.*, 2017). Intestinal presence of NTS has been interpreted as evidence of infection. However, NTS failed to cause disease following oral or immersion challenge of blue tilapia (*Oreochromis aureus*) (Baker and Smitherman, 1983; Baker *et al.*, 1983). It was not detected in the bloodstream or muscle tissue of positive tilapia (Baker *et al.*, 1983; Khalefa *et al.*, 2021), is commonly present in contaminated surface water or intentionally recycled human, poultry or pig waste (Baker *et al.*, 1983; Mhongole *et al.*, 2017; Khalefa *et al.*, 2021), and serovars detected in tilapia include the exclusively human pathogen *Salmonella* Typhi (Onyango *et al.*, 2009). All the above findings argue in favour of the presence of *Salmonella* spp. as an environmental contaminant rather than a tilapia pathogen. *Salmonella* Weltevreden deserves special mention because it is increasingly detected in human patients, aquatic environments and seafood, including in tilapia, in China and South-east Asia (Li *et al.*, 2017; Hounmanou *et al.*, 2020). In an aquatic version of the 'chicken-and-egg' problem, it is currently unknown whether humans are the main source of *S.* Weltevreden for fish or vice versa.

At retail, prevalence of NTS in tilapia ranges from undetected, e.g. 0% among tilapia bought from local retailers or online in the USA (Pao *et al.*, 2008) to high, e.g. 45.5% in fresh, nationally produced tilapia fillets from supermarkets in Brazil (Ferreira *et al.*,

2021) and 28–64% in imported tilapia purchased from supermarkets in Saudi Arabia (Elhadi, 2014). High levels of contamination may be due to the introduction of tilapia- or human-derived NTS during evisceration or other processing steps, the ability of NTS to grow rapidly on seafood (Kumar *et al.*, 2015) and survival of NTS in tilapia-processing environments, e.g. raw tilapia sashimi plants (Wang *et al.*, 2019). Traditional ceviche recipes, such as those that are popular in Latin America, 'cook' raw fish through lime juice marination but fail to control NTS loads in tilapia ceviche (Mathurand and Schaffner, 2013). With the rapid expansion of tilapia production in South America, it seems important to monitor the potential emergence of ceviche-borne salmonellosis or streptococcosis. In regard to *L. monocytogenes*, as for NTS, tilapia-processing facilities play an important role in the risk of product contamination: inadequate handling of tilapia in the slaughter room and inadequate sanitizing procedures, rather than contamination of aquaculture environments or transportation tanks, contribute to contamination of the final product (Chen *et al.*, 2016; Vázquez-Sánchez *et al.*, 2017).

Antimicrobial resistance

The presence of antimicrobial resistance (AMR) genes in aquaculture systems, including tilapia production, is potentially driven by two processes: the use of antimicrobials in aquaculture to control diseases of fish, and the exposure of fish to human and animal effluent. Effluent contains resistant bacteria, antimicrobial resistance genes (possibly on mobile genetic elements such as integrons or plasmids) and antimicrobial residues, all of which can end up in surface waters and aquatic species. FAO and other international agencies are investing in reducing the risk of AMR in aquaculture, e.g. through the promotion of disease prevention and prudent use of antimicrobials or alternatives such as vaccines. It is not clear, however, whether intentional treatment of fish with antimicrobials or intentional or unintentional exposure of fish to effluent contributes most to public health hazards. The examples described below suggest that human rather than tilapia pathogens may be of major concern, especially as major human pathogens, e.g. *Shigella* or *Klebsiella*, often don't cause disease in fish so that their presence is not noted at the production and processing stages of the supply chain.

In recent reviews, although methods for monitoring of antimicrobial use and trends in antimicrobial resistance in aquaculture in Asia are described (Narbonne *et al.*, 2021; Schar *et al.*, 2021), we could not find reviews specific to tilapia. Considering that tilapia is the second-most produced finfish species, its potential role in antimicrobial use and resistances statistics cannot be negligible. Regulation of production chains for food of terrestrial animal origin has contributed to reductions in antimicrobial use and antimicrobial resistance. Implementation of this approach in tilapia production in low- or middle-income countries and local, partly informal, supply chains could be challenging. This is not impossible, however, as demonstrated in Kenya, where regulated and unregulated tilapia value chains exist. The unregulated value chain had a significantly higher proportion of bacterial species that were resistant to antimicrobials than the regulated tilapia value chain (Onjong *et al.*, 2021). In that study, *Shigella*, a human rather than a tilapia pathogen, had the highest proportion of bacteria that were resistant to most of the antimicrobials used, hinting at contamination of water with human waste as the underlying mechanism rather than the use of antimicrobials for tilapia production (Onjong *et al.*, 2021). In an example from South Africa, numerous mobile genetic elements that contribute to the horizontal transfer of resistance genes within and between bacterial species were observed in *Aeromonas* spp. from tilapia and other farmed fishes. This was observed in the absence of selective pressure (antimicrobial use) in the aquaculture system itself, again hinting at a human origin, whilst also showing the potential for tilapia to serve as a vehicle for human exposure to resistance determinants (Jacobs and Chenia, 2007). Likewise, the gut of *Oreochromis mossambicus* (Mozambique tilapia) from polluted rivers and lakes near Pune in India harboured many human pathogens, including *Aeromonas hydrophila*, *Klebsiella pneumoniae*, *E. coli*, *Serratia marcescens*, *Enterobacter* spp. and *Shigella* spp., most of which are not known as tilapia pathogens. In those pathogens, the prevalence of class 1 and 2 integrons and various extended betalactamase genes ranged from 24 to 38% (Marathe *et al.*, 2016). Because there is no specific evidence to support human disease resulting from tilapia-borne resistant pathogens or resistance determinants, we think the cited examples suffice to illustrate the concept, even if many other examples would be available. To guide control of transmission of resistant bacteria or resistance determinants from tilapia to people, it would be useful to have greater insight into the drivers of their prevalence and in their contribution to human disease.

Protozoa

Two major protozoan genera that can be zoonotic are *Cryptosporidium* and *Giardia*. Both genera include multiple species and subtypes, and both can be excreted in human and mammalian faeces. This can result in contamination of surface water and cause exposure of aquatic species to the protozoa. Like *H. pylori* and various fish-borne trematodes, *Giardia* has been detected in Nile tilapia in Egypt. In humans, giardiasis is mainly caused by *Giardia duodenalis* assemblages A and B, which are incidentally found in other animals, whereas the other assemblages (C to H) are host specific. *G. duodenalis* assemblage A was detected at low prevalence in the faeces of two of 68 tilapia (Ghoneim *et al.*, 2012). In Papua New Guinea, the zoonotic *Cryptosporidium parvum* was detected in two of 83 Nile tilapia farmed in fresh water (Ghoneim *et al.*, 2012). Our current state of knowledge would suggest that these findings reflect intestinal passage following contamination of surface water with protozoa from human or animal faecal origin rather than true infection or an independent reservoir in tilapia.

Conclusions, controls and outlook

Proven or potential zoonotic pathogens from tilapia comprise mostly bacteria and trematodes. Case reports and evidence of transmission between tilapia and humans exist for some pathogen species, notably *S. iniae* and *V. vulnificus* pv. *piscis*, but data on the incidence and prevalence of human infection with tilapia-derived pathogens is generally scarce, even when the causative agent is a serious pathogen for both host species as in the case of *S. agalactiae*. This reflects a combination of factors including, but not limited to, the relatively rare occurrence in high-income countries with limited raw fish consumption; limited healthcare and diagnostic infrastructure in countries where raw fish consumption is more common; and limited clinical impact, particularly for tilapia-borne zoonotic trematodes. Interestingly, the strongest evidence exists for contact rather than foodborne zoonoses from tilapia. With the growth of the global human population there is a growing demand for protein of animal origin in inland areas, and a growing shortage of land or uncontaminated surface waters to produce such food. Tilapia are already a major aquaculture species and their ability to thrive in water that is nutrient rich through intentional or unintentional recycling of human and animal waste will support their continued or increased importance as food-fish species (Marathe *et al.*, 2016; FAO, 2021). As a result, enteric bacteria and antimicrobial resistance genes carried by tilapia may become more important public health risks than bacteria that cause disease in tilapia, with the notable exception of *S. agalactiae* ST283, which is already a food safety concern in Asia and may emerge as such in Brazil.

For pathogens that cause disease in both people and tilapia, the direct economic impact of disease on fish production or indirect economic impacts on sales – or export restrictions – may provide sufficient economic incentive to institute controls at the farm level. Ideally such controls would be based on prevention through adequate husbandry, breeding and vaccination rather than the use of antimicrobials. Although hygiene is often advocated, including the avoidance of contamination of ponds with human and animal faeces, this is difficult to implement in many countries where waste and wastewater disposal and infrastructure are poorly developed, or where recycling of nutrients from human or animal waste is implemented intentionally. Although vaccines are commercially available for some major tilapia pathogens, including *S. iniae* and *S. agalactiae*, their commercial viability is currently hampered by the need for handling, especially for injectable vaccines, and the low value of individual tilapia, but development of oral vaccination via feed might change that situation. Tilapia with enhanced streptococcosis resistance are also commercially available, but resistance is only partial. For pathogens or resistance determinants that can be transmitted by tilapia without causing disease, including human enteric pathogens, there is no fish health or production-based incentive for control. Rather, control would need to be at the level of water quality, which may be feasible in closed or recirculating water systems but is unlikely to be so at a large scale in the foreseeable future in low- or middle-income countries and emerging economies where river, lake, pond and other open-water production systems predominate. Impacts of climate change, including salination of estuaries and changes in water temperature, may affect and possibly increase the distribution of tilapia (because of its partial salt tolerance) as well its pathogens, *S. iniae*, *S. agalactiae* and *V. vulnificus*, which are associated with warm-water streptococcosis and vibriosis (Evans *et al.*, 2006; Amaro *et al.*, 2020).

Apart from zoonotic disease transmitted by contact or skin injuries, such as *S. iniae* and *V. vulnificus* infections, most tilapia-borne disease could be prevented by adequate heat treatment prior to fish consumption and, in the case of parasitic infestations, by freezing. Considering the popularity of raw fish dishes prepared as sashimi, yu-sheng, ceviche or salted fish, to name but a few, and the traditional importance of food to people's identity, such food safety advice may be culturally inappropriate as well as unlikely to be implemented, although it remains important to communicate the risks and preventive measures to reduce human infections. Indeed, a ban on the sale of ready-to-eat raw freshwater fish dishes helped to resolve a GBS outbreak in Singapore (Chau *et al.*, 2017). Improvement in healthcare and diagnostic infrastructure, for both humans and tilapia, combined with epidemiological investigations of suspected foodborne disease cases or outbreaks, should provide better insight into the occurrence and impact of tilapia-borne zoonoses. Only with sufficient evidence of human or tilapia health impacts will there be greater justification for investment in their ongoing monitoring and control of tilapia-borne zoonoses. There is a risk, however, that tilapia zoonoses may become stuck in the neglected tropical diseases loop of lack of evidence causing lack of investment, resulting in continuing lack of both evidence and investment.

References

Abdel-Moein, K.A., Saeed, H. and Samir, A. (2015) Novel detection of *Helicobacter pylori* in fish: A possible public health concern. *Acta Tropica* 152, 141–144. DOI: 10.1016/j.actatropica.2015.09.005.

Agnew, W. and Barnes, A.C. (2007) *Streptococcus iniae*: An aquatic pathogen of global veterinary significance and a challenging candidate for reliable vaccination. *Veterinary Microbiology* 122(1–2), 1–15. DOI: 10.1016/j.vetmic.2007.03.002.

Amaro, C., Sanjuán, E., Fouz, B., Pajuelo, D., Lee, C.T, *et al.* (2015) The fish pathogen *Vibrio vulnificus* biotype 2: Epidemiology, phylogeny, and virulence factors involved in warm-water vibriosis. *Microbiology Spectrum* 3(3). DOI: 10.1128/microbiolspec.VE-0005-2014.

Amaro, C., Fouz, B., Sanjuán, E. and Romalde, J. (2020) Chapter 10. vibriosis. In: Buchmann, J.A. K, (eds) *Climate Change and Infectious Fish Diseases, Woo, P.T.K., JLeong*. CABI Publishing, pp. 182–210.

Arguedas Cortés, D., Dolz, G., Romero Zúñiga, J.J., Jiménez Rocha, A.E. and León Alán, D. (2010) *Centrocestus formosanus* (Opisthorchiida: Heterophyidae) as a cause of death in gray tilapia fry *Oreochromis niloticus* (Perciforme: Cichlidae) in the dry Pacific of Costa Rica. *Revista de Biología Tropical* 58(4), 1453–1465.

Baker, D.A. and Smitherman, R.O. (1983) Immune response of *Tilapia aurea* exposed to *Salmonella* typhimurium. *Applied and Environmental Microbiology* 46(1), 28–31. DOI: 10.1128/aem.46.1.28-31.1983.

Baker, D.A., Smitherman, R.O. and McCaskey, T.A. (1983) Longevity of *Salmonella* typhimurium in *Tilapia aurea* and water from pools fertilized with swine waste. *Applied and Environmental Microbiology* 45(5), 1548–1554. DOI: 10.1128/aem.45.5.1548-1554.1983.

Barkham, T., Zadoks, R.N., Azmai, M.N.A., Baker, S., Bich, V.T.N, *et al.* (2019) One hypervirulent clone, sequence type 283, accounts for a large proportion of invasive *Streptococcus agalactiae* isolated from humans and diseased tilapia in Southeast Asia. *PLOS Neglected Tropical Diseases* 13(6), e0007421. DOI: 10.1371/journal.pntd.0007421.

Bisharat, N. (2002) *Vibrio vulnificus* infections can be avoided. *The Israel Medical Association Journal* 4(8), 631–633.

Burket, S.R., Sapozhnikova, Y., Zheng, J.S., Chung, S.S. and Brooks, B.W. (2018) At the intersection of urbanization, water, and food security: Determination of select contaminants of emerging concern in mussels and oysters from Hong Kong. *Journal of Agricultural and Food Chemistry* 66(20), 5009–5017. DOI: 10.1021/acs.jafc.7b05730.

Calif, E., Kaufman, B. and Stahl, S. (2003) *Vibrio vulnificus* infection of the lower limb after fish spine injuries. *Clinical Orthopaedics and Related Research* 411, 274–279. DOI: 10.1097/01.blo.0000069894.31220.a1.

Carmona-Salido, H., Fouz, B., Sanjuán, E., Carda-Diéguez, M., Delannoy, C.M.J, *et al.* (2021a) Draft genome sequences of *Vibrio vulnificus* strains recovered from moribund tilapia. *Microbiology Resource Announcements* 10(22), e0009421. DOI: 10.1128/MRA.00094-21.

Carmona-Salido, H., Fouz, B., Sanjuán, E., Carda, M., Delannoy, C.M.J, *et al.* (2021b) The widespread presence of a family of fish virulence plasmids in *Vibrio vulnificus* stresses its relevance as a zoonotic pathogen linked to fish farms. *Emerging Microbes and Infections* 10(1), 2128–2140. DOI: 10.1080/22221751.2021.1999177.

Centers for disease control and prevention (CDC) (1996) Invasive infection with *Streptococcus iniae*—Ontario, 1995–1996. *MMWR. Morbidity and Mortality Weekly Report* 45(30), 650–653.

Chai, J.Y. and Jung, B.K. (2020) Foodborne intestinal flukes: A brief review of epidemiology and geographical distribution. *Acta Tropica* 201, 105210. DOI: 10.1016/j.actatropica.2019.105210.

Chan, W.L., Chan, C.H. and Chan, T.Y. (1999) *Vibrio vulnificus* septicaemia and necrotizing fasciitis after a prick from the dorsal fin of a tilapia. *Transactions of the Royal Society of Tropical Medicine and Hygiene* 93(2), 174. DOI: 10.1016/s0035-9203(99)90298-5.

Chau, M.L., Chen, S.L., Yap, M., Hartantyo, S.H.P., Chiew, P.K.T, *et al.* (2017) Group B *Streptococcus* infections caused by improper sourcing and handling of fish for raw consumption, Singapore, 2015–2016. *Emerging Infectious Diseases* 23(12), 2002–2010. DOI: 10.3201/eid2312.170596.

Chen, B.Y., Wang, C.Y., Wang, C.L., Fan, Y.C., Weng, I.T, *et al.* (2016) Prevalence and persistence of *Listeria monocytogenes* in ready-to-eat tilapia sashimi processing plants. *Journal of Food Protection* 79(11), 1898–1903. DOI: 10.4315/0362-028X.JFP-16-149.

Chenia, H.Y. and Duma, S. (2017) Characterization of virulence, cell surface characteristics and biofilm-forming ability of *Aeromonas* spp. isolates from fish and sea water. *Journal of Fish Diseases* 40(3), 339–350. DOI: 10.1111/jfd.12516.

Chi, T.T., Dalsgaard, A., Turnbull, J.F., Tuan, P.A. and Murrell, K.D. (2008) Prevalence of zoonotic trematodes in fish from a Vietnamese fish-farming community. *The Journal of Parasitology* 94(2), 423–428. DOI: 10.1645/GE-1389.1.

Cho, S.T., Tsai, Y.M., Chen, C.Y. and Kuo, C.H. (2018) Draft genome sequence of *Vibrio vulnificus* 86573B, a bacterium isolated from diseased tilapia in Taiwan. *Microbiology Resource Announcements* 7(1). DOI: 10.1128/MRA.00813-18.

Choi, S., Kim, B.S., Hwang, J. and Kim, M.H. (2021) Reduced virulence of the MARTX toxin increases the persistence of outbreak-associated *Vibrio vulnificus* in host reservoirs. *The Journal of Biological Chemistry* 296, 100777. DOI: 10.1016/j.jbc.2021.100777.

Cortés-Sánchez, A.D.J. (2021) *Helicobacter pylori*, food, fish and tilapia. *Food Research* 5, 12.

Delannoy, C.M., Crumlish, M., Fontaine, M.C., Pollock, J., Foster, G, *et al.* (2013) Human *Streptococcus agalactiae* strains in aquatic mammals and fish. *BMC Microbiology* 13, 41. DOI: 10.1186/1471-2180-13-41.

Delannoy, C.M., Zadoks, R.N., Crumlish, M., Rodgers, D., Lainson, F.A, *et al.* (2016) Genomic comparison of virulent and non-virulent *Streptococcus agalactiae* in fish. *Journal of Fish Diseases* 39(1), 13–29. DOI: 10.1111/jfd.12319.

Elhadi, N. (2014) Prevalence and antimicrobial resistance of *Salmonella* spp. in raw retail frozen imported freshwater fish to eastern province of Saudi Arabia. *Asian Pacific Journal of Tropical Biomedicine* 4(3), 234–238. DOI: 10.1016/S2221-1691(14)60237-9.

Elsheikha, H.M. and Elshazly, A.M. (2008) Preliminary observations on infection of brackish and fresh water fish by heterophyid encysted metacercariae in Egypt. *Parasitology Research* 103(4), 971–977. DOI: 10.1007/s00436-008-1043-z.

Evans, J.J., Klesius, P.H. and Shoemaker, C.A. (2006) An overview streptococcus in warm-water fish. *Aquaculture Health International* 7, 10–14.

Evans, J.J., Bohnsack, J.F., Klesius, P.H., Whiting, A.A., Garcia, J.C, *et al.* (2008) Phylogenetic relationships among *Streptococcus agalactiae* isolated from piscine, dolphin, bovine and human sources: A dolphin and piscine lineage associated with a fish epidemic in Kuwait is also associated with human neonatal infections in Japan. *Journal of Medical Microbiology* 57(11), 1369–1376. DOI: 10.1099/jmm.0.47815-0.

Facklam, R., Elliott, J., Shewmaker, L. and Reingold, A. (2005) Identification and characterization of sporadic isolates of *Streptococcus iniae* isolated from humans. *Journal of Clinical Microbiology* 43(2), 933–937. DOI: 10.1128/JCM.43.2.933-937.2005.

Ferreira, A.C.A.O., Pavelquesi, S.L.S., Monteiro, E.D.S., Rodrigues, L.F.S., Silva, C.M.S, *et al.* (2021) Prevalence and antimicrobial resistance of

Salmonella spp. in aquacultured Nile tilapia (*Oreochromis niloticus*) commercialized in Federal District, Brazil. *Foodborne Pathogens and Disease* 18(11), 778–783. DOI: 10.1089/fpd.2021.0010.

Food and Agriculture Organization (2021) Risk profile – group B *Streptococcus* (GBS)/*Streptococcus agalactiae* sequence type (ST) 283 in freshwater fish. Bangkok. https://doi.org/10.4060/cb5067en

Food and Agriculture Organization/World Health Organization (2014) Multicriteria-based ranking for risk management of food-borne parasites. *Microbiological Risk Assessment Series*, p. 302. Rome.

Fouz, B., Llorens, A., Valiente, E. and Amaro, C. (2010) A comparative epizootiologic study of the two fish-pathogenic serovars of *Vibrio vulnificus* biotype 2. *Journal of Fish Diseases* 33(5), 383–390. DOI: 10.1111/j.1365-2761.2009.01130.x.

Gauthier, D.T. (2015) Bacterial zoonoses of fishes: A review and appraisal of evidence for linkages between fish and human infections. *Veterinary Journal* 203(1), 27–35. DOI: 10.1016/j.tvjl.2014.10.028.

Ghoneim, N.H., Abdel-Moein, K.A. and Saeed, H. (2012) Fish as a possible reservoir for zoonotic *Giardia duodenalis* assemblages. *Parasitology Research* 110(6), 2193–2196. DOI: 10.1007/s00436-011-2748-y.

Goh, S.H., Driedger, D., Gillett, S., Low, D.E., Hemmingsen, S.M, *et al.* (1998) *Streptococcus iniae*, A human and animal pathogen: Specific identification by the chaperonin 60 gene identification method. *Journal of Clinical Microbiology* 36(7), 2164–2166. DOI: 10.1128/JCM.36.7.2164-2166.1998.

Haenen, O.L., Evans, J.J. and Berthe, F. (2013) Bacterial infections from aquatic species: Potential for and prevention of contact zoonoses. *Revue Scientifique et Technique* 32(2), 497–507. DOI: 10.20506/rst.32.2.2245.

Haenen, O.L., Zanten, E., Jansen, R., Roozenburg, I., Engelsma, M.Y, *et al.* (2014) *Vibrio vulnificus* outbreaks in Dutch eel farms since 1996: Strain diversity and impact. *Diseases of Aquatic Organisms* 108(3), 201–209. DOI: 10.3354/dao02703.

Hounmanou, Y.M.G., Mdegela, R.H., Dougnon, T.V., Madsen, H., Withey, J.H, *et al.* (2019) Tilapia (*Oreochromis niloticus*) as a putative reservoir host for survival and transmission of *Vibrio cholerae* O1 Biotype El Tor in the aquatic environment. *Frontiers in Microbiology* 10, 1215. DOI: 10.3389/fmicb.2019.01215.

Hounmanou, Y.M.G., Dalsgaard, A., Sopacua, T.F., Uddin, G.M.N., Leekitcharoenphon, P, *et al.* (2020) Molecular characteristics and zoonotic potential of *Salmonella* Weltevreden from cultured shrimp and tilapia in Vietnam and China. *Frontiers in Microbiology* 11, 1985. DOI: 10.3389/fmicb.2020.01985.

Ip, M., Ang, I., Fung, K., Liyanapathirana, V., Luo, M.J, *et al.* (2016) Hypervirulent clone of Group B *Streptococcus* serotype III sequence type 283, Hong Kong, 1993–2012. *Emerging Infectious Diseases* 22(10), 1800–1803. DOI: 10.3201/eid2210.151436.

Jacobs, L. and Chenia, H.Y. (2007) Characterization of integrons and tetracycline resistance determinants in *Aeromonas* spp. isolated from South African aquaculture systems. *International Journal of Food Microbiology* 114(3), 295–306. DOI: 10.1016/j.ijfoodmicro.2006.09.030.

Kalimuddin, S., Chen, S.L., Lim, C.T.K., Koh, T.H., Tan, T.Y. *et al.* (2017) Epidemic of severe *Streptococcus agalactiae* sequence type 283 infections in Singapore associated with the consumption of raw freshwater fish: A detailed analysis of clinical, epidemiological, and bacterial sequencing data. *Clinical Infectious Diseases* 64, S145–s152.

Kawasaki, M., Delamare-Deboutteville, J., Bowater, R.O., Walker, M.J., Beatson, S. *et al.* (2018) Microevolution of *Streptococcus agalactiae* ST-261 from Australia indicates dissemination via imported tilapia and ongoing adaptation to marine hosts or environment. *Applied and Environmental Microbiology* 84, 16. DOI: https://doi.org/10.1128/AEM.00859-18.

Khalefa, H.S., Ahmed, Z.S., Abdel-Kader, F., Ismail, E.M. and Elshafiee, E.A. (2021) Sequencing and phylogenetic analysis of the *stn* gene of *Salmonella* species isolated from different environmental sources at Lake Qarun protectorate: the role of migratory birds and public health importance. *Veterinary World* 14(10), 2764–2772. DOI: 10.14202/vetworld.2021.2764-2772.

Koh, T.H., Kurup, A. and Chen, J. (2004) *Streptococcus iniae* discitis in Singapore. *Emerging Infectious Diseases* 10(9), 1694–1696. DOI: 10.3201/eid1009.040029.

Koh, T.H., Sng, L.H., Yuen, S.M., Thomas, C.K., Tan, P.L, *et al.* (2009) Streptococcal cellulitis following preparation of fresh raw seafood. *Zoonoses and Public Health* 56(4), 206–208. DOI: 10.1111/j.1863-2378.2008.01213.x.

Kumar, R., Datta, T.K. and Lalitha, K.V. (2015) *Salmonella* grows vigorously on seafood and expresses its virulence and stress genes at different temperature exposure. *BMC Microbiology* 15, 254. DOI: 10.1186/s12866-015-0579-1.

Lahey, T. (2003) Invasive *Mycobacterium marinum* infections. *Emerging Infectious Diseases* 9(11), 1496–1498. DOI: 10.3201/eid0911.030192.

Lara-Flores, M., Aguirre-Guzman, G., Balan-Zetina, S.B., Sonda-Santos, K.Y. and Zapata, A.A. (2014) Identification of *Mycobacterium* agent isolated from tissues of Nile tilapia (*Oreochromis niloticus*). *Turkish Journal of Fisheries and Aquatic Sciences* 14, 5.

Lau, S.K., Woo, P.C., Tse, H., Leung, K.W., Wong, S.S, *et al.* (2003) Invasive *Streptococcus iniae* infections outside North America. *Journal of Clinical Microbiology* 41(3), 1004–1009. DOI: 10.1128/JCM.41.3.1004-1009.2003.

Lau, S.K., Woo, P.C., Luk, W.K., Fung, A.M., Hui, W.T, *et al.* (2006) Clinical isolates of *Streptococcus iniae* from Asia are more mucoid and beta-hemolytic than those from North America. *Diagnostic Microbiology and Infectious Disease* 54(3), 177–181. DOI: 10.1016/j.diagmicrobio.2005.09.012.

Laviad-Shitrit, S., Izhaki, I. and Halpern, M. (2019) Accumulating evidence suggests that some waterbird species are potential vectors of *Vibrio cholerae. PLOS Pathogens* 15(8), e1007814. DOI: 10.1371/journal.ppat.1007814.

Leal, C.A.G., Queiroz, G.A., Pereira, F.L., Tavares, G.C. and Figueiredo, H.C.P. (2019) *Streptococcus agalactiae* Sequence Type 283 in farmed fish Brazil. *Emerging Infectious Diseases* 25(4), 776–779. DOI: 10.3201/eid2504.180543.

Li, K., Clausen, J.H., Murrell, K.D., Liu, L. and Dalsgaard, A. (2013) Risks for fishborne zoonotic trematodes in tilapia production systems in Guangdong province, China. *Veterinary Parasitology* 198(1–2), 223–229. DOI: 10.1016/j.vetpar.2013.08.011.

Li, K., Petersen, G., Barco, L., Hvidtfeldt, K., Liu, L, *et al.* (2017) *Salmonella* Weltevreden in integrated and non-integrated tilapia aquaculture systems in Guangdong, China. *Food Microbiology* 65, 19–24. DOI: 10.1016/j.fm.2017.01.014.

Li, Y.W., Liu, L., Huang, P.R., Fang, W., Luo, Z.P, *et al.* (2014) Chronic streptococcosis in Nile tilapia, *Oreochromis niloticus* (L.), caused by *Streptococcus agalactiae. Journal of Fish Diseases* 37(8), 757–763. DOI: 10.1111/jfd.12146.

Lobna, S.M., Metawea, Y.F. and Elsheikha, H.M. (2010) Prevalence of heterophyiosis in tilapia fish and humans in Northern Egypt. *Parasitology Research* 107(4), 1029–1034. DOI: 10.1007/s00436-010-1976-x.

Machimbirike, V.I., Uthaipaisanwong, P., Khunrae, P., Dong, H.T., Senapin, S, *et al.* (2021) Comparative genomics of *Edwarsiella ictaluri* revealed four distinct host-specific genotypes and thirteen potential vaccine candidates. *Genomics* 113(4), 1976–1987. DOI: 10.1016/j.ygeno.2021.04.016.

Mahmud, Z.H., Wright, A.C., Mandal, S.C., Dai, J., Jones, M.K, *et al.* (2010) Genetic characterization of *Vibrio vulnificus* strains from tilapia aquaculture in Bangladesh. *Applied and Environmental Microbiology* 76(14), 4890–4895. DOI: 10.1128/AEM.00636-10.

Majere, R.A. and Cortina, S. (2017) Necrotizing wound infection from a tilapia fish injury. *CMAJ* 189(14), E539–E541. DOI: 10.1503/cmaj.160663.

Manrique, W.G., Pereira Figueiredo, M.A., Charlie-Silva, I., Andrade Belo, M. and Dib, C.C. (2019) Spleen melanomacrophage centers response of Nile tilapia during *Aeromanas hydrophila* and *Mycobacterium marinum* infections. *Fish and Shellfish Immunology* 95, 514–518. DOI: 10.1016/j.fsi.2019.10.071.

Marathe, N.P., Gaikwad, S.S., Vaishampayan, A.A., Rasane, M.H., Shouche, Y.S, *et al.* (2016) Mossambicus tilapia (*Oreochromis mossambicus*) collected from water bodies impacted by urban waste carries extended-spectrum beta-lactamases and integron-bearing gut bacteria. *Journal of Biosciences* 41(3), 341–346. DOI: 10.1007/s12038-016-9620-2.

Mathurand, P. and Schaffner, D.W. (2013) Effect of lime juice on *Vibrio parahaemolyticus* and *Salmonella enterica* inactivation during the preparation of the raw fish dish ceviche. *Journal of Food Protection* 76(6), 1027–1030. DOI: 10.4315/0362-028X.JFP-12-526.

Mhongole, O.J., Mdegela, R.H., Kusiluka, L.J., Forslund, A. and Dalsgaard, A. (2017) Characterization of *Salmonella* spp. from wastewater used for food production in Morogoro, Tanzania. *World Journal of Microbiology and Biotechnology* 33(3), 42. DOI: 10.1007/s11274-017-2209-6.

Miniero Davies, Y., Oliveira, M.G., Paulo Vieira Cunha, M., Soares Franco, L., Pulecio Santos, S.L, *et al.* (2018) *Edwardsiella tarda* outbreak affecting fishes and aquatic birds in Brazil. *The Veterinary Quarterly* 38(1), 99–105. DOI: 10.1080/01652176.2018.1540070.

Narbonne, J.A., Radke, B.R., Price, D., Hanington, P.C., Babujee, A. *et al.* (2021) Antimicrobial use surveillance indicators for finfish aquaculture production: A review. *Frontiers in Veterinary Science* 8, 595152.

Nguyen, T.H., Nguyen, V.D., Murrell, D. and Dalsgaard, A. (2007) Occurrence and species distribution of fishborne zoonotic trematodes in wastewater-fed aquaculture in Northern Vietnam. *Tropical Medicine and International Health* 66–72. DOI: 10.1111/j.1365-3156.2007.01943.x.

Oh, W.T., Jun, J.W., Kim, H.J., Giri, S.S., Yun, S, *et al.* (2020) Characterization and pathological analysis of a virulent *Edwardsiella anguillarum* strain isolated from Nile tilapia (*Oreochromis niloticus*) in Korea. *Frontiers in Veterinary Science* 7, 14. DOI: 10.3389/fvets.2020.00014.

Onjong, H.A., Ngayo, M.O., Mwaniki, M., Wambui, J. and Njage, P.M.K. (2018) Microbiological safety of fresh tilapia (*Oreochromis niloticus*) from Kenyan fresh water fish value chains. *Journal of Food Protection* 81(12), 1973–1981. DOI: 10.4315/0362-028X.JFP-18-078.

Onjong, H.A., Ntuli, V., Wambui, J., Mwaniki, M. and Njage, P.M.K. (2021) Potential influence of regulation of the food value chain on prevalence and patterns of antimicrobial resistance: The case of tilapia (*Oreochromis niloticus*). *Applied and Environmental Microbiology* 87(23), e0094521. DOI: 10.1128/AEM.00945-21.

Onyango, D.M., Wandili, S., Kakai, R. and Waindi, E.N. (2009) Isolation of *Salmonella* and *Shigella* from fish harvested from the Winam Gulf of Lake Victoria, Kenya. *Journal of Infection in Developing Countries* 3(2), 99–104. DOI: 10.3855/jidc.56.

Pao, S., Ettinger, M.R., Khalid, M.F., Reid, A.O. and Nerrie, B.L. (2008) Microbial quality of raw aquacultured fish fillets procured from internet and local retail markets. *Journal of Food Protection* 71(8), 1544–1549. DOI: 10.4315/0362-028x-71.8.1544.

Pinto, H.A., Mati, V.L. and Melo, A.L. (2014) Metacercarial infection of wild Nile tilapia (*Oreochromis niloticus*) from Brazil. *The Scientific World Journal* 807492. DOI: 10.1155/2014/807492.

Rajendram, P., Mar Kyaw, W., Leo, Y.S., Ho, H., Chen, W.K, *et al.* (2016) Group B *Streptococcus* sequence type 283 disease linked to consumption of raw fish, Singapore. *Emerging Infectious Diseases* 22(11), 1974–1977. DOI: 10.3201/eid2211.160252.

Roig, F.J., González-Candelas, F., Sanjuán, E., Fouz, B., Feil, E.J, *et al.* (2017) Phylogeny of *Vibrio vulnificus* from the analysis of the core-genome: Implications for intra-species taxonomy. *Frontiers in Microbiology* 8, 2613. DOI: 10.3389/fmicb.2017.02613.

Rosinski-Chupin, I., Sauvage, E., Mairey, B., Mangenot, S., Ma, L, *et al.* (2013) Reductive evolution in *Streptococcus agalactiae* and the emergence of a host adapted lineage. *BMC Genomics* 14, 252. DOI: 10.1186/1471-2164-14-252.

Said, R., Volpin, G., Grimberg, B., Friedenstrom, S.R., Lefler, E, *et al.* (1998) Hand infections due to non-cholera *Vibrio* after injuries from St Peter's fish (*Tilapia zillii*). *Journal of Hand Surgery* 23(6), 808–810. Available at: https://doi.org/10.1016/s0266-7681(98)80104-6

Schar, D., Zhao, C., Wang, Y., Larsson, D.G.J., Gilbert, M, *et al.* (2021) Twenty-year trends in antimicrobial resistance from aquaculture and fisheries in Asia. *Nature Communications* 12(1), 5384. DOI: 10.1038/s41467-021-25655-8.

Seale, A.C., Bianchi-Jassir, F., Russell, N.J., Kohli-Lynch, M., Tann, C.J, *et al.* (2017) Estimates of the burden of Group B streptococcal disease worldwide for pregnant women, stillbirths, and children. *Clinical Infectious Diseases* S200–S219. DOI: 10.1093/cid/cix664.

Shirajum Monir, M., Yusoff, S.M., Mohamad, A. and Ina-Salwany, M.Y. (2020) Vaccination of tilapia against motile *Aeromonas* septicemia: A review. *Journal of Aquatic Animal Health* 32(2), 65–76. DOI: 10.1002/aah.10099.

Siddique, A.B., Moniruzzaman, M., Ali, S., Dewan, M.N., Islam, M.R, *et al.* (2021) Characterization of pathogenic *Vibrio parahaemolyticus* isolated from fish aquaculture of the southwest coastal area of Bangladesh. *Frontiers in Microbiology* 12, 635539. DOI: 10.3389/fmicb.2021.635539.

Son, M.S., Megli, C.J., Kovacikova, G., Qadri, F. and Taylor, R.K. (2011) Characterization of *Vibrio cholerae* O1 El Tor biotype variant clinical isolates from Bangladesh and Haiti, including a molecular genetic analysis of virulence genes. *Journal of Clinical Microbiology* 49(11), 3739–3749. DOI: 10.1128/JCM.01286-11.

Suezawa, C., Yasuda, M., Negayama, K., Kameyama, T., Hirauchi, M, *et al.* (2016) Identification of genes associated with the penetration activity of the human type of *Edwardsiella tarda* EdwGII through human colon epithelial cell monolayers. *Microbial Pathogenesis* 95, 148–156. DOI: 10.1016/j.micpath.2016.04.007.

Sumithra, T.G., Reshma, K.J., Anusree, V.N., Sayooj, P., Sharma, S.R.K, *et al.* (2019) Pathological investigations of *Vibrio vulnificus* infection in genetically improved farmed tilapia (*Oreochromis niloticus* L.) cultured at a floating cage farm of India. *Aquaculture* 511. DOI: 10.1016/j.aquaculture.2019.734217.

Sun, J.R., Yan, J.C., Yeh, C.Y., Lee, S.Y. and Lu, J.J. (2007) Invasive infection with *Streptococcus iniae* in Taiwan. *Journal of Medical Microbiology* 56(9), 1246–1249. DOI: 10.1099/jmm.0.47180-0.

Sun, Y., Chen, J., Liu, J., Huang, J., Ye, T, *et al.* (2020) The role of uhpA in *Edwardsiella piscicida* and the inflammatory cytokine response in tilapia. *Fish and Shellfish Immunology* 101, 192–197. DOI: 10.1016/j.fsi.2020.03.028.

Tan, S., Lin, Y., Foo, K., Koh, H.F., Tow, C, *et al.* (2016) Group B *Streptococcus* serotype III sequence type 283 bacteremia associated with consumption of raw fish, Singapore. *Emerging Infectious Diseases* 22(11), 1970–1973. DOI: 10.3201/eid2211.160210.

Tey, Y.H., Jong, K.J., Fen, S.Y. and Wong, H.C. (2015) Occurrence of *Vibrio parahaemolyticus, Vibrio cholerae*, and *Vibrio vulnificus* in the aquacultural environments of Taiwan. *Journal of Food Protection* 78(5), 969–976. DOI: 10.4315/0362-028X.JFP-14-405.

Van De, N., Le, T.H. and Murrell, K.D. (2012) Prevalence and intensity of fish-borne zoonotic trematodes in cultured freshwater fish from rural and urban areas of northern Vietnam. *The Journal of Parasitology* 98(5), 1023–1025. DOI: 10.1645/GE-3112.1.

Vázquez-Sánchez, D., Galvão, J.A. and Oetterer, M. (2017) Contamination sources, serogroups, biofilm-forming ability and biocide resistance of *Listeria monocytogenes* persistent in tilapia-processing facilities. *Journal of Food Science and Technology* 54(12), 3867–3879. DOI: 10.1007/s13197-017-2843-x.

Vinh, D.C., Mubareka, S., Fatoye, B., Plourde, P. and Orr, P. (2006) *Vibrio vulnificus* septicemia after handling tilapia species fish: A Canadian case report and review. *Canadian Journal of Infectious Diseases and Medical Microbiology* 17(2), 129–132. DOI: 10.1155/2006/164681.

Wang, C.L., Chen, B.Y., Hsu, C.T., Wu, R. and Chou, C.H. (2019) *Salmonella* contamination in ready-to-eat tilapia sashimi processing plants. *Journal of Food Protection* 82(2), 256–261. DOI: 10.4315/0362-028X.JFP-18-324.

Wang, F., Jiang, L., Yang, Q., Han, F., Chen, S, *et al.* (2011) Prevalence and antimicrobial susceptibility of major foodborne pathogens in imported seafood. *Journal of Food Protection* 74(9), 1451–1461. DOI: 10.4315/0362-028X.JFP-11-146.

Wang, Q., Yang, M., Xiao, J., Wu, H., Wang, X, *et al.* (2009) Genome sequence of the versatile fish pathogen *Edwardsiella tarda* provides insights into its adaptation to broad host ranges and intracellular niches. *PLOS ONE* 4(10), e7646. DOI: 10.1371/journal.pone.0007646.

Weinstein, M.R., Litt, M., Kertesz, D.A., Wyper, P., Rose, D, *et al.* (1997) Invasive infections due to a fish pathogen, *Streptococcus iniae*. *The*

New England Journal of Medicine 337(9), 589–594. DOI: 10.1056/NEJM199708283370902.

Wiriya, B., Clausen, J.H., Inpankaew, T., Thaenkham, U., Jittapalapong, S, *et al.* (2013) Fish-borne trematodes in cultured Nile tilapia (*Oreochromis niloticus*) and wild-caught fish from Thailand. *Veterinary Parasitology* 198(1–2), 230–234. DOI: 10.1016/j.vetpar.2013.08.008.

Wolf, J.C. and Smith, S.A. (1999) Comparative severity of experimentally induced mycobacteriosis in striped bass *Morone saxatilis* and hybrid tilapia *Oreochromis* spp. *Diseases of Aquatic Organisms* 38(3), 191–200. DOI: 10.3354/dao038191.

Yacisin, K., Hsieh, J.L., Weiss, D., Ackelsberg, J., Lee, E, *et al.* (2017) Outbreak of non-tuberculous mycobacteria skin or soft tissue infections associated with handling fish – New York City, 2013–2014. *Epidemiology and Infection* 145(11), 2269–2279. DOI: 10.1017/S0950268817001066.

Youssef, H., El-Timawy, A.K. and Ahmed, S. (1992) Role of aerobic intestinal pathogens of fresh water fish in transmission of human diseases. *Journal of Food Protection* 55(9), 739–740. DOI: 10.4315/0362-028X-55.9.739.

Zaidenstein, R., Sadik, C., Lerner, L., Valinsky, L., Kopelowitz, J, *et al.* (2008) Clinical characteristics and molecular subtyping of *Vibrio vulnificus* Illnesses, Israel. *Emerging Infectious Diseases* 14(12), 1875–1882. DOI: 10.3201/eid1412.080499.

Chapter 14

Histopathological maps of infectious diseases in tilapia

Ha Thanh Dong and Jorge del-Pozo

Introduction

The aim of this chapter is to provide a brief overview of the histopathological features of relevant infectious diseases of tilapia. It is our hope that this information may assist with the interpretation of histological field samples during disease outbreak investigations if access to the expertise of a qualified fish pathologist is not available. In addition, this information should also help interpret the output of pathology reports by such specialists. Histopathology entails the examination of microscopic disease features using, for example, haematoxylin-and-eosin staining. The focus of these examinations lies in identifying pathological tissue patterns resulting from disease mechanisms. While ideal cases involve pathognomonic lesions that indicate specific diseases, most lesions do not, necessitating further diagnostic tests. Samples, typically obtained from gills, hepatopancreas, spleen, kidney and intestine, are preserved in 10% formalin and sent to laboratories. Swift sampling post euthanasia is imperative due to the rapid autolysis process in fish. This chapter centres on commonly sampled organs, offering a practical reference for individuals in aquaculture involved in diagnosing diseases in tilapia.

Histopathology is the study of the microscopic features of disease using tissue samples that have been fixed, processed, embedded, sectioned, placed on glass slides and stained with a range of histochemical stains. The latter aims to underline the structural details of the tissue, and the stain combination most commonly used in the

Kim Thompson, Craig Shoemaker and David C. Little (eds) **Tilapia**
DOI: 10.1079/9781800629455.0014, © CAB International, 2025

diagnostic setting is haematoxylin and eosin (H&E). Haematoxylin is a base that is blue and binds to acids in the tissue (i.e. those parts of the tissue that are basophilic) whereas eosin is acidic and pink and binds to bases in the tissue (i.e. those parts of the tissue that are eosinophilic). In most situations these properties explain the blue staining of nuclei and pink staining of the extracellular matrix and cytoplasm. It is noteworthy that other histochemical stains can be used for highlighting specific structures/substances and/or detecting aetiological agents in tissue sections. However, in this chapter we will refer mostly to H&E-stained tissues.

One way to describe histopathological examinations is that they refer to the observation of pathological tissue patterns (or lesions) that are the result of one or more pathogenic mechanisms (i.e. the way in which normal function is deranged in disease – e.g. toxicity, ischaemia, etc.). The presence of one or a combination of these patterns confirms the presence of disease and is indicative (or suggestive) of one or more specific disease entities. This is where the usefulness of histopathology lies. Ideally, each lesion would indicate a specific disease entity and lesions that do so are referred to as pathognomonic. However, most lesions are not pathognomonic and may be present in different disease entities. A stereotypical lesion is one that is found in a wide range of disease entities. Histopathology is, therefore, an excellent tool for providing confirmation and classification of a diseased status and, frequently, it is a very good tool for suggesting a specific aetiology for it. However, there are limitations to this, especially in the presence of stereotypical lesions and the absence of visible aetiological agents. In these situations it is the combination of histopathological examination with additional diagnostic tests that will provide specific diagnostic information (e.g. bacteriology, virology, toxicology, etc.).

Samples for histopathology are frequently taken from the fish as tissue samples excised from dissected organs, which are then placed in a fixative (usually 10% neutral buffered formalin) and sent to a diagnostic laboratory. Both tissue sample thickness and tissue:fixative ratio are important factors in ensuring successful tissue fixation, with 5–10 mm thickness and 1:10 ratio being adequate parameters. Another relevant factor is the timing of sampling following euthanasia/death, because autolysis (i.e. post-mortem tissue self-digestion) occurs quickly in fish and will render histopathological examination difficult or even impossible. This is the reason why sampling immediately after euthanAsia is recommended, and sampling of moribund fish rather than sampling of dead fish is always preferred.

For practical reasons (this is a short and non-comprehensive chapter) we will focus on the lesions noted in commonly sampled organs (i.e. gills, hepatopancreas, spleen, kidney and intestine) in relevant infectious diseases of tilapia. Despite its limitations, we hope this chapter will provide a solid starting point and a reference for aquaculture workers working in tilapia disease diagnostics. To aid further with this, we also provide a glossary of commonly used terms at the end of the chapter.

Specific organ maps

Hepatopancreas

The tilapia hepatopancreas (HP, commonly referred to as the liver) is its largest digestive organ. It is an encapsulated, reddish-brown organ that extends posteriorly from the anterior coelomic cavity. When the head of the fish is placed to the left and the tail to the right during necropsy and the coelomic cavity is opened, the HP is clearly visible above the other internal organs. The HP comprises a thin connective tissue membrane and parenchymal cells including hepatocytes and exocrine pancreas, as well as vascular sinusoids and veins. The normal hepatocyte is polyhedral in shape and appears vacuolated (because it stores glycogen and fat). Hepatocyte nuclei typically have moderately stained chromatin margins and a prominent nucleolus (Fig. 14.1). Tilapia pancreatic tissue is distributed as islands within the HP, composed of glandular cells and bile ductal cells (Fig. 14.1). The HP is a vital organ in tilapia, acting as a storage location for fats and carbohydrates, a metabolic hub (e.g. lipid, protein and carbohydrate metabolism and protein synthesis) and a key player in homeostasis via regulation of blood chemistry and waste excretion.

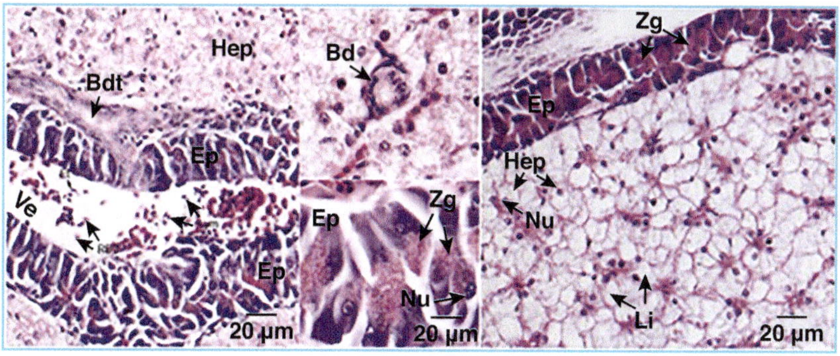

Figure 14.1 H&E-stained section of liver from a healthy Nile tilapia (*O. niloticus*). Normal hepatocytes have a polyhedral shape. Hep, hepatocyte; Ep, exocrine pancreas; Bdt, bile duct; Li, lipid/glycogen vacuole; Nu, nucleus; RBC, red blood cells; Si, sinusoid; Ve, vein; Zg, zymogen granules.

Pathognomonic lesions

The hepatopancreas is regarded as a useful organ for diagnosing infectious diseases in tilapia. And indeed, numerous viral infections in tilapia have reportedly resulted in pathognomonic histopathological lesions in this organ (Table 14.1). For instance, syncytial hepatitis in tilapia was described as a diagnostic histopathological feature of tilapia lake virus (TiLV) infection (syncytia – giant multinucleated hepatocytes) (Fig. 14.2 and Fig. 14.3) (Ferguson *et al.*, 2014; del-Pozo *et al.*, 2017; Dong *et al.*, 2017a; Behera *et al.*, 2018). Although the mechanism by which syncytia formation remains unknown, this histopathological feature is now recommended as one of the critical criteria for confirming a TiLV infection (Tang *et al.*, 2021). Regarding disease caused by *Megalocytivirus*, infectious spleen and kidney necrosis virus (ISKNV)-diseased fish typically exhibited basophilic hypertrophied cells or megalocytes (megalocytosis) in multiple organs, including the HP (Ramírez-Paredes *et al.*, 2021). Liu *et al.* (2020) described a novel virus, *Tilapia parvovirus* (TiPV) that infects farmed tilapia in China. Although the virus was detected in various organs of infected fish, there was no description of a pathognomonic histopathological lesion. We recently identified Cowdry type A inclusion bodies in the pancreas as a distinct

Table 14.1 Histopathological changes of common infectious diseases in tilapia. Data were collectively obtained from various publications and cases in the author's laboratory.

Disease	Causative agent	Histopathological changes						References
		Hepatopancreas	Kidney	Spleen	Gills	GIT	Secondary organs	
Syncytial hepatitis of tilapia (SHT) or tilapia lake virus disease (TiLVD)	*Tilapia tilapinevirus*	Syncytia formation (multinucleated giant cells); hepatocyte disassociation, foamy cytoplasm and necrosis	Increasing nos of MMCs, infiltration of inflammatory cells	Increasing cellular debris-laden macrophages	NA	Gastric mucosal epithelial necrosis	Meningeal inflammation, syncytia	Ferguson *et al.*, 2014; Eyngor *et al.*, 2014; de-Pozo *et al.*, 2017; Dong *et al.*, 2017; Behera *et al.*, 2018; Dinh-Hung *et al.*, 2021
Infectious spleen and kidney necrosis disease (ISKND)	*Megalocytivirus* ISKNV	Megalocytosis	Megalocytosis, necrosis of haematopoietic tissues and glomerulus	Megalocytosis, necrosis	Megalocytosis	Megalocytosis	Megalocytosis	Subramaniam *et al.*, 2016; Ramírez-Paredes *et al.*, 2021.
Viral nervous necrosis (VNN)	*Betanodavirus*	NA	NA	NA	NA	NA	Neuronal vacuolation (eyes/brain/spinal cord)	Keawcharoen *et al.*, 2015; Machimbirike *et al.*, 2019.

Streptococcosis	*Streptococcus agalactiae*	Granulomas, congestion, increasing numbers of MMCs	Haemorrhages and thrombosis in glomeruli and tubules, tubular degeneration, haematopoietic tissue atrophy	Granulomas, increasing nos of MMCs	NA	NA	Meningeal inflammation, infiltration of lymphocytes	Zamri-Saad *et al.*, 2010; Rodkhum *et al.*, 2011; Laith *et al.*, 2017.
Francisellosis	*Francisella orientalis*	Granulomas	Granulomas	Granulomas	Granulomas	Granulomas	NA	Soto *et al.*, 2009; Nguyen *et al.*, 2015.
Edwardsiellosis of tilapia	*Edwardsiella ictaluri*	Hepatomegaly with reduced fat reserve	Multifocal necrosis	Granulomatous splenitis, multifocal necrosis	Multifocal branchitis	NA	Encephalitis	Soto *et al.*, 2013; Dong *et al.*, 2019.
Columnaris	*Flavobacterium columnare* complex*	NA	NA	NA	NA	Severe necrosis of gill filaments	Dermal necrosis	

Continued

Table 14.1 (Continued)

		Histopathological changes						
Disease	Causative agent	Hepatopancreas	Kidney	Spleen	Gills	GIT	Secondary organs	References
Haemorrhagic septicaemia	*Aeromonas* spp.	Haemorrhages, congestion, necrosis, accumulation of haemosiderin	Collapsed kidney tubules and sloughing of epithelial cells into the lumen, hyaline degeneration	Haemorrhages	HyperplAsia and fusion of gill lamellae	Enteritis	Congestion in the brain	Dong *et al.*, 2017b; Korni *et al.*, 2017
Epitheliocystis	*Chlamydia*-like organism (CLO)	NA	NA	NA	Basophilic colonies of bacteria	NA	NA	Pádua *et al.*, 2015; Steckert *et al.*, 2018

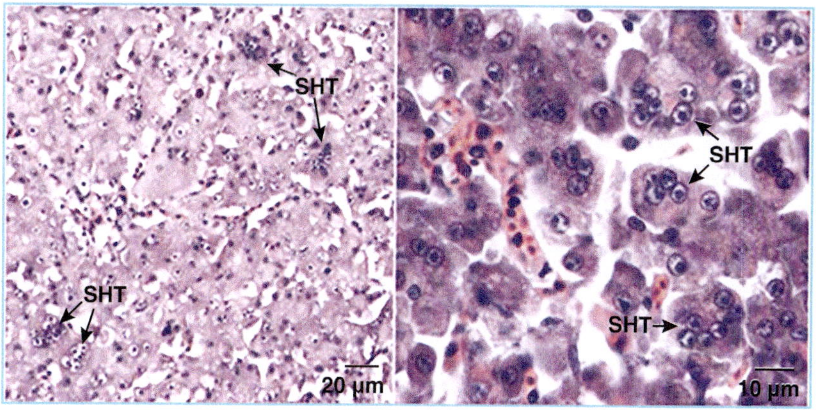

Figure 14.2 Microphotographs in H&E-stained liver from a TiLV-infected tilapia reveal typical syncytia formation (SHT, arrows), the pathognomonic lesion of TiLV infection.

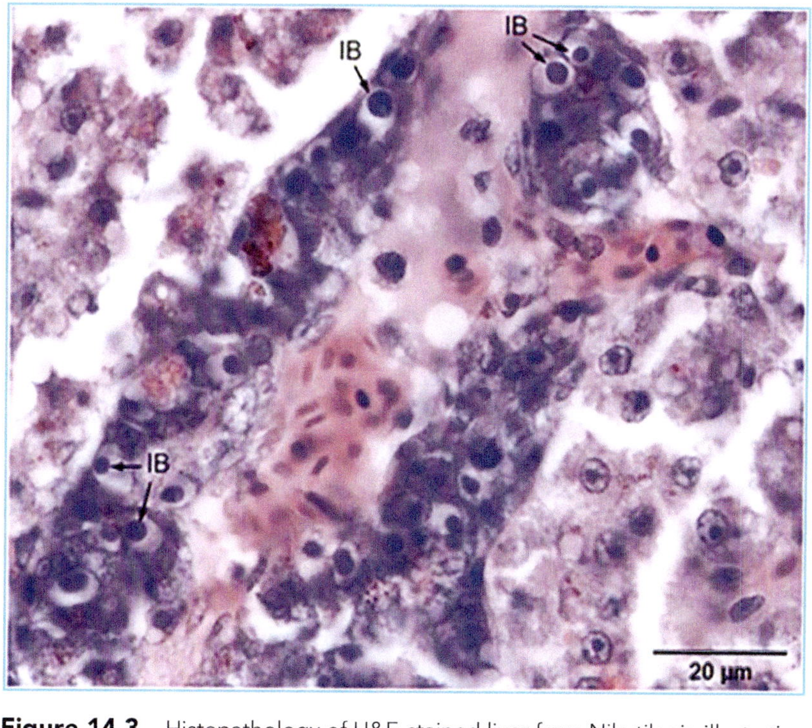

Figure 14.3 Histopathology of H&E-stained liver from Nile tilapia illustrating the presence of numerous basophilic intranuclear inclusion bodies (IBs; Cowdry type A) in the pancreatic cells, the pathognomonic lesion caused by *Tilapia parvovirus* (TiPV) infection.

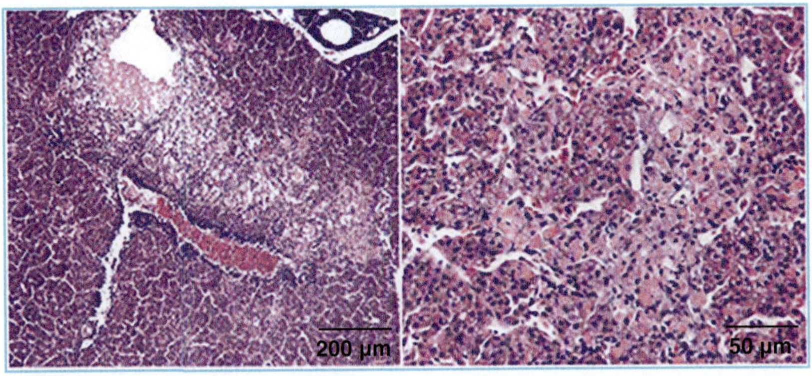

Figure 14.4 Liver of a TiLV-infected tilapia under low (left) and high (right) magnification reveals severe focal hepatocellular necrosis, pancreatic necrosis, vascular congestion and perivascular infiltration by lymphocytes. H&E stain.

pathological feature found in all clinical samples of adult Nile tilapia infected with this virus (Fig. 14.4). A Cowdry type A inclusion body is characterized by chromatin margination in the nucleus, with a clear halo around a central basophilic intranuclear inclusion body.

Non-pathognomonic lesions

Infectious diseases in tilapia usually present with non-pathognomonic histopathological patterns in the HP, either individually or in combination. These may include congestion, haemorrhage, hepatocyte degeneration/dissociation/necrosis/vacuolation/pigment accumulation and/or inflammatory infiltration (Figs. 14.4–14.8). For example, in TiLV-infected fish, typical syncytial formation is concurrent with focal/single hepatocyte necrosis, severe degeneration of pancreatic tissue with HP infiltration by inflammatory cells, and hepatic vacuolation (Fig. 14.4). Also, some hepatocytes can lose storage fats and have basophilic cytoplasm which contains eosinophilic inclusions or lipoprotein-like droplets (Ferguson *et al.*, 2014; del-Pozo *et al.*, 2017; Dong *et al.*, 2017a) (Fig. 14.4).

There are patterns highly indicative of a specific aetiology, but these can appear in other infections. For example, granulomas (occasionally necrotizing) form during chronic infection with *Francisella*

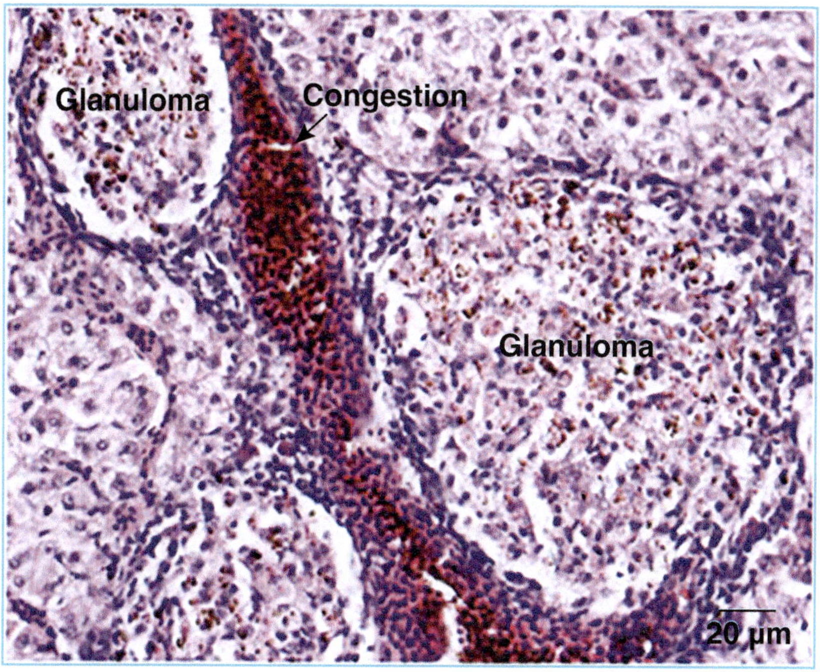

Figure 14.5 Liver histopathology of diseased tilapia caused by *Francisella orientalis* showing well-defined granulomas and severe congestion (i.e. intravascular packing of erythrocytes).

orientalis but are not exclusive to this infection (Fig. 14.5). In *Streptococcus* spp. infection there is a predominance of melanomacrophage centres within the exocrine pancreas and/or intra-lesional colonies of bacterial cocci during the chronic phase (Fig. 14.6). Hepatic necrosis, intra-sinusoidal congestion and intra-cytoplasmic accumulation of pigment (haemosiderin/ceroid/lipofuscin) frequently occur during infection with motile *Aeromonas* spp., but also with other bacterial species (Figs. 14.7 and 14.8).

Spleen

The spleen is a secondary lymphoid organ in tilapia. This organ is located alongside the stomach, beneath the HP and, in healthy fish, is dark red. Functionally the spleen is involved in 'filtering' blood, removing foreign agents (e.g. bacteria and viruses), recycling old red

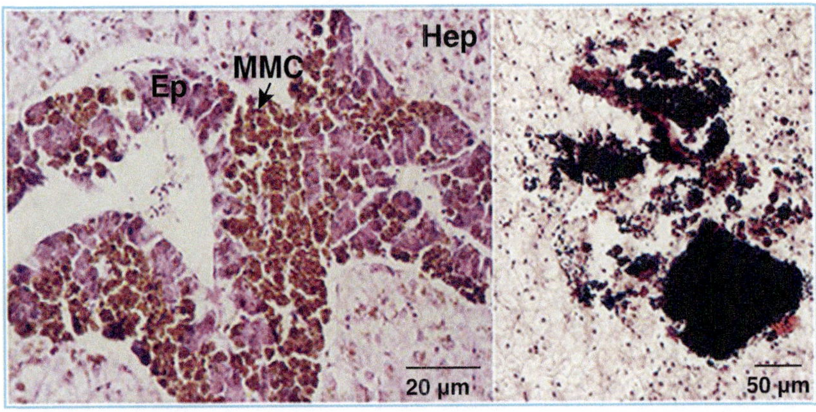

Figure 14.6 Photomicrographs of the H&E-stained liver from an adult Nile tilapia infected with *Streptococcus agalactiae* showing prominent melanomacrophage centres within exocrine pancreas islands (left). There are large Gram-positive bacterial coccal colonies in the liver (right, Gram stain). MMC, melanomacrophage centre; Ep, endocrine pancreas; Hep, hepatocyte.

blood cells and providing immune surveillance through its resident leukocytes. The spleen is composed of a thin capsule of fibrous connective tissue and parenchymal tissues that include red pulp, white pulp and melano-macrophage centres (MMCs), which are clusters of pigment-containing macrophages found in teleost tissues (Fig. 14.9). The red pulp is primarily composed of erythrocytes and thrombocytes within blood vessels and ellipsoids (which constitute the majority of the splenic parenchyma), whereas the white pulp is predominantly composed of leukocytes around ellipsoids and MMCs. Macrophages in MMCs contain pigments such as melanin (black), haemosiderin, ceroid or lipofuscin (yellow–pink to golden brown) within their cytoplasm. This cell type is abundant in the spleen but is also found in other organs (Agius & Roberts, 2003; Ferguson, 2006). MMCs serve as a repository for cell breakdown products (e.g. iron from senescent erythrocytes) and play a critical role in the capture, processing and presentation of antigens to lymphocytes. MMCs typically become larger and more numerous in response to chronic inflammatory lesions, but paradoxically they can also serve as foci for resistant intracellular bacteria (Agius & Roberts, 2003). The ellipsoids (a.k.a. periarterial macrophage sheaths) form

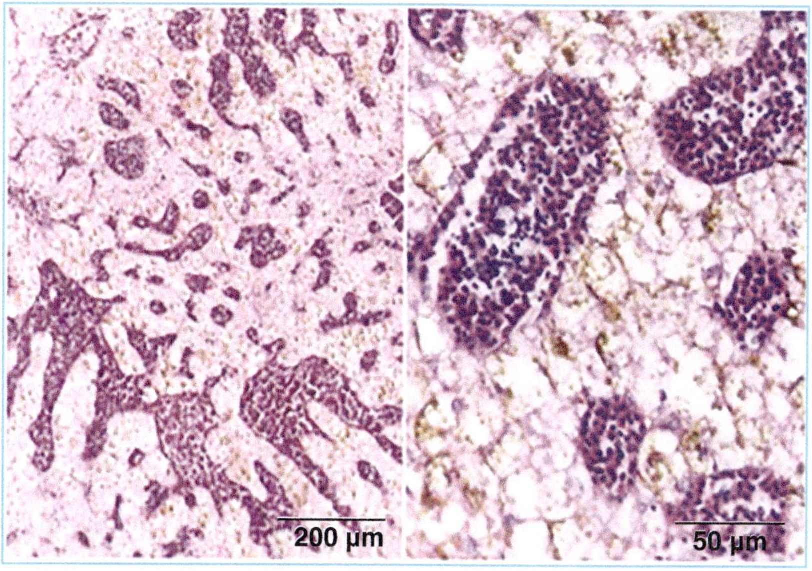

Figure 14.7 Photomicrographs of HP from a Nile tilapia concurrently infected with *S. agalactiae* and *Micrococcus* sp. There is severe intra-sinusoidal congestion and intra-cytoplasmic accumulation of brown pigment in hepatocytes.

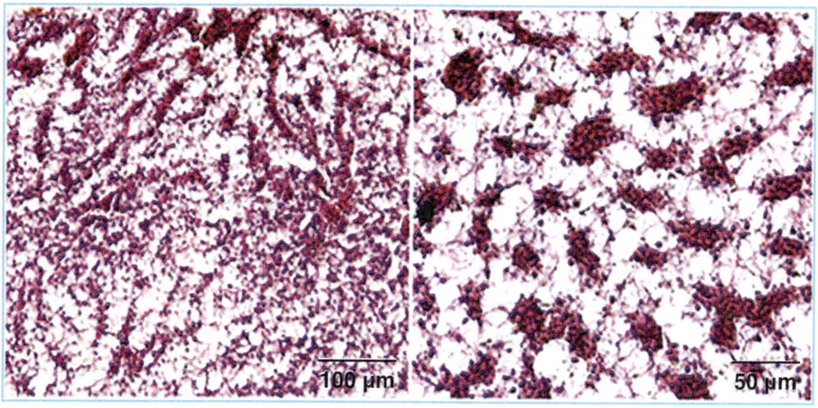

Figure 14.8 Liver of Nile tilapia experimentally infected with *Aeromonas veronii* showing severe sinusoidal congestion and intra-cytoplasmic accumulation of brown pigment (haemosiderin, suspected). Pictures were taken at low (left) and high (right) magnifications. H&E stain.

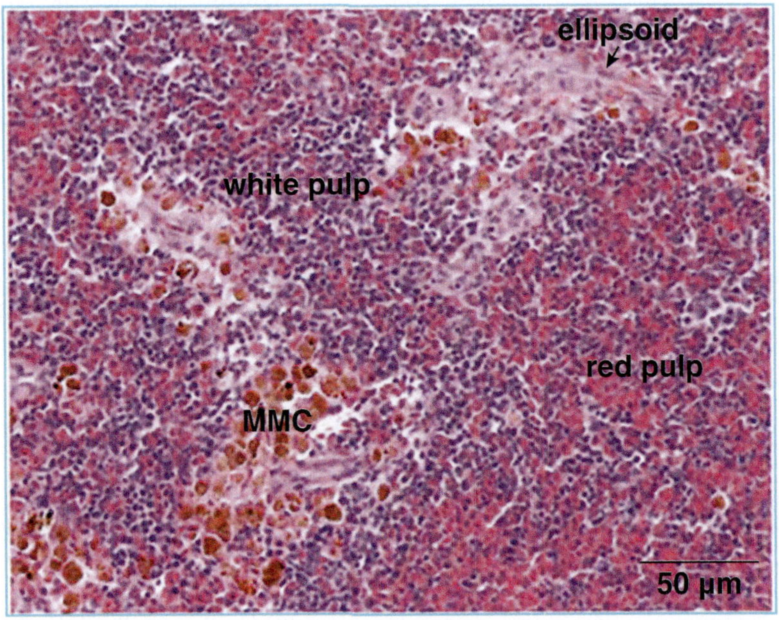

Figure 14.9 Photomicrograph of H&E-stained spleen from a normal Nile tilapia. Note that the boundary between red and white pulp is not well defined and, instead, these intersperse.

at the end of splenic arterioles, which are composed of a meshwork of reticular cells and fibres and macrophages. The ellipsoids capture foreign antigens (bacteria, viruses) and metabolic by-products, which are then transferred to MMCs.

Pathognomonic lesions

Megalocytivirus ISKNV infection usually results in the formation of variable numbers of splenic megalocytes (hypertrophied cells) superimposed on parenchymal necrosis. Megalocytes take on a deeply basophilic stain with H&E (Fig. 14.10).

Non-pathognomonic lesions

Unsurprisingly, various infections in tilapia cause severe lesions of the spleen, most of which are non-pathognomonic. For example, ISKNV

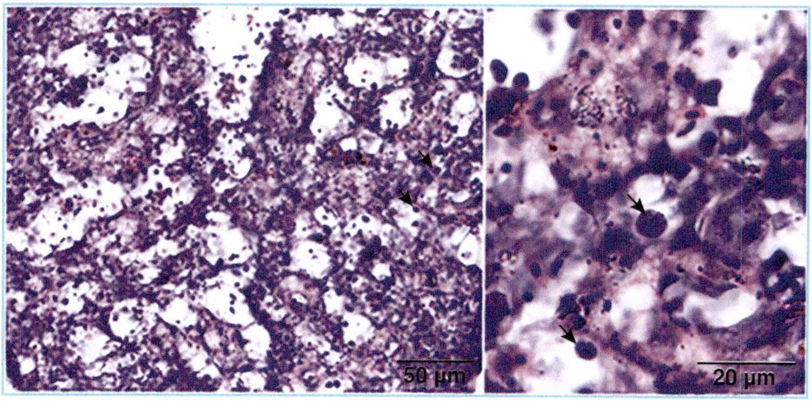

Figure 14.10 Histopathology of tilapia spleen naturally infected with *Megalocytivirus* ISKNV showing severe parenchymal necrosis with the presence of scattered basophilic hypertrophied cells (arrows).

infection causes severe splenic necrosis (Fig. 14.10) while TiLV infection presents with increased MMCs with melanin dispersion and cellular debris-laden macrophages (Fig. 14.11). Frequently, necrotic cells in viral infections present with condensation of chromatin in the nucleus (pyknosis) and fragmentation of the nucleus (karyorrhexis).

Several bacterial pathogens with an intracellular lifestyle cause diseases in tilapia, including *F. orientalis*, *E. ictaluri*, *S. agalactiae*, *Nocardia* sp. and *Mycobacterium* sp. (Soto *et al.*, 2010, 2013; Nguyen *et al.*, 2016; Dong *et al.*, 2019). These bacteria usually cause chronic infection with granulomas in the spleen and other organs. Granulomas are characterized by clusters of macrophages variably admixed with other inflammatory cell types, which are aimed at isolation of the pathogen and presenting it to the immune system. During *F. orientalis* infection these melanomas are mostly composed of epithelioid macrophages, can feature melanin-containing macrophages and frequently have a central necrotic core (Fig. 14.12). Increasing numbers of granulomas and hyperactivation of MMCs with severely overloaded melanophores are observed in the spleen of tilapia chronically infected with *S. agalactiae* (Fig. 14.13). In tilapia infected with *E. ictaluri*, histopathological lesions in the spleen include multifocal necrosis, pyogranulomas and occasional observation of clumps of rod-shaped bacteria within MMCs (Fig. 14.14).

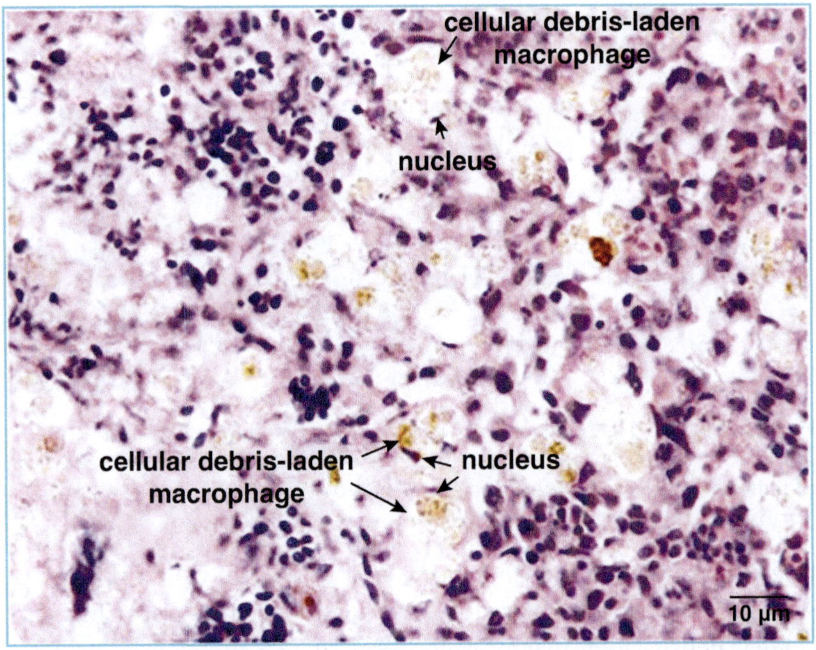

Figure 14.11 Histopathology of hybrid red tilapia spleen naturally infected with TiLV.

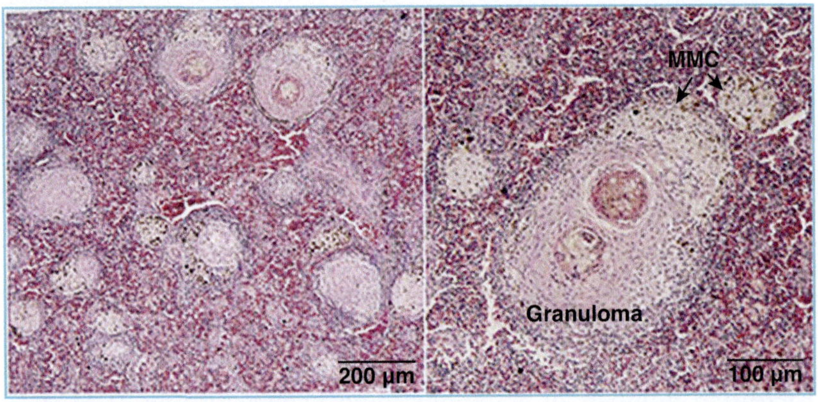

Figure 14.12 Histopathology of tilapia spleen naturally infected with *Francisella orientalis* under low (left) and high (right) magnification, showing multiple areas of granulomatous necrosis adjacent to MMCs.

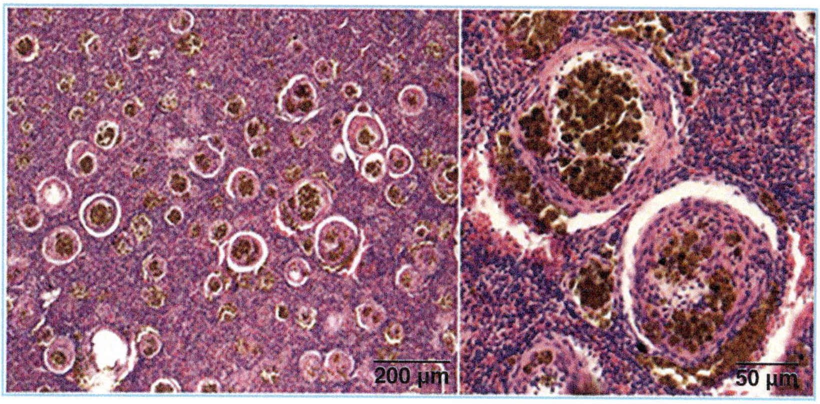

Figure 14.13 Histopathology of spleen from tilapia naturally infected with *S. agalactiae* and *Tilapia parvovirus* under low (left) and high (right) magnification, showing widespread prominence of MMCs and well-defined, frequently pigmented granulomas. H&E stain.

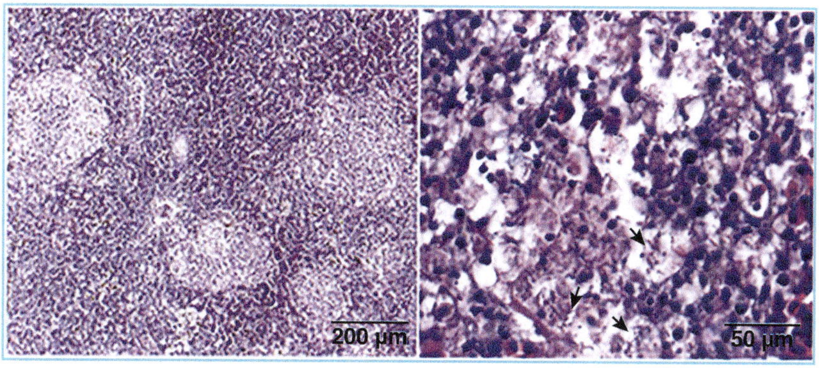

Figure 14.14 Histopathology of spleen from tilapia infected with *Edwardsiella ictaluri* under low magnification (left) showing severe multifocal pyogranulomas with associated necrosis. A necrotic area at higher magnification (right) showing necrotic cells, pyknosis and colonization of rod-shaped bacteria (arrows). H&E stain.

Kidney

Histology

The tilapia kidney is a complex organ that is involved in the maintenance of homeostasis through the clearance of metabolic waste and exogenous substances, as well as the production of white blood cells (haematopoiesis). Within and adjacent to the kidney there are two endocrine glands (the inter-renal gland and the corpuscles of Stannius), which are not described here. The tilapia kidney is located retroperitoneally (i.e. behind the peritoneal membrane) in the dorsal aspect of the coelomic cavity, and is an elongated, friable, dark tan organ that extends from the cranial to the caudal aspect of this cavity. This organ is divided into the anterior kidney (a.k.a. head kidney), the kidney body and the posterior kidney. These areas are distinct both functionally and histologically, whereby the anterior kidney is predominantly composed of haematopoietic tissue, the posterior is predominantly composed of nephrons (i.e. filtrating units) and the body is an even mixture of these two. Two ureters are found running below the peritoneum, along the lateral aspect of the kidney and converge caudally at the midline, where a small urinary bladder is present that empties into the cloaca. Histologically, haematopoietic tissue is composed of sheets of round cells within the renal interstitium (i.e. the space between nephrons). When examined closely, haematopoietic cells are of variable size. The details of haematopoiesis are beyond the scope of this review but, as a rule of thumb, it is useful to know that the size of these cells depends on their stage in the haematopoietic process (i.e. cells are larger at earlier stages of this process). Normally smaller cells tend to predominate (more mature), but there are situations where relatively larger haematopoietic cells predominate (less mature). For example, this can occur in a systemic inflammatory reaction where leukocytes are used more rapidly than they are produced (i.e. left shift in haematopoiesis). This is a stereotypical pattern of the renal interstitium, which can be noted in several diseases. The nephrons constitute the filtration unit of the kidney, and in tilapia they lack a Henlé loop (i.e. mesonephric kidney), as opposed to the mammalian nephron. This means that their kidney cannot concentrate urine beyond the level of the osmolarity of the blood. The nephron is composed of a glomerulus, proximal and distal tubules and collecting tubules, which are (apparently) haphazardly arranged in the kidney. The glomeruli are round to oval and are composed of a glomerular tuft (containing capillaries and mesangial cells) within

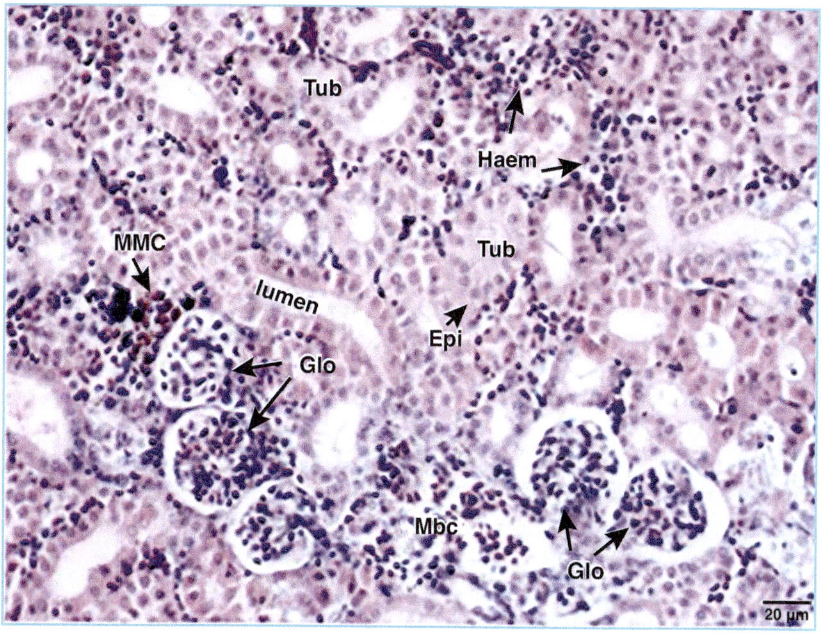

Figure 14.15 Histological section of posterior kidney from Nile tilapia. Epi, epithelial cell; Glo, glomerulus; Haem, haematopoietic tissue; Mbc, mature blood cells; Tub, tubules. H&E stain.

an encapsulated space (space of Bowman). The proximal tubules are involved in active excretion/resorption of metabolites, and can be recognized by a cuboidal/columnar epithelial lining featuring a brush border (a dense eosinophilic band at the luminal aspect of these cells). The distal tubules, which are involved in water resorption, are also lined by cuboidal/columnar epithelium but lack a brush border. Finally, the collecting tubules are much rarer and are lined by flattened epithelial cells (Fig. 14.15).

Pathognomonic lesions

There are very few infections in tilapia that cause pathognomonic lesions to the kidney. One of these lesions is megalocytosis caused by *Megalocytivirus* ISKNV, which can be observed in multiple organs, including the kidney. Figure 14.16 illustrates typical megalocytosis in the haematopoietic tissue of infected tilapia.

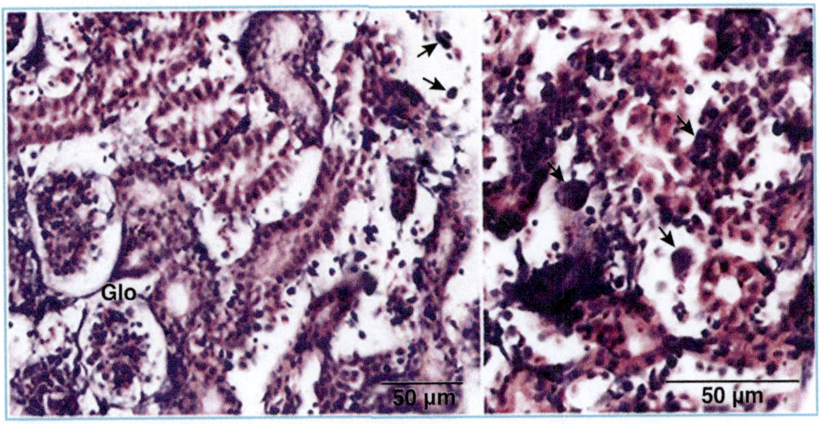

Figure 14.16 Photomicrographs of tilapia kidney infected with *Megalocytivirus* ISKNV taken under low (left) and high (right) magnifications showing the presence of typical megalocytes (arrows) and necrosis of haematopoietic tissue and glomeruli. H&E stain.

Non-pathognomonic lesions

As with the spleen, many bacterial and viral infections of tilapia cause non-pathognomonic lesions to the kidney. In particular, *Megalocytivirus* ISKNV infection causes severe necrosis and degeneration of haematopoietic tissue and glomeruli, superimposed on the presence of megalocytes (Fig. 14.16). TiLV infection usually results in an increase in numbers of melanin-containing macrophages and interstitial infiltration by lymphocytes, superimposed on the presence of hepatocellular syncytia (Fig. 14.17). Well-defined granulomas and interstitial infiltration of lymphocytes are commonly observed in the kidney of tilapia infected with *F. orientalis* (Fig. 14.18). In cases of motile *Aeromonas* spp. infection (i.e. *A. veronii*, *A. jandaei*) we sometimes observed attenuation of tubular epithelium with intra-luminal accumulation of necrotic epithelial cells, consistent with tubular epithelial necrosis possibly caused by bacterial toxins (Fig. 14.19). Another type of stereotypical pattern is hyaline droplets within the cytoplasm of tubular epithelial cells, which is frequently observed in the kidney of the fish following bacterial or viral infections, but can also be an incidental finding (Fig. 14.20).

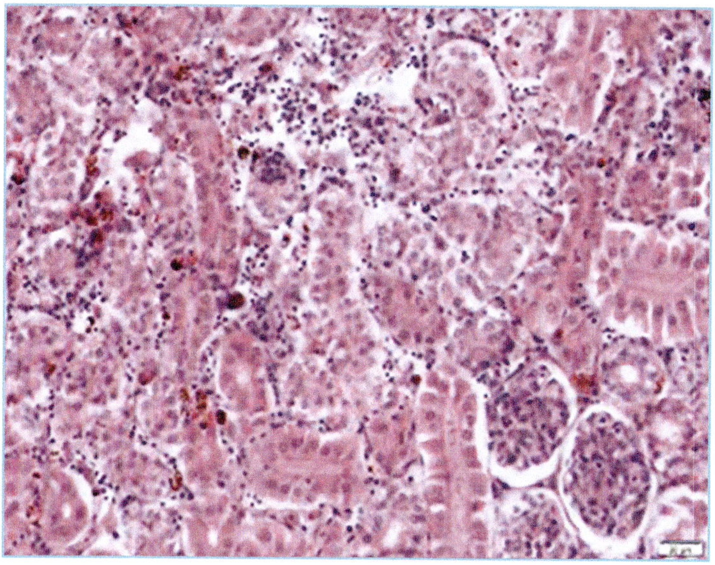

Figure 14.17 Photomicrograph of tilapia kidney infected with TiLV showing infiltration of basophilic inflammatory lymphocytes and melanin-containing cells. H&E stain.

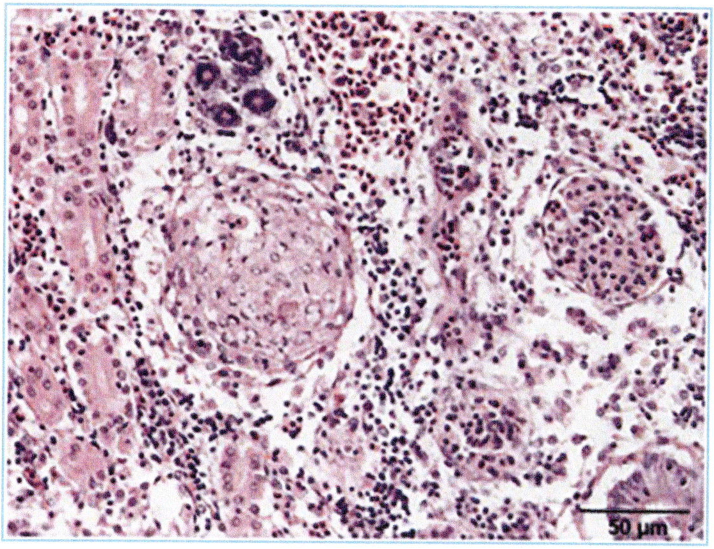

Figure 14.18 Photomicrograph of tilapia kidney infected with *F. orientalis* showing manifestation of typical granulomas and infiltration of basophilic inflammatory lymphocytes. H&E stain.

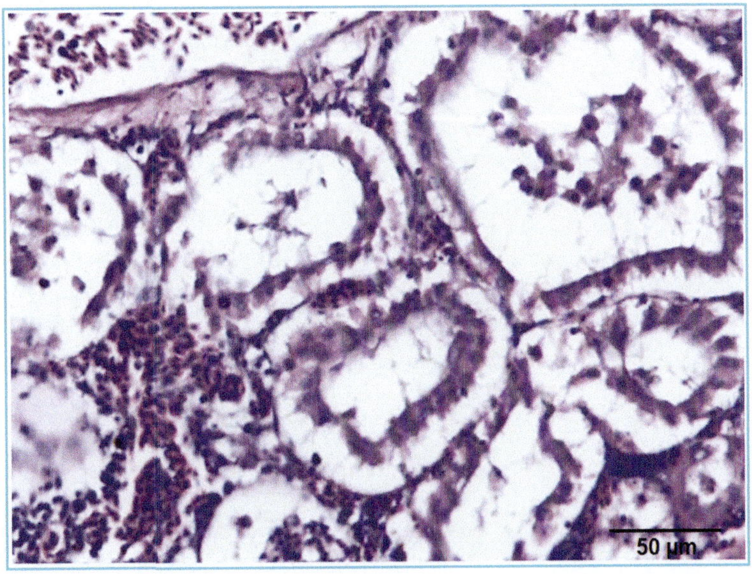

Figure 14.19 Histopathology of tilapia kidney infected with motile *Aeromonas* spp. revealing collapsed kidney tubules and sloughing of epithelial cells into the lumen. H&E stain.

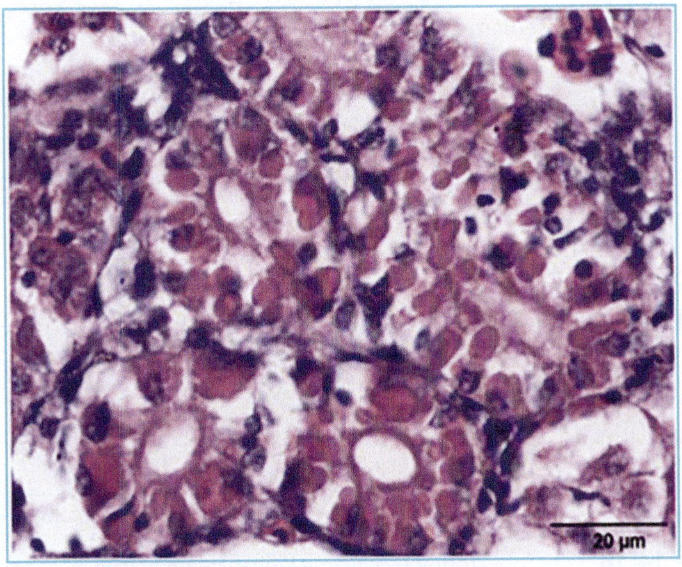

Figure 14.20 Histopathology of tilapia kidney showing hyaline droplets within tubular epithelial cells. H&E stain.

Gills

Histology

The gills of tilapia are involved not only in respiration but also in the maintenance of homeostasis (e.g. blood ion content and pH) through electrolyte transfer and excretion of metabolic nitrogenous waste (in the form of ammonia). They are found within both opercular cavities at the base of the head, where they comprise four gill arches on each side. The gills separate the oral and opercular cavities, and the gill arches have bony/cartilaginous processes cranially (i.e. gill rakers) that form a barrier to particulate matter from the oral cavity. The arches are supported by bone at their core and have a prominent vascular component (i.e. afferent and efferent branchial arteries). At their caudal aspect the arches support two lines of successive, slender, highly vascularized branchial filaments. These have a central cartilaginous core, have lymphoid aggregates at their base and are lined by stratified cuboidal to flattened epithelium. On the surface and extending along the length of the filaments there are two tightly packed rows of lamellae. These are the site of respiratory exchange and are thin, roughly semicircular structures continuous with the filament at their base. Histologically, lamella are overlaid by a single layer of flattened epithelial cells sitting on a basement membrane. This structure is supported at the centre by alternating pillar cells (which are part of the reticulo-endothelial system) and vascular spaces. Also overlaying the base of the lamella and the inter-lamellar space on the filament are chloride cells, which are involved in ion transfer. These cells are characterized by a relatively large eosinophilic cytoplasm. To avoid possible confusion, it is important to note that there is an alternative nomenclature to the above, where filaments are known as primary lamellae and lamellae are known as secondary lamellae (Fig. 14.21). The branchial system creates a large respiratory surface that is in contact with the external environment. This is particularly the case for the external arch of the gill, which can present with non-specific mechanical lesions that are not representative of the remainder of the gill. For this reason, sampling of the second gill arch is normally advocated to avoid misinterpretation of the sample.

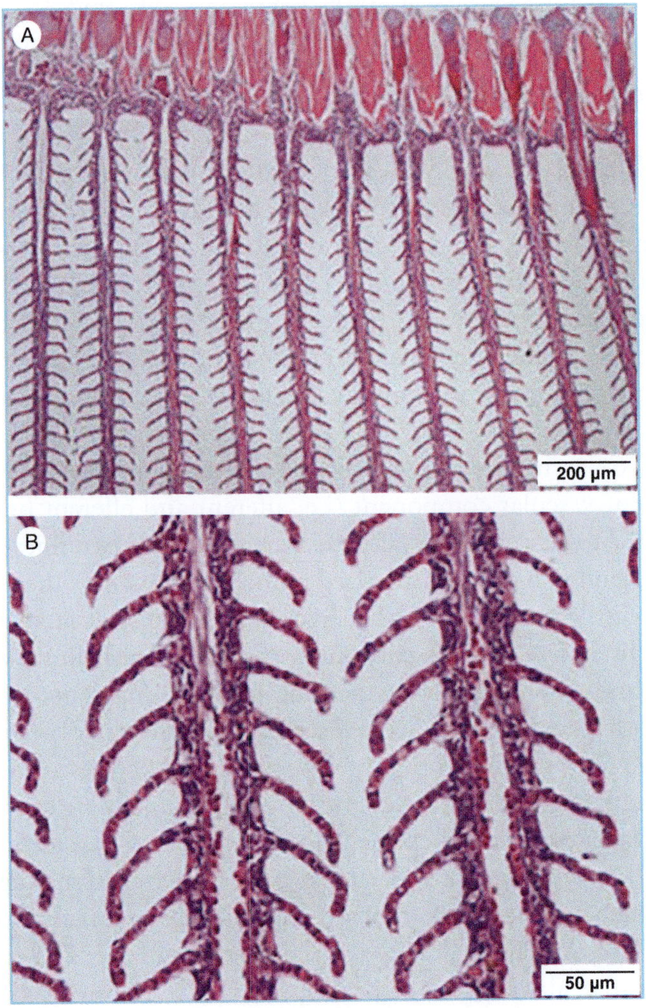

Figure 14.21 Photomicrographs of normal Nile tilapia gill arch with numerous gill filaments (a), and higher magnification of normal gill filaments and lamellae (b). H&E stain.

Pathognomonic lesions

Most parasitic infections caused by protozoa or monogeneans in tilapia have typical histopathological features due to the unique morphology of the parasites within the gill lesions. In particular, gill infection by the ciliated parasite *Trichodina* spp. usually

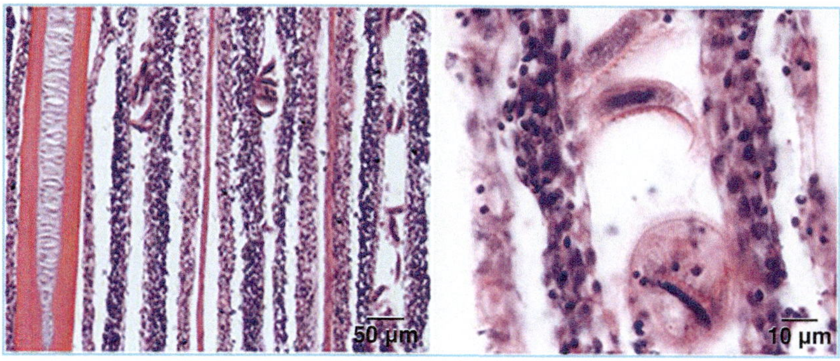

Figure 14.22 Histopathology of tilapia gill under low (left) and high (right) magnification showing *Trichodina* sp. infection free floating and attached to the gill, with severe infiltration of the gill filament by lymphocytes (inflammation). H&E stain.

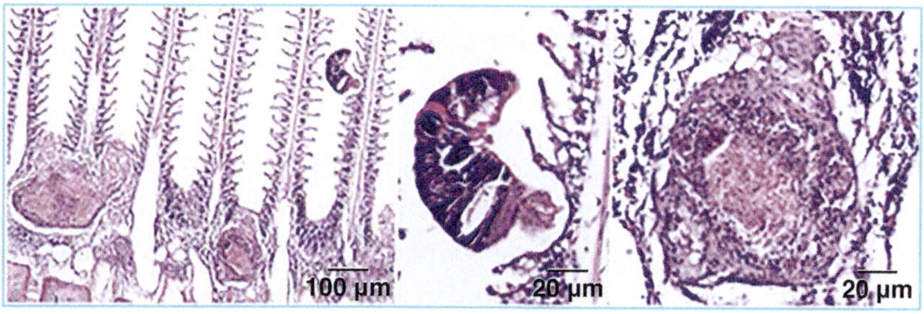

Figure 14.23 Histopathology of the gills of tilapia concurrently infected with *F. orientalis* and monogenean trematodes, taken under low (left) and high (middle and right) magnification. The middle image illustrates the typical morphology of the parasite attached to the gill filament by the opisthaptor; the right-hand image shows a necrotic granuloma. H&E stain.

presents with bell-shaped parasites located between the gill filaments (Fig. 14.22). In the case of monogenean trematode infection, parasites can be seen attached to the gill filament by their opisthaptor hooks (Fig. 14.23). Regarding bacterial infection caused by *Chlamydia*-like organisms, this features abundant, characteristic, cyst-like basophilic cytoplasmic inclusions within lamellar epithelial cells formed by colonization of bacteria, strikingly devoid of associated inflammation (Fig. 14.24).

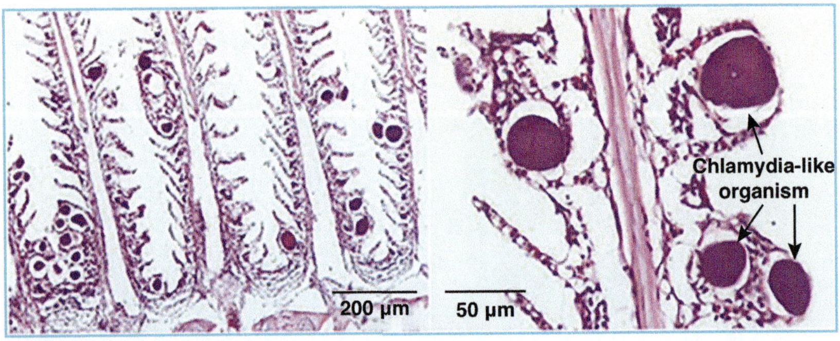

Figure 14.24 Photomicrographs of the gills taken under low (left) and high (right) magnifications showing pathognomonic lesion of epitheliocystis infection with abundant intra-epithelial cyst-like basophilic inclusions in scattered lamellar epithelial cells. H&E stain.

Non-pathognomonic lesions

The delicate structure of the gill – and in particular the lamella – is particularly sensitive to chemical and physical changes in the aquatic environment. These irritants result in lesions to the lamellar epithelium that may include swelling (hypertrophy), hydropic degeneration (vacuolation), detachment and even necrosis in acute exposures, with or without lamellar fusion. In chronic cases there is lamellar hyperplasia either in isolation or combined with acute lesions. This hyperplasia is characterized by the accumulation of haphazardly arranged epithelial cells that fill the inter-lamellar space to a variable degree. In more severe cases the entirety of the inter-lamellar space is obliterated by these cells and the filament adopts a club-like appearance. The presence/absence and severity of these lesions will depend on the severity of the insult itself and the timelines associated with it.

Gastrointestinal tract

The gastrointestinal tract (GIT) plays a critical role in food uptake, digestion and nutrient absorption. The GIT of tilapia consists of the mouth, pharynx, oesophagus, stomach (sac-like), intestine (long

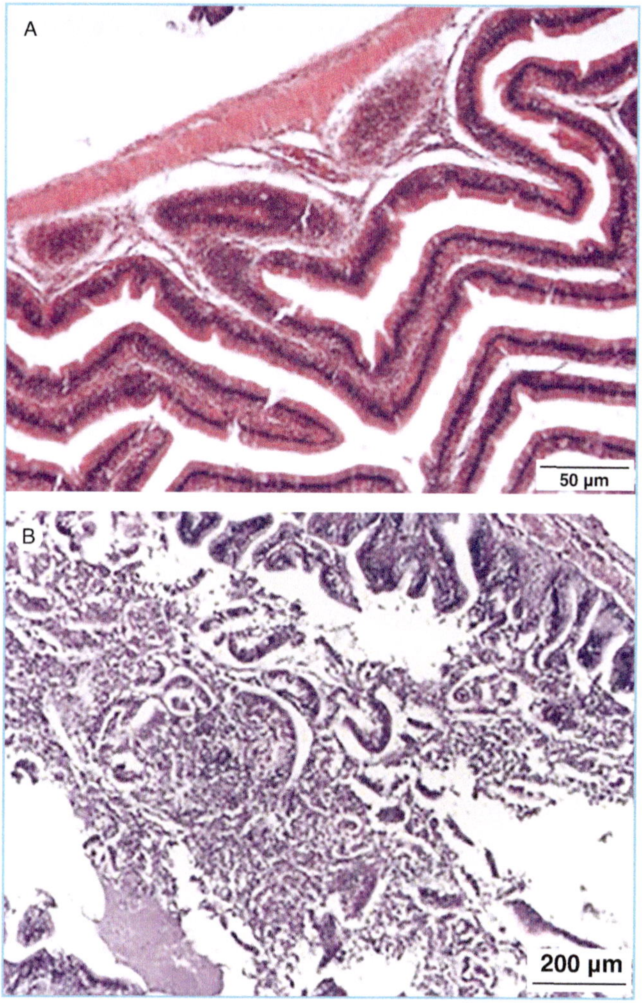

Figure 14.25 Photomicrographs of the normal intestine (a) and severe necrosis of microvilli due to motile *Aeromonas* infection (b). Note that the latter is difficult to differentiate from autolysis. H & E stain.

tubular) and anus. The intestine is an important part of the GIT for studying histopathological lesions due to infectious agents. Histologically, the GIT of tilapia consists of four layers, the mucosa, submucosa, muscularis and serosa, from the lumen to the outer aspect. Longitudinally the intestine of tilapia is divided into three segments: anterior (proximal), mid- (middle) and posterior (distal, including rectum). Structurally, the mucosa of the anterior intestine forms branched, zig-zag folds. The mid-intestine is made up of short longitudinal unbranched folds while the posterior intestine has shorter intestinal transverse folds. The epithelial layer of the entire intestine is composed of columnar epithelial cells with a prominent brush border (enterocytes – absorptive cells) and mucous cells with a basal nucleus (goblet cells – secrete mucin and create a protective mucus layer). Lymphocytes and rodlet cells are commonly present in the mucosa, interspersed with the above. The lamina propria consists of loose connective tissue supporting the microvilli. The lamina propria and submucosa usually contain eosinophilic granular cells (probably mucosal mast cells) and gut-associated lymphoid tissue (GALT – clusters of lymphoid tissue) (Phrompanya *et al.*, 2019; Okuthe and Bhomela, 2020). Notably, the GIT is prone to autolysis, which can look very similar to necrosis, leading to misinterpretation.

Pathognomonic lesions

As with its infection of other organs, *Megalocytivirus* ISKNV results in the formation of basophilic hypertrophied cells (megalocytosis) in the GIT. This pathognomonic pattern can be observed in both mucosal folds and the submucosa of the intestine (Ramírez-Paredes *et al.*, 2020).

Non-pathognomonic lesions

Several bacterial infections, especially motile *Aeromonas* spp. (*A. veronii*, *A. hydrophila*, *A. jandaei* and *A. dhakensis*), which usually produce extracellular toxins, cause widespread destruction of the mucosal folds/submucosa (Fig. 14.25) (Dong *et al.*, 2017b).

Other Organs

Other organs of tilapia are sometimes useful in disease diagnosis although they are less frequently sampled. Here we include the brain, eyes, skin and muscle.

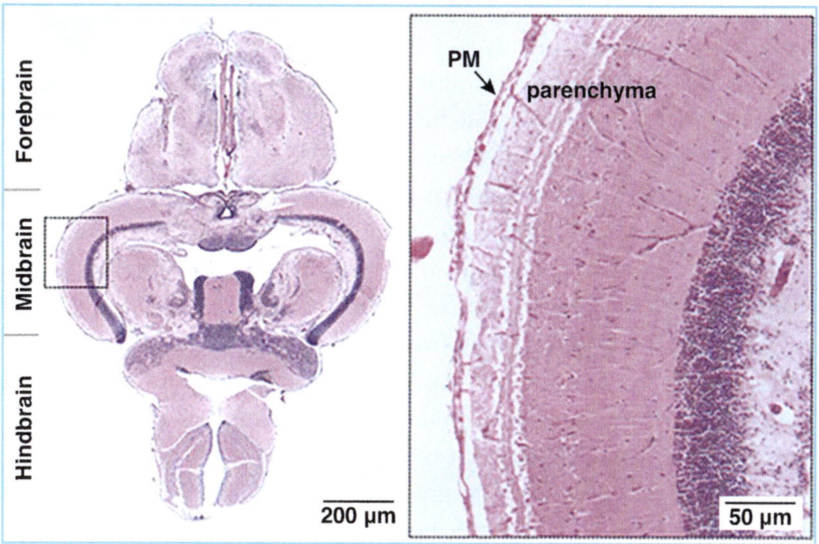

Figure 14.26 Histology of horizontal brain section from a healthy tilapia showing three main parts: forebrain, mid-brain and hindbrain (left). Histology of the optic tectum taken at a higher magnification (right). H&E stain. Images by N. Dinh-Hung.

Brain

The tilapia brain, like that of other teleosts, is divided into three parts: the forebrain, the midbrain and the hindbrain, which is protected by the primitive meninges (Fig. 14.26). Each brain region is responsible for different physiological and behavioural outputs (Dinh-Hung *et al.*, 2021). The forebrain contains the telencephalon and diencephalon. The telencephalon controls sensory, motor and cognitive tasks including memory, learning and emotion, while the diencephalon primarily serves as a centre of correlation for sensory inputs such as gustation and olfaction. The mid-brain (mesencephalon) is particularly involved in the reception and coordination of optic nerve inputs, such as motor and visual signal interpretation. The cerebellum and medulla oblongata, on the other hand, are located within the hindbrain (rhombencephalon) and are generally associated with the regulation of locomotion and balance stimuli as well as basal physiology (Roberts & Ellis, n.d.; Ferguson,

2006; Sado *et al.*, 2020; Dinh-Hung *et al.*, 2021). This is a highly vascularized organ and, as a consequence, several infections can enter it via the bloodstream and may cause neuronal/glial damage with resulting impairment of brain function. This explains abnormal behaviours during infection, such as lethargy, loss of appetite, erratic swimming and school suspension (Dinh-Hung *et al.*, 2021). Histopathologically the tilapia brain infected with *Betanodavirus* exhibits vacuolar degeneration and neuronal necrosis (Bigarré *et al.*, 2009; Keawcharoen *et al.*, 2015). Dinh-Hung *et al.* (2021) described the histopathological changes and distribution of *Tilapia tilapinevirus* in the brain where, superimposed on meningitis and widespread congestion, there was neuronal/glial degeneration within several areas (i.e. internal cellular layer of olfactory bulb, periventricular grey matter of optic tectum, granular layer of cerebellum and the vagal lobe of the medulla oblongata). Streptococcosis often results in lymphocytic meningitis with concurrent necrotic granulomas, as noted in other organs (Fig. 14.27).

Eyes

The tilapia eye is similar to that of other species, with a cornea, anterior chamber, lens, vitreous body and retina/optic nerve, from lateral to medial. The function of the eye is to project optical images onto the retina, which is stimulated and sends this information to the brain via the optic nerve for recognition and interpretation. The retina is a complex organ that is multi-layered and composed (from the conjunctival aspect into the lumen) of retinal pigment epithelium, rods and cones, the outer nuclear layer, the outer plexiform layer, the inner nuclear layer, the inner plexiform layer, the ganglion cell layer and, finally, the layer of ganglion cell axons. Keawcharoen *et al.* (2015) described necrotic retinitis with vacuolization in tilapia larvae as a characteristic lesion of *Betanodavirus* infection.

Skin and muscle

Tilapia skin is made up of two layers, the epidermis and the dermis. The thickness of the epidermis varies according to body part,

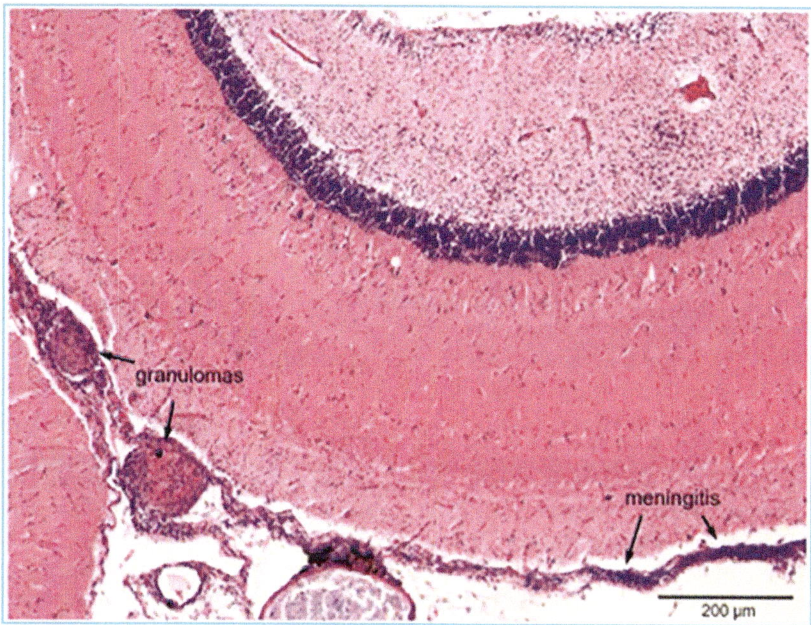

Figure 14.27 Lesions of streptococcosis in the the optic tectum of tilapia infected with *Streptococcus agalactiae*, including lymphocytic infiltration of the meninges (meningitis) and granuloma formation. H&E stain.

developmental stages, gender and environmental factors. The normal epidermis sits on a basement membrane and is mostly composed of cuboidal epithelial cells admixed with mucus-secreting cells. At the deepest aspect the epidermis has a layer of progenitor cells (basal layer), although it is necessary to highlight the fact that cell division can occur at all levels of the epidermis in fish. Other, less numerous, cell types in the epidermis include lymphocytes and macrophages, which may be more numerous during the course of infection. The dermis, which lies below the epidermis, is made up of two layers of intricate arrays of connective tissue. A superficial layer (stratum spongiosum) comprises a network of collagen, fibroblasts, pigment cells, phagocytic cells, scale bed cells and scales. The stratum compactum is the deeper layer, which is primarily a non-cellular layer with a few fibroblasts interspersed between orthogonal bands of collagen. Chromatophores (pigment-containing cells) are found in the dermis and play a role in colouring and protection. Below the

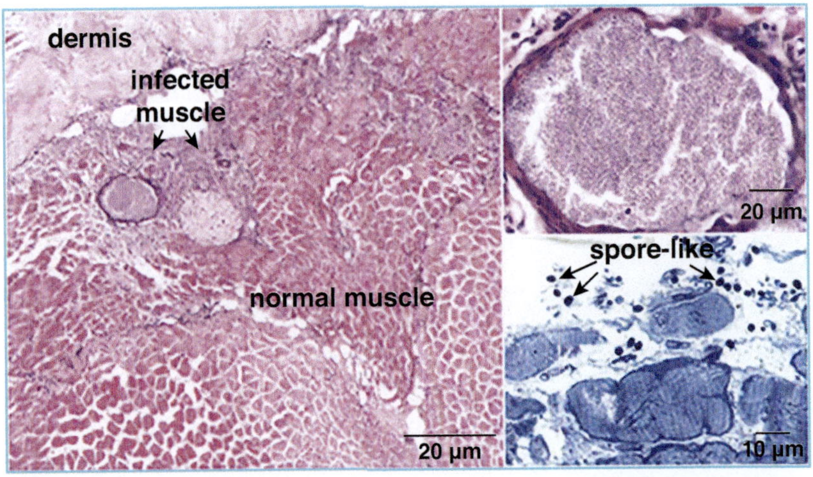

Figure 14.28 Histopathology of an unidentified microsporidian-like infection found in TiLV-infected tilapia in Thailand in 2017. There are intramuscular cysts containing basophilic spores, which were associated with areas of muscle necrosis and infiltration by macrophages. At higher magnification, toluidine blue stain highlighted extracellular spores, suggesting cyst rupture.

skin there is a very thin panniculus and a skeletal muscle. The latter is composed of multi-nucleated striated muscle cells. Examples of diseases that can lead to skin/muscle lesions are epizootic ulcerative syndrome, aeromonad septicaemia and microsporidiosis. Epizootic ulcerative syndrome results in skin/muscle necrosis with mild inflammation and innumerable intra-lesional fungal hyphae. Aeromonad septicaemia (and most other septicaemic diseases) result in multifocal areas of haemorrhage. Microsporidiosis features large cysts lined by a thick membrane that contains innumerable basophilic spores. If these cysts are broken they will result in infiltration by macrophages around the cyst, with secondary necrosis of skeletal myofibres (Fig. 14.28).

Glossary of relevant terms

Aetiology: The causative agent of a disease entity.

Aggregation: a group, body, or mass composed of many distinct parts or individuals.

Anterior (aka cranial): situated closer to the head or the front end of the animal.

Atypical: unusual and different from others of the same type.

Autolysis: destruction of cells or tissues by the body's own enzymes after death.

Basophilic: readily stained with basic dyes (e.g. haematoxylin, which is blue).

Chronic: protracted presentation that persists for a relatively long time.

Congestion: pathological accumulation of blood within veins.

Debris: the remains of tissue that has broken down or been destroyed.

Diagnosis: identification of the disease entity and/or aetiological agent causing a disease presentation.

Embed: in histology, this refers to fixing processed tissue by placing it within paraffin wax.

Encephalitis: inflammation of the brain.

Eosinophilic: readily stained by eosin, which is pink and an acid stain.

Erythrocyte: Red blood cell

Granuloma: an inflammatory cell aggregate composed of macrophages with/without other inflammatory cell types. Granulomas are chronic inflammatory lesions that are frequently associated with an aetiological agent (but not always). A necrotizing granuloma has a central area of necrosis.

H&E: hematoxylin and eosin stain.

Hematopoiesis: the formation of blood or of blood cells.

Hemorrhage: escape of blood from vessels (bleeding).

Hemosiderin: a yellowish-brown granular intracellular pigment derived from the metabolic degradation of hemoglobin.

Hepatitis: inflammation of the liver or hepatopancreas.

Infiltration: permeation of a liquid or a cell type into the tissue.

Karyorrhexis: fragmentation of the nucleus into unbound chromatin aggregates (basophilic), secondary to cell death.

Leukocyte: White blood cell.

Melanomacrophage centers (MMCs): clusters of macrophages containing melanin.

Necrosis: death of individual cells or groups of cells within a tissue.

Perivascular: situated or occurring around a blood vessel.

Posterior: further back in position or towards the hind end of the animal.

Post mortem: happening after death.

Pyknosis: shrinking and hyperbasophilia of the nucleus secondary to cellular degeneration/death.

Pyogranuloma: a granuloma that contains neutrophils within it.

Retroperitoneal: situated or occurring behind the peritoneum.

Septum: a structure that delimits spaces within a body cavity or an organ.

Splenic: relating to the spleen.

Syncytia: multinucleated cells resulting from previous cellular fusion.

Ventral: pertaining to the lower aspect of the body.

Viscera: the internal organs in the main cavities of the body.

References

Agius, C. and Roberts, R.J. (2003) Melano-macrophage centres and their role in fish pathology. *In Journal of Fish Diseases* 26(9), 499–509. Available at: https://doi.org/10.1046/j.1365-2761.2003.00485.x

Behera, B.K., Pradhan, P.K., Swaminathan, T.R., Sood, N., Paria, P. *et al.* (2018) Emergence of tilapia lake virus associated with mortalities of farmed Nile tilapia *Oreochromis niloticus* (linnaeus 1758) in India. *Aquaculture* 484, 168–174. Available at: https://doi.org/10.1016/j.aquaculture.2017.11.025

Bigarré, L., Cabon, J., Baud, M., Heimann, M., Body, A. *et al.* (2009) Outbreak of betanodavirus infection in tilapia, *Oreochromis niloticus* (L.), in fresh water. *Journal of Fish Diseases* 32(8), 667–673. Available at: https://doi.org/10.1111/j.1365-2761.2009.01037.x

del-Pozo, J., Mishra, N., Kabuusu, R., Cheetham, S., Eldar, A. *et al.* (2017) Syncytial hepatitis of tilapia (*Oreochromis niloticus L.*) is associated with orthomyxovirus-like virions in hepatocytes. *Veterinary Pathology* 54(1), 164–170. Available at: https://doi.org/10.1177/0300985816658100

Dinh-Hung, N., Sangpo, P., Kruangkum, T., Kayansamruaj, P., Rung-Ruangkijkrai, T. *et al.* (2021) Dissecting the localization of tilapia tilapinevirus in the brain of the experimentally infected Nile tilapia, *Oreochromis niloticus* (L). *Journal of Fish Diseases* 44(8), 1053–1064. Available at: https://doi.org/10.1111/jfd.13367

Dong, H.T., Siriroob, S., Meemetta, W., Santimanawong, W., Gangnonngiw, W. *et al.* (2017a) Emergence of tilapia lake virus in Thailand and an alternative semi-nested RT-PCR for detection. *Aquaculture* 476, 111–118. Available at: https://doi.org/10.1016/j.aquaculture.2017.04.019

Dong, H.T., Techatanakitarnan, C., Jindakittikul, P., Thaiprayoon, A., Taengphu, S. *et al.* (2017b) *Aeromonas jandaei* and *Aeromonas veronii* caused disease and mortality in Nile tilapia, *Oreochromis niloticus* (L). *Journal of Fish Diseases* 40(10), 1395–1403. Available at: https://doi.org/10.1111/jfd.12617

Dong, H.T., Senapin, S., Jeamkunakorn, C., Nguyen, V.V., Nguyen, N.T, *et al.* (2019) Natural occurrence of edwardsiellosis caused by edwardsiella ictaluri in farmed hybrid red tilapia (oreochromis sp.) in southeast Asia. *Aquaculture* 499, 17–23. Available at: https://doi.org/10.1016/j.aquaculture.2018.09.007

Ferguson, H.W. (2006) Systemic pathology of fish: A text and atlas of normal tissues in teleosts and their response in disease.

Ferguson, H.W., Kabuusu, R., Beltran, S., Reyes, E., Lince, J.A. *et al.* (2014) Syncytial hepatitis of farmed tilapia, *Oreochromis niloticus* (L.): A case report. *Journal of Fish Diseases* 37(6), 583–589. Available at: https://doi.org/10.1111/jfd.12142

Keawcharoen, J., Techangamsuwan, S., Ponpornpisit, A., Lombardini, E.D., Patchimasiri, T. *et al.* (2015) Genetic characterization of a betanodavirus isolated from a clinical disease outbreak in farm-raised tilapia *Oreochromis niloticus* (L.) in thailand. *Journal of Fish Diseases* 38(1), 49–54. Available at: https://doi.org/10.1111/jfd.12200

Korni, F.M.M., EL-Nahass, E.-S. and Ahmed, W.M.S. (2017) An outbreak of motile aeromonas septicemia in cultured Nile tilapia, *Oreochromis niloticus* with reference to hematological. *Biochemical and Histopathological Alterations Fish Pathology* 30, 11–24. Available at: https://doi.org/10.7847/JFP.2017.30.1.011

Laith, A.A., Ambak, M.A., Hassan, M., Sheriff, S.M., Nadirah, M. *et al.* (2017) Molecular identification and histopathological study of natural *Streptococcus agalactiae* infection in hybrid tilapia (*Oreochromis niloticus*). *Veterinary World* 10(1), 101–111. Available at: https://doi.org/10.14202/vetworld.2017.101-111

Liu, W., Zhang, Y., Ma, J., Jiang, N., Fan, Y. *et al.* (2020) Determination of a novel parvovirus pathogen associated with massive mortality in adult tilapia. *PLoS Pathogens* 16(9), e1008765. Available at: https://doi.org/10.1371/journal.ppat.1008765

Nguyen, V.V., Dong, H.T., Senapin, S., Pirarat, N. and Rodkhum, C. (2016) *Francisella noatunensis* subsp. orientalis, an emerging bacterial pathogen affecting cultured red tilapia (*Oreochromis sp.*) in thailand. *Aquaculture Research* 47(11), 3697–3702. Available at: https://doi.org/10.1111/are.12802

Okuthe, G.E. and Bhomela, B. (n.d.) Morphology, histology and histochemistry of the digestive tract of the Banded tilapia, *Tilapia sparrmanii* (Perciformes: Cichlidae). *Zoologia* 37, 1–14. Available at: https://doi.org/10.3897/zoologia.37.e51043

Pádua, S.B., Menezes-Filho, R.N., Martins, M.L., Belo, M.A.A., Ishikawa, M.M. *et al.* (2015) A survey of epitheliocystis disease in farmed Nile tilapia (*Oreochromis niloticus* linnaeus, 1758) in Brazil. *Journal of Applied Ichthyology* 31(5), 927–930. DOI: 10.1111/jai.12840.

Phrompanya, P., Saenphet, K. and Saenphet, S. (2019) Comparative histochemical study of the gastrointestinal tracts of the Nile tilapia

(*Oreochromis niloticus*) and the hybrid catfish (*Clarias batrachus* x *Clarias gariepinus*). *Acta Histochemica* 121(3), 261–267. Available at: https://doi.org/10.1016/j.acthis.2019.01.003

Ramírez-Paredes, J.G., Paley, R.K., Hunt, W., Feist, S.W., Stone, D.M. et al. (2021) First detection of infectious spleen and kidney necrosis virus (ISKNV) associated with massive mortalities in farmed tilapia in Africa. *Transboundary and Emerging Diseases* 68(3), 1550–1563. DOI: 10.1111/tbed.13825.

Ramírez-Paredes, J.G., Paley, R.K., Hunt, W., Feist, S.W., Stone, D.M, *et al.* (2020) First detection ofinfectious spleen and kidney necrosis virus (ISKNV) associated with massivemortalities in farmed tilapia in Africa. *Transboundary and Emerging Diseases* 68(3), 1550–1563. Available at: https://doi.org/10.1111/tbed.13825

Roberts, R.J. and Ellis, A.E. (n.d.) The anatomy and physiologyof teleosts. In: Rorberts, R. J. (ed.) *Fish Pathology (4th Edn)*. Wiley Blackwell, pp. 17–61.

Rodkhum, C., Kayansamruaj, P. and Pirarat, N. (2011) Effect of water temperature on susceptibility to *Streptococcus agalactiae* serotype ia infection in Nile tilapia (*Oreochromis niloticus*). *The Thai Journal of Veterinary Medicine* 41(3), 309–314. Available at: https://doi.org/10.56808/2985-1130.2314

Sado, R.Y., Souza, F.C., Behr, E.R., Mocha, P.R.E. and Baldisserotto, B. (2020) Chapter 2. anatomy of teleosts and elasmobranchs. In: Baldisserotto, B., Urbinati, E.C. and Cyrino, J.E.P. (eds) *Biology and Physiology of Freshwater Neotropical Fish.* Academic Press, pp. 21–47. Available at: https://doi.org/10.1016/B978–0-12–815872–2.00002–6

Soto, E., Bowles, K., Fernandez, D. and Hawke, J.P. (2010) Development of a real-time PCR assay for identification and quantification of the fish pathogen *Francisella noatunensis* subsp. orientalis. *Diseases of Aquatic Organisms* 89(3), 199–207. Available at: https://doi.org/10.3354/dao02204

Soto, E., Illanes, O., Revan, F., Griffin, M. and Riofrio, A. (2013) Bacterial distribution and tissue targets following experimental *Edwardsiella ictaluri* infection in Nile tilapia *Oreochromis niloticus. Diseases of Aquatic Organisms* 104(2), 105–112. Available at: https://doi.org/10.3354/dao02593

Steckert, L.D., Cardoso, L., Jerônimo, G.T., Pádua, S.B. and Martins, M.L. (2018) Investigation of farmed Nile tilapia health through histopathology. *Aquaculture* 486, 161–169. Available at: https://doi.org/10.1016/j.aquaculture.2017.12.021

Subramaniam, K., Gotesman, M., Smith, C.E., Steckler, N.K., Kelley, K.L. *et al.* (2016) Megalocytivirus infection in cultured Nile tilapia *Oreochromis niloticus*. *Diseases of Aquatic Organisms* 119(3), 253–258. Available at: https://doi.org/10.3354/dao02985

Tang, K.F.J., Bondad-Reantaso, M.G., Surachetpong, W., Dong, H.T., Fejzic, N, *et al.* (2021) Tilapia lake virus disease strategy manual. *FAO Fisheries and Aquaculture Circular*. Available at: https://doi.org/10.4060/cb7293en

Zamri-Saad, M., Amal, M.N.A. and Siti-Zahrah, A. (2010) Pathological changes in red tilapias (*Oreochromis* spp.) naturally infected by *Streptococcus agalactiae*. *Journal of Comparative Pathology* 143(2–3), 227–229. Available at: https://doi.org/10.1016/j.jcpa.2010.01.020

Chapter 15

Health management of tilapia from a farm perspective

Annette Simone Boerlage and Mona Dverdal Jansen

Introduction

The chapter discusses key components in tilapia health management systems with a farm-level focus. Following a short general introduction to fish health management, it introduces the reader to systems for collecting information and evidence for decision support. This includes common surveillance strategies, benchmarking and epidemiological studies. It subsequently takes the reader through major pathogen introduction pathways, focusing on common transmission pathways in tilapia farming systems. Farm-level mitigation measures and biosecurity are discussed, together with additional mitigation measures such as antimicrobials and chemotherapeutants. Vaccination, genetic strain selection and fish welfare are briefly mentioned, and the reader is referred to the other relevant chapters in the book for further information. Following this is a section on external support for fish health management, including the role of fish health services, diagnostics, best management practices and certification schemes. Finally, a short section on future perspectives on tilapia health management concludes the chapter.

Tilapia is the world's third-most produced aquatic species (FAO, 2020). The sustainable culture of tilapia is expected to be an important contributor to food security in the future (Béné *et al.*, 2015).

Kim Thompson, Craig Shoemaker and David C. Little(eds) **Tilapia**
DOI: 10.1079/9781800629455.0015, © CAB International, 2025

Further development and expansion of sustainable aquaculture are constrained by disease (FAO, 2020). Because suboptimal health management is one of the drivers of disease in aquaculture (FAO, 2019), a thorough understanding of health management is fundamental to a sustainable future of the sector.

Health management refers to a system designed to optimize the physical and behavioural health and welfare of animals. It is a broad term that encompasses the prevention, treatment and control of diseases and conditions affecting both individual fish and groups of fish (WOAH, 2019). The objectives of health management strategies are, for example, detecting the introduction of a specific pathogen, eradicating diseases to declare disease freedom or controlling and mitigating a disease when disease freedom seems impossible or deemed too costly (Peeler & Otte, 2016). The focus of a health management strategy can be, for example, a specific disease, a syndrome, mortality in general or fish health in general. Developing a good health management strategy requires understanding of the objective and focus and having institutional and technical abilities to implement the necessary measures (FAO, 2020). The most common limiting factors for effective health management strategies to be put in place include inadequate regulatory frameworks and guidelines for best practices, inadequate coordination between institutions involved, lack of adequate and well-implemented biosecurity strategies at different levels, and insufficient capacity for responses to emergencies (FAO, 2020). Although it may not be possible for many farmers to influence any of these limitations, awareness of these can help optimize farm-level strategies.

Health management strategies are designed and implemented at different levels, such as a farm, a cluster of farms (Kassam *et al.*, 2011), a sector, a watershed or a country. Interactions within, and characteristics of, these levels determine which health management measures are effective (Phillips, 1996). A country may aspire to eradicate disease *X* by implementing additional surveillance and biosecurity at the border, while at the farm level, measures to eradicate the same disease *X* may be part of a generic objective of minimizing pathogen emergence on the farm, and measures taken may not distinguish between stock obtained from a national or international source. It is, therefore, important to determine the level at which health management strategies are developed. Biosecurity plans are essential and integral components of such health management strategies.

Collecting information and evidence

Collecting information on the disease is an essential part of health management because it provides an evidence base on which to make decisions. Because it is impossible to sample all fish every day for every pathogen, a sampling strategy is usually developed, referred to as disease surveillance. The World Organization of Animal Health (WOAH) defines surveillance as 'a systematic series of investigations of a given population of aquatic animals to detect the occurrence of disease for control purposes, and which may involve testing samples of a population' (WOAH, 2019).

Simplified, surveillance activities fall into two broad categories, passive and active surveillance. There are many applications for surveillance activities, and only a brief overview is given below. For more detailed information on surveillance, a range of authoritative resources is available for consultation (e.g. Cameron, 2002; FAO, 2007; Corsin *et al.*, 2019).

Passive surveillance

Passive surveillance takes the form of a disease-reporting system where those who will use the information do not take any specific action to collect it. In the case of a country-wide programme, farmers typically report suspected or confirmed disease cases to the local or regional government officials, who file an official report in the national disease database of the competent authority. Where available, results from diagnostic laboratory investigations should additionally be reported. Depending on local routines and regulations, the farmer or the diagnostic laboratory might do this. Centralized disease databases should ensure data confidentiality and consistent data entry to allow retrieval of useful data for situation overview and decision support.

Because farmers are in close contact with their stock, they are best placed to detect unusual events and conditions. However, for the system to work efficiently they must be given clear information on when to report, what to report and how to report. In most systems, the suspicion of a notifiable disease or the observation of new clinical signs, which may suggest a new or exotic disease, should be reported as soon as possible to reduce the possibility of spread to other farms or areas. Farmers may additionally be requested to report more commonly occurring disease events, but these may have less stringent time requirements for reporting. With the increasing availability of smartphones and apps, digital reporting may be the most efficient reporting format.

Contributing farmers should receive useful information in return, such as feedback on the general disease situation in the country or their region, which can support farmer decision making regarding biosecurity measures. From the government's perspective, the system may aid detection of newly introduced, exotic or emerging diseases and the identification of local disease outbreaks. Because not all disease events will be reported, passive surveillance will not give a complete picture of the true number of disease cases or their geographical distribution. The system is therefore unsuited to inform about the real changes over time, because an increase in cases from one year to the next may, for example, occur due to increased farmer awareness following an information campaign and not due to an actual increase in disease occurrence. To answer such questions, an active surveillance programme must be initiated.

Active surveillance

As highlighted in the previous section, active surveillance may be able to provide the required information to fulfil some aims that passive surveillance was unable to do. In this case, high-quality data on a specific agent or disease are collected through a specific, structured disease survey designed to reflect the actual disease situation. Because it is aimed at gathering information on a particular (target) agent or disease, the surveillance activity will only be able to say something about this target and nothing about the range of other agents or diseases that may affect the same populations. To be scientifically valid, active surveillance needs to be carefully designed to ensure the representativeness of the surveyed population, thereby allowing inference of the results to the populations of farms in general. While expert epidemiology advice should be sought to instigate active surveillance activities, the step-by-step guide for active surveillance in aquatic animals may also be useful (Bondad-Reantaso *et al.*, 2021).

Syndromic surveillance

Syndromic surveillance is an emerging method to identify disease clusters early, before diagnoses are confirmed and reported, with the goal of detecting disease earlier and thus saving time and resources (Henning, 2004; Dórea *et al.*, 2011). There can be considerable variation between production cycles even within the company (Soares *et al.*, 2011); therefore, finding disease clusters early requires data-rich environments with (near)-real-time information, such as that collected by governments and research institutes. Developing a syndromic surveillance system requires a good understanding of relevant diseases and the sector and, although not generally led by farmers, farmer participation is essential to a good understanding (Brugere *et al.*, 2017; Muniesa *et al.*, 2018).

Benchmarking

Comparison of data from an individual unit (site, farm, company) to more/other data is called benchmarking. Within the unit, recent data can be compared to historical data. Between farms, data can be compared with a sector or region average (Soares *et al.*, 2011). Indices that standardize between farms, such as the production and management index proposed for tilapia (Muniesa *et al.*, 2016), can aid benchmarking.

Epidemiological studies

Collecting and recording production data can help provide an evidence base of trends and associations between disease and risk factors that human intuition might not be able to comprehend (Boerlage *et al.*, 2017; Opiyo *et al.*, 2018; Delphino *et al.*, 2019). Information on a variety of topics is useful to collect: for example, data on productivity such as daily count of mortalities and food conversion ratios. Furthermore, putative risk factors such as environmental factors like water temperature and husbandry practices like handling events and treatments can be recorded. Innovations in aquaculture technology increasingly allow for automatic data collection and recording (IDH, 2020), allowing for the display of interactive maps, such as www.barentswatch.no/fiskehelse, and leads to the next challenge: interpreting large amounts of data (Dórea and Revie, 2021). Observational studies are a type of epidemiological study that can help to provide insights into the long-term trends of disease. Outputs are associations that increase our understanding and help develop targeted studies that can find causation. For example, Boerlage *et al.* (2017) showed that the perceived mortality of tilapia was higher in provinces further downstream in the river. They devised possible reasons, such as the accumulation of contaminants in the water or a difference in local factors unrelated to water quality. In this example the study revealed an association and further investigations were needed to uncover causation, which could then lead to development of mitigation measures. One important barrier to such risk-factor studies is farmers' participation. Epidemiological studies are often not of immediate benefit to individual farmers, because associations arise when an appropriate amount of data is considered, often including data from many farms, and results will be presented long after the fish are harvested. These studies often rely on the voluntary provision of commercially sensitive information, and therefore communication about the benefits of these studies is important as is developing trust between farmers and the institution performing the study (Palmer *et al.*, 2009).

Farm-level pathogen introduction

Understanding how pathogens are introduced into the farm is essential to developing a health management strategy and biosecurity plans. Direct pathogen introduction pathways can be through (1) live fish movement, (2) dead fish movement, (3) environmental factors or (4) mechanical transmission and biosecurity breaches (Oidtmann *et al.*, 2011; Boerlage *et al.*, 2017). Different production systems vary widely in regard to opportunities for human control over these pathogen introduction pathways. For example, pathogen introduction pathways into closed recirculation systems can be controlled to a large extent but control over pathogen introduction into floating pens in rivers is limited due to uncontrolled water entering the farm and uncontrolled interaction between produced fish and wild fish species surrounding the farm. The presence of pathogens does not always lead to disease in fish, but pathogens can evolve and their presence may continuously pose a risk. For example, the pathogen *Francisella noatunensis* subsp. *orientalis* is considered non-infective when in a viable but non-culturable state but can become infective even after long periods without outbreaks (Delphino *et al.*, 2019).

Live fish movement

Live fish movement can be an important contributor to the spread of pathogens (Richard *et al.*, 2002; Dong *et al.*, 2017; Delphino *et al.*, 2019). One documented example is a case where live fish movement was most likely responsible for a *Francisella* sp. outbreak at a *Nile tilapia* recirculation farm in the UK (Jeffery *et al.*, 2010). Live fish movements are unavoidable in most production systems. Closed systems that produce broodstock, eggs, fry, fingerlings and grow-out fish within the same company often grow different life stages at different locations and therefore may still require live fish movement between locations. Most tilapia production systems are open and rely on the import of live input to produce and grow new batches.

The distance of travel can affect the risk of disease introduction. Globalization has led to transport opportunities for live fish across the globe – for example, to improve the genetic stock. Transport within countries or areas is more common, but distances can also be substantial if sectors rely on limited sources of fry or fingerlings. For example, in small-scale red-tilapia farming systems in Vietnam, about one quarter of grow-out farmers obtain their fry from outside their province (Boerlage *et al.*, 2017) and, in Egypt, most farmers, regardless of farm location, obtain their seed from the main production governorate (Ali *et al.*, 2020). There are different aspects to travel distance that may affect pathogen introduction. First, transport itself may affect the mortality burden of fish (Jia *et al.*, 2016), perhaps by suboptimal fish welfare during transportation resulting in disruption of the natural balance between putative pathogens and hosts, or pathogens may be introduced to live fish from transportation equipment that was insufficiently disinfected between transports. In both cases, durations may increase transmission chances. Secondly, transported live fish may carry (subclinical) infections native to the source farm but exotic to the recipient farm. This could lead to the emergence of either novel (to the recipient farm) pathogens or of pathogens already present but with (slightly) different genetic make-up and virulence (Kayansamruaj *et al.*, 2014). Biosecurity measures such as quarantine upon arrival may reduce the risk of disease transmission from new arrivals to pre-existing stock.

Dead fish movement

The effect of dead fish movement on the risk of introducing pathogens into a farm is through fish products for consumption or disposal of fish carcasses that perished during production.

Importing fish products for human consumption may pose a risk to human health when fish are eaten raw or partly uncooked (FDA, 2020). There is, to our knowledge, no evidence of disease introduction to tilapia farms linked to aquatic products sold for human consumption. This is similar to the shrimp sector, where, although aquaculture products for human consumption have tested positive for the disease under laboratory conditions (Nunan *et al.*, 1998), it was perceived more likely that the disease was transmitted through other means, such as live fish transport or ballast water (Flegel, 2009). In general, leftovers and by-products of food preparation should be disposed of appropriately without contact with farmed fish.

Dead fish should be disposed of frequently because they can be a source of pathogen transmission to live fish (Zamri-Saad *et al.*, 2014). Disposal methods of fish carcasses that may lead to pathogen introduction include selling them off as a food source for other fish species or leaving removed dead fish unattended on the ground near the fishpond (Boerlage *et al.*, 2017; Ali *et al.*, 2020). There are guidelines for better management practices that can help farmers develop low-risk practices; for example, the World Wildlife Fund, Inc. guidelines recommend procedures such as burying in compost piles with plastic-lined bottoms away from receiving waters, burning or using disposals as terrestrial fertilizer (WWF, 2011). Although some of these risks are for the introduction of disease on other farms and not on the source farm, an increase in infected farms in the area might indirectly affect the source farm, and thus it is advantageous to each farmer to correctly dispose of dead fish.

Environmental factors

Introduction of pathogens through water is a major risk to most tilapia production systems, because untreated water may contain pathogens, fomites with pathogens or reservoir hosts with pathogens. Systems with a degree of control to the water inlet, such as recirculation systems, tank systems, raceways and aquaponics, may have the ability to filter and disinfect water before it encounters fish. Pond systems may be able to disinfect water or use sedimentation ponds before the water gets pumped into the production pond. Farmers with ponds may be able to adjust production levels to water exchange rates that can be disinfected/treated. In some areas ponds are prone to flooding, which may introduce contaminated water or infected feral fish, thereby introducing pathogens. Flooding of ponds may additionally discharge pathogens into the environment from an infected farm, potentially spreading the infection to susceptible wild stocks or downstream farms. Cage structures in lakes and rivers have no control over water intake, and therefore pathogen introduction through water remains a high risk regarding pathogen introduction to the farm. Such pathogens may originate from upstream activities, such as fish farms, processing plants and recreational activities like angling or canoeing. In addition, the transmission of pathogens between wild and cultured fish might take place (Dong *et al.*, 2017; Delphino *et al.*, 2019).

Disease incidence often follows seasonal patterns, the knowledge of which can help in the design of a cost-effective health management strategy. For example, surveillance frequency for specific pathogens can be adjusted based on their expected seasonal occurrence (Surachetpong *et al.*, 2020).

Mechanical transmission

Mechanical transmission on farms can have a major impact on the introduction of pathogens from outside the farm. Because visitors may introduce pathogens, supplying visitors with protective clothing and boots can reduce the risk of pathogen transmission. Avian and non-human mammalian visitors also pose a risk: for example, predatorial types may spend considerable time at or around the pond, in which they may defaecate and contaminate the farm with trematodes (Sommerville, 1982) or bring with them infected material from other farms. Using fencing and/or nets may reduce the likelihood of contact between farmed fish and unwanted visitors with their pathogens, which may even lead to the elimination of pathogens in the fish host (Li *et al.*, 2013).

Equipment can also introduce pathogens. Sharing equipment between farms, such as harvesting equipment and aeration machines, is a common practice that can transmit pathogens (Boerlage *et al.*, 2017). If anglers are on site, angling equipment may be contaminated if not properly disinfected between use in different ponds. Vehicles and boats may also be contaminated if they have been at other farm sites, and disinfecting protocols should be in place. Fertilizers for the pond and unpasteurized fish food can also be contaminated, and both the source and risk of pathogen transmission need to be evaluated. If tilapia are fed an agricultural by-product, such as those derived from bran, oil cakes, vegetation, manure, inorganic fertilizer or feed, it is important to consider these as potential sources of pathogen transmission and evaluate risks using a risk assessment.

Farm-level mitigation

Fish farms rarely have the luxury of a pathogen-free environment and therefore must constantly mitigate against disease. Preventive mitigation approaches are often less costly than reactive approaches (FAO, 2020), but the effectiveness of preventive measures is less easy to quantify and therefore justify. Some mitigation activities may be anticipated – for example, when pathogens are known to be present and the risk of outbreak is consistent. Other disease outbreak events are unexpected emergencies and require research and unanticipated action. To help plan and prioritize mitigation measures, a risk assessment can be undertaken in which the likelihood of introduction is multiplied by the likely consequences to provide a measure of risk (Lind *et al.*, 2015).

General biosecurity concepts

According to the WOAH Aquatic Animal Health Code, biosecurity can be defined as 'a set of management and physical measures designed to reduce the risk of introduction, establishment and spread of pathogenic agents to, from and within an aquatic animal population' (WOAH, 2015). Other definitions of biosecurity include the broader definition used by the FAO, which also incorporates policy, regulatory frameworks and associated risks (FAO, 2007).

It is common to differentiate between so-called external biosecurity, also known as bio-exclusion, and biocontainment. External biosecurity comprises all measures adopted to prevent the introduction of infectious agents into a farm. Biocontainment, on the other hand, involves measures adopted to prevent the spread of infectious agents from an infected farm. Within an infected farm, internal biosecurity measures are adopted to prevent the spread of infectious agents, such as between different farm sections or fish groups (Dewulf & Immerseel, 2019). Several resources can assist with developing biosecurity strategies (Oidtmann *et al.*, 2011; Oidtmann *et al.*, 2013; Palić *et al.*, 2015). In addition, a Progressive Management Pathway for Aquaculture Biosecurity (FAO PMP/AB) is currently under development by the FAO (FAO, 2019).

An integral part of any biosecurity strategy is developing and adhering to biosecurity plans (Palić *et al.*, 2015). Such plans could be at either the national level, e.g. a national biosecurity strategy to prevent the introduction of exotic pathogens into the country, or the farm level, e.g. to prevent the entry of pathogens into a specific unit such as the hatchery. An assessment of required and available quarantine facilities is important when performing risk assessments and designing biosecurity strategies. This applies both at the national level, in the form of adequate quarantine units at major ports of entry, and at the farm level, with respect to the management of stock brought into the farm. Major mitigation strategies are typically determined and implemented by the competent authority. In its most severe form, mitigation measures may require infected farms to depopulate the farm of all fish. This is commonly used if an exotic disease (a disease not normally occurring in the area) is detected, but it has also been an essential component for controlling endemic and emerging diseases in salmon farming (Pettersen *et al.*, 2015). While reducing the risk of disease spread, and therefore being beneficial for the fish-farming community as a whole, depopulation measures are often very costly for affected farms because there is typically no economic compensation for lost fish. Therefore, only when the likelihood of recovery for the remaining fish is low and their welfare is severely affected might individual farmers consider depopulation of stock, perhaps in order to reduce the possibility of transmission to other stock.

The first and most important part of farm-level mitigation is the development of a biosecurity plan. A biosecurity plan is a plan that identifies important pathways for the introduction and spread of a pathogen. It describes the measures applied to mitigate such risks and detail how such measures should be audited (WOAH, 2019).

Farm-level biosecurity plans need to be carefully tailored to the hazards specific for every farm. Oidtmann *et al.* (2011) suggest that the following basic elements should be components in a farm-level biosecurity plan: (1) a description of best management practices used, (2) identification of the risks of contracting and spreading disease and subsequent risk-mitigation measures, (3) details of staff training in disease recognition and fish health management, (4) diagnostic and detection methods for infectious diseases and (5) disinfection and pathogen eradication methods used.

In order to achieve and maintain the desired level of biosecurity it is essential that all staff have a thorough understanding of the measures involved, the underlying reasons for these measures and the fact that a biosecurity level can be maintained only if everyone adheres to the same standard. As a result, staff training in biosecurity and general fish health management, including observations that should lead to reporting to the farm manager, are essential. The staff training system and auditing procedures should be an integral part of any farm-level biosecurity plan.

Antimicrobials and other chemotherapeutants

Even with many preventive measures in place, outbreaks of disease may occur, and chemical treatment may be the only effective reactive mitigation measure available. Administration of chemotherapeutants is usually by immersion or in-feed. Chemotherapeutants must be obtained through routes approved by the competent authorities and used with care. Doses and duration as indicated should be applied to prevent resistance issues or overdosing, and withdrawal times should be considered when planning harvests. Competent authorities and best management practice schemes often provide guidelines available to farmers in countries where veterinary involvement is not always mandatory, such as the guidelines provided by the Association of Southeast Asian Nations (ASEAN, 2013)

One category of chemotherapeutants, antimicrobials, deserves a special mention because of the increased worldwide attention they have been attracting due to the risk of antimicrobial resistance in both animal and human health (Scott *et al.*, 2018). Antimicrobials are used in tilapia husbandry both preventively as probiotics (Welker and Lim, 2011) and reactively as chemotherapeutants (Rico *et al.*, 2013). While worldwide antimicrobial use in livestock is reducing, there is still inappropriate use, particularly in less developed countries where veterinary services are less common or lacking and farmers have little to no access to diagnostics or guidance on chemical use (Li *et al.*, 2016). For example, one study showed that tilapia farmers prepared water-based antimicrobial solutions with their bare hands and increased doses on their own initiative (Li *et al.*, 2016). Misuse of antimicrobials is common in tilapia aquaculture (Watts *et al.*, 2017; Brunton *et al.*, 2019) and is a problem that needs to be addressed regionally. Solutions to reducing antimicrobial (mis) use include better monitoring of hotspots, training of farmers, improved access to veterinarians, regulatory guidance by government and certification bodies, alternative treatments and improved health management.

Alternatives to chemicals and antibiotics are becoming more popular in the search for tools that can help mitigate disease sustainably. One example is probiotics, i.e. micro-organisms that are introduced by immersion or in-feed to tilapia to enhance the immune response, disease resistance and survival and improve growth through different mechanisms (Hai, 2015; Tan *et al.*, 2019; Maas, 2021). Other alternatives that are common throughout Asia are herbal supplements. A wide variety of plant abstracts have been trialled to serve as immunostimulants or therapeutants for different diseases. Generally there is a lack of understanding of exactly how they work and which side-effects they may have (Gabriel and González-Redondo, 2019; Kuebutornye and Abarike, 2020)

Vaccination

Fish are vaccinated to produce immunity against a disease. This is a popular practice often incorporated into health management strategies (Delphino *et al.*, 2019). It is important to realize that vaccination is not always 100% effective, and a diverse health management strategy is key. See Chapter 16 for more detailed information.

Genetic strain selection

Genetic strain selection for specific, desired characteristics is common practice in both terrestrial and aquatic animal production systems. Work on genetic improvements of existing tilapia strains is a fast-evolving field, and modern methods and technologies are expected to enable rapid advances in the coming years (Yáñez *et al.*, 2020). In relation to fish health management, being able to stock fish that have an increased resistance against specific pathogens can be a valuable addition to general biosecurity measures. However, determining which characteristics may be desirable in each area requires some information on the existence of the current pathogen in the production area in question. Unfortunately, such information is currently lacking in many tilapia production areas, particularly in developing countries. Examples of recent work on genetic strain selection and disease resistance in relation to tilapia pathogens include that on Tilapia lake virus (Barría *et al.*, 2020), *Streptococcus* spp. (Barony *et al.*, 2017; LaFrentz *et al.*, 2020; Sukhavachana *et al.*, 2020), black spot disease (Charo-Karisa *et al.*, 2021) and *Flavobacterium* (Wonmongkol *et al.*, 2018).

Fish welfare

Fish health is closely linked to fish welfare. A good fish health management strategy includes ensuring good farmed fish welfare, as highlighted by many organizations including the Aquaculture Stewardship Council (Australian Sports Commission, 2017) and the Food and Agriculture Organization of the United Nations (Segner *et al.*, 2019). More information on the welfare of tilapia can be found in Chapter 8.

External support for farm-level health management

Fish health services and the role of diagnostics

The availability of fish health services and diagnostic laboratories varies widely between countries and regions. While this will impact the type of services available, there are some basic, general principles that apply almost irrespective of available resources. The regular inspection that farmers and farm workers automatically conduct when working on the farm (e.g. while feeding the fish) is a key, first-line activity to assess whether anything out of the ordinary is happening on the farm. As a result, farmers and farm workers are commonly the first to detect abnormalities that can raise suspicion of a disease event on a farm. In many countries, farms are additionally required to have regular inspections by fish health professionals or government extension workers who will be experienced in looking for, and identifying, potential signs of disease. Irrespective of who identifies an abnormality that may indicate a disease problem, it is important that all involved know what action should be taken. Even if the exact cause is unknown there may be requirements to report disease suspicions or events of increased mortality to official bodies and the procedures for such reporting should be known and easily accessible for the farm manager or person in charge. This may include required notification of increased mortality or notification of suspicion of certain diseases that the competent authority has classified as important. This may, for example, be diseases that can cause large economic losses, have the potential to cause severe harm to wild fish populations or diseases that are unwanted or not previously known to be present in the area. International and regional bodies like WOAH, the FAO and the Network of Aquaculture Centres in Asia-Pacific (NACA), as well as national/local bodies, have disease lists that may be used to set requirements in relation to trade or required disease control measures.

Even if there are no official requirements to report disease suspicions to an official body, it will be advantageous to do so in terms of disease control. For example, farmers sharing the same water body, or who have other links to each other that increase the probability that a pathogen from one farm might affect others, would benefit from being notified of disease suspicion. Such information sharing is facilitated in areas with open communication and general collaboration between farmers. It is important to communicate information about the advantages of early action to prevent a pathogen from establishing itself in an area (if a new introduction), or to prevent an unmanageable increase in infection pressure (if already present) on the farming community. The long-term goal should be to build sufficient trust between actors to facilitate such information sharing, allowing fish health management to be achieved via collective action and not solely by individual farmer efforts. A typical fish health collaboration network is shown in Fig. 15.1.

In many tilapia-producing countries, a lack of diagnostic services and facilities in relation to fish diseases frequently limits the possibility of thorough disease investigation. While there is a need to strengthen fish diagnostic services in many countries and regions, much can be done without highly specialized equipment. Utilizing what is available locally is important in regard to gathering baseline information on the current status allowing the building of local, regional and national competence. FAO/NACA (2001) have suggested three levels (levels I, II and III) of disease diagnosis, allowing a country to develop capacity and competence over time. Level I diagnostics include observations of the fish and its environment and gross clinical examination, and thus not requiring specialized equipment. This is the foundation of any diagnostic investigation, and its importance in yielding information should not be underestimated. Level III, at the other end of the scale, represents specialized diagnostic methods requiring specialist

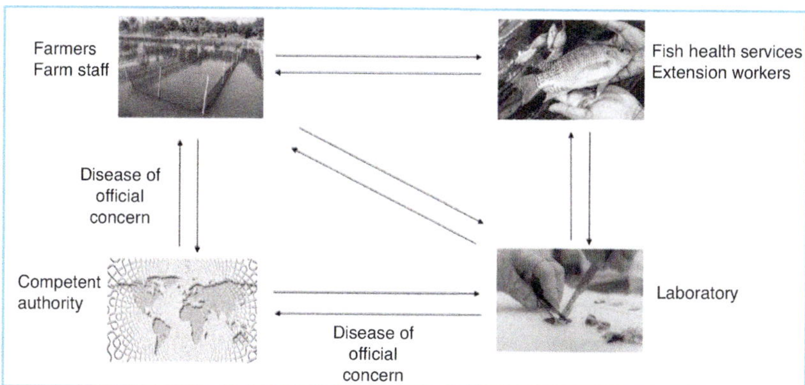

Figure 15.1 An example of collaborative action between the main actors involved in ensuring fish health. Photos by Mona Dverdal Jansen and Eivind Senneset, Norwegian Veterinary Institute and Gerd Altmann, Pixabay.

equipment, such as molecular biology methods, that can be essential to verifying a suspicion; however, the absence of such methods should not prevent basic information from being gathered.

Best management practice schemes

Best management practice (BMP) schemes for aquaculture typically aim at maximizing fish health and welfare and minimizing adverse environmental and social impacts while facilitating efficient and economically viable aquaculture operations, often with the objective to promote sustainability (FAO, 2020). An important aspect in relation to BMPs is the concept of optimal stocking density (Segner *et al.*, 2019; Zhou *et al.*, 2021). A range of BMPs for tilapia has been published, from general guidelines (e.g. ASC standards for responsible aquaculture; ASC, 2017) and best aquaculture practices (ASC, 2021) to more specific BMP examples for tilapia in specific countries/production technologies (e.g. Hoevenaars and Ng'ambi, 2019; IDH, 2020, 2021).

Certification

Best management practices may not always be immediately cost effective, but certification schemes can make them worthwhile. Tilapia is among the aquaculture species currently eligible for certification, which is a system in which organizations such as ASC provide a list of BMPs with which aquaculture operations must comply and, through a system of auditing, can become certified. Certification can lead to higher prices for products, or access to different markets. The process of obtaining and maintaining certification standards can be costly, and the technical capacity of smaller producers might not enable them to comply with the standards (FAO, 2020).

Future Perspectives

The importance of the different components of health management and how health management relates to biosecurity and surveillance were discussed previously. Once a system is established it should be considered dynamic, because frequent updates and ongoing evidence collection are essential. Knowledge towards (emerging) diseases is constantly evolving, and incorporation of up-to-date knowledge must be part of a health management strategy (Fig. 15.2). Review articles, fish health services, farmer organizations and governments may be able to provide sources that can help stay on top of new information.

Historically, health management has targeted specific diseases caused by specific pathogens, which will remain an important part of health management. However, novel strategies have emerged. Increasing evidence shows the importance of integrated health management and investing in mitigation methods that target a (genetically) diverse range of pathogens (Stentiford *et al.*, 2017; Hoa *et al.*, 2021), or diseases that are syndromic or complex and for which it is not possible to target a single pathogen. Another viewpoint is to move away from focusing on specific pathogens, but to focus mitigation on the pathobiome, i.e. the host-associated organisms linked to a reduced health status of fish resulting from interactions between those organisms, the hosts and the environment (Bass *et al.*, 2019). As more knowledge becomes available regarding the microbiome

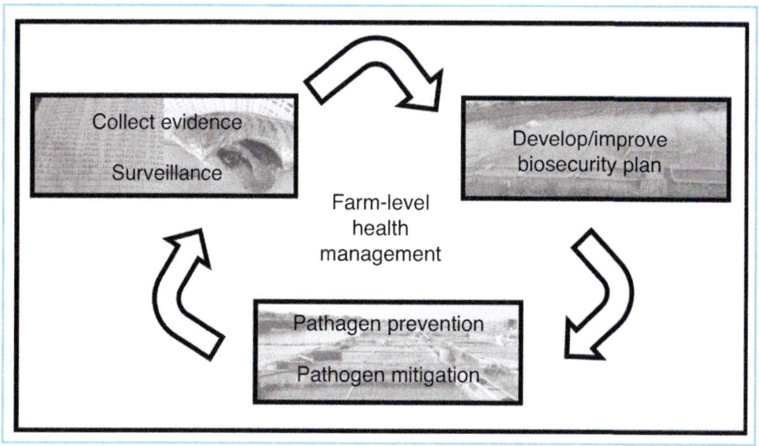

Figure 15.2 Dynamic nature of effective farm management requires frequent adjustments. Photos by Annette Boerlage, Scotland's Rural College.

and co-occurring pathogens with symbiont profiles, disease mitigation may develop towards a multi-organism focus.

As the world is gearing up for a 'green revolution', aquaculture will have to jump on board with a 'blue revolution'. What might such a future look like for health management? Long-term biosecurity measures focusing on prevention will be essential to improve sustainability and reduce economic and environmental losses (FAO, 2020). In many places the carrying capacity has been reached and, as a result, intensification in available habitats might be the only option for continuous growth. Intensification goes hand in hand with more human control; thus, health management will be of utmost importance. Technological developments such as apps, sensors and major data solutions could help modernize and automate production and may play a big part in modern health management. However, as animal production is a sensitive subject in some regions and aquaculture has had its fair share of negative publicity on subjects such as fish welfare and chemical use, modern health management would have to harmonize with public perception to maintain a social licence. The COVID-19 pandemic has shown how population-level disease control is essential to combat highly infectious diseases. This increased public and professional understanding must be utilized in the aquaculture sector to help develop strong health management strategies that deliver sustainable tilapia products with public acceptance.

References

Ali, S.E., Jansen, M.D., Mohan, C.V., Delamare-Deboutteville, J. and Charo-Karisa, H. (2020) Key risk factors, farming practices and economic losses associated with tilapia mortality in Egypt. *Aquaculture* 527, 735438. DOI: 10.1016/j.aquaculture.2020.735438.

Association of Southeast Asian Nations (2013) Guidelines for the use of chemicals in aquaculture and measures to eliminate the use of harmful chemicals. in association of Southeast Asian nations. Available at: https://www.asean.org/storage/images/resources/ASEANPublication/2013

Australian Sports Commission (2017) ASC tilapia. Retrieved April 2017, *The Standard Version 1.1*. Available at: w2w.asc-aqua.org

Australian Sports Commission. (2021) *Best aquaculture practices: Certification standards, guidelines (Issue 3.0)*.

Barony, G.M., Tavares, G.C., Pereira, F.L., Carvalho, A.F., Dorella, F.A, *et al.* (2017) Large-scale genomic analyses reveal the population structure and evolutionary trends of *Streptococcus agalactiae* strains in Brazilian fish farms. *Scientific Reports* 7(1), 1–10. DOI: 10.1038/s41598–017–13228-z.

Barría, A., Trinh, T.Q., Mahmuddin, M., Benzie, J.A.H., Chadag, V.M, *et al.* (2020) Genetic parameters for resistance to tilapia lake virus (TiLV) in *Nile tilapia* (*Oreochromis niloticus*). *Aquaculture* 522, 735126. DOI: 10.1016/j.aquaculture.2020.735126.

Bass, D., Stentiford, G.D., Wang, H.C., Koskella, B. and Tyler, C.R. (2019) The pathobiome in animal and plant diseases. *Trends in Ecology and Evolution* 34(11), 996–1008. DOI: 10.1016/j.tree.2019.07.012.

Béné, C., Barange, M., Subasinghe, R., Pinstrup-Andersen, P., Merino, G, *et al.* (2015) Feeding 9 billion by 2050 – putting fish back on the menu. *Food Security* 7(2), 261–274. DOI: 10.1007/s12571-015-0427-z.

Boerlage, A.S., Dung, T.T., Hoa, T.T.T., Davidson, J., Stryhn, H, *et al.* (2017) Production of red tilapia (*Oreochromis* spp.) in floating cages in the mekong delta, vietnam: Mortality and health management. *Diseases of Aquatic Organisms* 124(2), 131–144. DOI: 10.3354/dao03115.

Bondad-Reantaso, M.G., Fejzic, N., MacKinnon, B., Huchzermeyer, D., Seric-Haracic, S, *et al.* (2021) A 12-point checklist for surveillance of diseases of aquatic organisms: A novel approach to assist multidisciplinary teams in developing countries. *Reviews in Aquaculture* 13(3), 1469–1487. DOI: 10.1111/raq.12530.

Brugere, C., Onuigbo, D.M. and Morgan, K.L. (2017) People matter in animal disease surveillance: Challenges and opportunities for the aquaculture sector. *Aquaculture* 467, 158–169. DOI: 10.1016/j.aquaculture.2016.04.012.

Brunton, L.A., Desbois, A.P., Garza, M., Wieland, B., Mohan, C.V, *et al.* (2019) Identifying hotspots for antibiotic resistance emergence and selection, and elucidating pathways to human exposure: Application of a systems-thinking approach to aquaculture systems. *The Science of the Total Environment* 687, 1344–1356. DOI: 10.1016/j.scitotenv.2019.06.134.

Cameron, A. (2002) *Survey toolbox for aquatic animal diseases*. Available at: http://aciar.gov.au/files/node/2268/mn94.pdf

Charo-Karisa, H., Ali, S.E., Marijani, E., Ibrahim, N.A., Trinh, T.Q, *et al.* (2021) Genetic parameters for black spot disease (*diplopstomiasis*) caused by uvulifer sp. infection in *Nile tilapia* (*Oreochromis niloticus* L.). *Aquaculture* 532, 736039. DOI: 10.1016/j.aquaculture.2020.736039.

Corsin, F., Georgiadis, M., Hammell, L. and Hill, B. (2019) Guide for aquatic animal health surveillance. *In Autoimmunity* 1(3). DOI: https://www.oie.int/doc/ged/d6714.pdf.

Delphino, M.K.V.C., Leal, C.A.G., Gardner, I.A., Assis, G.B.N., Roriz, G.D, *et al.* (2019) Seasonal dynamics of bacterial pathogens of *Nile tilapia* farmed in a Brazilian reservoir. *Aquaculture* 498, 100–108. DOI: 10.1016/j.aquaculture.2018.08.023.

Dewulf, J. and Immerseel, F. (2019) *Biosecurity in animal production and veterinary medicine.*

Dong, H.T., Ataguba, G.A., Khunrae, P., Rattanarojpong, T. and Senapin, S. (2017) Evidence of tilv infection in tilapia hatcheries in Thailand from 2012 to 2017 reveals probable global spread of the disease. *Aquaculture* 479(June), 579–583. DOI: 10.1016/j.aquaculture.2017.06.035.

Dórea, F.C. and Revie, C.W. (2021) Data-driven surveillance: effective collection, integration, and interpretation of data to support decision making. *Frontiers in Veterinary Science* 8, 633977. DOI: 10.3389/fvets.2021.633977.

Dórea, F.C., Sanchez, J. and Revie, C.W. (2011) Veterinary syndromic surveillance: Current initiatives and potential for development. *Preventive Veterinary Medicine* 101(1–2), 1–17.

FDA (2020) Fish and fishery products hazards and controls guidance. In *Fish and Fishery Products Hazard and Control Guidance (4th Ed.) (Issue. March 2020*).

Flegel, T.W. (2009) Review of disease transmission risks from prawn products exported for human consumption. *Aquaculture* 290(3–4), 179–189. DOI: 10.1016/j.aquaculture.2009.02.036.

Food and Agriculture Organization (2007) *FAO biosecurity toolkit.* Portal, p. 128.

Food and Agriculture Organization (2019) Preventing and managing aquatic animal disease risks in aquaculture through a progressive management pathway, Retrieved August 23–27, 2019. Tenth Session of the Sub-Committee on Aquaculture of the Committee on Fisheries. Available at: http://www.fao.org/3/na265en/na265en.pdf

Food and Agriculture Organization (2020) Sustainability in action. Rome, *The State of World Fisheries and Aquaculture.* DOI: 10.4060/ca9229en.

Food and Agriculture Organization, & NACA (2001) Manual of procedures for the implementation of the Asia regional technical guidelines on health management for the responsible movement of live aquatic animals. *FAO Fisheries Technical Paper* 402(Suppl. 1), 106.

Gabriel, N.N. and González-Redondo, P. (2019) Review on the progress in the role of herbal extracts in tilapia culture. *Cogent Food and Agriculture* 5(1), 1619651. DOI: 10.1080/23311932.2019.1619651.

Henning, K.J. (2004) What is syndromic surveillance. *MMWR Supplements* 53(Suppl), 5–11.

Hoa, T.T.T., Boerlage, A.S., Duyen, T.T.M., Thy, D.T.M., Hang, N.T.T, *et al.* (2021) Nursing stages of striped catfish (*Pangasianodon hypophthalmus*) in vietnam: pathogens, diseases and husbandry practices. *Aquaculture* 533. DOI: 10.1016/j.aquaculture.2020.736114.

Hoevenaars, K. and Ng'ambi, J.W. (2019) Better management practices manual for smallholders farming tilapia in pond-based systems in zambia.

IDH (2020) *A Practical Guide for Integrating Data into Farmers' Decision-Making Lessons from Asia (Issue January)*.

IDH (2021) IDH Aquaculture program in China. Available at: https://www.idhsustainabletrade.com/uploaded/2019/04/2-Pager-China.pdf

Jeffery, K.R., Stone, D., Feist, S.W. and Verner-Jeffreys, D.W. (2010) An outbreak of disease caused by *Francisella sp.* in *Nile tilapia Oreochromis niloticus* at a recirculation fish farm in the UK. *Diseases of Aquatic Organisms* 91(2), 161–165. DOI: 10.3354/dao02260.

Jia, B., St-Hilaire, S., Stryhn, H., Yu, J., Groman, D.B, *et al.* (2016) Analysis of transaction records of live freshwater finfish in China: A case study of customers' claims of fish mortality using cross-classified modeling. *Aquaculture Reports* 4, 150–155. DOI: 10.1016/j.aqrep.2016.10.003.

Kassam, L., Subasinghe, R. and Phillips, M. (2011) Aquaculture farmer organizations and cluster management – concepts and experiences. *FAO Fisheries and Aquaculture Technical Paper* 563, 90pp.

Kayansamruaj, P., Pirarat, N., Katagiri, T., Hirono, I. and Rodkhum, C. (2014) Molecular characterization and virulence gene profiling of pathogenic *Streptococcus* agalactiae populations from tilapia (*Oreochromis sp.*) farms in Thailand. *Journal of Veterinary Diagnostic Investigation* 26(4), 488–495. DOI: 10.1177/1040638714534237.

Kuebutornye, F.K.A. and Abarike, E.D. (2020) The contribution of medicinal plants to tilapia aquaculture: A review. *Aquaculture International* 28(3), 965–983. DOI: 10.1007/s10499-020-00506-3.

LaFrentz, B.R., Lozano, C.A., Shoemaker, C.A., García, J.C., Ospina-Arango, J.F, *et al.* (2020) Genetic (co)variation between harvest weight and resistance to both *Streptococcus iniae* and *S. agalactiae* capsular type ib in *Nile tilapia* (*Oreochromis niloticus*). *Aquaculture* 529, 735726. DOI: 10.1016/j.aquaculture.2020.735726.

Li, K., Clausen, J.H., Murrell, K.D., Liu, L. and Dalsgaard, A. (2013) Risks for fishborne zoonotic trematodes in tilapia production systems in Guangdong Province, China. *Veterinary Parasitology* 198(1–2), 223–229. DOI: 10.1016/j.vetpar.2013.08.011.

Li, K., Liu, L., Clausen, J.H., Lu, M. and Dalsgaard, A. (2016) Management measures to control diseases reported by tilapia (*Oreochromis* spp.) and whiteleg shrimp (*Litopenaeus vannamei*) farmers in guangdong, China. *Aquaculture* 457, 91–99. DOI: 10.1016/j.aquaculture.2016.02.008.

Lind, C.E., Dana, G., Perera, R.P. and Phillips, M. (2015) *Risk Analysis in Aquaculture: A Step-by-Step Introduction with Worked Examples.* DOI: 10.13140/RG.2.1.2613.4560.

Maas, R. (2021) Upgrading low-quality feeds for tilapia by enzyme and probiotic supplementation.

Muniesa, A., Rey-Castaño, A.L., Ruiz-Zarzuela, I., Guarín, M., Iregui, C, *et al.* (2016) Proposal of a production and management index (PMI) for tilapia farms. *Journal of Animal Science* 94(11), 4872–4881. DOI: 10.2527/jas.2016-0605.

Muniesa, A., Ruiz-Zarzuela, I. and Blas, I. (2018) Design and implementation of a collaborative epidemiological surveillance system for aquaculture (VECA). *Reviews in Aquaculture* 10(2), 370–375. DOI: 10.1111/raq.12165.

Nunan, L.M., Poulos, B.T. and Lightner, D. (1998) The detection of white spot syndrome virus (WSSV) and yellow head virus (YHV) in imported commodity shrimp. *Aquaculture* 160(1–2), 19–30. DOI: 10.1016/S0044–8486(97)00222–6.

Oidtmann, B.C., Crane, C.N., Thrush, M.A., Hill, B.J. and Peeler, E.J. (2011) Ranking freshwater fish farms for the risk of pathogen introduction and spread. *Preventive Veterinary Medicine* 102(4), 329–340. DOI: 10.1016/j.prevetmed.2011.07.016.

Oidtmann, B.C., Peeler, E., Lyngstad, T., Brun, E., Bang Jensen, B, *et al.* (2013) Risk-based methods for fish and terrestrial animal disease surveillance. *Preventive Veterinary Medicine* 112(1–2), 13–26. DOI: 10.1016/j.prevetmed.2013.07.008.

Opiyo, M.A., Marijani, E., Muendo, P., Odede, R., Leschen, W, *et al.* (2018) A review of aquaculture production and health management practices of farmed fish in Kenya. *International Journal of Veterinary Science and Medicine* 6(2), 141–148. DOI: 10.1016/j.ijvsm.2018.07.001.

Palić, D., Scarfe, A.D. and Walster, C.I. (2015) A standardized approach for meeting national and international aquaculture biosecurity requirements for preventing, controlling, and eradicating infectious diseases. *Journal of Applied Aquaculture* 27(3), 185–219. DOI: 10.1080/10454438.2015.1084164.

Palmer, S., Fozdar, F. and Sully, M. (2009) The effect of trust on West Australian farmers' responses to infectious livestock diseases. *Sociologia Ruralis* 49(4), 360–374. DOI: 10.1111/j.1467-9523.2009.00495.x.

Peeler, E.J. and Otte, M.J. (2016) Epidemiology and economics support decisions about freedom from aquatic animal disease. *Transboundary and Emerging Diseases* 63(3), 266–277. DOI: 10.1111/tbed.12278.

Pettersen, J.M., Osmundsen, T., Aunsmo, A., Mardones, F.O. and Rich, K.M. (2015) Controlling emerging infectious diseases in salmon aquaculture. *In Revue Scientifique et Technique* 34(3), 923–938. DOI: 10.20506/rst.34.3.2406.

Phillips, M.J. (1996) Better health management in the Asia-pacific through systems management. In: Subasin, R.P. (ed.) *Health management in Asian aquaculture. Proceedings of the Regional Expert Consultation on Aquaculture Health Management in Asia and the Pacific.*

Richard, A.J., Phillips, M.J., Subasinghe, R.P., Reantaso, M.B. and MacRae, I.H. (2002) Primary aquatic animal health care in rural, small-scale, aquaculture development. In: Arthur, J.R.R., Phillips, M., Subasinghe, R.P., Bondad-Reantaso, M.G. and MacRae, I.H. (eds) *FAO Fisheries Technical Paper No.406.*

Rico, A., Phu, T.M., Satapornvanit, K., Min, J., Shahabuddin, A.M, *et al.* (2013) Use of veterinary medicines, feed additives and probiotics in four major internationally traded aquaculture species farmed in Asia. *Aquaculture* 412–413, 231–243. DOI: 10.1016/j.aquaculture.2013.07.028.

Scott, A.M., Beller, E., Glasziou, P., Clark, J., Ranakusuma, R.W, *et al.* (2018) Is antimicrobial administration to food animals a direct threat to human health? A rapid systematic review. *International Journal of Antimicrobial Agents* 52(3), 316–323. DOI: 10.1016/j.ijantimicag.2018.04.005.

Segner, H., Reiser, S., Ruane, N., Rösch, R., Steinhagen, D. *et al.* (2019) Welfare of fishes in aquaculture. In: *FAO Fisheries and Aquaculture Circular*. p. C1189.

Soares, S., Green, D.M., Turnbull, J.F., Crumlish, M. and Murray, A.G. (2011) A baseline method for benchmarking mortality losses in atlantic salmon (Salmo salar) production. *Aquaculture* 314(1–4), 7–12. DOI: 10.1016/j.aquaculture.2011.01.029.

Sommerville, C. (1982) The life history of *haplorchis pumilio* (looss, 1896) from cultured tilapias. *Journal of Fish Diseases* 5(3), 233–241. DOI: 10.1111/j.1365-2761.1982.tb00478.x.

Stentiford, G.D., Sritunyalucksana, K., Flegel, T.W., Williams, B.A.P., Withyachumnarnkul, B, *et al.* (2017) New paradigms to help solve the global aquaculture disease crisis. *PLOS Pathogens* 13(2), e1006160. DOI: 10.1371/journal.ppat.1006160.

Sukhavachana, S., Tongyoo, P., Massault, C., McMillan, N., Leungnaruemitchai, A, *et al.* (2020) Genome-wide association study and genomic prediction for resistance against *Streptococcus* agalactiae in hybrid red tilapia (*Oreochromis* spp.). *Aquaculture* 525, 735297. DOI: 10.1016/j.aquaculture.2020.735297.

Surachetpong, W., Roy, S.R.K. and Nicholson, P. (2020) Tilapia lake virus: The story so far. *Journal of Fish Diseases* 43(10), 1115–1132. DOI: 10.1111/jfd.13237.

Tan, H.Y., Chen, S.W. and Hu, S.Y. (2019) Improvements in the growth performance, immunity, disease resistance, and gut microbiota by the probiotic *Rummeliibacillus stabekisii* in *Nile tilapia* (*Oreochromis niloticus*). *Fish and Shellfish Immunology* 92, 265–275. DOI: 10.1016/j.fsi.2019.06.027.

van Hai, N.V. (2015) Research findings from the use of probiotics in tilapia aquaculture: A review. *Fish and Shellfish Immunology* 45(2), 592–597. DOI: 10.1016/j.fsi.2015.05.026.

Watts, J.E.M., Schreier, H.J., Lanska, L. and Hale, M.S. (2017) The rising tide of antimicrobial resistance in aquaculture: sources, sinks and solutions. *Marine Drugs* 15(6). DOI: 10.3390/md15060158.

Welker, T.L. and Lim, C. (2011) Use of probiotics in diets of tilapia. *Journal of Aquaculture Research and Development Special Issue* 1, 1–8. DOI: 10.4172/2155-9546.S1-014.

WOAH (2015) Glossary. Aquatic animal health code. Retrieved 2021/2/15. Available at: https://www.oie.int/fileadmin/Home/eng/Health_standards/aahc/2010/en_glossaire.htm

WOAH (2019) Glossary. Terrestrial animal health code. Retrieved 2021/2/15. Available at: https://www.oie.int/en/standard-setting/terrestrial-code/access-online

Wonmongkol, P., Sukhavachana, S., Ampolsak, K., Srisapoome, P., Suwanasopee, T, *et al.* (2018) Genetic parameters for resistance against *Flavobacterium columnare* in *Nile tilapia Oreochromis niloticus* (linnaeus, 1758). *Journal of Fish Diseases* 41(2), 321–328. DOI: 10.1111/jfd.12728.

World Wildlife Fund (2011) *Better Management Practices for Tilapia Aquaculture: A Tool to Assist with Compliance to the International Standards for Responsible Tilapia Aquaculture Version 1.0* 54.

Yáñez, J.M., Joshi, R. and Yoshida, G.M. (2020) Genomics to accelerate genetic improvement in tilapia. *Animal Genetics* 51(5), 658–674. DOI: 10.1111/age.12989.

Zamri-Saad, M., Amal, M.N.A., Siti-Zahrah, A. and Zulkafli, A.R. (2014) Control and prevention of streptococcosis in cultured tilapia in Malaysia: A review. *Pertanika Journal of Tropical Agricultural Science* 37(4), 389–410.

Zhou, A., Xie, S., Junaid, M., Sun, D., Tang, H, *et al.* (2021) First insight into the environmental microbial communities associated with potentially pathogenic strains in pond cultured tilapia (*Oreochromis niloticus*) at various growth stages based on 16S, 18S, and ITS2 rrna gene amplicons sequencing. *Aquaculture* 532, 736007. DOI: 10.1016/j.aquaculture.2020.736007.

Chapter 16

Progress and challenges in vaccine development for farmed tilapia

Hetron M. Munang'andu

Introduction

The expansion of tilapia production to become the second-largest farmed fish species in the world has led to an increase in the number of infectious diseases, which in turn has heightened the need for protective vaccines. The emergence of host-specific pathogens, such as tilapia lake virus disease (TiLVD) and tilapia parvovirus disease (TiPVD), that infect only tilapia calls for a good understanding of their pathogenic mechanisms before protective vaccines can be designed. Nevertheless, for pathogens that infect multiple fish species, vaccines developed for other fish species have the potential to be used for tilapia immunization, which could save resources needed for developing vaccines specific for tilapia. Thus, the availability of commercial vaccines for diseases such as streptococcosis, lactococcosis and columnaris used for the vaccination of various other fish species means that these vaccines can also be used for tilapia immunization. Despite this, there are several challenges, such as developing live vaccines able to evoke the cell-mediated immune responses needed to eliminate intracellularly replicating bacteria like *Francisella noatunensis orientalis* and *Edwardsiella* spp.; this is a challenge not only for tilapia but also for other fish species. The licensure of live attenuated vaccines (LAVs) poses the risk

Kim Thompson, Craig Shoemaker and David C. Little (eds) **Tilapia**
DOI: 10.1079/9781800629455.0016,

of reversion to virulence and also presents the danger of spreading LAVs into aquatic environments where they could cause disease in other aquatic organisms. Nonetheless, the ongoing expansion of the tilapia industry is also contributing to expansion of vaccine development in aquaculture aimed at coping with the growing demand for protective vaccines. Thus, ongoing efforts in tilapia vaccine development are bound to increase the number of vaccines licensed for use in aquaculture.

Tilapia is the second-largest farmed fish species globally after carp, being commercially cultured in >100 countries (Fitzsimmons and Watanabe, 2010). In 2014 global tilapia production was estimated at 4.5 million tons and is expected to increase to 7.3 million tons by 2030 (Bacharach *et al.*, 2016). In 2015, total tilapia production was estimated at 6.4 million metric tons (mt), with the highest-producing countries being China (1.78 million mt), Indonesia (1.12 million mt) and Egypt (0.88 million mt) (Food and Agriculture Organization of the United Nations (FAO), 2017). Other top tilapia-producing countries in the world include Bangladeshi, Brazil, India, Vietnam and the Philippines (FAO, 2017). Expansion of the tilapia industry in the last decade has brought with it the challenge of increased levels of infectious diseases (FAO, 2016). Diseases prevalent in the top tilapia-producing countries render the greatest threat to the growing tilapia industry. Vaccination is perceived as being the most environmentally friendly disease control strategy, which not only contributes to a reduction in infectious diseases but also has the potential to reduce antibiotic use in aquaculture.

Tilapia production has expanded globally due to its general hardiness and ready adaptation to different aquatic environments (Suresh and Lin, 1992). It is cultured in fresh water, brackish water and marine environments, which are also habitats for several pathogens that include members of the genera *Aeromonas*, *Vibrio*, *Edwardsiella* and *Flavobacteria* that infect various fish species (El-Sayed, 2006, 2019; de Godoy *et al.*, 2022). Thus, environmental changes that favour the replication of pathogens while increasing the susceptibility of fish to infectious diseases increase the occurrence of outbreaks. Moreover, high stocking densities brought about by intensive tilapia farming induce stress in fish thereby increasing their susceptibility to infection. Also, high stocking densities increase the transmission index of pathogens between fish, leading to disease outbreaks. Besides, water has been shown to have a higher transmission rate for pathogens than aerosol transmission in terrestrial environments (McCallum *et al.*, 2003, 2004). Taken together, these factors underscore the importance of developing vaccines for the control of fish diseases in aquaculture. Suffice to point out that tilapia are susceptible to pathogens that infect other fish species, which provides the opportunity to use vaccines developed for other fish species for tilapia vaccination. By learning from vaccine development strategies used for other fish species, considerable progress has been made in developing vaccines for tilapia despite its recent expansion into commercial production. The following is an outline of the progress and challenges faced in regard to vaccine development for tilapia based on ongoing research and existing commercial vaccines.

Vaccines for viral diseases of tilapia

Intensive tilapia farming has been linked to an increase in emerging viral diseases that include tilapia lake virus (TiLV) (Jansen *et al.*, 2018; Al-Hussinee *et al.*, 2019), tilapia parvovirus (TiPV) (Liu *et al.*, 2020) and infectious spleen and kidney disease virus (ISKV) (Dong *et al.*, 2015; Subramaniam *et al.*, 2016; Figueiredo *et al.*, 2021; Ramirez-Paredes *et al.*, 2021) (Table 16.1). These pathogens cause high levels of mortality, with the potential to adversely impact global tilapia production. Other viral pathogens reported to infect tilapia include viral nervous necrosis virus (VNNV) (Yanuhar *et al.*, 2017), infectious pancreatic necrosis virus (IPNV) (Hedrick *et al.*, 1983; Mulei *et al.*, 2018), lymphocytic disease virus (LCDV) (Paperna, 1973) and tilapia larvae encephalitis virus (TLEV) (Shlapobersky *et al.*, 2010), but with only a few isolated reports in tilapia (Table 16.1).

Table 16.1 Viruses infecting tilapia

Disease	Abbrev.		Genus	Country	Year	Ref
Lymphocystis disease virus	LCDV	DNA	*Lymphocystivirus*	Tanzania	1973	(Paperna, 1973)
Infectious pancreatic necrosis virus	IPNV	RNA	*Aquabirnavirus*	Taiwan	1983	(Hedrick *et al.*, 1983)
				Kenya	2018	(Mulei *et al.*, 2018)
Bohle virus	BV	DNA	*Ranavirus*	Australia	1997	(Ariel and Owens, 1997)
Iridovirus-like virus	ILV	DNA	*Iridovirus*	Canada	1998	
Viral nervous necrosis virus	VNN	RNA	*Betanodavirus*	France	1999	(Skliris and Richards, 1999)
				Thailand	2009	(Bigarré *et al.*, 2009)
				Indonesia	2015	(Keawcharoen *et al.*, 2015)
					2015	(Yanuhar *et al.*, 2017)
Tilapia larvae encephalitis virus	TLEV	DNA	*Herpesvirus*	Israel	2010	(Shlapobersky *et al.*, 2010)
Infectious spleen and kidney disease virus	ISKV	DNA	*Megalocytivirus*	USA	2015	(Subramaniam *et al.*, 2016)
				Thailand	2016	(Suebsing *et al.*, 2016)
				Ghana	2019	(Ramires *et al.*, 2019)
				Brazil	2021	(Figueiredo *et al.*, 2021)
Tilapia lake virus	TiLV	RNA	*Tilapinevirus*	Worldwide	2014	(Eyngor *et al.*, 2014)
Tilapia parvovirus	TiLP			China	2020	(Liu *et al.*, 2020)
				Thailand	2021	(Yamkasem *et al.*, 2021)
Megalocytivirus		DNA	*Iridovirus*	USA	2021	(Shahin *et al.*, 2021)

TiLV

Since its discovery in Israel in 2014 (Bacharach *et al.*, 2016), TiLV has been reported in Africa, Asia, America and the Middle East (Jansen *et al.*, 2018; Al-Hussinee *et al.*, 2019), including the top tilapia-producing countries globally. It causes mass mortality affecting different growth stages. The disease is mostly found in tropical and subtropical climates with temperatures > 20°C. As pointed out by Husseinee *et al.* (2018), the distribution of TiLV is still expanding as more cases continue being reported from different countries. Thus far the disease has attracted considerable interest in regard to the development of commercial vaccines, for which ongoing research suggests that protective vaccines can be developed against TiLV (Table 16.2). Challenges regarding TiLV vaccine development include the lack of information on whether isolates obtained from different geographical areas have different antigenic properties. It is not known whether there are antigenic differences among isolates that would challenge the design of vaccines with broad neutralizing ability against variant strains from different geographical areas. There is limited information on the structural proteins that form the viral capsid, making it difficult to identify immunogenic epitopes essential for the design of recombinant vaccines. Moreover, the functional roles and structural layout of the different viral segments are still under investigation, making it difficult to select key segments in the design of recombinant vaccines.

Only a few experimental vaccines have been tested, including whole-viral inactivated (WVI) and recombinant vaccines (Table 16.2). In regard to WVI vaccines, various inactivation methods have been used including formalin, β-propiolactone and heat 'killing' of TiLV to produce WVI vaccines (Table 16.2) (Mai *et al.*, 2022). Mai *et al.* (2022) showed that transfer of whole blood from fish vaccinated using WVI vaccines produced high antibody levels that led to protection in passively immunized fish. Zeng *et al.* (2021a) demonstrated an antigen dose-dependent increase in protection for tilapia vaccinated with doses of 10^8, 10^7 and 10^6 50% tissue culture infectious dose ($TCID_{50}$)/ml WVI vaccine, without adjuvant, that resulted in relative percentage survival (RPS) of

Table 16.2 Vaccine development for TiLV

Vaccine type	Antigen dose	Adjuvant	Mode of delivery	RPS (%)	Year	Ref
Formalin-killed	3.6×10^6 $TCID_{50}$ per fish		Intraperitoneal injection	86.7	2022	(Mai *et al.*, 2022)
Heat-killed	3.6×10^6 $TCID_{50}$ per fish		Intraperitoneal injection	80.0	2022	(Mai *et al.*, 2022)
Heat-killed	1.8×10^6 $TCID_{50}$ (100 µl/fish)		Intraperitoneal injection	71.3	2021	(Mai *et al.*, 2021)
Formalin-killed	1.8×10^6 $TCID_{50}$ (100 µl/fish)		Intraperitoneal injection	79.6	2021	(Mai *et al.*, 2021)
β-Propiolactone-inactivated	1×10^8 $TCID_{50}$/ml	Montanide IMS 1312 VG	Intramuscular injection	85.7	2021	(Zeng *et al.*, 2021a)
β-Propiolactone-inactivated	1×10^7 $TCID_{50}$/ml	Montanide IMS 1312 VG	Intramuscular injection	64.3	2021	(Zeng *et al.*, 2021a)
β-Propiolactone-inactivated	1×10^6 $TCID_{50}$/ml	Montanide IMS 1312 VG	Intramuscular injection	32.1	2021	(Zeng *et al.*, 2021a)
β-Propiolactone-inactivated	1×10^8 $TCID_{50}$/ml	None	Intramuscular injection	42.9	2021	(Zeng *et al.*, 2021b)
β-Propiolactone-inactivated	1×10^7 $TCID_{50}$/ml	None	Intramuscular injection	28.5	2021	(Zeng *et al.*, 2021b)
β-Propiolactone-inactivated	1×10^6 $TCID_{50}$/ml	None	Intramuscular injection	14.3	2021	(Zeng *et al.*, 2021b)

42.9, 28.5 and 14.3%, respectively. However, they observed that adjuvenated WVI vaccines for the same vaccine doses of 10^8, 10^7 and 10^6 $TCID_{50}$/ml yielded as twofold increase in protection resulting in RPS of 85.7, 64.3 and 32.1%, respectively. These findings underscore the importance of using adjuvants in the formulations of WVI vaccines. They also showed a positive correlation between increase in protection and antigen dose, indicating that high antigen doses will produce more highly protective vaccines than low antigen doses, as observed for other fish diseases (Munang'andu *et al.*, 2013a; Munang'andu and Evensen, 2019). In regard to recombinant vaccines, various TiLV proteins have been used to generate subunit and DNA vaccines, including segments 5, 6 and 10 TiLV proteins (N-t *et al.*, 2021; Lueangyangyuen *et al.*, 2022). Although no live vaccines have been developed against TiLV, Tattiyapong *et al.* (2020) showed that tilapia surviving TiLV infection developed high and long-lasting antibody levels. All fish first exposed to the virus survived the second challenge, suggesting that a single dose of a live vaccine could be protective over a long period.

In regard to the development of vaccination regimes, studies carried out in Israel (Bacharach *et al.*, 2016), Peru (Pulido *et al.*, 2019), Thailand (Surachetpong *et al.*, 2017) and Egypt (Fathi *et al.*, 2017) show that TiLV causes mortality during warm periods when water temperatures are high, suggesting that a specific timing for vaccination can be developed to ensure that vaccinated fish have protective immunity during the warm months when they are most susceptible to outbreaks. Apart from horizontal transmission from infected to non-infected fish (Mugimba *et al.*, 2019), vertical transmission through the eggs has also been reported (Yamkasem *et al.*, 2019; Dong *et al.*, 2020), implying that breeding stock in hatcheries should be vaccinated to prevent transmission of the virus through the eggs. Moreover, tilapia eggs stored over a long period were shown to contain the virus after several years, indicating that export of infected eggs could serve as a source of transboundary transmission of the virus to other countries (Dong *et al.*, 2017a). Thus, epidemiological data gathered to date could serve as a guide to the development of vaccination regimes and identification of populations that require vaccination at different stages of the tilapia production cycle.

ISKV

ISKV is the type species of the genus *Megalocytivirus* in the family Iridoviridae. It infects a variety of teleost species that include > 50 fish species (Wang *et al.*, 2007). It causes high mortality in tilapia and other fish species including barramundi (*Lates calcarifer*) (Dong *et al.*, 2017b), Chinese perch (*Siniperca chuatsi*) (Lin *et al.*, 2017), silver sweep (*Scorpis lineolate*) (Go and Whittington, 2019), pearl gentian grouper (*Epinephelus fuscoguttatus* × *E. lanceolatus*) (Huang *et al.*, 2020) and giant gourami (*Osphronemus goramy*) (Swaminathan *et al.*, 2021a). Although it has been reported from only Ghana, Thailand, the USA and Brazil in tilapia (Dong *et al.*, 2015a; Subramaniam *et al.*, 2016; Figueiredo *et al.*, 2021; Ramirez-Paredes *et al.*, 2021), its presence in other aquatic organisms expands its global distribution to include several tilapia-producing countries. Throngnumchai *et al.* (2021) developed a DNA vaccine using the capsid protein, yielding a strong immune response in vaccinated tilapia. Vaccines against ISKV tested in other fish species, including mandarin (*Siniperca chuatsi*) (Zhang *et al.*, 2018; Zhao *et al.*, 2019; Zhao *et al.*, 2020), Japanese flounder (*Epinephelus coioides*) (Huang *et al.*, 2012) and red sea bream (*Pagrus major*) (Shimmoto *et al.*, 2010), showed variable protection levels. For example, Zhao *et al.* (2020) developed a DNA vaccine encoding the major capsid protein (MCP) that yielded high protection (RPS 82.4%) while Zhao *et al.*, 2019 developed a subunit vaccine system encoding the *MCP* gene of ISKNV that yielded high protection (RPS 69.6%) in mandarin fish. Although the efficacy of these vaccines is not documented for tilapia, efforts aimed at developing protective vaccines against ISKV for different fish species might be beneficial for tilapia.

Other tilapia viral diseases

A recent emerging viral disease of tilapia is TiPV, which was first detected in 2015 in Hubei, China, where it caused severe mortality in adult tilapia (Liu *et al.*, 2020). To date TiPV has been reported only in China (Liu *et al.*, 2020) and Thailand (Yamkasem *et al.*,

2021). It is likely that the demand for TiPV vaccine development will be influenced by the increasing number of outbreaks, adverse economic impact on tilapia production and spread to other countries. Diseases caused by IPNV, NNV and TLEV in tilapia have been reported from only a few isolated cases (Table 16.1); it is thus unlikely that these would trigger the need for vaccine development specifically for tilapia. If these pathogens were to become a threat to tilapia production, however, then the need for vaccine development would increase. Besides, given that commercial vaccines for viral pathogens such as IPNV and NNV are available for other fish species (Brudeseth *et al.*, 2013; Salgado-Miranda *et al.*, 2013; Dhar *et al.*, 2014), the efficacy of these vaccines could be tested in tilapia.

Vaccines for bacterial diseases of tilapia

Bacteria account for the largest proportion of pathogens causing diseases linked to high economic losses in tilapia production. Common genera causing disease in tilapia include *Streptococcci*, *Aeromonas*, *Francisella*, *Flavobacteria*, *Lactococcus* and *Edwardsiella*.

Streptococcus agalactiae

Streptococcosis causes high mortality, leading to economic losses in tilapia production in several countries, including the top tilapia-producing countries such as China, India, Egypt, Bangladesh, Brazil, Thailand and Malaysia (Eldar *et al.*, 1995; Evans *et al.*, 2004; Kitancharoen *et al.*, 2006; Godoy *et al.*, 2013; Noraini *et al.*, 2013; Ismail *et al.*, 2016; Zhu *et al.*, 2017; Kawasaki *et al.*, 2018; Delphino *et al.*, 2019). There are several *S. agalactiae* serotypes of which the most prevalent in tilapia are serotypes 1a, 1b and III. Classification of serotypes 1a and 1b is based on the biochemical and haemolytic properties of which serotype 1a, also known as biotype-I, which is a β-haemolytic Group B Streptococcus (GBS) while 1b, also known as biotype-II, is a γ-haemolytic GBS (Li *et al.*, 2013; Soto *et al.*, 2015; Syuhada *et al.*, 2020). In Asia, serotype 1a is most prevalent in Thailand, Malaysia and Singapore while serotype 1b is mainly

Table 16.3 Commercially available vaccines against bacterial diseases of tilapia

Pathogen	Vaccine[a]	Reference
Streptococcus agalactiae *Streptococcus iniae*	AQUAVAC® Strep *Sa-Si* Strep *Sa-Si*	https://www.msd-animal-health.com/2019/11/26/msd-animal-health-launches-aquavac-strep-sa-si-vaccine-against-streptococcis-in-fish-in-latin-america/
Streptococcosis	ALPHA JECT® micro 1 Tila	https://www.pharmaq.com/en/pharmaq/our-products/
Motile aeromonad septicaemia	ALPHA JECT® Panga 2	https://www.pharmaq.com/en/pharmaq/our-products/
Aeromonas veronii	Autogenous Aeromonas veronii (Seabass)	https://www.pharmaq.com/en/pharmaq/our-products/
Streptococcus agalactiae *Lactococcus garvieae* *Enterococcus faecalis*	ME-VAC Aqua Strept, as a polyvalent inactivated vaccine	(Abu-Elala *et al.*, 2019)
Edwardisella ictaluri	AquaVac-ESC	(Shoemaker *et al.*, 2007; Abdelhamed *et al.*, 2018)
Flabobacterium columnare	AquaVac-COL	(Shoemaker *et al.*, 2007; Abdelhamed *et al.*, 2018)
Flavobacteria columnare	AQUAVAC-COL™	(Zhang *et al.*, 2006; Shoemaker *et al.*, 2007; Abdelhamed *et al.*, 2018)

[a]Not all vaccines are dedicated specifically for tilapia, but they are for other fish species

found in China and Indonesia (Li *et al.*, 2013; Soto *et al.*, 2015; Syuhada *et al.*, 2020). In South and Central America, serotype 1b is the most prevalent (Barony *et al.*, 2017; Chideroli *et al.*, 2017) whereas serotype III is widely distributed across all continents (Li *et al.*, 2013; Soto *et al.*, 2015; Barony *et al.*, 2017; Chideroli *et al.*, 2017; Syuhada *et al.*, 2020). Chen *et al.* (2012b) showed that the protective immunity against GBS is not only confined to serotypes but is also influenced by genotype within serotypes, suggesting that the choice of vaccine strains should be based on the prevalent strains where the vaccine will be used.

Vaccines licensed against *S. agalactiae* include the AQUAVAC® Strep *Sa-Si* Strep *Sa-Si*, produced by MSD Animal Health, and ALPHA JECT®, produced by Pharmaq (Zoetis) Ltd (Table 16.3).

In regard to experimental vaccines, various antigen delivery systems have been used including DNA, subunit, whole-cell inactivated (WCI) and live attenuated vaccines (LAVs) (Tables 16.4–16.7). Attenuation methods used for development of LAVs include continuous bacterial passage in growth media, selection for natural avirulent strains, selection of antibiotic resistance strains and mutational deletion (Table 16.4). As shown in Table 16.4, the general trend is that LAVs administered by intraperitoneal (IP) injection produce higher protection (RPS > 90%) than oral vaccination (RPS 70–82%), suggesting that the route of vaccination has a significant influence on the outcome of the immune response. In regard to WCI vaccines, various inactivation methods have been used including formalin, sonication and hydrogen peroxide treatment of bacterial cells (Table 16.5). For recombinant vaccines, various proteins have been used to produce DNA (Table 16.6) and subunit (Table 16.7) vaccines. Observations from experimental vaccines show that LAVs (Table 16.4) produce higher protection (>70%) than WCI (Table 16.5), DNA (Table 16.6) and subunit vaccines (Tables 16.6 and 16.7). However, the risks of attenuated bacteria reverting to virulence and of LAVs being transmitted to other aquatic organisms in the environment in which they could cause disease mean that LAVs have less likelihood of being licensed than WCI vaccines. This could be the reason why WCI vaccines have the largest share of licensed vaccines in aquaculture (Table 16.3). Even so, currently available commercial vaccines are contributing to the control of streptococcosis, not only in tilapia but in several other farmed fish species.

Streptococcus iniae *vaccines*

Streptococcus iniae causes high mortality in tilapia (Chen *et al.*, 2007; Anshary *et al.*, 2014) and has been reported in several countries, including top tilapia-producing countries such as China, Brazil, India, Bangladesh, Egypt, Malaysia and Thailand (Chen *et al.*, 2012a; Figueiredo *et al.*, 2012; Rodkhum *et al.*, 2012; Hossain *et al.*, 2014; Rahmatullah *et al.*, 2017; M Younes *et al.*, 2019; Dönmez and Cengizler, 2020). There are two *S. iniae* seroptyes, of

Table 16.4 *S. agalactiae* live vaccines for tilapia

Vaccine	Method of attenuation	Vaccination route	Challenge	Rps (%)	Ref.
Live attenuated YM001	Continuous passage	Oral	Intraperitoneal	75.44	(Huang *et al.*, 2019)
S. agalactiae HZAUSC001	Natural selection				(Zhang *et al.*, 2020)
Live attenuated	Δ*cps* In-frame deletion	Intraperitoneal	Intraperitoneal	90.47	(Zhang *et al.*, 2020)
Live attenuated capsule polysaccharide (CPS) non-encapsulated mutant (Δ*cps*)	Insertional mutagenesis	Intraperitoneal	Intraperitoneal	100	(Zhang *et al.*, 2019)
Live attenuated	Gossypol/novobiocin/ciprofloxan	Intraperitoneal	Intraperitoneal	88	(Pridgeon *et al.*, 2013)
Live attenuated vaccine LAV1,	Acriflavine dye	Oral	Intraperitoneal	81.58	(Laith *et al.*, 2019)
LAV2	Serial passage	Oral	Intraperitoneal	65.79	(Laith *et al.*, 2019)
S. agalactiae strain TFJ0901	Natural low-virulence strain	Intraperitoneal	Intraperitoneal	93.02–100	(Liu *et al.*, 2019)
S. agalactiae YM001	Continuous passage	Oral	Intraperitoneal	71.81	(Li *et al.*, 2015)
S. agalactiae YM001	Continuous passage	Immersion	Intraperitoneal	67.22	(Li *et al.*, 2018)
S. agalactiae YM001	Continuous passage	Injection	Intraperitoneal	96.88	(Li *et al.*, 2019)
Virulent strain HN016	Mutational deletion	Injection	Intraperitoneal	96.11	(Li *et al.*, 2019)
Virulent strain HN016	Mutational deletion	Oral	Intraperitoneal	74.80	(Li *et al.*, 2019)

Table 16.5 *Streptococcus* DNA vaccines for tilapia

Vaccine type	Immunogenic protein	Vaccination route	Challenge	RPS (%)	Ref.
DNA	Surface immunogenic protein (SIP)	Oral/lavage	Intraperitoneal	57–63	(Zhu *et al.*, 2017)
DNA	SIP	Oral	Intraperitoneal	47–53	(Huang *et al.*, 2014)
DNA	SIP	Intragastric lavage	Intraperitoneal	47–57	(Zhu *et al.*, 2017)
DNA	SIP	Mixed fodder	Intraperitoneal	37–63	(Zhu *et al.*, 2017)
DNA	45F2 and 42E2 proteins	Intramuscular	Intraperitoneal	76.19	(Pumchan *et al.*, 2020)
PLGA microspheres DNA	Surface immunogenic protein (SIP)-Nanoparticles	Lavage	Intraperitoneal	41.67–95.83	(Ma *et al.*, 2017)
DNA	Alpha-like protein1 (*alp1*) gene	Intramuscular	Intraperitoneal	51.4	(Inthasaeng *et al.*, 2018)

Table 16.6 *S. agalactiae* inactivated vaccines for tilapia

Vaccine type	Immunization	Challenge	RPS (%)	Ref.
Formalin-inactivated	Injection	Injection	92.3	(Wang *et al.*, 2020)
Formalin-killed	Injection	Injection	44.71–98.81	(Chen *et al.*, 2012)
Formalin-killed	Injection	Injection	83.4	(Evans *et al.*, 2005)
Formalin-killed	Injection	Injection	80.0	(Evans *et al.*, 2004)
Formalin-killed	Bath/immersion	Injection	35.0	(Evans *et al.*, 2004)
Formalin-killed *S. difficile* strains	Injection	Injection	25	(Eldar *et al.*, 1995)
Formalin-killed	Oral	Injection	70	(Ismail *et al.*, 2016)
Formalin-killed	Injection	Injection	62.22–98.89	(Wang *et al.*, 2018a; Wang *et al.*, 2018b)
Formalin-killed	Spray	Immersion	80	(Noraini and Sabri, 2013)
Formalin-killed	Spray	Injection	70	(Noraini and Sabri, 2013)
Formalin-killed	Oral	Injection	70	(Nur-Nazifah *et al.*, 2014)
Formalin-killed	Oral	Injection	75.1	(Ismail *et al.*, 2017)
Sonication	Intraperitoneal	Intraperitoneal	100	(Marcusso *et al.*, 2021)
Hydrogen peroxide inactivation	Intraperitoneal	Intraperitoneal	59.3 and 77.8	(Ramos-Espinoza *et al.*, 2020)

Table 16.7 *S. agalactiae* subunit vaccines for tilapia

Immunogenic protein	Immunization	Challenge	RPS (%)	Ref.
Immunogenic secreted protein (Isp)	Intraperitoneal	Intraperitoneal	48.61	(Li *et al.*, 2016)
S. agalactiae ghost vaccine	Intraperitoneal	Intraperitoneal	86.67	(Wang *et al.*, 2018a)
Phosphoglycerate kinase (PGK) and ornithine carbamoyl-transferase (OCT)	Intraperitoneal	Intraperitoneal	82.4	(Wang *et al.*, 2014)
Bacillus subtilis spores expressing SIP	Oral/gavage	Intraperitoneal	41.7	(Yao *et al.*, 2019)
SIP in PLGA	Oral	Intraperitoneal	100	(Zhang *et al.*, 2016)
Subunit glyceraldehyde-phosphate dehydrogenase protein (GapA)	Intraperitoneal	Intraperitoneal	76.7	(Zhang *et al.*, 2017)
Truncated recombinant proteins of cell wall surface anchor	Intraperitoneal	Intraperitoneal	62.5	(Liu *et al.*, 2016)
EAST-6-like secretion system proteins A (essA)	Intraperitoneal	Intraperitoneal	82.57	(Ma *et al.*, 2020)
EAST-6-like secretion system proteins B (essB)	Intraperitoneal	Intraperitoneal	85.29	(Ma *et al.*, 2020)
ABC transporter substrate-binding proteins (AP and LivK) AP	Intraperitoneal	Intraperitoneal	74.79	(Ma *et al.*, 2020)
ABC transporter substrate-binding proteins (AP and LivK) LivK	Intraperitoneal	Intraperitoneal	83.19	(Ma *et al.*, 2020)
Earlier secreted antigen target six protein (ESAT6)	Intraperitoneal	Intraperitoneal	83.19	(Ma *et al.*, 2020)
Subunit hormone-releasing peptide-6 (GHRP-6)	Intraperitoneal			(Martínez *et al.*, 2017)
Truncated SIP (tSIP)	Intraperitoneal	Intraperitoneal	90	(He *et al.*, 2014)
S. agalactiae ghost vaccines	Intraperitoneal	Intraperitoneal	73.1	(Wangkaghart *et al.*, 2021)

which serotype I isolates are mostly restricted to neural lesions while serotype II isolates produce generalized septicaemia and haemorrhages in several organs (Zlotkin *et al.*, 1998; Bachrach *et al.*, 2001; Barnes *et al.*, 2003). The two serotypes differ in their ability to react with arginine di-hydrolase (ADH) and ribose, for which serotype I is both ADH and ribose positive while serotype II is negative to both (Zlotkin *et al.*, 1998; Bachrach *et al.*, 2001; Barnes *et al.*, 2003). A vaccination programme against *S. iniae* reported from Israel on rainbow trout farms in the 1990s showed a significant reduction in mortality, from 50 to < 5%. Laboratory investigation showed that initial outbreaks were caused by serotype I (Eldar *et al.*, 1997; Bachrach *et al.*, 2001). Following vaccination, massive outbreaks followed that were caused by serotype II, with pathological lesions different from those linked with serotype I (Bachrach *et al.*, 2001; Barnes *et al.*, 2003). This difference could pose a challenge in the design of protective vaccines against different serotypes. Despite that, a multivalent commercial vaccine (ME-VAC Aqua), which is a polyvalent WCI vaccine containing inactivated *S. agalactiae*, *S. iniae*, *Lactococcus garvieae* and *Enterococcus faecalis* cells, has been reported to be protective against both *S. iniae* and *S. agalactiae* in tilapia (Abu-Elala *et al.*, 2019) (Table 16.3). In addition, the AQUAVAC® Strep *Sa-Si* multivalent vaccine containing *S. agalactiae* serotypes 1a, 1b and III together with *S. iniae*, a product of MSD Animal Health, is also available for tilapia vaccination (Table 16.3). Various approaches have been used in the design of experimental vaccines, including WCI, subunit, DNA and LAVs (Table 16.8). Attenuation to produce LAVs has been achieved by both serial passage of bacteria in growth media and selection of antibiotic resistance strains linked to loss of virulence (Table 16.8). For WCI vaccines, various inactivation methods that include formalin and heat treatment of bacterial cells have been used while different proteins have been utilized as antigens to produce DNA and subunit vaccines (Table 16.8). Overall, most experimental vaccines show protection in vaccinated fish (Table 16.8), suggesting that different approaches can be used to produce efficacious vaccines.

Table 16.8 *S. iniae* vaccines for tilapia

Vaccine type	Immunization	Challenge	RPS (%)	Ref.
Attenuated vaccine strain YM011	Oral		76.81	(Liu *et al.*, 2020)
Attenuated vaccine strain YM011	Intraperitoneal		93.25	(Liu *et al.*, 2020)
Attenuated live (ISET0901)	Intraperitoneal	Intraperitoneal	81–100	(Pridgeon and Klesius, 2011a)
Attenuated live (ISET0901)	Oral	Intraperitoneal	88	(Pridgeon and Klesius, 2011a)
Live attenuated (novobiocin resistance)	Intraperitoneal	Intraperitoneal	30	(Pridgeon *et al.*, 2013)
Live attenuated (ciprofloxacin resistance)	Intraperitoneal	Intraperitoneal	20	(Pridgeon *et al.*, 2013)
Inactivated bivalent vaccine *Streptococcus/Aeromonas*	Oral	Intraperitoneal	77.78	(Monir *et al.*, 2020; Monir *et al.*, 2021)
Modified *S. iniae* bacterin	Intraperitoneal	Intraperitoneal	100	(LaFrentz and Shoemaker, 2011)
Formalin-killed	Intraperitoneal	Intraperitoneal	82.5–97.2	(Klesius *et al.*, 2006)
α-Enolase-based DNA	Intramuscular	Intraperitoneal	72.5	(Kayansamruaj *et al.*, 2017a)
LAV rifampin-resistant strains	Intracoelomic	Intracoelomic	95.04	(Heckman *et al.*, 2022)
Formalin-killed (ARS-98–60) bacterin vaccine	Intraperitoneal	Intraperitoneal	79–100	(Shoemaker *et al.*, 2010)
Live attenuated vaccine mutant strain TBY-1ΔsrtA	Intraperitoneal	Intraperitoneal	79.4	(Wang *et al.*, 2014)
Subunit hydrophobic membrane protein (designated MtsB)	Intraperitoneal	Intraperitoneal	66.9	(Zou *et al.*, 2011)
Killed whole-cell vaccine	Intraperitoneal	Intraperitoneal	73.5	(Zou *et al.*, 2011)
Feed-based, formalin-killed vaccine	Oral	Injection	45–30 (80–55)	(Hayat *et al.*, 2021)
Formalin-killed	Injection	Injection	91.7	(Wang *et al.*, 2020)

L. Garvieae *vaccines*

Lactococcus garvieae is a member of the family Streptococcaceae. It has a wide host range that includes humans, animals, fish and shellfish. It is the aetiological agent of lactococcosis that causes high mortality in several fish species including tilapia, sea bass (*Centropristis striata*), eel (*Anguilla anguilla*), red-lipped mullet (*Chelon haematocheilus*), rainbow trout (*Oncorhynchus mykiss*), brook trout (*Salvelinus fontinalis*), red-lipped mullet (*Chelon haematocheilus*) and yellow-tail (*Amberjack seriola*) (Austin *et al.*, 2007; Anshary *et al.*, 2014; Meyburgh *et al.*, 2017; Rao *et al.*, 2020, 2021). Lactococcosis causes septicaemia and meningo-encephalitis in tilapia leading to high mortality (Vendrell *et al.*, 2006; Abu-Elala *et al.*, 2018). It has a worldwide distribution, including top tilapia-producing countries such as Brazil, Taiwan, Thailand, India, Egypt and Indonesia (Carson *et al.*, 1993; Avi Eldar *et al.*, 1999; Evans *et al.*, 2009; Itsaro *et al.*, 2012; Tsai *et al.*, 2012; Anshary *et al.*, 2014; Osman *et al.*, 2017; Swaminathan *et al.*, 2021b; Bwalya *et al.*, 2021). Several distinct serological variations have been identified based on differences in surface carbohydrates among *L. garvieae* isolates, although their impact on vaccine design has not been documented (Barnes and Ellis, 2004). Hence, serotyping would help identify *L. garvieae* strains prevalent in each ecosystem to guide the choice of vaccine strain (Ravelo *et al.*, 2003; Ture *et al.*, 2015).

A polyvalent commercial vaccine (ME-VAC Aqua) containing *S. agalactiae*, *S. iniae*, *L. garvieae* and *Enterococcus faecalis* has been reported to be protective against lactococcosis in tilapia (Abu-Elala *et al.*, 2019) (Table 16.3). Different strategies have been used in the design of *L. garvieae* vaccines including WCI, recombinant and autogenous vaccines. Bercovier *et al.* (1997) and Bwalya *et al.* (2020) developed autovaccines that showed protection against lactococcosis in tilapia. Recombinant vaccines tested include the glyceraldehyde-3-phosphate dehydrogenase (GAPDH) subunit vaccine from *L. garvieae*, which showed excellent protection in tilapia (Tsai *et al.*, 2013). Although oral delivery using chitosan coated with WCI vaccine produced a high level of protection in vaccinated tilapia (Halimi *et al.*, 2019), WCI vaccines administered IP have been shown to produce higher protection in several fish species including tilapia (Vendrell *et al.*, 2006). Given that lactococcosis causes losses in several farmed fish species of commercial importance (Austin *et al.*, 2007; Anshary *et al.*, 2014; Meyburgh *et al.*, 2017; Rao *et al.*, 2020; Rao *et al.*, 2021), ongoing efforts to produce vaccines for multiple host species are also bound to benefit the tilapia industry.

Flavobacterium columnare *vaccines*

Flavobacterium columnare is ubiquitous in aquatic environments (Bacharach *et al.*, 2016) and has been reported from > 30 fish species including tilapia, platies (*Xiphophorus maculatus*), goldfish (*Carassius auratus*), channel catfish (*Ictalurus punctatus*), common carp (*Cyprinus carpio*), goldfish (*Carassius auratus*), rainbow trout, Japanese eel (*Anguilla japonica*), piracanjuba (*Brycon orbignyanus*), pacu (*Piaractus mesopotamicus*), tambaqui (*Colossoma macropomum*) and cascudo (*Hypostomus plecostomus*) (Anderson and Conroy, 1969; Decostere *et al.*, 1998; Al-Harbi *et al.*, 2000; Figueiredo *et al.*, 2005; Verma *et al.*, 2015). It has a worldwide distribution, although in farmed tilapia it has been reported from only a few countries including Egypt (Eissa *et al.*, 2010), Brazil (Figueiredo *et al.*, 2005) and Thailand (Dong *et al.*, 2015b). Genetic diversity has classified *F. columnare* into four distinct genomovars (I, II, III and IV), with II, III and IV being reported in tilapia (LaFrentz *et al.*, 2018) while IV has been linked with highly virulent strains in tilapia (Dong *et al.*, 2015; Kayansamruaj *et al.*, 2017b). It infects fish through mucosal surfaces, mainly the skin and gills, that are in contact with the external aquatic environment. It is the aetiological agent for columnaris disease, also referred to as 'cotton wool disease' characterized by skin ulceration, erosion, fin rot and gill necrosis, leading to high mortality levels in warm-water-cultured fish (Schneck and Caslake, 2006). Columnaris outbreaks are influenced by changes in environmental factors such as temperature, and are stressful for the host resulting from high stocking density and elevated levels of ammonia and organic matter (Wakabayashi and Egusa, 1972; Wakabayashi, 1991).

A commercial vaccine known as AQUAVAC-COL™, which is an avirulent rifampicin-resistant mutant derived from genomovar I, confers protection in catfish and largemouth bass (Zhang *et al.*, 2006; Shoemaker *et al.*, 2007; Abdelhamed *et al.*, 2018) (Table 16.3). It is likely that this vaccine could also be protective against columnaris in tilapia. Experimental vaccines tested in tilapia include LAVs, WCI and muco-adhesive vaccines (Table 16.9). Grabowski *et al.* (2004) showed that tilapia immunized with WCI vaccines administered with Freund's complete adjuvant (FCA) produced high antibody levels, while Leal *et al.* (2010) found no correlation between increase in antibody levels with protection in tilapia vaccinated using WCI vaccines. Kitiyodom *et al.* (2019a, 2021) developed a muco-adhesive delivery system that increased adherence of the vaccine in the gills, resulting in increased protection in vaccinated fish. Mohammed *et al.* (Mohammed *et al.*, 2013) developed a LAV by selecting for rifampicin resistance using the genomovar II strain that produced a high level of protection in tilapia. They found core antigens shared by genomovars I and II that conferred protection against both genomovars in tilapia (Shoemaker *et al.*, 2011; Mohammed *et al.*, 2013). Zhu *et al.* (2012) developed a ghost vaccine that produced excellent protection in tilapia while Luo *et al.* (2016) evaluated the efficacy of three outer membrane proteins (OMPs) as antigens, namely the gliding motility lipoprotein (GldJ), lipoprotein (Lip), and outer membrane efflux protein precursor (Omep), which yielded excellent protection in grass carp (*Ctenopharyngodon idellus*). Overall, LAVs produce a high level of protection although the risk of reversion to virulence could limit their licensure. Given that *F. columnare* infects several warmwater fish species, the search for protective vaccines for multiple fish species is bound to benefit tilapia.

Francisella noatunensis *subspp.* orientalis *(Fno) vaccines)*

Francisella noatunensis is one of the major causes of high mortality in tilapia and other warmwater fish species (Soto *et al.*, 2009; Birkbeck *et al.*, 2011; Colquhoun and Duodu, 2011; Klinger-Bowen *et al.*, 2016; Ramírez-Paredes *et al.*, 2017; Pulpipat *et al.*, 2019). It causes multiple white granulomas in several internal organs including the gills, kidney, spleen, liver and gut (Birkbeck, 2011; Colquhoun and

Table 16.9 *Edwardsiella* and *Flavobacterium* vaccines for tilapia

Infectious agent	Vaccine type	Immunization	Challenge	RPS (%)	Ref.
Edwardsiella tarda FPC498	Avirulent mutant	Intraperitoneal	Intraperitoneal	100	(Igarashi and Iida , 2002)
Edwardsiella tarda FPC498	Formalin-killed	Intraperitoneal	Intraperitoneal	20	(Igarashi and Iida , 2002)
Edwardsiella tarda	Formalin-killed	Intraperitoneal	Intraperitoneal	92	(Kwon *et al*., 2006)
Edwardsiella ictaluri	Formalin-killed whole cells	Intraperitoneal	Intraperitoneal	28.6	(Trung Cao *et al*., 2014)
Edwardsiella tarda	Formalin-killed	Hyperosmotic infiltration	Intraperitoneal	0	(Lio-Po and Wakabayashi, 1986)
Edwardsiella ictaluri	Glyceraldehyde-3-phosphate dehydrogenase (GAPDH)	Intraperitoneal	Intraperitoneal	42.9	(Trung Cao *et al*., 2014)
Edwardsiella ictaluri	Recombinant EiCh protein	Intraperitoneal	Intraperitoneal	42	(Machimbirike *et al*., 2021)
Edwardsiella tarda	*E. tarda* ghosts (ETG)	Intraperitoneal	Intraperitoneal	100	(Kwon *et al*., 2006)
Flavobacterium columnare,	Attenuated	Immersion	Immersion	82.1	(Mohammed *et al*., 2013)
Flavobacterium columnare	Formalin-killed chitosan modification	Immersion	Immersion	83	(Kitiyodom *et al*., 2019b)
Flavobacterium columnare	Chitosan nanovaccine (CSNE)	Immersion	Immersion	78	(Kitiyodom *et al*., 2021)
Flavobacterium columnare	Muco-adhesive nanovaccine	Immersion	Immersion	60	(Kitiyodom *et al*., 2019a)

Duodu *et al.*, 2011). Outbreaks are mostly reported from intensive farming systems with high stocking densities when water temperature falls below 25°C (Soto *et al.*, 2009, 2012; Ramírez-Paredes *et al.*, 2017; Soto *et al.*, 2019). Soto *et al.* (2011) developed a LAV based on insertional mutation of the intracellular growth *locus* C gene (*iglC*), which produced a high level of protection in tilapia. However, due to the risk of reversion to virulence coupled with potential transmission of LAVs to other aquatic organisms in the environment in which the attenuated bacteria might cause disease, the mutant Fno vaccine was not licensed for commercial use (Soto *et al.*, 2011). Sahin *et al.* (2019) produced WCI vaccines of which the homologous vaccine strain with the challenge strain had higher protection than heterologous strains. Pulpipat *et al.* (2020) and Ramírez-Paredes *et al.* (2019) developed WCI vaccines that produced high protection in vaccinated fish, while Hoare *et al.* (2021) showed high antibody levels in serum and mucus induced by oral vaccination in tilapia (Table 16.10). Despite this, there are no commercial vaccines available against Fno (Ramírez-Paredes *et al.*, 2019). As pointed out previously (Munang'andu, 2018), Fno is a facultative intracellular bacterium that replicates in phagocytic cells such as macrophages and neutrophils, rendering it inaccessible to neutralization by antibodies; therefore, the most effective protective immune response would be cell-mediated immunity (CMI) that eliminates infected cells. However, licensure of LAVs that produce CMI remains a challenge due to the risk of reversion to virulence in some countries.

Edwardsiella *spp. vaccines*

Edwardsiella spp. causing disease in tilapia include *E. ictaruli* (Dong *et al.*, 2019; Nhinh *et al.*, 2021), *E. tarda* (Pirarat *et al.*, 2007; Abo El-Yazeed and Ibrahem, 2009), *E. piscicida* (Reichley *et al.*, 2015; Ye *et al.*, 2021) and *E. anguillarum* (Oh *et al.*, 2020). These bacteria cause high mortality in different fish species including channel catfish, largemouth bass, Japanese eel (*Anguilla japonica*), mullet (*Mugil cephalus*), crimson sea bream (*Evynnis japonicas*), freshwater catfish (*Tandanus tandanus*), chinook salmon (*Oncorhynchus tshawytscha*), red sea bream (*Chrysophrys major*), yellowtail (*Seriola

Table 16.10 Fno vaccines for tilapia

Infectious agent	Vaccine type	Immunizatiomn	Challenge	RPS (%)	Ref.
Fno	Membrane vesicles (MVs)	Injection	Cohabitation	<10	(Mertes *et al.*, 2010)
Fno	Formalin-killed	Intraperitoneal	Intraperitoneal	71	(Pulpipat *et al.*, 2020)
Fno	Formalin-killed	Intraperitoneal	Immersion	76	(Pulpipat *et al.*, 2020)
Francisella orientalis	Diatom-expressed IglC	Intracoelomic	Immersion	75	(Shahin *et al.*, 2020)
Francisella orientalis	GroEL recombinant	Intracoelomic	Immersion	53	(Shahin *et al.*, 2020)
Fno	Inactivated killed whole cell	Intraperitoneal	Intraperitoneal	82.3	(Shahin *et al.*, 2019)
Fno	Inactivated autogenous	Intraperitoneal	Intraperitoneal	42	(Ramírez-Paredes *et al.*, 2019)
Fno	Live attenuated	Immersion	Immersion	30–38	(Soto *et al.*, 2014)
Fno	Live attenuated	Immersion	Bath		(Soto *et al.*, 2013)
Francisella Asiatica	Live attenuated	Immersion	Immersion	90	(Soto *et al.*, 2011)
Lactococcus garviea	Recombinant GAPDH	Intraperitoneal			(Tsai *et al.*, 2013)
Lactococcus garviea	Formalin-killed	Intraperitoneal			(Tsai *et al.*, 2013)

gaingu eradiata), angel fish (*Pterophyllum scalare*), Japanese flounder (*Paralichthys olivaceus*), common carp (*Cyprinus carpio*), turbot (*Scophthalmus maximus*), Asian catfish (*Clarias batrachus*), rainbow trout, koi (*Anabas testudineus*), oyster toadfish (*Opsanus tau*) and European eel (*Anguilla anguilla*) (Mohanty and Sahoo, 2007; Park *et al.*, 2012; Dong *et al.*, 2019). Thus, vaccines are not only sought for tilapia, but several other farmed fish species.

Commercial vaccines used against edwardsiellosis include the bivalent live attenuated *E. ictaluri* (AquaVac-ESC) and *F. columnare* (AquaVac-COL) vaccines (Shoemaker *et al.*, 2007; Abdelhamed *et al.*, 2018), as well as an enteric septicaemia vaccine produced by Pharmaq/Zoetis Ltd for catfish (Ma *et al.*, 2019) (Table 16.3). Experimental vaccines tested for tilapia include LAVs, WCI, subunit and ghost vaccines (Table 16.9). Agarishi and Takaji (Igarashi and Iida , 2002) developed a LAV using strain SPM31 that yielded excellent protection in vaccinated fish, while Kwon *et al.* (2005, 2006) developed an *E. tarda* ghost (ETG) vaccine that produced higher protection than the formalin-killed *E. tarda* (FKC) vaccine. Cao *et al.* (2014) developed a recombinant vaccine using the glyceraldehyde-3-phosphate dehydrogenase (GAPDH) protein derived from *E. ictaruli*, which produced excellent protection in tilapia. Vaccine development methods used for other fish species include WCI (Swain and Nayak, 2003; Castro *et al.*, 2010), LAVs (Takano *et al.*, 2010), ghost cells (Kwon *et al.*, 2007), outer membrane proteins and vesicles (OMVs) (Maiti *et al.*, 2011; Park *et al.*, 2011), subunits (Park *et al.*, 2012) and DNA vaccines (Jiao, *et al.*, 2010). It is likely that vaccines developed for other fish species will also be protective for tilapia. Similar to Fno, *Edwardsiella* spp. replicate intracellularly in phagocytic cells like macrophages and monocytes, where they are inaccessible to neutralization by antibodies (Munang'andu, 2018). Thus, the CMI response produced by replicative vaccines like LAVs and DNA vaccines will be more protective, unlike antibody responses produced by WCI and subunit vaccines that neutralize extracellular replication bacteria. As with other LAVs, the risk of reverting to virulence and the challenge of producing DNA vaccines capable of evoking strong CMI responses may derail the development of protective vaccines against edwardsiellosis for tilapia and other fish species.

Aeromonas *spp. vaccines*

Aeromonads are opportunistic pathogens commonly found in freshwater, estuarine and marine environments (Holmes *et al.*, 1996). Although the genus *Aeromonas* comprises > 30 species (Chen *et al.*, 2016), the major species causing diseases in fish are *A. hydrophila*, *A. veronii*, *A. carvieae*, *A. sobria* and *A. jandaei* (Ashiru *et al.*, 2011; Li and Cai, 2011; Dong *et al.*, 2017c; Basri *et al.*, 2020). These *Aeromonas* spp. cause 'motile aeromonad septicaemia (MAS)' linked to high mortality and economic losses in aquaculture (Nielsen *et al.*, 2001; Pridgeon and Klesius, 2011b; Dharmaratnam *et al.*, 2019). Experimental vaccines developed include LAVs, WCI and subunit vaccines, of which WCI account for the largest proportion (Table 16.11). Inactivation methods used for producing WCI vaccines explored include formalin, autoclave and heat 'killing' of bacteria cells (Table 16.9) (Bactol *et al.*, 2018), while adjuvants used include lipopolysaccharide (LPS), turmeric, ascogen, incomplete Freund's adjuvant (IFA) and Freund's complete adjuvant (FCA) (Ruangpan *et al.*, 1986; Ramadan *et al.*, 1994; Ibrahem *et al.*, 2008; El Tantawy and Ayoub, 2016; Sukenda *et al.*, 2018). For LAVs, attenuation has been achieved using various methods such as serial passage of *A. hydrophila* in growth media (Mzula *et al.*, 2020), *Oreganum vulgare* herbal extract treatment of *A. hydrophila* and *A. putida* (Abdel-Hadi *et al.*, 2009) and selecting for novobiocin and rifampicin resistance (Pridgeon *et al.*, 2013). In regard to recombinant vaccines, different proteins have been used as antigens including *A. hydrophila* adhesion protein 1 (AHA1) (Fang *et al.*, 2004), various OMPs (Khushiramani *et al.*, 2007; Maiti *et al.*, 2011; Khushiramani *et al.*, 2012), surface (S)-layer protein (Poobalane *et al.*, 2010), EprJ1 (Ren *et al.*, 2006), ECPase54 (Qin *et al.*, 2000) and AsaP1 (Schwenteit *et al.*, 2013). The challenge is to find antigenic proteins that are cross-protective against different *Aeromonas* spp. Peepim *et al.* (Peepim *et al.*, 2016) showed that the protein designated as Epr3

Table 16.11 *Aeromonas* vaccines for tilapia

Pathogen	Vaccine type	Immunization	Challenge	RPS (%)	Ref.
Aeromonas hydrophila	Attenuated live	Intraperitoneal	Intraperitoneal	71.4	(Mzula *et al.*, 2020)
Aeromonas hydrophila	Attenuated live	Intramuscular	Intraperitoneal	82.3	(Mzula *et al.*, 2020)
Aeromonas putida	Attenuated live	Intraperitoneal	Intraperitoneal	80	(Abdel-Hadi *et al.*, 2009)
Aeromonas hydrophila	Attenuated live	Intraperitoneal	Intraperitoneal	60	(Abdel-Hadi *et al.*, 2009)
Aeromonas hydrophila	Attenuated live	Intraperitoneal	Intraperitoneal	100	(Pridgeon and Klesius, 2011c)
Aeromonas hydrophila	Recombinant LPS	Immersion	Intraperitoneal	27–45	(Fernandez *et al.*, 2014)
Aeromonas hydrophila	Autoclave-killed	Anal	Intramuscular	90	(Bactol *et al.*, 2018)
Aeromonas hydrophila	Heat-killed	Anal	Intramuscular	90	(Bactol *et al.*, 2018)
Aeromonas hydrophila	Heat-killed	Intraperitoneal	Intraperitoneal	90	(Abdel-Hadi *et al.*, 2009)
Aeromonas putida	Heat-killed	Intraperitoneal	Intraperitoneal	90	(Abdel-Hadi *et al.*, 2009)
Aeromonas hydrophila	Whole-cell-inactivated *Aeromonas*	Oral	Intraperitoneal	77.78	(Monir *et al.*, 2020; Monir *et al.*, 2021)
Aeromonas hydrophila	Formalin-killed	Immersion	Intramuscular	83–100	(Hardi *et al.*, 2019)
Aeromonas hydrophila	Formalin-killed	Oral	Intramuscular	83–87	(Hardi *et al.*, 2019)
Aeromonas hydrophila	Formalin-killed	Intraperitoneal	Intraperitoneal	53–61	(Ruangpan *et al.*, 1986)
Aeromonas hydrophila	Formalin-killed	Oral	Intraperitoneal	80–100%	(El Tantawy and Ayoub, 2016)
Aeromonas hydrophila	Formalin-killed	Intraperitoneal	Intraperitoneal	70.59–78.26	(Sukenda *et al.*, 2017)
Aeromonas hydrophila	Formalin-killed	Intraperitoneal	Intraperitoneal	53–61	(Ruangpan *et al.*, 1986)
Aeromonas hydrophila	Formalin-killed	Anal	Intramuscular	80	(Bactol *et al.*, 2018)
Aeromonas hydrophila	Formalin-killed	Intraperitoneal	Intraperitoneal	66.7	(Aly *et al.*, 2015)

Continued

Table 16.11 Continued

Pathogen	Vaccine type	Immunization	Challenge	RPS (%)	Ref.
Aeromonas hydrophila	Formalin-killed	Oral/immersion	Intraperitoneal	91.2	(Ismail *et al.*, 2010)
Aeromonas hydrophila	Formalin-killed	Immersion	Bath	90	(Osman *et al.*, 2009)
Aeromonas hydrophila	Formalin-killed	Oral	Bath	68	(Osman *et al.*, 2009)
Aeromonas sobria	Formalin-killed	Immersion	Bath	86	(Osman *et al.*, 2009)
Aeromonas sobria	Formalin-killed	Oral	Bath	90	(Osman *et al.*, 2009)
Aeromonas caviae	Formalin-killed	Immersion	Bath	84	(Osman *et al.*, 2009)
Aeromonas caviae	Formalin-killed	Oral	Bath	76	(Osman *et al.*, 2009)
Aeromonas hydrophila	Formalin-killed	Immersion	Immersion	76.67	(Prasad and Areechon, 2010)
Aeromonas hydrophila	Formalin-killed	Intramuscular		89.80	(Ramadan *et al.*, 1994)
Aeromonas hydrophila	Formalin-killed	Immersion		70.82	(Ramadan *et al.*, 1994)

is a conserved immunogenic protein among *Aeromonas* spp. able to induce antibody response in Nile tilapia.

Vaccination regimes

One of the key factors required for the design of effective vaccination programmes is correct timing, ensuring that vaccinated fish have protective immune responses during the period when they are most vulnerable to infection. For example, in the case of TiLV it is better for vaccinated fish to have high antibody levels during the warmer months when most fish are most susceptible to infection. However, for most vaccination programmes the duration of protective immunity induced by vaccination may not always be long enough to cover the period when they are most susceptible to infection. Vaccines used in aquaculture are broadly classified into (1) replicative vaccines, mainly comprising LAV and DNA vaccines, and (2) non-replicative vaccines, mainly comprising WCI and subunit vaccines. Replicative vaccines have the advantage of producing long-term protective immunity because of their replicative nature that continuously evokes immune responses in vaccinated individuals. A study carried out by Tattiyapong *et al.* (2020) showed that fish that survived TiLV infection had high antibody levels that lasted long after challenge, resulting in RPS of 100% following the second challenge. This finding suggests that a single dose of LAV against TiLV could be protective over a long period. By contrast, non-replicative vaccines do not induce long-term protective immunity on account of their innate or non-propagative state. To prolong the duration of protective immunity induced by non-replicative vaccines they require the inclusion of adjuvants that form a depot at the injection site essential for the slow release of antigens. For example, a study carried out by Zeng *et al.* (2021) showed that tilapia vaccinated with doses of 10^8, 10^7 and 10^6 $TCID_{50}$/ml IWV vaccines without adjuvant resulted in RPS of 42.9, 28.5 and 14.3%, respectively, whereas adjuvenated IWV vaccines for the same doses of vaccine showed a twofold increase in protection, resulting in RPS of 85.7, 64.3 and 32.1%, respectively. Thus, adjuvants are important for all non-replicative vaccines used in aquaculture. In most cases non-replicative vaccines also require a booster vaccination for continued activation of the immune response following waning of the host response to the prime vaccine. Ismail *et al.* (2017) showed a significant increase in antibody levels that remained high in tilapia that were given a second booster against *S. agalactiae*, in contrast to fish that were given only a single booster that showed a decline in antibody levels. They observed that fish given a second booster had higher survival (RPS 70%) than those given a single booster (RPS 45%). Similarly, Mai *et al.* (2021) showed significant increase in antibody responses following booster vaccination in tilapia immunized using WCI against TiLV that corresponded with increase in RPS. Ongoing studies show that various combinations of IP, immersion and oral prime booster vaccination regimes are now being used in the design of effective vaccination regimes for different pathogens infecting tilapia (Munang'andu *et al.*, 2016; Ismail *et al.*, 2017; Mo *et al.*, 2020).

Measures of vaccine efficacy

Establishing measures of vaccine efficacy is an important component of vaccine development aimed at ensuring that the vaccines produced are protective.

Challenge models

Challenge models are a vital part of evaluating vaccine efficacy. In aquaculture, three challenge methods are used namely IP, cohabitation and immersion. The advantages and disadvantages of these challenge methods have been discussed in detail in previous reviews (Munang'andu *et al.*, 2016; Munang'andu and Evensen, 2019). For a challenge model to be effective it is important to have a high discriminatory capacity between vaccinated and non-vaccinated fish (Munang'andu *et al.*, 2016). In addition, it must be reproducible to enable effective comparison between vaccine batches (Munang'andu *et al.*, 2016). Previous studies have shown that optimization of challenge models used for vaccine efficacy trials must include the selection of highly virulent pathogen strains, selection of a challenge dose able to cause high mortality and the use of highly susceptible fish able to exhibit the pathognomonic features of the disease (Munang'andu *et al.*, 2016). Studies carried out to date show that the most widely used challenge method for evaluating the efficacy of tilapia vaccines is IP, followed by immersion (Tables 16.4–16.11). However, there is a need to optimize these challenge methods to enable comparison between vaccine batches.

Relative percentage survival and Kaplan–Meyer survival analysis

The most widely used approach for evaluating the efficacy of vaccines against diseases that cause mortality in vaccinated fish is RPS (Tables 16.2–16.11) (Amend, 1981), which is based on quantifying the number of post-challenge survivors in vaccinated compared with unvaccinated control fish. Another approach used for evaluating the efficacy of fish vaccines based on post-challenge survival is Kaplan–Meyer survival analysis, which computes the post-challenge proportion of the survivors in both vaccinated and unvaccinated control fish (Dubey *et al.*, 2016; Liu *et al.*, 2016; Mugimba *et al.*, 2019; Joshi *et al.*, 2021). The advantage of this method is that it is useful in identifying vaccines at risk of not conferring protective immunity, by using Cox hazard risk analysis or other statistical methods (Munang'andu *et al.*, 2016). The major limitation for both RPS and Kaplan–Meyer survival analysis is that they evaluate protection based only on post-challenge survival and do not provide detail on the mechanisms of vaccine protection. They do not provide information as to whether vaccination prevents infection establishment, prevents pathogen dissemination to different organs or prevents pathology. Thus, for diseases that do not cause high mortality, both RPS and Kaplan–Meyer survival analysis may not be effective at determining vaccine efficacy. Therefore, to gain a better understanding of the mechanism of vaccine protection for non-lethal pathogens there is a need to use other measures of vaccine efficacy.

Reduction in pathogen levels

Several studies have shown that detection of low pathogen levels in vaccinated fish compared with high levels in unvaccinated fish post challenge can be used as a measure of vaccine protection (Dettleff *et al.*, 2015; Xing *et al.*, 2019). There are different methods used for evaluating the efficacy of vaccines based on reduction of pathogen levels in vaccinated fish. These include the use of sera from vaccinated fish to evaluate the inhibition of pathogen propagation via *in vitro* methods. For example, Zhang *et al.* (2020) used the serum inhibition test (SIT) to evaluate the ability of antibodies generated from tilapia vaccinated with the Sip protein to inhibit the replication of *S. agalactiae in vitro*. Similarly, Kwon *et al.* (2006) showed that sera from tilapia immunized against *E. tarda* using a ghost vaccine had significantly higher inhibition capacity in regard to bacteria growth *in vitro* that corresponded with the high RPS observed in vaccinated fish. Using the SIT method, Dubey *et al.* (2016) showed that sera from fish vaccinated with a high antigen dose (HiAg) of 8 ug/g of the recombinant outer membrane protein A (OmpA) had greater inhibition of *A. hydrophila* replication that corresponded with a high RPS (79.99%), unlike sera from fish vaccinated with a lower antigen dose (LoAg) of 4 ug/g that had a low RPS (37.33%). These studies show that sera from vaccinated fish can be used to evaluate the efficacy of different vaccines via *in vitro* methods.

Another approach for evaluation of vaccine efficacy is by quantifying post-challenge pathogen loads in organs of vaccinated fish compared with unvaccinated fish. For example, Gallage *et al.* (2017) found significantly higher *S. agalactiae* bacterial loads in the brain, head kidney and blood of unvaccinated tilapia ($>6.0 \times 10^4$ colony-forming units (CFU)/g), whereas vaccinated fish had no bacteria detected in the brain and blood but low levels in the head kidney (7.4×10^2 CFU/g). From their findings they observed that the absence of bacteria in the brain and blood, coupled with low levels detected in the head kidney, correlated with higher blood antibody levels in vaccinated fish. Similarly, Shahin *et al.* (2020) showed a correlation between reduction of bacterial load in the spleen of tilapia vaccinated against Fno using IglC and GroEL recombinant vaccines, while Kishimori *et al.* (2015) found a corresponding reduction in tissue damage in mucosal organs of tilapia exposed to the monogenean parasite *Neobenedenia melleni* with increase in antibody levels in vaccinated fish. Taken together, these studies show that reduction of pathogen loads in vaccinated fish correlates with increase in antibody levels and can be used as a measure of efficacy for tilapia vaccines.

Prevention of pathology

Several studies have shown that prevention of tissue damage in target organs can be used as a measure of protection in vaccinated fish (Munang'andu *et al.*, 2016; Munang'andu and Evensen, 2019; Mugimba *et al.*, 2020). For example, studies performed for infectious pancreatic necrosis virus (IPNV) in Atlantic salmon showed that viral titre $> 10^7$ $TCID_{490}$/ml correlated with establishment of pathology in the liver and pancreas that are target organs (Munang'andu *et al.*, 2013, 2014). On the other hand, vaccine efficacy studies on IPNV showed that antibody titres > 1.4 OD_{490} (diluted 1:50) correlated with prevention of pathology, resulting in RPS > 90% (Munang'andu *et al.*, 2013a, 2014). In tilapia, He *et al.* (2021) showed that increase in *S. agalactiae* bacterial loads corresponded with establishment of meningitis in the brain, which is a target organ for *S. agalactiae* infection in tilapia. Thus, an antibody titre can be established that correlates with prevention of meningitis in vaccinated tilapia, and used as a measure of vaccine efficacy. Mugimba *et al.* (2020) showed that the liver and brain were target organs for TiLV, in which an increase in viral loads corresponded with the establishment of pathology. Thus, virus quantification can be used to determine a cut-off antibody titre that correlate with prevention of pathology in target organs. So, by identifying antibody titres that prevent pathology linked to mortality, it is possible to establish a 'signature of protective immunity' based on antibody titres that prevents pathology, for use as a benchmark in developing protective vaccines for tilapia.

Correlates of protection

Developing correlates of protection is key to establishing the benchmarks needed for optimization of vaccine development.

Antibodies as correlates of protective immunity for tilapia vaccines

Antibodies are the most widely used correlates of protective immunity for mammalian and fish vaccines (Plotkin, 2010). As defined by Pulendran and Ahmed (2011), an antibody is considered a correlate of protection when it has reached a cut-off limit above which it confers protective immunity in vaccinated individuals. Thus, an antibody titre that reaches a protective threshold can serve as a signature of protective immunity for a given vaccine. Three independent studies carried out by Munang'andu *et al.* (Munang'andu *et al.*, 2012, 2013a; Munang'andu *et al.*, 2013b) on vaccination of Atlantic salmon against IPNV showed that antibody titres > 1.2 OD_{490} (serum dilution 1:50) corresponded to post-challenge survival proportion (PCSP) > 90% while titres > 1.4 OD_{490} produced PCSP > 92%, indicating that antibody titres can be established that correlate with > 90% protection (Munang'andu *et al.*, 2012; Munang'andu *et al.*, 2013a; Munang'andu *et al.*, 2013b). Similarly, Bricknell *et al.* (Bricknell *et al.*, 1999) showed that an antibody titre $\log_2 > 4$ produced RPS > 90% in Atlantic salmon vaccinated against *A. salmonicida*, indicating that a threshold antibody titre able to serve as a signature of protective immunity for RPS > 90% protection in vaccinated salmon can be established. In regard to tilapia vaccines, Pransik *et al.* (2005) showed high correlation ($r^2 = 0.883$) between RPS and antibody titre for tilapia vaccinated against *S. agalactiae*, while Ruangpan *et al.* (1986) found a significant correlation between increase in antibody levels and protection in tilapia vaccinated against *A. hydrophila*. Similarly, Eldar *et al.* (1995) showed a high correlation between antibody response and protection in tilapia vaccinated against S. *difficile*, while Shahin *et al.* (2020) showed a correlation between antibody response and protection in tilapia vaccinated against Fno using IglC and GroEL recombinant vaccines. In the case of recombinant vaccines, Wang *et al.* (2014) showed that increase in antibody levels correlated with protection in tilapia vaccinated against *S. agalactiae* using a subunit vaccine encoding the proteins phosphoglycerate kinase (PGK) and ornithine carbamoyl-transferase (OCT). LaFrentz and Shoemaker (2015) demonstrated that passively transferred antibodies showed correlation with protection in tilapia vaccinated with a *Vibrio vulnificus* vaccine. In regard to parasitic diseases, Kishimori *et al.* (2015) showed a significant correlation between reduction in the parasite load of the monogenean *Neobenedenia melleni* and increase in mucus and plasma antibody responses. Taken together, these studies show that an antibody titre can be established for each vaccine to serve as a correlate of protective immunity and thus a benchmark for vaccine production.

Antigen dose as a correlate of protection for tilapia vaccines

Studies in fish vaccinology have shown that antigen dose can be used as a correlate of vaccine protection (Munang'andu and Evensen, 2019). Research carried out in Atlantic salmon vaccinated against IPNV showed that an antigen dose of 10^{10} $TCID_{50}$/ml produced PCSP > 90%, while reducing the antigen dose by one log to 10^{9} $TCID_{50}$/ml resulted in a low PCSP of <42% (Munang'andu *et al.*, 2013, 2014). These findings indicate that an antigen dose of 10^{10} $TCID_{50}$/ml could serve as a correlate of protection for PCSP > 90%. Mo *et al.* (2020) showed that tilapia vaccinated against *S. agalactiae* using the LAV strain TFJ0901 with doses of 10^{2}, 10^{3}, 10^{4}, 10^{5}, 10^{6} and 10^{7} CFU/g resulted in RPS of 2.6, 20.5, 38.5, 56.4, 71.8 and 97.4%, respectively, indicating that a cut-off antigen dose of 10^{7} CFU/g could serve as a correlate protection for RSP > 90%. Zeng *et al.* (2021) showed that TiLV doses of 10^{6}, 10^{7} and 10^{8} $TCID_{50}$/ml resulted in RPS of 32.1, 64.3 and 85.7%, respectively, indicating that an antigen dose of 10^{6} $TCID_{50}$/ml could serve as correlate of protection for RPS > 80%. Similarly, Li *et al.* (2015) showed that antigen doses of 10^{5}, 10^{6}, 10^{7}, 10^{8} and 10^{9} CFU/fish in tilapia vaccinated against *S. agalactiae* using the LAV strain YM001 resulted in RPS of 10.2, 35.5, 50.7, 64.5 and 67.7%, respectively, giving a correlation of $r^2 = 0.965$ between antigen dose (CFU/fish) and protection (RPS). Taken together, these studies show that antigen dose can be used as a correlate of vaccine protection for the various pathogens infecting tilapia.

Conclusions

The expansion of global tilapia production has led to an increase in the number of diseases with adverse economic impacts on the tilapia industry, which has increased the need for protective vaccines. Although vaccines for emerging infectious diseases caused by pathogens such as TiLV that cause diseases only in tilapia are under research, ongoing studies show that protective vaccines can be developed against these pathogens (Table 16.2). For pathogens that infect multiple fish species, vaccines developed for other fish species have the potential to confer protective immunity to tilapia. Studies carried out to date show that WCI vaccines account for the largest proportion of vaccines developed for tilapia (Tables 16.4–16.11). In regard to LAVs, the risk of reversion to virulence is bound to hinder their use in aquaculture. Thus, intracellularly replicating bacteria such as Fno and *Edwardsiella* spp. that require LAVs able to evoke the cell-mediated immune response needed for elimination of intracellular bacteria are bound to remain 'hard-to-combat-pathogens' in aquaculture. Even so, the licensure of some vaccines for diseases such as streptococcosis, lactococcosis and columnaris used for the immunization of different fish species is bound to play an important role in the prevention and control of infections in the tilapia industry.

References

Abdel-Hadi, Y., Shamsudin, M., Yussof, K. and Zakaria, S. (2009) Indoor study on the immunization of red tilapia: *Oreohromis niloticus* × *O. mossambicus* against aeromonad and pseudomonad septicemias. *Journal of Fisheries International* 4(3), 45–51.

Abdelhamed, H., Lawrence, M.L. and Karsi, A. (2018) Development and characterization of a novel live attenuated vaccine against enteric septicemia of catfish. *Frontiers in Microbiology* 1819.

Abo El-Yazeed, H. and Ibrahem, M.D. (2009) Studies on *Edwardsiella tarda* infection in catfish and *Tilapia nilotica. Journal of Veterinary Medical Research* 19(1), 44–50.

Abu-Elala, N.M., Younis, N.A., AbuBakr, H.O., Ragaa, N.M., Borges, L.L. *et al.* (2018) Efficacy of dietary yeast cell wall supplementation on the nutrition and immune response of Nile tilapia. *The Egyptian Journal of Aquatic Research* 44(4), 333–341.

Abu-Elala, N.M., Samir, A., Wasfy, M. and Elsayed, M. (2019) Efficacy of injectable and immersion polyvalent vaccine against streptococcal infections in broodstock and offspring of Nile tilapia (*Oreochromis niloticus*). *Fish & Shellfish Immunology* 88, 293–300.

Al-Harbi, A.H., Truax, R. and Thune, R.L. (2000) Production and characterization of monoclonal antibodies against tilapia *Oreochromis niloticus* immunoglobulin. *Aquaculture* 188(3–4), 219–227.

Al-Hussinee, L., Subramaniam, K., Ahasan, M.S., Keleher, B. and Waltzek, T.B. (2018) Complete genome sequence of a tilapia lake virus isolate obtained from Nile tilapia (*Oreochromis niloticus*). *Genome Announcements* 6(26), e00580–00518.

Al-Hussinee, L., Kuttichantran, S., Win, S., Vsevolod, P., Kathleen, H. *et al.* (2019) *Tilapia Lake Virus (TiLV): A Globally Emerging Threat to Tilapia Aquaculture.* Universty of Florida IFAS Extension.

Aly, S.M., Albutti, A.S., Rahmani, A.H. and Atti, N.M.A. (2015) The response of new-season Nile tilapia to *Aeromonas hydrophila* vaccine. *International Journal of Clinical and Experimental Medicine* 8(3), 4508.

Amend, D.F. (1981) Potency testing of fish vaccines. *Fish Biologics: Serodiagnostics and Vaccines* 447–454.

Anderson, J. and Conroy, D. (1969) The pathogenic myxobacteria with special reference to fish diseases. Symposium on Myxobacteria and Flavobacteria: Paper VII, *Journal of Applied Bacteriology* 32(1), 30–39.

Anshary, H., Kurniawan, R.A., Sriwulan, S., Ramli, R. and Baxa, D.V. (2014) *Isolation and molecular identification of the etiological agents of streptococcosis in Nile tilapia* (*Oreochromis niloticus*) *cultured in net cages in Lake Sentani*, Vol. 3. SpringerPlus, Papua, Indonesia, pp. 1–11.

Ariel, E. and Owens, ljdoao (1997) Epizootic mortalities in tilapia *Oreochromis mossambicus* 29(1), 1–6.

Ashiru, A.W., Uaboi-Egbeni, P.O., Oguntowo, J.E. and Idika, C.N. (2011) Isolation and antibiotic profile of aeromonas species from tilapia fish (*Tilapia nilotica*) and catfish (*Clarias betrachus*). *Parkistan Journal of Nutrition* 10(10), 5.

Austin, B., Austin, D.A. and Munn, C. (2007) *Bacterial fish pathogens: Disease of farmed and wild fish*, Vol. 26. Springer.

Avi Eldar, M.G., Ghittino, C., Amir, zlotkin and Herve, B. (1999) Biodiversity of *Lactococcus garvieaestrains* isolated from fish in Europe, Asia, and Australia. *Appl Environ Microbiol* 65(3), 1005.

Bacharach, E.M.N, Zody, M.C., Kembou Tsofack, J.E., Zamostiano, R., Berkowitz, A. *et al.* (2016) Characterization of a novel orthomyxo-like virus causing mass die-offs of tilapia. *Mbio* 7(7(2)), 17.

Bachrach, G., Zlotkin, A., Hurvitz, A., Evans, D.L. and Eldar, A. (2001) Recovery of *Streptococcus iniae* from diseased fish previously vaccinated with a streptococcus vaccine. *Applied and Environmental Microbiology* 67(8), 3756–3758.

Bactol, I.D.C., Padilla, L.V. and Hilario, A.L. (2018) Immune response of tilapia (*Oreochromis niloticus*) after vaccination with autoclavekilled, heat-killed, and formalin-killed whole cell aeromonas hydrophila vaccines as possible serotype-independent vaccines. *International Journal of Agriculture and Biology* 20, 846–850.

Barnes, A.C. and Ellis, A.E. (2004) Role of capsule in serotypic differences and complement fixation by *Lactococcus garvieae*. *Fish & Shellfish Immunology* 16(2), 207–214.

Barnes, A.C., Young, F.M., Horne, M.T. and Ellis, A.E. (2003) *Streptococcus iniae*: Serological differences, presence of capsule and resistance to immune serum killing. *Diseases of Aquatic Organisms* 53(3), 241–247.

Barony, G.M., Tavares, G.C., Pereira, F.L., Carvalho, A.F., Dorella, F.A. *et al.* (2017) Large-scale genomic analyses reveal the population structure and evolutionary trends of *Streptococcus agalactiae* strains in Brazilian fish farms. *Scientific Reports* 7(1), 1–10.

Basri, L., Nor, R.M., Salleh, A., Saad, M.Z., Barkham, T, *et al.* (2020) Co-infections of tilapia lake virus, *Aeromonas hydrophila* and *Streptococcus agalactiae* in farmed red hybrid tilapia. *Animals* 10(11), 1.

Bercovier, H., Ghittino, C. and Eldar, A. (1997) Immunization with bacterial antigens: Infections with streptococci and related organisms. *Dev Biol Stand* 90, 153–160.

Bigarré, L., Cabon, J., Baud, M., Heimann, M., Body, A. *et al.* (2009) Outbreak of betanodavirus infection in tilapia, *Oreochromis niloticus* (L.), in fresh water. *Journal of Fish Diseases* 32(8), 667–673.

Birkbeck, T., Feist, S. and Verner–jeffreys, D. (2011) *Francisella* infections in fish and shellfish. *Journal of Fish Diseases* 34(3), 173–187.

Bricknell, I., King, J., Bowden, T. and Ellis, A. (1999) Duration of protective antibodies, and the correlation with protection in Atlantic salmon (*Salmo salar* L.), following vaccination with *Aeromonas salmonicida* vaccine containing iron-regulated outer membrane proteins and secretory polysaccharide. *Fish & Shellfish Immunology* 9(2), 139–151.

Brudeseth, B.E., Wiulsrød, R., Fredriksen, B.N., Lindmo, K., Løkling, K.-E. *et al.* (2013) Status and future perspectives of vaccines for industrialised fin-fish farming. *Fish & Shellfish Immunology* 35(6), 1759–1768.

Bwalya, P., Hang'ombe, B.M., Gamil, A.A., Munang'andu, H.M., Evensen, Ø. *et al.* (2020) A whole-cell *Lactococcus garvieae* autovaccine protects Nile tilapia against infection. *PLoS One* 15(3), e0230739.

Bwalya, P., Hang'ombe, B.M., Evensen, Ø. and Mutoloki, S. (2021) *Lactococcus garvieae* isolated from Lake Kariba (Zambia) has low invasive potential in Nile tilapia (*Oreochromis niloticus*). *Journal of Fish Diseases* 44(6), 721–727.

Cao, T.T., Tsai, M.-A., Yang, C.-D., Wang, P.-C., Kuo, T.-Y. *et al.* (2014) Vaccine efficacy of glyceraldehyde-3-phosphate dehydrogenase (GAPDH) from *Edwardsiella ictaluri* against *E. tarda* in tilapia. *The Journal of General and Applied Microbiology* 60(6), 241–250.

Carson, J., Gudkovs, N. and Austin, B. (1993) Characteristics of an enterococcus-like bacterium from Australia and South-Africa, pathogenic for rainbow trout, *Oncorhynchus mykiss* (Walbaum). *Journal of Fish Diseases* 16(4), 381–388.

Castro, N., Toranzo, A., Núñez, S., Osorio, C. and Magariños, B. (2010) Evaluation of four polymerase chain reaction primer pairs for the detection of *Edwardsiella tarda* in turbot. *Diseases of Aquatic Organisms* 90(1), 55–61.

Chen, C., Chao, C. and Bowser, P. (2007) Comparative histopathology of *Streptococcus iniae* and *Streptococcus agalactiae*-infected tilapia. *Bulletin-European Association of Fish Pathologists* 27(1), 2.

Chen, M., Li, L.-P., Wang, R., Liang, W.-W., Huang, Y. *et al.* (2012) PCR detection and PFGE genotype analyses of streptococcal clinical isolates from tilapia in China. *Veterinary Microbiology* 159(3–4), 526–530.

Chen, M., Wang, R., Li, L.P., Liang, W.W., Li, J. *et al.* (2012) Screening vaccine candidate strains against *Streptococcus agalactiae* of tilapia based on PFGE genotype. *Vaccine* 30(42), 6088–6092.

Chen, P.-L., Lamy, B. and Ko, W.-C. (2016) *Aeromonas dhakensis*, an increasingly recognized human pathogen. *Frontiers in Microbiology* 7, 793.

Chideroli, R.T., Amoroso, N., Mainardi, R.M., Suphoronski, S.A., Padua, S.B, *et al.* (2017) Emergence of a new multidrug-resistant and highly virulent serotype of *Streptococcus agalactiae* in fish farms from Brazil. *Aquaculture* 479, 45–51.

Colquhoun, D.J. and Duodu, S. (2011) Francisella infections in farmed and wild aquatic organisms. *Veterinary Research* 42(1), 1–15.

Decostere, A., Haesebrouck, F. and Devriese, L. (1998) Characterization of four flavobacterium columnare (*Flexbacter columnaris*) strains isolated from tropical fish. *Veterinary Microbiology* 62(1), 35–45.

de Godoy E.M., David, F.S., Fialho, N.S., Proença, D.C., Camargo, T.R. *et al.* (2022) Environmental sustainability of Nile tilapia production on rural family farms in the tropical Atlantic Forest region. *Aquaculture* 547, 737481.

Delphino, M.K., Barone, R.S., Leal, C.A., Figueiredo, H.C., Gardner, I.A. *et al.* (2019) Economic appraisal of vaccination against *streptoccocus agalactiae* in Nile tilapia farms in Brazil. *Preventive Veterinary Medicine* 162, 131–135.

Dettleff, P., Bravo, C., Patel, A. and Martinez, V. (2015) Patterns of *Cirickettsia salmonis* load in susceptible and resistant families of *Salmo salar*. *Fish & Shellfish Immunology* 45(1), 67–71.

Dhar, A.K., Manna, S.K. and Thomas allnutt, F. (2014) Viral vaccines for farmed finfish. *Virusdisease* 25(1), 1–17.

Dharmaratnam, A., Raja, S.A., Ramraj, D. and Lal, K.K. (2019) *Aeromonas veronii* caused bilateral exophthalmia and mass mortality in cultured Nile tilapia, *Oreochromis niloticus* (L.) in India. *Aquaculture* 15(512), 1.

Dong, H.T., Nguyen, V.V., Le, H.D., Sangsuriya, P., Jitrakorn, S. *et al.* (2015a) Naturally concurrent infections of bacterial and viral pathogens in disease outbreaks in cultured Nile tilapia (*Oreochromis niloticus*) farms. *Aquaculture* 448, 427–435.

Dong, H., LaFrentz, B., Pirarat, N. and Rodkhum, C. (2015b) Phenotypic characterization and genetic diversity of *flavobacterium columnare* isolated from red tilapia. *Oreochromis Sp., in T Hailand. Journal of Fish Diseases* 38(10), 901–913.

Dong, H., Ataguba, G., Khunrae, P., Rattanarojpong, T. and Senapin, S. (2017a) Evidence of TiLV infection in tilapia hatcheries from 2012 to 2017 reveals probable global spread of the disease. *Aquaculture* 479, 579–583.

Dong, H., Jitrakorn, S., Kayansamruaj, P., Pirarat, N., Rodkhum, C. *et al.* (2017b) Infectious spleen and kidney necrosis disease (ISKND) outbreaks in farmed barramundi (*Lates calcarifer*) in vietnam. *Fish & Shellfish Immunology* 68, 65–73.

Dong, H.T., Jindakittikul, P., Thaiprayoon, A. and Taengphu, S. (2017c) Char: *Aeromonas jandaei* and *Aeromonas veronii* caused disease and mortality in Nile tilapia, *Oreochromis niloticus* (L.). *Journal of Fish Diseases* 40(10), 9.

Dong, H., Senapin, S., Jeamkunakorn, C., Nguyen, V., Nguyen, N. *et al.* (2019) Natural occurrence of edwardsiellosis caused by *Edwardsiella ictaluri* in farmed hybrid red tilapia (*oreochromis sp.*) in southeast Asia. *Aquaculture* 499, 17–23.

Dong, HT, Senapin, S., Gangnonngiw, W., Rodkhum, C., Debnath, PP *et al.* (2020) Experimental infection reveals transmission of tilapia lake virus (TiLV) from tilapia broodstock to their reproductive organs and fertilized eggs. *Aquaculture* 515, 734541.

Dönmez, A. and Cengizler, I. (2020) Pathomorphology of experimental *Streptococcus iniae* infection in tilapia (*Oreochromis niloticus*). *Indian Journal of Animal Research* 54(2), 234–238.

Dubey, S., Avadhani, K., Mutalik, S., Sivadasan, S.M., Maiti, B. *et al.* (2016) *Aeromonas hydrophila* ompw PLGA nanoparticle oral vaccine shows a dose-dependent protective immunity in rohu (*Labeo rohita*). *Vaccines* 4(2), 21.

Eissa, A.E., Zaki, M.M. and Aziz, A.A. (2010) *Flavobacterium columnare/ Myxobolus tilapiae* concurrent infection in the earthen pond reared Nile tilapia (*Oreochromis niloticus*) during the early summer. *Interdisciplinary Bio Central* 2(2), 5.

Eldar, A., Shapiro, O., Bejerano, Y. and Bercovier, H. (1995) Vaccination with whole-cell vaccine and bacterial protein extract protects tilapia against *Streptococcus difficile* meningoencephalitis. *Vaccine* 13(9), 867–870.

Eldar, A., Horovitcz, A. and Bercovier, H. (1997) Development and efficacy of a vaccine against *Streptococcus iniae* infection in farmed rainbow trout. *Veterinary Immunology and Immunopathology* 56(1–2), 175–183.

El-Sayed, A.-F. (2019) *Tilapia culture*. Academic Press.

El Tantawy, M.M. and Ayoub, H.F. (2016) Efficiency of oral *Aeromonas hydrophila* vaccine and tumeric powder mixture on immune response of Nile tilapia (*Oreochromis niloticus*). *Egyptian Journal for Aquaculture* 6(1), 47–66.

El-Sayed, A.-F. (2006) Tilapia culture in salt water: Environmental requirements, nutritional implications and economic potentials. *Avances En Nutricion Acuicola.*

Evans, J.J., Klesius, P.H. and Shoemaker, C.A. (2004) Efficacy of *Streptococcus agalactiae* (Group B) vaccine in tilapia (*Oreochromis niloticus*) by intraperitoneal and bath immersion administration. *Vaccine* 22(27–28), 3769–3773.

Evans, J.J., Klesius, P.H., Shoemaker, C.A. and Fitzpatrick, B.T. (2005) *Streptococcus agalactiae* vaccination and infection stress in Nile tilapia, *Oreochromis niloticus*. *Journal of Applied Aquaculture* 16(3–4), 105–115.

Evans, J., Klesius, P. and Shoemaker, C. (2009) First isolation and characterization of *Lactococcus garvieae* from Brazilian Nile tilapia, *Oreochromis niloticus* (L.), and pintado, *Pseudoplathystoma corruscans* (Spix & Agassiz). *Journal of Fish Diseases* 32(11), 943–951.

Eyngor, M., Zamostiano, R., Tsofack, J.E.K., Berkowitz, A., Bercovier, H. *et al.* (2014) Identification of a novel RNA virus lethal to tilapia. *Journal of Clinical Microbiology* 52(12), 4137–4146.

Fang, H.-M., Ge, R. and Sin, Y.M. (2004) Cloning, characterisation and expression of *Aeromonas hydrophila* major adhesin. *Fish & Shellfish Immunology* 16(5), 645–658.

FAO: Food and Agriculture Organization of the United Nations (FAO) (2017) *The Future of Food and Agriculture: Trends and Challenges[Google Scholar].*

FAO (2016) *The state of world fisheries and aquaculture 2016. Contributing to Food Security and Nutrition for All.* FAO, Rome, 200pp.

Fathi, M., Dickson, C., Dickson, M., Leschen, W., Baily, J. *et al.* (2017) Identification of tilapia lake virus in Egypt in Nile tilapia affected by 'summer mortality'syndrome. *Aquaculture* 473, 430–432.

Fernandez, J.B., Yambot, A.V. and Almeria, O. (2014) Vaccination of Nile tilapia (*Oreochromis niloticus*) using Lipopolysaccharide (LPS) prepared from *Aeromonas hydrophila*. *International Journal of Fauna and Biological Studies* 1(4), 1–3.

Figueiredo, H., Klesius, P., Arias, C., Evans, J., Shoemaker, C, *et al.* (2005) Isolation and characterization of strains of *Flavobacterium columnare* from Brazil. *Journal of Fish Diseases* 28(4), 199–204.

Figueiredo, H., Nobrega Netto, L., Leal, C., Pereira, U.P. and Mian, G.F. (2012) *Streptococcus iniae* outbreaks in Brazilian Nile tilapia (*Oreochromis niloticus* L.) farms. *Brazilian Journal of Microbiology* 43, 576–580.

Figueiredo, H.C.P., Tavares, G.C., Dorella, F.A., Rosa, J.C.C., Marcelino, S.A.C. *et al.* (2021) First report of infectious spleen and kidney necrosis virus in Nile tilapia in Brazil. *Transboundary and Emerging Diseases.*

Fitzsimmons, K and Watanabe, W.O. (2010) Chapter 17: Tilapia (Family Cichlidae). In: *Finfish Aquaculture Diversification.* CABI, Cambridge, MA, USA, pp. 374–389.

Gallage, S., Katagiri, T., Endo, M. and Maita, M. (2017) Comprehensive evaluation of immunomodulation by moderate hypoxia in *S. agalactiae* vaccinated Nile tilapia. *Fish Shellfish Immunol* 66, 445–454.

Go, J. and Whittington, R. (2019) Experimental transmission of infectious spleen and kidney necrosis virus (ISKNV) from freshwater ornamental fish to silver sweep *Scorpis lineolata*, an Australian marine fish. *Diseases of Aquatic Organisms* 137(1), 1–21.

Godoy, D., Carvalho-Castro, G., Leal, C., Pereira, U., Leite, R. *et al.* (2013) Genetic diversity and new genotyping scheme for fish pathogenic *Streptococcus agalactiae*. *Letters in Applied Microbiology* 57(6), 476–483.

Grabowski, L., LaPatra, S. and Cain, K. (2004) Systemic and mucosal antibody response in tilapia, *Oreochromis niloticus* (L.), following immunization with *Flavobacterium columnare*. *Journal of Fish Diseases* 27(10), 573–581.

Halimi, M., Alishahi, M., Abbaspour, M.R., Ghorbanpoor, M. and Tabandeh, M.R. (2019) Valuable method for production of oral vaccine by using alginate and chitosan against *Lactococcus garvieae/ Streptococcus iniae* in rainbow trout (*Oncorhynchus mykiss*). *Fish & Shellfish Immunology* 90, 431–439.

Hardi, E.H., Nugroho, R.A., Isnansetyo, A., Agriandini, M., Kusuma, I.W. *et al.* (2019) Simultaneous Administration of *Boesenbergia pandurata* Extract and Vaccination to Stimulate Immune Response in Tilapia, *Oreochromis niloticus*. *Pakistan Journal of Biological Sciences* 22(9), 419–426.

Hayat, M., Mohd Yusoff, M.S., Samad, M.J., Abdul Razak, I.S., Md Yasin, I.S. *et al.* (2021) Efficacy of feed-based formalin-killed vaccine of *Streptococcus iniae* stimulates the gut-associated lymphoid tissues and immune response of red hybrid tilapia. *Vaccines* 9(1).

He, R., Li, Z., Li, S. and Li, A. (2021) Development of an immersion challenge model for *Streptococcus agalactiae* in Nile tilapia (*Oreochromis niloticus*). *Aquaculture* 531, 735877.

He, Y., Wang, K.Y., Xiao, D., Chen, D.F., Huang, L. *et al.* (2014) A recombinant truncated surface immunogenic protein (tSip) plus adjuvant FIA confers active protection against Group B streptococcus infection in tilapia. *Vaccine* 32(51), 7025–7032.

Heckman, T.I., Shahin, K., Henderson, E.E., Griffin, M.J. and Soto, E. (2022) Development and efficacy of *Streptococcus iniae* live-attenuated vaccines in Nile tilapia, *Oreochromis niloticus*. *Fish & Shellfish Immunology* 121, 152–162.

Hedrick, R., Fryer, J., Chen, S. and Kou, G. (1983) Characteristics of four birnaviruses isolated from fish in Taiwan. *Fish Pathology* 18(2), 91–97.

Hoare, R., Leigh, W., Limakom, T., Wongwaradechkul, R., Metselaar, M, *et al.* (2021) Oral vaccination of Nile tilapia (*Oreochromis niloticus*) against francisellosis elevates specific antibody titres in serum and mucus. *Fish & Shellfish Immunology* 1(113), 9.

Holmes, P., Niccolls, L.M. and Sartory, D.P. (1996) The ecology of mesophilic Aeromonas in the aquatic environment. *The Genus Aeromonas* 4.

Hossain, M.M.M., Ehsan, A., Rahman, M.A., Haq, M. and Chowdhury, M.B.R. (2014) Transmission and pathology of *Streptococcus inane* in monosex Nile tilapia (*Oreochromis niloticus*) in aquaculture of Bangladesh. *Journal of Fisheries* 2(1), 90–99.

Huang, L.Y., Wang, K.Y., Xiao, D., Chen, D.F., Geng, Y. *et al.* (2014) Safety and immunogenicity of an oral DNA vaccine encoding sip of *Streptococcus agalactiae* from Nile tilapia *Oreochromis niloticus* delivered by live attenuated *Salmonella typhimurium. Fish Shellfish Immunol* 38(1), 34–41.

Huang, T., Li, L.P., Liu, Y., Luo, Y.J., Wang, R. *et al.* (2019) Spatiotemporal distribution of *Streptococcus agalactiae* attenuated vaccine strain YM001 in the intestinal tract of tilapia and its effect on mucosal associated immune cells. *Fish and Shellfish Immunology* 87, 714–720.

Huang, Y., Cai, S., Jian, J., Liu, G. and Xu, L. (2020) Co-infection of infectious spleen and kidney necrosis virus and *Francisella* sp. in farmed pearl gentian grouper (♀ *Epinephelus fuscoguttatus* × ♂ *E. lanceolatus*) in China-A case report. *Aquaculture* 526, 735409.

Huang, Z., Tang, J., Li, M., Fu, Y., Dong, C. *et al.* (2012) Immunological evaluation of *Vibrio alginolyticus, Vibrio harveyi, Vibrio vulnificus* and infectious spleen and kidney necrosis virus (ISKNV) combined-vaccine efficacy in *Epinephelus coioides. Veterinary Immunology and Immunopathology* 150(1–2), 61–68.

Ibrahem, M.D., Arab, R., Mostafa, M. and Rezk, M.A. (2008) Evaluation of different vaccination strategies for control of (MAS) in Nile tilapia (*O. niloticus*) in Egypt.

Igarashi, A. and Iida, T. (2002) A vaccination trial using live cells of *Edwardsiella tarda* in tilapia. *Fish Pathology* 37(3), 145–148.

Inthasaeng, P., Unajak, S., Areechon, N., Hirono, I. and Surachetpong, W. (2018) Efficacy of pcDNA-Alp1 DNA vaccine against *Streptococcus agalactiae* in Nile tilapia (*Oreochromis niloticus*). *The Thai Journal of Veterinary Medicine* 48(2), 279–288.

Ismail, M.S., Siti-Zahrah, A., Syafiq, M.R., Amal, M.N., Firdaus-Nawi, M. *et al.* (2016) Feed-based vaccination regime against streptococcosis in red tilapia, *Oreochromis niloticus* × *Oreochromis mossambicus. BMC Veterinary Research* 12(1), 194.

Ismail, M.S., Syafiq, M.R., Siti-Zahrah, A., Fahmi, S., Shahidan, H. *et al.* (2017) The effect of feed-based vaccination on tilapia farm endemic for streptococcosis. *Fish and Shellfish Immunology* 60, 21–24.

Ismail, N., Atta, N., Aziz, A. and Ahmed, M. (2010) Oral vaccination of Nile tilapia (*Orechromis niloticus*) against motile aeromonas septicaemia. *Nature and Science* 6, 46–51.

Itsaro, A., Suanyuk, N. and Tantikitti, C. (2012) Multiplex PCR for simultaneous detection of *Streptococcus agalactiae, Streptococcus iniae* and *Lactococcus garvieae*: A case of S. agalactiae infection in cultured Nile tilapia (*Oreochromis niloticus*) and red tilapia (*Oreochromis niloticus* × *Oreochromis mossambicus*). *Songklanakarin Journal of Science & Technology* 34(5).

Jansen, M.D., Dong, H.T. and Mohan, C.V. (2018) Tilapia lake virus: A threat to the global tilapia industry? *Reviews in Aquaculture.*

Jiao, X. D., Hu, Y. H., & Sun, L. (2010). Dissection and localization of the immunostimulating domain of *Edwardsiella tarda* FliC. Vaccine, 28(34), 5635–5640.

Joshi, R., Skaaurd, A. and Tola Alvarez, A. (2021) Experimental validation of genetic selection for resistance against *Streptococcus agalactiae* via different routes of infection in the commercial Nile tilapia breeding programme. *Journal of Animal Breeding and Genetics* 138(3), 338–348.

Kawasaki, M., Delamare-Deboutteville, J., Bowater, R.O., Walker, M.J., Beatson, S. *et al.* (2018) Microevolution of *Streptococcus agalactiae* ST-261 from Australia indicates dissemination via imported tilapia and ongoing adaptation to marine hosts or environment. *Applied and Environmental Microbiology* 84(16), e00859–00818.

Kayansamruaj, P., Dong, H.T., Pirarat, N., Nilubol, D. and Rodkhum, C. (2017a) Efficacy of α-enolase-based DNA vaccine against pathogenic *Streptococcus iniae* in Nile tilapia (*Oreochromis niloticus*). *Aquaculture* 468, 102–106.

Kayansamruaj, P., Dong, H.T., Hirono, I., Kondo, H., Senapin, S. *et al.* (2017b) Comparative genome analysis of fish pathogen *Flavobacterium columnare* reveals extensive sequence diversity within the species. *Infection, Genetics and Evolution* 54, 7–17.

Keawcharoen, J., Techangamsuwan, S., Ponpornpisit, A., Lombardini, E., Patchimasiri, T. *et al.* (2015) Genetic characterization of a betanodavirus isolated from a clinical disease outbreak in farm-raised tilapia *Oreochromis niloticus* (L.) in thailand 38(1), 49–54.

Khushiramani, R., Girisha, S., Karunasagar, I. and Karunasagar, I. (2007) Protective efficacy of recombinant ompts protein of *Aeromonas hydrophila* in Indian major carp. *Vaccine* 25(7), 1157–1158.

Khushiramani, R.M., Maiti, B., Shekar, M., Girisha, S.K., Akash, N. *et al.* (2012) Recombinant *Aeromonas hydrophila* outer membrane protein 48 (Omp48) induces a protective immune response against *Aeromonas hydrophila* and *Edwardsiella tarda. Research in Microbiology* 163(4), 286–291.

Kishimori, J.M., Takemura, A. and Leong, J.-A. (2015) *Neobenedenia melleni*-specific antibodies are associated with protection after continuous exposure in mozambique tilapia. *Journal of Immunology Research.*

Kitancharoen, N., Hanjavanij, C. and Suwannapeng, N. (2006) Efficiency of vaccination with *Streptococcus agalactiae bacterin* on Streptococcosis prevention in Nile tilapia (Thai). *Asia-Pacific Journal of Science and Technology* 11(1), 53–61.

Kitiyodom, S., Kaewmalun, S., Nittayasut, N., Suktham, K., Surassmo, S. *et al.* (2019a) The potential of mucoadhesive polymer in enhancing efficacy of direct immersion vaccination against *Flavobacterium columnare* infection in tilapia. *Fish & Shellfish Immunology* 86, 635–640.

Kitiyodom, S., Yata, T., Yostawornkul, J., Kaewmalun, S., Nittayasut, N. *et al.* (2019b) Enhanced efficacy of immersion vaccination in tilapia against columnaris disease by chitosan-coated "pathogen-like" *mucoadhesive* nanovaccines. *Fish and Shellfish Immunology* 95, 213–219.

Kitiyodom, S., Trullàs, C., Rodkhum, C., Thompson, K.D., Katagiri, T. *et al.* (2021) Modulation of the mucosal immune response of red tilapia (*Oreochromis* sp.) against columnaris disease using a biomimetic-mucoadhesive nanovaccine. *Fish & Shellfish Immunology* 112, 81–91.

Klesius, P.H., Evans, J.J., Shoemaker, C.A. and Pasnik, D.J. (2006) A vaccination and challenge model using calcein marked fish. *Fish Shellfish Immunol* 20(1), 20–28.

Klinger-Bowen, R., Tamaru, C., McGovern-Hopkins, K., Li, Y., Sithole, F. *et al.* (2016) Dynamics of piscine francisellosis differs amongst tilapia species (*Oreochromis* spp.) in a controlled challenge with *Francisella noatunensis* subsp. orientalis. *Journal of Fish Diseases* 39(11), 1305–1312.

Kwon, S.R., Nam, Y.K., Kim, S.K., Kim, D.S. and Kim, K.H. (2005) Generation of *Edwardsiella tarda* ghosts by bacteriophage PhiX174 lysis gene E. *Aquaculture* 250(1–2), 16–21.

Kwon, S.R., Nam, Y.K., Kim, S.K. and Kim, K.H. (2006) Protection of tilapia (*Oreochromis mosambicus*) from edwardsiellosis by vaccination with *Edwardsiella tarda* ghosts. *Fish Shellfish Immunol* 20(4), 621–626.

Kwon, S.R., Lee, E.H., Nam, Y.K., Kim, S.K. and Kim, KH (2007) Efficacy of oral immunization with *Edwardsiella tarda* ghosts against edwardsiellosis in olive flounder (*Paralichthys olivaceus*). *Aquaculture* 269(1–4), 84–88.

LaFrentz, B.R. and Shoemaker, C.A. (2011) Klesius PH: immunoproteomic analysis of the antibody response obtained in Nile tilapia following vaccination with a *Streptococcus iniae* vaccine. *Veterinary Microbiology* 152(3–4), 346–352.

LaFrentz, B.R. and Shoemaker, C.A. (2015) Passive transfer of serum from tilapia vaccinated with a vibrio vulnificus vaccine provides protection from specific pathogen challenge. *Aquaculture* 442, 16–20.

LaFrentz, B.R., García, J.C., Waldbieser, G.C., Evenhuis, J.P., Loch, T.P. *et al.* (2018) Identification of four distinct phylogenetic groups in *Flavobacterium columnare* with fish host associations. *Frontiers in Microbiology* 9, 452.

Laith, A.A., Abdullah, M.A., Nurhafizah, W.W.I., Hussein, H.A., Aya, J. *et al.* (2019) Efficacy of live attenuated vaccine derived from the *Streptococcus agalactiae* on the immune responses of *Oreochromis niloticus*. *Fish and Shellfish Immunology* 90, 235–243.

Leal, C., Carvalho-Castro, G., Sacchetin, P., Lopes, C., Moraes, A. *et al.* (2010) Oral and parenteral vaccines against *Flavobacterium columnare*: Evaluation of humoral immune response by ELISA and in vivo efficiency in Nile tilapia (*Oreochromis niloticus*). *Aquaculture International* 18(4), 657–666.

Li, L., Wang, R., Liang, W., Gan, X., Huang, T. *et al.* (2013) Rare serotype occurrence and PFGE genotypic diversity of *Streptococcus agalactiae* isolated from tilapia in China. *Veterinary Microbiology* 167(3–4), 719–724.

Li, L., Huang, T., Liang, W. and Chen, M. (2019) Development of an attenuated oral vaccine strain of tilapia Group B Streptococci serotype Ia by gene knockout technology. *Fish & Shellfish Immunology* 93, 924–933.

Li, M., Li, L., Huang, T., Liu, Y., Lei, A, *et al.* (2018) On intestinal microbiota of tilapia are recoverable. *Frontiers in Microbiology* 9, 3251.

Li, W., Wang, H.Q., He, R.Z., Li, Y.W., Su, Y.L, *et al.* (2016) Major surfome and secretome profile of *Streptococcus agalactiae* from Nile tilapia (*Oreochromis niloticus*): Insight into vaccine development. *Fish and Shellfish Immunology* 55, 737–746.

Li, W., Li, Y., Hu, Y.Z., Mo, X.B., Xu, G.H. *et al.* (2019) GroEL, a novel vaccine candidate of piscine *Streptococcus agalactiae* identified by immunoproteome. *Fish and Shellfish Immunology* 84, 377–383.

Li, Y. and Cai, S.H. (2011) Identification and pathogenicity of *Aeromonas sobria* on tail-rot disease in juvenile tilapia *Oreochromis niloticus. Current Microbiology* 62(2), 5.

Lin, Q., Fu, X., Li, N., Wan, Q., Chen, W. *et al.* (2017) Co-infections of infectious spleen and kidney necrosis virus and *Siniperca chuatsi* rhabdovirus in Chinese perch (*Siniperca chuatsi*). *Microbial Pathogenesis* 111, 422–430.

Lio-Po, G. and Wakabayashi, H. (1986) Immuno-response in tilapia sarotherodon niloticus vaccinated with *Edwardsiella tarda* by hyperosmotic infiltration method. *Veterinary Immunology Immunopathology* 12(1–4), 351–357.

Liu, H., Zhang, S., Shen, Z., Ren, G., Liu, L. *et al.* (2016) Development of a vaccine against *Streptococcus agalactiae* in fish based on truncated cell wall surface anchor proteins. *Veterinary Record* 179(14), 359–359.

Liu, L., Lu, D.Q., Xu, J., Luo, H.L. and Li, A.X. (2019) Development of attenuated erythromycin-resistant *Streptococcus agalactiae* vaccine for tilapia (*Oreochromis niloticus*) culture. *Journal of Fish Disease* 42(5), 693–701.

Liu, W., Zhang, Y., Ma, J., Jiang, N., Fan, Y., Zhou, Y., Cain, K., Yi, M., Jia, K., Wen, H.: Determination of a novel parvovirus pathogen associated with massive mortality in adult tilapia. *PLoS Pathogens* 2020, 16(9):e1008765.

Liu, Y., Li, L., Yu, F., Luo, Y., Liang, W. *et al.* (2020) Genome-wide analysis revealed the virulence attenuation mechanism of the fish-derived oral attenuated *Streptococcus iniae* vaccine strain YM011. *Fish and Shellfish Immunology* 106, 546–554.

Lueangyangyuen, A., Senapin, S., Dong, H., Unajak, S., Wangkahart, E. *et al.* (2022) Expression and purification of S5196–272 and S6200–317 proteins from tilapia lake virus (TiLV) and their potential use as vaccines. *Protein Expression and Purification* 190, 106013.

Luo, Z., Fu, J., Li, N., Liu, Z., Qin, T. *et al.* (2016) Immunogenic proteins and their vaccine development potential evaluation in outer membrane proteins (Omps) of *Flavobacterium columnare. Aquaculture and Fisheries* 1, 1–8.

Ma, J., Bruce, T.J., Jones, E.M. and Cain, K.D. (2019) A review of fish vaccine development strategies: conventional methods and modern biotechnological approaches. *Microorganisms* 7(11), 569.

Ma, Y-p, Ke, H., Z-l, L., J-y, M., Hao, L. *et al.* (2017) Protective efficacy of cationic-PLGA microspheres loaded with DNA vaccine encoding the sip gene of *Streptococcus agalactiae* in tilapia. *Fish & Shellfish Immunology* 66, 345–353.

Ma, Y., Hao, L., Liang, Z., Ma, J., Ke, H. *et al.* (2020) Characterization of novel antigenic vaccine candidates for Nile tilapia (*Oreochromis niloticus*) against *Streptococcus agalactiae* infection. *Fish and Shellfish Immunology* 105, 405–414.

Machimbirike, V.I., Pornputtapong, N., Senapin, S., Wangkahart, E., Srisapoome, P. *et al.* (2021) A multi-epitope chimeric protein elicited a strong antibody response and partial protection against *Edwardsiella ictaluri* in Nile tilapia. *Journal of Fish Disease.*

Mai, T.T., Kayansamruaj, P., Taengphu, S., Senapin, S., Costa, J.Z. *et al.* (2021) Efficacy of heat-killed and formalin-killed vaccines against tilapia tilapinevirus in juvenile Nile tilapia (*Oreochromis niloticus*). *Journal of Fish Diseases* 44(12), 2097–2109.

Mai, T.T., Kayansamruaj, P., Soontara, C., Kerddee, P., Nguyen, D.-H. *et al.* (2022) Immunization of Nile tilapia (*Oreochromis niloticus*) broodstock with tilapia lake virus (TiLV) inactivated vaccines elicits protective antibody and passive maternal antibody transfer. *Vaccines* 10(2), 167.

Maiti, B., Shetty, M., Shekar, M., Karunasagar, I. and Karunasagar, I. (2011) Recombinant outer membrane protein A (OmpA) of *Edwardsiella tarda*, a potential vaccine candidate for fish, common carp. *Microbiological Research* 167(1), 1–7.

Marcusso, P.F., Silva Claudiano, G., Yunis-Aguinaga, J., Almeida Marinho-Neto, F., Eto, S.F. *et al.* (2021) Immunogenicity in *Oreochromis niloticus* vaccinated with sonicated antigens against streptococcosis. *Fish and Shellfish Immunology* 115, 134–141.

Martínez, R., Hernández, L., Gil, L., Carpio, Y., Morales, A, *et al.* (2017) Growth hormone releasing peptide-6 enhanced antibody titers against subunit antigens in mice (BALB/c), tilapia (*Oreochromis niloticus*) and African catfish (*clarias gariepinus*). *Vaccine* 35(42), 5722–5728.

McCallum, H., Harvell, D. and Dobson, A. (2003) Rates of spread of marine pathogens. *Ecology Letters* 6(12), 1062–1067.

McCallum, H.I., Kuris, A., Harvell, C.D., Lafferty, K.D., Smith, G.W. *et al.* (2004) Does terrestrial epidemiology apply to marine systems. *Trends in Ecology & Evolution* 19(11), 585–591.

Mertes, V., Bekkelund, A. K., Lagos, L., Ciani, E., Colquhoun, D., Haslene-Hox, H., Sletta, H., Sørum, H., & Winther-Larsen, H. C. (2021). The Use of Extracellular Membrane Vesicles for Immunization against Francisellosis in Nile tilapia (*Oreochromis niloticus*) and Atlantic Cod (*Gadus morhua* L.). *Vaccines*, 9(1), 34.

Meyburgh, C., Bragg, R. and Boucher, C. (2017) *Lactococcus garvieae*: An emerging bacterial pathogen of fish. *Diseases of Aquatic Organisms* 123(1), 67–79.

Mo, X.-B., Wang, J., Guo, S. and Li, A.-X. (2020) Potential of naturally attenuated *Streptococcus agalactiae* as a live vaccine in Nile tilapia (*Oreochromis niloticus*). *Aquaculture* 518, 734774.

Mohammed, H., Olivares-Fuster, O., LaFrentz, S. and Arias, C.R. (2013) New attenuated vaccine against columnaris disease in fish: Choosing the right parental strain is critical for vaccine efficacy. *Vaccine* 31(45), 5276–5280.

Mohanty, B. and Sahoo, P. (2007) Edwardsiellosis in fish: A brief review. *Journal of Biosciences* 32(3), 1331–1344.

Monir, M.S., Yusoff, S.B.M., Zulperi, Z.B.M., Hassim, H.B.A., Mohamad, A. *et al.* (2020) Haemato-immunological responses and effectiveness of feed-based bivalent vaccine against *Streptococcus iniae* and *Aeromonas hydrophila* infections in hybrid red tilapia (*Oreochromis mossambicus* × *O. niloticus*). *BMC Veterinary Research* 16(1), 226.

Monir, M. S., Yusoff, M. S. M., Zulperi, Z. M., Hassim, H. A., Zamri-Saad, M., Amal, M. N. A., Salleh, A., Mohamad, A., Yie, L. J., & Ina-Salwany, M. Y. (2021). Immuno-protective efficiency of feed-based whole-cell inactivated bivalent vaccine against *Streptococcus* and *Aeromonas* infections in red hybrid tilapia (*Oreochromis niloticus* × *Oreochromis mossambicus*). *Fish Shellfish Immunol* 113, 162–175.

Mugimba, K.K., Tal, S., Dubey, S., Mutoloki, S., Dishon, A, *et al.* (2019) Gray (*Oreochromis niloticus* × *O. aureus*) and red (*Oreochromis* spp.) tilapia show equal susceptibility and proinflammatory cytokine responses to experimental tilapia lake virus infection. *Viruses* 11(10), 893.

Mugimba, K.K., Lamkhannat, M., Dubey, S., Mutoloki, S., Munang'andu, H.M. *et al.* (2020) Tilapia lake virus downplays innate immune responses during early stage of infection in Nile tilapia (*Oreochromis niloticus*). *Scientific Reports* 10(1), 1–12.

Mulei, I., Nyaga, P., Mbuthia, P., Waruiru, R., Njagi, L. *et al.* (2018) Infectious pancreatic necrosis virus isolated from farmed rainbow trout and tilapia in Kenya is identical to European isolates. *Journal of Fish Diseases* 41(8), 1191–1200.

Munang'andu, H.M. (2018) Intracellular bacterial infections: A challenge for developing cellular mediated immunity vaccines for farmed fish. *Microorganisms* 6(2), 33.

Munang'andu, H.M. and Evensen, Ø. (2019) Correlates of protective immunity for fish vaccines. *Fish & Shellfish Immunology* 85, 132–140.

Munang'andu, H.M., Fredriksen, B.N., Mutoloki, S., Brudeseth, B., Kuo, T.-Y. *et al.* (2012) Comparison of vaccine efficacy for different antigen delivery systems for infectious pancreatic necrosis virus vaccines in Atlantic salmon (*Salmo salar* L.) in a cohabitation challenge model. *Vaccine* 30(27), 4007–4016.

Munang'andu, H.M., Fredriksen, B.N., Mutoloki, S., Dalmo, R.A. and Evensen, Ø. (2013a) Antigen dose and humoral immune response correspond with protection for inactivated infectious pancreatic necrosis virus vaccines in atlantic salmon (*Salmo salar* L). *Veterinary Research* 44(1), 1–16.

Munang'andu, H.M., Sandtrø, A., Mutoloki, S., Brudeseth, B.E., Santi, N. *et al.* (2013b) Immunogenicity and cross protective ability of the central VP2 amino acids of infectious pancreatic necrosis virus in Atlantic salmon (*Salmo salar* L). *PloS One* 8(1), e54263.

Munang'andu, H.M., Paul, J. and Evensen, Ø. (2016) An overview of vaccination strategies and antigen delivery systems for *Streptococcus agalactiae* vaccines in Nile tilapia (*Oreochromis niloticus*). *Vaccines* 4(4), 48.

Munang'andu, H.M., Mutoloki, S. and Evensen, Ø. (2014) Acquired immunity and vaccination against infectious pancreatic necrosis virus of salmon. *Developmental & Comparative Immunology* 43(2), 184–196.

Munang'andu, H.M., Santi, N., Fredriksen, B.N., Løkling, K.-E. and Evensen, Ø (2016) A systematic approach towards optimizing a cohabitation challenge model for infectious pancreatic necrosis virus in Atlantic salmon (*Salmo salar* L). *PLoS One* 11(2), e0148467.

Mzula, A., Wambura, P.N., Mdegela, R.H. and Shirima, G.M. (2020) Virulence pattern of circulating aeromonads isolated from farmed Nile tilapia in Tanzania and novel antibiotic free attenuation of *Aeromonas hydrophila* strain TZR7–2018. *Aquaculture Reports* 17, 100300.

Nhinh, D.T., Giang, N.T.H., Van, K., Dang, L.T., Dong, H, *et al.* (2021) The arrival, establishment and spread of a highly virulent *Edwardsiella ictaluri* strain in farmed tilapia, *Oreochromis* spp. *Authorea Preprints.*

Nielsen, M.E., Høi, L., Schmidt, A., Qian, D., Shimada, T. *et al.* (2001) Is *Aeromonas hydrophila* the dominant motile *Aeromonas* species that causes disease outbreaks in aquaculture production in the Zhejiang Province of China? *Diseases of Aquatic Organisms* 46(1), 23–29.

Noraini, O. and Sabri, M.Y. (2013) Siti-zahrah A: efficacy of spray administration of formalin-killed *Streptococcus agalactiae* in hybrid red tilapia. *Journal of Aquatic Animal Health* 25(2), 142–148.

Nur-Nazifah, M., Sabri, M.Y. and Siti-zahrah, A. (2014) Development and efficacy of feed-based recombinant vaccine encoding the cell wall surface anchor family protein of *Streptococcus agalactiae* against streptococcosis in *Oreochromis* sp. *Fish Shellfish Immunol* 37(1), 193–200.

Oh, W.T., Jun, J.W., Kim, H.J., Giri, S.S., Yun, S, *et al.* (2020) Characterization and pathological analysis of a virulent *Edwardsiella anguillarum* strain isolated from Nile tilapia (*Oreochromis niloticus*) in korea. *Frontiers in Veterinary Science* 14.

Osman, K.M., Mohamed, L.A., Rahman, E.H.A. and Soliman, W. (2009) Trials for vaccination of tilapia fish against *Aeromonas* and *Pseudomonas* infections using monovalent, bivalent and polyvalent vaccines. *World Journal of Fish and Marine Sciences* 1(4), 297–304.

Osman, K.M., Al-Maary, K.S., Mubarak, A.S., Dawoud, T.M., Moussa, I.M. *et al.* (2017) Characterization and susceptibility of streptococci and enterococci isolated from Nile tilapia (*Oreochromis niloticus*) showing septicaemia in aquaculture and wild sites in Egypt. *BMC Veterinary Research* 13(1), 1–10.

Paperna, I. (1973) Lymphocysits in fish from East African lakes. *Journal of Wildlife Diseases* 9(4), 331–335.

Park, S.B., Jang, H.B., Nho, S.W., Cha, I.S., J-i, H. *et al.* (2011) Outer membrane vesicles as a candidate vaccine against edwardsiellosis. *PloS One* 6(3), e17629.

Park, S.B., Aoki, T. and Jung, T.S. (2012) Pathogenesis of and strategies for preventing *Edwardsiella tarda* infection in fish. *Veterinary Research* 43(1), 1–11.

Pasnik, D.J., Evans, J.J. and Klesius, P.H. (2005) Duration of protective antibodies and correlation with survival in Nile tilapia *Oreochromis niloticus* following *Streptococcus agalactiae* vaccination. *Disease of Aquatic Organism* 66(2), 129–134.

Peepim, T., Dong, H.T., Senapin, S., Khunrae, P. and Rattanarojpong, T. (2016) Epr3 is a conserved immunogenic protein among *Aeromonas* species and able to induce antibody response in Nile tilapia. *Aquaculture* 464, 399–409.

Pirarat, N., Maita, M., Endo, M. and Katagiri, T. (2007) Lymphoid apoptosis in *Edwardsiella tarda* septicemia in tilapia, *Oreochromis niloticus*. *Fish & Shellfish Immunology* 22(6), 608–616.

Plotkin, S.A. (2010) Correlates of protection induced by vaccination. *Clinical and Vaccine Immunology* 17(7), 1055–1065.

Poobalane, S., Thompson, K.D., Ardó, L., Verjan, N., Han, H.-J. *et al.* (2010) Production and efficacy of an *Aeromonas hydrophila* recombinant S-layer protein vaccine for fish. *Vaccine* 28(20), 3540–3547.

Prasad, S. and Areechon, N. (2010) Efficacy of formalin-killed *Aeromonas hydrophila* and *Streptococcus* sp. vaccine in red tilapia. *Our Nature* 8(1), 231–240.

Pridgeon, J.W. and Klesius, P.H. (2011a) Development and efficacy of novobiocin and rifampicin-resistant *Aeromonas hydrophila* as novel vaccines in channel catfish and Nile tilapia. *Vaccine* 29(45), 7896–7904.

Pridgeon, J.W. and Klesius, P.H. (2011b) Identification and expression profile of multiple genes in Nile tilapia in response to formalin killed *Streptococcus iniae* vaccination. *Veterinary Immunology Immunopathology* 142(3–4), 201–206.

Pridgeon, J.W., Klesius, P.H. and Yildirim-Aksoy, M. (2013) Attempt to develop live attenuated bacterial vaccines by selecting resistance to gossypol, proflavine hemisulfate, novobiocin, or ciprofloxacin. *Vaccine* 31(18), 2222–2230.

Pulendran, B. and Ahmed, R. (2011) Immunological mechanisms of vaccination. *Nature Immunology* 12(6), 509–517.

Pulido, L.L.H., Mora, C.M., Hung, A.L., Dong, H.T. and Senapin, S. (2019) Tilapia lake virus (TiLV) from Peru is genetically close to the Israeli isolates. *Aquaculture* 510, 61–65.

Pulpipat, T., Lin, K.H., Chen, Y.H., Wang, P.C. and Chen, S.C. (2019) Molecular characterization and pathogenicity of *Francisella noatunensis* subsp. orientalis isolated from cultured tilapia (*Oreochromis* sp.) in taiwan. *Journal of Fish Diseases* 42(5), 643–655.

Pulpipat, T., Maekawa, S., Wang, P.C. and Chen, S.C. (2020) Immune responses and protective efficacy of a formalin-killed *Francisella noatunensis* subsp. *Orientalis* vaccine evaluated through intraperitoneal and immersion challenge methods in *Oreochromis niloticus*. *Vaccines (Basel)* 8(2).

Pumchan, A., Krobthong, S., Roytrakul, S., Sawatdichaikul, O., Kondo, H. *et al.* (2020) Novel chimeric multiepitope vaccine for streptococcosis disease in Nile tilapia (*Oreochromis niloticus* linn. *Sci Rep* 10(1), 603.

Qin, L., Yao, H. and Lu, C. (2000) Immune effect of conjugated *Aeromonas hydrophila* and *Edwardsiella tarda* subunit vaccines on mice. *Chinese Journal of Veterinary Science* 20(2), 163–166.

Rahmatullah, M., Ariff, M., Kahieshesfandiari, M., Daud, H., Zamri-Saad, M. *et al.* (2017) Isolation and pathogenicity of *Streptococcus iniae* in cultured red hybrid tilapia in Malaysia. *Journal of Aquatic Animal Health* 29(4), 208–213.

Ramadan, A., Afifi, N.A., Moustafa, M. and Samy, A. (1994) The effect of ascogen on the immune response of tilapia fish to *Aeromonas hydrophila* vaccine. *Fish & Shellfish Immunology* 4(3), 159–165.

Ramires, G., Paley, R.K., Hunt, W., Feist, S.W., Stone, D.M. *et al.* (2019) First detection of infectious spleen and kidney necrosis virus (ISKNV) associated with massive mortalities in farmed tilapia in Africa 680538.

Ramírez-Paredes, J. G., Thompson, K. D., Metselaar, M., Shahin, K., Soto, E., Richards, R. H., Penman, D. J., Colquhoun, D. J., & Adams, A. (2017) A polyphasic approach for phenotypic and genetic characterization of the fastidious aquatic pathogen *Francisella noatunensis* subsp. *Orientalis*. *Frontiers in Microbiology* 8, 2324.

Ramírez-Paredes, J.G., Mendoza-Roldan, M.A., Lopez-Jimena, B., Shahin, K., Metselaar, M. *et al.* (2019) Whole cell inactivated autogenous vaccine effectively protects red Nile tilapia (*Oreochromis niloticus*) against francisellosis via intraperitoneal injection. *Journal of Fish Diseases* 42(8), 1191–1200.

Ramírez-Paredes, J.G., Paley, R.K., Hunt, W., Feist, S.W., Stone, D.M. *et al.* (2021) First detection of infectious spleen and kidney necrosis virus (ISKNV) associated with massive mortalities in farmed tilapia in Africa. *Transboundary and Emerging Diseases* 68(3), 1550–1563.

Ramos-Espinoza, F.C., Cueva-Quiroz, V.A., Yunis-Aguinaga, J., Alvarez-Rubio, N.C., Mello, N. *et al.* (2020) Efficacy of two adjuvants administrated with a novel hydrogen peroxide-inactivated vaccine against *Streptococcus agalactiae* in Nile tilapia fingerlings. *Fish and Shellfish Immunology* 105, 350–358.

Rao, S., Byadgi, O., Pulpipat, T., Wang, P.C. and Chen, S.C. (2020) Efficacy of a formalin-inactivated *Lactococcus garvieae* vaccine in farmed grey mullet (*Mugil cephalus*). *Journal of Fish Diseases* 43(12), 1579–1589.

Rao, S., Pham, T.H., Poudyal, S., Cheng, L.W., Nazareth, S.C. *et al.* (2021) First report on genetic characterization, cell-surface properties and pathogenicity of *Lactococcus garvieae*, emerging pathogen isolated from cage-cultured cobia (*Rachycentron canadum*). *Transboundary and Emerging Diseases*.

Ravelo, C., Margarinos, B., Lopez-Romalde, S., Toranzo, A.E. and Romalde, J.L. (2003) Molecular fingerprinting of fish-pathogenic *Lactococcus garvieae* strains by random amplified polymorphic DNA analysis. *J Clin Microbiol* 41(2), 751–756.

Reichley, S.R., Waldbieser, G.C., Lawrence, M.L. and Griffin, M.J. (2015) Complete genome sequence of an *Edwardsiella* piscicida-like species, recovered from tilapia in the United States. *Genome Announcements* 3(5), e01004–01015.

Ren, Y., C-p, L.U. and Yao, H. (2006) Expression of the active fragment of an extracellular temperature-labile protease from *Aeromonas hydrophila* and its immune effect on mice. *Journal of Agricultural Biotechnology* 6, 009.

Rodkhum, C., Kayansamruaj, P., Pirarat, N. and Wongtawatchai, J. (2012) Duplex PCR for simultaneous and unambiguous detection of *Streptococcus iniae* and *Streptococcus agalactiae* associated with streptococcosis of cultured tilapia in Thailand. *The Thai Journal of Veterinary Medicine* 42(2), 153–158.

Ruangpan, L., Kitao, T. and Yoshida, T. (1986) Protective efficacy of *Aeromonas hydrophila* vaccines in Nile tilapia. *Veterinary Immunology and Immunopathology* 12(1–4), 345–350.

Salgado-Miranda, C., Loza-Rubio, E., Rojas-Anaya, E. and García-Espinosa, G. (2013) Viral vaccines for bony fish: Past, present and future. *Expert Review of Vaccines* 12(5), 567–578.

Schneck, J. and Caslake, L. (2006) Genetic diversity of flavobacterium columnare isolated from fish collected from warm and cold water. *Journal of Fish Diseases* 29(4), 245–248.

Schwenteit, J., Bogdanović, X., Fridjonsson, O.H., Aevarsson, A., Bornscheuer, U.T. *et al.* (2013) Toxoid construction of AsaP1, a lethal toxic aspzincin metalloendopeptidase of *Aeromonas salmonicida* subsp. achromogenes, and studies of its activity and processing. *Veterinary Microbiology* 162(2–4), 687–694.

Shahin, K., Shinn, A.P., Metselaar, M., Ramirez-Paredes, J.G., Monaghan, S.J. *et al.* (2019) Efficacy of an inactivated whole-cell injection vaccine for Nile tilapia, *Oreochromis niloticus* (L), against multiple isolates of *Francisella noatunensis* subsp. *Orientalis* from diverse geographical regions. *Fish and Shellfish Immunology* 89, 217–227.

Shahin, K., Pirezan, F., Rogge, M., LaFrentz, B.R., Shrestha, R.P. *et al.* (2020) Development of iglc and groel recombinant vaccines for francisellosis in Nile tilapia, *Oreochromis niloticus. Fish & Shellfish Immunology* 105, 341–349.

Shahin, K., Subramaniam, K., Camus, A.C., Yazdi, Z., Yun, S, *et al.* (2021) Isolation, identification and characterization of a novel megalocytivirus from cultured tilapia (*Oreochromis* spp.) from Southern California, USA. *Animals* 11(12), 3524.

Shimmoto, H., Kawai, K., Ikawa, T. and Oshima, S. (2010) Protection of red sea bream pagrus major against red sea bream iridovirus infection by vaccination with a recombinant viral protein. *Microbiology and Immunology* 54(3), 135–142.

Shlapobersky, M., Sinyakov, M.S., Katzenellenbogen, M., Sarid, R., Don, J. *et al.* (2010) Viral encephalitis of tilapia larvae: Primary characterization of a novel herpes-like virus. *Virology* 399(2), 239–247.

Shoemaker, C.A., Klesius, P.H. and Evans, J.J. (2007) Immunization of eyed channel catfish, *Ictalurus punctatus*, eggs with monovalent flavobacterium columnare vaccine and bivalent *F. columnare* and *Edwardsiella ictaluri* vaccine. *Vaccine* 25(6), 1126–1131.

Shoemaker, C.A., Lafrentz, B.R., Klesius, P.H. and Evans, J.J. (2010) Protection against heterologous *Streptococcus iniae* isolates using a modified bacterin vaccine in Nile tilapia, *Oreochromis niloticus* (L). *Journal of Fish Disease* 33(7), 537–544.

Shoemaker, C.A., Klesius, P.H., Drennan, J.D. and Evans, J.J. (2011) Efficacy of a modified live *Flavobacterium columnare* vaccine in fish. *Fish & Shellfish Immunology* 30(1), 304–308.

Skliris, G. and Richards, R. (1999) Nodavirus isolated from experimentally infected tilapia, *Oreochromis mossambicus* (peters). *Journal of Fish Diseases* 22(4), 315–318.

Soto, E., Fernandez, D. and Hawke, J.P. (2009) Attenuation of the fish pathogen francisella sp. by mutation of the iglC* gene. *Journal of Aquatic Animal Health* 21(3), 140–149.

Soto, E., Wiles, J., Elzer, P., Macaluso, K. and Hawke, J.P. (2011) Attenuated *Francisella Asiatica* iglC mutant induces protective immunity to francisellosis in tilapia. *Vaccine* 29(3), 593–598.

Soto, E., Abrams, S.B. and Revan, F. (2012) Effects of temperature and salt concentration on *Francisella noatunensis* subsp. orientalis infections in Nile tilapia *Oreochromis niloticus. Diseases of Aquatic Organisms* 101(3), 217–223.

Soto, E., Kidd, S., Mendez, S., Marancik, D., Revan, F. *et al.* (2013) *Francisella noatunensis* subsp. *Orientalis* pathogenesis analyzed by experimental immersion challenge in Nile tilapia, *Oreochromis niloticus* (L. *Veterinary Microbiology* 164(1–2), 77–84.

Soto, E., Brown, N., Gardenfors, Z.O., Yount, S., Revan, F. *et al.* (2014) Effect of size and temperature at vaccination on immunization and protection conferred by a live attenuated francisella noatunensis immersion vaccine in red hybrid tilapia. *Fish and Shellfish Immunology* 41(2), 593–599.

Soto, E., Wang, R., Wiles, J., Baumgartner, W., Green, C. *et al.* (2015) Characterization of isolates of *Streptococcus agalactiae* from diseased farmed and wild marine fish from the US Gulf Coast. *Latin America, and Thailand. Journal of Aquatic Animal Health* 27(2), 123–134.

Soto, E., Shahin, K., Talhami, J.J., Griffin, M.J., Adams, A. *et al.* (2019) Characterization of *Francisella noatunensis* subsp. *Orientalis* isolated from Nile tilapia *Oreochromis niloticus* farmed in Lake Yojoa, Honduras. *Diseases of Aquatic Organisms* 133(2), 141–145.

Subramaniam, K., Gotesman, M., Smith, C.E., Steckler, N.K., Kelley, K.L. *et al.* (2016) Megalocytivirus infection in cultured Nile tilapia *Oreochromis niloticus. Diseases of Aquatic Organisms* 119(3), 253–258.

Suebsing, R., Pradeep, P., Jitrakorn, S., Sirithammajak, S., Kampeera, J. *et al.* (2016) Detection of natural infection of infectious spleen and kidney necrosis virus in farmed tilapia by hydroxynapthol blue-loop-mediated isothermal amplification assay Journal of applied microbiology, 121(1), 55–67.

Sukenda, S., Carman, O., Rahman, R., Hidayatullah, D. and Yumaidawati, N.S. (2017) Vaccination in Nile tilapia broodstock with whole cell Journal and disease resistance in its fry against *Aeromonas hydrophila. Jurnal Akuakultur Indonesia* 16(2), 268–276.

Sukenda, S., Romadhona, E.I., Yuhana, M., Pasaribu, W. and Hidayatullah, D. (2018) Efficacy of whole-cell and lipopolysaccharide vaccine of *Aeromonas hydrophila* on juvenile tilapia *Oreochromis niloticus* against motile aeromonad septicemia. *Aquaculture, Aquarium, Conservation & Legislation* 11(5), 1456–1466.

Surachetpong, W., Janetanakit, T., Nonthabenjawan, N., Tattiyapong, P., Sirikanchana, K. *et al.* (2017) Outbreaks of tilapia lake virus infection, Thailand, 2015–2016. *Emerging Infectious Diseases* 23(6), 1031.

Suresh, A.V. and Lin, C K (1992) Tilapia culture in saline waters: A review. *Aquaculture* 106(3–4), 201–226.

Swain, P. and Nayak, S. (2003) Comparative sensitivity of different serological tests for seromonitoring and surveillance of *Edwardsiella tarda* infection of Indian major carps. *Fish & Shellfish Immunology* 15(4), 333–340.

Swaminathan, T.R., Nithyanantham, S.R., Narendrakumar, L., Dharmaratnam, A., Sood, N. *et al.* (2021a) Co-infection of lactococcus garvieae and tilapia lake virus (TiLV) in Nile tilapia *Oreochromis niloticus* cultured in India. *Diseases of Aquatic Organisms* 147, 127–140.

Swaminathan, T.R., Raj, N.S., Preena, P.G., Pradhan, P.K., Sood, N. *et al.* (2021b) Infectious spleen and kidney necrosis virus-associated large-scale mortality in farmed giant gourami, Osphronemus goramy, in India. India. *Journal of Fish Diseases* 44(12), 2043–2053.

Syuhada, R., Zamri-Saad, M., Ina-Salwany, M., Mustafa, M., Nasruddin, N, *et al.* (2020) Molecular characterization and pathogenicity of *Streptococcus agalactiae* serotypes ia ST7 and III ST283 isolated from cultured red hybrid tilapia in Malaysia. *Aquaculture* 515, 734543.

Takano, T., Matsuyama, T., Oseko, N., Sakai, T., Kamaishi, T. *et al.* (2010) The efficacy of five avirulent *Edwardsiella tarda* strains in a live vaccine against *Edwardsiellosis* in Japanese flounder, paralichthys olivaceus. *Fish & Shellfish Immunology* 29(4), 687–693.

Tattiyapong, P., Dechavichitlead, W., Waltzek, T.B. and Surachetpong, W. (2020) Tilapia develop protective immunity including a humoral response following exposure to tilapia lake virus. *Fish & Shellfish Immunology* 106, 666–674.

Throngnumchai, B., Jitrakorn, S., Sangsuriya, P., Unajak, S., Khunrae, P. *et al.* (2021) Refolded recombinant major capsid protein (MCP) from infectious spleen and kidney necrosis virus (ISKNV) effectively stimulates serum specific antibody and immune related genes response in Nile tilapia (*Oreochromis niloticus*). *Protein Expression and Purification* 184, 105876.

Trung Cao, T., Tsai, M.A., Yang, C.D., Wang, P.C., Kuo, T.Y. *et al.* (2014) Vaccine efficacy of glyceraldehyde-3-phosphate dehydrogenase (GAPDH) from *Edwardsiella ictaluri* against *E. tarda* in tilapia. *The Journal of General and Applied Microbiology* 60(6), 241–250.

Tsai, M.-A., Wang, P.-C., Liaw, L.-L., Yoshida, T. and Chen, S.-C. (2012) Comparison of genetic characteristics and pathogenicity of lactococcus garvieae isolated from aquatic animals in Taiwan. *Diseases of Aquatic Organisms* 102(1), 43–51.

Tsai, M.A., Wang, P.C., Cao, T.T., Liao, P.C., Liaw, L.L. *et al.* (2013) Immunoprotection of glyceraldehyde-3-phosphate dehydrogenase (GAPDH) from *Lactococcus garvieae* against lactococcosis in tilapia. *The Journal of General and Applied Microbiology* 59(6), 437–449.

Ture, M., Altinok, I. and Capkin, E. (2015) Comparison of pulsed-field gel electrophoresis and enterobacterial repetitive intergenic consensus

PCR and biochemical tests to characterize *Lactococcus garvieae*. *Journal of Fish Diseases* 38(1), 37–47.

Vendrell, D., Balcázar, J.L., Ruiz-Zarzuela, I., Blas, I., Gironés, O, *et al.* (2006) *Lactococcus garvieae* in fish: A review. *Comparative Immunology, Microbiology and Infectious Diseases* 29(4), 177–198.

Verma, D.K., Rathore, G., Pradhan, P.K., Sood, N. and Punia, P. (2015) Isolation and characterization of *Flavobacterium columnare* from freshwater ornamental goldfish carassius auratus. *Journal of Environmental Biology* 36(2), 433.

Wakabayashi, H. (1991) Effect of environmental conditions on the infectivity of flexibacter columnaris to fish. *Journal of Fish Diseases* 14(3), 279–290.

Wakabayashi, H. and Egusa, S. (1972) Preliminary experiments on environmental factors influencing the prevalence of columnaris disease. *Fish Pathology* 7(1), 58–63.

Wang, J., Zou, L.L. and Li, A.X. (2014) Construction of a *Streptococcus iniae* sortase A mutant and evaluation of its potential as an attenuated modified live vaccine in Nile tilapia (*Oreochromis niloticus*). *Fish and Shellfish Immunology* 40(2), 392–398.

Wang, J., He, R.Z., Lu, G.L., Luo, H.L., Lu, D.Q. *et al.* (2018a) Vaccine-induced antibody level as the parameter of the influence of environmental salinity on vaccine efficacy in Nile tilapia. *Fish Shellfish Immunol* 82, 522–530.

Wang, Q., Wang, X., Wang, X., Feng, R., Luo, Q, *et al.* (2018b) Generation of a novel *Streptococcus agalactiae* ghost vaccine and examination of its immunogenicity against virulent challenge in tilapia. *Fish and Shellfish Immunology* 81, 49–56.

Wang, J., Lu, D.Q., Jiang, B., Luo, H.L., Lu, G.L. *et al.* (2018c) The effect of intermittent hypoxia under different temperature on the immunomodulation in *Streptococcus agalactiae* vaccinated Nile tilapia (*Oreochromis niloticus*). *Fish Shellfish Immunol* 79, 181–192.

Wang, Q., Fu, T., Li, X., Luo, Q., Huang, J. *et al.* (2020) Cross-immunity in Nile tilapia vaccinated with *Streptococcus agalactiae* and *Streptococcus iniae* vaccines. *Fish Shellfish Immunol* 97, 382–389.

Wang, Y., Lü, L., Weng, S., Huang, J., Chan, S.-M. *et al.* (2007) Molecular epidemiology and phylogenetic analysis of a marine fish infectious spleen and kidney necrosis virus-like (ISKNV-like) virus. *Archives of Virology* 152(4), 763–773.

Wang, Y.-T., Huang, H.-Y., Tsai, M.-A., Wang, P.-C., Jiang, B.-H. *et al.* (2014) Phosphoglycerate kinase enhanced immunity of the whole cell of

Streptococcus agalactiae in tilapia, *Oreochromis niloticus. Fish & Shellfish Immunology* 41(2), 250–259.

Wangkaghart, E., Deville, S., Wang, B., Srisapoome, P., Wang, T. *et al.* (2021) Immune response and protective efficacy of two new adjuvants, montanide. *Fish and Shellfish Immunology* 116, 19–29.

Xing, J., Xu, H., Tang, X., Sheng, X. and Zhan, W. (2019) A DNA vaccine encoding the VAA gene of vibrio anguillarum induces A protective immune response in flounder. *Frontiers in Immunology* 10, 499.

Yamkasem, J., Tattiyapong, P., Kamlangdee, A. and Surachetpong, W. (2019) Evidence of potential vertical transmission of tilapia lake virus. *Journal of Fish Diseases* 42(9), 1293–1300.

Yamkasem, J., Tattiyapong, P., Gorgoglione, B. and Surachetpong, W. (2021) Uncovering the first occurrence of tilapia parvovirus in thailand in tilapia during co-infection with tilapia tilapinevirus. *Transboundary and Emerging Diseases* 68(6), 3136–3144.

Yanuhar, U., Christi, N. and Arfiati, D. (2017) Genetic characterization of viral nervous necrosis infects tilapia (*Oreochromis* sp.) in Indonesia. In: *1st International Conference in One Health (ICOH 2017)*, Atlantis Press, pp. 174–179.

Yao, Y.Y., Chen, D.D., Cui, Z.W., Zhang, X.Y., Zhou, Y.Y. *et al.* (2019) Oral vaccination of tilapia against *Streptococcus agalactiae* using bacillus subtilis spores expressing sip. *Fish and Shellfish Immunology* 86, 999–1008.

Ye, T., Wu, J., Mu, C., Pan, G., Chen, J. *et al.* (2021) FucP mediates fucose to regulate T3SS in edwardsiella piscicida and promotes intestinal colonization in tilapia. *Aquaculture International* 29(3), 1233–1243.

Younes, A.M., Gaafar, A.Y., El-Din, A. and Abu-Bryka, Z. (2019) Genotyping and pathogenicity of *Streptococcus iniae* strains recovered from cultured *Oreochromis niloticus* at Kafr El-Shaikh governorate, egypt. *Egyptian Journal of Aquatic Biology and Fisheries* 23(3), 467–474.

Yu, N., Zeng, W., Xiong, Z., Liu, Z. (2022) A High Efficacy DNA Vaccine against Tilapia Lake Virus in Nile tilapia (*Oreochromis niloticus*). Aquaculture Reports, 24, 101166.

Zeng, W., Wang, Y., Hu, H., Wang, Q., Bergmann, S.M. *et al.* (2021a) Cell culture-derived tilapia lake virus-inactivated vaccine containing montanide adjuvant provides high protection against viral challenge for tilapia. *Vaccines* 9(2), 86.

Zeng, W., Wang, Y., Chen, X., Wang, Q., Bergmann, S.M. *et al.* (2021b) Potency and efficacy of VP20-based vaccine against tilapia lake virus using different prime-boost vaccination regimens in tilapia. *Aquaculture* 539, 736654.

Zhang, C., Li, L.-H., Wang, J., Zhao, Z., Li, J. *et al.* (2018) Enhanced protective immunity against spring viremia of carp virus infection can be induced by recombinant subunit vaccine conjugated to single-walled carbon nanotubes. *Vaccine* 36(42), 6334–6344.

Zhang, D., Ke, X., Liu, Z., Cao, J., Su, Y. *et al.* (2019) Capsular polysaccharide of *Streptococcus agalactiae* is an essential virulence factor for infection in Nile tilapia (*Oreochromis niloticus* Linn.). *Journal of Fish Disease* 42(2), 293–302.

Zhang, D., Gao, Y., Li, Q., Ke, X., Liu, Z. *et al.* (2020) An effective live attenuated vaccine against *Streptococcus agalactiae* infection in farmed Nile tilapia (*Oreochromis niloticus*). *Fish Shellfish Immunol* 98, 853–859.

Zhang, L., Zeng, Z., Hu, C., Bellis, S.L., Yang, W. *et al.* (2016) Controlled and targeted release of antigens by intelligent shell for improving applicability of oral vaccines. *Biomaterials* 77, 307–319.

Zhang, Y., Arias, C., Shoemaker, C. and Klesius, P. (2006) Comparison of lipopolysaccharide and protein profiles between *Flavobacterium columnare* strains from different genomovars. *Journal of Fish Diseases* 29(11), 657–663.

Zhang, Z., Yu, A., Lan, J., Zhang, H., Hu, M. *et al.* (2017) GapA, a potential vaccine candidate antigen against *Streptococcus agalactiae* in Nile tilapia (*Oreochromis niloticus*). *Fish and Shellfish Immunology* 63, 255–260.

Zhang, Z., Li, Y., Hu, M. and Yu, A. (2020) Genome reanalysis to decipher resistome, virulome, and attenuated characters of attenuated *Streptococcus agalactiae* strain HZAUSC001. *Microbial Pathogenies* 147, 104416.

Zhao, Z., Zhang, C., Jia, Y.-J., Qiu, D.-K., Lin, Q. *et al.* (2019) Immersion vaccination of mandarin fish *Siniperca chuatsi* against infectious spleen and kidney necrosis virus with a swcnts-based subunit vaccine. *Fish & Shellfish Immunology* 92, 133–140.

Zhao, Z., Zhang, C., Lin, Q., N-q, L., Huang, Z.-B. *et al.* (2020) Single-walled carbon nanotubes as delivery vehicles enhance the immunoprotective effect of an immersion DNA vaccine against infectious spleen and kidney necrosis virus in mandarin fish. *Fish & Shellfish Immunology* 97, 432–439.

Zhu, L., Yang, Q., Huang, L., Wang, K., Wang, X. *et al.* (2017) Effectivity of oral recombinant DNA vaccine against *Streptococcus agalactiae* in Nile tilapia. *Developmental & Comparative Immunology* 77, 77–87.

Zhu, W., Yang, G., Zhang, Y., Yuan, J. and An, L. (2012) Generation of biotechnology-derived *Flavobacterium columnare* ghosts by phix174 gene E-mediated inactivation and the potential as vaccine candidates against infection in grass carp. *Journal of Biomedicine and Biotechnology.*

Zlotkin, A., Hershko, H. and Eldar, A. (1998) Possible transmission of *Streptococcus iniae* from wild fish to cultured marine fish. *Applied and Environmental Microbiology* 64(10), 4065–4067.

Zou, L., Wang, J., Huang, B., Xie, M. and Li, A. (2011) MtsB, a hydrophobic membrane protein of *Streptococcus iniae*, is an effective subunit vaccine candidate. *Vaccine* 29(3), 391–394.

Chapter 17

Certification in tilapia health management

Francis Murray

Introduction

Certified sustainability standards are voluntary, usually third-party-verified, norms relating to environmental, social, ethical and food safety issues. The most credible standards are based on normative international standards and guidelines and developed through multi-stakeholder consultation processes. Operating in the 'court of public opinion', voluntary standards help indemnify commercial brands against adverse reputational risks from civil society campaigns. Thus, most demand for tilapia originates from more affluent certification-centric markets (CCM) with strong civil society credentials. Demand for certified tilapia is most concentrated in the US market, and mainly provisioned for by two (of four) globally dominant aquaculture standards bodies, the Aquaculture Stewardship Council (ASC) and the Global Seafood Alliance (GSA).

Tilapia certified under these standards together accounted for up to 3.8% of a historic peak of 7.15 million tonnes of global production in 2021 (95% of it farmed). In turn, two countries producing over 1 million tonnes of production per year accounted for 64% of this certified volume: China (42%) and Indonesia (22%). Environmental and disease factors led to a subsequent 6% decline in global tilapia production to 6.72 million tonnes in 2023. Over the same period, the combined ASC, GSA–Better Aquaculture Practices (BAP) certified total fell by 30.3% from 363,455 tonnes to 238,223 tonnes, mainly due to falling production levels in lake-cage

Kim Thompson, Craig Shoemaker and David C. Little (eds) **Tilapia**
DOI: 10.1079/9781800629455.0017, © CAB International, 2025

systems facing interacting water quality (eutrophication) and disease challenges.

We also assess the complementary role of seafood guidance and rating schemes (SGRS) based on less costly (involuntary), third-party environmental assessments at national or regional sectoral scales. Monterey Bay's Seafood Watch (SFW) SGRS, the most influential such scheme in the US market, assessed tilapia sectors in six major exporting countries accounting for 1.6 million tonnes or 23% of global production in 2021. SGRS recognition of farm-level certification in otherwise poorly performing sectors rewards individual performance improvements, whilst unitary certification (variously of farms, hatcheries, feed-mills, processors and chain of custody) also supports wider value-chain performance improvements. Certification also reaches farms in some countries with low export volumes beyond the purview of SGRS. Standard benchmarking programmes such as the Global Seafood Sustainability Initiative (GSSI) help reduce consumer confusion and, potentially audit costs by facilitating mutual recognition of standard elements.

The influence of both unitary and SGRS systems is most fundamentally constrained by demand for certification stemming mainly from CCM. Between 2012 and 2024, this limited ASC and GSA farm certification to just 15 countries (mainly across S. America and S.E. Asia) of 126 countries with reported tilapia production. This excludes Egypt, with the third highest global production after China and Indonesia. Higher relative costs for certification also exclude many of the smaller enterprises that still dominate many of these sectors. Unitary certification also fails to adequately address cumulative interacting water quality and aquatic health impacts at the landscape or ecosystem level. Emergent responses include group certification, smallholder aquaculture improvement projects (AIPs) and area-based management (ABM) using certification to leverage coordinated sectoral responses to environmental pressures.

An analysis of 155 audits of 55 pond and cage farms in 13 countries (capacities ranging from 36 to 25,193 tonnes per year) against the ASC Tilapia Standard, between 2012 and 2019, confirmed the relative paucity of smallholder participation described above. Of a total of 354 nonconformities (NCs), 13% (46) were assigned to eight (of 11) indicators under a principle titled 'Manage fish health and welfare in an environmental responsible manner' (FHM; with primary intent to mitigate negative impacts on wild populations). FHM indicators were grouped under four sub-criteria addressing (1) stock recovery, i.e. survival rates; (2) use of therapeutants and chemicals, including prudent use of antibiotics; (3) removal and disposal of mortalities; and (4) implementation of a farm fish health plan. Some 40% (20) of all FHM NCs were assigned to a single indicator mandating a minimum recovery rate of 65% of fish stocked after attaining a size of 100 g, reflecting the wider global disease and water quality challenges described above. Unlike some other ASC standards, the Tilapia Standard includes no 'State' indicators reflecting health conditions of the environment itself (e.g. pathogen or disease prevalence rates), only Pressure and Response indicators.

Finally, we highlight longer-term sustainability challenges associated with the certified tilapia sector's dependence on export production from large oligotrophic and mesotrophic lake and reservoir systems in middle- and lower-income countries. These systems face eutrophication pressures associated with wider demographic transitions and biosecurity challenges associated with co-location with fragmented smallholder sectors in these open systems. ASC's decision to withdraw entirely from the Chinese market in March 2024 also highlights the influence of wider geopolitical risks and trends on global seafood value chains, including increased policy focus on achieving greater domestic food security.

In this chapter we assess the role of private environmental sustainability standards in the management of farmed tilapia health and welfare. Arising from perceived deficits in statutory regulation and operating in the 'court of public opinion', such voluntary standards help indemnify commercial brands against adverse reputational risks from civil society campaigns and other interest group pressures. Consequently, demand for this market-based form of regulation has been driven by a need to assure the quality of seafood commodities, often of uncertain regulatory provenance for consumption in more affluent certification-centric markets with strong civil society credentials. Demand growth has been further amplified by the enlargement of global seafood supply chains with net flows from lower- to higher-income economies.

Such demand will be sustained only where standards are perceived to be credible: that is, they are science-based and demonstrably impartial assurance systems are in place to assure their effective implementation. Thus, the most rigorous standards incorporate independent standard-setting and implementation processes including third-party conformity auditing of unitary production entities: farms, hatcheries, feed mills, processors and chain of custody (CoC) through these production stages as the basis for their certification.

The primary intent of animal health requirements in farm standards and sectoral rating schemes (see below) is to limit the negative externalities of production on local ecosystems and biodiversity. However, the higher costs for assurance of unitary farm-level standards must also be balanced against the economic sustainability of individual farms. This underpins a theory of change whereby standard requirements are subject to periodic revision based on median sectoral performance. Such revisions drive continuous improvement of farms, particularly those performing around the median threshold.

These attributes underpin two further implementation challenges. Firstly, whilst some requirements will also support improved economic performance, e.g. minimum survival rates, others simply impose additional compliance costs for farmers, the marginal cost of which is likely to be greater for smaller enterprises that still dominate tilapia and many other aquaculture sectors. Compliance with health management requirements can be especially challenging for smallholders. Secondly, the farm-level focus of third-party audited certification systems places limits on their ability to address cumulative impacts at a wider ecosystem 'landscape' level, with coordination efforts further challenged by smallholder exclusion (Fig. 17.1). This is especially critical for the management of two interacting environmental impacts, pathogen transmission and water quality. Below we review efforts by standard bodies to enable smallholder participation through lower-cost group certification and better management programmes (BMPS) and the intersection of these efforts with requirements for ABM.

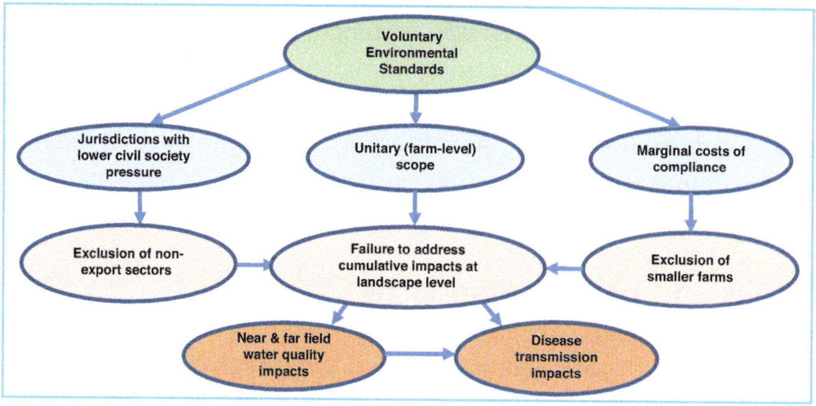

Figure 17.1 Challenges for implementation of farm-level environmental certifications systems incorporating animal health requirements.

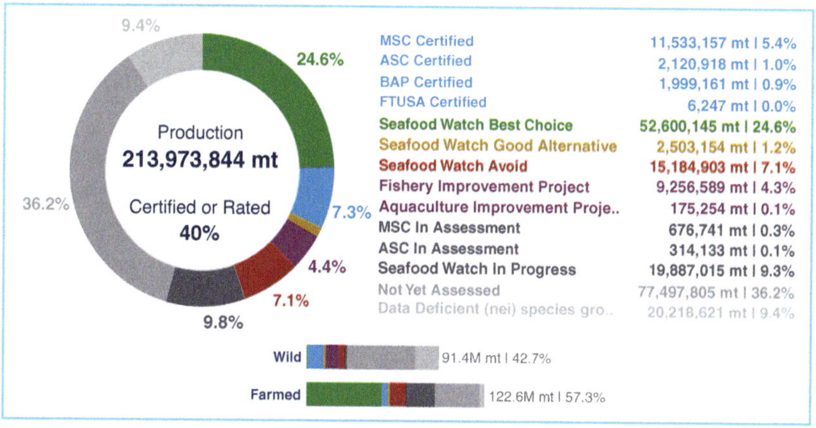

Figure 17.2 Global seafood volumes under environmental certification and rating schemes 2022 (source: certification and ratings collaboration data (CRC, 2023); global production data 2021 (FAO, 2023).

Below we review the complementary role of SGRS based on less costly, more generalized involuntary third-party assessments of species and system-specific (e.g. pond or cage) performance at national or regional sectoral scales. Figure 17.2 illustrates the marked difference in coverage between sectoral and more resource-intensive farm-level certification schemes. Monterey Bay Aquarium's Seafood Watch, one of the most influential rating schemes in the US market, covered 32.9% of global farmed seafood production in 2022. This contrasts with a combined total of 7.3%[1] by three of the largest voluntary environmental standards combined, one for wild capture, the Marine Stewardship Council (MSC; 5.4%) and two for farmed production, ASC and GSA–BAP; combined, 1.9%.

Public and private standards

Statutory and private standards are based on a subset of 'normative' public standards and guidelines established by intergovernmental agencies. Food safety standards established precedents for the more recent development of environmental, social and animal welfare standards.

Under the Sanitary and Phyto-sanitary Measures Agreement (SPS), the World Trade Organization established the joint FAO/WHO Codex Alimentarius as the standard-setting entity for food safety, whilst the World Organisation for Animal Health (WOAH, formally OIE) has responsibility for animal protection measures related to SPS. Health requirements in Environmental standards for aquaculture are grounded in the latter guidelines and more generally in the FAO Guidelines for Ecolabelling of Fish and Fishery Products from Marine/Inland Capture Fisheries (FAO, 2009) and the FAO Technical Guidelines on Aquaculture Certification (FAO, 2011). The development and implementation of food standards also relies on other private process and product standards such as those published by the International Organization for Standardization (ISO)[2] and codes of practice such as the International Social and Environmental Accreditation and Labelling alliance (ISEAL, 2023).

The proliferation of private voluntary food standards in global supply chains has also given rise to hybrid forms of public and private governance (FAO, 2011; FAO/UNEP, 2014). Private standards tend to be wider in scope and more dynamic and responsive to changing market demands than public standards. Influence can be bidirectional; for example, private organic standards have become the basis for an EU organic regulation (EC 835/07 and 889/08;

Commission, 2008), whilst private animal welfare standards continue to influence the (slower) development of public standards.

Environmental certification systems

Certified sustainability standards are voluntary, (usually) third party-verified norms relating to environmental, social, ethical and food safety issues, developed in consultation with a range of stakeholders and experts.

Private 'eco-labelling' schemes emerged in the 1980s as a response to societally perceived deficits in statutory environmental regulation. Agenda 21 of the Earth Summit 1992 (UNCED, 1992) went on to call for greater use of eco-labelling to help consumers make more informed choices.

A subset of the ISO 14000 series, 14020–14025, defines processes for creating eco-labels, differentiated as Types I, II and III based on their scope and conformity assessment requirements (Rusko and Koraus̆, 2013; OECD, 2017):

- Type I (ISO 14024) Eco-labels: requiring evaluation of defined, multi-attribute performance criteria by an independent, third-party (or external EMS registration), across the entire lifecycle of a product-group.
- Type II (ISO 14021) based on self-declarations according to general guiding principles, by any value-chain entity (producer, importer, distributor, retailer). Not third-party certified but expected to be verifiable and accurate.
- Type III (ISO 14025) Environmental labels and declarations (ELD): requiring quantification of a geographically expressed environmental load based on life-cycle analysis (LCA) for comparisons between products fulfilling the same function (usually for B2B or procurement purposes).

Despite its broad scope, the ISO 14000 series typology is not exhaustive. Table 17.1 lists attributes of four major certification schemes under which the greatest volumes of farmed seafood export

Table 17.1 Characteristics of major aquaculture sustainability certification schemes (adapted from Potts *et al.*, 2016).

Standard name-holder and consumer logo	Production system and provenance	Year scheme (initiated) effective	Finfish species scope	Geographic scope of certification	Type of certification	Value chain	Benchmarking[a]
Global Seafood Alliance Best Aquaculture Practices (GSA BAP)	Aquaculture (civil society, private sector, producers)	(2004) First tilapia certification 2008	Finfish standard 12 species certified. inc. tilapia	Asia, Australia Oceania, Central American Caribbean, Europe, N. America, S. America	Individual and multi-star production group along value chain	Seafood processors, farms, feed mills and hatcheries	GSSI: 4 Oct 2017 GFSI: 16 May 2013 GSCP: N/A SSCI: 8 Jul 20 (applied)
Aquaculture Stewardship Council (ASC)	Aquaculture (civil society)	(2010) Tilapia standard; first tilapia certification 2012	Standards for 11 species groups, inc. tilapia	Asia, Australia Oceania, Central American Caribbean, Europe, N. America, S. America	Individual and group	Farm and compound feed	GSSI: 2017 (4 of 7 spp. standards) GSFI: N/A GSCP: N/A
The Global Partnership for Good Agricultural Practice (GlobalGAP)	Aquaculture (private sector: industry, retailers)	(1997) 2004	All finfish spp. (crustaceans and molluscs)	30 Countries from North, Central and South America, Europe, Asia, Australia and Oceania	Individual and group	Aquaculture hatcheries and farms, compound feed	GFSI: 24 April 2013 GSSI: N/A GSCP: GlobalGAP Integrated Farm Assurance V5, Produce Safety Standard (PSS) V4 and Harmonized PSS

Friends of the Sea (FOS)	Fisheries and Aquaculture civil society)	(2008) First edition of wild catch and aquaculture standards, effective 2013	All species of fish	Africa, Asia, Australia and Oceania, Central America and Caribbean, Europe, N. America and S. America	Individual and group	Fisheries, fish farms, fish meal, fish oil	GFSI: N/A GSSI: N/A GSCP: N/A
Naturland (Organic)	Aquaculture (civil society, private sector, producers)	(1982) Aquaculture 1996 Wild Catch 2006	Carp, salmonids, whitefish, mussels, shrimp, tropical FW fish, perch-jack-cod like fish, macro-algae. Wild catch: all FW and marine finfish and invertebrates	Africa, Asia, Australia and Oceania, Central America and Caribbean, Europe and S. America	Individual and group	Producers, processors	GFSI: N/A GSSI: N/A GSCP: N/A

[a]Benchmarking schemes: GFSI, Global Food Safety Initiative; GSSI, Global Seafood Sustainability Initiative; GSCP, Global Social Compliance Programme; SSCI, Sustainable Supply Chain Initiative; N/A, not available

commodities have been certified, as well as a more niche example of organic certification. These all fall into a common subset of 'Other Type I (ISO 14024)-like Voluntary Sustainable Schemes (VSS)' (OECD, 2017); whilst they require third-party assessment against multiple environmental criteria, they are limited in life cycle scope with respect to the Type I definition.

The typology also lacks precision for third-party attributes beyond requirements for conformity assessment, a critical factor when perceptions regarding independence of the standard-holding entity can also impact the credibility and demand for market-based standards. Whilst industry-body 'second party' standards fall squarely under the Type II definition, national or state-operated standards could also be viewed as first party for internal markets, but 'second party' with respect to traded commodities such as seafood.

SGRS of the kind reviewed above also fall outwith this schema.

Voluntary standards are adopted by companies to complement their own internal and supply chain quality assurance systems, to demonstrate acceptable management of the negative externalities of business practices.

As an outsourced, third-party assurance system, certification schemes offer lead companies a credible means to demonstrate sustainable and ethical supply chain management. Thus, they have become a key tool in defending brand reputation against civil society campaigns linked to real or perceived deficits (Fig. 17.3). They also offer companies a strategic means to achieve other growth objectives requiring social licence (Olsen *et al.*, 2023). Contingent on business strategy and market conditions, standards may operate at business to business (B2B) and/or business to consumer (B2C) levels, the latter conferring the right for products to carry an eco-label.

Assurance also requires clear separation of standard setting, implementation, auditing and accreditation processes (Fig. 17.4). The demonstrable independence of involved parties is fundamental to the credibility of sustainability claims without which standards would lose market demand. Inter alia, transparent, 'third-party'-operated and -audited private standards offer greater

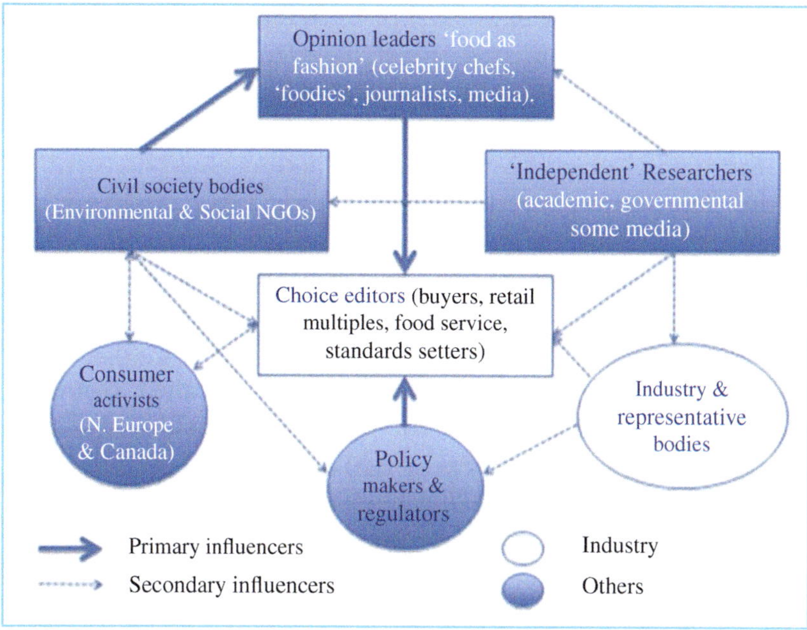

Figure 17.3 The relative influence of different stakeholder groups in driving demand for third-party certification by leading seafood brands (Murray and Taskov, 2018).

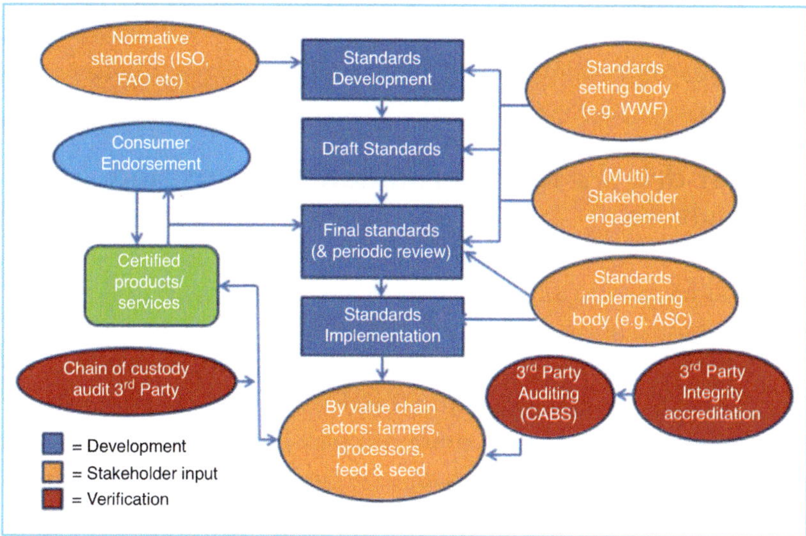

Figure 17.4 Sustainability standard development and certification processes (Murray and Taskov, 2018).

assurance compared with self-claims or industry body- or national state-operated 'second-party' standards, even where they include third-party auditing. The performance of third-party 'conformity assessment bodies' (CABs), along with standards-setting procedures, are themselves subject to formal accreditation processes based on further tiers of process standards.

National Standards: To support smallholders, and domestic producers more generally, many exporting countries have attempted to establish their own national standards. However, as second-party systems they lack the level of credible impartiality and assurance brought to market by the established dominant mainstream voluntary standards with their provenance in European and North American CCM (Table 17.1). Second-party standards developed by producer organizations (POs) have seen application in more consolidated sectors capable of underwriting their implementation (e.g. salmon), but face similar challenges in more fragmented (e.g. tilapia) sectors, where assurance needs are often greatest. Redundant national standards may be maintained as entry points for mainstream certification under equivalence assessments, or as elements of public–private smallholder AIPs. One national scheme, ThaiGAP, has morphed into a traceability system serving the Thai shrimp export sector (Yamprayoon and Sukhumparnich, 2010).

Market drivers for environmentally certified aquaculture

Demand for sustainably certified foods is concentrated in the global north, where commercial brands are likely to be exposed to greater civil society scrutiny of their environmental and social performance (Fig. 17.3). Conversely, assurance needs are greater for regions with less effective regulation and more complex fragmented value chains (OECD, 2013a; FAO, 2014) more typical of lower income countries. This encourages seafood companies to engage in ethical supply chain management of commercial activities beyond their own direct ownership and legal jurisdictions.

Certification systems must also engage with the supply chain to assure responsible sourcing and origin of certified products. This typically includes a separate CoC certification to verify that standard requirements are being implemented for all production determined to be eligible for certification and to prevent mislabelled products entering supply chains. Although 'voluntary' in principle, sustainability certification has become a necessity for continued access to sizeable market segments. This is especially the case for seafood commodities sold through large retail multiples. There has been less penetration of the more fragmented food service section, with the notable exception of some large restaurant chains.

Theoretical ways for producers to offset certification costs and incentivize adoption include (1) achieving a price premium for certified products; (2) reduction of production costs through greater eco-efficiency; (3) access to otherwise inaccessible markets; and (4) social licence for wider strategic objectives (e.g. new site licences). The significance of individual drivers is contested and contingent on supply chain characteristics, including the degree of sectoral consolidation.

Unless a premium is explicitly guaranteed, e.g. as in the Fairtrade model, prices remain subject to market forces. Thus premiums are more likely to accrue to niche differentiated, e.g. organic, than more mainstream aquaculture standards for high-volume commodity flows. Certification costs increase further where multiple compliance audits for different standards are required to meet the demands of individual or multiple target market segments. Larger companies in more consolidated sectors are also better positioned to engage in precompetitive collaborations, such as the Global Salmon Initiative (GSSI, 2023) using certification to achieve social licence for strategic industry sectoral growth objectives (Murray and Taskov, 2018).

Certification can create barriers to smallholder participation in increasingly globalized markets (FAO, 2014; CFC, 2018). In these instances, achieving certification may require significant investment in upgrading of farm facilities, management practices, auditing and certification costs. Lack of diagnostic capacity and access to professional animal health services is particularly challenging for many smallholders, from which an estimated 70% of all global aquaculture production originated in 2015 (Vince & Haward, 2017).

An associated antiglobalization critique positions certification bodies and environmental non-governmental organizations (eNGOs) as complicit in enabling this exclusionary process (Belton *et al.*, 2010; Vandergeest & Unno, 2012; Marschke & Wilkings, 2014; Saha, 2022). From this perspective, lead seafood companies (e.g. retail multiples, processers and vertically integrated farming companies) accrue commercial benefits whilst smaller producers incur a disproportionate share of costs, further accelerating consolidation of export value chains. Intrinsic sustainability attributes of semi-intensive polyculture systems practised by many smallholders are also ignored or undervalued. Rather than 'lifting all boats', periodic revision of standards to drive 'continuous improvement' then only reinforces a performance dichotomy, as there will inevitably always be a lower percentile of producers, principally smallholders, who will remain 'beyond certification'. One response has been for standard holders to develop lower cost group certification options, in conjunction with transitional cluster-based AIPs, the latter co-sponsored with public-and private-sector and NGO partners.

The importance of low-cost tilapia production to domestic food security in the global south also explains the relatively low percentage of global production certified under environmental standards, compared with more export-orientated, higher-value commodities such as salmon or shrimp. In China the large proportion of small and medium enterprises supplying domestic or export markets that lack demand for environmental and social certification is a particular distortion.

More fundamentally, the extent to which standards can improve environmental management has also been questioned (Jacquet and Pauly, 2007), or how impacts can even be reliably measured (Tlusty, 2011; Bridson *et al.*, 2020). Below we assess impacts of the ASC Tilapia Standard on environmental performance (with a focus on animal health requirements), through an analysis of farm conformance audits conducted between 2012 and 2019. This was selected (1) as the only tilapia-specific mainstream standard at time of writing, multi-species standards being the industry norm and (2) ASCs transparency policy of positioning of audit reports in the public domain is consistent with its own compliance with ISEAL Codes of Good Practice.

In the next section we compare the global scope of tilapia certification and rating systems in the light of these exclusionary challenges.

Trends in environmentally certified and rated tilapia production

Seven species groups – salmon, shrimp, trout and sea bream, mussels, pangasius and tilapia – accounted for 97% of a modest 6.3% share (4,828 million tonnes) of global aquaculture production (excluding seaweeds) certified under environmental standards in 2015 (Potts *et al.*, 2016). This reflects the relatively limited nature of demand for these export commodity species, stemming from more certification-centric markets (Bush *et al.*, 2013). Four 'mainstream' aquaculture standards accounted for most of this production: (1) GSA–BAP (formerly Global Aquaculture Alliance), (2) ASC, (3) GlobalGAP (GG) and (4) Friend of the Sea (FoS) (Table 17.1).

In 2021, a total of 363,455 tonnes of tilapia was certified under the first two standards, nominally accounting for 5.1% (with some overlap) of a historically high total of 7,145,207 tonnes of recorded tilapia production (94.6% farmed) recorded by 126 countries in the same year (Figs. 17.5 and 17.6). Two standards, GSA–BAP and ASC, dominate tilapia certification, consistent with demand for these eco-labels in the USA as the largest market for environmentally certified tilapia. FoS and GG mainly cater for the European market, where pangasius has supplanted tilapia as a low-cost whitefish substitute. Only ten tilapia producers were GG certified in 2021, with no data available for FoS (one Vietnamese supplier of certified tilapia was listed on their website).

Farm-level certification under ASC and GSA standards was limited to 12 countries in 2021, with shares of the 363,455 tonne

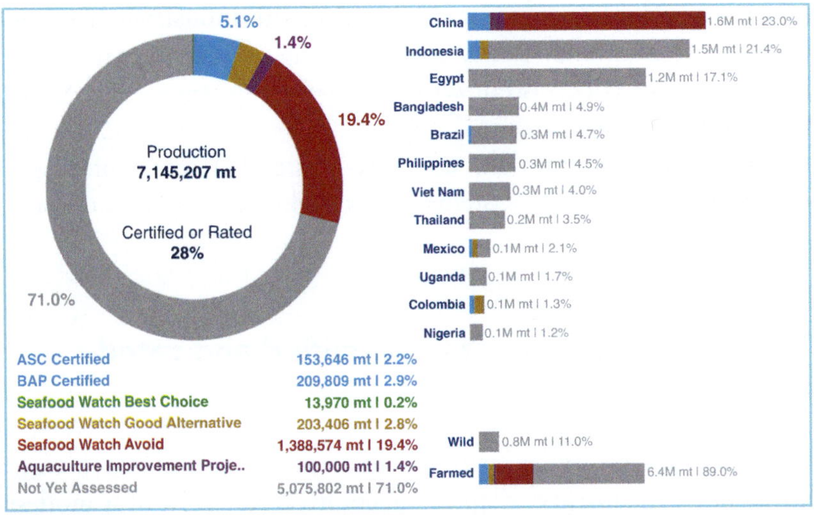

Figure 17.5 Certification and rating of global farmed tilapia production (L) by the top 12 producing countries (R) in 2021 (source: CRC, 2023; global production data 2021 (FAO, 2023).

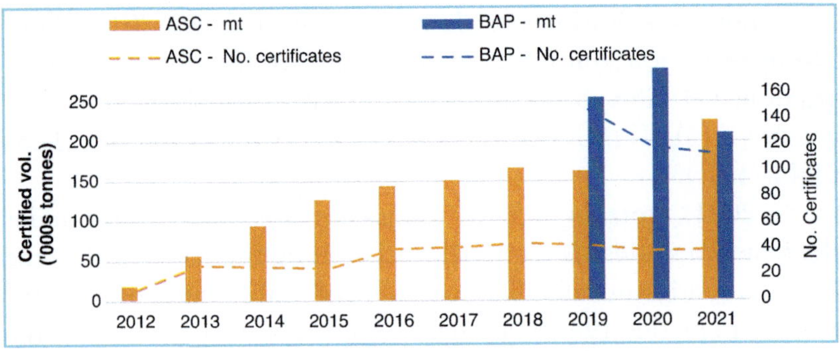

Figure 17.6 ASC- and GSA–BAP-certified tilapia volume and number of certificates 2012–2021 (sources: ASC, 2023b; GSA–BAP, 2023).

certified total as follows: China (42%), Indonesia (22%), Honduras (9%), Colombia (9%), Brazil (7%), Mexico (6%), Costa Rica (3%), Malaysia (1%), Taiwan (0.2%), Ecuador (<1%), Belize (<1%) and Poland (one recirculating aquaculture system (RAS)).

Monterey Bay's SFW sectoral ratings programme covered 1.6 million tonnes of production, nearly 4.5 times the certified volume in the same year. However, they were more limited in scope, to just six major exporters to the USA: China (87% of the rated total), Colombia (4%), Indonesia (4%), Taiwan (3%), Mexico (2%) and Honduras (<1%). SFW rated all Chinese tilapia red (avoid) whilst acknowledging that farms accounting for an additional 0.1 million tonnes of Chinese production were participants in AIPs implemented by the Sustainable Seafood Partnership (SSP). Thus, globally in 2021, SFW rated a total volume of 303,406 tonnes as either a 'good equivalent' (orange) or operating under an AIP, in addition to its acknowledgement of the 363,455 tonnes of blue-rated certified production as at least 'good equivalent' (again with some potential for group overlap).

These patterns reflect the relatively low global demand for tilapia production from CCM compared with commodities such as salmon and shrimp. Most demand for certified and rated tilapia is thereby limited to the US market which, in turn, is often supplied by specialized export production (particularly lake/reservoir cage) sectors. Major producers prioritizing domestic and/or less demanding export markets are almost entirely excluded. This includes most Chinese and African production (including the world's third-largest producer, Egypt, which alone produced 1,081,202 tonnes of tilapia in 2021; FAO, 2023) and many Asian and South American countries, where low-cost tilapia makes a significant contribution to national food security. Consequently in 2021 tilapia production, under the certification and rating schemes described above, totalled 28% of global production compared with totals of 94 and 78% for the more export-orientated farmed Atlantic salmon and shrimp commodity sectors, respectively (CRC, 2023).

US and EU tilapia markets: The USA imported 96% of its tilapia supply in 2020, with imports totalling 193,230t (Table 17.2). Frozen fillets from Asia and frozen whole tilapia, mainly from China, accounted for 87% of total imports. Lake-grown fresh tilapia fillets from Central and South America, which achieved a 38% premium on frozen fillets, accounted for 12% of imports. The EU imported just 33,000t (live weight equivalent; AIPCE-CEP, 2022) in 2021 (82% from China), a significant share of which was processed for re-export.

Table 17.2 US tilapia import volume (tonnes) and value by product, 2020 (source, MSU, 2020).

Product	Import vol. (t)	Imports (%)	Imports value (US$ mill.)	US$/kg	Source/notes
Frozen fillets	126,824	65.6	493.2	3.89	Asia
Fresh fillets	23,405	12.1	151.2	6.46	& Central America (mainly Honduras, Colombia)
Frozen whole (HoG)	41,368	21.4	73.2	1.77	Mainly China
Other	1633	0.8	8.4	5.14	
Total	**193,230**	**100**	**726**	**4.23**	Weighted average, US$/kg

Note: Fillet size ranges from 150 to 850g

Recent trends in GSA–BAP and ASC certification: The first tilapia farms achieved GSA–BAP and ASC certification in 2008 and 2012, respectively. Certified volumes achieved near equivalence at 209,809t (GSA–BAP) and 225,820t (ASC, 2023a) in 2021 (Fig. 17.6)[3]. Of the combined volume of 435,629t (5.1% of global production), 76% (331,078t) originated from farms certified under both schemes. Thus, a corrected total of 270,090t was certified under one or both schemes, equivalent to 3.78% of the total 7.15 million tonnes of tilapia production recorded in 2021. As a result, costs for multiple compliance audits are incurred by many farms.

GSA–BAP volume and certificate numbers plateaued in 2021, averaging 1857t across 113 certificates, whilst ASC average certificate volume rose to 5790t across just 39 certificates (Fig. 17.7). This corresponded with an increase in multi-site (MS) group certification, accounting for six of the 39 ASC certificates with two to six farms per certificate. The resulting average of 4428t per farm site (51 in total) was 2.4 times larger than the average GSA–BAP certificate volume.

Reasons for this divergence are unclear but may reflect capacities of different production groups to meet respective compliance requirements, different audit cost models or appeal to different market segments associated with the provenance of the standards

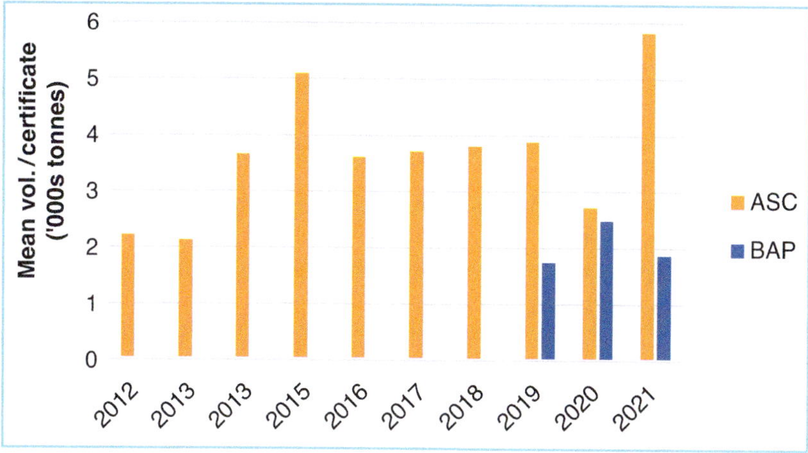

Figure 17.7 ASC and GSA–BAP mean tilapia volume per certificate, 2012–2021 (sources: ASC, 2023b; GSA–BAP, 2023).

(eNGO and industry; Table 17.1). Some 81% of 118 GSA–BAP certificates in 2021 were issued for farms in China compared with just 18% of ASC certificates (Fig. 17.8). Thus, GSA–BAP appeared to be certifying a larger volume of lower-cost frozen whole fish (Table 17.2) from smaller pond and cage farms as raw material for value-added processing and reprocessing. Separate GSA–BAP hatchery and processor (and feed-mill) standards also offer a path to full product certification for the many vertically integrated tilapia operations in China.

In contrast, ASC certified six of the largest tilapia farms globally, all adopting MS certification for cage systems with a combined volume of nearly 120,000 t. This accounted for 53% of ASC-certified volume in 2021 and 20% of all recorded Indonesian tilapia production in 2021 (FAO, 2022). Five of these farms were located on Lake Toba, a 505-m-deep, 1,130 km^2 volcanic lake in central Sumatra and the largest freshwater body in S. Asia (Fig. 17.5 and Fig. 17.9). Four were owned by a single operator, PT Aquafarm Nusantara (Regal Springs Indonesia). GSA–BAP certified their individually quick-frozen processing plant in Medan, ASC lacking a processing standard.

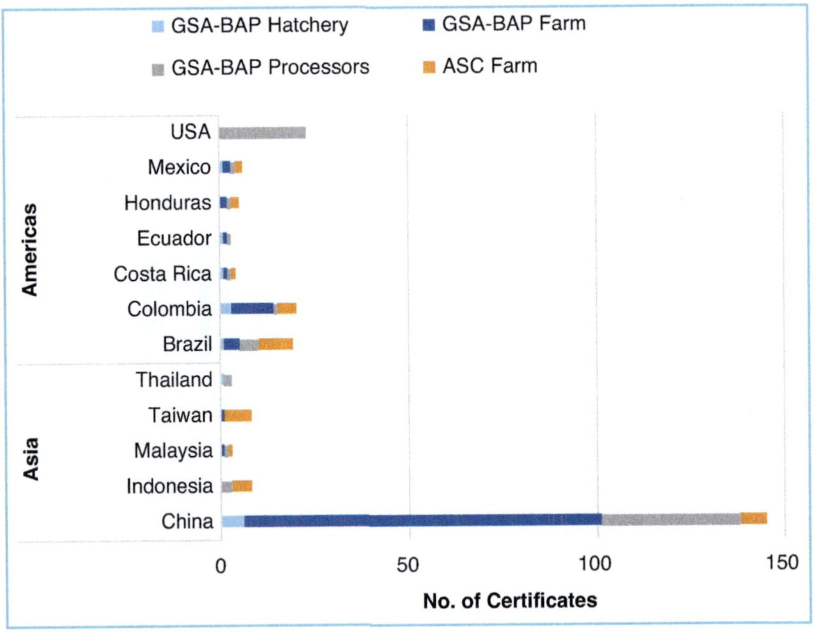

Figure 17.8 Number of GSA–BAP and ASC tilapia-entity certificates and types by country, December 2021 (sources: ASC, 2023b; GSA–BAP, 2023)[4].

In March 2024, ASC announced it would phase out all its certification operations in China (mainly for tilapia, shrimp and bivalve farms; after 'Ocean Outlaw', a US investigative journalism advocacy project, exposed state-sponsored forced labour conditions in seafood-processing plants (Scalia-Bruce, 2024; Urbina, 2019, 2023). Lacking a processing standard, ASC relies on third-party audits of a joint MSC–ASC CoC standard to provide such assurance. Detection failures were attributed to over-reliance on self-assessments in audits under three social standard systems benchmarked as equivalent by the Sustainable Supply Chain Initiative (SSCI) and introduced in 2019: Sedex Members Ethical Trade Audit (SMETA), AMFORI Business Social Compliance Initiative (BSCI) and the SA8000 Standard (White, 2023). The issue highlights a fine strategic balance to be drawn between retention of (1) influence on environmental and social conditions in challenging markets and (2) credibility in the wider market for certified environmental and social standards.

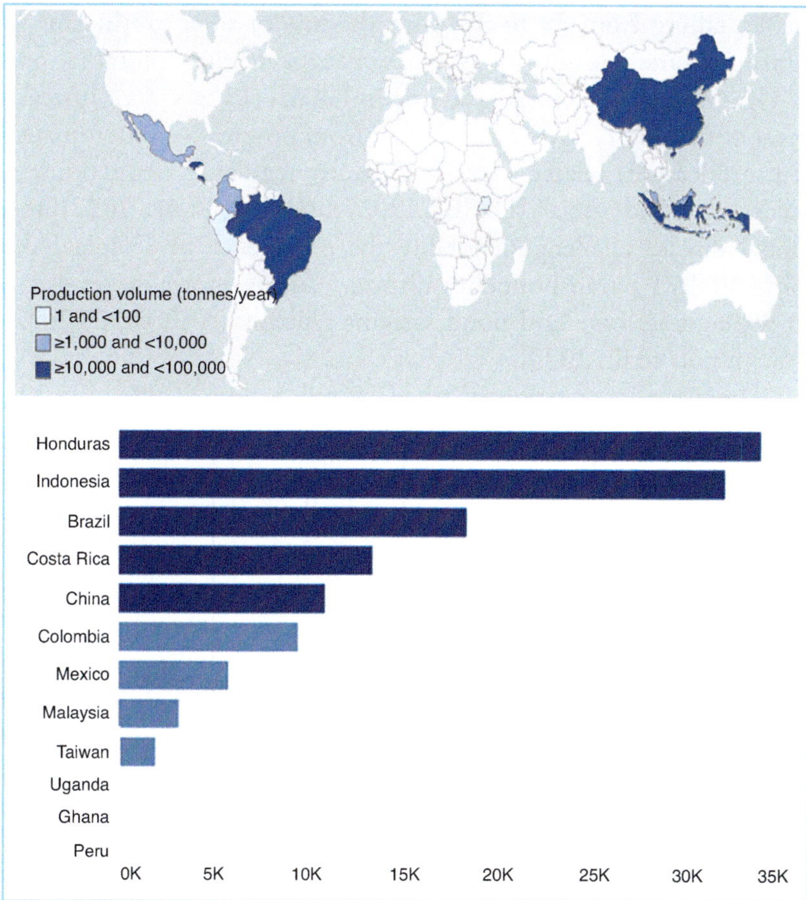

Figure 17.9 ASC certified tilapia production volume (000s t) by country, December 2021 (source: ASC, 2023b).

The ASC decision resulted in cancellation of six Chinese certificates accounting for 12,062 tonnes of tilapia production in 2023 (nearly double the 6482 tonnes certified in 2022). This makes a relatively small contribution to a much greater 63% reduction from the peak of 225,820 tonnes of ASC certified tilapia production in 2021 (Fig. 17.7), down to 96,088 tonnes in 2023. Most of the decline resulted from a 76% reduction (of 91,660 tonnes) in Indonesian certified capacity, augmented by a combined reduction of 30,852 tonnes in Mexican, Brazilian and Malaysian certified production. Over the same period, certificate numbers remained relatively stable, falling from 39 to 38, with the loss of three certificates in Brazil and the addition of two new certificates in Columbia, one in Uganda (Lake Victoria) and one in Ghana (Lake Volta). Instead, most of the observed decline resulted from progressive water quality (eutrophication)-related carrying capacity regulatory reductions in Lake Toba, Indonesia[5] (USSEC, 2017; Tanjung *et al.*, 2022) and global disease challenges, notably the emergence of Tilapia Lake Virus (TiLV) a transboundary disease, causing mortality levels up to 90% across cage and pond systems globally (Aich *et al.*, 2022; Mackinnon et al., 2023).

These factors contributed to an overall decline of 6% (425,293 tonnes) in recorded global tilapia production from its peak of 7,145,207 tonnes in 2021 to 6,721,614 tonnes in 2023. Whilst the coverage of SFW ratings (in just six countries) rose by 9.7% (from 22.7 to 24.9% of global production) over the same period, the combined ASC, GSA–BAP certified total (12–14 countries) fell by 30.3% (125,232 tonnes), from 5.1% (363,455 tonnes) to 3.5% (238,223 tonnes) of global production[6]. With its greater exposure to cage production, ASC accounted for a disproportionate share of this loss, further compounded by its subsequent withdrawal from China in 2024.

Seafood environmental ratings and guidance schemes

Seafood ratings and guidance schemes (SRGS) aim to drive improved performance by increasing consumer awareness of the environmental sustainability of their seafood choices in retail or food service settings, i.e. like certification but aiming to drive ethical consumerism. SGRS generally operate through publication of regularly updated consumer guides based on involuntary assessments of production systems at broader sectoral and regional scales, conducted in-house or by expert consultants. This contrasts with the detailed farm-level, third-party assessments central to elective certification schemes. SGRS recommendations are typically represented as easy-to-understand traffic-light/numerical scales. Overall ratings are further disaggregated across individual sustainability criteria, accompanied by short qualitative summaries for species and culture system combinations.

Rating schemes lack the more direct business (B2B or B2C) focus of certification schemes, aiming to leverage improved farm performance through more advocacy-based influence on consumer behaviour. Unlike certification systems, SGRS also communicate the relative environmental performance of different species and production systems through assignment of rating scores against the same criteria at the sectoral level. Unlike certification systems, this also encourages consumers to make sustainability choices between species and production methods.

Conversely, certification schemes typically assess farm-level compliance against a far more expansive set of process and/or metric performance compliance requirements. To achieve certification, time-bound corrective actions must be made against non-compliances against standard requirements detected at audit. Such requirements are, to varying degrees species- and culture-system specific. GSA–BAP has separate salmon, mollusc and hatchery/nursery standards alongside its general farm standard (including tilapia). ASC currently have 11 species-/system-specific standards (including tilapia) pending completion of their own alignment process for a single farm standard. Standards-setting requirements based on metric limits will often set species and/or system-specific thresholds, even within a single multi-species standard.

As a more process-orientated scheme, i.e. avoiding specification of metric limits, GlobalGAP has the most integrated set of requirements. Its flagship Integrated Farm Assurance (IFA) Standards include base requirements applicable to both agricultural and aquaculture systems, along with more specialized modules and options. These include an 'integrated farm assurance for aquaculture' IFA AQ Standard (covering hatchery and grow-out systems; with '2.5 million tonnes' of '35 species of finfish, crustaceans, and molluscs/seaweed currently under certification'), a Compound Feed Manufacturing Standard (CFM) and CoC standard (GlobalGAP, 2023).

SGRS aggregation of scores across multiple criteria also necessitates subjective weighting of the contribution of individual criteria, e.g. across regulatory, environmental, animal health and welfare or food safety groupings. Although this simplifies communication based on overall scores or traffic-light indications, it also risks the generation of overly generalized and reductive outcomes. Set against this shortcoming, as involuntary systems SGRS also place greater reliance on their advocacy attributes to drive change. Consequently they also lack the multiple assurance levels (Fig. 17.4) required to sustain the credibility of farm-level certification schemes.

Whilst being less costly to implement, the SGRS sectoral focus can fail to adequately acknowledge and incentivize individual farm-level improvements. This is particularly important in fragmented production contexts with high variability in performance levels, i.e. typical of many tilapia sectors. Regional SGRS assessments also typically rely on assessment revisions at a minimum of every 2–3 years, compared with an annual frequency for most certification audits. SGRS also lack any consideration of wider supply chain risks. Consequently, it is increasingly common for rating schemes to acknowledge certified farms in otherwise poorly performing sectors as having, at a minimum, an intermediate amber rating. SGRS tend to rate closed production systems most highly, as best-choice (green) based on their biosecurity attributes, limiting disease and escape risk and their greater ability to concentrate and manage solid wastes.

The SGRS sectoral approach allows for some consideration of the cumulative impacts of aquaculture, e.g. through disease transmission or on water quality at a landscape level, which has proved particularly challenging for farm-level certification. Cooperation is also being extended to AIPs targeting small-scale farming being implemented by many certification schemes. Thus, the two approaches are becoming more complementary hybrid governance components (Vince & Haward, 2019).

Like voluntary certification schemes, SGRS work by influencing demand in markets that are subject to strong civil society pressure, i.e. principally more affluent markets of the global north (Table 17.3).

The attributes of three of the most influential schemes in North American and European markets, the Monterey Bay Aquarium Seafood Watch (SFW), Marine Conservancy Council (MCS) and Sustainable Fisheries Partnership (SFP), are further summarized below.

Table 17.3 Examples of SGRS incorporating aquaculture.

SGRS by region	Scheme output/characteristics
N. America	
Monterey Bay Aquarium Seafood Watch (SFW)	Wild/farmed ratings and advisories
Monterey Fish Market Seafood advisory list	West Coast, USA
Audubon Society's National Seafood Wallet Card (USA)	
Sustainable Fisheries Partnership (SFP)	Wild/farmed ratings (FishSource) & advisories, AIPs
The Environmental Justice Foundation (EJF)	Consumer Guide to Prawns
Sustainable Seafood Canada	*Canada's Seafood Guide* (SeaChoice),
Europe	
Marine Conservation Society (MCS)	*Good Fish Guide*: Wild/ farmed advisories
Good Fish Foundation & WWF Europe	*The Seafood Guide*
National and International	
Greenpeace	*Seafood Red list*
S. Asia	
Private scheme (wild fish)	*In Season Fish: A sustainable seafood guide for India*
SE Asia	
Australian Marine Conservation Society (AMCS)	*Australia's Sustainable Seafood Guide*
Royal Forest & Bird Protection Society of New Zealand	*Best Fish Guide*

Monterey bay aquarium seafood watch (SFW, 2020, 2023)

- Ratings are based on ten separately scored sustainability criteria and sub-factors. Those most directly related to animal health are disease pathogens and parasite interactions, chemical use, escapes and wildlife mortalities.
- The disease criterion offers two assessment options: (1) evidence- or (2) risk-based. The former, which assesses known (or demonstrated lack of) impacts on ecosystems, is allowed only when a separate 'data' score (considering availability, quality and confidence metrics) is above a threshold level. The risk-based option relies on (1) evidence of disease outbreaks on a 'typical'

farm and (2) openness of the culture system as proxies for impact on wild populations.

- Aggregate results are communicated on a three-point scale and traffic-light system: best (green, ≥6.6651 and ≤10); good alternative (orange, ≥3.335 and ≤6.664); avoid (red, ≥0 and ≤3.334). Classifications are further moderated by limits on the frequency 'red' criteria-level scores (≤3.334) and 'critical' scores (very severe impact) on criterion sub-factors.
- Farm-level certification under recognized schemes is positioned as a fourth category (blue), equivalent to at least 'good alternative' (orange) or higher rating.
- Cage systems are conceived as 'worst-case' options relative to pond or raceway systems because of their greater propensity for escapes. Thus, all non-certified open freshwater net pen systems receive at maximum an orange rating. All tilapia systems in China are rated red (avoid) unless third-party certified. Closed-production systems with wastewater treatment are amongst the few farm systems (of any species) to routinely merit a green rating.
- Certification schemes eligible for blue rating (for any production system) include: Naturland Organic (for tropical freshwater fish including tilapia), GSA–BAP (2, 3 and 4 star) and ASC standards.
- Of an estimated 30.5% of globally farmed tilapia rated in 2021 (Fig. 17.5), 5.9, 3.2% and 22.4% were rated blue (i.e. certified), orange and red, respectively (citing evidence that banned antibiotics and antimicrobials are used in the Chinese industry, leading to antibiotic resistance).

Marine conservation society (MCSUK, 2022, 2023)

- The MCS guide rates species and production systems on four criteria: feed, environment, management and fish health & welfare (FHW). Aggregate results are communicated on a five-point scale: 'sustainable choice' (1–2/green), 'needs improvement' (3–4/orange) and 'avoid' (5/red).

- Farmed tilapia (all species and production systems) with ASC certification is rated 2 (good choice). Other certification schemes are not explicitly covered, although ASC is likely to account for much of the very low-certified volume entering MCS's focal UK market. Otherwise, all open net pen and freshwater pond production in Asia (including China) is rated 4 (needs improvement).
- Humane slaughter is, as yet not an explicit welfare requirement within ASC standards. Thus, any production from countries lacking national animal welfare legislation currently elicits a red rating on the MCS FHW criteria.
- Risk of (fish-borne zoonotic) parasite transference is cited in guidance as an issue in China, along with overuse of antibiotics (including those critical for human health), linked to antimicrobial resistance, and risk of escapes in flood-prone areas.

Sustainable Fisheries Partnership (SFP, 2023)

- SFP provides a rating of the status of fisheries, fish stocks and aquaculture in their online 'FishSource' database (based on assessment of a subset of MSC standards and sectoral reports using publicly available scientific and technical information). The primary focus is on seafood entering US and European markets.
- The database includes assessments of 'aquaculture management units' (AMUs) for tilapia culture – currently two in Guandong and Hainan Provinces of China and six in Indonesia; also in central and west Java, north, south and west Sumatra and north Sulawesi.
- AMUs are operationally defined as 'the primary unit within which aquaculture management practices should be coordinated across a group of farms to mitigate against cumulative impacts and shared risks'.
- AMUs level recommendations are further linked to SFP AIPs incorporating area-based management and smallholder group certification.

Table 17.4 Global volume of farmed tilapia and percentage by species in 2021 (FAO, 2023).

SN	Species	Latin name	Volume (t)	%
1	Nile tilapia	*Oreochromis niloticus*	5,540,836	77.5
2	Tilapias nei	*Oreochromis* (=*Tilapia*) spp.	1,101,560	15
3	Blue Nile tilapia, hybrid	*Oreochromis aureus x O. niloticus*	410,553	6.6
4	Mozambique tilapia	*Oreochromis mossambicus*	74,435	1.2
5	Three-spotted tilapia	*Oreochromis andersonii*	4793	0.08
6	Tilapia shiranus	*Oreochromis shiranus*	4711	0.08
7	Blue tilapia	*Oreochromis aureus*	3100	0.05
8	Redbreast tilapia	*Tilapia rendalli*	2999	0.05
9	Longfin tilapia	*Oreochromis macrochir*	1800	0.03
10	Sabaki tilapia	*Oreochromis spilurus*	300	0.005
11	Blackchin tilapia	*Saratherdon*	95	0.002
12	Mango tilapia	*Oreochromis galilaeus*	19	0.0003
13	Redbelly tilapia	*Tilapia zilli*	6	0.0001
	Total		**7,145,207**	**100**

Environmentally certified and rated tilapia species

A total of 12 different tilapia species were recorded as being farmed in 126 countries by FAO in 2021 (Table 17.4). One species, Nile tilapia (*O. niloticus*), accounted for at least 75.5% of recorded volume. All production certified by ASC and GSA–BAP in 2021 with species indication (omitted in 14% of ASC and 5% of GSA–BAP certificates) was exclusively for Nile tilapia (Table 17.5). These two standards accounted for 3.2 and 2.9% of global farmed tilapia volume, respectively, and together up to 3.78% of global volume farmed in 12 countries in 2021.

In the same year, the SFW rating scheme encompassed 23% of global tilapia volume. In 2023, SFW assessments resulted in 29 recommendations differentiated by species and culture methods and comprehensive sectoral ratings in six countries. Closed production systems (with wastewater treatment) worldwide are rated green (best choice, pond and net cage production in Colombia, Honduras, Indonesia, Mexico and Taiwan are rated amber (good

Table 17.5 Total volume (tonnes) of certified and rated farmed tilapia by scheme and as a percentage of global farmed volume in 2021 (ASC, 2003b; CRC, 2023; FAO, 2023).

Species	ASC (t)	ASC %[a]	GSA-BAP	BAP %[a]	SFW (t)	SFW %[a]	SFP/ AIP[b]	SFP/ AIP[b] %[a]
Nile tilapia	194,046	2.7	199,146	2.7	1,081,221	18.9	100,000	1.6
Tilapias nei	31,774	0.5	10,663	0.2	127,812	2.4		
Blue Nile tilapia, hybrid					414,000	5.7		
Total	**225,820**	**3.2**	**209,809**	**2.9**	**1,623,033**	**23**	**100,000**	**1.4**

[a]Percentage of total global farmed production of all species: 7,145,207 mt
[b]Sustainable Fisheries Partnership, Aquaculture Improvement projects

alternative) and all Chinese production is rated red (avoid), with the exception of production certified under specific standards (see below)). Despite its higher volume coverage compared with voluntary certification, the SFW focus on production destined for the US market also precludes major global tilapia sectors from any assessment (including eight countries with ASC- or GSA-certified farms in 2023).

Species covered include *O. niloticus*, *O. mossambicus*, *Oreochromis niloticus* × *Oreochromis aureus* hybrids, *O. spilurus*, *O. urolepsis* as well a general *Oreochromis* spp. category. The Chinese farmed tilapia sector, mostly concentrated in the four southern provinces of Guangdong, Hainan, Fujian and Guangxi, also includes 410,553t of low-temperature-tolerant Blue Nile tilapia hybrid (*O. aureus* × *O. niloticus*), *O. niloticus* (largely genetically improved farmed tilapia, GIFT) and its hybrids.

SFW recommendations currently acknowledge three 'robust' aquaculture certification schemes, internally benchmarked for equivalence against SFW criteria: Naturland Organic standards (with separate standards for ponds and cages) and ASC and GSA–BAP farm standards, i.e. also the principal schemes operating in the US market. Recognition is for all production systems and species certified under these schemes worldwide.

Development of a separate (more agnostic) SFW blue 'certification' category superseded an earlier approach of directly benchmarking standards against the SFW criteria with overall assignment to one of the three SFW traffic-light categories. Previously cage systems were more likely to be downgraded due to the higher inherent risk of escapes, together with associated chemical use and disease transmission risks.

All production from China not currently certified under one of these three certification schemes is assigned an overall red (avoid) sectoral rating, with red assignments against two criteria, habitats and disease, and critical assignments against two further criteria, escapes and chemicals. Many of these farms operate low-lying repurposed shrimp ponds, abandoned in the 1990s due to disease and having water eutrophication problems, and thus exposed to ongoing flooding risk (SFW, 2018).

Benchmarking, equivalence and harmonization

A range of 'benchmarking' schemes has emerged to support and promote certification as a means of improving seafood sustainability. These are designed to calibrate equivalence between third-party audited standards and foster greater harmonization against the intent of normative guidelines underpinning their collective development. They are framed as a means of building consumer confidence in response to confusion associated with the proliferation of eco-labels and/or as a mechanism for reducing barriers to trade by lowering the costs of entry to markets requiring certification under multiple overlapping schemes.

Benchmark initiatives take various forms associated with different objectives including: (1) internal transitions from individual to more 'aligned' multi-species and system farm standards; (2) recognition between third-party audited standards, usually partial, e.g. covering feed sourcing, CoC requirements, etc.; (3) recognition between third- and second-party (e.g. national or producer organization) standards enabling the latter to become a gateway to third-party certification with greater market appeal; and (4) independent benchmarking schemes for environmental, social and food safety standards (Table 17.1).

Table 17.6. Animal health ECs and SCs in the GSSI Global Benchmark Tool v.2.0.

Type	Essential or supplementary component
EC	The decision to treat with antimicrobial agents, and their subsequent application, is consistent with the Principles for Responsible & Prudent Use of Antimicrobial Agents in Aquatic Animals and other guidance of the OIE Aquatic Animal Health Code (2015 Articles 6.2.7 and 6.2.8), i.e. by the aquatic animal health professional or other relevant competent authority and in response to a diagnosed disease. (C1.01)
SC	The standard prohibits use of antimicrobials listed by the World Health Organization (WHO) as highly and critically important to human health. (C1.01.01)
SC	The standard prohibits aquatic animals treated with antimicrobials from being labelled with its standard; however, applications deemed necessary by an aquatic health professional cannot be withheld solely to preserve certification. (C1.01.02)
EC	Workers with responsibilities in aquatic animal husbandry have been adequately trained and are aware of their responsibilities in aquatic animal health management. (C1.02)
EC	Aquatic animals are kept under conditions suitable for the species being farmed. (C1.03)
EC	The facility establishes, implements and maintains appropriate procedures to respond to disease outbreaks. (C1.04)
EC	The facility establishes, implements and maintains appropriate procedures for the early detection of aquatic animal health issues, including routine monitoring of stocks and the environment. (C1.05)
EC	Mortalities and moribund aquatic animals are routinely collected, where collection is feasible. (C1.06)
EC	The facility has operational fish health management practices addressing the following elements (where relevant to the species, scale and production system covered by the Standard's scope): (1) effective biosecurity; (2) identification and use of suitable available vaccines; (3) introductions and transfers of farmed animals, overseen by an aquatic animal health professional. (C1.07)
SC	The facility reviews the Aquatic Health Management Plan. (SC1.07.01)
SC	The facility determines cause of death when losses are significantly greater than expected and cause is unclear, and to use laboratory analysis where feasible. (SC1.07.02)
SC	The facility establishes, implements and maintains a written Aquatic Animal Health Management Plan (AAHMP), overseen by an aquatic animal health professional, at a minimum, that is compliant with the following Essential Components: C.1.01–C.1.10. (SC1.07.02)

Continued

Table 17.6 Continued

Type	Essential or supplementary component
SC	The written health plan (SC1.07.03) includes an emergency response protocol in the event of an invasive disease, which includes depopulation where appropriate; a written list of all diseases that the aquatic animals are likely to face during production; and annual/end of production review and failure analysis. Where multiple effective chemical treatments are available, and while maintaining compliance with the OIE Prudent Use guidance, there is a rotation to reduce the risk of resistance. (C1.07.04)
SC	Aquatic animals are vaccinated against all relevant/important diseases for which vaccines are both available and effective. (SC1.07.05)
SC	There are suitable performance-based metric limits on survival rate (or a similar system that incorporates survival rates, e.g. recovery rate) or similar criteria that demonstrate that aquatic health management practices are effective. (C1.07.06)
SC	There is a legally binding, appropriately defined and operational area management system in place that ensures that all participant aquaculture facilities use common and, where applicable, coordinated, practices for the shared management of aquatic animal disease risk. (C1.07.07)
EC	The facility establishes and implements procedures for the disposal of mortalities using appropriate methods that prevent the spread of disease. (C1.08)
EC	The facility establishes, implements and maintains appropriate procedures and/or systems to reduce the likelihood of disease and parasite transmission within the aquaculture facility and between it and natural aquatic fauna. (C1.09)
SC	Where the production system allows the discharge of parasites of known concern to local wildlife, the facility monitors and adapts farming practices based on parasite prevalence on wild fish. (C1.09.01)
EC	The facility maintains records on veterinary drug and chemical usage and the rationale for their use [C1.10]
EC	Establishment, implementation and maintenance of an appropriate system for the application of chemicals and veterinary drugs. (C2.01)
EC	Appropriate controls for all chemicals, including veterinary drugs, that enter the environment during or after use (whether covered by GSSI EC or not) to minimize adverse impacts on environmental quality. Manufacturer's guidance or equivalent directions should be followed and, where appropriate, relevant examples of chemicals that pose a high risk of adverse impacts to environmental quality should be specifically defined by the standard. (C2.02)
SC	The standard prohibits the use of chemicals within the aquaculture facility that may enter the local environment due to farming practices that are listed as highly polluting by relevant organizations or other justification. (C2.02.01)
SC	The standard requires that chemicals used on the aquaculture facility, and that may enter the local environment, are restricted to identified environmentally benign products (e.g. rapidly denaturing chemicals), with a suitable justification for their listing as benign. (C2.02.02)

Continued

Table 17.6 Continued

Type	Essential or supplementary component
EC	The standard requires that the aquaculture facility and its daily operations ensure that good culture and hygienic conditions are maintained. Relevant aspects include proper management of all chemicals, fuels and feeds including their safe storage. (C3.01)

Examples of the latter include the Global Seafood Sustainability Initiative (GSSI), the Global Food Safety Initiative (GFSI), the Global Social Compliance Programme (GSCP) and the Sustainable Supply Chain Initiative (SSCI). These variously address the environmental, social, food safety and audit process requirements of voluntary certification systems.

Established in 2013, the GSSI is an industry-sponsored, public–private partnership that has developed the leading benchmark scheme for (capture and farmed) seafood environmental certification schemes (UN DESA, 2023). Based on normative internationally negotiated and agreed (FAO) codes and guidelines (FAO, 1995, 2009, 2011), the initiative was designed to 'support greater alignment of a growing number of certification schemes, accelerate sustainability commitments and to reduce market failures and duplicate compliance costs associated with expansion of the sector'. The GSSI Global Benchmark Tool (GBT; GSSI, 2023a) covers fisheries and aquaculture production standards as well as scheme governance and operational management systems. More recently GSSI collaborated with the Consumer Goods Forum SSCI on the development of a Social Benchmark Tool.

The GSSI benchmarking approach was designed as a pre-competitive initiative, with the intent to limit scope for scheme differentiation. Recognition (against v.2.0) is granted based on alignment with a core of 175 'Essential components' (ECs), including 40 aquaculture environmental ECs, grouped under nine performance areas including animal health (Table 17.6). Alignment decisions against a further suite of 40 generally more demanding 'Supplementary Components (SCs)' are documented in a downloadable benchmark report but are not itemized by GSSI for any further recognition purpose. In early applications, scheme operators were inclined to omit evidence for compliance against ECs, although more recent submissions for determination of 'on-going alignment' suggest that schemes are becoming more focused on their attainment as a basis for self-claims around their differentiation from competing standards. This may lead to greater discrimination between more 'process'- and 'metric'-orientated approaches, as the ECs are home to most of the more quantitative benchmark requirements.

Although GSSI does not aim to differentiate the performance of aligned standards, founder seafood industry partners initially hoped to bolster the competitiveness of emergent national fisheries certification schemes, particularly those for major fisheries in Alaska and Iceland through benchmarking alongside the dominant MSC eco-label. Established in 1997, MSC has achieved significant market share, aided by its first mover advantage and eNGO provenance. This has proved less of an issue in the aquaculture sector, where the three or four dominant third-party schemes emerged more contemporaneously, with ASC as the latest arrival.

Over the decade since the launch of GSSI, nine voluntary environmental schemes have been successfully benchmarked against the GBT (GSSI, 2023b). These include the more robust and better-resourced voluntary third-party schemes capable of achieving recognition, providing them with further opportunity to secure their market position. Six are for third-party schemes: three for aquaculture (GSA–BAP, ASC and GlobalGAP) and three for fisheries (including MSC), with the balance a combination of second-party national schemes for fisheries (including Alaskan and Icelandic schemes) and/or aquaculture. In 2020 the Alaskan Responsible Fisheries Management (AKRFM) scheme transitioned from a PO second-party to a private not-for-profit third-party standard as a strategic means of broadening its market appeal (ASMI, 2020).

As involuntary schemes, SRGS have (as yet) not been subject to formal benchmarking themselves. However, they play an important role in shaping equivalence and harmonization of voluntary certification schemes in at least two ways. First, SRGS are typically commissioned by eNGOs. Consistent with their advocacy mandates, they are also some of the most active and influential stakeholders in the ongoing revision of voluntary private standards, particularly those with strong commitment to multi-stakeholder engagement (Fig. 17.3). Second, whilst their sectoral scope encompasses significant volume, they are unable to discriminate between good performers in otherwise poorly performing sectors. Thus, recognition of farm-level certification within these domains (such as the SFW recognition of ASC, GSA–BAP and Naturland schemes) provides a missing feedback loop for incentivization of individual improvement. Larger standards bodies are also collaborating on recognition of equivalence between elements of their own standards, e.g. on sourcing of sustainable feed ingredients.

Animal health and welfare requirements in aquaculture certification schemes

Risk of disease transmission has been a primary driver in the development of aquaculture environmental standards. Animal health requirements are primarily intended to mitigate the impacts of farms on wild populations and ecosystems, as well as contributing to food safety and human health assurance. Any economic benefits to farmers then accrue indirectly. Some specialist standards such as the Royal Society for the Prevention of Cruelty to Animals (RSPCA) welfare standards for farmed Atlantic salmon (RSPCA, 2021) do position health and welfare of the cultured species more directly as a primary concern. Mainstream standards are also following this trend, facilitated by considerable overlap with requirements focused on the health of wild species, making these objectives mutually re-enforcing. Requirements under different environmental criteria can also work in an integrated manner. For example, those addressing farm escapes, primarily intended to limit the risk of hybrid introgression or displacement or habitat modifications of wild species, also serve to limit pathogen amplification and transmission to and from farms.

In this section we summarize the animal health-related content of the mainstream aquaculture environmental standard schemes through a review of relevant components of the GSSI Global

Benchmark Tool (GSSI, 2023a). Of the nine certification schemes to be recognized by GSSI to date, five include farm standards: ASC (salmon and shrimp only; ASC, 2023c), GSA–BAP (farm, salmon, mollusc; GSA–GSA–BAP, 2023), GlobalGAP (IFA Aquaculture; GlobalGAP, 2023), Bord Iascaigh Mhara (BIM) Ireland (CQA salmon and shellfish; BIM, 2022) and MEL Japan (Aquaculture and Fisheries; MEL, 2018). Of these only the GSA–BAP multi-species 'Farm' standard covers tilapia environmental requirements. ASC with its multi-species standards elected to limit assessments to their high-certification-volume salmon and shrimp standards pending release of their 'aligned' (multi-species) standard.

GSSI Benchmark environmental components are differentiated as 34 essential (EC) and 46 more demanding supplementary (sub) components (SCs). GSSI recognition requires, at a minimum, alignment with all ECs. Table 17.6 lists the ECs and SCs of most direct relevance to animal health and welfare. These are positioned in three of a total of nine GSSI performance areas: (1) aquatic animal health management (AAHM), (2) chemical and veterinary drug use and (3) environmentally responsible construction, waste disposal and storage.

With 20 components (ten ECs and ten SCs), AAHM is also the largest performance area. Components cover prudent use of antimicrobials to ensure animal welfare whilst minimizing antimicrobial resistance risk, health management standard operating procedures (SOPs) and linked training covering disease prevention (including vaccination, diagnostics and prevention of on- and off-farm pathogen and parasite transmission), surveillance and response measures.

ECs are predominantly process based, acknowledging the split between more process-oriented (e.g. GlobalGAP) and metric-based standards (e.g. ASC, GSA–BAP). Only one supplementary health component (C1.07.06) imposes a requirement for performance-based metric limits on survival rates. Area-based (health) management requirements are also limited to one SC, SC1.07.07 (Table 17.6).

Requirements under the other two performance areas cover systems for application and control of chemicals and veterinary drugs and exclusion of more environmentally harmful and persistent chemicals.

The ASC Tilapia Standard (v.1.2 ASC, 2019), the only mainstream standard focused entirely on tilapia (in the genera *Oreochromis* and *Tilapia*) contains 11 indicators under a principle titled 'Manage fish health and welfare in an environmentally responsible manner.' These are grouped under four criteria (A–D) itemized below.

Using the OECD PSR framework (OECD, 2013b), environmental indicators can be further classified as indicators of anthropogenic 'pressures' (P) on the environment, the 'state' (S) of the environment itself or societal 'responses' (R) designed to modify and limit the potential impact of pressures (Bridson *et al.*, 2020). In this schema, a state might be a low pathogen or disease prevalence rate or water-body trophic status to be conserved, a pressure might be a survival rate (as an indicator of disease risk) or nutrient load with associated control limits, whilst response indicators, such as a farm health plan or waste treatment system, reduce the risk and/or severity of a pressure impact in a non-specific manner.

PSR indicators can also operate at near- or far-field scales, at the site or wider ecosystem 'resource' level. Of the ASC FHM indicators listed above, those addressing prudent use of antibiotics operate at the widest, potentially global ecosystem level.

Thus the 11 ASC FHM indicators (prefixed P, S or R) listed below include six pressure and five response, but no state, indicators. This contrasts with the ASC Salmon Standard, which requires adaption of farming practices in response to monitoring of sea louse prevalence on wild fish. Unlike most other ASC standards, beyond the requirement for general legal compliance, the Tilapia Standard also lacks a specific requirement for juveniles or broodstock to come from sources meeting regional, national and international importation guidelines (e.g. WOAH and ICES) with appropriate disease-free status. Greater focus is placed on preventing further introductions of tilapias outside their native range, with the most conservative transboundary restrictions set at watershed level in Africa.

Under a separate principle, 'Conserve species diversity and wild populations', the Tilapia Standard also specifies a minimum 95% of males or sterile fish in a culture unit (i.e. cages or ponds), to mitigate the risk and impact of escapes. This effectively limits certification to more intensive production systems, precluding all mixed-sex production, which remains a prevalent method for smallholders practising semi-intensive polyculture in fertilized ponds in many lower-income countries or countries with emergent tilapia sectors. These low-input out systems have many favourable sustainability traits that differentiate them from more intensive systems (Belton *et al.*, 2010).

Health requirements in the ASC Tilapia Standard (v.1.2) classified as P, R or S indicators:

A. Stock recovery (one indicator):

1. (P) ≥ 65% of fish stocked in production stages after attaining a size of 100 g.

B. Chemicals (seven indicators):

2. (R) Health records proving that all therapeutants used were prescribed by a veterinary or accredited fish health professional.
3. (R) No use of chemicals or therapeutants banned in the importing or producing country.
4. (P) No prophylactic use of antibiotics.
5. (P) No use of antibiotics critical for human medicine, as categorized by the WHO.
6. (P) Calculation and verification of the total weight (kg) of each antibiotic (active ingredient) used per tonne of fish produced per year (no limit currently imposed).
7. (P) A maximum of ≤3 separate antibiotic treatments (per diagnosed disease) over the most recent production cycle.
8. (P) A minimum hold time of ≥48 hour before any water in which fish have been fed with feed containing methyl or ethyl testosterone can be released.

C. Mortalities (two indicators):

9. (R) Records demonstrating prudent removal of mortalities (minimum daily).
10. (R) Records demonstrating prudent disposal of mortalities.

D. Health management (one indicator):
11. (R) Implementation of a site-specific fish health plan.

Performance of farms certified under the ASC tilapia standard against animal health requirements

The evidence base for assessing the impacts of voluntary standards remains limited. Further context-specific empirical studies are required to draw robust generalizations (FAO, 2014; Naylor *et al.*, 2021). The strongest proof of impact occurs where quantitative data are available to evaluate state indicators (at the site or ecosystem level), although no such indicators were included in the ASC FHM Principle. Management (pressure and response) indicators are more amenable to qualitative assessment (Bridson *et al.*, 2020).

In this section we analyse 354 non-conformities detected against environmental and animal health-specific indicators in the ASC Tilapia Standard over the course of 155 audits of 55 export-orientated tilapia farms between August 2012 and March 2019. A total of 13 (22%) of these sites were audited under six group certificates, with membership ranging from two to a maximum of four farms.

Data on indicator non-compliances, along with farm characteristics including location, production system and scale as ex post explanatory independent variables, were extracted from published audit reports (ASC, 2023b), cleaned and coded for analysis in an ACCESS relational database. To contextualize health management within wider environmental performance, NCs were first grouped across the seven Tilapia Standard 'Principles' (Fig. 17.12) including FHM. This was followed by a more granular analysis of NCs on eight of the 11 indicators under the FHM principle, four of these addressing antibiotic use (Fig. 17.14).

Countries, production systems and volumes: Audits took place in 11 countries: China, Taiwan, Indonesia, Malaysia, Brazil, Colombia, Mexico, Ecuador, Belize, Costa Rica and Honduras. Of these, eight were classed as upper-middle-income countries (UMICs), two as lower-middle-income (LMICs; Honduras and

Indonesia) and two as higher-income (HICs; Taiwan and Poland) (World Bank, 2023). Production systems were mainly ponds (51% of sites) and reservoir cages (42% of sites). The balance of four farms operated nursery ponds, reservoir pens, river cages and a single RAS system (in Poland). Thirty-seven farms in five South Asian countries accounted for 61% of sites and 38% of total production capacity. A further 22 farms in seven Central and South American countries accounted for 37% of sites and 61% of capacity (the RAS accounting for the balance). Three countries with large reservoir cage sites – Honduras (two farms), Indonesia (eight farms) and Mexico (three farms) – together accounted for 50% of an estimated mean total annual production capacity of 127,442 tonnes for the audited sites (Fig. 17.10).

All ASC certification of Vietnamese tilapia production ceased following the loss of six certificates between 2019 and 2021, for reasons that are unclear.

Farm scale: Five reservoir cage farms in Honduras (two farms), Mexico (three farms) and a single pond farm in Costa Rica with an area of 197 ha had the largest mean annual production capacities,

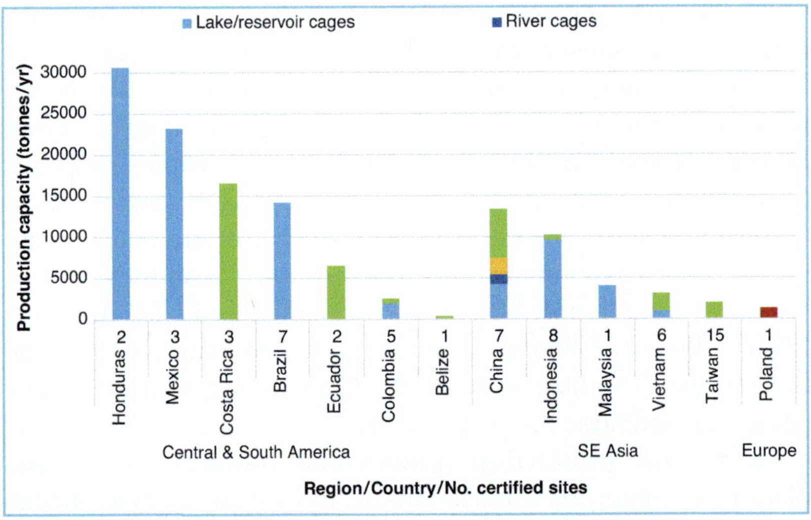

Figure 17.10 Total production capacity of ASC-certified tilapia farms by country and culture method; mean tonnes per year, August 2012–December 2019.

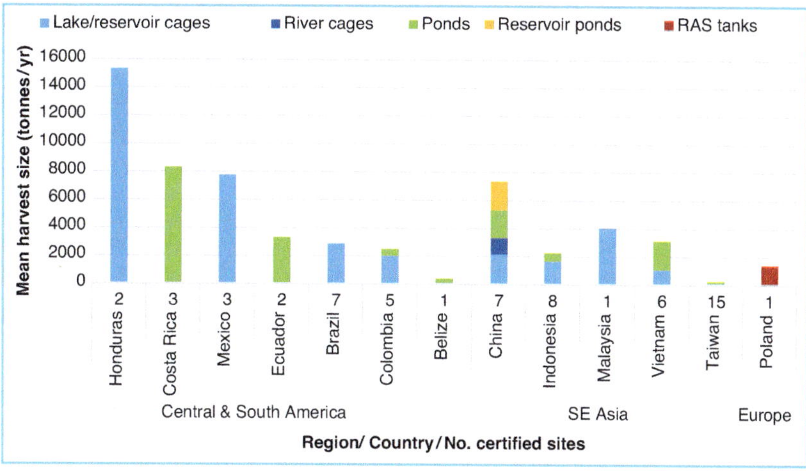

Figure 17.11 Mean annual production capacity of ASC-certified tilapia farms by country and culture method; mean tonnes per year, August 2012–December 2019.

ranging from 7708 to 15,296 tonnes per year (Fig. 17.11). The smallest sites to be certified were pond farms in Taiwan (15 farms, mean 194 tonnes per year, SD 116 tonnes per year), Belize (one farm, 335 tonnes per year) and Indonesia (one farm, 593 tonnes per year, part of a larger reservoir-cage operation on Lake Toba). The smallest reservoir cage site was in Colombia (Betania Reservoir, 484 tonnes per year). China had the greatest diversity of system types, but all were at intermediate 'medium' production scales ranging from 1149 to 2100 tonnes per year, i.e. lacking any smallholder certification.

The smallest farms in the sample were pond sites operated by members of Taiwan Nan Ying Aquaculture Association (NYAA), a community non-profit organization established in the 1970s to support small-scale producers through 'technical and management assistance and production control and monitoring' (Taiwan Today, 2013). The farms are co-located within a dense zone of continuous ponds culturing tilapia, milkfish, shrimp, carp, mullet and sea bass, some 12 km from the south-west coast in Syuejia District adjacent to Tainan Municipality. They are contracted as outgrowers to supply two processors in Tainan City (also NYAA members) mainly producing frozen whole tilapia for export to the USA. The farms, which have also held GlobalGAP certification, are mostly family owned and operated. Most operators were older locals following a wider trend in economic out-migration of younger generations from rural areas. Farm sizes ranged from three to 28 ponds of 3.3–21.2 ha culture area, operated by between one and seven workers as semi-closed monocultures with partial recirculation.

NCs against ASC indicator requirements: Smaller and medium-scale farms in Taiwan (ponds), China (mixed systems) and Malaysia (reservoir cages) recorded the highest mean number of NCs per audit across the seven environmental principles, with 6.6, 4.3 and 3.2 NCs per audit, respectively (Fig. 17.12), and 1.2, 0.47 and 0.83 NCs against the FHM principle alone (Figs. 17.12 and 17.14). Taiwan and Malaysia also recorded the highest share of FHM NCs as a proportion of all NCs for any country – 18 and 26%, respectively. For the remaining countries the proportion ranged from 0% in Honduras (reservoir cages), Costa Rica (ponds)

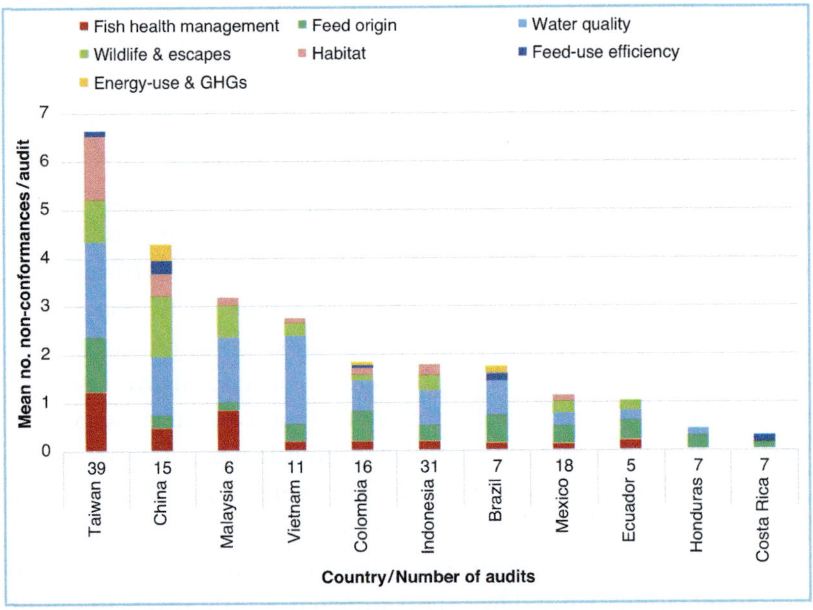

Figure 17.12 Mean NCs per audit by country and ASC environmental principle, August 2012–Decembedr 2019 (155 audits and 354 NCs).

and Poland (RAS) to 11% in China, Colombia and Indonesia (Fig. 17.12).

These findings indicate a stronger correlation between health outcomes and farm scale relative to other independent variables, e.g. production system type, country/region or outcomes against other environmental principles such as water quality. These deficits also meant greater scope for improvement in the performance of smaller relative to larger farms. By their second year of ASC certification, the first NYAA members to be certified in 2013 had achieved a marked reduction in the large number of NCs detected in the initial audits (Fig. 17.12). At the time of writing, this performance level had been maintained following a fourth consecutive 3-year cycle of certification.

Prevalence of antimicrobial treatments: Some 16% of farms (nine of 55) reported treatment with antibiotics in 12% of all annual audits (18 of 155). This was limited to farms in only four of 11 countries: Malaysia (one cage farm), Costa Rica (three pond farms), Taiwan (three pond farms) and Ecuador (two pond farms). Country differences may be indicative of effective regulatory and national residue-testing programmes and/or undetected farm under-reporting.

Most audit reports failed to record the pathogen responsible for triggering antimicrobial treatments. The single cage farm in Malaysia reporting the highest recurrent use rates (see below) cited *Streptococcus* infection.

The type of antibiotic active ingredient(s) used was specified in 74% (14) of audit reports for farms reporting treatments. Four different active ingredients were applied in the following combinations: oxytetracycline (seven audits in Costa Rica and one in Malaysia), oxytetracycline and amoxicillin (three in Taiwan) and amoxicillin, oxytetracycline, enrofloxacin and florfenicol (one in Malaysia). All the above are broad-spectrum antibiotics routinely used in aquaculture, subject to country-specific regulation including appropriate withdrawal periods.

Mean application rates per unit harvest ranged from 0.022g per tonne (Ecuador) to 2.11 kg per tonne (SD 3.34 kg per tonne, Malaysia), with the rate in Malaysian cages more than double the highest pond application rate recorded in Taiwan (Fig. 17.13). The

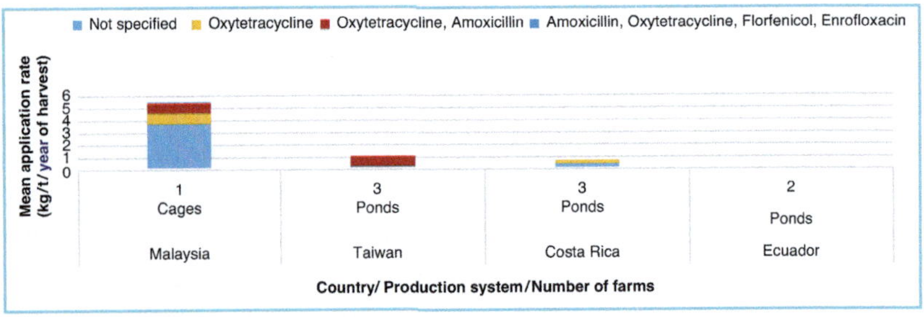

Figure 17.13 Mean antibiotic application rates by country and year, as grams active ingredient per tonne tilapia harvest per year, August 2012–December 2019.

Malay result was skewed by a single outlier of 8.89 kg per tonne recorded in 2015, but was also responsible for recorded inputs of 0.85–1.14 kg per tonne over six consecutive audits. Lower survival rates during severe disease episodes triggering treatments will also further amplify the resulting metric ratio.

Imprudent antimicrobial use: Of a total of 46 NCs assigned to indicators under the FHM principle, 15% (seven NCs) related to the imprudent use of antibiotics (Fig. 17.14). Of these, three were due to failure to calculate the quantity of active ingredient used per unit harvest as per requirement (Taiwan ponds),

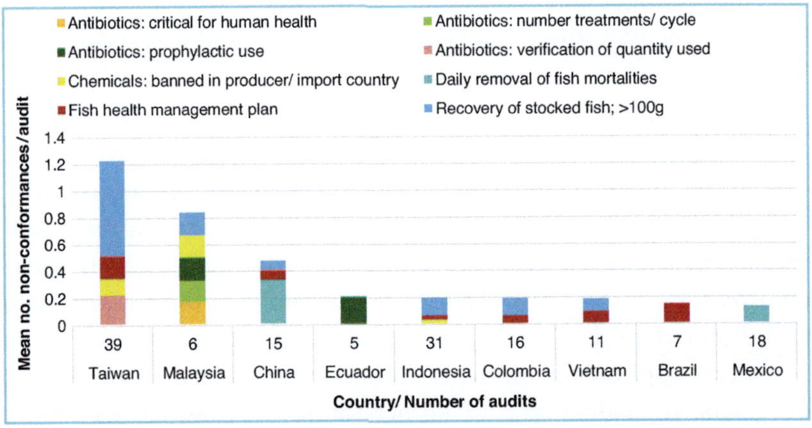

Figure 17.14 Mean NCs per ASC Tilapia Standard audit on FHM criteria by country, August 2012–December 2019 (155 audits and 46 NCs).

two for prophylactic applications (Ecuador ponds and Malaysia cages), one for exceeding the maximum limit for three treatments allowed per cycle (Malaysia cages) and one for use of amoxicillin trihydrate (Malaysia cages) mistakenly interpreted by the auditor as belonging to one of the classes of antimicrobials classed as critical to human health by WHO (FAO, 2008; WHO, 2018). The critical list refers more specifically to 'aminopenicillins with betalactamase inhibitors' (e.g. amoxicillin-clavulanic-acid). A correct conformance decision was assigned to the amoxicillin treatments reported in Taiwan. Inevitable variability in the capacities of auditors, particularly in such specialist areas, highlights the concomitant need for effective internal and external scheme assurance mechanisms.

A further seven NCs were assigned against the use of unsanctioned therapeutants (types not specified), five of these for pond farms in Taiwan. As participants in Taiwan GaP (a national voluntary certification scheme), members of the NYAA routinely underwent pre-harvest testing of fish samples by the National Aquaculture Products Inspection Centre, hosted by National Taiwan Oceanography University. Samples were tested for the following panel of antibiotics and their metabolites: chloramphenicol, thiamphebnicol, florphenicol, florphenicol-amine, semicarbazide (SC), 1-aminohydantoin (AH), 3-amino-2-oxazolidinone (AOZ) and 3-amino-5-morpholinomethyl-2-oxazolidinone (AMOZ), as well as malachite green and leuco-malachite green.

Six further NCs across multiple countries were assigned to deficiencies detected in the farms' fish health management plan, and a further six were assigned due to insufficiently frequent removal of mortalities, five of these for pond farms in China (Fig. 17.14).

Recovery rates: Overall, the largest number of FHM NCs (20; 44% of the total) was assigned to recovery rates below the minimum requirement of 65% of fish stocked after attaining a size of 100 g. Of these, 11 were assigned to pond farms in Taiwan followed by four for reservoir cage farms in Indonesia.

Mean reported recovery rates for fish >100g ranged from maxima of 97–98% for ponds in Taiwan (in contrast with the poor performance of individual farms under the group certificate in early

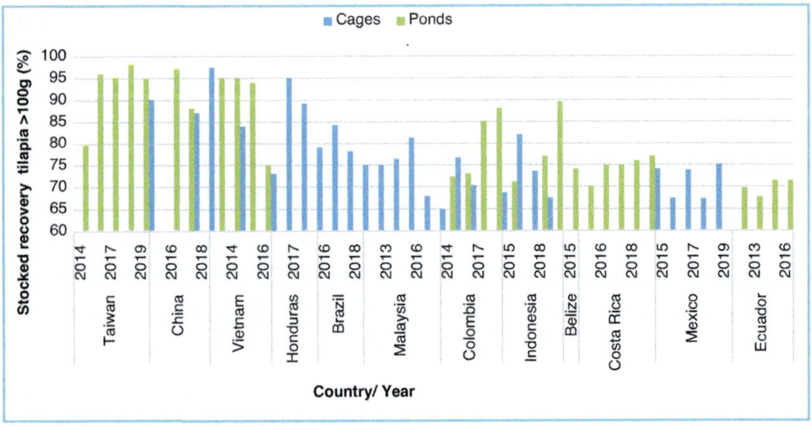

Figure 17.15 ASC FHM metrics; mean percentage stocked recovery, August 2012–December 2019.

audits), China and Vietnam to minima of 65–67.7% for cages in Colombia, Indonesia, Mexico and Ecuador (Fig. 17.15). Results indicate marginal progressive improvements for cages in Ecuador and ponds in Costa Rica and Taiwan with more variable trends in other cases. Further overall declines in survival rates are likely to have accompanied water quality- and disease-related declines in ASC certification levels between 2021 and 2023 (particularly in cage systems).

Sex reversal rates: Mean sex-reversal rates (data not shown) were lowest for cage systems in Malaysia, averaging just 95%, followed by Vietnam (96.4%) and Indonesia (97.6%). Superior levels were reported for pond production systems, i.e. where routine achievement of rates > 98% is also most critical to economic performance. The highest such (pond) rates were achieved in Vietnam (98.1%), followed by Indonesia (98%) and Taiwan (97.8%).

Counterfactual analysis: Assurance programmes should also evaluate reasons for indicators with consistently low NC rates; over extended periods, variable production conditions and geographies. Such deficits may also be indicative of auditability factors rather than superior performance.

In this analysis, no NCs were assigned against three of the 11 FHM indicators: (2) prescription of therapeutants by a health professional, (10) mortality disposal and (8) hold time for effluents containing methyl or ethyl testosterone. Mortality disposal and effluent hold time could present detection challenges for spot-audits. Sex-reversal is also likely to be implemented off-farm by specialist hatcheries supplying most smaller and medium-sized operations and is thus beyond the purview of farm-level standards.

A more comprehensive assessment of progressive improvement should also incorporate analysis of NC repeat frequencies for individual sites, NC root cause analysis and derived corrective actions (proposed by the client and sanctioned by the auditor).

Group certification and smallholder inclusion

In Asia, most aquaculture production still originates from medium- and small-scale producers lacking any environmental certification. The cumulative environmental impact of such production is unknown, though inter alia, individual farm environmental footprints averaged per unit of production are likely to be greater relative to those of larger-scale operations at similar production intensities. Key smallholder challenges include (1) limited health management capacity, particularly in disease diagnosis; (2) capacity for effective record keeping and data evaluation; and (3) higher marginal costs for quality assurance and production assembly logistics associated with fragmented production geographies.

In response, standard holders have developed 'cluster' or 'group' certification options to spread and lower auditing costs for individual farms. Two options meet the needs of entities at different production scales: (1) without or (2) with requirement for an "internal "management" system' (IMS), also sometimes referred to as an 'internal control system' (ICS).

Group certification 'without IMS' is typically adopted by larger horizontally and/or vertically integrated companies operating multiple sites under an existing common management process. Although there is no cap, in most cases participation involves relatively small numbers of farms (two to six) at intermediate scales. Lead supply chain companies, usually exporting processors, may also assume responsibility for coordinating the management of contracted outgrowers, as the certificate holder, a model that is most prevalent in the shrimp sector.

In the previous ASC Tilapia Standard performance case study, all six group certifications were classed as multi-site 'without IMS'. Five of these were for multiple (two to four) cage sites in individual reservoirs (in Indonesia, Brazil and Mexico) and one for three pond sites in Costa Rica. All were operated by vertically integrated companies as the certificate holders.

Group certification 'with IMS' provides an option for independent farms, including smallholders, to achieve certification in the absence of an external certificate holder providing management oversight. This model has proved most challenging to implement because costs of coordination, including responsibility for internal audits, must be met by the group (a sample of which will then be reviewed by the third-party CAB auditor). A common expectation that larger farming entities in the membership will take a leading role ignores the potential for free-rider effects. The potentially large number of smallholders required to assemble commercially viable volumes further increases free-riding as well as assurance risks. Furthermore, the highly perishable nature of seafoods elevates food safety risks relative to other agricultural commodity sectors that have achieved greater progress with smallholder group certification.

These factors lower the incentive for lead companies to contract the smallest producers, contributing to their exclusion from certification-centric export markets. This is a key reason for the failure of many donor-funded AIPs variously implemented by governments, NGOs and scheme-holders, to enable sustained access to such markets beyond project lifespans. Scale thresholds are also context specific, with greater incentive for inclusion of more geographically clustered and intensive producers of higher-unit-value species. Unsurprisingly then, most pilot smallholder AIPs have been implemented with shrimp farmers or for more niche, e.g. organic certified, premium production. Here too, left to market forces alone, the smallest family-operated farms in emerging economies generally remain beyond certification. For example, in Bangladesh group-certification projects predicated on organic and Fairtrade certification for value-added shrimp exports stalled for these reasons, combined with high costs for CoC assurance of micro-harvests. Where external support is ongoing, it is very likely to be the result of wider donor development and social responsibility objectives. Lessons learned include the need for 'value-chain shortening' eliminating intermediate value-chain product-handling nodes, through the creation of decentralized processing capacity within production concentrations.

Conditions are likely to be more enabling in well-regulated jurisdictions with a history of private and public sector cooperation and support. The Taiwanese NYAA group is one such example. However, having established an internal management system to assure compliance with statutory environmental regulations and traceability requirements, members still opted for ASC single-site rather than group certification. The reasons for this are unclear, although Taiwan remains one of only two high-income countries (along with Poland) with tilapia producers seeking ASC certification where audit costs may be less of an impediment to certification.

Area-based management of biosecurity

Farm-level certification schemes have been critiqued for their failure to adequately address the cumulative environmental impacts of aquaculture at a landscape, a.k.a. 'zonal' or 'area-based', level (Rector *et al.*, 2023). The ecosystem approach to aquaculture (EAA) provides a practical strategy for integrating activities within the social and biophysical dimensions of ecosystems (FAO, 2010; Aguilar-Manjarrez *et al.*, 2017). Such an approach is critical for mitigation of the interacting impacts of aquaculture on disease (pathogen and parasite) amplification, transmission or re-transmission to wild species and water quality.

To maintain wider ecosystem function and services, ABM requires a transition from farm-level to coordinated management of shared environmental risk by producer organizations and regulators within appropriate agro-ecological system boundaries. Certification alone is unlikely to drive institutional changes and reforms; rather, the aim should be to build on and complement existing legal instruments and practices, particularly where deficits exist. Emphasis should be on appeal to the enlightened self-interests of the often larger and better-resourced entities seeking certification to leverage shared benefits. Joint actions may also be required with water users in other sectors as well as aquaculture operators. Although this may be possible only through primary legislation, there may be opportunities to integrate area-based aquaculture organizations where (1) governance objectives and (2) ecosystem and administrative boundaries are aligned. As with group certification, building consensus based on understanding of incentives and disincentives for participation, and critically who will take on coordination roles, is key to operationalizing ABM.

An industry code of good practice (CoGP; SSPO, 2023), established by the Scottish producer organization 'Salmon Scotland' in 2006, is a well-recognized example of industry taking a lead in ABM with mitigation of sea louse transmission as the primary incentive for mutually beneficial collective action. CoGP certification is a requirement for membership of the industry body initiating the scheme, which is consequently adhered to by most almost the entire growout sector (although a second-party standard, like third-party standards the CoGP requires verification through regular independent farm-level audits). However, the highly consolidated nature of the Scottish salmon industry dramatically simplifies the challenge for zonal coordination compared with more fragmented sectors with high levels of smallholder participation. In 2023 the Scottish sector consisted of just 12 active companies, of which four accounted for >90% of output. Consolidation has also allowed companies to gain exclusive rights to operate in individual loch systems.

A voluntary disease notification system sponsored by larger member members of the Surat Shrimp Club in southern Thailand (Boromthanarat and Nissapa, 2000, Yamprayoon and Sukhumparnich, 2010) parallels the Salmon Scotland example. Driven by recurrent epizootic disease outbreaks over recent decades, coordination is also enabled by a relatively high degree of sectoral consolidation in the Thai shrimp sector.

Mainstream environmental standards are attempting to leverage industry cooperation through at least three complementary approaches: group certification and zonal requirements, either as a stand-alone standard or as an integrated component of farm-level standards. Options for, and challenges to, smallholder participation in group certification (approach 1) are outlined above; the other two approaches are discussed below.

Stand-alone zonal standards: In 2014 GSA (then GAA) initiated the development of a 'Zone Management Standard' (ZMS) focused on biosecurity management alongside its existing farm standards as an optional add-on (Marshall, 2015). Becoming effective in 2018, it was piloted with tilapia, shrimp and salmon producers in China, Honduras and Canada. Requirements were organized under six headings: (1) committee formation and participation; (2) risk-based boundary setting of a management area (MA); (3) a biosecurity management plan; (4) risk management indicators; (5) data collection and sharing; and (6) regular iterative performance review.

The standard defined an MA as an area 'containing an aquatic animal population with a distinct health status with respect to a specific disease or diseases, for which required surveillance and control measures are applied and basic biosecurity conditions are met under a common biosecurity management system'[7]. 'Most' farms within a zone were required to adhere to biosecurity requirements in farm-level standards, though not all needed to be individually certified against the base standard. The standard also mandated a competent health professional to act as zone manager to assure farm-level compliance and ensure that the proposed zone and actions offer effective disease control. The standard was withdrawn after the piloting phase, for reasons that are unclear, although it is likely that similar implementation challenges were encountered as described for 'group certification with IMS', including resourcing and retention of the zone manager.

Integrated zonal requirements: The third approach, integration of zonal management requirements as a component of general farm-level environmental standards, eliminates the need for a dedicated zone manager by making coordination a joint responsibility for all certified farms and their affiliates. Initial efforts have focused on a mature, highly consolidated cage salmon sector where sea louse transmission has become a major environmental and economic concern. The ASC salmon standard requires farms accounting for at least 80% of farmed production in a zone to participate in ABM of disease and parasite transmission and prevention of resistance to therapeutic treatments. This requires the participation of non-certified farms where necessary. Where not already determined by a regulatory body, boundary setting must reflect cumulative impacts on wild populations based on salmonid migration patterns, water movement and wider ecosystem structure and function. ABM members must participate in (1) the development and documentation of shared, disease management objectives for reducing risks to wild fish, subject to regular performance related revision; and (2) information sharing and coordination of stocking and fallowing (all fish in the ABM must be single-year class), on-farm disease and parasite monitoring, application and rotation of therapeutic treatments, evidence of increased resistance and any suspicions of an unidentifiable transmissible agent. The ABM must also commit to public disclosure of sea louse loads, demonstrate proactive engagement with public research bodies and respond to concerns raised by local communities and wild fish interests.

Integrating health and water quality requirements: Potential exists for greater future integration of biosecurity and water quality management requirements at local and/or regional zonal scales. Water quality can play a key role in pathogen–host–environment interactions whilst consideration of overlapping stakeholder interests can preclude conflicts and enhance coordination.

Boundary differences will be a key challenge. For water quality, boundary setting is most fundamentally linked to receiving water hydraulic residence time (HRT) and nutrient flushing rates whilst for health management, boundaries will reflect the distribution and migration routes of host species and vectors within connected waterways. Thus, health management boundaries will often extend beyond water-quality boundaries encompassing more discrete lentic settings. Greater consideration of far-field, downstream water quality impacts could enhance boundary congruence.

For health management, non-participation of just a small number of farms at any scale could compromise biosecurity objectives, whilst for water quality it may only be necessary to mobilize stakeholders accounting for the greatest share of nutrient inputs to preclude shifts in trophic status based on zonal assimilative capacity. However, this is likely to encompass a more diverse range of water-user groups. It may also be more challenging to hold to account those responsible for diffuse source nutrient effluents, e.g. from agriculture, compared with those responsible for more localized or point sources of effluents, including aquaculture operators. Regulatory and infrastructural disparities between lower- and higher-income countries will further compound such challenges.

An integrating approach of this kind is also highly consistent with the One Health concept (Stentiford *et al.*, 2020; Militzer *et al.*, 2023; WOAH *et al.*, 2023). One Health aims to integrate management of animal, human and environmental health to upscale disease surveillance, preparedness and response measures (from the farm- to the wider ecosystem-scale), cognisant of boundary problems of the kind described above.

Summary and conclusions

The global scope for farm-level environmental certification is most fundamentally constrained by demand for this form of market-based governance being limited to certification-centric markets (CCM) in affluent jurisdictions with strong civil society credentials. Such demand stems foremost from the need of consumer-facing brands to manage reputational risk, rather than more direct consumer pressure. Moreover, whether for B2B (with no consumer-facing label) or B2C models, demand tends to be most concentrated in retail supply chains, with less penetration of more fragmented food service sectors.

Global demand for environmentally certified tilapia originates primarily from just one CCM, the USA, also the world's largest importer, with much smaller flows into the EU. Certification is limited to export sectors capable of supplying the demanding environmental and other quality standards required by these markets. Respective US and EU imports of 193,230 tonnes (2020) and 33,000 tonnes (live weight equivalent, 2021) together accounted for just 3.17% of global tilapia production in 2021 (live weight equivalent), whilst total production certified under the two major environmental farm standards for tilapia, ASC and GSA–BAP together accounted for an estimated 3.15–3.78% of global production. This originated from just 12 of 126 tilapia-producing countries: seven in Central and South America, five in SE Asia and one in Europe. Of these, eight were classed as upper-middle-, two as lower-middle- and two as high-income countries (World Bank, 2023).

This relatively low share contrasts with an estimated 14% of total seafood supply, including capture fisheries under environmental certification in 2015 (Potts *et al.*, 2016). The tilapia deficit reflects the primacy of production for domestic consumption in emerging economies, placing entire production geographies – that together accounted for 71% of global production in 2022 – beyond the scope of voluntary schemes. Major exclusions include most production in the Middle East and most of Africa[8], where tilapias are also native, including nearly 1.2 million tonnes of Egyptian production in 2021 (FAO, 2023) as well many lower-income countries worldwide where low-cost tilapias make a significant contribution to national food security.

Synergies between certification and ratings and SGRS: When ASC and GSA–BAP farm-level certification is combined with SFW SGRS, the most influential such scheme in the US market, coverage rises from 38 to 40% of global seafood production and 27 to 25% of tilapia production between 2020 and 2023 (CRC, 2023)[9]. However, whilst the involuntary and sectoral nature of SGRS enables significantly higher volume coverage than farm-level schemes, the sectoral generalization of rating outcomes fails to acknowledge individual farm improvements in otherwise poorly performing sectors. Certification systems also have a much broader value chain scope.

Benchmarking of third-party farm-level standards against normative, e.g. FAO environmental and other codes, supports their adoption in various ways. Public–private benchmark initiatives, notably the Global Seafood Sustainability Initiative (GSSI, 2023), have helped consolidate demand for more robust schemes, enhancing consumer confidence and lowering compliance costs for producers supplying markets requiring certification under multiple standards.

SFWs internal benchmarking of ASC-, GSA–BAP- and Naturland-certified tilapia farms as the only acceptable ('blue') choice from sectors otherwise rated as 'red' (avoid) further leverages demand for these voluntary standards in CCM. In 2022 SFW red-rated (avoid) all Chinese tilapia production accounting for up to 19.4% of global production. Farms certified under the above schemes, accounting for up to 15.4% of Chinese production, were the only exceptions to this recommendation (Fig. 17.5).

ASCs decision to withdraw from the Chinese market in 2024, following the detection of forced labour in some seafood-processing plants, highlights wider strategic challenges facing standard holders vis-à-vis trade-offs between retention of influence and wider market credibility.

Smallholder participation: Voluntary standards do not always provide positive social and economic benefits for smallholders. Smallholders face greater challenges and higher marginal costs to implement the levels of quality assurance demanded by more affluent import markets. Moreover, mainstream commodity standards do not guarantee a price premium, only continued access to these markets. Statutory food safety and phytosanitary measures constitute the most significant and longest established non-tariff barriers to have accelerated sectoral consolidation and ever-more horizontally and vertically integrated supply chains serving major CCMs. The more recent introduction of environmental and social standards has only amplified this trend. The increasing share of certified tilapia production from larger, vertically integrated enterprises includes global operators of cage systems in large pristine lakes, better able to guarantee premium off-flavour-free, fresh fillets to the US market.

Co-sponsored by certification schemes, SGRS, NGOs, buyers and governments, AIPs are conceived as a means of enabling smallholders to enter certification through capacity-building support in conjunction with lower cost options for group certification (Kassam *et al.*, 2011). Smallholders responsible for an estimated 100,000 tonnes, or 1.4%, of global tilapia production, principally in Indonesia and China, the world's two largest producers and exporters, were participants in AIPs in 2021 (Fig. 17.5). However, evidence of the effectiveness of AIPs as a route to certification is still lacking. Furthermore, there will inevitably be a minimum farm size threshold below which the implementation costs for group certification (e.g. for harvest assembly logistics, management systems and chain of custody assurance, etc.) will preclude group certification as an economically viable model for supportive buyers. Recognizing this limitation, most certification bodies now differentiate between transitional group programmes and more philanthropic AIPs simply supporting smallholder capacity building.

Group certification, AIPs and ABM: Group certification models that aim to lower entry barriers for certification take two main forms: with or without requirement for an IMS. The latter, in which the certificate holder assumes management responsibility, has seen widespread adoption by horizontally integrated companies as a 'multi-site' option, or by processors contracting small numbers of (typically) medium-sized third-party producers. Group certification with IMS has proved a more challenging path to certification because internal management becomes the responsibility of the group, which may include large numbers of smallholders. Both group models have also been the focus of capacity-building AIPs variously co-sponsored by certification bodies, SGRS, eNGOs and governments.

Farm-level 'unitary' standards have also been critiqued for their failure to adequately address cumulative environmental impacts at the landscape (ecosystem) level, particularly with respect to disease transmission and water quality. Implementation challenges become much greater in fragmented sectors with large numbers of smallholders. Consequently, coordinated ABM must also address the problems of smallholder exclusion described above. Integration of group certification and AIPs within ABM models offers a potential solution. Further integration of ABM requirements within existing standards (rather than as an optional add-on) increases the ability to leverage cooperation of entities seeking farm-level certification.

Impacts of certification on tilapia health performance: Our analysis of 155 audits of 55 farms against the ASC Tilapia Standard between 2012 and 2019 confirmed the relative paucity of small-holder participation described above. The audited farms, mainly pond or lake-reservoir cage systems, were distributed across seven South American, five South-east Asian and one European country. Production from pond farms ranged from 36 (Taiwan) to 11,600 tonnes per year (Honduras), whilst cage farms ranged from 1100 (Vietnam) to 25,193 tonnes (Honduras). The smallest farms, pond systems averaging 194 tonnes per year, were operated by 15 members of a Taiwanese producer association, i.e. in a high-income country. This exclusion notwithstanding, analysis of NCs still revealed the strongest correlation between the frequency of indicator NCs and farm scale.

Of a total of 354 NCs, 13% (46) were assigned to eight (of 11) indicators under a principle titled 'Manage fish health and welfare in an environmental responsible manner' (FHM; with primary intent to mitigate negative impacts on wild populations). Indicators were grouped under four sub-criteria addressing (1) stock recovery, i.e. survival rates; (2) use of therapeutants and chemicals; (3) removal and disposal of mortalities; and (4) implementation of a farm fish health plan. Four indicators under criterion two addressed the prudent use of antibiotics linked to food safety (residue) and public health (antimicrobial resistance) concerns. Unlike some other ASC standards, the Tilapia Standard includes no 'state' indicators reflecting health conditions of the environment itself (e.g. pathogen or disease prevalence rates), only pressure and response indicators (OECD, 2013b).

Eleven of the 15 Taiwanese pond farms and a single 2,264-tonne reservoir-cage farm in Malaysia together accounted for 44% of all NCs against FHM indicators, with an average of 1.2 and 0.83 NCs per audit, whilst no FHM NCs were assigned against the largest cage and pond farms in Honduras and Costa Rica or a single RAS farm in Poland. With the largest number of NCs assigned during initial audits, this also meant greater scope for continuing improvement by these farms. All had subsequently progressed through at least three complete 3-year cycles of certification with much lower rates of NC detection following implementation of corrective actions.

Some 40% (20) of all FHM NCs were assigned to a single indicator mandating a minimum recovery rate of 65% of fish stocked after attaining a size of 100 g. Cage farms in Colombia, Indonesia and Mexico recorded some of the lowest and most erratic recovery rates, with recurrent country averages of 65–67%. A further 30% (14) of FHM NCs resulted from the use of unsanctioned therapeutants and imprudent use of antimicrobials, including prophylactic applications, exceedance of a maximum limit of three treatments per cycle and failure to calculate the quantity of active ingredient used per tonne of fish harvest.

An incorrect NC assignment detected against one indicator (use of antimicrobials determined critical to human health by WHO) highlights the potential significance of auditor capacity as an uncontrolled independent variable in the preceding analysis. Additionally, differences in the intrinsic 'auditability' of different requirements, as well as audit timing and duration, will influence detection rates and thereby account for some of the difference in NC frequencies observed between indicators. This highlights the absolute necessity for robust assurance capacity, both internal and external, to preserve the credibility of any certification scheme. 'Auditability' must also be a practical consideration in development of feasible indicator requirements during standard-setting processes.

Global value chains and future certification supply and demand: Rising demand for seafoods by the emergent middle classes in developing economies could increase demand for voluntary certification. However, embedded food safety attributes are likely to be primary drivers in markets with weaker CCM attributes. This has also tended to stimulate greater demand for niche organic certification above the more mainstream eco-labels in such settings. However, multinational brands and larger vertically integrated companies operating across value chains serving both strong and weaker CCM domains also remain incentivized to manage spillover of reputational risk through mainstream certification. This is further enabled by the lower marginal costs of certification for larger entities (often implemented as a B2B model with no requirement for a consumer-facing label).

A greater challenge stems from the rising share of seafood moving through channels exclusively dedicated to serving these emergent affluent non-CCM markets. A 'deglobalization' trend following recent major global value chain (GVC) disruptions is driving increased on- or 're-shoring' as companies and governments prioritize resilience and food security above GVC benefits (WEF, 2023). China, for example, became a net seafood importer in 2022 (Fletcher, 2023), most notably for shrimp, spurred by a post-Covid-19 rebound in food service demand. This on-shoring trend may see further spillover into lower-value seafood market segments more critical to basic food security, i.e. including tilapia production. ASC's recent withdrawal from the Chinese market provides a further example of how global geopolitical factors might amplify this trend.

Interaction of environmental and disease factors led to a 6% decline in recorded global tilapia production, from a historic peak of 7.15 million tonnes in 2021 (95% of it farmed) to 6.72 million tonnes in 2023. Over the same period, the combined ASC-, GSA–BAP-certified total fell by 30.3% from 363,455 to 238,223 tonnes, mainly due to falling production levels particularly in lake-cage systems, rather than to declining certificate numbers (coverage of SFW ratings rose by 9.7% over the same period, reflecting its sectoral scope). Findings highlight longer-term sustainability challenges associated with the certified tilapia sector's dependence on export production from large oligotrophic and mesotrophic lake and reservoir systems in middle- and lower-income countries. These water bodies face growing intersectoral water quality and eutrophication pressures associated with wider demographic transitions. Co-location with fragmented smallholder sectors in these open systems elevates biosecurity challenges and future risk of transboundary epizootics, as demonstrated by the recent emergence of TiLV.

Notes

1 6.5%, adjusting for farms certified under both major aquaculture certification schemes (ASC and GSA-BAP).

2 The ISO 14000 series of voluntary standards define processes required to set up and implement Environmental Management Systems (EMS).

The series are also integral components of (the more demanding) Eco-management and Audit Scheme (EMAS) of the European Union.

3 The CRC website reporting of 153,646t in 2021 is 32% lower than the 225,820t reported directly in an ASC 'Certification Update'. This would lower the corrected global percentage to 3.15% of total tilapia production.
A review of production capacities declared for 55 sites audited between 2012 and 2019 totalled 116,213 t.

4 Fig.17.8 excludes one ASC-certified RAS tilapia farm located in Poland.

5 Similar plans to limit tilapia farming in Honduras' largest natural lake, Lake Yojoa, were announced by the Honduran Government in June 2023 (The FishSite, 2023).

6 These aggregate percentages may double-count some overlapping ASC and GSA–BAP certification (CRC, 2023).

7 https://www.globalseafood.org/wp-content/uploads/2016/08/Day0_PeterMarshall_GOAL2015-copy.pdf

8 Two large lake cage farms in Ghana and Uganda achieved ASC certification in 2023.

9 While only Nile tilapia (*O. niloticus*) has formally entered ASC and GSA–BAP certification, SFW guidance encompasses more marginal tilapia species, including those farmed only in their African native range.

References

Aguilar-Manjarrez, J., Soto, D. and Brummett, R.E. (2017) Aquaculture zoning, site selection and area management under the ecosystem approach to aquaculture. *Full Document. Report ACS113536. Food and Agriculture Organization.* Available at: http://www.fao.org/3/a-i6992e.pdf

Aich, N., Paul, A., Choudhury, T.G. and Saha, H. (2022) Tilapia lake virus (tilv) disease: current status of understanding. *Aquaculture and Fisheries* 7(1), 7–17. Available at: https://doi.org/10.1016/j.aaf.2021.04.007

AIPCE-CEP (2022) Finfish Study 2022 AIPCE-CEP EU fish processors and traders association brussels September 2022. Brussels. Available at: https://www.aipce-cep.org/wp-content/uploads/2022/09/Finfish-Study-2022-final.pdf

ASC (2023b) ASC impacts dashboard. Available online. https://us.asc-aqua.org/homepage/asc-impacts-dashboard/. https://us.asc-aqua.org/homepage/asc-impacts-dashboard/ (accessed on 3 July 2023)

ASMI (2020) Certified Seafood Collaborative (SCS) takes ownership of RFM certification program. Announcements, press releases, RFM certification. Available at: https://www.alaskaseafood.org/news/for-release-certified-seafood-collaborative

Australian Sports Commission (2019) ASC tilapia Standard version 1.2. Available at: https://www.asc-aqua.org/wp-content/uploads/2019/03/ASC-Tilapia-Standard_v1.2_Final.pdf

Australian Sports Commission (2023a) Find a farm. Retrieved August 2, 2022. Available at: https://www.asc-aqua.org/find-a-farm/

Australian Sports Commission (2023b) ASC impacts dashboard. Retrieved July 3, 2023. Available at: https://us.asc-aqua.org/homepage/asc-impacts-dashboard/. https://us.asc-aqua.org/homepage/asc-impacts-dashboard/

Australian Sports Commission (2023c) ASC farm standards. Available at: https://asc-aqua.org/producers/farm-standards/

Belton, B., Murray, F., Young, J., Telfer, T. and Little, D.C. (2010) Passing the panda standards: A TAD off the mark. *Ambio* 39(1), 2–13. Available at: https://doi.org/10.1007/s13280-009-0009-4

BIM (2022) Certified quality aquaculture (CQA) standards (farm and processing) Issue 2. Available at: https://bim.ie/wp-content/uploads/2021/01/BIM-CQA-Farm-Standards-Issue-2.pdf

Boromthanarat, S. and Nissapa, A. (2000) Shrimp farming experiences in thailand. – A continued pathway for sustainable coastal aquaculture. NACA report submitted by Prince of Songkla University 1–109.

Bridson, P.B., Fransen, M.H. and Stoner, J.M.S. (2020) The aquaculture sustainability continuum – defining an environmental performance framework. *Environmental and Sustainability Indicators* 8(December), 100050.

Bush, S.R., Belton, B., Hall, D., Vandergeest, P., Murray, F.J. *et al.* (2013) Certify sustainable aquaculture. *Science* 341, 1067–1068.

CFC (2018) The impact of voluntary sustainability standards on small-scale farmers in global commodity chains. Common Fund for Commodities annual report. Available at: https://www.commonfund.org/sites/default/files/Publications/CFC_AR_18_paper_Certification.pdf

Commission (2008) Commission Regulation (EC) No 889/2008 of 5 September 2008 laying down detailed rules for the implementation of Council Regulation (EC) No 834/2007 on organic production and labelling of organic products with regard to organic production, labelling and control. Commission regulation (EC). Available at: https://faolex.fao.org/docs/pdf/eur82157.pdf

CRC (2023) Certification and Ratings Collaboration, Sustainable Seafood Data Tool. Available online Data Tool – C&R. 2023. certificationandratings.org Retrieved July 2, 2023.

Fletcher, R. (2023) Rabobank/Gorjan Nikolik reportsA seismic shift in the global seafood trade. The fish site. Available at: https://thefishsite.com/articles/a-seismic-shift-in-the-global-seafood-trade-china-rabobank

Food and Agriculture Organization (1995) Code of conduct for responsible fisheries Rome (CCRF). Food and Agriculture Organization. Available at: https://www.fao.org/3/v9878e/v9878e.pdf

Food and Agriculture Organization (2008) Joint FAO/WHO/OIE expert meeting on critically important antimicrobials report of the FAO/WHO/OIE expert meeting FAO headquarters. Rome Retrieved November 26–30, 2007. Available at: https://www.fao.org/3/i0204e/i0204e.pdf

Food and Agriculture Organization. (2009) Food and Agriculture Organization, *Guidelines for the Ecolabelling of Fish and Fishery Products from Marine Capture Fisheries (1st Revision)*. Available at: https://www.fao.org/3/i1119t/i1119t.pdf

Food and Agriculture Organization. (2010) Aquaculture development. 4. Ecosystem approach to aquaculture. FAO technical guidelines for responsible fisheries. no.5. Food and Agriculture Organization. Available at: https://www.fao.org/3/i1750e/i1750e.pdf, Suppl. 4

Food and Agriculture Organization (2011) The FAO technical guidelines on aquaculture certification. Rome 122. Available at: https://www.fao.org/3/i2296t/i2296t00.htm

Food and Agriculture Organization (2014) Impact of international voluntary standards on smallholder market participation in developing countries – A review of the literature. Rome. Available at: https://www.fao.org/3/i3682e/i3682e.pdf

Food and Agriculture Organization (2022) The state of world fisheries and aquaculture, towards blue transformation. Available at: https://doi.org/10.4060/cc0461en

Food and Agriculture Organization (2023) FishStatJ - Software for fishery and aquaculture statistical time series. Retrieved July 2, 2023. Available at: https://www.fao.org/fishery/en/topic/166235/en

Food and Agriculture Organization/UN Environmental Program (2014) *Conference Proceedings Workshop on Voluntary Standards for Sustainable Food Systems: Challenges and Opportunities*. Available at: http://www.fao.org/3/a-i3421e.pdf

GlobalGAP (2023) Integrated farm assurance for Aquaculture Standard IFA v6. Available at: https://www.globalgap.org/uk_en/for-producers/globalg.a.p./integrated-farm-assurance-ifa/aquaculture/

GSA–BAP (2023) Global seafood alliances better aquaculture practices, standards and guidelines. Available at: https://www.bapcertification.org/Standards

GSSI (2023a) GSSI Global Benchmark Tool v2.0. Available online https://www.ourgssi.org/wp-content/uploads/2021/11/Benchmark-Framework-version-2.0.xlsx

GSSI (2023b) GSSI recognised certification. Available online https://www.ourgssi.org/gssi-recognized-certifcation/ (accessed on 10 Jul 2023)

ISEAL (2023) Defining credible practice. Available at: https://www.isealalliance.org/defining-credible-practice

Jacquet, J.L. and Pauly, D. (2007) The rise of seafood awareness campaigns in an era of collapsing fisheries. *Marine Policy* 31(3), 308–313. Available at: https://doi.org/10.1016/j.marpol.2006.09.003

Kassam, L., Subasinghe, R. and Phillips, M. (2011) Aquaculture farmer organizations and cluster management: concepts and experiences. FAO fisheries and aquaculture technical paper. *Food and Agriculture Organization* 563.

MacKinnon, B., Debnath, P.B., Bondad-Reantaso, M.G., Fridman, S., Bin, H.,Nekouei, O. (2023) Improving tilapia biosecurity through a value chain approach. In *Reviews in Aquaculture, Tilapia health: quo vadis?* Eds. Bartley, D., Phillips, M., Fitzsimmons, K., Arthur, R. Bondad-Reantaso, M. Volume 15, Supplement 1, February 2023. https://doi.org/10.1111/raq.12776

Marschke, M. and Wilkings, A. (2014) Is certification a viable option for small producer fish farmers in the global south? insights from vietnam. *Marine Policy* 50 A, 197–206. Available at: https://doi.org/10.1016/j.marpol.2014.06.010

Marshall, P. (2015) Zone Management Workshop. GOAL Vancouver Cananda, 26 Oct. Available at: https://www.globalseafood.org/wp-content/uploads/2016/08/Day0_PeterMarshall_GOAL2015-copy.pdf

MCSUK (2022) Marine Conservation Society; aquaculture ratings methodology handbook version 4. Available at: https://media.

mcsuk.org/documents/MCS_Aquaculture_Methodology_FULL_V4_FINAL_72022.pdf

MCSUK (2023) Marine conservation society: Good fish guide. Retrieved May 10, 2023. Available at: https://www.mcsuk.org/goodfishguide/ratings/aquaculture/373/

MEL (2018) Marine eco-label Japan aquaculture management Standard. Available at: http://melj.jp/eng/wp-content/uploads/2018/07/%E3%80%90Standard%E3%80%91-Aquaculture-FINAL.180730.pdf

Militzer, N., McLaws, M., Rozstalnyy, A., Li, Y., Dhingra, M. *et al.* (2023) Characterising biosecurity initiatives globally to support the development of A progressive management pathway for terrestrial animals: A scoping review. *Animals: An Open Access Journal from MDPI* 13(16), 2672. Available at: https://doi.org/10.3390/ani13162672

MSU (2020) Mississippi market maker. *Newsletter*. U. S. Tilapia Production, 10(9), and Wholesale Prices (9th ed., vol. 10, pp. 11). Available at: http://extension.msstate.edu/newsletters/mississippi-marketmaker/2020/vol-10-no-9-us-tilapia-consumption-production-imports-and. MSU Extension Service

Murray, F. and Taskov, D. (2018) Report on costs and benefits of compliance with voluntary market-based labelling and certification schemes. Developing Innovative Market Orientated Prediction Toolbox to Strengthen the Economic Sustainability and Competitiveness of European Seafood on Local and global market. Grant agreement No: 635761 EU Horizon 2020 research and innovation program. Available at: https://cordis.europa.eu/project/id/635761/results

Naylor, R.L., Hardy, R.W., Buschmann, A.H., Bush, S.R., Cao, L. *et al.* (2021) A 20-year retrospective review of global aquaculture. *Nature* 591(7851), 551–563. Available at: https://doi.org/10.1038/s41586-021-03308-6

Olsen, M.S., Amundsen, V.S. and Osmundsen, T.C. (2023) Exploring public perceptions and expectations of the salmon aquaculture industry in norway: A social license to operate. *Aquaculture* 574, 739632. Available at: https://doi.org/10.1016/j.aquaculture.2023.739632

Organization for Economic Co-operation and Development (2013a) Interconnected economies: Benefitting from global value chains. Meeting of the OECD Council at Ministerial level. Paris 29-20 May 2013. Available at: https://www.oecd.org/mcm/C-MIN(2013)15-ENG.pdf

Organization for Economic Co-operation and Development (2013b) Framework of OECD work on environmental data and indicators. In:

In Environment at a Glance 2013: OECD Indicators. OECD Publishing, pp. 9–10. Available at: https://doi.org/10.1787/9789264185715-3-en

Organization for Economic Co-operation and Development (2017) *Environmental labelling and information schemes; policy perspectives. OECD Environment working papers on which this synthesis report.* Available at: https://www.oecd.org/env/policy-persectives-environmental-labelling-and-information-schemes.pdfPotts

Potts, J., Wilkings, M., Lynch, M. and McFatridge, S. (2016) State of sustainability initiatives. Review: Standards and the blue economy. International Institute for Sustainable Development. Available at: https://www.iisd.org/publications/state-sustainability-initiatives-review-standards-and-blue-economy

Rector, M.E., Filgueira, R. and Grant, J. (2023) From farm sustainability to ecosystem sustainability: exploring the limitations of farm-applied aquaculture eco-certification schemes. ISSN 0301-4797, *Journal of Environmental Management* 339, 117869. Available at: https://doi.org/10.1016/j.jenvman.2023.117869

Royal Society for the Prevention of Cruelty to Animals (2021) RSPCA welfare standards for farmed Atlantic. Salmon Publishing. Available at: https://science.rspca.org.uk/sciencegroup/farmanimals/standards/salmon

Rusko, M. and Korauš, A. (2013) Types I, II and III of ecolabels. *Journal of Environmental Protection, Safety, Education and Management* 2(1). Available at: https://www.sszp.eu/wp-content/uploads/2013_No2-Vol_Journal-JEPSEM__p-1__Rusko-Koraus_f4.pdf

Saha, C.K. (2022) Emergence and evolution of aquaculture sustainability certification schemes. *Marine Policy* 143. Available at: https://doi.org/10.1016/j.marpol.2022.105196

Scalia-Bruce (2024) ASC initiates phaseout of Chinese operations; GSA responds to BAP Criticism. Seafood source. Published in environment and sustainability. Retrieved March 28, 2024. Available at: https://www.seafoodsource.com/news/environment-sustainability/asc-ceases-operations-with-china-pedro-bank-jamaica-queen-conch-industrial-dive-fishery-achieves-msc-certification

SFP (2023) Sustainable Fisheries Partnership. Retrieved July 3, 2023. Available at: https://sustainablefish.org/tools-science-services/fishsource/

SFW (2018) *Tilapia* (*Orechromis spp.*) China Ponds aquaculture Standard version a3.2 Seafood Watch Consulting Researchers. Available at: https://www.seafoodwatch.org/globalassets/sfwdatablocks/reports/T/MBA_SeafoodWatch_TilapiaChinaReport.pdf

SFW (2020) Monterey Bay Aquarium seafood watch. Standard for aquaculture version a4.0, 2020. Available at: https://www.seafoodwatch.org/globalassets/sfw/pdf/standards/aquaculture/seafood-watch-aquaculture-standard-version-a4.pdf

SFW (2023) Monterey Bay seafood watch. Retrieved May 10, 2023. Available at: https://www.seafoodwatch.org/recommendations/search?query=%3Afree%3BTilapia%7Cspecies

SSPO (2023) Code of Good Practice for Scottish Finfish Aquaculture (CoGP). Available at: https://thecodeofgoodpractice.co.uk/

Stentiford, G.D., Bateman, I.J., Hinchliffe, S.J., Bass, D., Hartnell, R. *et al.* (2020) Sustainable aquaculture through the one health lens. *Nature Food* 1(8), 468–474. Available at: https://doi.org/10.1038/s43016-020-0127-5

Taiwan Today (2013) Taiwan tilapia farms get ASC certification. (April 22, 2013). Available at: https://taiwantoday.tw/news.php?unit=6&post=11579

Tanjung, D., Hutagaol, M.KriswantriyonoA.Hasututi, Y., Nirmala, K. *et al.* (2022) Regulatory impact assessment analysis of regulations on pollution load capacity and carrying capacity of lake toba for aquaculture fisheries. *ICAS 2021 E3S Web of Conferences 348* 00037. Available at: https://doi.org/10.1051/e3sconf/202234800037

Tlusty, H. (2011) Environmental improvement of seafood through certification and ecolabelling: Theory and analysis. *Fish and Fisheries* 13(1), 1–13. DOI: 10.1111/j.1467–2979.2011.00404.xUN.

UN DESA (2023) A Global Partnership supporting the operationalization of internationally agreed codes and guidelines for the seafood sector. The global sustainable seafood initiative (GSSI). United Nations Department of Economic and Social Affairs. Available at: https://sdgs.un.org/partnerships/global-partnershipsupportingoperationalizationinternationallyagreedcodesand#:~:text=Since%20its%20founding%20in%202013,and%20agreed%20codes%20and%20guidelines

United Nations Conference on Environment and Development. (1992) Agenda 21 of the rio declaration on environment and development. In: *UN.Org. Agenda. UN Conference on Environment and Development Rio de Janerio*, Vol. 21 [Doc]. Brazil, pp. 3–14.

Urbina, I. (2019) *The outlaw ocean: Journeys across the last untamed frontier.* Knopf Doubleday Publishing Group. 9780451492951

Urbina, I. (2023) Print and online. The shadow armada. The New Yorker. Available at: https://www.newyorker.com/magazine/2023/10/16/the-crimes-behind-the-seafood-you-eat

USSEC (2017) Carrying capacity of Lake Toba in north Sumatra is being determined. United States Soybean Export Council. Available at: https://ussec.org/carrying-capacity-lake-toba-north-sumatra-determined/

Vandergeest, P. and Unno, A. (2012) A new extraterritoriality? aquaculture certification. *Sovereignty, and emPolitical Geography* 31, 358–367.

Vince, J. and Haward, M. (2017) Hybrid governance of aquaculture: opportunities and challenges. *Journal of Environmental Management* 201, 138–144. Available at: https://doi.org/10.1016/j.jenvman.2017.06.039

Vince, J. and Haward, M. (2019) Hybrid governance in aquaculture: certification schemes and third-party accreditation. *Aquaculture* 507, 322–328. Available at: https://doi.org/10.1016/j.aquaculture.2019.04.041

White (2023) Social audits for MSC, ASC, BRC certification likely missed evidence of Uyghur forced labor. Seafood source. Supply and trade, Retrieved October 16, 2023. Available at: https://www.seafoodsource.com/news/premium/supply-trade/social-audits-for-msc-asc-brc-certification-likely-missed-evidence-of-uyghur-forced-labor

WOAH (2023) One Health. Retrieved August 1, 2023. Available at: https://www.woah.org/en/what-we-do/global-initiatives/one-health/

World Bank. (2023) The world by income and region. Retrieved August 1, 2023. Available at: https://datatopics.worldbank.org/world-development-indicators/the-world-by-income-and-region.html

World Economic Forum (2023) Deglobalisation: What you need to know. Retrieved January 17, 2023. Available at: https://www.weforum.org/agenda/2023/01/deglobalisation-what-you-need-to-know-wef23

World Health Organization (2018) *Critically important antimicrobials for human medicine* (6th revision) ISBN 9789241515528 World Health Organization 2019. Available at: https://apps.who.int/iris/bitstream/handle/10665/312266/9789241515528-eng.pdf

Yamprayoon, J. and Sukhumparnich, K. (2010) Thai aquaculture: achieving quality and safety through management and sustainability. *Journal of the World Aquaculture Society* 41(2), 274–280. Available at: https://doi.org/10.1111/j.1749-7345.2010.00355.x

Index

Printed and bound by CPI Group (UK) Ltd, Croydon, CR0 4YY

13/09/2026

14961083-0002